FRACTURE MECHANICS OF CONCRETE AND CONCRETE STRUCTURES

BALKEMA – Proceedings and Monographs
in Engineering, Water and Earth Sciences

PROCEEDINGS OF THE 6TH INTERNATIONAL CONFERENCE ON FRACTURE MECHANICS OF
CONCRETE AND CONCRETE STRUCTURES, CATANIA, ITALY, 17–22 JUNE 2007

Fracture Mechanics of Concrete and Concrete Structures

VOLUME 3: High-Performance Concrete, Brick-Masonry and Environmental Aspects

Editors

Alberto Carpinteri
Politecnico di Torino, Department of Structural Engineering and Geotechnics, Italy

Pietro G. Gambarova
Politecnico di Milano, Department of Structural Engineering, Italy

Giuseppe Ferro
*Politecnico di Torino, Department of Structural Engineering and Geotechnics –
Italian Group of Fracture, Italy*

Giovanni A. Plizzari
*University of Brescia, Department of Civil Engineering, Architecture,
Land and Environment, Italy*

Taylor & Francis
Taylor & Francis Group

LONDON / LEIDEN / NEW YORK / PHILADELPHIA / SINGAPORE

Cover illustrations
Courtesy of S. Jox, C. Becker & G. Meschke

Taylor & Francis is an imprint of the Taylor & Francis Group, an informa business

© 2007 Taylor & Francis Group, London, UK

Typeset by Charon Tec Ltd (A Macmillan Company), Chennai, India
Printed and bound in Great Britain by Bath Press Ltd (a CPI-group company), Bath

Published by: Taylor & Francis/Balkema
                P.O. Box 447, 2300 AK Leiden, The Netherlands
                e-mail: Pub.NL@tandf.co.uk
                www.balkema.nl, www.taylorandfrancis.co.uk, www.crcpress.com

ISBN 13 Set (3 volumes):   978-0-415-44066-0
ISBN 13 Vol-1:   978-0-415-44065-3
ISBN 13 Vol-2:   978-0-415-44616-7
ISBN 13 Vol-3:   978-0-415-44617-4

*Fracture Mechanics of Concrete and Concrete Structures – High-Performance Concrete,
Brick-Masonry and Environmental Aspects – Carpinteri, et al. (eds)
© 2007 Taylor & Francis Group, London, ISBN 978-0-415-44617-4*

Table of contents

VOLUME 2 – Design, Assessment and Retrofitting of RC Structures

XIII

VOLUME 3 – High-Performance Concrete, Brick-Masonry and Environmental Aspects

Preface

The present volume is the third part of the Proceedings of the 6th International Conference on Fracture Mechanics of Concrete and Concrete Structures (FraMCoS-6), co-organized by the Politecnico di Torino, the Politecnico di Milano, the University of Brescia, and the Italian Group of Fracture (IGF), under the auspices of the International Association of Fracture Mechanics for Concrete and Concrete Structures (IA-FraMCoS). FraMCoS-6 was also scientifically supported by FIB, ICF, RILEM, ACI, ESIS, and JCI, as well as by the National Research Council of Italy (CNR). The venue was the Conference Centre of Catania Sheraton Hotel, in Catania (Italy), a very attractive place on the island of Sicily, not far from the well-known archaeological sites of Syracuse and Taormina.

Since the IA-FraMCoS' establishment by Professor Zdenek Bazant, Founding and Honorary President of our Scientific Society, the FraMCoS series have taken place in different parts of the World on a triennial basis. The previous conferences were held in:

- Breckenridge (Colorado, USA) – FraMCoS-1, organised in 1992 by Z.P. Bazant;
- Zurich (Switzerland) – FraMCoS-2, organised in 1995 by F.H. Wittmann;
- Gifu (Japan) – FraMCoS-3, organised in 1998 by H. Mihashi and K. Rokugo;
- Cachan (France) – FraMCoS-4, organised in 2001 by R. de Borst, J. Mazars, G. Pijaudier-Cabot, and J.G.M. van Mier;
- Vail (Colorado, USA) – FraMCoS-5, organised in 2004 by V.C. Li, C.K.Y. Leung, K.J. Willam, and S.L. Billington.

It should be reminded that, before the formal foundation of IA-FraMCoS in 1992, significant and pioneering conferences on the same subject had been held since 1984, in Evanston, Lausanne, Houston, Vienna, Cardiff, Torino, and Noordwijk, organised by the same Scientific Community, and, in particular, by S.P. Shah, F.H. Wittmann, S. Swartz, H.P. Rossmanith, B.I.G. Barr, A. Carpinteri, and J.G.M. van Mier.

As always, the organisers were responsible for pointing out the main themes and the open problems on which our Scientific Community should work and debate. These themes and research directions usually change in time, and represent the genetic mutations that are necessary for the natural selection of ideas and for the scientific evolution of the IA-FraMCoS. This Community, on the other hand, has produced important ideas, that have been appreciated by other Communities also interested in Material Strength and Structural Integrity. Among the many possible examples, the Fictitious Crack Model by Arne Hillerborg, proposed in the Seventies, may be cited, since it is nowadays utilised even for materials other than concrete, under the more general denomination of "Cohesive Crack Model".

The present volume, titled "High-Performance Concrete, Brick-Masonry and Environmental Aspects", is divided into four Parts: (1) High-Performance Concrete; (2) Fiber Reinforced Concrete; (3) Brick-Masonry and other Quasi-Brittle Materials; and (4) Environmental Issues.

Concrete technology has developed at a fast pace indeed during the last two decades and material performance has been significantly improved; as a consequence, high-performance concrete (HPC) is now a reality. Initially the attention was mostly focused on the compressive strength and the enhanced concrete was named "high-strength concrete" (HSC). Later, other issues came up, such as workability and durability. As a consequence, enhanced rheology (in terms of flowability and cohesion, i.e. no segregation effects) at the fresh state and compactness at the hardened state were increasingly requested. The researchers' response was the development of self-consolidating concrete (SCC), that allows a reduction in manpower, especially when the reinforcement is highly congested.

Since higher strength generally implies higher brittleness, fibre-reinforced concrete (FRC) has raised considerable interest for its enhanced toughness under both static and dynamic loading, as well as for its ability to control concrete cracking. Nowadays, many types of fibres are available on the market, with different materials and geometry. By taking into account the remarkable toughness of FRC through its fracture energy, and by modelling the structural behaviour within nonlinear fracture-mechanics, the advantages ensuing from the introduction of the fibres can be fully exploited. Furthermore, by adopting optimized mix-designs (in terms of fibre content and type, and of pozzolanic or hydraulically-active adjuncts) the increasingly important requirements of durability can be met, even under the most severe environmental conditions (high and low temperatures, corrosive atmosphere, etc.).

Recently, fracture mechanics has been extended also to other brittle or quasi-brittle materials, such as brick-masonry, glass, polymers and ice, and a more realistic evaluation of the actual safety level of the structures has been obtained.

Finally, as at the end of the two previous volumes, the Editors would like to express their sincere and warm thanks to all the Sponsors of the FraMCoS-6, whose generous financial support was instrumental in making this Event feasible, as well as to all the actors of the Event: the Board of Directors, the Advisory Board, the International Scientific Committee, the Local Organising Committee, the Chairmen of the various Sessions, the Plenary Lecturers, the Participants and, in particular, the Authors, for their excellent contributions.

The Editors of the Volume
Alberto Carpinteri, Pietro G. Gambarova,
Giuseppe Ferro, and Giovanni A. Plizzari

Fracture Mechanics of Concrete and Concrete Structures – High-Performance Concrete,
Brick-Masonry and Environmental Aspects – Carpinteri, et al. (eds)
© 2007 Taylor & Francis Group, London, ISBN 978-0-415-44617-4

Sponsors

Organized by

Politecnico di Torino

University of Brescia

Politecnico di Milano

Italian Group of Fracture (IGF)

Under the auspices of

International Association of Fracture Mechanics
for Concrete and Concrete Structures (IA-FraMCoS)

With the scientific support of

Fédération Internationale du Béton (fib)

International Congress on Fracture (ICF)

International Union of Laboratories and Experts in Construction
Materials, Systems and Structures (RILEM)

American Concrete Institute (ACI)

European Structural Integrity Society (ESIS)

Japan Concrete Institute (JCI)

National Research Council of Italy (CNR)

Italian National Agency for New Technologies, Energy and the Environment (ENEA)

University of Applied Science of Southern Switzerland, Lugano, Switzerland

University of Bologna

aicap

Italian Association for Reinforced and Prestressed Concrete (AICAP)

Italian Society of Building Experts (CTE)

With the support of

City of Catania

Provincia Regionale di Catania

XXII

This Volume was edited in the framework of the ILTOF Project (EU Leonardo da Vinci Programme): Innovative Learning and Training On Fracture – www.iltof.org.

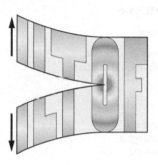

Main sponsors

ANAS S.p.A.   Italian Agency of National Roads (ANAS)

MAPEI, Milan (Italy)

Saint-Gobain Vetrotex Espana, Madrid (Spain)

Technochem, Barzana (BG, Italy)

Board Members of IA-FraMCoS

Part IX
High-performance concrete

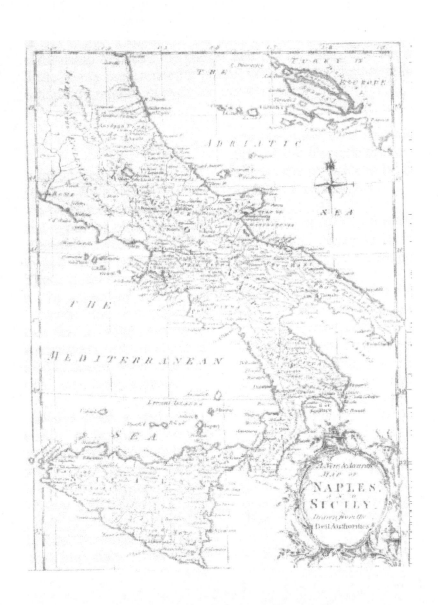

Fracture Mechanics of Concrete and Concrete Structures – High-Performance Concrete,
Brick-Masonry and Environmental Aspects – Carpinteri, et al. (eds)
© 2007 Taylor & Francis Group, London, ISBN 978-0-415-44617-4

Fracture mechanics of concrete and its role in explaining structural behaviour

J.C. Walraven

Delft University of Technology, The Netherlands

ABSTRACT: Major steps forward in the history of the development of fracture mechanics, as a tool for describing or predicting the behaviour of concrete structures, have been the fictitious crack model, introduced by Hillerborg, and the compressive damage zone model, introduced by Markeset. In structural elements, however, often the conditions are less clear than in concentric tension or compression tests. The behaviour of concrete beams subjected to shear and slabs subjected to punching are illustrating examples, especially when size effects are concerned. Another phenomenon that required further study is the rotation capacity of slabs at intermediate supports. Since mostly the ultimate rotation is reached at failure of the concrete in the compression area, the question is justified whether rotation capacity is subject to size effects as well. Since concrete is a brittle material, it is necessary to design structures in such a way that the structural behaviour becomes ductile nevertheless. One way is to adequately reinforce the structure. Another way is to provide the material concrete itself with ductility. This can be done by adding steel (or other) fibers to the concrete mixture. Indeed the fracture toughness of the concrete is considerably enhanced: however further questions can be raised with regard to the influence of the production process of fiber reinforced structural members on the mechanical properties and the most appropriate method of determining the mechanical properties. An essential question is furthermore whether improving the fracture toughness of concrete by adding fibers by definition leads to an improved behaviour of the structural members.

1 INTRODUCTION

The introduction of the fictitious crack model by Hillerborg and his co-workers (1976), and Petersson (1981) has been a major step forward in understanding and modeling the behaviour of concrete structures. By providing tensile stress–crack opening relations further to stress–strain relations, it became possible to better describe the behaviour of structures, especially those which exhibit a brittle failure behaviour. The first time that this behaviour was described in a structural design code was in the CEB-FIP Model Code for Concrete Structures in 1990. Fig 1 shows the relation as given in the MC'90. The corresponding expressions for the crack widths w_1 and w_c are:

$$w_1 = 2\frac{G_F}{f_{ctm}} - 0.15w_c \qquad (1)$$

$$w_c = \alpha_F \frac{G_F}{f_{ctm}} \qquad (2)$$

In the equations G_F is the fracture energy which is a function of the concrete compressive strength and the maximum aggregate size. The latter influence is justified since there is a relation between the crack band

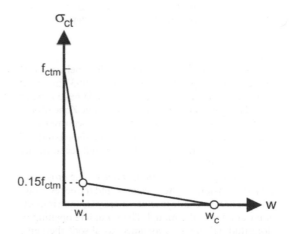

Figure 1. Stress–crack opening relation according to CEB-FIP Model Code 1990.

width and the maximum particle diameter. Van Mier (1996) showed, that within the band of microcracks "crack bridging" occurs: this phenomenon is the mechanism behind the residual stresses, transmitted across the crack faces during crack widening, Fig. 2.

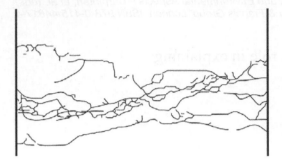

Figure 2. Crack band in a concrete with $d_{max} = 16$ mm (van Mier, 1996).

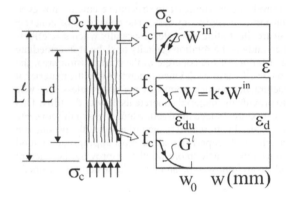

Figure 3. CDZ-model according to Markeset (1993) illustrated for a specimen loaded in concentric compression.

Of similar significance is the failure behaviour of concrete in compression. Markeset (1993) assumed that compressive failure localizes in a damage zone of limited length. Her Compressive Damage Zone Model (CDZ-Model) takes into account the occurrence of longitudinal splitting cracks as well as a shear band and can be applied both to concentrically and eccentrically loaded concrete, Fig. 3.

The damage zone length depends on the cross sectional dimensions of the specimen and the eccentricity of the load. The tensile fracture energy G_F is an important parameter in the model. The complete opening of a longitudinal crack is assumed to absorb the same amount of energy as the opening of a pure tensile crack.

Both in the case of tension and compression, the failure is mostly an integral part of a larger and more complex system.

In the following this will be illustrated for the cases of shear in slender and short members not reinforced in shear, slabs subjected to punching and the rotation capacity of statically indeterminate slabs.

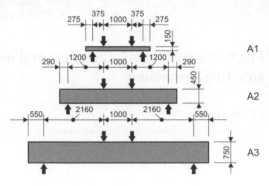

Figure 4. Size effect tests on slender beams, series A (Walraven, 1978).

2 FRACTURE MECHANICS CONSIDERATIONS FOR SLENDER AND SHORT CONCRETE MEMBERS FAILING IN SHEAR

2.1 Slender shear critical beams

The first time that a relation was found between the shear capacity of non shear reinforced concrete beams and crack propagation was in a research project at TU Delft (Walraven, 1978). Three series of three beams were tested in shear. The beams in any series differed in size. Fig. 4 shows the dimensions of the beams in series A. The heights of the beams were 150 mm, 450 mm and 750 mm respectively. The reinforcement ratio was 0.8%. The beams contained no shear reinforcement. The loads were applied at a distance 3d from the supports, in order to have a shear critical loading geometry. The tests were carried out in order to investigate the existence of a size effect in shear and to find an appropriate explanation. In those days researchers were convinced that the reason for the size effect was the size dependency of aggregate interlock in the cracks. They argued that the cracks in larger specimens would by wider for the same crack pattern. If the same concrete mixture would be used, with the same size of the aggregates, the effect of aggregate interlock in the cracks should consequently be smaller in the larger beams, which would explain the size effect. In order to verify this hypothesis, a series of beams was tested similar to those shown in Fig. 4, but now with lightweight concrete in stead of normal weight concrete. If the hypothesis of aggregate interlock would be valid, this would mean that no size effect would be observed in the lightweight concrete series, because the lightweight particles fracture at cracking.

A comparison of the nominal ultimate shear stress ($v_u = V_u/bd$) clearly showed, that also in the lightweight concrete beams a clear size effect occurred, Fig. 5.

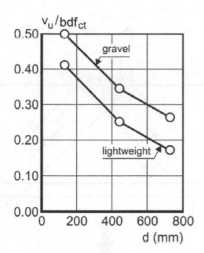

Figure 5. Nominal shear strengths as a function of the effective cross sectional depth, for gravel – and lightweight concrete (Walraven, 1978).

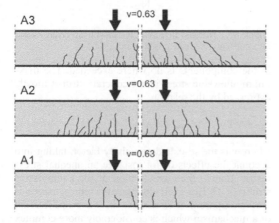

Figure 6. Crack pattern development in shear critical beams (a/d = 3), with cross sectional depths varying from 750 mm (beam A_3) to 150 mm (beam A_1), represented equally large, at the same nominal shear stress V/bd = 0.63 MPa.

A comparison of the crack patterns at similar values of the nominal shear stress v = V/bd showed, that the progress of the cracks in the large specimens is significantly faster than in small specimens. This is shown in Fig. 6.

In this figure the dimensions of the beams are graphically represented at the same size, in order to facilitate a good comparison of the state of development of the crack pattern. It is seen that the highest beam (A3, with h = 750 mm) shows a crack pattern indicating that the beam is near to failure (which occurred at a nominal shear stress of 0.70 MPa). Failure of the beams always

occurred abruptly, when one of the inclined bending cracks suddenly developed into an unstable shear crack. The failure was violent because no alternative bearing mode was available.

The bearing mechanism of the shear capacity in such type of beams is complex. In building codes with regard to the ultimate shear stress v_u normally three dominating influencing factors are distinguished, namely the concrete compressive strength f_c, the longitudinal reinforcement ratio $\rho_l = A_{sl}/bd$ and the effective member depth. E.g. in the new Eurocode for Concrete Structures the expression

$$v_d = C \cdot k \, (100\rho_l \cdot f_{ck})^{1/3} \tag{3}$$

is given, where ρ_l is the longitudinal reinforcement ratio as defined previously, f_{ck} is the characteristic cylinder compressive strength of the concrete, C is a coefficient (advisory value 0.12) and k is a size factor according to

$$k = 1 + \sqrt{200/d} \leq 2.0 \tag{4}$$

with the effective cross sectional depth d in mm.

König et al (1993) showed on the basis of fracture mechanics considerations, that it would be scientifically more correct to directly involve the fracture energy G_F and the concrete tensile strength f_{ct}. This can be done by introducing the characteristic length l_{ch}, which is defined as

$$l_{ch} = E \cdot G_F / f_{ct}^2 \tag{5}$$

The mean ultimate nominal shear strength can then be formulated as:

$$v_u = C \cdot f_{ct} \cdot \sqrt[3]{(l_{ch} \rho_l / d)} \tag{6}$$

Such a formulation will prove its value in future, since many innovative concretes will be introduced, for which the compressive strength, the tensile strength and the fracture energy cannot be related anymore by straightforward expressions.

E.g, if a concrete gets artificial ductility by steel fibres, a formula of the type given by Eq. 5/6 may be a better basis than that of Eq. 4.

A very interesting new area for which well based expertise in the field of shear is asked, is the determination of the shear capacity of existing concrete bridges. Recently in The Netherlands investigations on existing bridges were carried out, with the aim to determine the actual shear bearing capacity. This was necessary because the Dutch government had decided to extend a large number of existing highroads with an additional lane. This includes, that also the bridges have to be extended with an additional lane, which

Figure 7. Shear test on strip, sawn out of an existing slab viaduct, with an unusual relation of concrete tensile to concrete compressive strength.

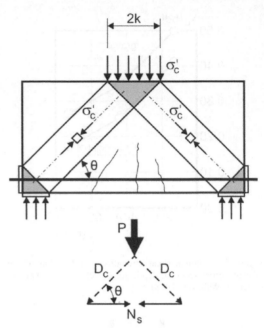

Figure 8. Strut and tie model for short deep beam loaded by a concentrated load at mid-span.

raises the question of the actual bearing capacity. Tests on drilled cylinders showed, that the centric tensile strength of the concrete was unexpectedly low. For a compressive cylinder strength of 60–65 MPa a centric tensile strength in the order of only 1–1.5 MPa was found. It is clear that here new considerations with regard to the shear capacity have to be developed. Fig. 7 shows a shear test on a strip, sawn out of an existing slab bridge, investigated in the laboratory of TU Delft.

2.2 Short deep beams loaded in shear

Detailing of concrete structures has always been a point of concern, because damage to concrete structures could often be traced back to insufficient knowledge of the designing engineer with regard to appropriate detailing. Especially the work of Schlaich and Schäfer (1987), developing strut and tie models for detailing regions, contributed to the development of rational detailing models. The principle of such models is the following:

– orient the compression struts to the compressive stress trajectories in the uncracked loading stage
– apply tensile ties to obtain an equilibrium system, considering that the shortest tensile ties offer the best solution, since they guarantee the smallest deformation and the lowest degree of redistribution of forces
– connect the compression struts and the tensile ties by appropriate nodes, striving at confining the concrete at the nodes.

Fig. 8 gives and example of the application of the strut and tie model to a short deep beam with a load at midspan. The model shown is, in terms of the theory of plasticity, a lower bound solution if the allowable stress

in the components is no where exceeded. The maximum allowable stress in the concrete strut is mostly expressed by the relation

$$\sigma_{cu} = \nu \cdot f_c' \qquad (7)$$

where ν is the so-called effectivity factor, taking into account the effects of for instance an unequal stress distribution in the cross section of the compressive strut ($\nu < 1$). However, although this representation is very transparent at first sight, it is a simplification of a mechanism which is considerably more complex in reality. A shortcoming which is detected quickly is that the shear bearing capacity, if the longitudinal reinforcement does not yield, is equal to

$$V_u = \nu \cdot f_c' bk \qquad (8)$$

where k is the length of the bearing plate and b is the width of the specimen. So, the bearing capacity is independent of the angle θ. This is contrary to all experimental findings, which confirm a strong dependence on the angle θ (or the ratio a/d). An important question is further whether in such a case a size effect occurs or not. The simplified model, shown in Fig. 8, suggests that this is not to be expected. In order to investigate this question a series of 5 tests was carried out on short deep beams with different dimensions (Lehwalter, 1988). The specimens were geometrically similar, but differed in size, where the cross sectional depth

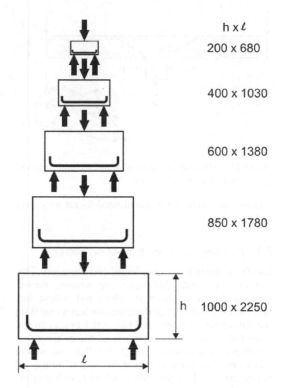

h x ℓ
200 x 680
400 x 1030
600 x 1380
850 x 1780
1000 x 2250

Figure 9. Shear tests on small beams with the same geometry but different size, Lehwalter, 1988.

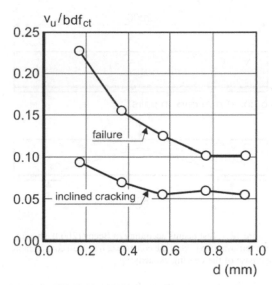

Figure 10. Relative shear stress at inclined cracking and at failure for short deep beams with a/d = 1.

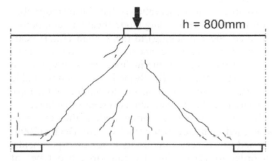

Figure 11. Failure of compression strut at the junction with the loading area.

varied from a lowest value of 200 mm to a highest value of 1000 mm, Fig. 9.

The concrete strength was about 20 MPa and the longitudinal reinforcement ratio ρ_l was in all cases about 1.1%, which means that they were all over reinforced in bending. The behaviour was remarkable in a number of respects. Like in the case of slender beams, the progress of the cracks in the first stage of loading strongly depended on the size of the beam. In the largest specimens the development of the crack pattern was much faster than in the smaller specimens, in analogy with the slender beams (Fig. 6). However, in this case the occurrence of an inclined shear crack did not lead to failure, contrary to the slender beams. The explanation to this is that a redistribution of forces is possible to another bearing mode, in a qualitative sense resembling the model shown in Fig. 8. Fig. 10 shows that there is still a significant reserve capacity after inclined cracking. The figure shows that, with regard to the bearing capacity, a significant size effect exists, in spite of the different bearing mode. With regards to crack propagation and failure two further observations are done, see fig 11:

– The inclined crack intersects the compression strut near to the loaded area

– Failure occurs at the upper part of the compression strut, adjacent to the node area.

This suggests that a static system with hinges, like shown at the bottom of fig. 8 is actually not a good description of the real behaviour.

In order to refine the description of the behaviour Asin (2000), introduced a damage localization zone. Recalculating Lehwalter's tests he found that a reasonable agreement could be obtained by introducing a localization zone with a length s = 200 mm and a failure strain of $4 \cdot 10^{-3}$ at the compressed side of the localization area. The localization area is subjected to a combination of normal force and bending, due to the fixed node between the struts just below the load introduction area. With this relatively simple model, which was not worked out further, it was shown that a reasonable description of the behaviour is obtained,

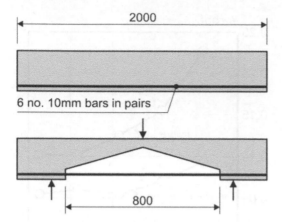

6 no. 10mm bars in pairs

Figure 12. Test beams according to Beeby (2000), showing that removal of a part of the concrete can cause a considerable increase of the bearing capacity.

including the size effect. It would be worthwhile to see if an improved formulation, for instance based on Markeset's CDZ-model, would further improve the agreement.

An interesting observation is that, during redistribution from the one bearing mode (beam action, before inclined shear cracking) to the second (final) bearing mode (strut and tie model), the compression struts are first damaged by the inclined crack, progressing through the strut, and finally loaded in an unfavourable way (non-uniform loading, due to bending at the top). Considering these observations, it seems that the redistribution from the one bearing mode to the other goes along with damage of the final system. Beeby (2000) carried out a very interesting series of experiments, in order to verify this. He compared two types of beams. The reference type of beam is a traditional beam, in which a process of redistribution, as sketched before, can occur. The second type of beam was made in the same mould and had the same reinforcement but much of the tension zone was blocked out with polystyrene. The objective was to induce the beam to behave as a truss rather than as a beam. Fig. 12 shows the beams, which had a compressive strength of around 50 MPa. The longitudinal reinforcement ratio was 0.87% and the a/d ratio was 2.5. The regular beams A and B failed in shear at loads of 135 and 123 kN respectively. Beam C, where a triangular part of the concrete was removed, failed at a load of 202 kN. So, the removal of a part of the concrete generated an increase of the bearing capacity of about 60%. The failure of beam C was not shear but, finally, a bond failure over one support. Nevertheless, the beam was carrying a load at failure in excess of the calculated flexural strength and so the reinforcement was likely to have started to yield. The reason of the much better behaviour of beam C is that, by adapting its shape, the

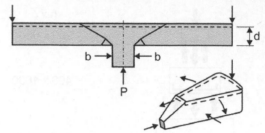

Figure 13. Mechanical model by Kinnunen and Nylander (1960) decribing punching shear.

member is much better conditioned to act as a strut and tie system.

2.3 Punching shear and fracture mechanics

Punching shear is a complicated phenomenon. This explains that still PhD-theses are written, further exploring the mechanism of sudden and violent failure. The only models that describes the behaviour from the formation of radial and tangential cracks to the final failure of the concrete around the column was developed by Kinnunen et al. (1960). With this model Kinnunen succeeded to model the role of nearly all parameters, such as the radial and tangential reinforcing ratio, the punching shear reinforcing ratio, the concrete strength and all geometrical effects. Unfortunately the model did not include any size effect. The model was based on the development of a kinematic mechanism, shown in Fig. 13. The only weakness was the failure criterion: the structure was supposed to fail when the tangential strain ε_{cT} in the bottom of the slab around the column, reached a limit value, determined experimentally.

For B/d $\leq$ 2

$$\varepsilon_{cT,} = 0.0035(1-0.22 \, B/d) \qquad (9)$$

and for B/d > 2 as

$$\varepsilon_{cT,} = 0.0019 \qquad (10)$$

where B is the diameter of the cross section of a circular column and d is the effective slab depth.

36 Years later Hallgren (1996) found the key to the complete description of the phenomenon, completing Kinnunen's model with a fracture mechanics component, for which he based himself on the Multifractal Scaling Law by Carpinteri and Chiaia, (1995). The structural size, which is the main variable of the law, was set equal to x, the depth of the compression zone. The Multifractal Scaling Law gives:

$$G_F = G_F^{\infty}(1 + \frac{\alpha_F \cdot d_a}{x})^{-1/2}$$

where G_F^∞ is the fracture energy for an infinitely large structural size, d_a is the maximum aggregate size and α_F is an empirical factor, which was determined to be 13. A FEM analysis showed that the radial length of the zone with tensile strains at about $r = B/2 + y$ is approximately equal to the compression zone depth x (y is the length of the truncated wedge in the model of Kinnunen). After an extensive FEM analysis a modified criterion for the ultimate tangential strain was formulated, according to:

$$\varepsilon_{cTu} = \frac{3.6 \cdot G_F^\infty}{x \cdot f_{ct}}(1 + \frac{13 \cdot d_a}{x})^{-1/2}$$

where x is the depth of the compression zone and f_{ct} = concrete tensile strength. With this formulation the size effect could be well described, which made the model complete.

2.4 Rotation capacity and fracture mechanics

The rotation capacity of plastic hinges in reinforced concrete slabs is a tool that allows the designer to make optimum use of the potential of redistribution of bending moments by virtue of the yielding capacity of reinforcing steel. In slabs with low reinforcement ratios, fracture of the steel limits the rotation capacity. For higher reinforcement ratio's the rotation capacity is reached by crushing of the concrete in the compression zone of the element. Considering the effect of localization in compression, see also Fig. 2, Hillerborg, 1989, supposed that the rotational capacity of plastic hinges would be size dependent in the latter case. In order to investigate this phenomenon, a series of tests was carried out in order to investigate this hypothesis, Bigaj (1999). Tests were carried out for two longitudinal reinforcement ratio's: 0.28% and 1.12%. For every reinforcement ratio a series of three beams was tested. The beams were geometrically similar but of different size. The cross sectional depths were 90 mm, 180 mm and 450 mm respectively. Fig. 14 shows the final crack patterns for the beams with a reinforcement ratio of 1.12%. The plastic rotation as observed in the tests is shown in Fig. 15.

The analysis of the results showed, that the composition of the longitudinal reinforcement is even more important than the effect of localization of damage in the compression area. The tensile reinforcement is hard to correctly scale, because it is not possible to proportionally increase both the bar diameter and the bar distances at the same time. The bond properties of the bars determine the crack distance and as such the number of cracks that is involved in plastic deformation. Furthermore the bar distances and the concrete cover play a role, because of the occurrence of eventual splitting cracks around the reinforcing bars and

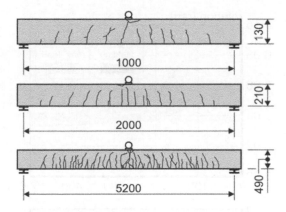

Figure 14. Crack patterns for rotational capacity tests on beams with different depths and a reinforcement ratio of 1.12%, after Bigaj, 1999.

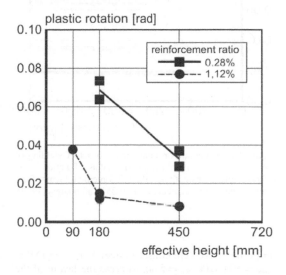

Figure 15. Observed plastic rotations at peak load versus effective depth of the beams (Bigaj, 1999).

their influence on the yielding length of the bar. Moreover the type of steel (f_u/f_y and ε_u) plays a role. In order to fundamentally investigate the behaviour of plastic hinges, at first an extensive study was carried out with regard to the bond of reinforcing bars, including the yielding stage of the steel. This resulted in complete bond-slip relations, including the effect of steel strain and magnitude of the concrete cover, for low, medium and high strength concrete. The relations observed were implemented in a plastic hinge model, where for the compression zone Markeset's CDZ-model was used. A substantial number of simulations was carried out. Fig. 16 shows one of them. The steel

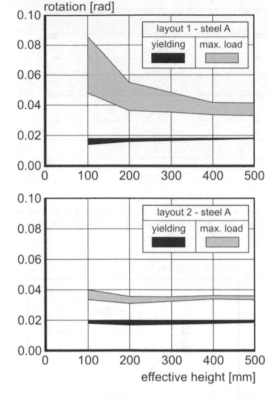

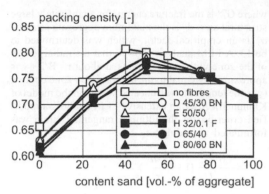

Figure 17. Effect of the sand content and the type of steel fibres (at 1.5 Vol.%) on the packing density (fibre types: first index: L_f/d_f: second index: L_f), according to Grünewald, 2004.

3 INCREASING THE DUCTILITY OF CONCRETE BY ADDING STEEL FIBRES: ACHIEVEMENTS AND ANOMALIES

3.1 Optimizing the properties of fiber concrete by defined performance mix design

The brittleness of concrete and its relatively low tensile strength have always required special design skill in order to create non-brittle structures. Providing fibers to concrete makes the material much more ductile. Looking back to three decades of development in fiber concrete, it can be concluded that the major steps in the development have been made during the last few years. At first it has finally been understood that a fibre concrete should be "designed" in order to get appropriate properties. Whereas in the past fibers were simply added to conventional concrete mixtures, without further considerations, now it has been understood that fibers and the aggregate particle skeleton interact and that an optimum combination of those components has to be found in order to get the most appropriate mixtures. Fig. 17 shows an experiment carried out by Grünewald (2004). He mixed fibres with combinations of aggregate particles, consisting of different volumes of sand and gravel (no other components like cement or water were added). He then vibrated this mixture until the highest packing density was obtained. Fig. 17 shows the maximum packing density values, obtained for volumes of 1.5 Vol.% of different steel fibres, as a function of the ratio sand to gravel. It can be seen that by adding fibers, keeping the sand to total aggregate ratio constant, the packing density decreases. This means basically, that the fibers disturb the aggregate skeleton. The maximum packing density moves to higher values of the ratio sand/total aggregate. This shows that, to compensate for the effect of the fibers, the mixture composition has to be adjusted

Figure 16. Calculated rotation at the onset of yielding of the reinforcement and at maximum load as a function of the effective member depth and the reinforcement layout, for a longitudinal reinforcement ratio of 0,25%, after Bigaj (1999).

properties were kept constant (steel A: $f_y = 550$ MPa, $f_{max} = 594$ MPa, $\varepsilon_u = 5\%$), whereas the layout of the tensile tie was varied:

– Layout I: constant bar diameter $d_s = 16$ mm, bar spacing adjusted to the actual member size and to the required reinforcement ratio
– Layout II: constant bar spacing $s = 50$ mm, bar diameter adjusted to the actual member size and to the required reinforcement ratio

Fig 16, simulating the behaviour for a reinforcement ratio of 0.25% and the two reinforcement layouts, shows that the layout determines whether there is a size effect or not. The figure shows for any combination a band, rather than a line. This is to take account of the probabilistic aspect of crack formation: the number of cracks contributing to the rotation capacity depends on the accidental position of the cracks, adjacent to the load. Whether one or two cracks contribute means a significant difference.

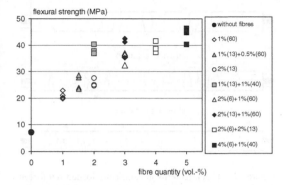

Figure 18. Flexural tensile strengths of various types of hybrid fiber concrete, Markovic, 2006.

by increasing the content of grains that are relatively small compared with the fiber length (cement, fillers or small aggregate grains).

Grünewald showed that, based on such type of considerations, high performance self compacting fiber reinforced concrete mixtures with fiber volumes over 125 kg/m³ are possible.

Another interesting technology is that of combining short and long fibers in the same concrete mixture. Markovic (2006) showed the potential of this method. The short fibers are activated as soon as microcracks occur, so that the behaviour remains quasi elastic until the formation of macrocracks. Then the long fibres take over. Fig. 18 shows, that flexural tensile strengths of 40 MPa and more can be achieved by appropriate combinations of fibers. The figure shows moreover, that by a suitable choice of the fibers optimization is well possible: with 1 Vol.% of fibers 13 mm and 1 Vol.% of fibers 40 mm, about the same flexural tensile strength is obtained as with 4 Vol.% of fibers 6 mm in combination with 1 Vol.% of fibers 40 mm.

3.2 Measuring the mechanical properties of high performance fiber concrete

A distinct advantage of those new, high performance fiber concretes is that they show a pronounced hardening behaviour, which makes them to an appropriate material for structural design.

Any material that is used for structural design purposes should be qualified in terms of mechanical behaviour. The question which is the most appropriate test, has been a subject of intensive discussions. RILEM defined a standard tests, by which the load deflection relation is established on a specimen like shown in Fig. 19.

This load deflection relation is the basis for the stress–crack opening relation. Since it is known that the result of the RILEM bending test is very sensitive to the production of the specimen and the execution

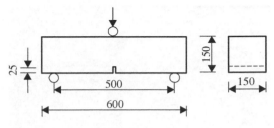

Figure 19. Rilem standard test for fiber concrete.

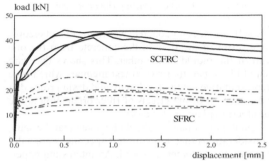

Figure 20. Test results of three point bending tests on self compacting concrete and conventional concrete, with the same concrete strength and the same amount of fibers, according to Grünewald, 2004.

of the test, recommendations are given on how to fill the mould and to compact the specimen. Filling the mould is done in a certain, well described sequence: at first a batch of concrete is placed in the middle of the mould, and subsequently at the two ends. After filling the mould in this way, the concrete is compacted with a prescribed intensity and time. However, modern high performance fiber concretes have, as an advantage, mostly a very good workability. Many of them are even self compacting. This means that filling the mould in the RILEM way is not an option.

Grünewald, 2004, showed that considerable differences are obtained between two mixtures C55/67, which both contained 60 kg/m³ hooked end steel fibers Dramix 80/60 BP. One was a traditional mixture, which allowed filling of the mould in the RILEM way, and one was self compacting. In the last case the concrete was poured into the mould from one side. Fig. 20 shows the load deflection relations obtained: the difference is remarkable. The bending capacity of the self compacting concrete is about twice as large as that of the conventional fiber concrete with the same compressive strength. Moreover the scatter in the results is smaller. The analysis showed that there are two reasons for the better behaviour. The pull-out resistance of the fibers in the self compacting concrete is better, because of the better embedment of the fibers in the matrix. Furthermore the orientation of the fibers in

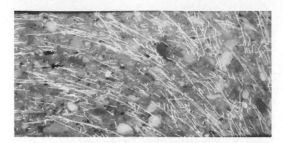

Figure 21. X-Ray photo of fiber orientation in tunnel element, cast with self compacting fiber concrete.

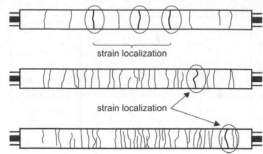

Figure 22. Crack patterns in centrically loaded reinforced tensile bars. From top to bottom 0, 0.8 and 1.6 Vol.% short steel fibers: (Shionaga, 2006).

the self compacting concrete was more advantageous, as a result of the flow of the concrete from the one end of the mould to the other. This shows that care should be taken to arrive at test methods which give representative results for design, and design rules that take account of the effects of fiber orientation at the production plant or at the site.

On the other hand, it should be noted that the same considerations that apply for test specimens are valid for structural elements as well. An interesting experiment was done with self compacting fiber concrete. A self compacting fiber mixture was used to cast a tunnel lining element. This curved element was cast through a window at the centre of the curved side. The concrete was delivered from a truck mixer: it flew through a half pipe into the mould for the tunnel element, were it spread until the whole mould was filled. After hardening cylinders were drilled at various locations and in various directions. From those cylinders discs were sawn, which were subjected to X-ray photography.

Fig. 21 shows clearly that there is a strong orientation, due to the flow of the concrete. The fiber efficiency varied between the extreme values 0.24 and 0.91. So, the fiber efficiency may vary substantially throughout the specimen. Up to now this has not been considered in building codes. Therefore general provisions should be formulated to take account of this phenomenon. On the other hand profit should be taken from the possibility to orient fibers intentionally.

3.3 Combining steel fibers with traditional reinforcement

An interesting possibility for structural design is to combine fibers with traditional reinforcement. In this way very thin walled structures can be obtained. The rebars take care of the main bearing function, whereas the fibers give the concrete a large ductility, which is favourable in any region where stress concentrations can be expected (anchorage regions, regions with lapped bars, impact loading, vibrations). Moreover the fibers contribute to an advantageous cracking behaviour, with fine cracks at small distances. An

important question is still which is the best ratio of fibers to classic steel. In order to answer this question at TU Delft a research program has been started in which concentric tensile tests on prismatic concrete bars, reinforced with different ratio's of classic steel and fibers, are carried out. The concrete types investigated had compressive strengths of 130 and 180 MPa respectively. The fiber content was 0 Vol.%, 0.8 Vol.% and 1.6 Vol.%, which corresponds to 0, 60 and 120 kg/m^3 steel fibers. Fig. 22 shows the cracking patterns that were obtained. The role of the fibers on the cracking pattern is clearly recognizable.

However, also another interesting phenomenon is observed: if fibers are added, the plastic deformation localizes finally in one single crack, whereas the specimen without fibers shows localization in several cracks. This may have consequences for cases where design for ductility is concerned, as shown in the following case.

3.4 Means adding fibers to a concrete always an improvement of the structural behaviour?

It was argued already before, that just adding fibers to a concrete does not necessarily mean an improvement of the material properties. A good design of the material should make sure that fibers indeed generate improved properties in strength and ductility. However the question can be raised if a well designed fiber concrete by definition improves the structural behaviour. In order to answer this question the phenomenon of rotation capacity is treated again, now in combination with fibers.

In Fig. 22 it was shown that the combination fibers and conventional reinforcement leads to strain localization in a single crack. The reason for this is that there is a certain scatter, both in the homogeneity of fiber distribution and in fiber orientation. So, if cracks have been formed, the capacity of transmission of forces across the crack varies from crack to crack, because of the scatter of the fiber capacity in the subsequent

Figure 23. Plastic hinge in plain concrete (top) and fiber reinforced concrete (bottom) (Schumacher, 2006).

sections. This explains the localization in one crack after yielding of the reinforcing bars. A similar observation was done in recent research on the rotation capacity of reinforced concrete hinges, with and without fibres (Schumacher, 2006), Fig. 23 shows the difference between concrete with and without fibers.

The figure shows for the plain concrete a more brittle behaviour of the compression zone, but a localization of steel yielding in more cracks. The fiber reinforced concrete shows a more ductile compression zone but localization of bar yielding in one crack. The rotation capacity of fiber reinforced concrete was finally smaller than that of plain concrete.

4 CONCLUSIONS

1. Fracture mechanics still learns us a lot with regard to material and structural behaviour.
2. Initiatives should be taken to implement fracture mechanics parameters in structural design recommendations. This will be useful for the design of structures with defined performance materials in future.

REFERENCES

Asin, M., 2000, The behaviour of reinforced concrete continuous deep beams, PhD-Thesis, Delft University of Technology, The Netherlands

Bigaj, A.J.,1999, Structural dependence of rotation capacity of plastic hinges in RC beams and slabs, PhD Thesis, Deflt University of technology, The Netherlands

Bigaj, A., J, Walraven, J.C., Size effects in plastic hinges in RC members, Heron, Vol. 47, 2002, No. 1, pp. 53–75 Beeby, A.W., The behaviour of reinforced concrete beams without shear reinforcement in shear, Draft paper, School of Civil Engineering, The University of Leeds, Aug. 2000.

Carpinteri, A., Chiaia, B., Multifractal Scaling Law for the-Fracture Energy Variation of Concrete Structures, Fracture Mechanics of Concrete Structures, ed. F.H. Wittmann, Proc. of the Second International Conference FRAMCOS 2, ETH Zürich, Switzerland, July 25–28, Vol. 1. pp. 581–596

Grünewald, S., 2002, Performance based design of self compacting fibre concrete, PhD Thesis, Delft University of Technology, The Netherlands

Hallgren, M., Punching shear capacity of reinforced high-strength concrete slabs, Trita-Bkn Bulletin 23, 1996, Kungl Tekniska Högskolan, Institutionen för Byggkonstruktion, Stockholm, PhD-Thesis

Hillerborg, A., Modeer, M., Petterson, P.E., 1976, Analysis of crack formation and growth in concrete by means of fracture mechanics and finite elements, Cement and Concrete Research, Vol. 6m Pergamon Press Inc. pp. 773–782

Kinnunen, S., Nylander, H, 1960, Punching of concrete slabs without shear reinforcement, Transactions of the royal Institute of Technkology, No. 158, Stockholm,

König. G., Shear behaviour of longitudinally reinforced concrete members of HSC, Proceedings of the JCI International Workshop on Size Effect in Concrete Structures, Oct.31–Nov. 2, 1993, Sendai, Japan, pp. 63–74

Lehwalter, N., 1988 The bearing capacity of concrete compression struts in strut and tie models for the example of short deep beams, PhD Thesis, Darmstadt University of Technology, Germany

Markeset, G., 1993, Failure of concrete under compressive strain gradients, PhD Thesis, The Norwegian Institute of Technology, Trondheim

Markovic, I., 2006, High performance hybrid fiber concrete: development and utilization, PhD-Thesis, TU Delft

Mier, J.G.M. van, 1996, Fracture Processes in Concrete, CRC Press, ISBN 0849-391237

Petersson, P.E., 1981, Crack growth and development of fracture zones in plain concrete and similar materials, PhD Thesis, Technical University of Lund, Sweden

Schlaich, J., Schäfer, K., Jennewein, M. 1987, Towards a consistent design of reinforced concrete structures, PCI Journal 32 (3): pp. 74–150

Schumacher, P., 2006, Rotation Capacity of Self-Compacting Steel Fiber Reinforced Concrete, PhD-Thesis, Delft University of Technology

Shionaga, R., 2006, Structural behaviour of high performance fiber reinforced concrete in tension and bending, Proceedings of the 6th international PhD Symposium in Civil Engineering, Zürich, Aug. 23–26, 2006

Walraven, J.C., Lehwalter, N., Size effects in short bemas loaded in shear, ACI Structural Journal, Vol. 91, No. 5, Sept.–Oct. 1994, pp. 585–595

Walraven, J.C., 1978, The influence of cross sectional depth on the shear strength of normal weight and light weight beams without shear reinforcement

Leung, without, main reinforcement. Journal, Dr. flexure, Journal of Civil Engineering, The University, Intl. Eeds, Aug. 2000.

Kompuntu, A., Chapre, B., Mathematical modeling Law for the flexure, Energy, Structural, concrete Structures, concrete Structure, of Concrete Structures, (ed.) H.W. Reinhardt, Proc. for the Second International Conference, FRA, A. CPN, J.I. Taerwe, Eade, Ghent, July 25, vol. 1, pp. 451-460.

Finite, Peoky, 2002, Experimental based design structure, journey, flute concrete, PhD Thesis, Delft University of Technology, The Netherlands.

Kullar, J. M., Fracture, sheet, energy of reinforced high strength concrete edges, Trin-Dev, Studies, TU, 1999, Kunzl, Test, on, flag, beam, Mechanisms for the Experimental, Stockholm, Royal Inst.

Thibaseen, A., Muchete, A., Patterson, P.L., 1996, Total strain restraint analysis in placement of monotonic structure incongruity and Stress elements, Fracture and Concrete Research, Vol. 6th, Pergamon Press, Sweden, 771-782.

Vonmingen, Reidyman, H. Proff, Proshtling of concrete slabs without shear, reinforcement, Heron, Journ., of the Royal Institute, Technology, Pat. 5th, Stockholm.

Weigle, P., Structure, verum of Bridge, analytically, explored concrete members of HSC, Proceedings of the 3rd International Workshop on Shear Effect in Conc. for Structures, Or, H. Noro, 1993, Sendai, Japan, pp. 67-67.

Labouthe, H., 1998, The Exp. type, type, of concrete compression, stress, in, structured, the, stress for, the, exp. type of short bar, beam, child, Thesis, Darmstadt, University of Technology, Germany.

Flutche, P., New, Failure, of, sheer, under, compressive concrete, columns, PhD Thesis, The New Reg. University of Technical Un. University.

Moerman, Emm, H., Shear, under stress and flux concrete demos, concr. and structures, FRA, Eeds, TU Delft.

Nori, Eduard, 1974, Flexure, Freeman, Francisco, San CPN.

Nori et al., 1981, Crack growth, mechanism of fracture, and cause, in placement, and sensor, sensed, PhD Thesis, Technical Univ, of Delft, Delft, Royal.

Schlaich, J., Schafer, K., Jennewein, M., 1987, Towards a consistent design of structural concrete structures, PCI Journal, 32 (3) pp. 74-150.

Schumaker, P., Plain concrete placement, of, structure plate, reinforcement, concrete, conc., 6th Inter, Delft, The Netherlands.

Surface, and stress Structural in Concrete, Fracture, structure, fracture, of the structure for a concrete structure, frac., of, the stress, placement International FRA, Structure en Tech.

Vakhtoor, B.I., Subotto, R., New, effect, in, steel, column, to, concrete design, M. Structural Journal, Vol. 40, No. 5, 1994, pp. 555-575.

Wakuoka, J.C., 1977, The influence of main structure upon the shear strength of normal weight and light weight concrete beams, without shear reinforcement.

sections. This explain, the localization in one crack, after, yielding of the, reinforcing, bar. A similar obser-vation was done in, recent research on the, behaviour, capacity, of reinforced concrete beams, with and without, fibre, reinforcement, Street, Fig. 2. Shows the, difference, between, concrete, with, and, without, fibre. This figure, shows that the, plain, concrete, a, more, brittle, behaviour, of the, concrete, can, also, be shown, behaviour, if fibre, cracks. To, reduce, cracks. The, fibre reinforced, concrete, shows, a, more, ductile, compres-sion, zone, but, localization, of, the, yielding, in, one, crack. The, rotation, capacity, of, fibre, reinforced, concrete, was, finally, reached, in, the, test of, plain, concrete.

4 CONCLUSIONS

1. Fracture, mechanics, still, is, not, taken, for, with, it, and, in, material and, structural, behaviour.

Influences, should, be, taken, to, make, sure, fracture, mechanics, parameters, in, structural, design, recom-mendations. This, will, be, useful, for, the, designing, practices, point, of, field, reinforcement, movements, in, future.

REFERENCES

Van, M., The, new, et al., concrete, reinforcement, of, concrete, steel, design, Stuttgart, placement, The, University, of, Toughness, of, the, M., struct.

Bakir, A.I., 1993, Structural, design, model, structure, aggre., concrete, concrete, shear, based, on, fracture, aggr., University of, technology, The, Netherlands.

Reinhardt, H.W., Weerheijm, J.C., Structure, size, in, plate, in, Mechanics, struct., Vol. 12, 2002, 555, pp.

Reinhardt, A.W., Flow, Reinforced, Reinforced, concrete.

*Fracture Mechanics of Concrete and Concrete Structures – High-Performance Concrete,
Brick-Masonry and Environmental Aspects – Carpinteri, et al. (eds)
© 2007 Taylor & Francis Group, London, ISBN 978-0-415-44617-4*

Microtensile testing and 3D imaging of hydrated Portland cement

P. Trtik & P. Stähli
Institute for Building Materials (IfB), ETH Zurich, Zurich, Switzerland

E.N. Landis
Department of Civil and Environmental Engineering, University of Maine, Orono, Maine, USA

M. Stampanoni
Swiss Light Source, Paul Scherrer Insitute, Villigen, Switzerland

J.G.M. van Mier
Institute for Building Materials (IfB), ETH Zurich, Zurich, Switzerland

ABSTRACT: Portland cement has an extremely complex microstructure that defies simple characterizations. While electron microscopy has provided valuable insight on microstructure development, the 2D nature of the data and the required specimen preparation steps continues to provide controversy as to what is real and what is artifact (see for example Diamond (2004a, b) and Scrivener (2004)). Here we present the results of synchrotron-based microtomographic investigations of 130 µm diameter cement specimens subject to uniaxial tension. With the quantitative 3D image analysis of the tomographic data we are able to isolate the different phases, and identify the relationship between the phases and the resulting fracture path. In particular, we see that the crack path shows little if any preference for phase, undermining the notion that at small scales, unhydrated cement grains act as hard aggregates in hydrated cement matrix.

1 INTRODUCTION

At micro-scale cement is composed of various hydrate phases (Calcium Silicate Hydrates (CSH) of various densities, etc.), un-hydrated cement grains (built-up from the different types of clinkers), Calcium Hydroxide (CH), (latently hydraulic) fillers like micro-silica at [nm] and [µm] size, pores of varying size and water. These phases arrange themselves in a random array. In assessing mechanical properties of hydrated cement, the heterogeneity leads us to use test methods that average over representative volumes, or, alternatively, extremely local test methods should be used at scales where the constituting phases can be considered homogenous.

Traditionally, there has always been a trade-off in microstructural measurement techniques. Electron microscopy, in its various forms, has been applied to problems of cement microstructure for more than 25 years (Scrivener (1984)). Especially when combined

with energy dispersive X-ray spectroscopy (EDX) (Xu and Sarkar (1991)), insight into the distribution of phases has been invaluable. However, the 2D nature of the data means that important structural information is lost. Recent advances in synchrotron-based X-ray microtomography (SRμCT), see Stampanoni et al. (2002), have allowed us to produce 3D images at a spatial resolution beginning to approach that of the electron microscope. In addition, SRμCT allows the samples to be tested non-invasively in an ambient environment.

Figure 1 illustrates the exploitation of the different techniques. Figure 1a shows a micrograph of cement paste that clearly shows not only distinction between the hydrated (grey) and unhydrated (white) cement phases, and pore space (black), but subtle variations within the unhydrated cement particles are a result of the different compounds that make up the cement clinker. Figure 1b shows a representative slice from a 3D microtomographic image of a similar cement paste. The image also presents the hydrated and unhydrated cement phases, as well as pore space at spatial resolution of 0.7 μm. However, through simple threshold-based image segmentation, we are able to quantify and visualize the distributions of phases in three

Pavel Trtik moved in 2006 to the Institute of Mechanical Systems at ETH-Zurich; Eric N. Landis was on sabatical leave from the University of Maine, and collaborated in the experiments presented in this paper during his stay at IfB.

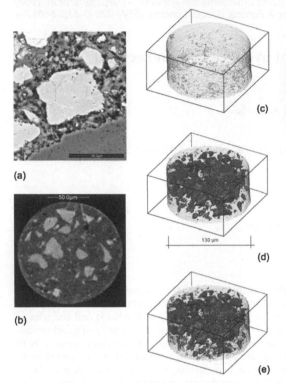

(a)

(b)

(c)

(d)

130 μm

(e)

Figure 1. Comparison of ESEM and SRμCT: (a) Micrograph of the polished surface of hydrated Portland cement. (b) Axial slice from 3D SRμCT reconstruction of a cylindrical specimen of hydrated Portland cement. 3D-reconstruction of voids (c), unhydrated cement grains (d), and the relationship between the two (e) in part of the volume of 130 μm-diameter specimen.

dimensions. Voids (Figure 1c), unhydrated cement grains (Figure 1d), and the relationship between the two (Figure 1e) are all visible in the volume.

In the work presented here, microtomographic imaging was exploited to examine fracture characteristics of cement pastes at the micron scale. The important issue to be addressed is tensile fracture under a properly defined state of stress. This approach is counter to the current popularity of indentation testing (Shuh et al. (2005), Velez et al. (2001) and Constantinides and Ulm (2004)), which while suitable for homogeneous materials or homogeneous phases, appears to be ill-suited for materials like cement and concrete, where the size of indents may be at the scale of material disorder. The high stress concentration of the indenter can lead to compaction of pore space, pop-out of material at indentation edges, and a continuously changing state of stress due to local material changes. Tensile testing results tend to be easier to interpret. Following extensive past experience in the field of tensile testing of concrete and rock at the macro-scale (van

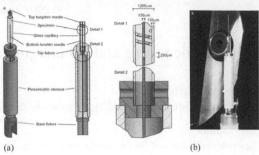

(a)   (b)

Figure 2. In-situ SRμCT microtensile testing device: (a) Schematic view of the microtensile device based on piezo-electric actuation, and (b) Photograph of the device in front of the SRμCT detector.

Vliet and van Mier (2000), using sample sizes in the range of 50–1600 mm), additional insight was sought at a scale 1000 times smaller. At this scale cement is highly heterogeneous, with "aggregates" consisting of approximately 5–30 μm unhydrated cement grains. This can be contrasted with the concrete samples tested at the macro-scale where aggregates are in the range of 4–32 mm. Of interest here is the preference of the crack growth towards different phases, and whether there are weak interfaces to direct the crack growth.

2   TEST PROCEDURE

In order to test tensile specimens at a scale approaching 100 μm, a small loading device was built based on piezoelectric actuation. As illustrated in Figure 2a, a 15 mm tubular piezoelectric element was mounted to a base fixture. Above this element was a 1.2 mm diameter glass capillary. Tungsten needles are glued to the top and bottom of the 130 μm diameter cylindrical specimen. This 'tungsten-cement' assembly is inserted through the capillary and piezo stack, and glued to the base of the device. As a final step, the top of the top tungsten needle is glued to the top of the capillary. When voltage is applied to the piezoelectric element its expansion results in a tensile loading of the 'tungsten-cement' assembly. Prior to assembly, the cement specimen was circumferentially notched at its centre to ensure that cracks would propagate through the material and not at the glue line where stress concentrations occur. The notch was made using femtosecond laser pulses while the specimen was rotating about its principal axis. Figure 2b shows a photograph of the entire tensile device mounted in front of the X-ray detector.

Details of the specimen preparation, the synchrotron X-ray computed microtomography and the applied image analysis techniques are given in the appendix at the end of the paper.

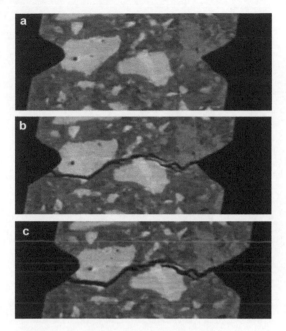

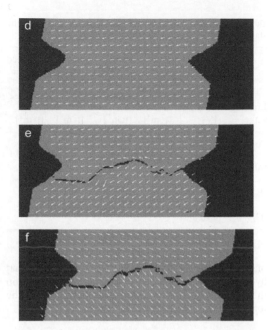

Figure 3(a–c). 2D crack reconstructions, displacement vector fields and crack bridging: (a–c) 2D vertical slices from 3D SRμCT reconstruction of the notched area of the same specimen for three subsequent loading stages ('A', 'B' and 'C').

Figure 3(d–f). Vector fields showing the displacement between the respective loading stage ('A', 'B' or 'C') and the unloaded state of the specimen.

3 FRACTURE PROCESS

Figures 3a to 3c provide visual comparison of tomographic data by showing the equivalent 2D cross-sections of one of the tested specimen at different stages of loading. The corresponding binary images are fitted with the results of the displacement vector field analyses. These fields provide information about the internal displacement fields between the respective loading stage and the specimen in an unloaded state. In stage 'A' the specimen is still in the elastic regime showing fairly uniform displacement vector field indicating in particular the occurrence of rigid body motion of the entire specimen between the unloaded stage and stage 'A' (see Figure 3d). The further stages ('B' and 'C') show cross-section of the crack exhibiting bridging and branching phenomena. The corresponding displacement vector fields for the loading stages 'B' and 'C' show a distinct difference in the direction of the displacement vectors in the regions above and below the crack due to the crack opening (see Figures 3e–3f).

Also, clear non-uniformity of the displacement vector fields in the vicinity of the crack is observed that can be possibly explained by the existence of strains in the vicinity of crack bridges. The thickness of some of the bridges and the widths of some crack branches is

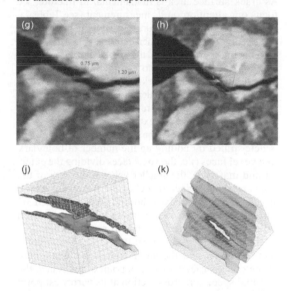

Figure 3(g–k). Detail of one of the vertical slices showing one of the bridging events with the sizes of the bridge and crack in micrometers; (h–k) 3D reconstruction of the bridge shown in Figure 3g.

apparently below 1 micrometer (see Figure 3g). Also, the size and shape of the crack bridges can be assessed and visualized in 3D (see Figures 3h–3k). As similar bridging mechanisms were observed in concrete

Table 1. Crack volumes and crack surface areas for the stages shown in Figures 3a–3c.

Loading stage	Voltage [V]	Calculation based on count of voxel faces		Calculation based on 3D triangular model	
		ΔV [μm^3]	ΔA [μm^3]	ΔV [μm^3]	ΔA [μm^3]
O	0.0				
		77	178	101	157
'A'	71.0				
		29798	23673	29805	18727
'B'	138.0				
		17182	2425	17161	394
'C'	290.0				

on the macroscale (van Mier (1991)), we suggest that hydrated Portland cement exhibits – when tested on this scale – quasi-brittle post-peak behaviour in the load versus displacement diagram.

4 CRACK SURFACE AREA

As crack surface area is fundamental in fracture, we focus our attention on its measurement (see Table 1). In our binary images (Figures 3d–3f), the largest zero (black) object is the specimen exterior. When the specimen becomes cracked, and the crack connects with the exterior, they combine into a single object. Thus we can follow changes in cracking by tracking the size of the largest object. Two independent measurements of the crack surface area were carried out (for identical threshold values between exterior and sample interior). While the larger value of the surface area measurement is determined by adding up the number of boundary free voxel faces (i.e. the voxel faces dividing the exterior and material), the smaller value is the area of the respective 3D surface model of the specimen. As triangular elements are basic building blocks of such 3D surface models, the ratio between the two presented values for the surface area has to remain below $\sqrt{3}$. At the same time we suggest that these values represent the upper and the lower bounds for assessment of the crack surface area. As a point of reference, the nominal specimen cross section at its narrowest point is 7700 μm^2.

5 THREE-DIMENSIONAL RECONSTRUCTION

Figure 4a shows the total volume of void in the specimen. The 3D reconstructions confirmed that no detectable fracture occurred outside the notch area. Even though it may appear that the majority of crack path shown in two dimensional cross-sections in

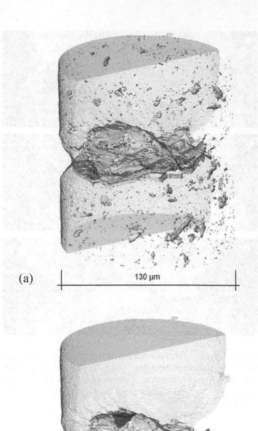

(a)

130 μm

(b)

Figure 4. 3D reconstruction of crack with respect to cement microstructure: (a) 3D reconstruction of the crack surface area and porosity in the microspecimen in the stage 'B' shown in 2D in Figure 3b; (b) 3D reconstruction of the major tensile crack in the microspecimen in the stage 'B' showing the areas (dark purple colour) in which the crack intersects the unhydrated cement particles.

Figures 3b and 3c followed the interfaces between the unhydrated cement particles and the hydration products, the full reconstruction of the crack surface (Figures 4c–4d) clearly proves that this is not the case in 3D. At the same time, the results suggest that the positions of the crack bridges seem to have no apparent preferential occurrence with regard to the cement microstructure.

It was recently shown that the fracture process in disordered quasi-brittle materials can be simulated by

1280

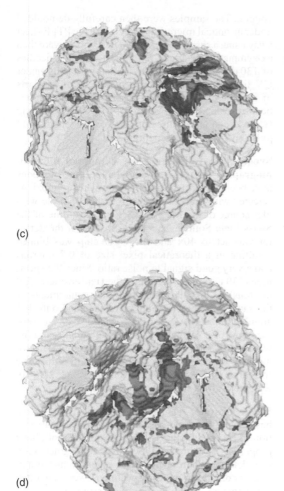

(c)

(d)

Figure 4(c–d). Top-down and bottom-up views of the 3D reconstruction of the crack shown in Figure 4b. The parts of crack surfaces contacting the unhydrated cement particles made as partially translucent. *Yellow* – top crack surface contacting hydration products, *Light purple* – top crack surface in contact with unhydrated cement particle, bottom crack surface in contact with hydration product (i.e. crack through interface of the hydration products and unhydrated particles), *Light pink* – both crack surfaces in contact with unhydrated cement particle, *White islands* – crack bridges connecting the two specimen halves.

simple models (Herrmann et al. (1989)), in which the fracturing occurs at those locations where stress over strength is most critical in the structure of the material. Such models effectively describe the behaviour of concrete and sandstone (Lilliu and van Mier (2003)) provided that a realistic material structure is indicated instead of a statistical distribution of element properties. Based on the presented results, it seems to be

likely that should the same modelling methodology be extended to smaller scales, i.e. fracture in hydrated Portland cement, the internal microstructural characteristics of unhydrated cement grains will have to be taken into account since they are in the same size range as other elements in the microstructure.

6 CONCLUSION

In this paper we present the results of some preliminary microtensile experiments. Synchrotron-based microtomography in combination with quantitative 3D image analysis revealed details of the fracture process in cement samples of extremely smalls size. Due to the constraints in the microtomography set-up, the experiments were not stable, and no load-displacement diagrams were measured. However, the experiments revealed what happened before and after critical crack growth. Since similar bridging and branching phenomena were observed comparable to those found in concrete at the meso- and macro-scale, it can be deduced that hardened cement paste is a quasi-brittle material (i.e. shows softening) at the scale of observation ([μm]-scale). Moreover, it can be concluded that the unhydrated kernels of partially hydrated cement grains do not act like rigid stiff sand and gravel particles in concrete at the meso- and macro-scale. Due to the clearly visible internal structure of some of the un-hydrated cement kernels, cracks were found to propagate through all phases of the cement. Indentification of the properties of all relevant phases is essential for developing micro-structural mechanics models like we did in the past for concrete (Lilliu & van Mier (2003)).

ACKNOWLEDGEMENTS

The authors would like to acknowledge contributions of Messrs Cornelius Senn, Jürg Inhelder, Martin Suter and Heinz Richner – technicians at the Institute for Building Materials, ETH Zurich. Mr Christoph Hauri, Institute for Quantum Electronics, ETH Zurich is thanked for help with femtosecond laser part of the experiment. Contributions of Dr Paul Beaud and Dr Amela Grošo (both Paul Scherrer Institute, Villigen) are also gratefully acknowledged.

REFERENCES

Constantinides, G. & Ulm, F.J. 2004. The effect of two types of C-S-H on the elasticity of cement-based materials: Results from nanoindentation and micromechanical modeling. *Cem. & Concr. Res.* 34, 67–80
Diamond, S. 2004a. Special issue on scanning electron microscopy of cements and concretes. *Cem. Conc. Comp.* 26: 917–918.

Diamond, S. 2004b. The microstructure of cement paste and concrete -a visual primer. *Cem. Conc. Comp.* 26: 919–933.

Gonzalez, R.C. & Woods, R.E. 2002. *Digital Image Processing*. Upper Saddle River, NJ, USA, Prentice Hall.

Herrmann, H.J., Hansen, A. & Roux, S. 1989. Fracture of disordered, elastic lattices in 2 dimensions. *Phys. Rev. B* 39: 637–648.

Lilliu, G. & van Mier, J.G.M. 2003. 3D-lattice type fracture model for concrete. *Eng. Fract. Mech.* 70: 927–941.

Schuh, C.A., Mason J.K. & Lund A.C. 2005. Quantitative insight into dislocation nucleation from high-temperature nanoindentation experiments. *Nature Mater.* 4: 617–621.

Scrivener, K.L. 2004. Backscattered electron imaging of cementitious microstructures: understanding and quantification. *Cem. Conc. Comp.* 26: 935–945.

Scrivener, K.L. 1984. *The Development of Microstructure during the Hydration of Portland Cement*. PhD-thesis, Imperial College, London.

Snigirev, A., Snigireva, I., Kohn, V., Kuznetsov, S. & Schelokov, I. 1999. On the possibilities of X-ray phase contrast micro-imaging by coherent high-energy synchrotron radiation. *Rev. Sci. Inst.* 66: 5486–5492.

Stampanoni, M., Borchert, G., Wyss, P., Abela, R., Patterson, B., Hunt, S., Vermeulen, D. & Ruegsegger 2002. High resolution X-ray detector for synchrotron-based microtomography. *Nucl. Inst. & Meth.* A 491: 291–301.

van Mier, J.G.M. 1991. Mode I fracture of concrete: Discontinuous crack growth and crack interface grain bridging. *Cem. Concr. Res.* 21: 1–15.

van Vliet, M.R.A. & van Mier, J.G.M. 2000. Experimental investigation of size effect in concrete and sandstone under uniaxial tension. *Eng. Fract. Mech.* 65: 165–188.

Velez, K., Maximilien, S., Damidot, D., Fantozzi, G. & Sorrentino, F. 2001. Determination by nanoindentation of elastic modulus and hardness of pure constituents of Portland cement clinker. *Cem. Concr. Res.* 31: 555–561.

Xu, A.M. & Sarkar, S.L. 1991. Microstructural study of gypsum activated fly-ash hydration in cement paste. *Cem. Conc. Res.* 21: 1137–1147.

APPENDIX: METHODS

Sample preparation. As a cylinder is the most convenient specimen shape from the tomography point of view, all the samples used were microcylinders produced from Portland cement CEM I 42.5 N with water/cement ratio equal to 0.33. Special coupons, made of polytetrafluorethylen (PTFE) sheets of 250 μm in thickness with apertures of 130 μm in diameter, were developed for their production. The samples were cast in such a manner that the whole coupon was submerged into the fresh cement paste, while being held in tweezers. Having been removed from the fresh paste, the coupons were then placed in a small plastic bag that was stored in a climate box at 25°C and 75% R.H. After curing the specimens for 6 days, the thin layers of material on the top and the bottom of the coupon was carefully removed, leaving the microcylinders of 130 μm in diameter and approximately 250 μm in height embedded in the PTFE coupon. The samples were then carefully demoulded under an optical microscope by cutting the PTFE sheet away using a sharp razor blade. The samples were then axisymmetrically glued onto short tungsten needles of 130 μm in diameter using a glue hardening under UV light. This needle later became one of the parts of the microtensile device. In this state a notch has been machined in the centre of the specimen using femtosecond laser pulses of 100 fs pulse length, 4 μJ pulse energy, 100 Hz repetition rate.

Synchrotron radiation x-ray computed microtomography (SRμCT). Tomographic investigations have been performed at an X-ray wavelength of 1 Å, selected with a Si111 monochromator from the wiggler source of the Materials Science beamline of the Swiss Light Source. The magnification of the detector was set to 40x and the CCD chip was binned resulting in a theoretical pixel size of 0.7 microns and a very good signal to noise ratio. Since the specimen (130 microns) was filling the field-of-view (720 microns) by less that one third, the requirements of the sampling theorem could be relaxed and only 401 angular projections have been acquired. With an exposure time of 2 seconds per image, the total scan time was approximately 20 minutes. Sample-to-detector distance has been minimized in order to reduce the formation of Fresnel fringes (Snigirev et al. (1999)), which can affect the spatial resolution.

Image Analyses. 3D image analysis consisted of three basic steps: threshold-based segmentation, identification of connected objects, and measurement of object properties. The threshold level used for segmentation was determined from the voxel intensity histogram. The histograms for these specimens were trimodal, consisting of voxel distributions representing void (darkest), hydrated cement phases (darker grey), and unhydrated cement particles (lighter grey). The threshold used to separate void space from solid is then taken as the local minimum between the two adjacent peaks on the histogram, minimizing the error associated with mislabelling a voxel, see Gonzalez & Woods (2002). The remaining steps are to identify and measure all the resulting void objects in the 3D space. This was done using an efficient connected components algorithm, which labels and measures the surface area and volume of each void object. Surface area is calculated by counting the number of free (not connected) voxel faces and multiplying by the unit voxel area. 3D displacement vectors were estimated using a block-matching method for intensity-based registration. Displacement vectors are calculated by finding the maximum cross correlation between a segment in the reference image (unloaded stage of the specimen) and a same-sized region in a larger search window in subsequent images (respective loaded stages of the specimen).

Fracture Mechanics of Concrete and Concrete Structures – High-Performance Concrete,
Brick-Masonry and Environmental Aspects – Carpinteri, et al. (eds)
© *2007 Taylor & Francis Group, London, ISBN 978-0-415-44617-4*

Time-dependent characteristics of a Self-Consolidating Concrete mix for PC concrete bridge girders

K. Larson, R.J. Peterman & A. Esmaeily
Dept. of Civil Engineering, Kansas State University, Manhattan, KS, USA

ABSTRACT: Creep and shrinkage test results of the proposed SCC mix for bridge girders (in the state of Kansas) are presented. Four bridge girders with an inverted tee profile were used to measure the creep and shrinkage in typical bridge girders. In two of the girder specimens the strands were tensioned to seventy-five percent of ultimate tensile strength. The strands of the other two girder specimens used were left un-tensioned to evaluate the shrinkage effect of the concrete. In addition, the fully tensioned girder specimens were used to determine the transfer length of the pre-stressing strand.

1 INTRODUCTION

Self-Consolidating Concrete (SCC) has rapidly become a widely used material in the construction industry. The primary reasons for the increased use of SCC are the economical advantages that SCC has over normal conventional concrete (NC). The Interim Guidelines for the use of Self-Consolidating Concrete in PCI Member Plants defines SCC as "a highly workable concrete that can flow through densely reinforced or complex structural elements under its own weight and adequately fill voids without segregation or excessive bleeding without need for vibration."

The Interim Guidelines also state in the commentary that the hardened properties of SCC may be different than those of NC. Where modulus, creep, and shrinkage are important design guidelines, it states that the mix should be properly investigated before using in design. When designing pre-stressed concrete members, these properties are very important for an accurate estimation of time-dependent losses.

2 PROBLEM STATEMENT

Departments of Transportation including the Kansas Department of Transportation (KDOT) would like to use SCC in pre-tensioned bridge members to enhance the aesthetics and improve consolidation in congested areas. However, before allowing the use of SCC in state bridges, KDOT needed to investigate the time-dependent losses of an SCC mix proposed by the local pre-caster. A previous study ("Bond Characteristics of an SCC Mix for Pre-stressed Concrete Bridge Girders," companion paper at the 2005 National

Bridge Conference) using SCC was conducted to investigate the bond between the strand and the concrete. At the time of this study the American Concrete Institute (ACI) and American Association of State Highway Transportation Officials (AASHTO) did not address the issue of members cast with SCC. KDOT wanted a thorough investigation of long term creep and shrinkage properties of a proposed SCC mix before allowing it to be used in state projects.

3 BACKGROUND

American Concrete Institute Building Code Requirements, AASHTO LRFD Bridge Design Specifications, KDOT, and The PCI Design Handbook, Sixth Edition all have slight differences in determining the losses of pre-stressed members. They are listed in the appendix section.

4 EXPERIMENTAL PROGRAM

4.1 *Prestress loss determination*

To determine the time-dependent losses, KDOT funded an experimental program to evaluate the long-term performance of inverted T-shape (IT) members cast with SCC. Four IT's, twelve foot in length, were cast in order to determine the time-dependent losses of actual bridge girders. A twelve foot length was considered to be an adequate length for this SCC mix because previous tests concluded that a six-foot development length was all that was needed to achieve full bond. The girder type used was the IT600. The cross-section

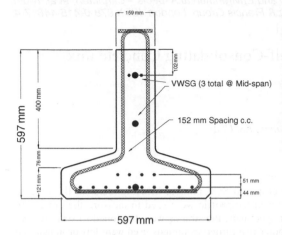

Figure 1. Cross section of IT600.

Table 1. List of measured geometric properties of IT600.

$A = 165,096 \, mm^2$	$I = 8,272,242 \, mm^2$	$E_{ps} = 196,501 \, MPa$
$E_{ci} = 24,821 \, Mpa$	$Y_{bot} = 215 \, mm$	$e = 98 \, mm$
$M_{sw} = 498 \, KN.mm$	$RH = 65\%$	$A_{ps} = 1579 \, mm^2$
$f_{pj} = 1363 \, Mpa$		$V/S = 73 \, mm$

Figure 2. Location of VWSG's.

of this shape can be seen in Figure 1. Table 1 presents the geometric properties of the IT600. Dimensions of the cast specimens were measured and then used for all calculations. Of the four girders, two of them had the pre-stressing strand jacked to seventy-five percent of the ultimate guaranteed tensile stress (f_{pu}) and were designated as FT #1 and FT #2, where FT stands for fully-tensioned. These two girders were used to determine the combined effects of creep, shrinkage, and relaxation. In addition, the two fully tensioned specimens were used to evaluate the transfer lengths, as discussed in the following section. The remaining two specimens, used to determine the effect of shrinkage alone, had the pre-stressing strand pulled to a "hand-tight" condition. The designations of these specimens were UT #1 and UT #2, where UT stands for un-tensioned. The strand was placed in these shrinkage specimens to match the transformed section properties of the two specimens that had the strand jacked to seventy-five percent of f_{pu}.

Elastic shortening losses occur at the time the pre-stress force is transferred to the concrete and thus can be eliminated from long term calculations. The increase in relaxation losses from transfer to final loss calculations are very small compared to the creep and shrinkage losses and are not included in calculations. Elastic shortening and relaxation were determined from the change in strain just after detensioning and subtracted from the measurements that were recorded

from the two fully-tensioned specimens. Then the values of shrinkage, obtained from the two un-tensioned specimens, was subtracted from the above mentioned fully-tensioned specimens along with the elastic shortening and relaxation to determine the effects of creep alone. To record the long term strains that all the specimens had undergone, Vibrating Wire Strain Gages (VWSG) were used. Gages with a 152.4 mm gage length were chosen for this project. All the gages were connected to a data-logger that could take data readings at any desired time interval. Readings were taken every five minutes for the first day and then the collection interval was changed to two hours. Gages were placed at the height of the top strand (101.6 mm from the top), the neutral axis of the section (208.5 mm from the bottom and in one case this value changed slightly and thus the recorded value was used), and the last gage placed at the bottom strand height (50.8 mm from the bottom). The setup of the gages is shown in Figure 2. The gage at the bottom was primarily used to gather the long term strains. To determine the stress at the strand height, Hooke's Law was used.

$$\sigma = E_{ps}\varepsilon \qquad (1)$$

where E_{ps} = modulus of elasticity (MPa); ε = recorded strain value (mm/mm).

Strains were zeroed just prior to detensioning. Therefore, the subsequent strain changes were due to pre-stress losses. A load cell was placed at the "dead" end of the pre-stressing bed in order to get an exact pre-stress force at detension. The nominal value of the jacking stress, f_{pj}, was calculated to be 1396 MPa.

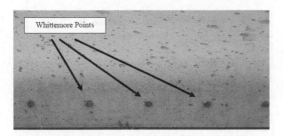

Figure 3. Whittemore points adhered to the bottom the specimen.

Figure 4. Measuring concrete surface displacements with a Whittemore gage.

However experimentally it was found to be 1365 MPa and this value will be used for all calculations using code and experimental data. For the methods that were compared to the experimental data, no intermediate days are calculated, just final values. In order to estimate creep and shrinkage values for periods less than two years, the expression by Corley and Sozen (1966) was used. The following equation made it possible to compare creep and shrinkage values for periods less than two years.

$$R = 0.13 \ln(t+1) \tag{2}$$

where R = the total time-dependent proportion; and t = time (days).

4.2 Transfer length setup

Measurements of the transfer length were accomplished by measuring concrete surface strains with a mechanical strain gage. Whittemore points, stainless steel discs with a machined hole in the center, were adhered along the bottom flange of the specimen of both fully-tensioned specimens prior to detensioning, Figure 3. The location chosen for the line of points was the center of gravity between the bottom two layers of pre-stressing strand, 68 mm from the bottom. Readings were then taken just prior to detensioning and after detensioning, shown in Figure 4. The procedure outlined by Russell & Burns (1993) was used to evaluate the data. Measured changes in strain were then plotted against the length of the specimen. The concrete strains were the numerical difference between the initial reading and the final reading (just after

detensioning). To eliminate any anomalies, measured strains were smoothed by averaging the data over three gage lengths. The equation used to smooth the data is shown as follows:

$$(strain)_i = \frac{(strain)_{i-1} + (strain)_i + (strain)_{i+1}}{3} \tag{3}$$

where i = the current strain reading.

So at any given strain point, that strain and the values just ahead and behind were averaged to obtain the "smoothed" average.

Transfer lengths were then determined by plotting the strains versus the specimen length. The method known as the "95% Average Maximum Strain," also detailed in Russell and Burns (1993), was used to estimate the transfer lengths. The first step is to plot the "smoothed" strain profile. Then the maximum strain that the specimen underwent is determined. Next, 95% of this maximum strain value is computed and a line corresponding to this value is drawn. Lastly, the transfer length is determined by taking the intersection of the 95% maximum strain line and the "smoothed" strain profile line.

Transfer length is implied by both ACI and AASHTO to be expressed as:

$$L_{tr} = \frac{f_{se}}{3} d_b \tag{4}$$

where d_b = diameter of strand (in) f_{se} = effective stress is pre-stressing strand after allowance of pre-stress losses (ksi).

For the IT specimen geometry and strand configuration used in this study, the transfer length was calculated (using equation 4) to be 24 inches.

5 RESULTS

5.1 Prestress losses

By using the methods for determining pre-stress losses as described earlier and presented in the appendix, all the values for elastic shortening, creep, shrinkage, and relaxation were calculated, see Table 1. It was found that the ACI and PCI methods gave the same results; therefore they are presented in the same row. The experimental values for elastic shortening, creep, shrinkage, and relaxation are also given. However, it must be noted that data collection for this paper was stopped at 144 days and collection of strains will continue for a much longer time in order to obtain more complete data.

Strains have been recorded throughout the life of the specimens. By using equation 1, time-dependent losses were estimated at several days and compared to the experimentally-determined values. It must be noted that the strains from FT #1 and FT #2 were

Table 2. Summary of time-dependent losses.

Method	Elastic short-ening	Creep	Shrink-age	Relax-ation	Effective prestress Loss
AASHTO	121.3	183.4	50.3	13.1	992.8
ACI / PCI	110.3	239.2	46.9	18.6	951.5
KDOT	119.3	183.4	50.3	11.0	999.7
Experimental	122.0	129.6	3.4	7.6	1096.3

* All values in MPa, data recorded at 144 days.

Table 3. Strand stress at various ages (MPa).

Time (days)	AASHTO	ACI/PCI	KDOT	Experimental
Transfer	1234.2	1213.5	1213.5	1241.1
25	1137.6	1117.0	1137.6	1179.0
50	1117.0	1089.4	1117.0	1158.3
75	1103.2	1075.6	1103.2	1130.7
96	1096.3	1068.7	1096.3	1123.8
120	1089.4	1054.9	1089.4	1117.0
144	1082.5	1048.0	1082.5	1096.3
Long term	992.8	951.5	999.7	1034.2*

* Estimated by extrapolating curve to 2 years.

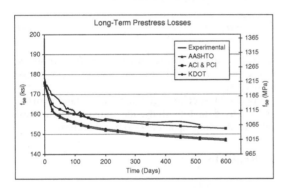

Figure 5. Long term effective pre-stress losses.

averaged and then converted to stresses. The results are shown in Table 2. Figure 5 presents the long term losses in graphical form.

5.2 Transfer length results

The concrete strain profile verses length for both FT #1 and FT #2 are shown below. FT #1, seen in Figure 6, had transfer lengths of 813 mm on one end and 584 mm on the other. FT #2, seen in Figure 7, had transfer lengths of 610 mm on one end and 711 mm on the other. The ends of the specimens that had the greatest transfer lengths both were flame-cut at detensioning while the ends with the smaller values underwent a gradual release during detensioning.

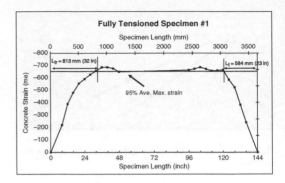

Figure 6. Transfer length profile for FT #1.

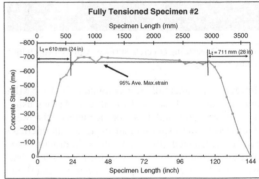

Figure 7. Transfer length profile for FT #2.

6 CONCLUSIONS

Total observed losses for the experimental specimens were slightly less than those predicted by the current AASHTO, ACI/PCI, and KDOT design expressions.

Measured transfer lengths for the proposed SCC mix and specimen geometry were in general accordance with the current AASHTO and ACI/PCI design assumptions, with the maximum measured transfer lengths being 33 percent larger than the calculated value.

Using the proposed SCC mix, it was found that effective prestress losses were in general accordance with current AASHTO, ACI/PCI, and KDOT code equations and no special design considerations need to be taken.

REFERENCES

AASHTO. 2004. *LRFD bridge design specifications, 3rd edition.* Washington, DC. American Association of State Highway and Transportation Officials.

ACI Committee 318. 2002. *Building code requirements for structural concrete (ACI 318-02) and commentary (ACI 318R-02)*. Farmington Hills, MI. American Concrete Institute.

Corley, W.G. & Sozen, M.A. 1966. Time dependent deflections of reinforced concrete beams. *Journal of the American Concrete Institute, Proceedings* 63(3): 373–386.

Interim guidelines for the use of self-consolidating concrete in precast/prestressed concrete institute member plants, first edition. 2003. Chicago, IL.

KDOT. 2003. *Kansas department of transportation design manual, bridge section(3)*. Kansas Department of Transportation.

PCI. 1999. *PCI design handbook, fifth edition*. Chicago, IL Precast/Prestressed Concrete Institute.

Russell, B.W. & Burns, N.H. 1993. Design guidelines for transfer, development and debonding of large diameter seven wire strands in pretensioned concrete girders. *Texas Department of Transportation, Research Project 3-5-89/2-1210*:286

Zia, P., Preston, H.K., Scott, N.L., & Workman, E.B. 1979. Estimating prestress losses. *Concrete International* 1(3): 32–38.

APPENDIX

ACI (ACI Committee 318. 2002) and PCI methods (PCI 1999)

Both use the findings of Zia et al. (1979) for calculations.

Elastic Shortening of Concrete (ES)
For members with bonded tendons,

$$ES = K_{es} E_s \frac{f_{cir}}{E_{ci}}$$ (1)

where $K_{es} = 1.0$ for pre-tensioned members; $K_{es} = 0.5$ for post-tensioned members when tendons are tensioned in sequential order to the same tension. With other post-tensioning procedures, the value for K_{es} may vary from 0 to 0.5.

$$f_{cir} = K_{cir} f_{cpi} - f_g$$ (2)

where $K_{cir} = 1.0$ for post-tensioned members; $K_{cir} = 0.9$ for pretensioned members.

Creep of Concrete (CR)
For members with bonded tendons,

$$CR = K_{cr} \frac{E_s}{E_c} (f_{cir} - f_{cds})$$ (3)

where $K_{cr} = 2.0$ for pre-tensioned members; $K_{cr} = 1.6$ for post-tensioned members.

Shrinkage of Concrete (SH)

$$SH = 8.2x10^{-6} K_{sh} E_s \left(1 - 0.06 \frac{V}{S}\right)(100 - RH)$$

in which $K_{sh} = 1.0$ for pre-tensioned members or K_{sh} is taken from Table 1 (Russell et al. 1993) for post-tensioned members.

Relaxation of Tendons (RE)

$$RE = \left[K_{re} - J(SH + CR + ES) \right] C$$

where the values of K_{re}, J, and C are taken from Tables 2 and 3 (Russell et al. 1993).

AASHTO method

Taken from the Third Edition (AASHTO 2004) for pre-tensioned members

$$\Delta f_{pT} = \Delta f_{pES} + \Delta f_{pSR} + \Delta f_{pCR} + \Delta f_{pR2}$$

where Δf_{pT} = total loss (ksi); Δf_{pES} = loss due to elastic shortening (ksi); Δf_{pSR} = loss due to shrinkage (ksi); Δf_{pCR} = loss due to creep of concrete (ksi); and Δf_{pR2} = loss due to relaxation of steel after transfer (ksi).

Elastic shortening (Δf_{pES})

$$\Delta f_{pES} = \frac{E_p}{E_{ci}} f_{cgp}$$ (4)

where f_{cgp} = sum of concrete stresses at the center of gravity of pre-stressing tendons due to the pre-stressing force at transfer and the self-weight of the member at the sections of maximum moment (ksi); E_p = modulus of elasticity of pre-stressing steel (ksi); and E_{ci} = modulus of elasticity of concrete at transfer (ksi)

Shrinkage (Δf_{pSR})

$$\Delta f_{pSR} = (17.0 - 0.150H)$$ (5)

where H = the average annual ambient relative humidity (percent)

Creep (Δf_{pCR})

$$\Delta f_{pCR} = 12.0 f_{cgp} - 7.0 \Delta f_{cdp} \geq 0$$

where f_{cgp} = concrete stress at center of gravity of pre-stressing steel at transfer (ksi); Δf_{cdp} = change in concrete stress at center of gravity of pre-stressing steel due to permanent loads with the exception of the load acting at the time the pre-stressing force is applied. Values of Δf_{cdp} should be calculated at the same section or at sections for which f_{cgp} is calculated (ksi).

Relaxation (Δf_{pR2}

At transfer

In pre-tensioned members, the relaxation loss in pre-stressing steel, initially stressed in excess of $0.50 f_{pu}$, may be taken as:

For stress-relieved strand:

$$\Delta f_{pR1} = \frac{\log(24.0t)}{10.0}\left[\frac{f_{pj}}{f_{py}} - 0.55\right] f_{pj}$$

For low-relaxation strand:

$$\Delta f_{pR1} = \frac{\log(24.0t)}{40.0}\left[\frac{f_{pj}}{f_{py}} - 0.55\right] f_{pj}$$

where t = time estimated in days from stressing to transfer (days); f_{pj} = initial stress in the tendon at the end of the stressing (ksi); and f_{py} = specified yield strength of pre-stressing steel (ksi).

After Transfer

Losses due to relaxation of pre-stressing steel may be taken as:

For pre-tensioning with stress-relieved strands:

$$\Delta f_{pR2} = 20.0 - 0.4\Delta f_{pES} - 0.2\left(\Delta f_{pSR} + \Delta f_{pCR}\right)$$

where Δf_{pES} = loss due to elastic shortening (ksi); Δf_{pSR} = loss due to shrinkage (ksi); and Δf_{pCR} = loss due to creep of concrete (ksi).

For pre-stressing steels with low relaxation properties conforming to AASHTO M 203 (ASTM A 416 or E 328): Use 30 percent of Δf_{pR2} given by equation 12.

KDOT method

As described in the 2003 release (KDOT 2003), the loss of stress in the pre-stressing steel is as follows:

$\Delta f_S = SH + ES + CR_C + CR_S$
Δf_S = Loss of stress, psi
SH = Loss due to concrete shrinkage, psi
ES = Loss due to elastic shortening, psi
CR_C = Loss due to creep of concrete, psi
CR_S = Loss due to relaxation of steel, psi

Shrinkage

The shrinkage loss is computed as follows,

$$SH = 17,000 - 150RH$$

where RH is the average relative humidity in percent. For Kansas, the humidity may be assumed at 65 percent.

Elastic Shortening

Elastic shortening is computed as follows:

$$ES = \left(E_s \big/ E_{ci}\right) f_{cir} \qquad (6)$$

where $E_s = 28 \times 10^6$ psi; E_{ci} = Modulus of elasticity of concrete at transfer of stress ($33W^{3/2}\sqrt{f'_{ci}}$ psi.); W = 145 pcf for normal weight concrete; and f_{cir} = Concrete stress at the center of gravity of the pre-stressing steel due top pre-stressing force and dead load of the beam immediately after transfer. (At this stage, the initial stress in the tendon has been reduced by elastic shortening of the concrete and tendon relaxation during placing and curing of the concrete.)

Creep of concrete:

For pre-tensioned and post-tensioned members

$$CR_C = 12 f_{cir} - 7 f_{cds}$$

Where f_{cds} is the concrete stress at the center of gravity of pre-stressing steel due to all dead loads except the dead loads present at the time pre-stressing force is applied.

Relaxation of pre-stressing steel – (Low relaxation strand)

$$CR_S = 5,000 - 0.10ES - 0.05\left(SH + CR_C\right)$$

The values of ES, SH, and CR_C are those computed previously,

Total losses: $\Delta f = SH + ES + CR_C + CR_S$

The minimum loss of pre-stress to be used when computing service load stresses shall be 35,000 psi.

Fracture Mechanics of Concrete and Concrete Structures – High-Performance Concrete,
Brick-Masonry and Environmental Aspects – Carpinteri, et al. (eds)
© 2007 Taylor & Francis Group, London, ISBN 978-0-415-44617-4

Bond characteristics of a Self-Consolidating mix for PC bridge girders

K. Larson, R.J. Peterman & A. Esmaeily

Dept. of Civil Engineering, Kansas State University, Manhattan, KS, USA

ABSTRACT: Results from tests used to determine the material and bond characteristics of a proposed SCC mix for bridge girders in the state of Kansas are presented. Eleven full-scale, pre-tensioned SCC flexural specimens were tested to evaluate the transfer and development lengths. These specimens were single-strand specimens that included specimens designed to evaluate the so-called "top-strand" effect. These top-strand specimens, with more than twenty inches of concrete below the strand, were tested to evaluate the current AASHTO requirement of a thirty percent increase in the development length when more than the twelve inches of concrete is cast below the strand. Prior to casting the beams, the pre-stressing strand was pre-qualified using the Large Block Pullout Test procedure. Strand end-slip measurements, used to estimate the transfer lengths, indicated that the proposed SCC mix meets the ACI and AASHTO requirements. In addition, flexural tests on the same specimens, confirmed that the SCC mix also meets the current code requirements for development length. Furthermore, the test results indicated that a thirty percent increase in development length was not necessary to achieve full tensile capacity of the strand in the "top-strand" specimens.

1 INTRODUCTION

Self-Consolidating Concrete (SCC) has rapidly become a widely used material in the construction industry. SCC is defined as "a highly workable concrete that can flow through densely reinforced or geometrically complex structural elements under its own weight and adequately fill voids without segregation or excessive bleeding without need for vibration.[1]"

The Interim Guidelines for the use of Self-Consolidating Concrete in PCI Member Plants recommend that "strand bond tests shall be run with new SCC mixes to verify that the bond with SCC is equivalent or better than a conventional concrete of similar design when using similar strand." Furthermore, these guidelines state that "this can be done using a flexural development length test or by direct load testing." Since SCC does not require any external vibration during placement, there has been concern by some design engineers about the ability to achieve adequate bond between the SCC and the pre-stressing strand.

Departments of Transportation, including the Kansas Department of Transportation (KDOT) would like to use SCC in pretensioned bridge members to enhance aesthetics and improve consolidation in congested areas. A drawback with conventional concrete is that in hard-to-vibrate areas, air is trapped at the surface of the form producing "bug" holes (Fig. 1). SCC will help ensure proper consolidation and a smooth finish on these surfaces.

Figure 1. "Bug" holes in bottom flange of IT.

Before allowing the use of SCC in state bridge girders KDOT wanted to investigate the bond and flexural characteristics of an SCC mix proposed by the local precaster. Since SCC is placed without external vibration, KDOT was concerned that the bond between the SCC and strand may not be as strong as that achieved with a conventional concrete mix.

Moreover, at the time of this study, information about the transfer and development length of prestressing steel in SCC and the applicability of the American Concrete Institute (ACI) and American Association of State Highway Transportation Officials (AASHTO) equations to these members, were essentially absent from the literature.

Transfer length is the distance required to transfer the fully effective prestressing force from the strand to the concrete. Development length is the bond length required to anchor the strand as it resists external loads on a member (PCI 1999). As external loads are applied to a flexural member, the member resists the increased moment demand through increased internal tensile and compressive forces. The increased tension in the strand is achieved through anchorage to the surrounding concrete (Khayat et al. 2004).

Current ACI and AASHTO design requirements do not address the use of SCC in prestressing applications. The ACI code expressions for transfer and development lengths are based on tests performed with conventional concrete and are shown below.

Transfer length (L_{tr}):

$$L_{tr} = f_{se}d_b / 3 \qquad (1)$$

Development length (L_{dev}):

$$L_{dev} = f_{se}d_b / 3 + (f_{ps} - f_{se})d_b \qquad (2)$$

where d_b = diameter of strand (in.); f_{se} = effective stress is prestressing strand after allowance of prestress losses (ksi); and f_{ps} = stress in prestressing strand at calculated ultimate capacity of section (ksi).

The AASHTO specifications are similar but require an additional 1.6 multiplier to equation 2 for precast, prestressed beams.

2 BACKGROUND

KDOT funded an initial investigation in which Large Block Pullout Tests (LBPTs) were performed at Kansas State University (KSU) using both the standard mix recommended by Logan (1997) and the proposed SCC mix. The results with SCC had both lower first-slip and ultimate load values compared to those values when conventional concrete was used (Tables 1, 2). Both of the LBPTs used strand from the same un-weathered reel and which had exhibited satisfactory bond performance in flexural beam tests.

3 TEST PROGRAM

Based on these early findings it was then determined that full-scale development length girder tests were necessary to further investigate the bond between SCC and the pre-stressing strand. Therefore, KDOT funded an experimental program to evaluate the flexural performance of pre-tensioned concrete members with the proposed SCC mix.

Table 1. LBPTs conducted with SCC.

	SCC Block with control strand	
Specimen	Max load (kips)*	Load at 1st slip (kips)*
#1	21.8	11.8
#2	21.4	12.5
#3	19.7	12.4
#4	27.5	10.7
#5	23.2	12.7
#6	21.4	10.7
Average	22.5	11.8

* 1 Kips = 4.448 KN.

Table 2. LBPTs conducted with control mix.

	Control mix with control strand	
Specimen	Max load (kip)	Load at 1st slip (kips)
#1	42.0	28.2
#2	41.7	27.8
#3	40.4	27.3
#4	36.5	24.9
#5	36.9	24.2
#6	39.9	25.0
Average	39.5	26.2

3.1 Material properties

3.1.1 Large block pullout tests

Prior to casting any flexural test specimens, the prestressing strand that would be used for all test girder specimens was pre-qualified using the LBPTs. Standard LBPT procedures, as stipulated by Logan (1997), were followed while performing these tests. These strand qualification tests were performed with the standard mix proposed by Logan6 and not with SCC. The average first-observed slip was 96.1 KN (21.6 kips) and the average ultimate was 176.1 KN (39.6 kips). The values are both above the minimum values recommended by Logan (1997) of 71.2 KN (16 kips) and 160.1 (36 kips), respectively. Thus, the strand reel was deemed acceptable for use in this study. This reel was then covered to prevent weathering and used for all flexural beams reported herein.

3.1.2 Mix design

Casting of test specimens was performed at Prestressed Concrete Inc, in Newton, Kansas (PCIN), which is a PCI certified plant that produces bridge members. PCIN developed their proposed SCC mix design with the help of their admixture supplier. The SCC mix used in this study along with the conventional concrete mix that this plant uses is presented in Table 3. It should be noted that, both mixes use a 19.05 mm (3/4-inch) maximum aggregate size and have a 0.30 and 0.41 water-to-cementicious materials ratio for the SCC and the conventional concrete mix, respectively.

Table 3. SCC and conventional concrete mix design.

Materials	SCC quantity per m^3	Conventional quantity per m^3
Cement (Type III)	338 kg	293 kg
Fine aggregate (MA1 sand)	675 kg	666 kg
Coarse aggregate (CA-6, 1″ −#67)	612 kg	656 kg
Air entrainment	148 mL	178 mL
Viscosity modifying agent	0 mL	0 mL
Water	102 L	120 L
W/C ratio	0.30	0.41

Figure 3. J-Ring test for SCC.

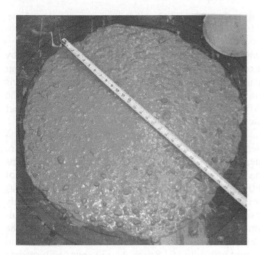

Figure 2. Spread Test for SCC.

Also note that a different high range water reducer is used for the SCC and conventional concrete mix.

3.1.3 Fresh concrete evaluation

During the casting of the specimens, the SCC mix was tested to determine its rheological properties. At the time of casting, there were no existing ASTM standards for testing SCC, but the PCI Interim Guidelines document many test methods to evaluate the plastic properties of SCC for production qualifications. In this study, Inverted Slump Flow (Fig. 2) VSI, J-Ring (Fig. 3) and L-Box (Fig. 4) tests were all performed on the concrete during casting. The Inverted Slump Flow measures the flow separation resistance, stability/settlement resistance, air migration, and relative viscosity. The J-Ring and L-Box are both tests that measure the passing ability and blocking resistance of the SCC mix.

Figure 4. L-Box test for SCC.

3.1.4 Hardened concrete properties

The compressive strength and modulus of elasticity of the concrete were measured for future use in analytical computations. Standard ASTM procedures were followed for compressive strength and modulus of elasticity testing. In addition to measuring one-day (release) strengths; compressive strengths were determined just prior to loading the flexural specimens to failure. A set of three 101.6 × 203.2 mm (4 × 8 in.) cylinders were tested for each flexural specimen and the average values were recorded.

3.2 Transfer length measurements

Mast's strand slip theory as presented by Logan (1997) was used to determine the transfer length of the girders experimentally. End-slip values were obtained by measuring the distance that the strand slipped into the beam at the ends. Prior to detensioning, a mark was made on the strand with a saw blade at a distance approximately 25.4 mm from the specimen end. A steel block having a width of exactly 12.7 mm was then held against the concrete at the strand location. The distance between this machined block and the mark on the strand was then measured using a digital caliper having a precision of 0.0254 mm. This value was then used as the baseline for measurements taken after detensioning to determine the amount of end-slip that occurred. Subsequent measurements were taken up to the time of testing of the specimen. The following equations were used to determine the implied transfer length values from the end-slip measurement data.

$$\Delta = \frac{avg\, f_{si} L_{tr}}{E_{ps}} \qquad (3)$$

where Δ = end slip (in.), equal to measured length between steel block and strand minus elastic shortening between the mark on the strand and specimen end; avg f_{si} = average initial strand stress over the transfer length after release of pre-stress (ksi); L_{tr} = transfer length (in.); and E_{ps} = elastic modulus of strand (ksi).

Assume straight line variation in the strand stress from zero at the end of the beam to full pre-stress:

$$\Delta = \frac{0.5 f_{si} L_{tr}}{E_{ps}} \qquad (4)$$

thus

$$L_{tr} = \frac{\Delta E_{ps}}{(0.5 f_{si})} \qquad (5)$$

4 FLEXURAL SPECIMEN TYPES

4.1 Single-strand development length specimens

Twelve single-strand development length specimens with different embedment lengths were fabricated and tested in this investigation. However, due to a handling error with one of the specimens only eleven were tested to failure. The single-strand specimens were used to evaluate two different embedment lengths. Two different cross-sections were utilized in order to evaluate the so-called "top strand" effect, having 304.8 mm or more of concrete cast below the reinforcement. ACI requires a 1.3 multiplier on development length for "horizontal reinforcement so placed that more than 304.8 mm (12 in.) of fresh concrete is cast in the member below the development length or splice," (ACI 12.2.4). AASHTO uses a similar 1.3 multiplier for strand development length when using an Alternate Development Length Equation (AASHTO 5.11.4.2-2).

The first cross-section cast was the standard 203.2×304.8 mm (8×12 in.) section that was used by Peterman et al. (2000). The nomenclature used for these specimens was Single-Strand Beams (SSB). This section contained a single pre-stressing strand at a depth d_p of 254 mm (Fig. 5). The section chosen

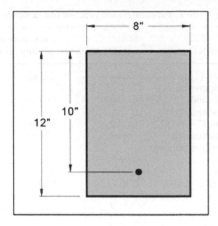

Figure 5. Cross section of bottom strand girders. Note: 1 inch = 25.4 mm.

was slightly larger than the 165.1 mm wide tested by Logan (1997) in order to provide an increased shear capacity. This was desirable since these specimens did not have any shear reinforcement. Refer to the Appendix for shear capacity calculations and other sample calculations.

Specimens with the second single-strand cross-section used to evaluate the "top-strand" effect, are denoted Top-Strand Beams (TSB). These specimens had a width of 203.2 mm and an overall height of 609.6 mm (Fig. 6). The strand in these specimens was located 558.8 mm from the bottom, and thus greatly exceeded the 304.8 mm height requiring a 1.3 multiplier for development length by AASHTO. At the center portion of these specimens, however, the strand height was only 304.8 mm. At mid-span, a Styrofoam blockout was used to reduce the height from 609.6 mm to 304.8 mm (Fig. 7). These specimens were inverted prior to testing. Note at mid-span, which is the critical section; these specimens had an identical cross-section to the SSB specimens. Therefore, direct comparisons between results are possible.

4.2 Embedment lengths

At the outset of this experimental program the researchers decided to evaluate two different embedment lengths. Crack formers (Fig. 8) were cast at the embedment length to insure that during loading the first cracks would open at his location. The first set of specimens was tested at an embedment length equal to 100% of the calculated development length (l_{dev}). The second set of specimens was tested at either 80% l_{dev} or 120% l_{dev}, depending on the results obtained from the 100% l_{dev} specimen tests. The second set of specimens was specifically designed to allow for testing at either embedment length as explained below.

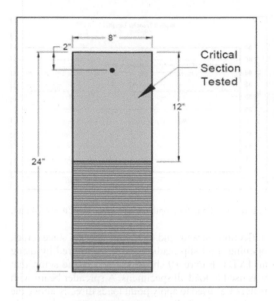

Figure 6. Cross section of top strand girders. Note: 1 inch = 25.4 mm.

Figure 7. Blockouts for top strand beams.

Figure 8. Crack formers.

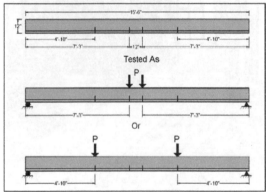

Figure 9. Test setup for 4'-10" embedment length for bottom strand beams. Note: 1 inch = 25.4 mm.

If the 100% l_{dev} specimens failed (by flexure) at a moment greater than or equal to the calculated nominal moment capacity, M_n, then the second set of specimens would be tested at an embedment length equal to 80% l_{dev}. However, if the 100% l_{dev} specimens failed (by bond) at a moment less than the calculated nominal moment capacity, M_n, then the second set of specimens would be tested at an embedment length equal to 120% l_{dev}. Since all of the 100% l_{dev} specimens failed by flexure (as later discussed in results section of this manuscript), the second set of specimens were tested at an embedment length equal to 80% l_{dev}.

The different embedment length testing of the second set of specimens was made possible by utilizing four crack formers per beam (Fig. 9). As shown in this figure, the 80% l_{dev} tests required the use of the spreader beam with loading points directly above the outer-most crack former.

4.3 Loading conditions

Three types of loading rate conditions were used for evaluating the different embedment lengths. The first loading condition was designated as the SLOW test and was targeted to take about ten hours. During a SLOW test, the specimen was loaded at 444.8 N/min until cracking. Then the loading rate was reduced to 44.48 N/min until failure. This slow loading rate was used in order to accurately measure the amount of strand slip, if any, occurring prior to failure. For the second loading condition, designated as 76.5% M_n, the specimen was loaded at 444.8 N/min up to 76.5% of nominal capacity of the specimen, and then this

Table 4. Loading conditions for beams tests.

	Beam	Embedment length*	Loading condition
Bottom strand	SSB A	6'-1"	76.5% M_n
	SSB C	6'-1"	SLOW
	SSB D	4'-10'	100% M_n
	SSB E	4'-10"	SLOW
	SSB F	4'-10"	76.5% M_n
Top strand	TSB A	4'-10"	76.5% M_n
	TSB C	4'-10"	100% M_n
	TSB D	6'-1"	SLOW
	TSB E	6'-1"	76.5% M_n
	TSB F	6'-1"	SLOW

* 1 inch = 25.4 mm.

Figure 10. Beam setup.

load was maintained for twenty-four hours. This load condition was modeled after ACI 20.3.2 for the testing and evaluation of existing structures. If the specimen successfully withstood the load for 24 hours, it was then loaded at 44.48 N/min to failure. The final loading condition, designated as 100% M_n, was similar to the 76.5% M_n procedure, except that load was maintained at 100% M_n for 24 hours. Table 4 shows the loading condition of each specimen tested along with the corresponding development length.

4.4 Test setup

All specimens were tested using a 97.86 KN (22 kips) MTS servo-controlled actuator in the KSU Civil Engineering Department Mechanics of Materials Laboratory. Data was collected for load, mid-span

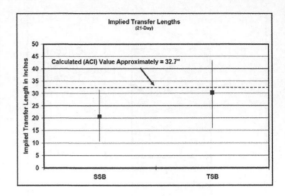

Figure 11. Transfer length results. Note: 1 inch = 25.4 mm.

deflection, strand end-slip, and tension face crack opening. End slip readings were monitored by using an LVDT. Figure 10 shows the test frame setup that was used to load all specimens. A spreader beam with rollers was used to apply point loads directly above the crack formers. Roller connections were used to apply the point load at these locations.

5 RESULTS

5.1 Transfer length

As described earlier, end-slip measurements were used to estimate the transfer length of each girder. In these calculations, f_{si} was assumed to be 1352.75 MPa (196.2 ksi) for all single strand specimens. For all bottom strand beams, none had a longer implied transfer length (21-day) than assumed by the ACI code. The average implied transfer length was 533.4 mm for the SSB specimens and 812.8 mm. for the TSB specimens. Figure 11 presents the range and average implied transfer lengths for all the specimens tested, along with the ACI code assumptions.

5.2 Flexural test results

Flexural failure by strand rupture was the failure mode of all specimens tested in this study. In each case, the experimental moment exceeded the calculated nominal moment capacities by 10%–20% (Table 5). In Table 5 the spread column refers to the test referred to in Figure 2. Furthermore, the maximum end-slip recorded for all specimens during testing was less than 0.01 in.

6 CONCLUSIONS

Transfer lengths estimated from 21-day strand end-slip measurements were in general accordance with the values assumed by the AASHTO and ACI specifications. The average implied transfer lengths for

Table 5. Results of specimens tested.

Beam		Spread (in)*	% M_n achieved	Strand rupture	Strand slip >0.01″
SSB	A	21	110.9	Yes	No
	C	21	115.4	Yes	No
	D	22	117.2	Yes	No
	E	22	113.3	Yes	No
	F	22	115.7	Yes	No
TSB	A	28	116.2	Yes	No
	B	28	116.4	Yes	No
	C	28	115.3	Yes	No
	D	28	112.6	Yes	No
	E	28	116.6	Yes	No
	F	28	114.0	Yes	No

* 1 inch = 25.4 mm.

the top-strand beams were approximately 50% greater than those for the corresponding bottom-strand beams.

Flexural tests indicated that the current ACI (and thus also the AASHTO) equations for strand development length were conservative for the SCC mix and specimen geometry used in this study. Moreover, all of the load tests conducted on specimens with an embedment length equal to eighty percent of the ACI development length, including those with more than 304.8 mm of concrete below the strand, failed in flexure by strand rupture.

REFERENCES

AASHTO. 2004. *LRFD bridge design specifications, 3rd edition*. Washington, DC. American Association of State Highway and Transportation Officials.

ACI Committee 318. 2002. *Building code requirements for structural concrete (ACI 318-02) and commentary (ACI 318R-02)*. Farmington Hills, MI. American Concrete Institute.

Interim guidelines for the use of self-consolidating concrete in precast/prestressed concrete institute member plants, first edition. 2003. Chicago, IL.

Khayat, K.H., Assaad, J. & Daczko, J. 2004. Comparison of field-Oriented test methods to assess dynamic stability of self-consolidating concrete. *ACI Materials Journal* 101(2):168–176.

Logan, D.R. 1997. Acceptance criteria for bond quality of strand for pretensioned prestressed concrete applications. *PCI JOURNAL* 42(2):52–90.

PCI. 1999. *PCI design handbook, fifth edition*. Chicago, IL Precast/Prestressed Concrete Institute.

Peterman, R.J., Ramirez, J.A. & Olek, J. 2000. Influence of flexure-shear cracking on strand development length in prestressed concrete members. *PCI JOURNAL* 45(5): 76–94.

NOTATION

AASHTO	= American Association of State Highway Transportation Officials;
ACI	= American Concrete Institute;
CC	= Conventional concrete;
db	= Diameter of prestressing strand;
dp	= Depth of prestressing strand;
Ec	= 28-day modulus of elasticity of concrete;
Eci	= Release modulus of elasticity of concrete;
Eps	= Modulus of elasticity of prestressing strand;
fpj	= Stress in strand after tensioning;
fps	= Stress in prestressing strand at failure;
fse	= Effective prestress, after all time – dependent deformations;
fsi	= Prestress after transfer before time-dependent losses;
FT	= Fully tensioned;
KDOT	= Kansas Department of Transportation;
LBPT	= Large-block pullout test;
IT	= Inverted-T;
Ldev	= Development length;
Le	= Embedment length;
Ltr	= Transfer length;
Mn	= Nominal moment;
PCI	= Prestressed/Precast Concrete Institute;
SCC	= Self-consolidating concrete;
SSB	= Single-strand beam;
TB	= T-beam;
TSB	= Top-strand beam;
UT	= Untensioned;
VWSG	= Vibrating wire strain gage;
□	= Measured end slip;
□	= Strain; and
□	= Stress.

APPENDIX

Prestress Losses and Nominal Calculations for SSB specimen

Assume that

$$\varepsilon_c = 0.003 \qquad f_c' = 8,000\,\text{psi} \qquad f_{pu} = 270\,\text{ksi}$$

See Figure 4 for rectangular prestressed concrete beam with following properties:

$$b = 8\,\text{in} \quad h = 12\,\text{in} \quad d_p = 10\,\text{in}$$

$$A = 96\,\text{in}^2 \quad I = 1152\,\text{in}^4 \quad y_b = 6\,\text{in}$$

$$e = 4\,\text{in} \quad \beta_1 = 0.85 - [0.05(f_c' - 4,000)/1,000] = 0.65$$

$$E_{ps} = 28,500\,\text{ksi} \quad A_{ps} = 0.153\,\text{in}^2 \quad f_{pj} = 0.75*(270) = 202.5\,\text{ksi}$$

$$P = 202.5*0.153 = 30.983\,\text{kips} \qquad E_{ci} = 3,600\,\text{ksi}$$

$$E_c = 5,000\,\text{ksi} \quad \text{Self Weight} = 93.33\,\text{lb/ft}$$

$$L_b = 13.17\,\text{ft} \quad M_{sw} = \frac{93.33*(13.17)^2}{8}*12 = 24,376\,\text{lb} - \text{in}$$

V/S = 2.33 RH = 65%

Loss Calculations (Based on PCI Handbook 5th Edition)

$ES = K_{es} * E_{ps} * f_{cir} / E_{ci}$

$$f_{cir} = K_{cir}\left[\frac{P}{A} + \frac{P*e^2}{I}\right] - \frac{M_{sw}*e}{I}$$

$$= 0.9\left[\frac{30,983}{96} + \frac{30,983*4^2}{1152}\right] - \frac{24,276*4}{1152}$$

$$594 \text{ psi}$$

with $K_{cir} = 0.9$ for pretensioned members

$\qquad K_{es} = 1.0$ for pretensioned members

$ES = 1.0 * 28,500 * 0.594 / 3600 = 4.70$ ksi

Creep (CR)

$CR = K_{cr} \times (E_p / E_c) \times (f_{cir} - f_{cds})$

with

$CR = 2.0 \times (28,500/5000) \times 0.594 = 6.77$ ksi

Shrinkage (SH)

$SH = (8.2x10^{-6}) \times K_{SH} \times E_p \times (1 - 0.06 \times (V/S)) \times (100 - RH)$

with

$SH = (8.2x10^{-6}) \times 1.0 \times 28,500 \times (1 - 0.06 \times 2.33) \times (100 - 65) = 7.04$ ksi

Relaxation (RE)

$RE_L = [K_{RE} - J*(SH + CR + ES)]*C$

with

$RE_L = [5.0 - 0.04*(4.7 + 6.77 + 7.04)]*1.0 = 4.26$

$RE_i = f_{st}\left\{[\log 24t - \log 24t_1]/24\right\}*\left[f_{st}/f_{py} - .55\right]$

$\quad = 202,500*[\log 18]/45*[202,500/243,000 - .55] = 1.60$ ksi

Total Losses

$f_{si} = f_{pj} - ES - RE_i$

$\quad = 202.5 - 4.7 - 1.6 \approx 196$ ksi

$f_{se} = f_{pj} - ES - CR - SH - RE_L$

$\quad = 202.5 - 4.7 - 6.77 - 7.04 - 4.26 \approx 180$ ksi

Calculated Transfer Length (using equation 1)

$L_{tr} = f_{se} d_b / 3$

$\quad 196*.5/3 = 32.67$ inches

Experimental Implied Transfer Length (Sample Calculation)

$\Delta_{end-slip} = .060 - \left(\frac{PL}{AE}\right)_{elastic\,shortening} = .060 - \left(\frac{30.98*1}{.153*28,500}\right) = .055$ inch

$L_{tr} = \frac{\Delta E_{ps}}{.5 f_{si}} = \frac{.055*28,500}{.5*196} = 16$ inch

Calculated Development Length (using Equation 2)

$L_{dev} = f_{se} d_b / 3 + (f_{ps} - f_{se}) d_b$

$\quad = 180*.5/3 + (268.2 - 180)*.5 = 74$ inch

Nominal Capacity using Strain Compatibility

$P_e = f_{se} * A_{ps} = 180*0.153 = 27.54$ kips

$\varepsilon_1 = f_{se}/E_p = 180/28,500 = 0.00623$

$\varepsilon_2 = \frac{1}{E_c}\left[\frac{P_e}{A} + \frac{P_e*e^2}{I}\right] = \frac{1}{5000}\left[\frac{27.54}{96} + \frac{27.54*4^2}{1152}\right] = .000134$

Assume $f_{ps} = 268.2$ ksi

$a = \frac{A_{ps}*f_{ps}}{0.85*f_c'*b} = \frac{0.153*268.2}{0.85*8*8} = 0.754$

$c = a/\beta_1 = 0.754/0.65 = 1.16$

$\varepsilon_3 = \left(\frac{d_p - c}{c}\right)*\varepsilon_c = \left(\frac{10 - 1.16}{1.16}\right)*0.003 = 0.0229$

$\varepsilon_{ps} = \varepsilon_1 + \varepsilon_2 + \varepsilon_3 = 0.00623 + 0.000134 + 0.0229 = 0.0294$

From Curve in Handbook

$f_{ps} = 270 - \frac{0.04}{\varepsilon_{ps} - 0.007} = 268.2$ ksi *(Equals assumed value)*

$M_n = A_{ps}*f_{ps}*\left(d_p - \frac{a}{2}\right) = 0.153*268.2*\left(10 - \frac{.754}{1}\right) = 394.9$ kip-in

$\qquad\qquad 32.9$ kip-ft

Shear Capacity

Test Span $\qquad L_{test} = 12.83$ *ft*

Shear Span $\qquad a_{test} = 5.92$ *ft*

$M_D + M_L = 32.9$ *kip* $-$ *ft*

$\frac{(0.0933)*(12.83)^2}{8} + \frac{P_F}{2}(5.92) = 32.9$

$P_F = 10.5$ *kips*

$V_{max} = 0.0933*(6) + \frac{10.5}{2} = 5.8$ *kips*

$V_c = 2*\sqrt{f_c'}*b*d_p = 2*\sqrt{8,000}*8*10 = 14.3$ *kips*

$V_c > V_{max}$ *(Good)*

Fracture Mechanics of Concrete and Concrete Structures – High-Performance Concrete,
Brick-Masonry and Environmental Aspects – Carpinteri, et al. (eds)
© 2007 Taylor & Francis Group, London, ISBN 978-0-415-44617-4

Size and shape effects on the compressive strength of high-strength concrete

J.R. del Viso, J.R. Carmona & G. Ruiz

E.T.S. de Ingenieros de Caminos, Canales y Puertos, Universidad de Castilla-La Mancha

ABSTRACT: In this paper we investigate the influence of the shape and of the size of the specimens on the compressive strength of high strength concrete. We use cylinders and cubes of different sizes for performing stable stress-strain tests. The tests were performed at a single axial strain rate, $10^{-6}\,s^{-1}$. This value was kept constant throughout the experimental program. Our results show that the post-peak behavior of the cubes is milder than that of the cylinders, which results in a strong energy consumption after the peak. This is consistent with the observation of the crack pattern: the extent of micro-cracking throughout the specimen is denser in the cubes than in the cylinders. Indeed, a main inclined fracture surface is nucleated in cylinders, whereas in cubes we find that lateral sides get spalled and that there is a dense columnar cracking in the bulk of the specimen. Finally, we investigate the relationship between the compressive strength given by both types of specimen for several specimen sizes.

1 INTRODUCTION

By far the most common test carried out on concrete is the compressive strength test. The main reason to understand this fact is that this kind of test is easy and relatively inexpensive to carry out (Mindess et al., 2003). Testing Standard requirements use different geometries of specimens to determine the compressive concrete strength, f_c. The most used geometries are cylinders with a slenderness equal to two and cubes. Shape effect on compression strength has been widely studied and different relationships between compression strength obtained for these geometries have been proposed, mainly from a technological standpoint. Such approach eludes the fact that there is a direct relation between the nucleation and propagation of fracture processes and the failure of the specimen. Indeed, experimental observations confirm that a localized micro-cracked area develops at peak stress Shah and Sankar, 1987 or just prior to the peak stress (Torreti et al., 1993). For this reason compressive failure is suitable to be analyzed by means of Fracture Mechanics (Van Mier, 1984).

Some recent experimental works based on Fracture Mechanics to study the compressive behavior of concrete are especially noteworthy. Jansen and Shah (Jansen and Shah, 1997) planned an experimental program aimed at analyzing the effect of the specimen slenderness on the compressive strength. Also notable is the work by Choi et al. (Choi et al., 1996), on the

strain softening in compression under different end constraints. Borges et al. (Borges et al., 2004) studied the ductility of concrete in uniaxial and flexural compression. The results of these experimental programs suggest that the compressive test may be considered as an structural test, because the results depend not only on the actual mechanical properties but on the geometry of the specimen and on the boundary conditions, like end constraints, feedback signal or specimen capping (Shah et al., 1995).

In the last decades concrete technology has made it easier to reach higher strengths and the so-called High-Strength Concrete (HSC) has appeared as a new construction material. This amount of the compressive strength provokes that the normalized specimens for normal strength concrete, e.g. $150 \times 300\,mm$ cylinder (diameter × height), may surpass the capacity of standard laboratory equipment. To overcome this drawback HSC mixtures are often evaluated using $100 \times 200\,mm$ cylinders, which also meet the requirements of ASTM C39. The strength obtained with this specimen is higher in average than that obtained with $150 \times 300\,mm$ cylinders. This size effect on the compressive strength manifests that the specimen is not behaving as simply as it may look. Indeed, size effect is understood as the dependence of the nominal strength of a structure on its size (dimension) when compared to another geometrically similar structure (Bažant and Planas, 1998). For quasi-brittle materials as concrete, the presence of a non-negligible fracture process zone

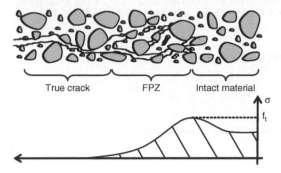

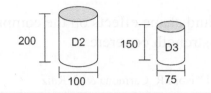

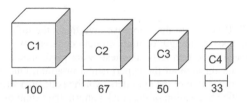

Figure 1. Fracture process zone (FPZ).

Dimensions in mm.

Figure 2. Specimen geometry.

at the front of the crack, which is commensurate with an internal characteristic length and with the typical size of the material inhomogeneities, provokes a deterministic size effect (please, see Figure 1).

In this paper we investigate the mechanical behavior of HSC (around 100 MPa) tested in a compressive setup in axial strain control. We are particularly interested in the influence of the shape and of the size of the specimens on the compressive strength, f_c, of the material. According to our results we propose a new relationship between cylinder and cube specimens.

The article is organized as follows. An outline of the experimental program is given in Section 2. Concrete production and specimens are described in Section 3. In Section 4 we describe the testing procedures. The experimental results are presented and discussed in Section 5. Section 6 includes a simple analysis of size effect and establishes a relationship between the compressive strength measured from standard cylinders and the one obtained from cubes of different sizes. Finally, in Section 7 some conclusions are extracted.

2 OVERVIEW OF THE EXPERIMENTAL PROGRAM

The experimental program was designed to study the size and shape effect on compressive tests performed on High-Strength Concrete (HSC) specimens. Specifically, we wanted to disclose the influence of the size in cubes and the relationship between compressive strength obtained from standard cylinders, ASTM C39 (100 × 200 mm) and the strength got from cubes. We also intended to analyze the variations in the crack pattern and in the mechanical behavior due to the size and shape of the specimens.

With these intentions in mind, we chose the specimens sketched in Figure 2. We use cylinders and cubes of different sizes: the dimensions of the cylinders are 75 × 150 and 100 × 200 mm (diameter × height); the edges of the cubes are 33, 50, 67 and 100 mm long. The dimensions of the cylinders were scaled to the

diameter D, while the dimensions of the cylinders were scaled to the edge L, please see Figure 2. Each specimen was named by a letter indicating the shape, D for cylinders and C for cubes, and one number which indicates the size. For example, D2 names a cylinder of 100 × 200 mm. We performed at least four tests for each type of specimen.

An important feature of our experimental program —unlike most of the tests available in the scientific literature— consists of providing complete material characterization obtained by independent tests, in many cases performed according to Standard recommendations. Specifically, we measured the compressive strength, the tensile strength, the elastic modulus and the fracture energy of the concrete.

3 MATERIAL AND SPECIMENS

A single high-strength concrete mix was used throughout the experimental program. It was made with a andesite aggregate of 12 mm maximum size and ASTM type I cement. Microsilica fume slurry and superplasticizer (B-255, BASF) were used in the concrete composition. The water to cement ratio (w/c) was fixed to 0.28.

There was a strict control of the specimen-making process, to minimize scatter in test results. In the case of cylinders the concrete was poured into steel molds in three layers and compacted on a vibration table. In the case of the cubes the specimens were taken by wet-sawing from 100 × 100 × 420 mm concrete prisms. The prisms were also made using steel molds and a

Table 1. High strength concrete mechanical properties.

	$f_c^{(a)}$ MPa	$f_{ts}^{(b)}$ MPa	E_c GPa	G_F N/m	$\ell_{ch}^{(c)}$ mm
mean	89.6	5.4	36.1	119.0	147.3
std.dev.	7.13	0.6	1.1	13.5	–

[a] Cylinder, compression tests, D2.
[b] Cylinder, splitting tests.
[c] $\ell_{ch} = \frac{E_c G_F}{f_t^2}$.

vibration table. Cylinders and prisms were wrap-cured for 24 hours, demolded and stored in a moist chamber at 20°C and 96% relative humidity. Right after making the cubes from the prisms their sides, as well as the bases of the cylinders, were surface-ground with a wet diamond wheel to ensure perfectly-plane and parallel surfaces. Of course, the specimens were immediately stored again in the chamber until testing time (six months from concrete making). Just before testing the length, diameter and weight of the cylinders and the height and weight of cubes were measured.

Table 1 shows the characteristic mechanical parameters of the micro-concrete determined in the various characterization tests. In the table it is also shown the characteristic length of the concrete, ℓ_{ch}, (Petersson, 1981), which is defined as $\ell_{ch} = E_c G_F/f_t^2$, being E_c the elastic modulus, G_F the fracture energy and f_t the tensile strength. Please, notice that the characteristic length, ℓ_{ch}, of this concrete is roughly 150 mm, closely half the ℓ_{ch} of normal concrete ($\ell_{ch} = 300$ mm). This means that, from a Fracture Mechanics standpoint, we expect that high-strength concrete (HSC) may present a more brittle behavior than NSC for the same specimen size. As mentioned above, the characteristic length is related with the extension of the fracture processes zone (FPZ), see Figure 1.

4 EXPERIMENTAL PROCEDURES

4.1 Characterization tests

Compressive tests were carried out on 4 cylindrical specimens according to ASTM C-39 on 100 × 200 mm (diameter × height), except for a reduction of the axial displacement rate of the machine actuator, which was 0.018 mm/min.

The Elastic modulus was obtained according to ASTM C-469 on cylindrical specimens of 150 × 300 mm. The strain was measured over a 50 mm gage length by means of two inductive extensometer placed symmetrically, on a similar setup that it is shown in Figure 3. The tests were run under displacement control, at a rate of 0.3 mm/min.

Brazilian tests were also carried out on cylindrical specimens following the procedures recommended by

Figure 3. Testing setup for D3 specimen.

ASTM C496 on 150 × 300 mm cylinders (diameter × height). The specimen was loaded through plywood strips with a width of 1/6 of the specimen diameter. The velocity of displacement of the machine actuator was 0.18 mm/min.

Stable three-point bend tests on 100 × 100 × 420 mm notched beams were carried out to obtain the fracture properties of concrete following the procedures devised by Elices, Guinea and Planas (Guinea et al., 1992; Elices et al., 1992; Planas et al., 1992). During the tests the beams rested on two rigid-steel semi-cylinders laid on two supports permitting rotation out of the plane of the beam and rolling along the beam's longitudinal axis with negligible friction. These supports roll on the upper face of a very stiff steel beam fastened to the machine actuator.

Table 1 shows the characteristic mechanical parameters of the micro-concrete determined in the various characterization tests.

4.2 Testing procedure to obtain stable $\sigma - \varepsilon$ curves

All the specimens were tested on a compressive setup as illustrated in Figure 3. The experiments were performed on an INSTRON 8800 closed-loop servo-hydraulic dynamic testing machine, with a capacity of 1000 kN. The loading platens are made of hardened steel and have polished surfaces. Two linear variable differential transformers (LVDTs), having a calibrated range of ±2.5 mm, were placed opposed to the specimen in a symmetric fashion. These LVDTs were used to measure axial platen-to-platen displacement, δ. The axial deformation referred to in this paper is the average value of the two LVDTs.

For controlling the tests, the signal of the average axial displacement from the LVDTs was chosen. This average signal represents the axial displacement velocity at the center of the loading platen. This type of control leads to stable tests for our particular material and specimens geometry. The axial strain rate is

Table 2. Axial displacement rate for each of the specimens.

Denomination	Axial displacement rate, $\dot{\delta}$ $mm \times min^{-1}$
D2	1.8×10^{-2}
D3	1.2×10^{-2}
C1	0.9×10^{-2}
C2	0.6×10^{-2}
C3	0.4×10^{-2}
C4	1.98×10^{-3}

Table 3. Strength and strain at peak and ultimate load.

Specimen		σ_c MPa	ϵ_c %	σ_u MPa	ε_u %
D2	mean	89.6	0.37	31.4	0.71
	desv. std.	7.11	0.02	23.31	0.51
D3	mean	89.9	0.34	41.6	0.22
	desv. std.	4.65	0.02	22.08	0.42
C1	mean	96.1	0.57	12.7	1.53
	desv. std.	1.63	0.01	1.55	0.12
C2	mean	102.4	0.61	20.9	1.59
	desv. std.	9.93	0.03	1.81	0.19
C3	mean	104.2	0.66	38.5	1.46
	desv. std.	2.19	0.06	6.63	0.13
C4	mean	110.0	0.85	33.8	1.94
	desv. std.	7.34	0.09	5.09	0.35

defined as $\dot{\varepsilon} = \dot{\delta}/L$, where $\dot{\delta}$ is the axial platen-to-platen displacement rate and L is the initial distance between the top and bottom plane surfaces over the height of the specimen. This axial strain rate was kept constant for all tests having a value of $10^{-6} s^{-1}$. The axial displacement rates selected for each specimen are showed in Table 2.

An elastic band was stuck around the specimen, see Figure 3, (1) to preserve the fragments together for a posterior analysis, and (2) to avoid a interchange of humidity with the environment during the test. The transversal force provoked by this elastic band is negligible.

The load, P, and the axial displacement, δ, were continually monitored and recorded. We used a pair of clip extensometer centered on the specimen in D2 and D3 to measure the displacement in the central part of the specimen in order to get the elastic modulus. To complete the experimental information, we also took pictures of the crack pattern resulting from each test.

5 RESULTS AND DISCUSSION

The discussion of the results is organized as follows. First we present and describe the $\sigma - \varepsilon$ curves obtained from the tests. Then we proceed to discuss size effect

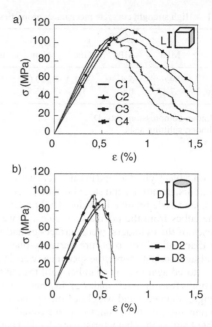

Figure 4. $\sigma - \varepsilon$ Curves corresponding to: (a) Cubes; (b) Cylinders.

in cubes and cylinders. Finally the crack patterns are analyzed. Let us emphasize again that cylinders and cubes were tested in strain control keeping constant the axial strain rate throughout the experimental program.

5.1 $\sigma - \varepsilon$ curves

Table 3 lists the average value and the standard deviation for the peak stresses, σ_c, the strain at peak stress, ε_c, the ultimate stress recorded, σ_u, and the ultimate strain recorded, ε_u for all specimens. As expected, σ_c increases noticeably as the size of the specimen decreases. The strain at peak load increases also as size decreases.

Figure 4a and 4b compare $\sigma - \varepsilon$ curves for selected sizes. Specifically, Figure 4a shows the $\sigma - \varepsilon$ obtained for cubes and Figure 4b for cylinders. The x-axis corresponds to the average strain, i.e. the displacement between the top and bottom plane surfaces over the height of the specimen. The y-axis corresponds to the concrete stress ($\sigma_c = \frac{P}{Area}$). A typical $\sigma - \varepsilon$ curve starts with a linear ramp-up. The initial slope shows no significant variations due to the size in the case of cubes while in cylinders the shorter specimen shows a higher slope. In both cases there is a loss of linearity before reaching the peak load, which indicates the initiation of fracture processes.

The post peak behavior depends on the specimen shape. In the case of cylinders, Figure 4b, a sudden drop in the load takes place after the peak load and no significant changes are reported in the softening

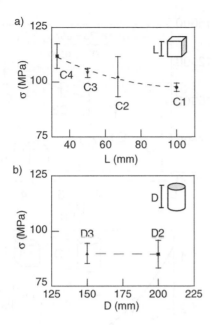

Figure 5. Size effect on the peak strength: (a) Cubes; (b) Cylinders.

branch between the scaled tested cylinders. In the case of cubes, Figure 4a, $\sigma - \varepsilon$ curves show a not so steep softening branch which is associated to a milder failure localization and to a large amount of volumetric energy dissipation, in contrast with the strong failure localization observed in cylinders.

5.2 Size effect

Figures 5a and b plot the peak loads or maximum strength obtained in the $\sigma - \varepsilon$ curves for the different tested specimens, against a characteristic dimension of the specimen. In Figure 4a is drawn the case of cubes specimens. It is clearly observed that large specimens resist less in terms of stress than the smaller ones. A interpolation line has been plotted attached to the tests results to facilitate the comparison. Size effect is less noticeable as size increases.

By contrast, the average stress, σ, obtained from cylinders is roughly constant within the experimental size interval planned for this research and approximately 10% lower than the horizontal asymptote for the cubes as it can be observed in Figure 4b.

5.3 Crack pattern

The crack pattern observed after the test is sensitive to the shape of the specimen as Figure 6 shows. The extent of micro-cracking throughout the specimen is denser in the cubes than in the cylinders. Indeed, a main inclined fracture surface is nucleated in cylinders, whereas in cubes we find that the lateral sides

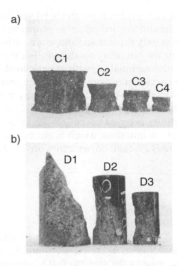

Figure 6. Crack pattern for: (a) Cubes; (b) Cylinders.

get spalled and that there is a dense columnar cracking in the bulk of the specimen. This is consistent with the observation made in Section 5.1 about the differences in the $\sigma - \varepsilon$ curves between cubes and cylinders. The steeper response observed in the $\sigma - \varepsilon$ curves for cylinders may be associated to the inclined fracture surface or cone type failure, see Figure 6b, with a initial localization failure in the central part of the specimen. In the case of cubes the fracture process is provoked by a stress concentration near the cube corners. Inclined micro-cracks appear and coalesce near the corners and provoke the crack pattern observed. The stress state inside the cubes provokes vertical cracks that en-up forming column-like fragments. The extent of the micro-cracking in this fracture process leads to a higher energy consumption than in the case of cylinders.

Another interesting point is that the failure pattern observed is almost independent for scaled elements in the size interval planned for the experimental program. Cylinders broke in all cases by a diagonal fracture plane and cubes present a bursting rupture combined with a dense columnar cracking. We have also included the crack pattern obtained for a cylinder of 150×300 mm, called D1, in Figure 6b.

6 RELATIONSHIP BETWEEN CYLINDER AND CUBE STRENGTH BASED ON FRACTURE MECHANICS THEORIES

In this section we analyze the test results showed in section 5 within a Fracture Mechanics frame. The aim of this analysis is to get a relationship between cylinder and cube strength. According to Bazant and Planas (Bažant and Planas, 1998) the maximum load P_c is a function of the specimen geometry (shape and size), boundary conditions and concrete properties. In this investigation only the size and shape are varied, keeping constant the boundary conditions. For the sake of simplicity, the nominal strength at peak load, σ_N, for a concrete specimen is represented by the generalized deterministic energy-based formula for size effect at crack initiation proposed by Bažant (Bažant, 1998). Its applicability is based on the fact the failure is provoked by crack initiation, which is our case as shown in the previous section. So, we can write σ_N as:

$$\sigma_N = \sigma_\infty \left(1 + \frac{B}{\beta_H}\right)^{\frac{1}{r}} \qquad (1)$$

$$\sigma_\infty = \kappa \sigma_0$$

where σ_∞ is the theoretical strength for a specimen of a infinite size and σ_0 is a reference strength, which in our case we take as the strength that is obtained from the standard test (ASTM C-39 on 100×200 mm cylinders). B and κ are empirical constants dependent on the specimen shape and fracture properties of the material, but not on the structural size. Both constants can be obtained from test results. β_H is the denominated Hillerborg's brittleness number (Bažant and Planas, 1998). It is defined as the ratio between the representative size of the specimen —represented by the depth D in the case of cylinders, and L in the case of cubes— and the characteristic length of the concrete. The values of the exponent r using in concrete ranges from 1 to 2 (Bažant, 1998). The formulation when the exponent r is equal to 2, is identical to the formula proposed on the basis of strictly geometric arguments proposed by Carpinteri and coworkers (Carpinteri et al., 1998). The size effect law can be rewritten as:

$$\sigma_N = \sigma_\infty \sqrt{1 + \frac{B}{\beta_H}} \qquad (2)$$

Applicability of Eqs. 1 and 2 requires not only that the specimens are scaled to each other, but also that the shape of cracks patterns must be similar. In our case the crack patterns for cubes are very similar, as we showed in Section 5.3.

Figure 7a shows the linear regression made with the cubes to get the constants κ and B. Please note that the Pearson's correlation coefficient, R in Figure 7a is close to 1. Figure 7b shows the test results compared to the obtained size effect law in a non-dimensional fashion. It may be pointed out that size effect tends to disappear when $D \to \infty$. The compressive strength converges to a value that in our case is quite similar to 1, that is, for big sizes the compressive strength in cubes converges to the compressive strength obtained with the standard procedure. Further analysis would be necessary to evaluate the influence of the displacement control platen-to-platen velocity. The strength

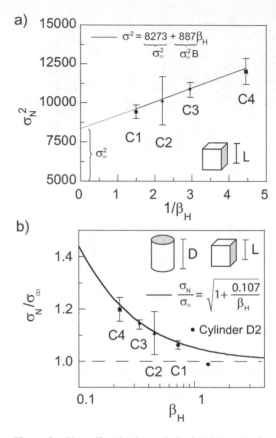

Figure 7. Size effect in the peak load: (a) Results for the regression to calibrate σ_∞ and B coefficients for cubes specimens; (b) Size effect law.

for a specimen of infinite size can be evaluated as $\sigma_\infty = \sigma_0 \kappa$. In our case σ_∞ is equal to 88.7 MPa.

Based on the obtained law a correlation between compressive strength obtained for different cubes specimens, σ_{cub}, and for Standard tests, f_c, can be derived. In a general way for concrete which compression standard is around 100 MPa and have a characteristic length of 150 mm, the next simplified relationship arises. For HSC like the one used in this research $f_c = \sigma_0 \approx 100$ MPa, $\ell_{ch} \approx 150$ mm and $\sigma_{cub} = \sigma_N$, the expression is:

$$f_c = \sigma_{cub} \sqrt{\frac{L}{L + L_0}} \qquad (3)$$

where L is the side of the cube and L_0 is a empirical constant equal to 20 mm in this case. The fact that cubes are somewhat easier to test, since cubes provide directly two couples of perfectly-plane parallel surfaces and they not require capping, together with the necessity of using specimens whose strength do not surpass the capacity of standard equipment, suggest that cubes would be a solution for compressive test characterization. The tendency to prefer cylinder to cubes, in our opinion does not have a solid foundation due to the structural character of the compressive test. There is not a reason to think that cylinders catch concrete compressive behavior better than cubes, since compressive tests depend strongly on many factors other than the actual material properties. The standard compressive strength of HSC could be obtained from cubes using the simplified expression showed in Eq. 3.

With this example we want to show how that using Fracture Mechanics theories provide solutions for concrete technological problems. These solutions may help to understand concrete behavior and structural failures better.

7 CONCLUSIONS

In this paper we investigate the mechanical behavior of high-strength concrete (around 100 MPa) in compression and tested in strain-control. We are particularly interested in the influence of the shape and the size of the specimens on the compressive strength, f_c, of the material. Cylinders and cubes were tested in strain control at a rate that was kept constant throughout the experimental program. Concrete making and testing procedures were closely controlled to reduce experimental scatter. The following conclusions can be drawn from the study:

1. The prepeak and post-peak behavior in the $\sigma - \varepsilon$ curves are dependent on the specimen size and shape. Cubes show a mild failure localization in contrast with the strong failure localization observed in cylinders.
2. Test results show size effect. Large specimens resist less in terms of stress than the small ones. The size effect in the cubes is quite stronger than in cylinders, where the average strength obtained is roughly constant within the size interval planned for this research.
3. The crack pattern observed after the test is sensitive to the shape of the specimen. A main inclined fracture surface is nucleated in cylinders, whereas in cubes we find that the lateral sides get spalled and that there is a dense columnar cracking in the bulk of the specimen. This is consistent with the differences observed in the $\sigma - \varepsilon$ curves between cubes and cylinders. Interestingly, the failure pattern does not change with the size.
4. The size effect in the compressive strength of cubes is described by a simple model based on Fracture Mechanics concepts. A relationship between

standard cylinder strength and strength obtained from cubes of any size is derived. It discloses the influence of the cube dimension and may help to deduce relations between the compressive strength determined from cylinders and cubes, inside a theoretical Fracture Mechanics frame.

ACKNOWLEDGEMENT

The authors gratefully acknowledge financial support for this research provided by the *Ministerio de Fomento*, Spain, under grant BOE305/2003, and the *Junta de Comunidades de Castilla -La Mancha*, Spain, under grant PAI05-028. Javier R. del Viso and Jacinto R. Carmona also thank the *Junta de Comunidades de Castilla -La Mancha*, Spain, and the *Fondo Social Europeo* for the fellowships that support their research activity.

REFERENCES

Bažant, Z. P. (1998). Size effect in tensile and compression fracture of concrete structures: Computational modeling and designing. In *Fracture Mechanics of Concrete Structures*, pages 1905–1922, Freiburg, Germany. F.H. Wittmann, Ed., Aedificatio Publishers.

Bažant, Z. P., and Planas, J. (1998). *Fracture Size Effect in Concrete and Other Quasibrittle Materials*. CRC Press, Boca Raton.

Borges, J. U. A., Subramaniam, K. V., Weiss, W. J., Shah, S. P., and Bittencourt, T. (2004). Length effect on ductility of concrete in uniaxial and flexural compression. *ACI Structural Journal*, 101(6):765–772.

Carpinteri, A., Chiaia, B., and Ferro, G. (1998). Size effect on nominal tensile strength of concrete structures: Multifractality of material ligaments and dimensional transition from order to disorder. *Materials and Structures*, 28:311–317.

Choi, S., Thienel, K. C., and Shah, S. P. (1996). Strain softening of concrete in compression under different end constraints. *Magazine of Concrete Research*, 48(175): 103–115.

Elices, M., Guinea, G. V., and Planas, J. (1992). Measurement of the fracture energy using three-point bend tests. 3. Influence of the cutting the $P - \delta$ tail. *Materials and Structures*, 25:327–334.

Guinea, G. V., Planas, J., and Elices, M. (1992). Measurement of the fracture energy using three-point bend tests. 1. Influence of experimental procedures. *Materials and Structures*, 25:121–128.

Jansen, D. C., and Shah, S. P. (1997). Effect on length on compressive strain softening of concrete. *Journal of Engineering Mechanics-ASCE*, 123(1):25–35.

Mindess, S., Young, J. F., and Darwin, D. (2003). *Concrete*. Prentice Hall, Pearson Education, Inc. United States of America.

Petersson, P. E. (1981). *Crack Growth and Development of Fracture Zones in Plain Concrete and Similar Materials*. Report No. TVBM-1006, Division of Building Materials, Lund Institute of Technology, Lund, Sweden.

Planas, J., Elices, M., and Guinea, G. V. (1992). Measurement of the fracture energy using three-point bend tests. 2. Influence of bulk energy dissipation. *Materials and Structures*, 25:305–312.

Shah, S. P., and Sankar, R. (1987). Internal cracking and strain-softening response of concrete under uniaxial compression. *ACI Materials Journal*, 84(3):200–212.

Shah, S. P., Swartz, S. E., and Ouyang, C. (1995). *Fracture Mechanics of Concrete*. Wiley, New York.

Torreti, J. M., Benaija, E. H., and Boulay, C. (1993). Influence of boundary conditions on strain softening in concrete compression tests. *Journal of Engineering Mechanics-ASCE*, 119(12):2369–2384.

Van Mier, J. G. M. (1984). *Strain-softening of concrete under multiaxial loading conditions*. PhD thesis, Eindhoven University of Technology, Eindhoven, The Netherlands.

Fracture Mechanics of Concrete and Concrete Structures – High-Performance Concrete,
Brick-Masonry and Environmental Aspects – Carpinteri, et al. (eds)
© 2007 Taylor & Francis Group, London, ISBN 978-0-415-44617-4

Relationship between compressive strength and modulus of elasticity of high-strength concrete

T. Noguchi
University of Tokyo, Japan

K.M. Nemati
University of Washington, Seattle, Washington, USA

ABSTRACT: Modulus of elasticity of concrete is frequently expressed in terms of compressive strength. While many empirical equations for predicting modulus of elasticity have been proposed by many investigators, few equations are considered to cover the entire data. The reason is considered to be that the mechanical properties of concrete are highly dependent on the properties and proportions of binders and aggregates. This investigation was carried out as a part of the work of the Research Committee on High-strength Concrete of the Architectural Institute of Japan (AIJ) and National Research and Development Project, called New RC Project, sponsored by the Ministry of Construction. More than 3,000 data, obtained by many investigators using various materials, on the relationship between compressive strengths and modulus of elasticity were collected and analyzed statistically. The compressive strength of investigated concretes ranged from 20 to 160 MPa. As a result, a practical and universal equation is proposed, which takes into consideration types of coarse aggregates and types of mineral admixtures.

1 INTRODUCTION

Modulus of elasticity of concrete is a key factor for estimating the deformation of structural elements, as well as a fundamental factor for determining modular ratio, n, which is used for the design of structural members subjected to flexure. Based on the relationship of modulus of elasticity of concrete that it is proportional to the square root of compressive strength in the range of normal concrete strength, AIJ specifies the following equation to estimate modulus of elasticity of concrete.

$$E = 2.1 \times 10^5 (\gamma/2.3)^{1.5} (f_c/200)^{1/2} \qquad (1)$$

where E = modulus of elasticity (kgf/cm^2)
γ = unit weight of concrete (t/m^3)
f_c = specified design strength of concrete (kgf/cm^2)

Eq. 1 is applied to concrete of a specified design strength of 36 MPa or less which is defined as normal strength concrete, a number of experiments have revealed that Eq. 1 over-estimates the modulus of elasticity as the compressive strength increases (Figure 1). This study aims to derive a practical and universal equation which is applicable to high-strength concretes with compressive strengths greater than 36 MPa,

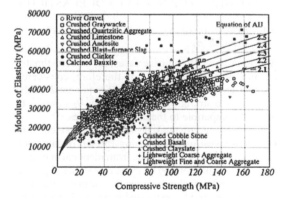

Figure 1.

using regress ional analysis of numerous results of experiments published in Japan. The outline of this study was published in 1990.

2 REGRESSIONAL ANALYSIS

Before performing any analysis, it was necessary to create a basic form of the equation for modulus of elasticity. In this study the authors adopted the conventional form, in which modulus of elasticity, E, is expressed as a function of compressive strength, σ_B, and unit weight, γ. Since it is self-evident that the concrete with a compressive strength of 0 MPa has modulus of elasticity of 0 MPa, the basic form of the equation is expressed as Eq. 2.

$$E = a\sigma_B{}^b \gamma^c \qquad (2)$$

The parameters examined are compressive strength, modulus of elasticity, and unit weight of concrete at the time of compression test, as well as types and mechanical properties of materials for producing concrete, mix proportioning, unit weight and air content of fresh concrete, method and temperature of curing, and age.

2.1 Estimation of the unit weight

Out of the 3000 experimental data collected, only one third included measured unit weight of specimens, γ. In order to express modulus of elasticity as a function of compressive strength and unit weight, the unit weights of hardened concrete had to be estimated when measured unit weights were not available from the data on materials used, mix proportioning, curing conditions, and age.

3 EQUATION FOR MODULUS OF ELASTICITY

3.1 Evaluation of exponent b of compressive strength, σ_B

As compressive strength increases, Eq. 1 overestimates the modulus of elasticity. It is therefore, considered appropriate to reduce the value of exponent b of the compressive strength, σ_B, to less than $^1/_2$ in order to make it compatible to the measured values.

Firstly, range of possible values of exponent b in Eq. 2 was investigated evaluating 166 sets of data, each set of which had been obtained from identical materials and curing conditions by the same researcher. Figure. 2 shows the relationship between the ultimate compressive strengths and the estimated exponent b. Similarly, Figure. 3 shows the relationship between the exponent b and the ranges of compressive strengths in the sets of data. In Figures. 2 and 3, while the estimated values of exponent b vary widely, the values show a tendency to decrease from around 0.5 to around 0.3, as the maximum compressive strengths increase and the ranges of compressive strength widens. In other words, whereas modulus of elasticity of normal-strength concrete has been predictable from the compressive strength with exponent b of 0.4 to 0.5, the values of 0.3 to 0.4 are more appropriate a general-purpose equation to estimate modulus of elasticity for a wide range of

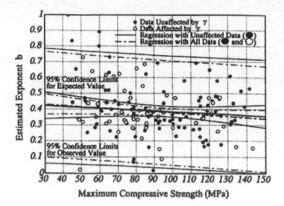

Figure 2.

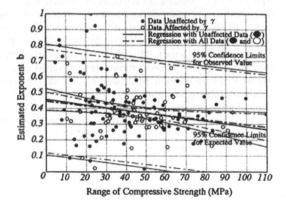

Figure 3.

concretes from normal to high-strength. Consequently, the 1/3 is proposed as the value of exponent **b**.

3.2 Evaluation of exponent c of unit weight, γ

Secondly, by fixing exponent **c** at 1/3, as mentioned above, exponent **c** of the unit weight, γ, was investigated. The relationship between the unit weight, γ, and the value obtained by dividing modulus of elasticity by compressive strength to the 1/3 power, $E/\sigma_B^{1/3}$ is shown in Figure. 4 with a regression equation (Eq. 3) that was obtained from data on all aggregates as shown below.

$$E = 1630\sigma_B^{1/3}\gamma^{1.89} \qquad (3)$$

From Figure. 4, it can bee seen that Eq. 3 clearly shows the effect of the unit weight on modulus of elasticity, for concretes made with lightweight, normal weight, and heavy weight aggregates (bauxite, for example). The concretes made with normal weight aggregate, however, are scattered over a rather wide

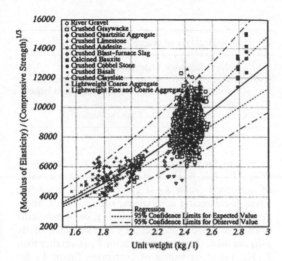

Figure 4.

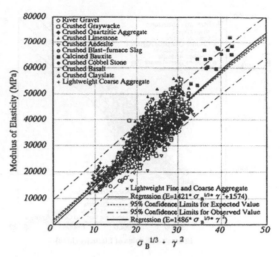

Figure 5.

range of 6000 to 12000 of $E/\sigma_B^{1/3}$, while they gather in a relatively small unit weight range of 2.3 to 2.5. This suggests differences in the effects of lithological types of aggregates on modulus of elasticity, which will be discussed later in this paper. Whereas 1.5 has been used conventionally as the value for exponent **c**, as indicated in Eq. 1, 1.89 was obtained from the regression analysis as the value for exponent **c** that is applicable for a wide range of concretes from normal to high-strength concretes. In consideration of the utility of the equation, a value of 2 is proposed as the value for exponent **c**.

3.3 Evaluation of coefficient a

Thirdly, after fixing exponent **b** and exponent **c** at 1/3 and 2, respectively, the value for coefficient **a** was investigated. The relationship between modulus of elasticity, E, and the product of compressive strength to the 1/3 power and unit weight to the second power, $\sigma_B^{1/3}\gamma^2$, is shown in Figure. 5, together with regression equation (Eq. 4) obtained from the data on all aggregates as shown below.

$$E = 1486\sigma_B^{1/3}\gamma^2 \qquad (4)$$

The coefficient of determination is as high as 0.769, and the 95% confidence interval of modulus of elasticity is within the range of ±8000 MPa, as shown in Figure. 5. The relationship between modulus of elasticity and $\sigma_B^{1/3}\gamma^2$ can therefore be virtually expressed by Eq. 4.

3.4 Evaluation of correction factor k

In the conventional equation for modulus of elasticity, Eq. 1, the only difference in the type of coarse

aggregate taken into account is the difference in the specific gravity, the effect of which is represented by the unit weight of concrete, γ. However, use of a wide variety of crushed stone has revealed that the difference in unit weight is not the only factor to account for the differences in moduli of elasticity of concretes of the same compressive strength. Lithological type should also be considered as a parameter of coarse aggregate. Besides, it has also been pointed out by many researchers that modulus of elasticity cannot be expected to increase with an increase in compressive strength, when the concrete contains a mineral admixture for high-strength, such as silica fume. This suggests the need to include the type of admixtures as another factor affecting modulus of elasticity. Thus, the type of coarse aggregate, as well as type and amount of mineral admixtures should be considered on the investigation of the values of correction factor, k.

$$E = k \cdot 1486\sigma_B^{1/3}\gamma^{1/2} = k_1 k_2 \cdot 1486\sigma_B^{1/3}\gamma^2 \qquad (5)$$

Where $k = k_1 \cdot k_2$
k_1 = correction factor corresponding to coarse aggregates
k_2 = correction factor corresponding to mineral admixtures.

3.4.1 Evaluation of correction factor k_1 for coarse aggregate

Figure 6. shows the relationship between the values estimated by Eq. 4 and the measured values of modulus of elasticity of concretes without admixtures. According to Figure. 6, most of the measured values/the calculated values, i.e. values of k_1 in Eq. 5, fall in the range of 0.9 to 1.2, indicating that each lithological type of coarse aggregate tends to have

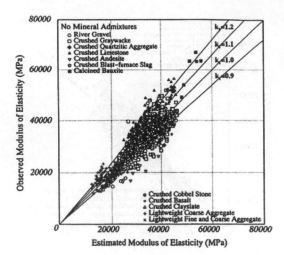

Figure 6.

Table 1.

Aggregate	k_1
River Gravel	1.005
Crushed Graywacke	1.002
Crushed Quartzitic Aggregate	0.931
Crushed Limestone	1.207
Crushed Andesite	0.902
Crushed Basalt	0.922
Crushed Clayslate	0.928
Crushed Cobbel Stone	0.955
Blast-furnace Slag	0.987
Calcined Bauxite	1.163
Lightweight Coarse Aggregate	1.035
Lightweight Fine and Coarse Aggregate	0.989

an inherent k_1. The correction factor k_1 for each coarse aggregate is presented in Table 1. According to Table 1, the effects of coarse aggregate on modulus of elasticity are classified into three groups. The first group, which requires no correction factor, includes river gravel, crushed greywacke, etc.; the second group, which requires correction factors of greater than 1, includes crushed limestone and calcined bauxite; and the third group, which requires correction factors smaller than 1, includes crushed quartzitic aggregate, crushed andesite, crushed cobble stone, crushed basalt, and crushed clayslate. Consequently, the value for each type of coarse aggregate is proposed as shown in Table 2, in consideration of the utility of the equation.

3.4.2 Evaluation of Correction Factor k_2 for Admixtures

Table 3 presents the averages of correction factor k_2 obtained for each lithological type of coarse aggregate

Table 2.

k_1	Lithological type of coarse aggregate
1.20	Crushed Limestone, Calcined Bauxite
0.95	Crushed Quartzitic Aggregate, Crushed Andesite, Crushed Basalt, Crushed Claystate, Crushed Cobbel Stone
1.00	Coarse Aggregate Other Than the Above

as well as for each type and amount of admixtures. When fly ash is used as an admixture the value of correction factor k_2 is greater than 1, but when strength-enhancing admixtures, such as silica fume, ground granulated blast furnace slag, or fly ash fume (ultra fine powder produced by condensation of fly ash) are used, the correction factor k_2 is smaller than 1. The proposed values of correction factor k_2 for admixtures are shown in Table 4, in consideration of the utility of the equation.

3.5 Practical Equation for Modulus of Elasticity

Eq. 5 was derived as an equation for modulus of elasticity. Meanwhile, conventional equations such as Eq. 1 have been convenient in such a way that standard moduli of elasticity can be obtained simply by substituting standard values of compressive strength and unit weight in the equation. In this study, Eq. 6 is proposed as the equation to be used for modulus of elasticity calculations. The equation is based on 60 MPa, a typical compressive strength of high-strength concrete, and uses a unit weight of 2.4, which leads to the compressive strength of 60 MPa.

$$E = k_1 k_2 \cdot 3.35 \times 10^4 (\gamma/2.4)^2 (\sigma_B/60)^{1/3} \qquad (6)$$

4 COMPARISON OF EQUATIONS

Figures 7–10 show the accuracy of estimation by Eq. 1 and Eq. 6 as well as by ACI 363R and CEB-FIP equations, which are presented in Table 5.

As pointed out by a number of researchers, the equation by the Architectural Institute of Japan (AIJ)(Figure. 7) tends to overestimate moduli of elasticity in the range of compressive strength over 40 MPa, except in the cases where crushed limestone or calcined bauxite is used as the coarse aggregate. The residuals also tend to increase in relation to the compressive strength.

The equation by ACI 363R (Figure. 8) slightly underestimates moduli of elasticity when crushed limestone or calcined bauxite is used as the coarse aggregate, regardless of the compressive strength.

Table 3.

Coarse aggregate	Silica fume			Glanulated blast-furnace slag		Fly Ash	
	<10%	10–20%	20–30%	<30%	30%<	Fume	Fly ash
River Gravel	1.045	0.995	0.818	1.047	1.118	–	1.110
Crushed Graywacke	0.961	0.949	0.923	0.949	0.942	0.927	–
Crushed Quartzitic Aggregate	0.957	0.956	–	0.942	0.961	–	–
Crushed Limestone	0.968	0.913	–	–	–	–	–
Crushed Andesite	–	1.072	0.959	–	–	–	–
Crushed Basalt	–	–	–	–	–	–	1.087
Calcined Bauxite	–	0.942	–	–	–	–	–
Lightweight Coarse Aggregate	1.026	–	–	–	–	–	–
Lightweight Fine and Coarse Aggregate	1.143	–	–	–	–	–	–

Table 4.

k_2	Type of addition
0.95	Silica Fume, Ground Glanulated Blast-furnace Slag, Fly Ash Fume
1.10	Fly Ash
1.00	Addition Other Than the Above

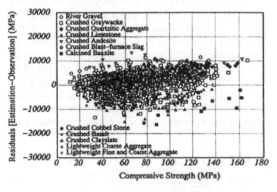

Figure 8.

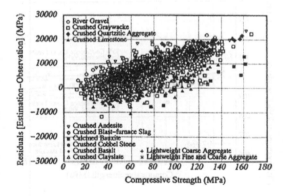

Figure 7.

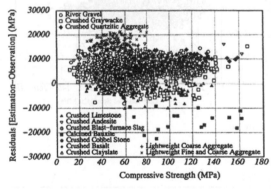

Figure 9.

In the case of other aggregates, the equation tends to overestimate the moduli, though marginally, as compressive strength increases.

The CEB-FIP equation (Figure. 9) leads to clear differences in residuals depending on the lithological type of coarse aggregate. When lightweight aggregate is used, the equation overestimates the moduli, and the value of the residuals tends to decrease as the specific gravity of coarse aggregate increases from crushed quartzitic aggregate to crashed graywacke, crushed limestone, and calcined bauxite.

The residuals by the new equation (Figure. 10), as a whole, fall in the range of ±5000 MPa, regardless of the compressive strength levels, although a portion of data display residuals near ±10000 MPa. The new equation is therefore assumed to be capable of estimating moduli of elasticity for a wide range of concretes from normal to high-strength.

5 EVALUATION OF 95% CONFIDENCE INTERVALS

The accuracy of the equation being enhanced by incorporating the correction factors, 95% confidence intervals should be indicated, because the reliability of the estimated values is required in structural design and is used when determining materials and mix proportioning so as to ensure safety.

Excluding the ease of using fly ash as an admixture, only five values of the product of the correction factors, k_1 and k_2, are possible, i.e. 1.2, 1.14, 1.0, 0.95, and 0.9025.

A regression analysis of Eq. 2 was conducted for the combinations of a coarse aggregate and an admixture corresponding to each of the five values of $k_1 \cdot k_2$, to obtain 95% confidence intervals of both estimated and measured moduli of elasticity. The results are shown in Figures. 11–15. The curves indicating the upper and lower limits of 95% confidence of the expected values for all $k_1 \cdot k_2$, are within the range of approximately ±5% of the estimated values, regardless of compressive strength and unit weight. The curves indicating those for the observed values are also within the range of approximately ±20% of the estimated values. Consequently, the 95% confidence limits of the new equation (Eq. 6) are expressed in a simple form as Eq. 7, and the 95% confidence limits of measured modulus of elasticity can be expressed as Eq. 8.

$$E_{e95} = (1 \pm 0.05)E \tag{7}$$

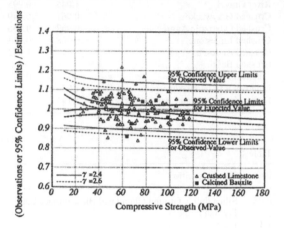

Figure 11.

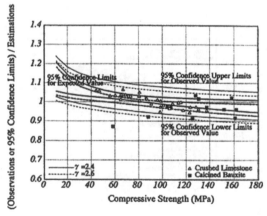

Figure 12.

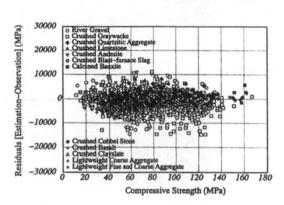

Figure 10.

Table 5.

ACI 363R Equation	$E + (40000 \cdot \sigma_B^{0.5} + 1.0 \cdot 10^6) \cdot (\gamma/2.346)^{1.5}$	(psi)	State-of-the-Art Report on High-Strength Concrete ACI JOURNAL, July–August 1984
CEB-FIP Equation	$E = \alpha \cdot 21500 \cdot (\sigma_{28}/10)^{1/3}$ $\alpha = 1.2$: Basalt, Dense Limestone Aggregates $\quad$ 1.0 : Quartzitic Agregates $\quad$ 0.9 : Limestone Aggregates $\quad$ 0.7 : Sandstone Aggregates	(MPa)	CEB–FIP MODEL CODE 1990 Final Draft Contribution a la 28e Session Pleniere du CEB Vienne – September 1991

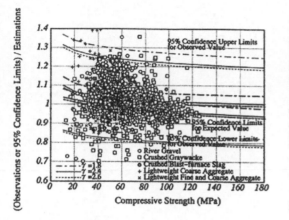

Figure 13.

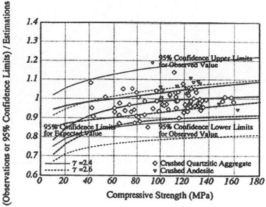

Figure 15.

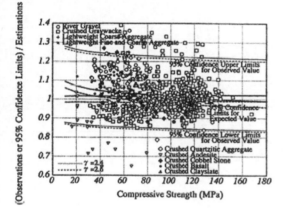

Figure 14.

$$E_{o95} = (1 \pm 0.2)E \tag{8}$$

where $E_{e95} = 95\%$ confidence limits of expected modulus of elasticity.
$E_{o95} = 95\%$ confidence limits of observed modulus of elasticity.

6 CONCLUSIONS

Multiple regression analyses were conducted using a great deal of data published in Japan regarding the relationship between compressive strength and modulus of elasticity of concrete, by assuming compressive strength and unit weight as explanatory variables and modulus of elasticity as the target variable. As a result, a new equation (Eq. 6) is proposed as a practical and universal equation for modulus of elasticity. It is applicable to a wide range of concretes from normal to high-strength. The 95% confidence limits of the new equation were also examined, and Eq. 7 and 8 were proposed as the equations to indicate the 95% confidence limits for the expected and observed values, respectively.

REFERENCE

Tomosawa, F., Noguchi, T. and Onoyama, K. "Investigation of Fundamental Mechanical Properties of High-strength Concrete", Summaries of Technical Papers of Annual Meeting of Architectural Institute of Japan, pp.497–498, October 1990.

Fracture Mechanics of Concrete and Concrete Structures – High-Performance Concrete,
Brick-Masonry and Environmental Aspects – Carpinteri, et al. (eds)
© *2007 Taylor & Francis Group, London, ISBN 978-0-415-44617-4*

Fracture characteristics of high-strength light-weight cement mortar composites reinforced with the waste products of alluminium processing

J. Hemanth

Department of Mechanical Engineering, Siddaganga Institute of Technology (S.I.T.), Karnataka, India

ABSTRACT: An investigation was carried out to summarize the results of the mechanical properties of high strength lightweight eloxal reinforced cement mortar subject to short term loads. Eloxal (in the solid slag form) is a waste obtained during the production of aluminum. It is mainly of aluminum oxide, SiO_2, CaO, MgO and other substances. It's a hard substance, having sufficient strength with additive properties and bonds very rapidly. Eloxal reinforced cement mortars in the present investigation are tested for their compression and fracture behavior. Data were obtained pertaining to compressive strength, fracture behavior (using dogbone tension tests and double cantilever beam tests), role of moisture and drying effects. Deformation properties under load were studied to provide insight into the internal behavior and failure mechanism of light weight eloxal reinforced cement mortar. To analyze the mode of failure, distribution of eloxal particles in cement mortar and the deformation behavior, several optical and Scanning Electron Microscope (SEM) photographs were taken to study the mechanism. Results of the tests of eloxal reinforced cement mortar are compared with unreinforced cement mortar and information obtained else where in earlier tests of normal weight cement mortar. Structural composites materials offer an excellent opportunity to produce components that achieve weight savings and improved structural properties. The eloxal particles (dispersoid) added to cement mortar in the present investigation is varied from 20 to 40 wt%. In steps of 10 wt.%. The resulting composite blocks cast were tested for their properties.

1 INTRODUCTION

Lightweight cement mortar has been used successfully for many years for structural members and systems in buildings and bridges. One of the earliest applications in North America was in the construction of cement mortar ships during World War I (C. Wilson). Through the years, by judicious selection of the lightweight aggregate and careful proportioning, semi-lightweight cement mortars having high compressive strengths have been made (B.H. Spratt). Although such strength is not necessary in many structural applications, there are advantages to the use of very high strength lightweight concrete in such applications as offshore drilling platforms. Such concrete has greater buoyancy and thus is easier to tow in shallow waters, and less excavation is required in construction of the dry dock compared to heavier structures. There is one instance where such cement concrete have been used for oil drilling platforms in the Arctic (G. Wischers and W. Manns). In addition to its lighter weight, which permits savings in dead load and so reduces the cost of both super structure and foundations, this cement concrete is more resistant to fire and provides better heat

and sound insulation than cement mortar of normal density (R.L. Carrasquillo, J.J. Shideler and J.A. Hanson). For lightweight cement mortar structures, as for structures of normal weight cement concrete, there is a well-established trend toward using higher compressive strengths. This permits the use of smaller member sizes, which in turn permits further reduction in dead load with attendant cost savings, and extends the practical range of span as well (P.H. Kaar, P.T. Wang and R.L. Carrasquillo).

Little useful information has been available to the structural engineer on the engineering properties of high-strength lightweight cement mortar. The main purposes of the work described in this paper were a) to gain insight into the difference in the internal behavior of high-strength light weight cement mortar compared with normal cement mortar and b) to establish its mechanical properties and compare these with the properties of cement mortar of normal density. The work of this paper is aimed at establishing the fracture characteristics of eloxal particles to the one by the author of this paper for the fracture study of chilled aluminum alloy-quartz particulate composite (Joel Hemanth). The salient feature of this model is the presence of a zone, called the pseudoplastic zone, in which the aluminum matrix is cracked, but reinforcing fibers continue to provide resistance to the crack opening. Such a behavior requires use of a test specimen allowing sufficient length of crack development in order that a complete crack resistance and estimate of the fracture toughness may be obtained.

Advanced composite materials such as fiber-reinforced polymers have the potential to revolutionize engineering technology. In order to use these advanced composite materials, a detailed knowledge of their mechanical behavior is imperative. Moreover, structurally efficient design using these advanced composites mandates a detailed knowledge of their mechanical properties. The mechanical properties of composites have received much attention for decades, but their performance in environments simulating hostile service and their long-term durability are only beginning to be studied. These composites are particularly suitable because they are lightweight, presumably durable, corrosion resistant and have high compression strength. Their lower density is important not only because it adds less weight to the existing structures but also because it is very important during construction. Heavy equipment is needed for construction with steel; however, it is not needed for lightweight composites. The fact that these composites are electrically non-conductive and have impact resistance also helps in certain applications. Extensive research has been carried out on using these composites for strengthening; however, information about their durability is still lacking (Y. Xiao, T.Harmon, P.Labossier, A.Nanni and H. Toutanji). Eloxal, the waste is obtained during the production of aluminum in the plant produced by means of electrolysis or limited oxidation processing of metal plates in the electrolyte solution. The present composite developed using eloxal as the dispersoid has the properties of rapid hardening, light weight, high sulphate resistance, low thermal coefficient of expansion along with good mechanical properties (J. Piasta and A.N. Scian).

2 EXPERIMENTAL PROCEDURE

2.1 Composition of the matrix material and the dispersoid

Chemical composition of the matrix (cement mortar) is given in Table 1.

In this investigation eloxal particles from 20 to 40 wt.% in steps of 10 wt.% were dispersed in the matrix. Chemical composition and some properties of eloxal (dispersoid) are as follows: Density: 1.3 gm/cc, Hardness: BHN 310, Melting point: 680°C and Youngs Modulus 75 GPa, Size distribution of the dispersoid: 3.72 to 7.8 mm, Chemical composition of the dispersoid is indicated in Table 2.

Table 1. Chemical composition of the cement (Birla super 53 Grade, normal portland cement, ASTM type I).

S_iO_2	CaO	Al_2O_3	Fe_2O_3	MgO
22.02	62.26	3.99	2.26	3.30

(Balance of the composition contains SO_3, Na_2O and K_2O).

Table 2. Chemical composition of the dispersoid (eloxal).

Al_2O_3	S_iO_2	MgO	CaO	Na_2O
48.21	29.82	3.32	15.83	1.21

(Balance of the composition contains Al-Hydroxide).

Table 3. Sand aggregate gradation.

Passing sieve no.	Retained sieve no.	Aggregate size (mm)	Proportion %
8	16	1.18–2.37	16
16	30	0.7–1.19	30
30	50	0.3–0.6	40
50	100	0.15–0.3	10
100	Pan	Less than 0.15	5

Eloxal available from aluminum industry will be in the form of a hard solid slag and this is reduced to the required size into the granular form using a ball mill. Before adding, eloxal was prepared by washing it with water and thus the soluble NaOH and other substances were removed. Finally the solid eloxal was dried in an oven at 105°C for 2 hours.

2.2 Mix proportions and casting procedure

The cement-aggregate ratios and water-cement ratios used in the present investigation were 1:3 respectively. The aggregate grading is shown in Table 3.

Normal portland cement (by Birla, India) ASTM Type 1, was used. The casting procedure consists of mixing dry, predetermined quantities of fine graded aggregate and cement for about three minutes. Water was then added slowly and the mixing continued for another minute. Dry eloxal particles (size 3.72 to 7.8 mm) were slowly and continuously added as the mixing continued. Mixing was done by a medium sized Hobart mixing machine of planetary mixing action was used. The resulting mix was used to prepare compression (152*305 mm cylinders, ASTM C 109 standards), tension (dogbone tension test specimen with a notch) and Double Cantilever Beam (DCB) testing specimens. Finally the prepared specimens were compacted and leveled and left to cure. The specimens were de-molded after 24 hours and moist cured in a water bath at room temperature for 7, 14, 21 28, 35 and 42 days before testing. A series of four cement mortar mixtures were made, containing 20%, 30% and 40% eloxal particles (dispersoid) respectively.

2.3 The Notched Tensile Specimen (NTS) testing and Double Cantilever Beam (DCB) Testing

Direct tension test on the notched specimens were performed in an Instron machine. The dogbone tension test specimen of 175 mm effective length with right angled notch 3 mm deep cast on opposite sides to resist the crack in one plane was used in the present investigation. The casting procedure, mixes and curing history were identical to those used for DCB specimens. Similar to DCB testing, a tensile load of 2.0 KN and cross head speed of 2 mm per minute were used in the tests. An extensometer to measure the displacement at the notch was fixed on the specimen, each arm a distance of 25 mm from the notch.

The dimensions of the Double Cantilever Beam (DCB) specimen used in this study were adopted according to findings available elsewhere (K. Visalvanich). On each side of the specimen, a groove was cast which was 12.7 mm wide at the surface and 7.6 mm deep before sloping into an apex angle of 60 degrees. To prevent the crack from deviating from the assigned path along the groove, two 2 mm diameter plain steel wires, placed as close to the groove as possible, were used to reinforce each arm of the beam. The initial notch depth was constant throughout the experiments and set at 101.6 mm. The mould was made such that the cement mortar could be cast in a horizontal position. Each cover had a pre-molded notch attached to it, to produce the side grooves in the specimen. On the bottom cover was also attached a removable pre-molded notch to produce the required initial notch in the specimen.

The DCB specimens were tested in an Instron machine in vertical position, the vertical load being transmitted to the arms through a loading wedge and a roller bearing system. Details of wedge and roller bearing system may be found in other findings (Y.N. Zibra).

The full load cell scale chosen for the DCB specimen testing was 2 KN and a crosshead speed of 2 mm/min was used throughout the investigation. An extensometer with a range of 25 mm was used to measure the Arm Opening Displacement (AOD). The arms of the extensometer attached to the axles of the roller bearings by small screws. The signal from the extensometer was amplified by the strain data unit and used to control the chart servo-drive mechanism. This way, the crack mouth displacement and the load could be plotted as the abscissa and ordinate, respectively, on am X-Y recorder. Calibration of the extensometer, carried out at the beginning of each test, was done with the help of a micrometer screw gauge having an accuracy of 10^{-3} mm. Crack lengths were observed and measured through a vertical axis traveling microscope with an accuracy of ±0.1 mm. The specimen was loaded at a rate of 2 mm per minute until crack extension was observed to occur after which the specimen was unloaded at the same rate, having noted the crack length at which the crack began to extend. This process of loading and unloading was repeated until the crack length was developed to the zero zone or until the specimen failed. The process of loading and unloading eliminates the influence of permanent deformation on the compliance of the DCB specimen.

Microscopic examination was conducted on the specimens using a Scanning Electron Microscope (SEM) and the optical microscope (Neophot–21) under different magnifications to study the distribution, orientation, bonding of eloxal particles in cement mortar and the mode of fracture.

3 RESEARCH SIGNIFICANCE

The research shows that high-strength lightweight cement mortar possesses properties that are significantly different form normal cement mortar and also different from high-strength normal weight cement mortar. Such cement mortar as used in this investigation can be used in buildings and bridges mainly to reduce dead load and increase the practical range of spans. Eloxal in the form of scrap which is cheaply and abundantly available can be recycled to produce cement mortar-eloxal light weight and high strength composite. Thus, this paper provides important new information on the behavior and engineering properties of this material.

4 RESULTS AND DISCUSSION

Results of the micro and macro examination indicate that the shape of pores present, distribution and orientation of eloxal particles determines the mechanical properties of the composite developed. It is observed in the present investigation that, the maximum mechanical properties are obtained for composites containing 40% dispersoid and hence the discussion is based on this composition.

4.1 Effect of adding eloxal to cement mortar

The effect of adding eloxal to cement mortar as a dispersoid is that, the setting times of the mixture is gradually decreased and the strength is increased as compared against other cement composites. Microstructural studies reveal that, this decreasing effect of setting time and increase in the strength are attributed to solving of eloxal near the boundray

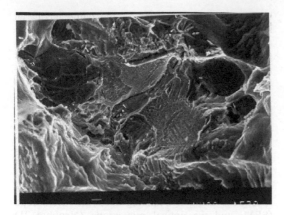

Figure 1. SEM photograph showing the microstructure of composite containing different phases (dispersoid content 40%).

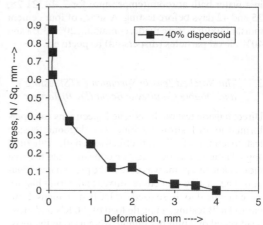

Figure 2. Stress-deformation plot of NTS specimen containing 40% dispersoid.

Table 4. Summary of mechanical properties of the composite.

% Dispersoid	Compression Strength, MPa (152*305 mm cylinders) Days			Splitting Tensile Strength (MPa) of 152*305 mm cylinders (42 days)
	28	35	42	
00	32.1	41.8	50.2	2.8
20	49.2	56.6	60.3	3.9
30	49.8	58.2	62.7	3.7
40	52.7	56.2	68.9	3.3

(between eloxal and cement mortar) forming ettringite and hydrogranate ($3Ca(Fe_2O_3, Al_2O_3, H_2O)$) phase, calcium aluminate hydrates and calcium silicate hydrates that occurs by reaction of cement mortar and eloxal, which forms a strong bond between the two (matrix and the dispersoid). The setting time of all cement mortar – eloxal mixes is in accordance with Indian Standards (IS) limits. SEM photograph (Fig. 1) shows the structure which contains calcium silicates (fibrous phase), portlandite [$Ca(OH)_2$], ettringite (dimple form) and hydrogranate phase (dark phase) (A. Benthur).

Results of the microstructural analysis also supports that, hydration yields a white cover appearance and thus the main structure was formed as a growth of calcium hydroxide and hydrogranate phase among calcium silicate hydrate gels. A summary of the mechanical properties of the composite developed are indicated in Table 4.

4.2 Fracture characteristics of the composite

A typical stress-deformation plot of one NTS specimen (containing 40wt.% dispersoid) is shown in Fig. 2.

Fracture of cement mortar reinforced with eloxal particles can be adequately determined using a stress release law that can be observed form the dispersoid pullout curve of the notched dogbone specimens during the tension testing. SEM analysis reveals that the crack opening of the eloxal particles (dispersoid) reinforced in cement mortar depends on its volume and orientation. Microstructural studies also reveal that the shape of the dispersoid and its distribution in the test specimen greatly affect the fracture behavior. It is observed form Fig. 3 (Fracture energy with crack extension curve) that, the crack opening at which dispersoid offer no resistance to pullout is relatively small. This means that steady state crack propagation may be expected to occur at shorter crack lengths. SEM analysis shows that, failure response of all the DCB specimens tested was stable one throughout the test expect for a couple of specimens where a series of cracks were developed due to mis-orientation of eloxal particles.

The mechanisms which control the fracture of the composite are dependent upon both microstructure and the deformation. The manner in which the stress response varies is an important feature of the fracture process. The stress required for cracking the specimen provides an useful information pertaining to the mechanical stability of the intrinsic microstructural features during straining and coupled with an ability of the material to distribute the strain over the entire volume are key factors governing the fracture of the composite. It is observed in the present investigation that the eloxal content of the composite is the most significant factor that affects fracture. Further, it is observed that, the eloxal content beyond 30% by wt., the fracture energy values register a decreasing trend. The possible micro-mechanisms controlling the

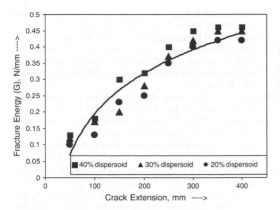

Figure 3. Plot of fracture energy Vs crack extension.

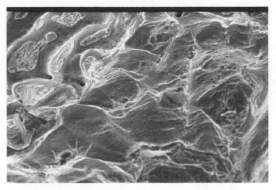

Figure 4. SEM structure of fractured surface of unreinforced composite.

fracture behavior during loading are ascribed to the following synergistic influences.

(a) Load transfer between the cement mortar matrix and the hard and brittle eloxal particle reinforcement.

(b) Hardening arising from constrained deformation and triaxiability in the cement mortar matrix due to the presence of the brittle eloxal reinforcements. As a direct result of the particles resisting the deformation of the matrix, an average internal stress or back stress is created.

(c) Residual stresses are also generated in the matrix (cement mortar) due to mismatch of eloxal particles.

During deformation it seems possible that the mismatch that exist between the brittle reinforcing particle and the matrix favors concentration of stress at and near the particle/matrix interface, causing the matrix in the immediate vicinity to fail permanently or the particle to separate from the matrix.

Conversely, fracture of the unreinforced composite on the microscopic scale, exhibited limited fracture energy and the fracture is brittle (Fig. 4) and is normal to the major stress axis. Thus the presence of eloxal in the composite as the reinforcement has a pronounced effect on the fracture.

4.3 Microsrtuctural observation of the composite

Optical microstructural studies (Fig. 5) reveal good bonding and distribution of eloxal particles throughout the matrix (cement mortar) with non-affective interfacial reaction. This may be one of the reasons for increase in strength and soundness of the composite developed.

Examination of the fracture surface features in the SEM indicate that, specimens at low magnification to identify the final fracture regions, and at higher magnification to identify regions of micro-crack initiation,

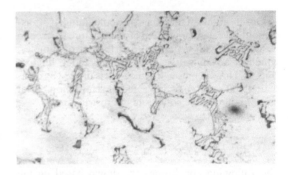

Figure 5. Optical microstructure of the composite containing 40% dispersoid.

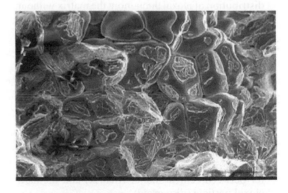

Figure 6. SEM structure of fractured surface of composite containing 20% dispersoid.

early crack growth and final scale fracture features. Fracture surfaces revealed different topographics for the composite containing different weight percent of eloxal particles. Fracture of the composites containing 20% dispersoid on macroscopic and microscopic scales exhibited brittle fracture with isolated cracks in the matrix (Fig. 6).

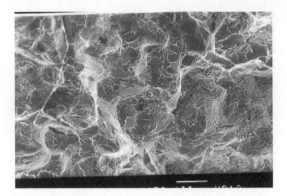

Figure 7. SEM structure of fractured surface of composite containing 40% dispersoid.

Observations of the composite containing 40% dispersoid revealed large areas of the fracture surface to be covered with a bimodal distribution of dimples, which is an indication of the mixed mode fracture (Fig. 7). However growth of the void is limited by competing and synergistic influence of reinforcing eloxal particles and in the composite microstructure.

5 CONCLUSIONS

The following conclusions are drawn based on the experimental results presented.

1. Microstructural studies reveal good bonding and distribution of eloxal throughout the matrix without any interfacial reaction. This is one of the reasons for increase in the strength of the composite.
2. High strengths can be achieved with various admixtures of minerals but the use of eloxal as the dispersoid is mandatory to obtain light weight and high strength cement mortar composite.
3. The high strength light weight cement mortar composite investigated has satisfactory resistance to cyclic loading and this is conveniently studied from the tensile tests of notched dogbone specimens.
4. The major mechanism that controls the fracture characteristics of the composite are the load transfer between cement mortar (matrix) and the hard eloxal particle (dispersoid).
5. The composite developed exhibited steady state crack propagation at shorter crack lengths.

REFERENCES

Benthur, A and Cohen, M.D. 1991. *Advances in Cement Concrete Research. ACI journal.* 4(3): 47–55.
Carrasquillo, R.L. Nilson A.H. and Slate, F.O. 1982. Short term mechanical properties of high-strength lightweight cement mortar. *Research report, Department of structural engineering, Cornell university,* Ithaca,: 98–111.
Harmon, T and Ramakrishnan, K. 1999. The effect of confined cement mortar. *Proceedings of II International RILEM symposium (FRPRCS-2) on non metallic (FRP) reinforced cement mortar structures, RILEM proceedings,* 29(19): 584–597.
Hanson, J.A. 1958. Shear strength of lightweight reinforced cement mortar beams. *ACI journal,* 55(14): 387–394.
Joel Hemanth, 2001. Production and Mechanical properties of Al-quartz particulate composite. *Journal of Materials Engineering and Performance, American Society of Metals (ASM) International,* Ohio, USA, 10(12): 143–152.
Kaar, P.H., Hanson, N.W. and Capell, H.T. 1993. Stress-strain characteristics of high strength cement mortar. *International symposium on cement mortar and cement mortar structures, American cement mortar institute.* Detroit. 6(2): 161–175.
Labossier, P. Neale, K and Demers, M. 1994. Repair of reinforcement cement mortar columns withadvanced composite materials confinement. *Proceedings of the conformance/workshop on repair and rehabilitation of infrastructure in the Americans,* Mayaguez, Puetro Rico, 14(4): 153–164.
Nanni, A. Norris, M.S and Bradord, N.M. 1983. Lateral confinement of cement mortar using FRP reinforcement. *American cement mortar institute special publication",* Detroit, 138(19): 193–208.
Piasta, J. Sawicz Z and Piasta, W.G. 1989. Recent developments in light weight composites. *Journal of Cement Concrete Research,* 19(16): 103–113.
Spratt, B.H, 1974. The structural use of lightweight aggregate cement mortar. *Cement cement mortar Association* 23(4): 19–25.
Shideler, J.J. 1957. Lightweight-Aggregate cement mortar for structural use. *ACI journal,* 54 (19): 299–314.
Scian, A.N. Porto, J.M and Pereira, E. 1991. Cement mortar composites. *Journal of Cement Concrete Research,* 51(11): 51–64.
Toutanji, H and Rey, F. 1978. Stress-strain characteristics of cement mortar columns externally confined with advanced fiber composite sheets. *ACI journal,* 78(32): 786–794.
Visalvanich, K and Naaman, A.E. 1983. Fracture model for fiber reinforced cement mortar. *Journal of the American cement mortar Institute, Proceedings,* 80(13): 128–137.
Wilson, C, 1954. Cement mortar ship resists sea water thirty-four years. *Cement mortar* 62(5): 45–52.
Wischers, G and Manns, M. 1974. Technology of structural lightweight cement mortar. *Lightweight aggregate cement mortar Technology and World applications,* CEM Bureau, 12(3): 16–22.
Wang, P.T. Shah, S.P and A.E. Naaman 1978. Stress-strain curves of normal and lightweight concrete in compression. *ACI journal,* 75(22): 603–614.
Xiao, Y, Martin, G.R and Yin, Z. 1996. Seismic retrofit of existing reinforced cement mortar bridge using a prefabricated composite wrapping system. *proceedings of I International conference on composites in infrastructure (ICCI'96),* Tucson, AZ, 6(3): 903–909.
Zibra, Y.N. 1984. Material and fracture characterization of sisal fiber cement mortar. *M.S. Thesis, Department of Civil Engineering, University of Petroleum and Minerals,* Dharan, Saudi Arabia.

Mitigating autogenous shrinkage of hardening concrete

H.W.M. van der Ham, E.A.B. Koenders & K. van Breugel
Delft University of Technology, Delft, The Netherlands

ABSTRACT: The probability of early age cracking in hardening concrete can be affected by using additives that reduce the total shrinkage of a mixture with emphasis on the reduction of autogenous shrinkage. In order to become more insight in the effectiveness of shrinkage reducing additives, an experimental programme is currently running at TU Delft. Experiments are performed for water/cement ratios varying from 0.3 to 0.5, for Portland cement as well as Blast Furnace Slag cement. All mixtures are tested isothermally at 20°C. The early-age shrinkage deformations are measured over a testing period of 90 days. The development of compressive strength, tensile splitting strength and elastic modulus will be determined from respectively cubes and prisms that harden under similar conditions. Results obtained from this experimental programme provide the effectiveness of shrinkage reducing additives and their influence on development of the mechanical properties. The autogenous shrinkage, being a single contribution of total shrinkage, will be addressed explicitly.

1 GENERAL

Autogenous shrinkage is one of the causes of cracking of hardening concrete elements. Autogenous shrinkage occurs if less water is present in the mixture than required to hydrate all binder material and that, as a result of this, capillary forces develop. During the hardening process of a so called low-water-binder ratio concretes, the internal demand for water is higher than the amount of water available. This is called 'internal' drying and causes the concrete to shrink ('autogenous shrinkage'). Mitigating autogenous shrinkage will reduce the probability of cracking of the hardening concrete elements.

Super Absorbent Polymers can be used to mitigate autogenous shrinkage by internal curing. To become more insight in the effectiveness of those Super Absorbent Polymers, an experimental programme (figure 1) is currently running at TU Delft. First two mixtures are compared, one mixture without polymers and one mixture with polymers. Preliminary results of those tests are presented are presented in this paper.

2 MIX DESIGN

Two mixtures have been tested, as presented in table 1. Firstly, a C68/85 concrete without polymers and secondly, the same concrete with polymer addition. First mixture has also been used in previously research by Sule (2003), Lokhorst (2001), and Koenders (1997). The polymer dosage was $2 \, kg/m^3$. The water-cement ratio was increased from 0.28 to 0.31 to compensate for the influence of the polymer on the workability.

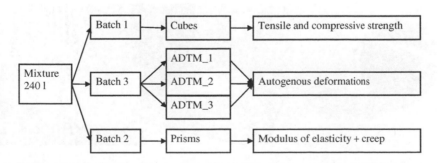

Figure 1. Testing program.

3 SUPER ABSORBEND POLYMERS

Super absorbent polymers (figure 2) and extra water are added to the concrete mixture separately, in order to mitigate autogenous deformations. The polymers are considered to absorb all the extra added mixing water which will release during cement hydration. The release of the extra water during hardening will affect the internal drying and, as a result of this, the autogenous deformations as well.

Mixing procedure:

– prepare mixer;
– mix sand and gravel for 1 minute;
– add dry polymers and mix for 2 minutes;
– add cement and mix for 1 minute;
– add water and additives and mix for 1.5 min.

In order to achieve a homogenous distribution of the polymers and avoid clusters of polymers in the mixture, the polymers are added to the sand-gravel mixture in a dry configuration.

4 TESTING PROGRAM AND METHODS

The early-age shrinkage tests were carried out using three ADTM (Autogenous Deformation Testing Machine) apparatus. It measures the deformation of a specimen that is free to deform. Figure 5a gives an

Table 1. Mix proportions (kg/1000 l).

Mixture	32REF	32A133
Water	125.4	144.9
CEM III/B 42.5 LH HS	237	232.6
CEM I 52.5 R	238	233.6
Sand 0–4 mm	755	740.7
Aggregate 4–16 mm	1001	981.9
Addiment BV1	1	1
Addiment FM951	9.5	9.3
Silica fume slurry	50	49.1
Polymer Powder	0	2

overview of its functioning. Three insulated moulds are used (figure 5b); two of 1000 mm long and one of 850 mm long. All three are 150 mm wide and 100 mm high. The walls of the mould contain plastic tubes

Prisms   Cubes

Figure 3.

Figure 4. Long term shrinkage measurements.

Figure 2. From left to right two polymers, one unsaturated and one saturated [4], 25 grams of unsaturated polymers and 630 grams of water absorbed by 25 gram polymers.

Figure 5a. ADTM1, 2 and 3.

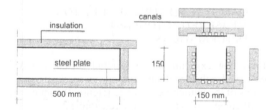

Figure 5b. Schematically drawing of ADTM-setup Lokhorst (2001).

that are connected to cryostats units. The temperature of the specimen is measured by means of three thermocouples, one in the middle and two in the ends of the specimen. The cryostats control the water flow through the tubes to maintain the required temperature regime in the specimen. All tests are performed under isothermal conditions, at 20°C. Directly after casting, all moulds are covered with plastic foil and covered with an insulated plate.

Two small steel bars were embedded in the three specimens, 750 mm apart from each other. These bars protrude through the two holes in the long side faces of the mould, while not making contact with the mould itself. Thus, the ends of the steel bars could move freely over a certain deformation range. As soon as the fresh concrete starts to develop some strength and stiffness,

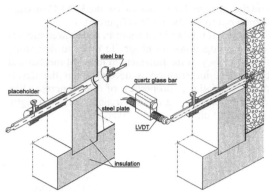

Figure 5c. Schematically drawing of LVDT placement Lokhorst (2001).

the LVDT's (linear voltage displacement transducers) positioned along the long side faces of the mould were activated and attached to the small steel bars that protrude through the mould (figure 5c). From that moment on (t-zero [8]), the deformation of the specimen over a 750 mm measuring length was registered, both at the front and rear of the specimen.

5 TEST RESULTS

5.1 *Adiabatic temperature development*

With respect to the adiabatic temperature development of the two tested mixtures (figure 7), slight differences are expected in terms of the degrees of hydration. It is likely that a prolonged hydration process will be observed due to the addition of the release of extra water from the SAP.

5.2 *Short term shrinkage – ADTM-tests*

Autogenous deformations are measured from three sealed specimens. During testing, similar conditions were applied, i.e. temperature of 20°C and a relative humidity of 50%. The measured autogenous deformations are presented in figure 6 and represent the mean values of the three specimens. Variation of the results, which turned out to be 2% for the concrete without SAP and 23% for the mixture with SAP, is calculated by dividing the standard deviation by the mean value. This high variation of the concrete with SAP is due to the mean value fluctuates around zero.

The deformations of the 32A133 mixture are corrected for the expansion in the first 20 hrs (figure 6) because this will lead to relative small compressive stresses which will relax, compared to the tensile stresses due to the shrinkage after the expansion. The reduction of shrinkage between the reference concrete and the corrected deformations of the 32A133 mixture turned out to be about 65% (figure 6).

Note that this will not result in 65% lower stresses due to autogenous shrinkage. In order to determine the efficiency of the polymers in terms of mechanical properties, information is needed about the development of these properties of the concrete, like development of compressive strength, tensile strength, modulus of elasticity and relaxation.

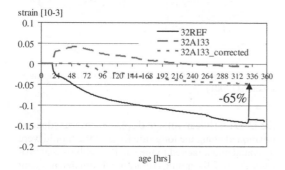

Figure 6.   Measured autogenous deformations.

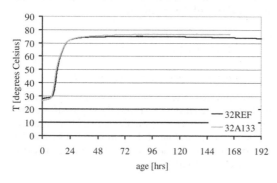

Figure 7.   Adiabatic temperature development.

5.3   Long term shrinkage – ADTM-tests

Because the extra water added to the mix to obtain a similar workability as the reference concrete it is to be expected that the hydration process will proceed for a longer period. This might result in an ongoing refinement of the pore structure and, hence, a possible higher shrinkage. It is therefore that the shrinkage of the 32A133 mixture is measured for a longer period of at least 90 days (figure 4). The results up to now are presented in figure 8. After correcting the expansion in the first 20 hours, the difference is still 50%.

5.4   Tensile strength

For the tensile strength, even bigger reductions (−14%) are observed (measured from cubes, figure 3) due to the addition of SAP, with a maximum reduction of almost 20% after 14 days of hardening (figure 9 middle).

5.5   Modulus of elasticity

Reductions in measured modulus of elasticity (measured from prisms, figure 3) are less (−6%) compared to differences in compressive strength and tensile strength. The results of the measurements are provided in figure 9 (right).

6   CALCULATED STRESSES DUE TO MEASURED AUTOGENOUS DEFORMATIONS

From the experiments, substantial differences are observed in the development of the visco-elastic properties of the hardening concrete (figure 9) containing SAP compared to the ref. concrete (table 2).

To calculate the actual stresses due to the autogenous deformations, a trend line is fitted into the test

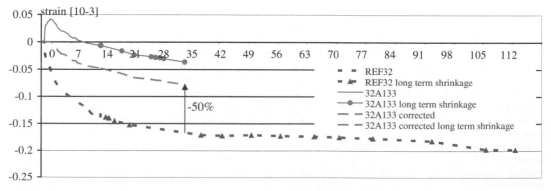

Figure 8.   Measured and corrected long term shrinkage.

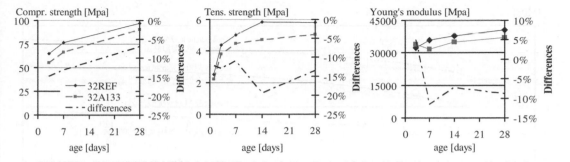

Figure 9. Development concrete properties and influence of SAP.

Table 2. Average effect SAP on properties and probability of failure P_f.

32A133 compared to reference mixture	Effect on P_f
Short term shrinkage (2 weeks) – 65%	Positive
Long term shrinkage (5 weeks) – 50%	Positive
Compressive strength – 12%	Negative
Tensile strength – 14%	Negative
Modulus of elasticity – 6%	Positive

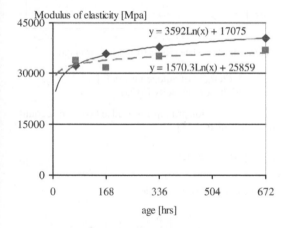

Figure 10. Measured and modeled modulus of elasticity.

results of the modulus of elasticity, both for the mixture with and without polymers. The development of the modulus of elasticity can be described by:

$$E = 3592 \cdot Ln(t) + 17075 \qquad (1)$$

for the mixture without polymers (32REF), and

$$E = 1570{,}3 \cdot Ln(t) + 25859 \qquad (2)$$

for the mixture with polymers (32A133) in which t is the time after casting in hrs (figure 10).

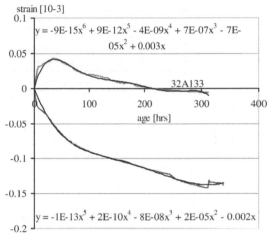

Figure 11. Measured and modeled autogenous deformations of both mixtures.

Measured autogenous deformations of both mixtures are modeled by polynomial formulations, as presented in figure 11. To correct for the starting time of the measurements (17 hours after casting, see figure 6), the polynomial formulations are shifted for 17 hours.

The adiabatic degree of hydration is calculated from the adiabatic temperature development by:

$$\alpha_h(t) = \frac{(T_a(t) - T_a(0)) \cdot \frac{\rho \cdot c_c}{C}}{Q_{max}} \qquad (3)$$

in which:
T_a = adiabatic temperature [K]
ρ = volumetric density of concrete [kg/m^3]
c_c = specific heat of concrete [kJ/kg/K]
C = amount of cement [kg]
Q_{max} = maximum heat production [kJ/kg]

Because the experiments are performed at isothermal conditions of 20°C, the degree of hydration in

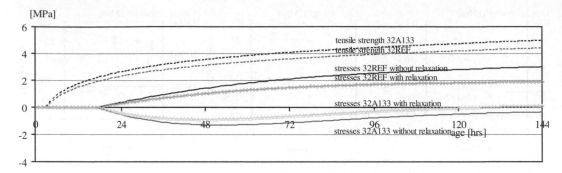

Figure 12.   Calculated stress development due to measured autogenous deformations.

Degree of Hydration

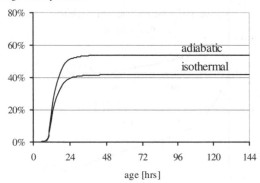

Figure 13.   Adiabatic and isothermal degree of hydration.

the specimen will be lower compared to the adiabatic degree of hydration. To calculate the process degree of hydration, the heat production in the specimen can be calculated by:

$$\Delta Q_{p;j+1} = \Delta Q_{a;j+1} \cdot e^{-\frac{E_A(T_{p;j})}{R} \frac{T_{a;j} - T_{p;j}}{T_{a;j} \cdot T_{p;j}}}$$ (4)

In which:
E_a = apparently activation energy [kJ/mol]
R = universal gas constant [R = 8,31 J/mol/K]
$T_{a;j}$ and $T_{p;j}$ are the temperatures [K]
$\Delta Q_{p;j+1}$ = produced process heat in timestep j
$\Delta Q_{p;j+1}$ = produced adiabatic heat in timestep j

The calculation of the stress development is based on elastic and time-dependent deformational behavior of concrete. The elastic stress increments are calculated according to:

$$\Delta \sigma_e(\tau) = \Delta \varepsilon_a(\tau) \cdot E(\tau)$$ (5)

Where:
$\Delta \sigma$ = elastic stress increment [MPa]
τ = age at loading [hrs]

stress/(0.75*tensile_strength-ratio) [-]

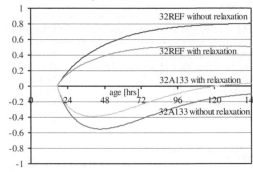

Figure 14.   Calculated stress/(0.75*strength)-ratio.

$\Delta \varepsilon_a$ = increment of autogenous def. [-]
E = modulus of elasticity [MPa]

The superposition principle is used to take into account the stress history [5]:

$$\sigma(t) = \sum_{i=0}^{j} \psi(t, \tau) \cdot \Delta \sigma_e(\tau)$$ (6)

With for the relaxation factor:

$$\Psi(\tau_i, t, \alpha_{\tau,i}) = \exp\left(-\left[\frac{\frac{\alpha_h(t)}{\alpha_h(\tau_i)} - 1 + 1.34 * \omega^{1.65} *}{* \tau_i^{-d} * (t - \tau_i)^n * \frac{\alpha_h(t)}{\alpha_h(\tau_i)}}\right]\right)$$ (7)

And
α_h = degree of hydration [-]
τ = age at loading [hrs]
ω = water/cement ratio [-]
d = 0.3 (slow) and 0.4 (rapid cement) [-]
n = 0.3 [-]

The results of the calculations with and without relaxation are shown in figure 12. The reference concrete without Super Absorbent Polymers will reach a tensile stress of about 2 MPa, while the concrete with Super Absorbent Polymers develops almost no tensile stresses during the first 144 hours. After 144 hours, no extra tensile stresses will develop due to more autogenous shrinkage because those extra stresses will relax immediately.

For the calculations with relaxation, both calculations are performed using equal relaxation. For the concrete containing Super Absorbent Polymers, bigger relaxation is expected compared to the reference concrete, because of the bigger pores in the microstructure. The total pore volume will not change, but the amount of small pores will decrease Mönnig (2005). To determine the influence of Super Absorbent Polymers on creep and relaxation of hardening concrete, creep experiments under a constant compressive load will be performed soon. This effect is not taken into account yet.

Stress/strength-ratios are calculated for both mixtures for the first 144 hours after casting. From experiments, a cracking criterion is proposed by Lokhorst (2001) to be stress >0.75*tensile strength. If relaxation is taken into account, a maximum stress/strength-ratio due to only autogenous shrinkage is observed to be almost 0.5 for the reference concrete without Super Absorbent Polymers. For the concrete which contains Super Absorbent Polymers, The maximum stress/strength-ratio in the first 144 hours is almost 0. This positive effect of the SAP might be less at higher ages, due to faster shrinkage of the specimen what contains SAP, compared to the specimen without SAP (compare figure 6 and 8).

7 CONCLUSIONS

– Adding 2 kg/1000l Super Absorbent Polymers to the tested mixture results in a reduction of autogenous shrinkage after 336 hours of 65%. After 800 hours, this is reduced to 50%.

– Adding 2 kg/1000l Super Absorbent Polymers to the tested mixture results in an average reduction of the compressive strength of 12%, the tensile strength of 14% and the modulus of elasticity of 6%.
– Adding 2 kg/1000l Super Absorbent Polymers to the tested mixture, more relaxation is expected.
– Adding 2 kg/1000l Super Absorbent Polymers results to lower stresses in the hardening concrete (2 MPa after 144 hours), which results in a lower stress/strength-ratio due to only autogenous deformations from 0.5 to 0.04 after 144 hours. At higher ages, the effect of SAP might be less, due to a faster shrinkage after 144 hours of the mixture containing SAP compared to the mixture without SAP.

REFERENCES

Bjøntegaard, Ø. and Hammer, T.A. 2006. RILEM TC 195-DTD: Motive and Technical content. *Volume changes of Hardening Concrete: Testing and Mitigation. RILEM Proceedings PRO 52*: 357–366.

Breugel, K. van. 1991. Relaxation of young concrete, *Delft University of Technology*.

Breugel, K. van. 1997. Simulation of hydration and formation of structure in hardening cement-based materials (second edition). *Delft University of technology*.

Jensen, O.H., Hansen, P.F. 2001. Water-entrained cement-based materials, I. Principles and theoretical background, *Cement and Concrete Research 31*: 647–654.

Koenders, E.A.B. 1997. Simulation of volume changes in hardening cement-based materials. *Delft University of Technology*.

Lokhorst, S.J. 2001. Deformational behavior of concrete influenced by hydration related changes of the microstructure, *Delft University of Technology*.

Mönnig. 2005. Water saturated super-absorbent polymers used in high strength concrete, *Otto-Graf Journal Vol. 16*: 193–202.

Sule, M.S. 2003. Effect of Reinforcement on Early-Age Cracking in High Strength Concrete. *Delft University of Technology*.

Part X
Fibre reinforced concrete

Fracture Mechanics of Concrete and Concrete Structures – High-Performance Concrete,
Brick-Masonry and Environmental Aspects – Carpinteri, et al. (eds)
© 2007 Taylor & Francis Group, London, ISBN 978-0-415-44617-4

Toughening mechanism of hybrid fiber reinforced cement composites

H. Mihashi & Y. Kohno
Tohoku University, Sendai, Japan

ABSTRACT: A new type of fiber reinforced cement composites (FRCC) has been developed by means of hybrid fiber reinforcement in which randomly distributed short fibers of two types rationally bridge crack surfaces. They are steel cord and polyethylene fibers. In this paper, toughening mechanisms of this newly developed cement composite material are experimentally studied. Focused on the bridging mechanism of steel cord on the crack surface, pull-out tests on a single fiber embedded in cement-based matrixes are carried out to clarify the toughening mechanisms taking into account influences of the inclination angle, surface property and length of the fiber, including mechanical property of the matrix.

1 INTRODUCTION

Concrete is widely used in the world as the most common and important construction material. Although it is good at resisting under compression, plain concrete is a brittle material and the strength and strain capacity under tension are much lower than that under compression. From ancient times, it was well-known that poor crack resistance of brittle materials such as brick under tension is improved by fiber reinforcement. In the field of concrete engineering, extensive research works have been carried out on developing FRCC technology since the early 1960s and a wide range of practical applications was developed. Especially in these two decades, new types of fibers and material design methodologies have been developed. Processes of the development are reviewed in several books such as Bentur & Mindess (1990), Balaguru & Shah (1992) and Brandt (1995). Now several high performance fiber reinforced cement composites (HPFRCC) are available in practice. They can be found, for example, in the special issues of Journal of Advanced Concrete Technology (Japan Concrete Institute 2003 and 2006), an international journal published by Japan Concrete Institute.

Mechanical properties of brittle cement-based materials such as energy absorption and ductility can be significantly improved by reinforcing with fibers. Although fiber reinforcement generally doesn't improve the crack initiation strength, it does change the post-cracking behavior by bridging across cracked surfaces. The crack bridging efficiency of a fiber depends on its length, interfacial properties, geometry, diameter, mechanical properties such as Young's modulus and strength of the fiber as well as mechanical properties of the matrix.

Mechanisms of the crack bridging are usually studied with fiber pull-out tests. Since short fibers in FRCC are randomly distributed in the matrix, the orientation effect of fibers crossing the crack surface is also taken into consideration. There are many papers published on fiber pull-out tests. Pull-out tests on steel fibers with different profile and the inclination angle to the crack surface were carried out to study the influence of fiber geometry and the mix proportion of the matrix on the bond properties (e.g. Banthia et al. 1994, Naaman 1999).

Generally speaking, increasing the lateral surface of a fiber increases frictional and adhesive bond forces along the fiber to increase the pull-out resistance. However, deformed fiber such as screw may have a too strong grip to achieve a ductile behavior, though it may increase the pull-out strength. Naaman (1999) carried pull-out tests on smooth, hooked and twisted fibers and concluded that twisting is the best way to improve the mechanical component of bond of fibers.

Pull-out tests on inclined fibers were carried out by various researchers, too. Morton & Groves (1974) clearly showed that substantial stress concentration occurred in the matrix near the fiber exit point where the matrix always yields in compression. They also introduced a theoretical model based on elementary beam theory in which interfacial shear stress was neglected and the reaction stress from the matrix on the fiber was equal to the matrix yield strength. Brandt (1985) extended the analytical model in which the fracture energy was given by summing the energy absorbed in the following five processes as a function of the inclination angle: (1) debonding of the fibers from the matrix; (2) pulling-out of fibers against the interfacial friction; (3) plastic deformation of the fibers; (4) yielding of the matrix in compression near the exit points of the fibers; (5) complementary friction between the fibers and the matrix due to local compression at the bending points.

Naaman & Shah (1976) carried out an experimental study on pull-out mechanism in steel fiber-reinforced concrete and concluded that spalling and disruption of the mortar matrix lead to a substantial reduction in the pull-out resistance. They also reported that as the number of fibers increases, the contribution of randomly oriented fibers relative to that of parallel fibers decreases. On the basis of a set of experiments with synthetic fibers, Li et al. (1990) reported that the increase of pull-out resistance is due to the additional friction stress caused by snubbing effect. Leung et al. (1992, 1995) developed a model to analyze the fiber bending/matrix spalling mechanisms by a micro mechanical approach, in which beam was bent on an elastic foundation with variable stiffness and with possibility of spalling. The theoretical analyses showed that the presence of an optimal yield strength for inclined fiber is due to the increased matrix spalling with increasing fiber yield strength. It was proved by an experimental work (Leung & Shapiro 1999).

In spite of the large number of studies, only a limited number of FRCC have achieved the high performance that accompanies a pseudo strain-hardening property and multiple cracking (Reinhardt and Naaman 1995). For example, Li & Leung (1992) developed Engineered Cementitious Composite (ECC) with synthetic fibers on the basis of a micro mechanical approach, in which the volume content of the synthetic fiber is less than 2%. In their work, it was clarified that bridging and snubbing friction of fibers on cracked surfaces under a certain condition are the key to achieve the multiple cracking which increases the ductility. On the other hand, Naaman (1999) developed a new type of deformed fiber by twisting triangular or rectangular steel fibers, which can achieve the high performance with a fiber volume content of less than 2% (Naaman 2003).

Besides changing the type and properties of the fiber, other various attempts have been made to develop new and improved FRCC systems, too. One of such attempts is hybrid fiber reinforcement systems. The combination of different types of fibers has been studied. For example, Rossi et al. (1987) presented an idea that fibers play a role at two different levels: at the material level, fibers can improve the strength and the ductility of concrete if a high percentage of fibers with a small diameter is used; at the structural level, fibers can improve the load bearing capacity and the ductility

of the structure by using a low percentage of fibers long enough to allow sufficient anchorage. In the multiscale concept (Rossi 1997, 2000), it is conceptually supposed that a large number of short and thinner metal fibers can "sow up" microcracks during the first stage of loading and then lower percentages of long metal fibers can also "sow up" the macrocracks during the stage of macrocrack propagation. Lawler et al. (2005) also suggested that microfibers delay the formation of macrocracks to increase the strength and toughness of the composites. However, there are quite few studies on the toughening mechanisms in the hybrid FRCCs.

Kawamata et al. (2000, 2003) developed a new hybrid FRCC using steel cord and polyethylene fibers. Bending tests on notched beams clearly showed that the hybrid fiber reinforcement substantially improved the ductility, though the peak load was almost same as that of the virtual composite (Fig. 1). Cracking patterns observed in FRCC specimens containing only steel cord were very different from those in specimens reinforced with both of steel cord and polyethylene fibers together (Fig. 2). A parametric study was also carried out about the influence of materials and mixproportions on the mechanical properties. Length and volume content of the steel cord and water-cement ratio of the matrix were the dominant influencing factors. Uniaxial tension tests on cylindrical specimens made with an optimized mixproportion of this type of hybrid FRCC showed the pseudo strain-hardening until the strain of about 1.5% and multiple cracking.

The main objective of the present paper is to clarify the toughening mechanism of the hybrid FRCC using steel cord and polyethylene fibers. An experimental program of pull-out tests is carried and the toughening mechanism is discussed. Besides a single steel cord embedded in hardened cement paste (hcp) reinforced with short polyethylene fiber, a single straight steel fiber embedded in plain hcp as a reference as well as a single steel cord embedded in plain hcp are pulled out. The inclination angle of the fiber and the embedded length are varied, too.

2 EXPERIMENTAL PROGRAM

2.1 Test groups

There were three groups of pull-out tests, each of which has the different combination of the pulled-out fiber and the matrix. Two types of steel fiber (steel cord: SC and straight fiber: SF) and two types of matrix (plain hardened cement paste: PLM and hardened cement paste reinforced with polyethylene fiber: PEM) were employed. In all series, the influence of the angle of fiber orientation on the pull-out behavior was investigated and the inclination angle was changed to be 0°, 15°, 30°, 45° and 60°. Three specimens were tested for each test condition.

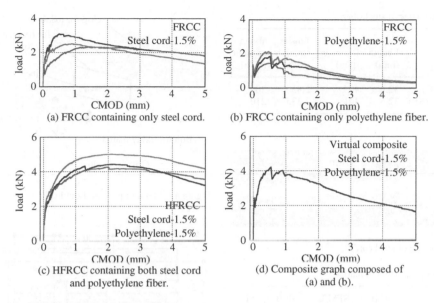

(a) FRCC containing only steel cord.

(b) FRCC containing only polyethylene fiber.

(c) HFRCC containing both steel cord and polyethylene fiber.

(d) Composite graph composed of (a) and (b).

Figure 1.   Load-CMOD curves of bending tests (Kawamata et al. 2003).

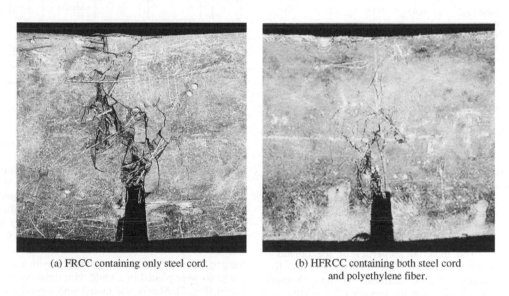

(a) FRCC containing only steel cord.

(b) HFRCC containing both steel cord and polyethylene fiber.

Figure 2.   Cracking near notch of specimens (Kawamata et al. 2003).

The first group was aimed at assessing the influence of the geometry of steel fiber embedded in PLM: one is the straight fiber and the other is the steel cord. All fibers were used without any treatment on the surface.

The second group of pull-out tests was conducted to assess the influence of the toughness of the matrix. Two different types of material for the matrix were used for this comparison. While PLM was very brittle, PEM was much more ductile than PLM.

The third group of experiments was aimed at assessing the influence of the embedded length of the fiber. A steel cord of four different length was embedded in PEM.

2.2   Materials

Properties of two types of steel fibers are shown in Table 1. Since the tensile strength of both steel fibers

Table 1. Properties of steel fiber.

	Length (mm)	Diameter (μm)	Tensile strength (MPa)	Young's modulus (GPa)
Steel cord (SC)	32.0	405	2,850	206
Straight fiber (SF)	32.5	401	2,335	210

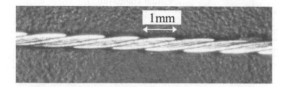

Figure 3. Profile of steel cord.

Table 2. Mix proportion of matrix.

	W/B wt %	SFM/B wt %	SP/B wt %	PE vol %
Plain hardened cement paste (PLM)	0.45	0.15	2	0
Hardened cement paste reinforced with polyethylene (PEM)	0.45	0.15	2	1

W: Water, C: High early strength Portland cement,
SFM: Silica fume, B = C + SFM, SP: Superplasiticizer,
PE: Polyethylene fiber.

2.3 Specimen preparation

Fiber pull-out specimens with a steel fiber inclined at 0°, 15°, 30°, 45° and 60° were prepared with each type of fiber in Table 1 and matrix in Table 2, that is PLM-SF, PLM-SC and PEM-SC. Generally the embedded length was 16 mm. In case of PEM-SC, however, the embedded length of the fiber was changed for four levels such as 16 mm, 12 mm, 8 m, and 4 mm. They are described as PEM-SC16,

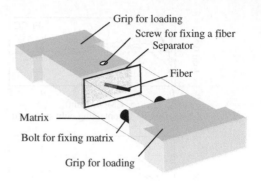

Figure 4. Metal grip for pull-out loading.

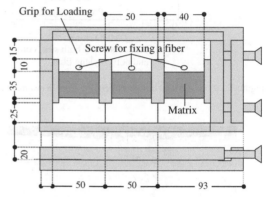

Figure 5. Mold for specimen preparation (unit: mm).

PEM-SC12, PEM-SC08 and PEM-SC04, respectively. The size of the pull-out specimen was 40 mm × 35 mm × 20 mm. Before setting the mold, the fiber was first inserted into a small metal grip with a hole drilled at the appropriate angle in the middle (Fig. 4). A thin plastic sheet was glued with silicon grease to be a separator between the metal grip and the matrix, which works as the artificial crack surface. The embedded length of the fiber was left outside of the grip. After fixing the fiber in the grip with a screw, the grips were placed in a mold with three compartments (Fig. 5). Matrix was mixed with a bowl-mixer whose capacity was one liter. After casting and curing the specimens in a moisture room of 20°C and about 95% RH for seven days, they were loaded.

2.4 Pull-out testing

Outline of the loading device for the pull-out test is shown in Figure 6. After the specimen was carefully placed in the apparatus, a LVDT was placed between the grips for measuring the pull-out displacement. Loading was conducted by the displacement control with the rate of 0.2 mm/sec.

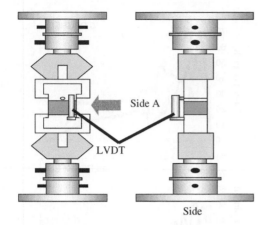

Figure 6. Outline of loading device.

3 EXPERIMENTAL RESULTS

While cylindrical specimens of 50ø × 100 mm were tested, mean values of the compressive strength and the Young's modulus of the matrix were 42.4 MPa and 24.7 GPa for PLM, and 54.6 MPa and 27.5 GPa for PEM, respectively.

Figures 7–9 show the pull-out curves obtained for each experimental series. In PLM-SF series (Fig. 7), the adhesive bond strength was negligible as obviously shown in the case of zero inclination angle (0°, i.e. the fiber is embedded perpendicular to the crack surface) probably because of oil on the surface. Except in the case of 0°, there was a very steady ascending part between 15% and 60% of the embedded length and the peak load was obtained only after a large slip representing around 70% of the embedded length. The load level and the slope were almost proportional to the inclination angle. It might be caused by the frictional bond due to the compression stress at the fiber exit point since a fine stick-slip behavior of high frequency is observed on the curve.

Figure 8 shows the pull-out behavior of PLM-SC series. Even in the case of 0°, the peak load was very high and it was obtained in the early part of the pull-out curve. It means the peak load in this case is corresponding to the adhesive bond strength and the first peak represents the end of debonding along the fiber. Pull-out process after the debonding was composed of many small peaks whose cycle was about 1 mm that is corresponding to the pitch (Fig. 3). Although the pull-out curve in the case of 15° descended monotonically after the peak, generally a high level of resistance was held. It is rather curious to find that some secondary peaks were observed around 60 to 70% of the embedded length in the case of 0°. It may mean that some untwisting of the steel cord occurs to increase the friction of the fiber. Although the load level was lower

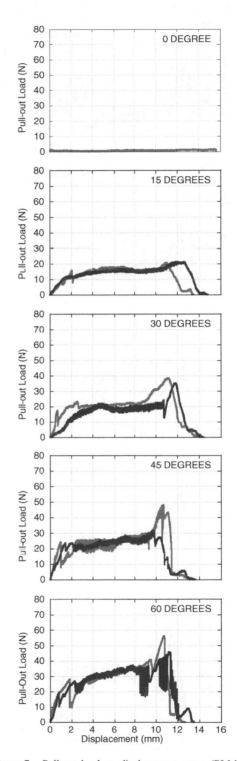

Figure 7. Pull-out load vs. displacement curve (PLM-SF series).

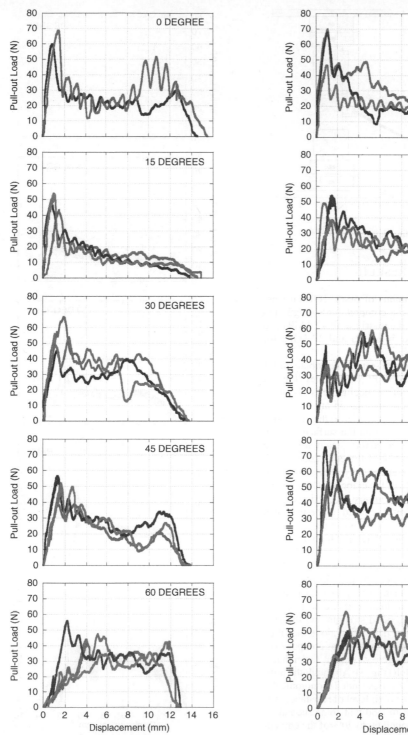

Figure 8. Pull-out load vs. displacement curve (PLM-SC series).

Figure 9. Pull-out load vs. displacement curve (PEM-SC16 series).

than that of the first peak, the last peak was observed around 75% of the embedded length in several curves. This tendency is similar to that in PLM-SF series.

Pull-out load versus displacement curves of PEM-SC series are shown in Figure 9, which shows a very high strength and ductility. As same as the curves in PLM-SC series, many small peaks were observed after the first peak. Especially over 30° of the inclination angle, the load levels were maintained at very high values which are almost equivalent to those of the first peaks.

Figure 10 shows the mean curves of PEM-SC series with the different embedded length of SC. As the inclination angle increases, the influence of the fiber length becomes significant.

4 DISCUSSION

4.1 *Pull-out load and displacement curve*

Figure 11 shows each mean curve of the relation between pull-out load and displacement in the three series. It is very obvious that the shape of the curve in case of PLM-SF is different from those in cases of PLM-SC and PEM-SC. Since the difference between two series of PLM-SF and PLM-SC is only properties of the steel fiber, the contributions of SC to the bridging can be found from the comparison. Significant characteristics of PLM-SC series are the high stiffness of the first ascending curve, the high load level of the first peak, and the periodic behavior after the first peak. These characteristics are observed in PEM-SC series, too. The slope of the first ascending part is very different between SF series and SC series. While the slope of SF series becomes higher as the inclination angle increases, the changing process of SC series is not so obvious up to 30°. However, the slope remarkably decreases from 60°. These different tendencies may be related to the difference in the bridging mechanisms. The main mechanism of SF series is bending of the fiber and the frictional bond at the fiber exit point due to the compressive stress which increases as the inclination angle increases. On the other hand, the fiber of the SC series can resist directly to the pull-out load due to the high adhesive bond strength in the early stage of loading. As the inclination angle increases, the bending component of the stress for the local matrix at the fiber exit point becomes dominant and it causes a local failure of the matrix. It may be the reason why the slope in SC series suddenly decreases when the inclination angle becomes larger.

In PLM-SF series, the peak load was obtained just before the final pull-out. It may be because the fiber rotated as a short rigid cylinder and cut into the matrix (Leung & Shapio 1999). The post peak descending behavior of PLM-SF series is much steeper than that in other series. In several curves in case of PLM-SC

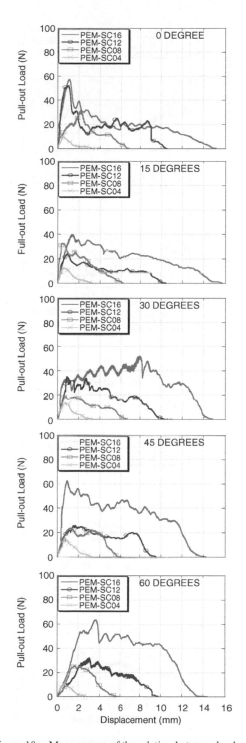

Figure 10. Mean curves of the relation between load and displacement (influence of embedded length in PEM-SC series).

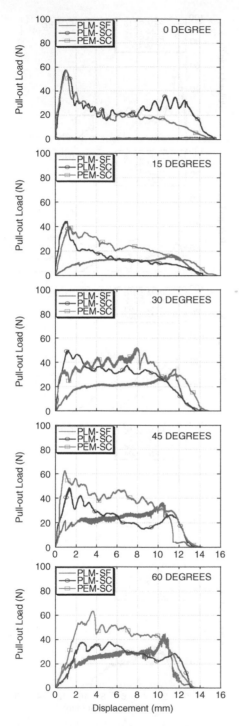

Figure 11. Mean curves of the relation between pull-out load and displacement in three series.

series, the last peak similar to that in PLM-SF series was also observed around the displacement of 75% of the embedded length, though the load level was lower than that of the first peak. In PEM-SC16 series, however, the final peak was not observed as done in PLM-SF and PLM-SC series.

Since the crack width of multiple cracking observed in cylindrical specimen of hybrid FRCC was much less than 1 mm (Kawamata et al. 2003), the most important parts in these curves are the first ascending part, the slope of this part, the first peak load and the post-peak behavior. From these view points, PEM-SC16 series achieves the highest performance especially for the cases of inclination angles higher than 45°. The peak loads and resistance levels after the first peak was obviously higher in case of PEM-SC16 series than those of PLM-SC series except in some limited number of portions.

4.2 Influence of the fiber length

In case of shorter embedded length (PEM-SC12, PEM-SC08, PEM-SC04 series), the peak loads and maintained resistance levels were mostly lower than those in PEM-SC16 series as shown in Figure 10. It was recognized that the influence of the fiber length is very significant from the view point of both strength and ductility. Now it is obvious that shorter steel cords can't sufficiently toughen the composite material even if the matrix is reinforced with synthetic fiber. These results are corresponding to the previous ones (Kawamata et al. 2003).

4.3 Matrix spalling

As shown in previous studies, matrix failure at the fiber exit point is always observed in pull-out tests on inclined fibers. Typical examples observed in this study are shown in Figure 12. While cone type of the matrix failure was observed in cases of both PLM series, a much smaller portion damaged with many cracks was observed in case of PEM series. Since the difference of the test condition between PLM-SC series and PEM-SC16 series was only the matrix, the difference in the pull-out curves could be caused by the different cracking behavior around the fiber exit point.

Size of the matrix spalling was measured by the diameter of an envelope circle as shown in Figure 13. Relationship between the diameter and the inclined angle is significantly influenced by the type of matrix, fiber and the embedded length as shown in Figures 14–15. Because of the rough surface of SC, bond of the fiber might be too strong for hcp to prevent the cone failure in PLM-SC series. On the other hand, the higher crack resistance of the matrix at the fiber exit point prevents a large spalling of the matrix that is cone failure, though some microcracks were accumulated

| PLM-SC series | PLM-SF series | PEM-SC series |
| (Magnification:2.0) | (Magnification:2.0) | (Magnification:14.0) |

Figure 12.   Matrix spalling from the fiber exit point (60 degree).

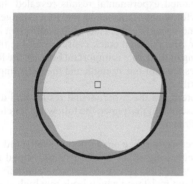

Figure 13.   Measurement of diameter of matrix spalling.

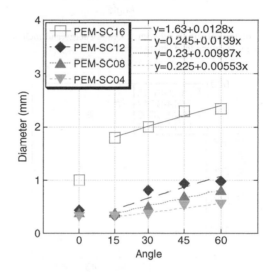

Figure 15.   Diameter of matrix spalling and inclination angle (influence of embedded length).

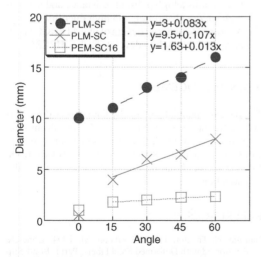

Figure 14.   Diameter of matrix spalling and inclination angle (embedded length = 16 mm).

around the fiber (i.e. bond cracks) during the pull-out process. The small diameter in PEM-SC series may reflect the anchorage force of the fiber in the matrix on the pull-out resistance.

While Figure 14 shows the diameter is proportional to the inclination angle between 15° and 60°, the diameter of PEM series is about 1/6 smaller than that of PLM series for the case of SC series. When the matrix is same, the diameter of PLM-SC series is always much larger than that of PLM-SF series. It is due to the very strong bond strength of the steel cord and it means that the critical condition to control the pull-out behavior of PLM-SC series is the cone creation in the matrix. However, even such a strong bond strength can't be sufficiently showed because of a limited space of the matrix if the fibers are densely distributed (Naaman & Shah 1976).

While Figure 15 shows the influence of the embedded length on the diameter of the matrix spalling is significant, the diameter is much smaller than that in case of PLM-SC series.

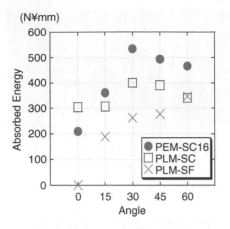

(N¥mm)

Figure 16. Relation between absorbed energy and inclination angle (embedded length = 16 mm).

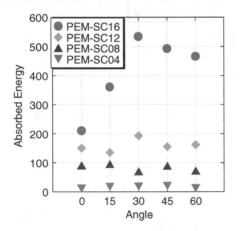

Figure 17. Relation between absorbed energy and inclination angle (influence of embedded length).

4.4 *Absorbed energy*

Absorbed energy during the pull-out test was calculated from the area under each load-displacement curve shown in Figures 7–9. Figure 16 shows the influence of the type of fiber and the matrix on the relation between the absorbed energy and the inclination angle. From this figure, it is obvious that the most dominant factor for the absorbed energy is the type of matrix on the condition with the sufficient fiber length. Especially when the inclination angle is over 30°, the absorbed energy in case of PEM-SC16 series is much higher than that in the corresponding case of PLM-SC series. It means that higher crack resistance of the matrix is the key to maintain the pull-out load up to the large displacement at least in the range of the inclination angle over 30°. If the size of the matrix spalling at the fiber exit point is small, the

fiber can achieve a sufficient bridging up to a large displacement.

Figure 17 shows the absorbed energy remarkably decreases as the embedded length is shorter. Although the highest values were recorded in case of 30° for SC16 and SC12, the influence of the inclination angle on the absorbed energy was negligible for shorter fibers such as SC08 and SC04. Because of the shorter bridging length, friction due to the rough surface of the fiber can't work sufficiently for enhancing the load nor the energy absorption.

5 CONCLUSIONS

The obtained experimental results revealed that the high performance of the new hybrid FRCC is due to the high bond strength of the fiber with a sufficient length together with the high crack resistance of the matrix. Providing both factors is important because the former increases the pull-out strength and the latter improves the ductility.

According to the experimental results and discussion presented in this paper, the following conclusions can be drawn:

(1) Increased surface area of the steel cord leads to increasing adhesive and frictional bond forces along the fiber. It causes a significant increase in pull-out resistance which leads to a high strength.
(2) Increased fracture energy of the matrix prevents a large spalling in the matrix at the fiber exit point. It maintains a high pull-out resistance and leads to high energy absorption and high ductility.
(3) Appropriate material design of hybrid FRCCs is an efficient approach to develop further HPFRCCs.

ACKNOWLEDGEMENT

The partial financial support by Grant-in-Aid for Scientific Research of Japan Society for the Promotion of Science is gratefully acknowledged. The authors would like to express their thanks to Tokyo Rope MFG. Co. for donating the steel fibers used in the investigation.

REFERENCES

Banthia, N. Trottier, J.-F. and Pigeon, M. 1994. Concrete Reinforced with Deformed Steel Fibers, Part I: Bond-Slip Mechanisms, *ACI Material Journal*, 91(5): 435–446.
Balaguru, P.N. and Shah, S.P. 1992. *Fiber Reinforced Cement Composites*. New York: McGraw-Hill, Inc.
Bentur, A. and Mindess, S. 1990. *Fiber Reinforced Cementitious Composites*. London and New York: Elsevier Applied Science.
Brandt, A.M. 1995. *Cement Based Composites*. London: E & FN Spon.

Japan Concrete Institute 2003. Ductile Fiber Reinforced Cementitious Composites. *Journal of Advanced Concrete Technology*, 1(3): 211–340.

Japan Concrete Institute 2006. *High-Performance Fiber Reinforced Cementitious Composites*, 4(1): 3–78.

Kawamata, A. Mihashi, H. and Fukuyama, H. 2000. Flexural failure properties of hybrid fiber reinforced cementitious composites, *Proceedings of AIJ Tohoku Chapter Architectural Research Meeting*, (63): 69–72. (in Japanese).

Kawamata, A., Mihashi, H. and Fukuyama, H. 2003. Properties of Hybrid Fiber Reinforced Cement-based Composites. *Journal of Advanced Concrete Technology*, 1(3): 283–290.

Leung, C.K.Y. and Chi, J. 1995. Crack-Bridging Force in Random Ductile Fiber Brittle Matrix Composites. *Journal of Engineering Mechanics*, ASCE, 121(12): 1315–1324.

Leung, C.K.Y. and Li, V.C. 1992. Effect of Fiber Inclination on Crack Bridging Stress in Brittle Fiber Reinforced Brittle Matrix Composites. *Journal of Mech. Phys. Solids*, 40(6): 1333–1362.

Leung, C.K.Y. and Shapiro, N. 1999. Optimal Steel Fiber Strength for Reinforcement of Cementitious Materials. *Journal of Materials in Civil Engineering*, 11(2): 116–123.

Leung, C.K.Y. and Ybanes, N. 1997. Pullout of Inclined Flexible Fiber in Cementitious Composite. *Journal of Engineering Mechanics*: 239–246.

Li, V.C., Wang, Y. and Backer, S. 1990. Effect of Inclining Angle, Bundling and Surface Treatment on Synthetic Fiber Pull-out from a Cement Matrix. *Composites*, 21(2): 132–140.

Morton, J. and Groves, G.W. 1974. The Cracking of Composites Consisting of Discontinuous Ductile Fibers in a Brittle Matrix – Effect of Fiber Orientation, *Journal of Material Science*, 9: 1436–1445.

Naaman, A.E. 1999. Fibers with Slip-Hardening Bond. In Reinhardt H.W. and Naaman A.E. (eds.), *High Performance Fiber Reinforced Cement Composites (HPFRCC 3)*, Pro6, RILEM Publication S.A.R.L.: 371–385.

Naaman, A.E. 2003. Engineered Steel Fibers with Optimal Properties for Reinforcement of Cement Composites, *Journal of Advanced Concrete Technology*, 1(3): 241–252.

Naaman, A.E. and Reinhardt, H.W. (eds.). 1996. *High Performance Fiber Reinforced Cement Composites 2 (HPFRCC 2)*. London: E & FN Spon.

Naaman, A.E. and Shah, S.P. 1976. Pull-out Mechanism in Steel Fiber-Reinforced Concrete. *Journal of the Structural Division*, ASCE, 102(8):1537–1548.

Rossi, P. 1997. High Performance Mutimodal Fiber Reinforced Composite (HPMFRCC): The LCPC experience. *ACI Material Journal*, 94(6): 478–483.

Rossi, P. 2000. Ultra-High Performance Fiber Reinforced Concretes (HUPFRC): An overview. In Rossi, P. and Chanvillard, G. (eds.), *Fiber-Reinforced Concrete* (FRC) – BEFIB 2000, RILEM Publications S.A.R.L.: 87–100.

Rossi, P. Acker, P. and Mailer, P. 1987. Effect of Steel Fibers at Two Different Stages: *The Material and the Structure, Materials and Structures*, 20: 436–439.

Fracture Mechanics of Concrete and Concrete Structures – High-Performance Concrete,
Brick-Masonry and Environmental Aspects – Carpinteri, et al. (eds)
© 2007 Taylor & Francis Group, London, ISBN 978-0-415-44617-4

Toughness properties and fiber dispersion in vibrated and self-consolidating fiber reinforced concrete

L. Ferrara*
Department of Structural Engineering, Politecnico di Milano, Italy

Y.D. Park[†]
Faculty of Construction and Architectural Design, Daegu Haany University, South Korea

S.P. Shah
Center for Advanced Cement Based Materials, ACBM, Northwestern University, Evanston, IL, USA

ABSTRACT: The connections between mechanical properties of fiber reinforced concrete and fiber dispersion, as influenced by fresh state properties of the material and casting modalities, are investigated in this work with reference to mostly two-dimensional structural geometry, selected as representative of significant potential applications in precast construction industry (folded plate roof elements, structural topping in precast slabs etc.). Square plates 600 mm wide and 60 mm thick were cast with three different fiber reinforced concretes featured by different fresh state properties. Specimens were cast in different ways, allowing for either a radial spreading or a prevalent one-directional flow of the fresh mix in the moulds. After analyzing fiber dispersion through Alternate-Current Impedance Spectroscopy, beam specimens have been cut from the plates with their axis differently oriented with respect to the flow direction, and tested in 4 point bending, either according and upside-down with respect to the casting. The thus measured mechanical properties are going to be correlated to the detected fiber dispersion.

1 INTRODUCTION

The benefits of incorporating (steel) fibers in concrete and other cement based materials have been widely investigated over the past forty years (Shah et al., 2004) leading to the current thorough knowledge of the multifold behavior of Fiber Reinforced Concrete (FRC). Nevertheless structural applications have not followed a similar development trend, only in the very last years the use of FRC having been extended from partially load-carrying structures, such as pavements, facade and wall panels, to more demanding structural elements e.g. precast roof elements (di Prisco M. & Plizzari, 2004), tunnel linings (Falkner & Henke, 2004; Gettu, 2004) and girders for slope stabilization (di Prisco C. et al., 2006). The partial substitution of conventional reinforcement with fibers, or even its complete replacement, such as for transverse shear

reinforcement, represents a surely attractive characteristics of this material from the point of view of (precast) construction industry.

The effectiveness of what above said relies, more or less explicitly, on the assumption of a uniform dispersion of fibers within the elements, since poor dispersion may lead to poor fresh and hardened state properties, thus affecting the resulting structural performance (Ferrara & Meda, 2006). Fiber dispersion related issues hence stand as a crucial point to be tackled for a wide promotion of safe and reliable structural applications of FRCs. Various techniques have been developed and assessed for non destructive and time effective monitoring of fiber dispersion (Franchois et al., 2004; Ozyurt et al., 2006a,b).

The use of self consolidating concrete may be helpful at guaranteeing a more uniform dispersion of fibers, thanks to both its self placeability, which lead to the elimination of compaction by vibration, and rheological stability in the fresh state. The possibility of even driving, through a suitable balance of fresh state properties, the orientation of fibers along the anticipated stress pattern within the element when in service,

---

* formerly Fulbright Visiting Scholar at Center for ACBM, Northwestern University, Evanston, IL, USA.

† formerly Visiting Scholar at Center for ACBM, Northwestern UNiversity, Evanston, IL, USA.

e.g. under dead and long term live loads, may represent an interesting opportunity to be exploited. The compactness of the SCC matrix, due to the higher amount of fine and extra-fine particles, may improve interface zone properties (Corinaldesi & Moriconi, 2004), and consequently also the fiber-matrix bond, leading to enhanced post-cracking toughness and energy absorption capacity. The synergy between self compacting and fiber reinforced technologies, thanks to the elimination of vibration and the reduction or even the complete substitution of conventional reinforcement with fibers, is likely to improve the economic efficiency of the construction process. Increased speed of construction, reduction or suitably focused rearrangement of labor resources, costs and energy consumption, better working environment, with reduced noise and health hazards, also contribute toward the automation and reliability of quality control.

In this work, toughness properties of steel fiber reinforced concrete are investigated into a multifaceted framework, aiming at establishing connections with fresh state properties and fiber dispersion related issues. Three fiber reinforced concrete mixes were designed for different fresh state properties (vibrated, self consolidating and highly fluid and hence prone to segregation) and three square plates 600 mm wide and 60 mm thick were cast for the self consolidating and segregating mixes, while two were cast for vibrated concrete. For self consolidating steel fiber reinforced concrete – SCSFRC – and segregating SFRC, plates were cast either from the center of the formwork, allowing a radial spread of fresh concrete, and from one side, an almost unidirectional flow of the fluid mixture filling the moulds in this case. After non destructive measurement of fiber dispersion through Alternate Current Impedance Spectroscopy, AC-IS, the plates were cut into five beams 120 mm wide, to be tested in 4-point bending. For plates cast from the side, beams were cut with their longitudinal axis either parallel or vertical to the casting flow direction.

The material performance in the hardened state is strongly influenced by fresh state properties and fiber dispersion, which are on their hand dependant on each other. An attempt of assessing the above said correlations in the framework of an omni-comprehensive approach is performed in this study.

2 MIX DESIGN OF SFRCS AND FRESH STATE CHARACTERIZATION

The mix proportions of the three different steel fiber reinforced concrete investigated in this study are shown in Table 1. Cement type I-ASTM and class C fly ashes were employed, together with a polycarboxilate superplasticizer. In all cases fiber reinforcement consisted of hooked end steel fibers 35 mm long and with

an aspect ratio equal to 65. Maximum aggregate diameter was equal to 10 mm. For vibrated SFRC the fresh state performance was characterized only by means of the slump test: a slump height loss equal to 210 mm was measured for both batches, each batch corresponding to one plate. In the other cases, the slump flow test was performed, measuring the final spread diameter and the time to reach a 500 mm spread, T_{50} (Table 2). A Visual Segregation Index – VSI, ranging from 0 to 3, respectively for very stable and loose mixes, (Khayat et al., 2000) was also assigned from the observation of the flown patty as an indicator of the stability of the mix.

As illustrated in detail in Ferrara et al. (2007a), the fresh state performance is related to an average spacing of solid particles, d_{ss}, which is an indicator of their degree of suspension in the fluid cement paste:

$$d_{ss} = d_{av} \left[\sqrt[3]{1 + \frac{V_{paste} - V_{void}}{V_{concrete} - V_{paste}}} - 1 \right] \quad (1)$$

where V_{paste} is the volume of paste, V_{void} is the void ratio of the solid particle skeleton (aggregates and fibers), determined according to ASTM C29/29. The average diameter of solid particles, d_{av}, is computed from the grading of aggregates and fibers, these being handled as a further "aggregate fraction" with 100% passing fraction at an equivalent diameter, $d_{eq,fibers}$, calculated on the basis of the equivalence with respect to specific surface area:

$$d_{eq,fiber} = \frac{3 L_f}{1 + 2 \dfrac{L_f}{d_f}} \frac{\gamma_{fiber}}{\gamma_{aggregate}} \quad (2)$$

with L_f and d_f fiber length and diameter, γ_{fiber} and $\gamma_{aggregate}$ specificy gravity of fibers and aggregates (average of coarse and fine) respectively.

For the fibers herein employed $d_{eq,fibers} = 2.37$ mm and $d_{av} = 3.54$ mm, irrespective of the different mix proportions.

The role of the average spacing of solid particles in justifying the fresh state performance is evident from the data in Tables 1 and 2: for the given grading of aggregates and fibers a higher value of d_{ss} corresponds to a more dilute suspension in the fluid cement paste and hence to a highly flowable concrete, whereas a tighter suspension (lower d_{ss}) yields a less workable concrete. As shown in Table 1, different values of d_{ss}, and hence different levels of fresh state performance, are obtained in this study increasing the past volume ratio, and correspondingly decreasing the volume ratio of aggregates.

Table 1. Mix proportions of SFRCs and model parameters.

Material	Vibrated kg/m^3	Self cons. kg/m^3	Segregating kg/m^3
Cement	355	400	440
Fly ash 1.3	132	148	164
Water	166	186	205
SP	24	27	30
Fine aggr.	975	918	861
Coarse aggr.	795	749	703
Fibers	50	50	50
Relevant data for mix-design model			
V_{paste}	0.33	0.37	0.41
V_{void}	0.23	0.23	0.23
d_{ss} (mm)	0.27	0.36	0.46

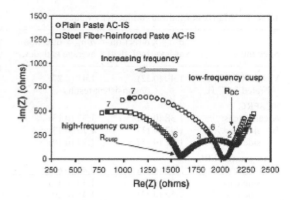

Figure 1. Sample Nyquist plot curves for plain and FRC cement paste (w/c = 0.40, fiber volume ratio = 0.35% – from Ozyurt et al., 2006a).

Table 2. Fresh state characterization of self consolidating and segregating SFRC.

	Self cons.	Segregating
Slump flow diameter (mm)	630	775
	665	800
	650	760
T_{50} (sec)	4, 4, 3	2, 2, 1
VSI	1.5, 1.5,	1 3, 3, 3

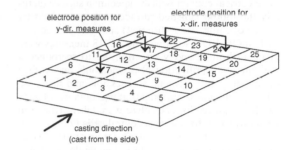

Figure 2. Scheme for AC-IS measurements.

3 NON DESTRUCTIVE MONITORING OF FIBER DISPERSION THROUGH AC-IS

Alternate Current Impedance Spectroscopy AC-IS is an electrical characterization technique to monitor characteristics of cement based materials (Peled et al., 2001) and, as recently shown, can be usefully employed to monitor dispersion of conductive fibers in concrete (Ozyurt et al., 2006a). It involves the application of an excitation voltage (1 V in this study) over a certain range of frequencies (10 MHz–10 Hz in the present case, stepped down according to a log-scale) and the measurement of the magnitude and phase of the current.

When the obtained data are converted into the real and imaginary part of the impedance Z and presented on the so-called Nyquist plot (-Im(Z) vs. Re(Z)) the so-called dual-arc behavior occurs (Figure 1) due to the frequency dependent behavior of conductive fibers. They are insulating under direct current (DC) and low frequencies of AC, because of, a thin oxide film that forms at the fiber-electrolyte interface, whereas they are conductive under high frequencies of AC. The low-frequency that can be detected on the Nyquist plot hence corresponds the resistance of the matrix, R_m, whereas the leftmost cusp represents the one of the composite, R. The following relationship holds for highly conductive fibers, for which the ratio of the fiber conductivity to the matrix conductivity can be considered infinite (see Ozyurt et al. 2006b for theoretical framework):

$$\frac{R_m}{R} = \frac{\sigma}{\sigma_m} = 1 + [\sigma]_{fibers} \, V_{fibers} \qquad (3)$$

where σ and σ_m represent the conductivity (inverse of resistance) of the composite and of the matrix respectively, $[\sigma]_{fibers}$ is the intrinsic conductivity of the fibers and V_{fibers} is the fiber volume ratio (0.64% in the present investigation). The conductivity of the fibers can be calculated knowing only their aspect ratio $AR = L_f/d_f$.

$$[\sigma]_{fibers} = \frac{1}{3}\left[\frac{2(AR)^2}{3\ln(4AR)-7}+4\right] \qquad (4)$$

The experimental set-up for the present investigation is shown in Figure 2. A grid has been drawn on the plate surface (actually the mould-finished one) consisting of 5×5 120 mm squares to locate measuring points. Stainless steel square plates (80 mm side) were

Table 3. Matrix normalized conductivity values.

Specimen	σ/σ_m X direction average (std. dev)	σ/σ_m Y direction average (std. dev.)
Vibrated FRC A	1.40 (0.22)	1.41 (0.27)
Vibrated FRC B	inconsistent results	
SCSFRC		
A–center cast	1.38 (0.14)	1.32 (0.13)
B – side cast	1.42 (0.11)	1.73 (0.13)
C – side cast	1.52 (0.15)	1.74 (0.19)
Segregating SFRC		
A – center cast	2.35 (0.55)	2.45 (0.69)
B – side cast	1.36 (0.22)	1.33 (0.19)
C – side cast	1.52 (0.23)	1.48 (0.20)

employed as electrodes, constantly spaced at 240 mm; the electrical contact with the specimen surface (actually the mould finished surface) was guaranteed by means of same size (when wet) sponges, soaked in a 1 M NaOH-KOH solution. The AC-IS measures were taken in both x and y direction, so to reproduce a real "in-situ" application, also when only one surface of the structural element is accessible and hence no "through thickness" information was obtained. Thirty AC-IS measures for each plate were taken, fifteen in each direction: for the nine central cells (7–9, 12–14, 17–19) it was possible to obtain information on fiber dispersion in both x and y direction, while for twelve edge cells information on either the x (2–4, 22–24) or y (6,11,16,10,15,20) direction was gathered. No measure was taken for the four corner cells (1,5,21,25).

The average values and standard deviations of the matrix normalized conductivity $R_m/R = \sigma/\sigma_m$, in both x and y direction, and the related standard deviations are listed in Table 3. The measured values of the R_m/R ratio have also been plot for some representative situations. The influence of the fresh state properties of the concrete is evident also from this bulk information.

Self consolidating concrete is effective in guaranteeing a more uniform dispersion of fibers within the specimen, as well as in effectively driving their orientation along the casting direction (Figures 3a–b; for plates cast from the side the y direction is always the direction of casting), the exceptions along the edges of the center cast plate being consistently attributable to the boundary effect of the formwork.

Vibrated SFRC plates show, with respect to analogously cast SCSFRC ones, a dispersion of fibers almost twice as much scattered. The dispersion of matrix normalized conductivity values is shown, for the only plate which gave consistent measures, in Figure 4. The points are likely to be arranged according to a shape which resembles the filling modality of the moulds as driven by vibration, with peak values in the center and a radial decreasing spread, still with some exceptions along the edges, justified as above.

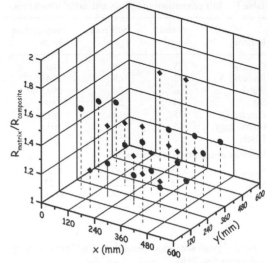

SCSFRC (cast from the center)
◆ x direction
● y direction

(a)

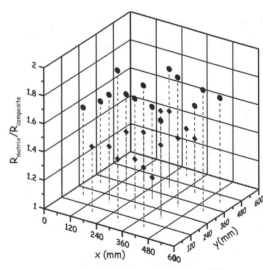

SCSFRC (cast from the side)
◆ x direction (transverse to casting)
● y direction (parallel to casting)

(b)

Figure 3. Plots of matrix normalized conductivity for SCSFRC plates cast from the center (a) and from the side (b).

The same holds with reference to plates cast from the side with segregating SFRC (Figure 5), the strong decrease in the matrix normalized conductivity representing a significant impoverishment in the content of

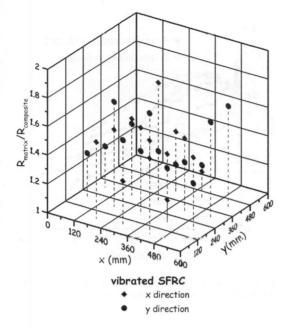

vibrated SFRC

◆ x direction

● y direction

Figure 4. Plots of matrix normalized conductivity for vibrated SFRC plate.

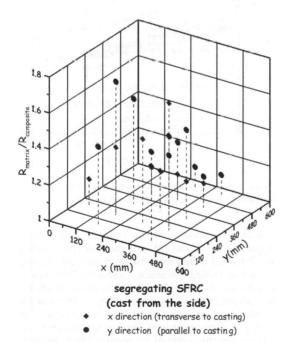

segregating SFRC
(cast from the side)

◆ x direction (transverse to casting)

● y direction (parallel to casting)

Figure 5. Plots of matrix normalized conductivity for segregating SFRC plate cast from the side.

Table 4. Fractional intrinsic conductivity of fibers.

Specimen	f_x	f_y	f_x/f_y
Vibrated FRC A	0.50	0.50	1.0
Vibrated FRC B		inconsistent results	
SCSFRC			
A-center cast	0.56	0.44	1.27
B – side cast	0.37	0.63	0.59
C – side cast	0.40	0.60	0.67
Segregating			
A – center cast	0.52	0.48	1.08
B – side cast	0.39	0.61	0.64
C – side cast	0.44	0.55	0.8

fibers along the flow direction. High scattering was detected also in these cases, whereas the effectiveness in governing the orientation of the fibers being lower than for SCSFRC due to the poor viscosity of fresh concrete. The measured high values of normalized matrix conductivity, with likewise high scattering, for segregating concrete plates cast from the center are, could be attributed to the strong segregation of fibers which may have affected the current path during AC-IS measures.

The issue of fiber orientation has been further investigated. From the measured values of the matrix normalized conductivities, the intrinsic conductivity of fibers in both x and y direction can be calculated:

$$[\sigma]_{fibers\ x,y} = \frac{\left(\dfrac{\sigma}{\sigma_m}\right)_{x,y} - 1}{V_{fibers}} \qquad (5)$$

and then normalized to the sum of both to calculate a fractional effective conductivity

$$f_{fibers\ x,y} = \frac{[\sigma]_{fibers\ x,y}}{[\sigma]_{fibers\ x} + [\sigma]_{fibers\ y}} \qquad (6)$$

Results, listed in Table 4, confirm the above said statement with reference to the possibility of effectively driving fiber orientation along the casting direction thanks to the suitably "selected" fresh state performance of the concrete. It has to be remarked, for the sake of clarity, that results in Table 3 have been obtained, for each direction, from all 15 AC-IS taken, whereas in Table 4 refer only to the nine central cells of the grid for which data in both x and y directions were gathered.

Since a linear correlation holds between the matrix normalized conductivity and the orientation number of fibers (Ozyurt et al., 2006a), the f_x/f_y ratio can be related to the ratio between orientation factors transverse and parallel to the casting direction. Results

(Table 4) are consistent with recent findings by Ferrara et al. (2007 in press), which measured, through fiber counting, an orientation factor ratio equal to 0.56 with reference to SCSFRC plates, 500 mm wide and 1 m long, cast parallel to the long side, the specimen geometry (width/length ratio) being likely to exert some influence.

4 TOUGHNESS PROPERTIES OF SFRC AND FIBER DISPERSION

In order to assess the connections existing between fresh state behavior, fiber dispersion and mechanical properties of SFRC in the hardened state, the 600 mm square plates were cut into five 120 mm wide beams, which were tested in 4-point bending (loading span 128 mm). For plates cast from the side the beams were cut with their axis either parallel or transverse to the casting direction. Tests were performed in displacement control and the crack opening displacement across the midspan section at the beam intrados was measured (base length 150 mm). One of the beams, always cut from the same position inside the plate (4–24 for plates A and C and 16–20o for plates C – see Figure 2), was tested upside down with respect to casting, to check the influence of downward settlement of fibers, due, if any, to vibration of poor stability of the fresh concrete.

Results are shown, in terms of dimensionless load vs. COD in Figures 6–8, the current load value being normalized to the cracking load P_{crack}, defined as the maximum load in the COD range 0–0.1 mm. Edge beams have not been considered, due to the boundary effect of formworks on fiber dispersion and hence on the measured mechanical properties.

For vibrated SFRC beams (Figure 6) a quite high scattering affects experimental results, even for beams cut from the same plate; the downward settlement of fibers, due to vibration, justifies the worst performance of the beam tested upside down to casting, when compared to the one of companion beams cut from the same plate and tested according to it.

In the case of SCSFRC (Figure 7) results are, for beams cut from the same plate, fairly less scattered; a higher dispersion has been detected for beams cut from the center-cast plate. The effectiveness in avoiding the downward settlement of fibers is remarkable: results for beams tested upside down to casting are absolutely comparable with the ones of companion beams tested according to it. A certain sensitivity, even if lower than expectable, of the response to the relative orientation of the beam axis to the direction of casting, also appears. This may be justified considering that beams cut transverse to the direction of casting belonged to plate 3, for which a higher fractional conductivity in that direction was measured (Table 4).

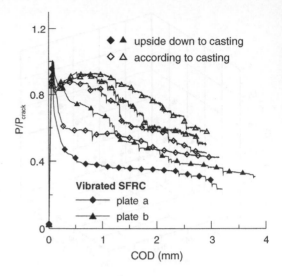

Figure 6. 4pb tests on vibrated SFRC beams – dimensionless load vs. crack opening displacement.

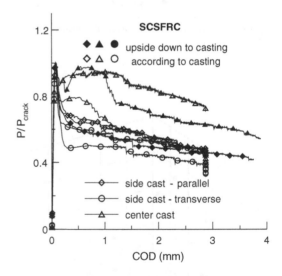

Figure 7. 4pb tests on self consolidating SFRC beams – dimensionless load vs. crack opening displacement.

When the fluidity of the concrete is too high (Figure 8) the downward settlement of fibers yields to a remarkable difference between the performance of the beam tested upside down and according to casting, mainly for beams cut from the side-cast plates. A moderate post-cracking hardening is observed for beams tested according to casting, reasonably due to the higher concentration of fibers in the tension zone. The good performance of the beam cut from the centrally cast plate and tested upside down to casting can be attributed to a "late supply", during casting, of the

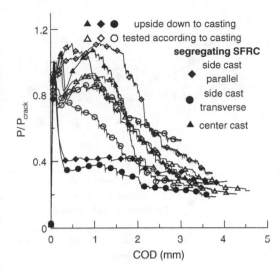

upside down to casting
tested according to casting
segregating SFRC
◆ side cast parallel
● side cast transverse
▲ center cast

Figure 8. 4pb tests on segregating SFRC beams – dimensionless load vs. crack opening displacement.

Table 5. First cracking strength, post cracking strengths and their ratios for beams tested in 4 point bending (beam 3 tested upside down to casting).

	f_{ctf}	$\sigma_{eq,I}$	$\sigma_{eq,II}$	$\sigma_{eq,I}/f_{ctf}$	$\sigma_{eq,I}/\sigma_{eq,II}$
Vibrated SFRC					
Plate A					
Beam 1	6.98	6.05	5.17	0.87	0.85
Beam 2	7.92	4.81	4.43	0.61	0.92
Beam 3	7.28	3.69	2.66	0.51	0.72
Plate B					
Beam 1	8.37	6.77	5.04	0.81	0.74
Beam 2	8.11	7.12	6.68	0.88	0.94
Beam 3	7.55	6.53	6.73	0.86	1.03
Self consolidating SFRC					
Plate A – cast from the center					
Beam 1	11.11	8.64	6.92	0.78	0.80
Beam 2	10.04	9.25	9.33	0.92	1.01
Beam 3	9.90	8.51	7.97	0.86	0.94
Plate B – cast from the side – parallel tested					
Beam 1	7.04	5.23	4.30	0.74	0.82
Beam 2	7.24	5.07	4.32	0.70	0.85
Beam 3	6.95	4.60	3.79	0.66	0.84
Plate C – cast from the side – transverse tested					
Beam 1	8.69	5.07	4.27	0.59	0.84
Beam 2	8.52	5.28	4.71	0.61	0.89
Beam 3		meaningless results			
Segregating SFRC					
Plate A – cast from the center					
Beam 1	8.62	8.24	7.77	0.96	0.94
Beam 2	8.44	6.64	7.07	0.79	1.06
Beam 3	8.88	7.60	8.84	0.87	1.16
Plate B – cast from the side – parallel tested					
Beam 1	8.15	8.60	8.96	1.06	1.04
Beam 2	8.78	8.78	7.16	1.00	0.82
Beam 3	7.08	3.30	2.95	0.46	0.90
Plate C – cast from the side – transverse tested					
Beam 1	8.02	5.96	6.61	0.74	1.11
Beam 2	8.53	6.66	5.15	0.78	0.77
Beam 3	7.50	2.64	2.77	0.35	1.05

fibers settled down in the mixing bowl right in the center of the specimen, i.e. in the zone where the crack is likely to form, which may have compensated the segregation of fibers themselves.

Results of 4pb tests have been processed according to the prescriptions of the recently issued Italian guidelines for the design of SFRC structures (CNR-DT204). In Table 5 are listed for each beam:

– the first cracking strength f_{ctf}, calculated from the first cracking load, defined as above;
– the first equivalent post-cracking strength, $\sigma_{eq,I}$, averaged in the crack opening range 3–5 w_I, where w_I is the crack opening at first cracking;
– the second equivalent post-cracking strength, defined as the average stress in the crack opening range 0.8 w_u – 1.2 w_u, where w_u is assumed equal to 0.02 h, with h depth of the specimen (h = 60 mm, w_u = 1.2 mm in the present case).

The ratios $\sigma_{eq,I}/f_{ctf}$ and $\sigma_{eq,II}/\sigma_{eq,I}$ can be assumed as toughness indicators, and will be thus referred to in the following. Data in Table 5 confirm what above said, qualitatively, from the observation of experimental curves.

The correlation existing between the mechanical performance of the composite and the dispersion of the fibers, as monitored through AC-IS and influenced by the fresh state performance of concrete, has been hereafter tentatively assessed also from the quantitative point of view. Data from beams tested upside down to casting have not been considered in this stage of the investigation. The toughness indicators are plot in Figure 9 vs. the normalized matrix conductivity in

the direction of the beam axis, measured in the cell where fracture occurred during 4pb tests. Despite a common trend hardly could be found, both indicators are likely to increase with the matrix normalized conductivity,. A not clear trend of the $\sigma_{eq,I}/f_{ctf}$ ratio is observed for lower values of the matrix normalized conductivity; this deserves further investigation. For the $\sigma_{eq,II}/\sigma_{eq,I}$ ratio a reverse in the increasing trend is observed for matrix normalized conductivities higher than 1.7. These higher values correspond to segregating SFRC plates cast from the center: this may be due to the downward settlement of fibers, which alterates the the current pathin the AC-IS measures and also affects the mechanical performance of the composite.

In the literature more or less successful attempts of correlating the mechanical performance of the

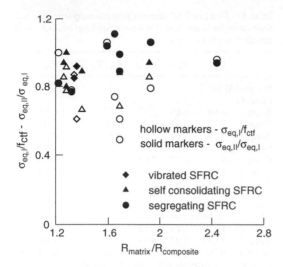

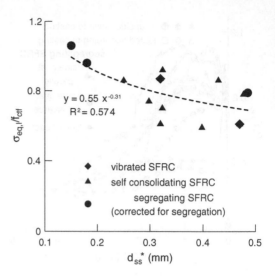

Figure 9. Toughness indicators of SFRCs vs. matrix normalized conductivity.

Figure 10. $\sigma_{eq,I}/f_{ctf}$ vs. average spacing of solid particles modified to take into account fiber dispersion and orientation.

hardened fiber reinforced cement composites to an "average fiber spacing" have been found (Soroushian & Lee, 1990; Voigt et al., 2004). In this work, also consistently with the framework in which the mix-design and fresh state characterization was performed, the average spacing of solid particles d_{ss} [Eq (1)] has been chosen at the purpose (see also Ferrara et al., 2007 in press). The value of this parameter (Table 1) would correspond to a perfectly isotropic and uniformly dispersed case. To account for fiber dispersion and orientation it has been corrected through two factors:

- a dispersion factor, equal to the ratio between the average value of the matrix normalized conductivity parallel to the beam axis, measured for the whole plate (Table 3), and the one measured in the fracture cell in the same direction;
- an orientation factor equal to $0.5/f_{//}$, where $f_{//}$ is the fractional intrinsic conductivity [Eq. (6)] for the cell where fracture occurred in the direction of the beam axis (either f_y for plates A and C and f_x for plates C). The value 0.5 is to the fractional intrinsic conductivity for perfectly planar isotropic distribution of fibers in a 2D specimen

Both correction term takes into account that a higher concentration of fibers, furthermore better aligned along one direction, correspond to a tighter spacing.

For segregating SFRC the actual spacing of aggregates and fibers is tighter than the theoretical one. Visual observation showed that almost all fibers and coarser aggregates settled down in the half bottom part of the specimen: An attempt to consistently process also these data has been done herein, through a further rough correction factor equal to 0.5. As an exmaple the

values of $\sigma_{eq,I}/f_{ctf}$ have been plot vs. the thus modified average spacing of solid particles, d_{ss}^* (Figure 10).

The trend, despite the somewhat not excellent correlation, is consistent with the physics of the problem: a better mechanical performance of the composite corresponds to a tighter spaced fiber reinforcement. This can be achieved through the mix design and suitably balanced fresh state performance, leading to a more uniform and effectively oriented dispersion of fibers.

5 CONCLUSIONS AND FURTHER WORK

In this work the correlations between toughness properties of steel fiber reinforced concrete and fiber dispersion related issues, as influenced by the fresh state performance, have been investigated. A mixdesign model recently proposed for SCSFRC has been employed to analyze the performance in the fresh state. It is based on an average spacing of the solid particles, which accounts for the grading of aggregates and fibers; these are regarded as an equivalent aggregate with respect to the specific surface area. The average spacing of solid particles, d_{ss}, is an indicator of the degree of suspension of the solid particles into the fluid cement paste.

The effectiveness of the self consolidating concrete in driving the orientation of fibers along the casting direction and in guaranteeing a uniform dispersion of the fibers, furthermore preventing their downward settlement, has been proved.

The performance of the composite in the hardened state, measured through 4 pb tests, has been shown to be strongly affected by the dispersion and orientation of fibers and have been tentatively correlated to the same parameter, d_{ss}, employed for mix design and fresh state characterization. A correction factor has been suitably defined to include in this parameter the effect of the dispersion and orientation of fibers, as influenced by the casting process, fresh state behavior of concrete and specimen geometry. The parameter d_{ss}, once modified as above, can be hence regarded also as an indicator of the binding efficacy of aggregates and fibers by the hardened cement paste.

Further investigation to improve the approach, including e.g. more explicitly the interaction of fibers with corase aggregates, and assess its reliability with reference to a broader range of cases is currently on going.

ACKNOWLEDGMENTS

This work has been partially performed during period spent by the first author at ACBM, Northwestern University, in the framework of a Fulbright Program, whose support is gratefully acknowledged. The authors also thank prof. T. O. Mason and dr. S. Wansom, Department of Materials Science and Engineering, Northwestern University, for their kind availability during the AC-IS investigations.

REFERENCES

Corinaldesi, V. & Moriconi, G. 2004 Durable fiber reinforced self-compacting concrete, *Cement and Concrete Research*, 34: 249 254.

CNR DT-204: Guidelines for design and execution of SFRC structures (in Italian).

di Prisco, M. & Plizzari, G.A. 2004. Precast SFRC elements: from material properties to structural applications. In M. di Prisco et al. (eds.) *BEFIB 2004, Proc. 6th Int. RILEM Symp., Varenna, Italy, 20–22 Sept. 2004*, Rilem Pubs.: 81–100.

di Prisco, C., di Prisco, M., Mauri, M. & Scola, M. 2006. A new design for stabilizing ground slopes. In *Proc. 2nd fib Congress, Napoli (Italy), June 5–8, 2006*, ID 4-1-CDRom.

Falkner, H. & Henke, V. 2004. SFRC – shear load bearing capacity and tunnel linings. In M. di Prisco et al. (eds.) *BEFIB 2004, Proc. 6th Int. RILEM Symp., Varenna, Italy, 20–22 Sept. 2004*, Rilem Pubs.: 111–124.

Ferrara, L. & Meda, A. 2006. Relationships between fibre distribution, workability and the mechanical properties of SFRC applied to precast roof elements, *Materials and Structures, 39*: 411–420.

Ferrara, L., Park, Y.D. & Shah, S.P., 2007a. A method for mix-design of fiber reinforced self consolidating concrete, submitted for publication to *Cement and Concrete Research*.

Ferrara, L., Dozio, D. & di Prisco, M. 2007 in press. On the connections between fresh state behavior, fiber dispersion and toughness properties of steel fiber reinforced concrete, *Proc.* HPFRCC5, Mainz, 4–7 July 2007.

Franchois, A., Taerwe, L. & Van Damme, S. 2004. A microwave probe for the non-destructive determination of the steel fiber content in concrete slabs', In M. di Prisco et al. (eds.) *BEFIB 2004, Proc. 6th Int. RILEM Symp., Varenna, Italy, 20–22 Sept. 2004*, Rilem Pubs.: 249–256.

Gettu, R., Barragàn, B., Garcia, T., ramos, G., Fernàndez C. & Oliver, R. 2004. Steel fiber reinforced concrete for the Barcelona metro line 9 tunnel lining. In M. di Prisco et al. (eds.) *BEFIB 2004, Proc. 6th Int. RILEM Symp., Varenna, Italy, 20–22 Sept. 2004*, Rilem Pubs.: 141–156.

Khayat, K.H., Ghezal, A. & Hadriche, M.S. 2000. Utility of statistical models in proportioning self consolidating concrete, Materials and Structures, 33: 338–344.

Ozyurt, N., Mason, T.O. & Shah, S.P. 2006a. Non destructive monitoring of fiber orientation using AC-IS: an industrial scale application. *Cement and Concrete Research*, 36: 1653–1660.

Ozyurt, N., Woo, L.Y., Mason, T.O. & Shah, S.P. 2006b. Monitoring fiber dispersion in fiber reinforced cementitious materials: comparison of AC-Impedance Spectroscopy and Image Analysis, *ACI Materials Journal*, 103: 340–347.

Peled, A., Torrents, J.M., Mason, T.O., Shah, S.P. & Garboczi, E.J. 2001. Electrical impedance spectra to monitor damage during tensile loading of cement composites, *ACI Materials Journal*, 98: 311–316.

Shah, S.P., Kuder, K.G. & Mu, B. 2004. Fiber-reinforced cement-based composites: a forty year odyssey. In M. di Prisco et al. (eds.) *BEFIB 2004, Proc. 6th Int. RILEM Symp., Varenna, Italy, 20–22 Sept. 2004*, Rilem Pubs.: 3–30.

Soroushian, P. & Lee, C.D. 1990. Tensile strength of steel fiber reinforced concrete: correlation with some measures of fiber spacing *ACI Materials Journal*, 87: 541–546.

Voigt, T., Van Bui, K. & Shah, S.P. 2004. Drying shrinkage of concrete reinforced with fibers and welded-wire fabric, *ACI Materials Journal*, 101: 233–241.

Effect of manufacturing methods on tensile properties of fibre concrete

P. Stähli & J.G.M. van Mier

Institute for Building Materials (IfB), ETH Zurich, Switzerland

ABSTRACT: This Paper presents the first results of the experiments of fibre reinforced concrete tested using a newly developed uniaxial tensile test. The aim was to develop a tensile test set-up in which the specimen supports have no restraind freedom of movements. This goal has been achieved by means of an arrangement based on 'pendulum-bars'. Different filling/casting methods were used for optimizing the fibre distribution and orientation aimed at an increase of the tensile strength will increase. Results of a test series, where the filling method, the type and the amount of fibres are varied is presented in this paper.

1 INTRODUCTION

Compaction may have a significant influence on the properties of fibre reinforced concrete. In particular fibre orientation and fibre distribution may be affected, especially when vibration needles are inserted in the fresh concrete. With the development of self compacting concrete, the use of vibrational energy for compaction has become obsolete, and with current generation of super-plasticizers it is possible to develop self-compacting fibre concrete as well (Grünewald (2004), Markovic (2006), Stähli (2004), Stähli (2007), Rossi (1996)). Increasing the amount of fibres may have a positive effect on the mechanical properties, but because fibres are not all necessarily aligned in the direction of stress, the effectivity is debatable. Better would be to align fibres in the direction of stress, which might lead to improved performance of fibre reinforced concrete (FRC) and hybrid fibre concrete (HFC) in a structure, probably against lower cost. Not only strength should be considered, but also ductility.

Aligning fibres has been tried in the past under a variety of circumstances. Recently a method based on magnetic fields was proposed by Linsel (2005). For SIFCON, fibre alignment can be achieved by sprinkling fibres in a narrow space, or in very thin elements (Van Mier (1992), van Mier (1996)). Moreover, during extrusion of FRC, fibres align in the longitudinal direction as well (see for example Shao (1997)), leading to quite some anisotropy with, probably the related differences of properties in the various directions.

As mentioned, the development of self-compacting concrete leads to easier placement of the fresh material, and the fibre distribution and orientation is not affected since compaction becomes obsolete. An interesting idea is to investigate to what extent the filling method of the fresh material can be used to affect the fibre distribution and orientation, and to see if a possible influence on the mechanical properties emerges. In this paper a test series is reported that confirms that the filling method has an influence on the mechanical properties of fibre reinforced materials. These and earlier experiments also show that material properties for fibre reinforced materials are dependant on the geometry of the tested specimen. Therefore, parameters, such as the tensile strength, are not constant for the whole specimen. They change with the flow and the geometry of concrete in the specimen and that means that structures can not be conventionally designed. Moreover, the whole filling process has to be taken into account to design structures. This is also a chance for optimizing a structure by controlling the flow of the material so that fibres are oriented most favourably and perhaps fewer fibres need to be used.

This paper also introduces a newly developed tensile test using pendulum bars. This test set-up was specially designed for materials like hybrid fibre concrete with a maximum uniaxial tensile strength of 20 MPa and a maximum fibre length of 30 mm.

2 FILLING METHODS AND MATERIAL

In order to optimize the filling process and the resulting fibre orientation and distribution three different filling methods were performed. The 'conventional' method (C), where the concrete was filled from the top and the material did not flow, but falls in the mould (Fig 1); the 'filling' method (F), where the concrete was filled in the mould in such a way that the material flowed a little (Fig. 2) and the 'climbing' method (CL), where the material was filled/pumped from the bottom to the top of the mould (Fig 3).

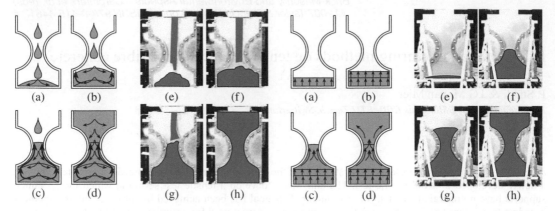

Figure 1. Sketch (a to d) and images (e to f) of the filling sequence of a tensile specimen with the 'conventional' method. The arrows show the direction of the flow of the material during the filling process.

Figure 3. Sketch (a to d) and images (e to f) of the filling sequence of a tensile specimen with the 'climbing' method. The arrows show the direction of the flow of the material during the filling process.

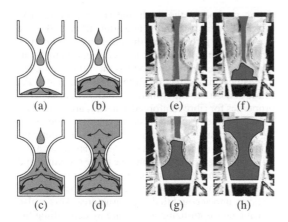

Figure 2. Sketch (a to d) and images (e to f) of the filling sequence of a tensile specimen with the 'filling' method. The arrows show the direction of the flow of the material during the filling process.

Figure 4. Sketch and image of the 'U-Shaped' mould for the tensile test specimens.

Due to the available equipment in the laboratory (no concrete pump was available) the 'climbing' method was performed in such a way, that the material first was filled in the 'filling' method specimen and there, the material flowed trough a connecting part and climbed up into the 'climbing' method specimen (Fig. 4). The difference between the conventional and the filling method is in the flow of the material. Figure 1b shows that as soon as the material starts to fill the mould, the material starts to flow upwards and concrete swirls develop. This also happens when the mould is almost filled (Fig. 1c). Figure 2b and c show that the material starts to flow downwards the more the mould is filled. No swirls develop and the fibres can properly align in the direction of the material flow.

To ensure that the fibres did not segregate and the material was capable to self-level the diameter of the small slump flow of the fresh material had to be in-between 20 and 26 cm. The concrete matrices only differed in the amount of super-plasticizer because of the aim to have equal rheological properties for all the mixtures. In most of the mixtures additional super-plasticizer was added to increase the flow ability of the material. Three mixtures with only one type of steel fibre and two hybrid fibre concretes with three different types of steel fibres were investigated. An overview of all the mixtures and their respective rheological properties is given in Table 1.

The matrix was composed as follows: $1000 \, kg/m^3$ CEM I 52.5R, $200 \, kg/m^3$ fly-ash, $100 \, kg/m^3$ Micro-silica and $800 \, kg/m^3$ sand with a maximum diameter of 1 mm. This composition of the materials guaranteed a compact material structure where the fibres were quite perfectly embedded. The water-binder-ratio

Table 1. Overview of the concrete mixtures.

	3 0 0	0 3 0	0 0 3	1 1 1	3 1 1
Super-plasticizer [%]	2.0	2.0	2.6	2.0	2.3
Additional super-plasticizer [%]	0.3	0.3	–	0.4	0.3
0.15 × 0.6 mm fibre [Vol.-%]	3	–	–	1	3
0.20 × 12 mm fibre [Vol.-%]	–	3	–	1	1
0.60 × 27 mm fibre [Vol.-%]	–	–	3	1	1
Small slump flow [mm]	20.5	23	23	26	27
Large slump flow [mm]	62.5	67	78	76	84
Fresh concrete density [kg/m³]	2428	2370	2378	2415	2503
Air content [%]	4	3.5	3.8	3.8	2.8

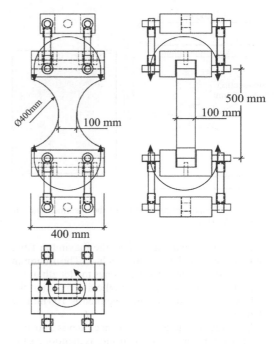

Figure 5. Sketch of the 'pendulum-bar' test set-up. The arrows shows the ability to rotate around the centre of rotation, e.g. around the interface between aluminium plate and specimen.

(w/b-ratio) was taken as low as possible to improve the strength, i.e. the w/b-ratio was 0.18. To keep the mixture self-compacting Glenium® super-plasticizer was between 2.3% and 2.6% of the cement weight. The used fibres were Stratec steel fibres with different shapes and geometries. Straight steel wire fibres with a length of 6 and 12 mm and a diameter of 0.15 and 0.2 mm respectively and undulated steel fibres with a length of 30 mm and a diameter of 0.6 mm were used. The steel fibres had a tensile strength of 2400 MPa. Fresh concrete properties were tested using slump flow tests Stähli (2004) and the air contend was measured using the test according to EN 12350-7.

3 'PENDULUM-BAR' TENSILE TEST

3.1 Operational principle

The reason for using pendulum bars is based on the idea that during a tensile test the forces should remain centric and the supports should be able to rotate (Fig. 5 and 6). This becomes important especially when a material of increased ductility, such as HFC, is used. Van Mier (1994) showed that the boundary conditions in a uniaxial tensile test have a significant influence on the result of the test. The tensile strength but also the tensile strength obtained from uniaxial tensile tests between fixed boundaries are higher than the values received from tests with rotation supports. The centre of rotation of the supports is positioned at the centre of the interface between the specimen and the glued aluminium plates (Fig. 5).

The deformations are measured using four LVDT's, two at each side of the specimen as shown in Figure 7.

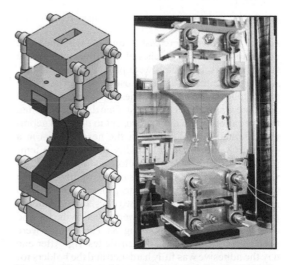

Figure 6. Sketch and image of the 'pendulum-bar' tensile test set-up.

3.2 Specimen geometry

To determine the uniaxial tensile strength of concrete, dog-bone shaped specimens are well proven (Markovic (2006), Van Vliet (2000), van Mier (1994), Carpinteri (1994)). In the presented experiments, the geometry

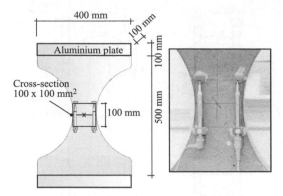

Figure 7. Sketch and photograph of the specimen and the attached LVDT's.

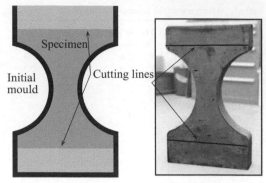

Figure 8. (a): An image of a tensile test specimen with its cutting lines. (b): The same specimen after gluing the aluminium plates.

of the specimen was derived by the maximum fibre length, the glue and the testing machine. The minimum cross section was at least three times the length of the largest fibre (30 mm). Therefore the dimension of the smallest cross section of the specimen is 100 mm by 100 mm.

The height of the tensile test specimen is 500 mm. The maximum width is 400 mm and, as mentioned, the minimum width is 100 mm. The chosen depth of the specimen was 100 mm. The radius of the curved sides is 200 mm. The capacity of the load cell used in the testing machine is 200 kN.

3.3 Specimen preparation

All the specimens were cast on one day with the same concrete batch. Two batches of 300 litres each of the same mixture per batch were produced on one day. All the presented results are those from the first of the two batches. For the whole series 3 m³ fibre reinforced concrete was produced and cast. One day after casting, the specimens were demoulded, cut to the final geometry (Fig. 7, 8) and stored for the next 24 days in a climate chamber with a humidity of 95% and a temperature of 20°C. Three days before testing the samples were prepared for gluing. The surfaces of the specimen and the aluminium plates were ground, sand blasted and cleaned properly. Within two days, two aluminium plates with four M20 threads were glued to the top and the bottom of the specimen (Fig. 8). These plates were used to fix the sample in the tensile test rig. After one day the adhesive was fully hardened and the holders for the LVDT's were glued to each side of the specimen, after which the sample was ready to be tested.

4 RESULTS AND DISCUSSION

Figures 9–11 show results derived from the tensile test. It can be seen that the characteristics of the load displacement curves and the crack patterns are

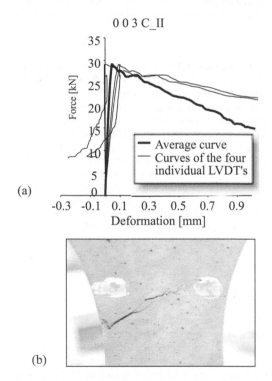

Figure 9. (a): Curves derived by the pendulum-bar tensile test with the specimen 0 0 3 C_II. (b): Crack pattern of the above mentioned specimen.

related. The figures also show that the curves from the four LVDT's can have completely different shapes. In Figure 9 the peak is very sharp. The same 'sharpness' can be seen in the crack pattern of this specimen. This crack propagated from the left to the right. After the peak, the deformations on the right side of the

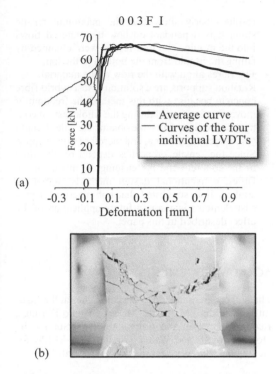

(a)

(b)

Figure 10. (a): Curves derived by the pendulum-bar tensile test with the specimen 0 0 3 F_I. (b): Crack pattern of the above mentioned specimen.

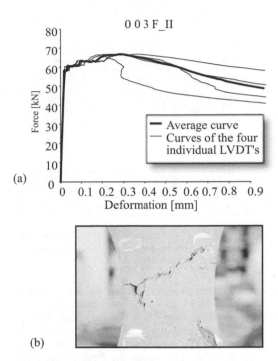

(a)

(b)

Figure 11. Upper: Curves derived by the pendulum-bar tensile test with the specimen 0 0 3 F_II. Lower: Crack pattern of the above mentioned specimen.

specimen decreases as expected. This behaviour is typical for a test set-up with freely rotating supports.

Figure 10 shows a similar characteristic, but the deformation on the backside decreases now there the peak is very smooth. The crack propagated from the front to the backside of the specimen and its pattern is completely different. On the image of the crack pattern in Figure 10, it can be seen that there is not one single crack, but there is, typical for fibre reinforced materials, a crack zone with multiple cracking.

Figure 11 shows a third type of behaviour. The deformations of all sensors always increase during the tensile test. Here the whole cross-section is pulled symmetrically. The crack propagated from the left to the right and showed two branches. After cutting the aluminium plates from the sample with a water cooled diamond saw, white lines appeared on the surface of the specimen in the cracking zone. It seems that these white lines are micro cracks which can not be detected by the naked eye. Such phenomena also appeared on other specimens which are not reported in this paper.

The influence of the filling method for mixtures with one fibre only is significant. Figure 12 gives an overview of the results for the mixtures 3 0 0, 0 3 0 and 0 0 3. It can be seen that difference in the tensile

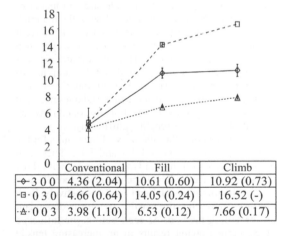

	Conventional	Fill	Climb
3 0 0	4.36 (2.04)	10.61 (0.60)	10.92 (0.73)
0 3 0	4.66 (0.64)	14.05 (0.24)	16.52 (-)
0 0 3	3.98 (1.10)	6.53 (0.12)	7.66 (0.17)

Figure 12. Summery of the results in [MPa] derived from tensile test from the mixtures with only one type of fibre. The standard deviation is given in brackets.

strength for the conventional filling method is negligible. The tensile strengths are within the standard deviations. These standard deviations are relatively large in comparison to the ones from the two other filling methods. The fibre amount of these three mixtures was constant at 3%. The diagram also shows that

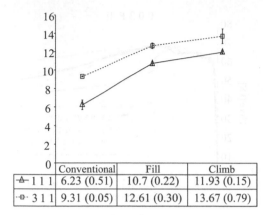

	Conventional	Fill	Climb
—▲— 1 1 1	6.23 (0.51)	10.7 (0.22)	11.93 (0.15)
···▣··· 3 1 1	9.31 (0.05)	12.61 (0.30)	13.67 (0.79)

Figure 13. Summery of the results in [MPa] derived from tensile test from the mixtures with HFC. The standard deviation is given in brackets.

the tensile strength for the 'filling' and the 'climbing' method is higher than for the 'conventional' method. This leads to the assumption that the tensile strength increases with the ability of flowing of the fresh concrete and with that the alignment of the fibres to the tensile direction. The longer the fresh concrete can flow, the higher the tensile strength is. This can be seen for all types of fibres, even for the small 0.15/6 mm fibres. But why does the tensile strength increase? To the author's opinion, the tensile strength increases because the fibres align with the flow of the material. This fact would explain the results from the current tensile tests. As seen before, when casting a specimen with the 'filling' method, the flow of the material is less than with the 'climbing' method, but the jump in tensile strength between 'filling' and 'climbing' method is not that large. It seems that the material does not need to flow that long; only a short, but controlled flow can increase the tensile strength significantly.

Figure 13 shows the summary of the results for the experiments with mixture 1 1 1 and 3 1 1. These two mixtures were both hybrid fibre concretes with three types of steel fibres. The fibre amounts were 3% and 5% for the mixtures 1 1 1 and 3 1 1 respectively. Both mixtures show that the tensile strength is dependant on the filling method. As seen before, the influence of concrete flowing results in an increasing tensile strength. Figure 13 also shows that the tensile strength for the mixture with 5% of fibres is higher then the one with 3% only. That means, by increasing the amount of fibres, the tensile strength increases as well. This was seen before in several studies, e.g. Markovic (2006).

5 CONCLUSIONS

– The filling method has a significant influence on the tensile strength (Fig 12 and 13). The presented results clearly show that the maximum tensile strength is dependant on how the material flows into the mould. The best results were obtained by filling the mould from the bottom to the top.
– The fibres align with the flow of the material.
– Rotation supports are a solution to test hybrid fibre concrete because with this method the freedom of movement is ensured during the whole test. In contrary to fixed supports, the obtained tensile strength results are real minima, and therefore safe to apply.
– Since the tensile properties depend on the flow of the concrete and the ensuing alignment of the fibres, conventional design methods cannot be applied. It is suggested that optimization of precast elements is a sound future application of the effect described in this paper.

ACKNOWLEDGEMENTS

The authors are indebted to the colleagues at the Institute of Building Materials, Messrs Claudio Derungs, Leonardo Bressan and Antoni Amerana. Support by Holcim, BASF and Walter Trindler vom EMPA for sieving the sand is gratefully acknowledged.

REFERENCES

Carpinteri, A. and Ferro, G. (1994), Size effects on tensile fracture properties: a unified explanation based on disorder and fractality of concrete microstructure, Materials and Structure, Volume 10, Number 2, 563–571
Grünewald, S. (2004), Performance-Based Design of Self-Compacting Fibre Reinforced Concrete, PhD thesis, Delft University of Technology.
Linsel, S. (2005), Magnetische Positionierung von Stahlfasern in zementösen Medien (Magnetic positioning of steel fibres in cementitious medias), PhD thesis, Technische Universität Berlin
Markovic, I. (2006), High-Performance Hybrid-Fibre Concrete, PhD thesis, Delft University of Technology.
Rossi, P. and Renwez, S. (1996), High performance multimodal fiber reinforced cement composites (HPMFRCC), Presses de l'Ecole nationale des ponts et chausses, Paris, 29.–31. May 1996
Shao, Y. and Shah, S.P. (1997), Mechanical Properties of PVA Reinforced Cement Composites Fabricated by Extrusion Processing, ACI Mater. J., 94(6), 555–564.
Stähli, P., Van Mier, J.G.M. (2004), Three-Fibre-Type Hybrid Fibre Concrete, In Proceedings 5th International Conference on 'Fracture Mechanics of Concrete and Concrete Structures' (FraMCoS-V), April 12–16, 2004, Vail, Colorado, ed. V. C. Li et al, pp. 1105–1112.
Stähli, P., Van Mier, J.G.M. (2007), Manufacturing, Fibre Anisotropy and Fracture of Hybrid Fibre Concrete, Eng.Frac.Mech. 74, 223–242.
Van Mier, J.G.M., Timmers, G. (1992), Shear Fracture in Slurry Infiltrated Fibre Concrete (SIFCON), in Proceedings RILEM/ACI Workshop on High Performance Fibre

Reinforced Cement Composites (H.W. Reinhardt and A.E. Naaman eds.), Mainz, June 24–26, 1991, E&FN Spon/Chapman & Hall, London/New York, 348–360.

Van Mier, J.G.M., Vernuurt A., Schlangen E. (1994), Effect of loading rate on the strength of concrete subjected to uniaxial tension, Materials and Structure, Volume 27, Number 5, 260–264

Van Mier, J.G.M., Timmers, G. (1996), Werkwijze en Inrichting voor de Vervaardiging van een Gewapend Constructie-Element en een Dergelijk Constructiedeel (Device for Manufacturing a Reinforced Construction Element), Dutch patent No. 1000285, submitted July 25, 1995, granted November 6, 1996 (in Dutch).

Van Vliet, M.R.A., (2000), Size Effect in Tensile Fracture of concrete and rock, PhD thesis, Delft University of Technology.

Fracture Mechanics of Concrete and Concrete Structures – High-Performance Concrete,
Brick-Masonry and Environmental Aspects – Carpinteri, et al. (eds)
© 2007 Taylor & Francis Group, London, ISBN 978-0-415-44617-4

Influence of different PVA fibres on the crack behaviour of foamed cement paste

D. Meyer & J.G.M. van Mier
Materials Research Centre, Institute for Building Materials, Department of Civil,
Environmental and Geomatic Engineering (D-BAUG), ETH Zurich

ABSTRACT: Foamed cementitious materials are highly porous materials with low density, low compressive and bending strengths, caused by their high porosity. Due to the small strength and the scatter of the properties the material is not commonly considered as a load bearing material.

The aim of this work is to improve the mechanical properties and to reduce the variability of the properties of foamed concrete while keeping its favourable physical properties. To achieve the main goal of this research fibres are added to the foamed cement paste to improve the mechanical properties, specifically the compressive and bending strengths at the macroscopic level.

The crack patterns are analyzed by sectioning after impregnation with fluorescent epoxy resin. Emphasis is on fracture mechanisms in various foamed concretes. The varied parameters are the material density of the foamed concrete (with various amounts of foam), the w/c-ratio, fibre content and fibre size. The main result is that, in particular the flexural strength and ductility of foamed concrete can be improved significantly by adding fibres, achieving a bending strength a factor 5 higher than commonly observed values for plain mixtures without fibres.

1 INTRODUCTION

The idea of adding fibres to a cementitious matrix, to improve the mechanical properties, is well known from different applications in steel fibre concretes.

In the field of lightweight concretes, especially not autoclaved foamed materials, fibres are not common, as not autoclaved foamed cementitious materials are not common as load bearing materials.

In this preliminary test series, we are interested to see whether mechanical properties of foamed concrete can be improved by adding fibres. Moreover, do fibres affect the foam structure, also in the hardened cement, and to what extend are the mechanical properties affected? Do fibres cause different fracture mechanisms of the foamed concrete; if yes, is the influence favourable or not.

The foamed cementitious materials that are investigated are produced with protein-based pre-formed foam that is mixed with cement paste. Density, pore structure and the final water-cement ratio are related to the amount of foam that is mixed with the cement paste. At present no aggregates are added. The amount of foam that can be added is between 0 and 80% of the volume of the final mixture. The used fibres are made of PVA (poly vinyl alcohol). For these test series three different lengths were used (3, 6 and 12 mm). All fibres have the same diameter of 0.2 mm and a density

of 1300 kg/m^3 with a Young's modulus of 30 GPa and a tensile strength of 1000 MPa.

PVA was chosen, because steel fibres are much heavier and according to the matrix density (in-between 600 and 1600 kg/m^3) they lead to sagging and segregation. A further consideration is the fact that durability of steel in cementitious materials with air void contents of more than 40% is rather low. Another reason for choosing PVA was the fact that it is well known in our group from earlier projects see Bäumel (2002) and Meyer and van Mier (2004).

The effect of foam behaviour and the amount of foam on the reproducibility was investigated before, and published in Meyer et al (2005) and Meyer & van Mier (2006).

2 TEST PROGRAM

The test program contains ten different mixtures containing fibres of varying length, amount of fibres and different amounts of foam. Additionally, four reference mixtures without fibres, but variable amounts of foam, were made. The details of the mixtures are given in Table 1.

The fibre contents were defined after preliminary tests where the mix-ability of mixtures with different

Table 1. Mixture overview.

Mixture [–]	Amount of foam [V%]	Fibre length [mm]	Amount of fibres [V%]
Ref 1	0	No fibres	
Ref 2	20		
Ref 3	40		
Ref 4	60		
M 1.1	40	3	7
M 1.2	40		1.4
M 2.1	40	6	4.2
M 2.2	40		0.84
M 3.1	20	12	2.5
M 3.2	20		0.5
M 4.1	40	12	2
M 4.2	40		0.4
M 5.1	60		1
M 5.2	60	12	0.2

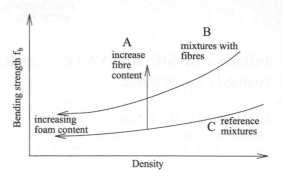

Figure 1. Idea behind the mixture composition.

Table 2. Test program mechanical tests.

Test type		Sample size [mm]	Nr. of samples
f_b	3-point bending*	150 × 50 × 50	5
$f_{c,cube}$	ID-compression	50 × 50 × 50 (1/2 Prism)	5
$f_{c,prism}$	ID-compression	150 × 50 × 50	5

* Span = 120 mm

Table 3. Test parameters for mechanical tests.

Test	Starting load	Loading velocity	Deformation measurement
3 point bending	50 N	0.005 mm/s	Yes
Compression ½ prisms	2000 N	0.5 kN/s	No
Compression standing prisms	2000 N	0.5 kN/s	Yes

fibres and various foam ratios had been investigated. In Table 1 the upper value of the fibre content corresponds to the maximum amount that can be added to a given mixture, without sagging, segregation or balling to occur. Results from these preliminary mixing experiments are gathered in Section 4.1. The lower fibre ratio in Table 1 represents 20% of the maximum value. With this, a first-order approximation of the effect of the fibre addition on the mechanical parameters and the crack behaviour is obtained.

With this test program two effects are to be separated mainly (see also Figure 1), one is the influence of the fibre length and the fibre content on mixing, rheology, pore structure and mechanical properties (A) and the second the influence of the amount of foam on the allowable fibre content and the ensuing mechanical properties (B) in comparison to the reference mixtures without fibres (C).

3 SAMPLE PREPARATION AND TEST SETUP

All samples were manufactured following the same procedure, which was defined during the preliminary tests. The following list shows the chronological order of the steps during the mixing and production of the samples.

– Preparation of test instruments, moulds and materials for plain cement paste
– Mixing of plain cement paste
– Rheology test on plain cement paste
– Weighing out plain cement paste and preparation of tests for cement paste with fibres
– Mixing fibres and cement paste
– Rheology tests on cement paste with fibres

– Weighing out cement paste with fibres, preparation of tests for foamed cement paste with fibres (new mixer)
– Production of foam
– Mixing of foam and cement paste with fibres
– Rheology tests on foamed cement paste with fibres
– Filling of moulds (surfaces of moulds are treated with silicon spray)
– Cleaning

After casting the samples were stored in controlled climate (20°C and 95% RH). After one day the samples were de-moulded. About one week prior to testing, the top and bottom surfaces were ground plan-parallel. The top surface during casting was also ground in order to obtain samples of regular size and shape. Before the samples were tested, geometry and the weight were measured. The mechanical tests carried out are summarized in Tables 2 and 3.

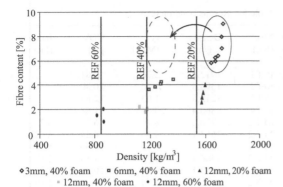

♦ 3mm, 40% foam   ▫ 6mm, 40% foam   ▲ 12mm, 20% foam
▪ 12mm, 40% foam   • 12mm, 60% foam

Figure 2.   Context of fibre type, fibre amount and fresh density.

4   RESULTS

4.1   Preliminary test (mixing)

With the preliminary experiments on mixing, the main interest was to evaluate the maximum amount of fibres that can be added to a cement paste containing a pre-defined amount of foam, without sagging, segregation or balling of the added fibres.

During the preliminary test it was found that a certain amount of foam was destroyed during mixing and placing, see Figure 2. Short fibres (3–6 mm) lead to a higher loss of foam than the long fibres (12 mm) causing an increase of the fresh density in comparison to the reference mixtures without fibres but containing the same amount of foam. A possible explanation therefore could be the number of fibre-tips, that is much higher in the case of short fibres at constant fibre volume, and so the probability that fibre tips are piercing foam bubbles.

4.2   Pore structure and fibre distribution

The pore structure of foamed cement paste can be influenced by the foam properties (amount of protein concentrate and the production method) and by the total amount of foam that is added to the cement paste. For these tests on fibre reinforced foamed cement paste we focused on the comparison of the macroscopic material (bubble) structure of samples with the same amount of foam, with and without fibres.
The effects of fibres of different sizes on the material/pore structure can be seen in Figures 3 and 4. The different fibre types (different length) lead to different loss of foam volume during mixing (see Figure 4), filling of moulds and hardening and as a result they have different effects on the pore structure. Smaller fibre ratios, more exactly smaller numbers of fibres, lead to a smaller loss of foam. The more foam is added,

Figure 3.   Comparison of pore structure of M 1.2 (left) with 1.4% 3 mm fibres and M1.1 (right) with 7% 3 mm fibres, both produced with 40% foam.

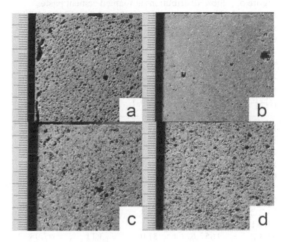

Figure 4.   Comparison of pore structure of mixtures produced with an amount of 40% foam by volume. (a) Reference mixture without fibres, (b) M1.1 7% 3 mm fibres, (c) M2.1 4.2% 6 mm fibres and (d) M4.1 2% 12 mm fibres.

the greater this effect is, based on the probability of contact between foam bubbles and fibres. This perception leads us to the model, that greater numbers of fibres, and so great numbers of fibre tips, destroy increasingly more foam bubbles. We think the foam bubbles are pierced by the fibres. The fact that no fibres were found reaching an air void in hardened material confirms this assumption.

One of the main interests of the preliminary experiments was to know if the fibres are distributed regularly over the whole cross section of the samples, or if there are effects such as balling and segregation, which have not been recognised in the preliminary tests or the rheology tests.

We benefit from the fact that the PVA fibres are sensitive to UV light. Figure 5 shows images, where the fibres were made visible with UV light. The images then were optimised by using an image editing software. As can be seen, the fibre distribution is quite regular over the whole cross sections.

Figure 5. Fibre distributions in foamed cement paste. Left: M 1.1 7% 3 mm and right: M 5.2 1% 12 mm White spots represent the fibres, black and grey shades are matrix material.

Table 4. Results compression tests.

Mix [Nr.]	Density [kg/m³]	f_c [MPa]	COV [%]	E-Modul [GPa]
Ref 1	1912.6	47.8	0.89	11.54
Ref 2	1595.8	27.6	2.49	7.14
Ref 3	1215.5	13.2	2.85	4.55
Ref 4	868.4	7.1	2.76	2.96
M 1.1	1770.2	42.4	0.64	9.51
M 1.2	1653.1	32.2	5.60	8.29
M 2.1	1279.3	14.0	7.17	4.47
M 2.2	1304.4	14.8	4.18	5.45
M 3.1	1602.1	30.1	4.76	7.41
M 3.2	1652.6	31.0	3.11	7.95
M 4.1	1232.6	13.2	3.26	4.79
M 4.2	1251.4	12.2	7.64	4.76
M 5.1	871.3	5.6	3.77	5.33
M 5.2	922.7	6.4	8.78	4.34

COV ....Coefficient of variation

4.3 Mechanical tests

As mentioned before, compression tests and bending tests were done to determine the mechanical properties. The main results are shown in Table 4 and 5. PVA fibres showed no appreciable increase of the ultimate compressive strength (Figure 6). The only improved parameters were the post peak behaviour.

The bending strength was significantly affected by the addition of fibres, see Figures 7–10. The increase of the bending strength and ductility is surprising. With the 12 mm fibres the difference in density between the reference mixture and mixtures containing fibres is small. The fact, that the scatter of all parameters is smaller than the reference mixtures, besides the mixture with 60% foam and the low fibre content (M 5.1), confirms the idea that adding fibres to foamed cement paste improves the mechanical parameters, in particular the behaviour in bending significantly.

Table 5. Results bending tests.

Mix [Nr.]	Density [kg/m³]	f_b [MPa]	COV [%]	Increase [factor]
Ref 1	1912.6	2.70	7.81	–
Ref 2	1595.8	1.76	1054	–
Ref 3	1215.5	1.03	12.65	–
Ref 4	868.4	0.74	21.88	–
M 1.1	1770.2	8.70	6.79	3.22
M 1.2	1653.1	4.43	4.24	2.52
M 2.1	1279.3	5.52	10.56	5.37
M 2.2	1304.4	2.23	7.18	2.17
M 3.1	1602.1	9.50	10.87	5.41
M 3.2	1652.6	3.74	11.30	2.13
M 4.1	1232.6	5.00	16.04	4.86
M 4.2	1251.4	2.10	5.89	2.05
M 5.1	871.3	1.92	38.14	2.61
M 5.2	922.7	1.03	6.41	1.40

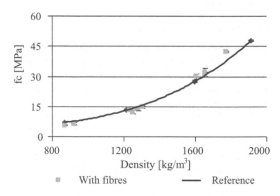

Figure 6. Relation of density and compressive strength.

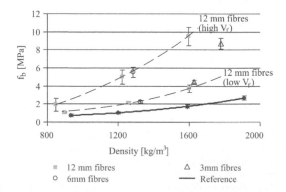

Figure 7. Influence of fibre addition to the bending strength.

As can be seen in Figure 9 all mixtures with high fibre content lead to a very high bending strength compared to the references (Figure 8) and the mixtures with the low fibre content. Moreover the mixtures with

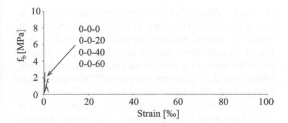

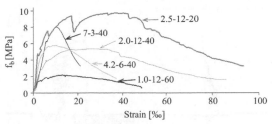

Figure 8. Stress-strain diagrams for the reference mixtures from bending tests. In the caption of the lines the first number means the fibre content [%], the second the fibre length [mm] and the last number represents the percentage of foam that is added to the total foam volume.

Figure 10. Flexural stress-strain diagrams for mixtures with low fibre content.

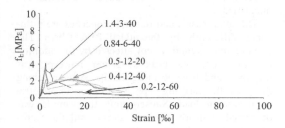

Figure 9. Flexural stress-strain diagrams for mixtures with high fibre content. Explanation of the captions 1.4-3-40 stand for 1.4% of 3 mm fibres in a mixture where 40% foam is added.

Figure 11. Crack pattern for mixture M3.1 (20% foam, 2.5% 12 mm PVA fibres), showing multiple cracking.

12 mm fibres and high foam amount show strain hardening and a high ductility. When only 20% foam is added to the mixture strains up to 4.0% are observed before peak-stress is reached. In the mixtures with only 20% foam fibres appear to be anchored better in the matrix, so the strength and the ductility are more favourable compared to samples with smaller amounts of foam.

The mixtures with the lower fibre content (20% of maximum mixable amount of fibres) show no strain hardening. After the peak the applicable load drops immediately to a level between 40 to 75% of the peak load (depending on the fibre length). After that drop all mixtures show a light secondary strain hardening and a long trail in the softening diagram except the mixture with the short 3 mm fibres (see Figure 10).

The diagrams kept on the same scale in order to show the differences between plain foamed cement paste and fibre reinforced foamed cement paste as clearly as possible.

4.4 Crack patterns

As described in Section 4.3 the PVA fibres had a cause effect on the strength and the ductility. Also cracking behaviour is changing by adding fibres to foamed cement paste. Depending on the fibre type and the

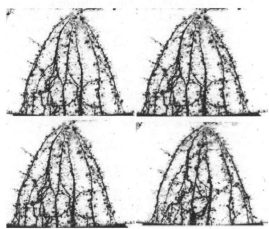

Figure 12. Crack pattern of longitudinal cuts of the same sample as shown in Figure 12. The sample was impregnated with epoxy resin and photos were taken under UV light. The contrast of the crack was optimized with image editing software.

amount of fibres that is added to the mixture, but also influenced by the amount of air voids that a sample contains, the crack behaviour leads to multiple cracking or not. Multiple cracking was recognized in the mixtures M 3.1 and M 4.1; all the other mixtures lead to a single crack or had only in some samples more than one crack, as it was recognized on series M 1.1, M 2.1 and M 5.1. Some selected photographs of crack patterns are shown in Figures 11–13 to illustrate the crack behaviour.

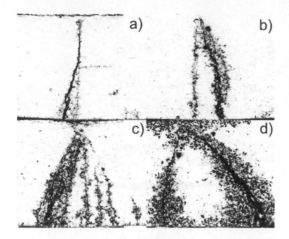

Figure 13. Crack patterns from different mixtures. (a) M 1.1 7% 3 mm fibres, 40% foam; (b) M 2.1 4.2% 6 mm fibres, 40% foam; (c) M 4.1 2% 12 mm fibres, 40% foam & (d) M 5.1 1% 12 mm fibres, 60% foam.

Figure 12 shows the devolution of the crack from the sample from Figure 11 in four longitudinal sections. It can bee seen that over the whole section multiple cracking takes place; all major cracks grow through the entire depth of the beam.

In Figure 13 typical crack patterns from different mixtures are shown, indicating the various crack modes from a single localized macro crack (Figure 13a) to multiple cracking (Figure 13c) and apparent shear fracture (Figure 13d).

5 DISCUSSION AND CONCLUSIONS

The main interest of the presented test series was to evaluate the influence of PVA fibres on the mechanical properties, pore structure and fracture mechanisms of foamed cement paste, in comparison to the behaviour of plain reference mixtures without fibres. Of additional interest was the reproducibility of the results.

The tests showed that it is important that plain cement paste and fibres are mixed properly before the pre-formed foam is added. The amount of foam that is destroyed during mixing, placing and hardening differs strongly, depending on the size of the fibres used and the amount of fibres added to the mixture. The short 3 mm fibres, where a high volume can be added to foamed cement paste, lead to the largest loss of the added foam. Even if only a low amount of such short fibres is added the volume of foam that is destroyed is considerable, so the difference between planned and final density is considerable. With longer fibres this difference is smaller. With the longest used fibres (12 mm) the difference was irrelevant. The fact

that high volume of short fibres, and thus a large number of fibres and fibre tips, lead to a larger volume of foam loss during mixing and placing, guided us to a possible explanation that the foam bubbles are pierced when the mixture is in motion. So a greater number of fibres, and thus a greater number of fibre tips, lead to a higher loss of foam volume.

In all mixtures, independent of fibre type, the fibres were distributed homogeneously over the whole cross section in the observed samples. The fibre orientation, however, differed in the observed samples, influenced by the way the mould was filled. If the material could flow, fibres tended to align in the flow direction. Note that this is a qualitative observation only since the effect was not studied systematically, and was not quantified.

The compressive and flexural properties are influenced differently by the added PVA fibres. The compressive strength is not affected by the addition of fibres. The post peak behaviour, however, was improved and became more ductile. After reaching the peak, the carrying capacity dropped to 30–50% of maximum load. Subsequently the sample was deformed at nearly constant stress until the fibres bridging the cracks were pulled out.

The bending strength f_b can increase up to a factor of five compared to plain foamed cement, depending on the fibre volume, fibre type and amount of foam. The ductility also is improved by adding PVA fibres to foamed cement paste mixtures and a high amount of fibres lead to multiple cracking. Mixtures with the long 12 mm fibres reach remarkable strains. Some mixtures with longer fibres and a high fibre volume lead to multiple cracking. Other mixtures, as mentioned in the text before, only show multiple cracking in individual samples and not over the whole test series.

PVA fibres are a good and easy way to improve the bending strength and the ductility of foamed cement paste while keeping density and reproducibility under control. The effects of fibres to the foam structure during mixing, placing and hardening have to be taken in consideration. Based on the experiments that have been done PVA fibres can be added easily to foamed cement pastes to reach a homogeneous mixture and a relatively small scatter of the parameters.

From a practical point of view, the addition of PVA fibres brings us one step closer to foamed cement paste as a load-carrying material. The low weight of the material, in combination with a high bending strength seems to defy common response observed in porous media.

ACKNOWLEDGMENT

Parts of the preliminary tests carried out in the bachelor thesis of A. Bucher and Y. Yao.

REFERENCES

Bäumel, M. 2002. Steigerung der Dauerhaftigkeit selbstverdichtender Betone durch den Einsatz von Polymerfaserkurzschnitt. Dissertation. Zurich: ETH Zurich.

Gibson, L.J. & Ashby, M.F. 2001. *Cellular solids: structure and properties – second edition.* Cambridge: Cambridge University press.

Meyer, D. Trtik, p. & van Mier, J.G.M. 2004. Foamed cementitious materials. Internal report. Institute for building materials. http://e-collection.ethbib.ethz.ch/

Meyer, D. Trtik, P. Rindlisbacher, M. Voide, R. Müller, R. & van Mier, J.G.M. 2005. Preliminary study on the properties of protein foam and hardened foamed cement paste. In Proc. Intern. Conf. ConMat 05, Vancouver, 22–24. August 2005. Vancouver: University of British Columbia.

Meyer, D. & van Mier, J.G.M. 2006. The influence of the foam behaviour on the properties of foamed cement paste. In *6th international PhD symposium in Civil Engineering; Proc. intern. Symp., Zurich, 23–26. August 2006.* Zurich: Institute for structural engineering.

van Mier, J.G.M. 1997. *Fracture processes of concrete.* London: CRC Press.

Sieghart, K. 1979. *Leichtzuschlag-Schaumbeton als Konstruktionsleichtbeton mit abgeminderter Rohdichte.* Dissertation. Darmstadt: TU Darmstadt.

Postcracking behaviour of hybrid steel fiber reinforced concrete

L. Vandewalle

K.U.Leuven, Department of Civil Engineering, Leuven, Belgium

ABSTRACT: This papers presents research on the mechanical performance (compressive strength, bending behaviour according to RILEM TC162-TDF) of both normal and hybrid steel fiber reinforced concrete. The investigated parameters were fiber type (shape, length) and fiber dosage. The results show that very short and short fibers affect the cracking process at small crack openings while the longer fibers provide a good ductility at larger deformations. For the mixtures with one type of fiber the absolute scatter on the results of the bending tests is much smaller than for the long fibers. In this research the synergetic effect of the combination of different types of steel fiber is ambiguous.

1 INTRODUCTION

Cementitious composites are typically characterized as brittle, with a low tensile strength and strain capacity. Fibers are incorporated into cementitious materials to overcome this weakness, producing materials with increased tensile strength, ductility and toughness and improved durability (Balaguru and Shah, 1992).

Fibers can differ from each other by size, shape and material (steel, carbon, synthetics, glass, natural fibers, …). However, for most structural and non-structural purposes, steel fibers are the most used of all fiber types. Synthetic fibers on the other hand are mainly used to control early cracks in slabs on grade and to avoid spalling of high strength concrete during fire.

A more recent development in fiber reinforced cement-based composites is the use of fiber hybridization to optimize material performance based on the intended use. Two or more fiber types are combined so that the material can achieve the beneficial performance characteristics of each fiber. Typically this improvement is attained by using fibers that will affect the cracking process during different stages of loading, which often involves using fibers of varying sizes and moduli (Shah and Kuder, 2004).

The idea is that the short fibers can bridge efficiently the microcracks, which develop in the first phases of the tensile loading (Fig. 1a). When microcracks join with each other into larger cracks, long fibers may be activated as bridging mechanism (Fig. 1b). As a result the ductility enhancement depends mainly on the long fibers (Markovic, 2006).

This papers presents research on the mechanical performance (compressive strength, bending

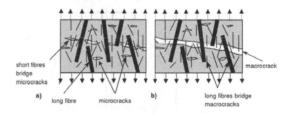

Figure 1. Action of the different fiber types in concrete (Markovic, 2006).

behaviour according to RILEM TC162-TDF) of both normal and hybrid steel fiber reinforced concrete. The investigated parameters were fiber type (shape, length) and fiber dosage. The results show that very short and short fibers affect the cracking process at small crack openings while the longer fibers provide a good ductility at larger deformations. For the mixtures with one type of fiber the absolute scatter on the results of the bending tests is much smaller than for the long fibers. In this research the synergetic effect of the combination of different types of steel fiber is ambiguous.

2 RESEARCH PROGRAM

The test program, executed at the Department of Civil Engineering of the K.U.Leuven, involved RILEM 3-point bending tests (Vandewalle et al, 2002) to measure the postcracking tensile behaviour of fiber concrete. The test set-up is shown in Figure 2. The test specimen is a beam of 550 to 600 mm length and a cross section with a width and depth of 150 mm. The span of the

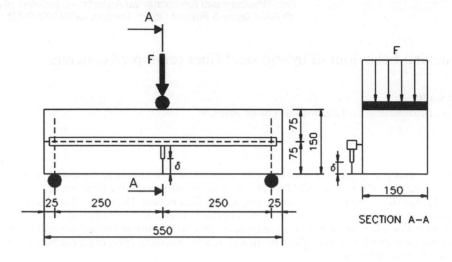

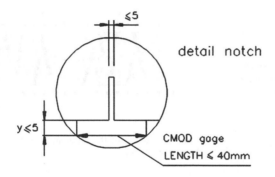

Figure 2. Test set-up (Vandewalle et al., 2004).

test specimen is equal to 500 mm. In the middle of the span the specimen is notched, the depth of the notch is equal to 25 mm. The test is performed under CMOD (crack mouth opening displacement) control. Besides the CMOD and the load, also the relative deflection at midspan at both sides of the prism can be measured optionally.

The compressive strength ($f_{cm,cube}$) is measured on cubes with side = 150 mm.

Three types of steel fibers are applied, i.e. one very short straight steel fiber (SK : length = 6 mm, diameter = 0.16 mm – OL6/.16), one short straight steel fiber (K: length = 13 mm, diameter = 0.16 mm – OL13/.16) and one long hooked-end steel fiber (L : length = 35 mm, diameter = 0.55 mm – RC65/35BN).

The total fiber content ranges from 0 (reference mix) to 90 kg/m³. Fifteen mixtures in total were tested as shown in Table 1. LxKySKz means a mixture with a dosage of x kg/m³ long hooked-end fibers, y kg/m³ short fibers and z kg/m³ very short fibers.

The concrete composition is identical for all mixtures (see Table 2). Only the dosage of superplasticizer changed since the presence of steel fibers in concrete decreases its workability.

After casting, the specimens were cured at +20°C and 95–100 % relative humidity for 4 weeks.

3 TEST RESULTS

3.1 Compressive strength

The mean cube compressive strength of the different mixes is given in Table 1. $f_{cm,cube}$ ranges between 54.5 MPa (reference mix) and 76.9 MPa (L00K30SK30). The addition of steel fibers results in a higher compressive strength. Moreover, the test results show also that very short and short steel fibers provide a higher improvement of $f_{cm,cube}$ than the long hooked-end steel fibers do. The reason for this is probably the

Table 1. Overview of test program and mechanical results.

Series	$f_{cm,cube}$ (MPa)	$f_{fct,L}$ (MPa)	$f_{R,1}$ (MPa)	$f_{R,4}$ (MPa)
L00K00SK00	54.5	–	–	–
L00K00SK30	59.6	4.45 (0.33 – 7.42)[(x)]	1.81 (0.30 – 16.57)	0.30 (0.07 – 23.33)
L00K00SK60	62.1	5.45 (0.27 – 4.95)	2.63 (0.28 – 10.65)	0.49 (0.10 – 20.41)
L00K30SK00	62.8	5.19 (0.31 – 5.97)	2.78 (0.29 – 10.43)	1.26 (0.13 – 10.32)
L00K60SK00	66.9	5.98 (0.34 – 5.69)	4.40 (0.38 – 8.64)	1.98 (0.18 – 9.09)
L30K00SK00	55.9	5.11 (0.30 – 5.87)	3.45 (0.92 – 26.67)	2.64 (0.75 – 28.41)
L60K00SK00	57.2	6.15 (0.73 – 11.87)	5.96 (0.76 – 12.75)	4.24 (0.59 – 13.92)
L00K30SK30	76.9	5.34 (0.36 – 6.74)	3.46 (0.53 – 15.32)	0.80 (0.24 – 30.00)
L30K30SK00	65.2	5.37 (0.34 – 6.33)	4.77 (0.39 – 8.18)	2.95 (0.25 – 8.47)
L20K40SK00	67.2	5.69 (0.49 – 8.61)	4.85 (0.70 – 14.43)	2.70 (0.48 – 17.78)
L40K20SK00	61.6	6.06 (0.62 – 10.23)	5.75 (0.72 – 12.52)	3.47 (0.39 – 11.24)
L20K20SK20	64.8	6.10 (0.64 – 10.49)	5.30 (0.82 – 15.47)	3.03 (0.62 – 20.46)
L30K30SK30	69.2	6.63 (0.70 – 10.56)	6.37 (0.72 – 11.30)	3.76 (0.36 – 9.57)
L30K60SK00	67.2	7.07 (0.56 – 7.92)	6.94 (0.53 – 7.64)	4.48 (0.36 – 8.04)
L60K30SK00	58.6	6.36 (0.48 – 7.55)	6.30 (0.46 – 7.30)	4.47 (0.50 – 11.19)

(x) : value (MPa) (absolute scatter (MPa) – relative scatter (%)).

fact that the very short and short steel fibers can bridge efficiently the microcracks.

3.2 Tensile behaviour

Each fiber concrete mixture consists of 6 prisms for the bending test. The mean value and the scatter (absolute – relative) of the bending test results of all series are mentioned in Table 1. $f_{fct,L}$ is the limit of proportionality while $f_{R,1}$ and $f_{R,4}$ are the residual flexural tensile strength at CMOD = 0.5 mm, CMOD = 3.5 mm respectively.

3.2.1 Scatter

The load-CMOD-curves of series L00K00SK60 are shown in Figure 3, of series L00K60SK00 in Figure 4, of series L60K00SK00 in Figure 5 and of the hybrid mixes L60K30SK00 and L30K60SK00 in Figures 6 and 7 respectively. The detailed results of the other mixtures can be found elsewhere (De Smedt and Rolies, 2005, Fevrier and Vangoidsenhoven, 2006).

The absolute scatter of the results of the concrete with only very short steel fibers (L00K00SK60), short steel fibers (L00K60SK00) respectively is much smaller than of the steel fiber concrete with only long hooked-end fibers (L60K00SK00). However, when looking to the relative scatter (see Table 1), the short fibers give the best results if only one type of steel fiber is used.

The number of fibers in one kg increases when its shape decreases. Fiber counts have shown that the thoughness parameters were directly related to the number of fibers intersecting the fracture surface. A small variation or difference in number of fibers has a direct and relatively large influence on

Table 2. Concrete composition.

	kg/m³
Gravel 4/16	1012
Sand 0/5	865
Cement CEM I 52,5N	350
Water	175
W/C	0,5

the postcracking behaviour of the materials tested. This is particularly important for low fiber dosages and relatively small cross sections (Barr et al., 2003). As a result the mixtures with only one fiber type (see Table 1) confirm the statement that the scatter would be more pronounced in specimens with a lower absolute number of fibers.

However, the same tendency can not always be found for the hybrid steel fiber concretes. For instance, although the absolute number of fibers in L30K60SK00 is higher than in L60K30SK00 both the absolute and relative scatter are almost the same in the two mixtures.

3.2.2 Postcracking behaviour

The mean-load-CMOD-curves for concrete reinforced with 60 kg/m³ of fibers are shown in Figure 8 (0–0.5 mm) and 9 (0–4.5 mm).

For concrete reinforced with one type of steel fiber, it can be seen that the overall postcracking behaviour in the CMOD-region 0 to 0.15 mm is the best for concrete reinforced with short steel fibers. The length of the very short steel fibers is probably. too small in comparison with the size of the used aggregates (gavel 4/16)

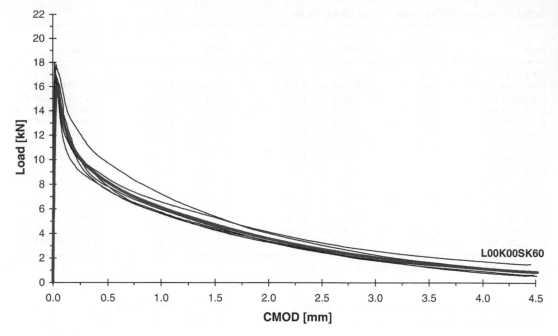

Figure 3.   Load-CMOD-curves for L00K00SK60.

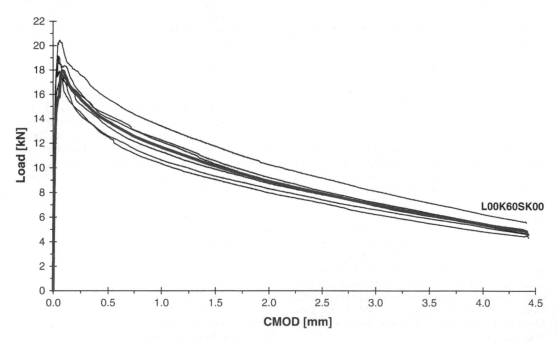

Figure 4.   Load-CMOD-curves for L00K60SK00.

to bridge efficiently the microcracks. On the other hand, the tensile behaviour for CMOD-values larger than 0.15 mm is much better for concrete reinforced with long hooked-end fibers.

Short fibers can bridge microcracks more efficiently because they are very thin and their number in concrete is much higher than that of the long thick fibers for the same fiber volume quantity. However,

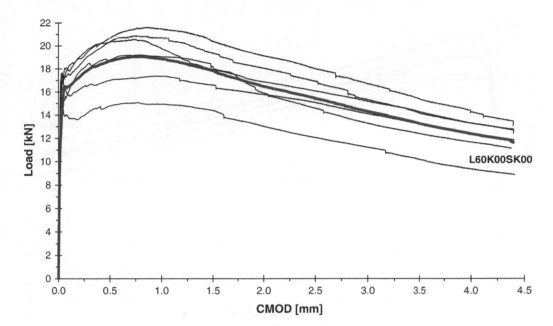

Figure 5.   Load-CMOD-curves for L60K00SK00.

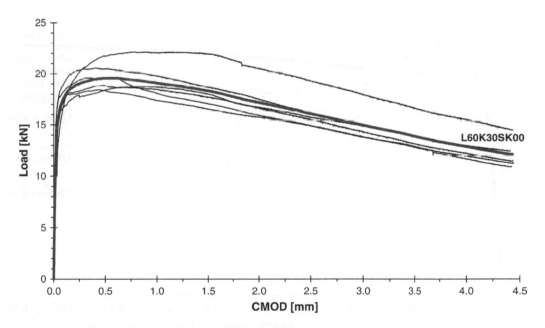

Figure 6.   Load-CMOD-curves for L60K30SK00.

for larger crack widths the ductility of the mixtures with long fibers is much better than that of the corresponding mixtures with the short fibers. As the microcracks grow and join into larger macrocracks, the long hooked-end fibers become more and more active in crack bridging. The origin of the higher residual forces for long hooked-end fibers at larger CMOD-values is twofold:

– presence of a hooked-end
– long embedded length (anchorage length).

1371

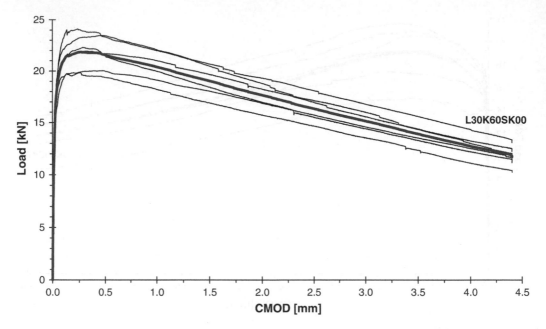

Figure 7. Load-CMOD-curves for L30K60SK00.

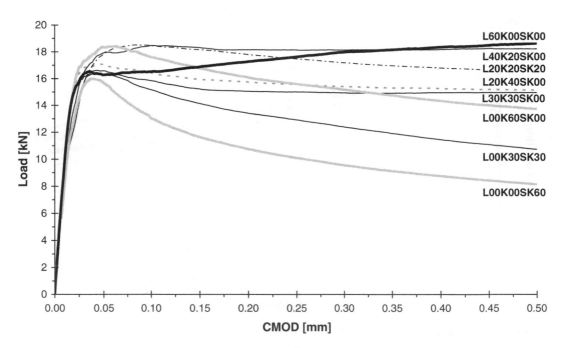

Figure 8. Mean load-CMOD-curves for concretes with 60 kg/m^3 of fibers (CMOD: 0–0.5 mm).

Both aspects provide a higher pull-out force for long hooked-end fibers in comparison with short fibers, particularly at larger crack widths. Long fibers can therefore provide a stable post-peak response. Short straight fibers will be less active because they are being pulled out more and more as the crack increases.

The postcracking behaviour in the CMOD-region of 0 to 0.15 mm is for the hybrid mixes relatively

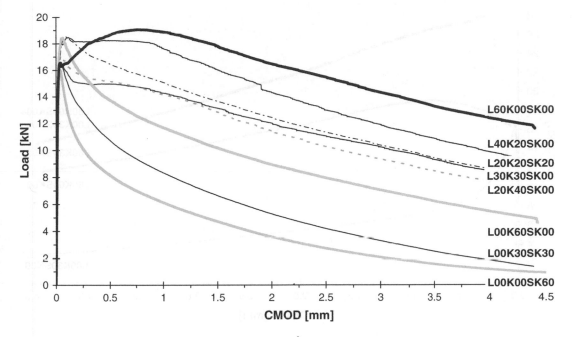

Figure 9. Mean load-CMOD-curves for concretes with 60 kg/m³ of fibers (CMOD: 0–4.5 mm).

similar when taking the scatter of the individual results within a series into account. For larger CMOD-values, however, the ductility increases when a higher volume percentage of long hooked-end fibers in the mixture is used. The series with only very short fibers and/or a combination of very short and short fibers show the worst overall postcracking behaviour.

3.2.3 Synergetic effect

Synergy is the phenomenon where acting of two or more subjects together leads to a better result than the action of the same subjects independently of each other. Translated to hybrid fiber concrete, the synergy of very short or short fibers and long fibers should lead to an improved tensile response of the hybrid fiber concrete, compared to the arithmetic sum of tensile responses of two concretes, one of which contains only long and another only very short or short fibers (in the same quantity as the hybrid fiber concrete) (Markovic, 2006).

In this paper the synergetic effect has been investigated for two hybrid fiber concretes, i.e. L60K30SK00 as shown in Figure 10 and L30K60SK00 as shown in Figure 11. The calculated curve in both figures starts from a CMOD of 0.3 mm in order to account only the effect of the fibers.

As shown in Figure 10 no synergetic effect at all can be observed for L60K30SK00. On the contrary the sum of the individual curves gives a smaller stress at a certain CMOD than the experimental curve

does. With regard to L30K60SK00 it can be seen that the arithmetic sum of the two individual curves for CMOD-values larger than 2 mm is almost identical to the experimental curve. Again no synergy can be recorded.

Markovic (Markovic, 2006), however, did find synergetic effects. The reason for this is perhaps due to the fact that the maximum grain size of the used concrete in his research was only 1 mm and the fiber dosages used were rather high, i.e. 1 to 2 Vol.%. Further research with regard to this phenomenon is necessary.

4 CONCLUSIONS

Fifteen series of steel fiber reinforced concrete have been investigated at the Department of Civil Engineering of the K.U.Leuven. The following conclusions can be drawn:

– for a certain applied fiber volume percentage, the number of fibers crossing a cracked section increases as the size of the used steel fiber decreases. This results in a lower absolute scatter on the test results of steel fiber concrete which contains only very short or short fibers. An analogous conclusion can not be drawn for the hybrid fiber concrete mixes;
– for the used concrete containing aggregates with a maximum size of 16 mm, the efficiency of the very

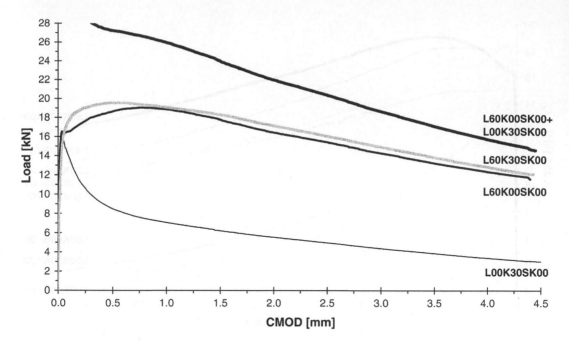

Figure 10. Synergetic effect of L60K30SK00.

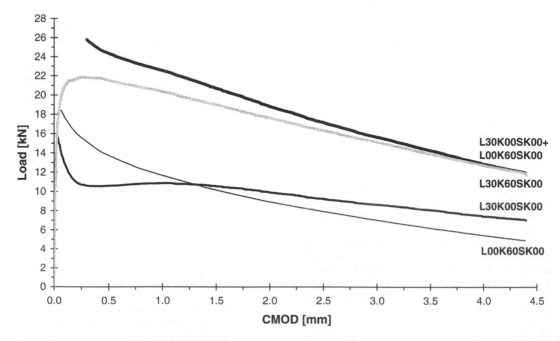

Figure 11. Synergetic effect of L30K60SK00.

short fibers is worse than of the short fibers even in the CMOD-region of 0 to 0.15 mm;
- the overall ductility of the fiber concrete is dominated by the volume of the long hooked-end steel fibers: the higher is the dosage of the long hooked-end steel fibers, the better is the postcracking performance of the fiber concrete;
- no synergetic effect has been found in the hybrid mixes.

REFERENCES

Balagaru, P.N. & Shah, S.P., 1992, *Fiber-Reinforced Cement Composites*, New York, McGrax-Hill Inc.

Shah, S.P. & Kuder, K.G., 2004, *Proceedings of the International Workshop on Advances in Fiber Reinforced Concrete*, Bergamo, Italy : 83–92.

Vandewalle, L. & al., 2002, Recommendation of RILEM TC162-TDF: Test and design methods for steel fiber reinforced concrete: final recommendation for bending test, *Materials and Structures,* Vol.35: 579–582.

Barr, B.I.G., Lee, M.K., De Place Hansen, E.J., Dupont, D., Erdem, E., Schaerlaekens, S., Schnütgen, B., Stang, H. & Vandewalle, L., 2003, Round-robin analysis of the RILEM TC162-TDF bending test – Part 3 – Fibre distribution, *Materials and Structures,* Vol.36 : 631–635.

De Smedt, K. & Rolies, K., 2005, Onderzoek naar de fysische en mechanische eigenschappen van hybride staalvezelbeton (in dutch: Investigation of the physical and mechanical properties of hybrid steel fiber reinforced concrete), *Master thesis K.I.H. De Nayer Belgium.*

Fevrier, B. & Vangoidsenhoven, G., 2006, Onderzoek naar de fysische en mechanische eigenschappen van hybride staalvezelbeton (in dutch: Investigation of the physical and mechanical properties of hybrid steel fiber reinforced concrete), *Master thesis K.I.H. De Nayer Belgium.*

Markovic, I., 2006, High-performance Hybrid-Fibre Concrete – Development and Utilisation, *Ph.D.-thesis T.U.Delft.*

Fracture Mechanics of Concrete and Concrete Structures – High-Performance Concrete,
Brick-Masonry and Environmental Aspects – Carpinteri, et al. (eds)
© 2007 Taylor & Francis Group, London, ISBN 978-0-415-44617-4

Curing effects on fracture of high-performance cement based composites with hybrid steel fibers

F. Ozalp, Y. Akkaya, C. Sengul, B. Akcay, & M.A. Tasdemir
Istanbul Technical University, Civil Engineering Faculty, Istanbul, Turkey

A.N. Kocaturk
Iston, Prefabricated Concrete Products and RMC Co., Istanbul, Turkey

ABSTRACT: In this work, effects of fiber strength and curing conditions on the mechanical behaviour and fracture properties of high performance cement based composites with hybrid steel fibers were investigated. Depending on the high temperature curing conditions, concretes with hybrid steel fibers showed behaviour of enhanced toughness and ductility. Fracture energy of the matrix increased up to 96 times owing to normal strength steel fibers, while in concretes with high strength steel fibers the increase in fracture energy due to steel fibers was 168 times. Petrographic examinations indicated that a connected crack system occurred which was filled with calcium hydroxide at 200°C curing condition and no indication of delayed ettringite formation was observed. Furthermore, the prefabricated elements successfully produced with these materials, such as gully tops and manhole covers used in the city infrastructure, are presented along with selected examples from several applications.

1 INTRODUCTION

There have been incredible advances in concrete technology over the years. Not more than 45 years ago the maximum compressive strength obtained at the construction site was about 40 MPa, such a concrete is now considered quite low strength when compared to the modern very-high strength concretes with cube compressive strengths between 200 MPa and 800 MPa, tensile strengths between 25 MPa and 150 MPa, and fracture energies of about 30000 J/m^2 (Alexander 1993, Richard & Cheyrezy 1995, Dugat et al. 1996, Bonneau et al. 1997). In recent years, a number of specific developments have occurred in parallel to the evolution of what is commonly described as high performance concretes. Such concretes have high strength combined with highly enhanced durability. The brittleness of concrete, however, increases with an increase in its strength. In other words, the higher the strength of concrete, the lower is its ductility. This inverse relation between strength and ductility is a serious drawback and limits the use of high strength concrete (Balaguru et al. 1992). Their potential use is yet to be established and more work is required, in particular, to evaluate their long-term performance. These materials have already successfully been laboratory tested for strengthening structures and their potential use in areas subjected to earthquakes needs to be further investigated (Ilki et al. 2004).

The fracture energies of Reactive Powder Concretes (RPCs), thus, are about 300 times that of normal strength concrete or even 1350 times for Slurry Infiltrated Fiber Reinforced Concrete (SIFCON) (Fritz 1991). The low porosity of RPC gives it important durability and low transport properties and makes it potentially suitable for some structures exposed to harsh environmental conditions (Feylessoufi et al. 1996, Matte & Moranville 1999).

The interface between cement paste and aggregate particles is the weakest zone in concrete, and the use of ultra-fine particles, such as silica fume, is important for densification and for the improvement of the stability of fresh concrete, thus, enhancing the overall durability and strength. Micro-silica or silica fume is a by product of silicon and ferrosilicon industries; it has been used since 1950s to improve the properties of concrete. To gain benefits from these particles, good dispersion within the concrete system is necessary and can be provided by means of a superplasticizer (Tasdemir et al. 1998, Tasdemir et al. 1999). The internal structure of RPC is optimized by precise gradation of all particles in the mixture, to add short steel fibers for improved ductility and to allow the resulting concrete to harden under pressure and increased temperatures (Walraven 1999).

The main objective of the work presented here is to demonstrate the combined effects of curing conditions, and the hybrid steel fibers and their strengths on the mechanical and fracture properties of high perform-ance steel fiber reinforced composites.

2 EXPERIMENTAL WORK

2.1 Materials and mix proportions

A total of 6 composite mixtures were cast for this investigation. In all the mixtures, the volume fractions of cement, siliceous sands (0–2 mm and 0–0.5 mm), silica fume and water were kept constant. Cement used was ordinary Portland cement with a density of $3.20\,g/cm^3$ and its nominal content in the composite was $1000\,kg/m^3$. The maximum particle size of aggregate was 2 mm and the density of siliceous sand was $2.65\,g/cm^3$. The amount of high-range water reducing admixture (HRWRA) varied between 11% and 12% by weight of cement for different composite mixtures to maintain the same workability. Three different steel fibers with and without hooked ends were added. The short ones without hooked ends were straight high strength steel fibers coated with brass, 6 mm in length, 0.16 mm in diameter and the aspect ratio (L/d) was 40. The tensile strength of normal and high strength hooked end steel fibers were 1150 and 2250 MPa, respectively, while their aspect ratios were the same ($l/d = 55$). The total volume fractions of fibers were kept constant at 3%. The mixture proportions of the matrix were as follows; cement: silica fume: water: siliceous sand (0.5–2 mm): siliceous powder (0–0.5 mm): HRWRA = 1: 0.250: 0.114: 0.325: 0.493: 0.120. Partial replacement of aggregate by steel fiber was based on one to one volume basis. The properties of the steel fibers are given in Table 1.

The composite mixtures were designated using the following codes: P = matrix, R = composite with high strength straight steel fibers, S = composite with normal strength hooked-end steel fibers, T = composite with high strength hooked-end steel fibers. RS and RT contain equal percentages of steel fibers; 1.5% HSSSF + 1.5% NSHSF and 1.5% HSSSF + 1.5% HSHSF as shown in Table 2.

2.2 Specimen preparation

In all mixtures, because the efficiency of HRWRA was low, over dosage was required. For reaching the high values of strength, cement and silica fume content were significantly increased, thus the water/binder ratio decreased. Because of the fact that 60% of the HRWRA is composed of water, water/cement and water/binder ratios were re-calculated and expressed with the names of the total water/cement and total water/binder ratios. Hence, the total water/binder ratio in the mixtures was 0.17.

In mixing, cement, silica fume, siliceous sand, and siliceous powder were blended first in dry condition. Halves of the HRWRA and the water were mixed in a pan and added to the mixture. The remaining halves of the HRWRA and the water were added to the mixture gradually to provide homogeneity in the mixture. Steel fibers were scattered in the mixture and carefully mixed to achieve a uniform distribution. The specimens were cast in steel moulds and compacted on a vibration table. Details of the tests and dimensions of the specimens are given in Table 3.

All specimens were demoulded after 24 hours, and then three different curing regimes were applied: i) Steam curing was as follows; an initial moisture curing for one day was followed by steam curing of 65°C at atmospheric pressure for 3 days and further in a water tank saturated with lime at 20°C until testing day. ii) The period of hot curing regime was similar to the steam curing regime; however, a higher temperature of 200°C was applied to the specimens. During this high temperature curing regime, specimens were wrapped by the two layered wrapping material. The first layer wrapping material was a high temperature resistant

Table 1. The properties of straight and hooked-end steel fibers.

	Type of steel fiber		
	High strength straight steel fiber (HSSSF)	Normal strength hooked-end steel fiber (NSHSF)	High strength hooked-end steel fiber (HSHSF)
Length (L) mm	6	30	30
Diameter (d) mm	0.15	0.55	0.55
Aspect ratio (L/d)	40	55	55
Tensile strength (N/mm²)	2200	1100	2200

Table 2. The designation of the mixtures.

Mix Code			Steel fiber percentage		
Normal curing	Steam curing at 65°C	Hot curing at 200°C	HSSSF	NSHSF	HSHSF
PN	PV	PZ	0	0	0
RN	RV	RZ	3	0	0
SN	SV	SZ	0	3	0
TN	TV	TZ	0	0	3
RSN	RSV	RSZ	1.5	1.5	0
RTN	RTV	RTZ	1.5	0	1.5

plastic sheet and the second one was an aluminium foil. iii)The third curing regime involved standard water curing in a water tank saturated with lime at 20°C prior to testing.

2.3 Test procedure

The tests for determining the fracture energy (G_F) were performed in accordance with the recommendation of RILEM 50-FMC Technical Committee. As schematically shown in Figure 1, the deflection was measured using a linear variable displacement transducer (LVDT). The load was applied using a closed-loop testing machine (Instron 5500R) with a maximum capacity of 100 kN. The beams prepared for the fracture energy tests were 280 mm in length and 70 mm × 70 mm in cross section. The notch to depth ratios (a/D) of specimens were 0.40 and the notches were formed using a diamond saw. The effective cross section was reduced to 70 mm × 42 mm to accommodate long fibers in more abundance, and the length of support span was 200 mm. The crack mouth opening displacement (CMOD) was used as a feedback control variable to obtain stable tests. Thus, load versus CMOD and load versus displacement at the midspan (δ) curves were obtained for each specimen.

The fracture energy was determined using the following expression

Table 3. Test methods and specimen size.

Test type	Specimen	Dimensions (mm)	Parameters
Compression	cylinder	Ø100, h200	f'_c (MPa), E (GPa)
Splitting	disc	Ø150, h60	f_{st} (MPa)
Bending	beam	70 × 70 × 280	G_F(N/m), f_{flex} (MPa)

f'_c = compressive strength, E = modulus of elasticity, f_{st} = splitting tensile strength, G_F = fracture energy, f_{flex} = flexural strength.

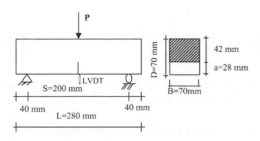

Figure 1. Schematic representation of the three point bending test.

$$G_F = \frac{W_0 + mg\dfrac{S}{U}\delta_s}{B(D-a)} \qquad (1)$$

where, B, D, a, S, U, m, δ_s, and g are the width, depth, notch depth, span, length, mass, specified deflection of the beam and gravitational acceleration, respectively. At least five specimens of each mixture were tested at the 56th day. All the beams were loaded at a constant rate of 0.02 mm/minute. The fracture energy of the matrix is based on the area under the complete load versus displacement at mid span curve. The results for composites with steel fibers are based on the area under the load versus displacement curve up to a specified displacement. A cut-off point was chosen at 10 mm displacement. It is seen from Figure 2 that the energy at this displacement (i.e. 10 mm), however, is not totally dissipated.

In order to determine the ductility of the composites, characteristic length (l_{ch}) was calculated using the measured G_F, E, and f'_t, according to the Eq(2) introduced in fictitious crack model by Hillerborg et al. (1976).

$$l_{ch} = \frac{EG_F}{f'^2_t} \qquad (2)$$

where f'_t, E, and G_F are direct tensile strength, modulus of elasticity, and fracture energy, respectively.

In this study, the term "direct tensile strength (f'_t)" in this relation was replaced by "splitting tensile strength (f_{st})".

3 EXPERIMENTAL RESULTS AND DISCUSSION

Table 4 summarizes the mechanical test results of composites produced and their fracture parameters.

3.1 Compressive strength and modulus of elasticity

A significant increase is observed in compressive strength of composites with increasing curing temperature. The composites containing high strength straight fibers have the highest compressive strength for all curing conditions. In addition, compressive strengths of composites with high strength hooked-end steel fibers are higher than those with normal strength hooked-end steel fibers. As seen in Table 4, it is clear that high temperatures activate the pozzolanic reaction between the calcium hydroxide formed during the normal cement hydration and silica fume contained in the mixture. On the other hand, the reaction may take place between the very fine ground silicious sand and calcium hydroxide product at hot curing conditions

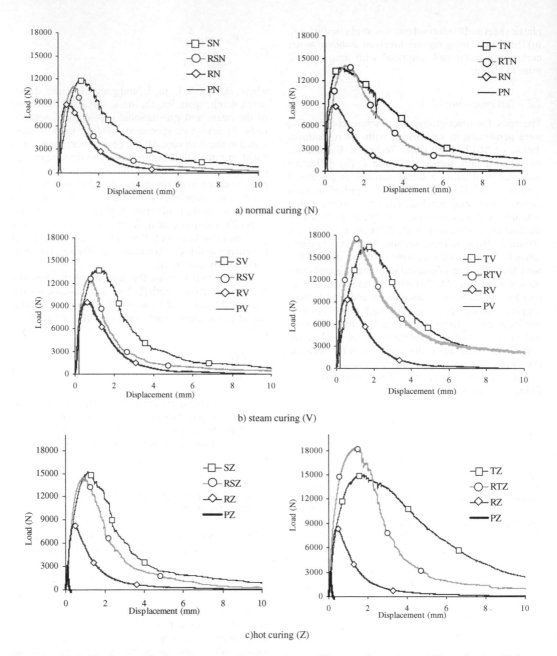

Figure 2. Typical load versus displacement curves for composites at different curing regimes: a) Normal curing, b) Steam curing c) Hot curing.

(Alaee 2002, Massaza & Costa 1986). It is also clearly shown that the high temperature curing conditions may provide a shorter curing period compared to the standard curing regime.

There is no significant effect of the type of steel fibers and their strengths, and curing conditions on the modulus of elasticity. In composites with hooked-end fibers, however, the fiber strength causes little increase in the modulus of elasticity.

3.2 *Splitting tensile and flexural strengths*

As seen in Table 4, the splitting tensile and flex-ural strengths of composites notably increased with

Table 4. Fracture and strength properties of composites.

	Mix code	f'_c (MPa)	E (GPa)	f_{st} (MPa)	G_F (N/m)	f_{flex} (MPa)	l_{ch}(mm)
Normal	PN	131.2	44.1	8.7	181	10.2	105
curing	RN	146.7	45.4	12.4	6043	23.8	1784
	SN	113.4	41.6	13.5	14254	35.2	3254
	TN	130.0	45.9	15.6	21617	37.7	4077
	RSN	143.7	44.6	15.0	8067	28.1	1599
	RTN	133.1	48.1	15.1	19829	38.5	4183
Steam	PV	136.3	44.1	9.4	201	12.8	100
curing at	RV	175.1	45.1	15.5	6778	25.4	1272
65°C	SV	133.8	38.5	16.0	14079	37.4	2117
	TV	149.8	45.1	17.0	24821	44.9	3873
	RSV	144.3	42.8	15.4	10343	35.7	1867
	RTV	149.4	41.3	17.6	20140	45.8	2685
Hot	PZ	145.2	41.5	6.2	179	8.6	193
curing at	RZ	177.0	45.6	16.0	5421	23.2	966
200°C	SZ	136.0	38.8	16.4	17212	41.1	2483
	TZ	156.4	41.3	16.7	30169	42.0	4468
	RSZ	163.4	34.8	16.9	11336	36.6	1381
	RTZ	168.2	43.9	17.3	20235	46.8	2968

addition of fibers as expected. In addition, composites containing longer fibers (S and T) have significantly greater splitting tensile strength than those containing short fibers (R), while the increasing tensile strength of steel fibers increased the splitting tensile strength of composites.

The use of hybrid fibers enhanced the splitting tensile strength of the composites. Depending on the type and strength of the fibers, and curing conditions, splitting and flexural strengths increased up to 5 times compared to those of the matrix. It is also indicated in Table 4 that the curing temperature increased the splitting and flexural strengths of composites. Since hot curing regimes activate pozzolanic reaction and enhance the microstructure of the composites, both splitting and flexural strengths can be greatly improved. On the other hand, the hot curing condition significantly reduced these strengths of plain composite. This is due to the fact that higher curing condition caused the formation of the microcracks in the matrix phase and this in turn, caused reduction in the splitting tensile strength. In addition, the composites containing both straight and high strength hooked-end steel fibers (RT) have the highest splitting and flexural strengths. Along the fracture plane of the composites in both splitting and bending tests, the opening and propagation of the crack are controlled by the steel fibers. During crack propagation some fibers are broken, but some are pulled-out of the matrix. After completion of the splitting and bending tests, the fracture surfaces were examined. In most cases the high strength hooked-end fibers did not break, but were pulled-out of the matrix.

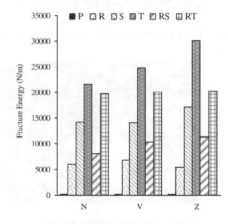

Figure 3. The variation of fracture energy with curing condition and type of steel fibers.

In some other cases the normal strength hooked-end fibers were broken into two parts. Thus, in the latter case, the mechanical mismatch may have played a role on these two mechanical properties. As indicated earlier, high strength steel fiber with a tensile strength of 2000 MPa is suggested for the high strength concretes (Vandewalle 1996, Grünewald & Walraven 2002).

3.3 *Fracture energy*

The test results have shown that the most significant effects of both fiber addition and curing condition are observed on the fracture energy of composites. Figures 2 and 3 show the effects of curing conditions, and the type and strength of the steel fibers on the mechanical behaviour of the composites. In Figure 2, each composite shows the brittle behaviour of a typical matrix. It is seen that the use of high strength steel fibers improves the mechanical performance and increases the fracture energy of the composite. Especially after the first crack, the formation of strain hardening in the ascending branch of the curve is a typical indication of high performance cementitious composites. The composites with normal strength

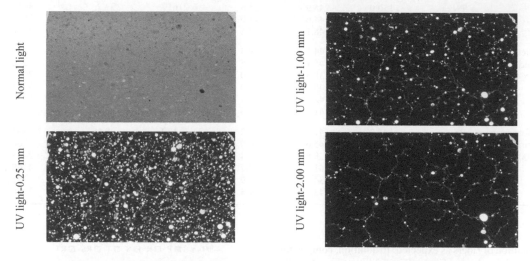

Figure 4. Plane section views of the matrix at 200°C hot curing condition.

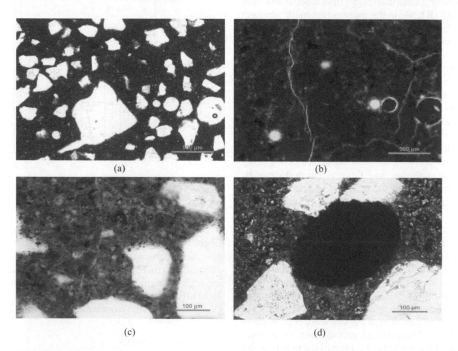

Figure 5. View of microcracks in thin section view of the specimen subjected to 200°C (a): plane polarized view at 50x, (b): fluorescence light view at 50x, (c): Microcracks filled with Ca(OH)₂, No sign of delayed ettringite formation is observed (plane polarized view at 200x), (d): bond between fiber and cement matrix, the dark circle shows steel fiber (plane polarized view at 200x).

hooked-end steel fibers have lower peak loads and steeper gradients of the softening branch compared to the composites with high strength hooked-end steel fibers and the highest fracture energies were obtained from these composites.

As seen in Table 4, curing regimes enhance the microstructure of the composites and as a result the fracture energy has been greatly improved. It can be concluded that increasing the curing temperature increased the peak load significantly and this resulted in an increase in fracture energy, while the slope of the ascending branch of the curves remained almost constant.

The increase in the fiber aspect ratio and also the fiber tensile strength significantly increased the fracture energy of composites. The short fibers have a limited effect on the post-peak response of load versus displacement at midspan of beam, while there is a substantial effect of long fibers on the post peak response part of curve, which results in high value of fracture energy.

It is also shown that, depending on the high temperature curing condition, fracture energy of the matrix increased up to 168 times compared to composites with high strength steel fibers; while in composites with normal strength steel fibers the increase in fracture energy due to the steel fibers was 96 times.

In all curing regimes, while the composite hybrid mixture of high strength straight and high strength hooked-end steel fibers (i.e. mixture RT) have the highest peak load, the fracture energy of composites with high strength hooked-end steel fibers (i.e. mixture T) have been found greater than those of other series.

3.4 Characteristic length

Characteristic length (l_{ch}) should be taken into consideration in the design of cementitious composites, because it controls the nominal strength, failure mode, and crack pattern (Lange-Kornbak & Karihaloo, 1998). The variation of l_{ch} with curing conditions, and the types and the strengths of steel fibers are shown in Table 4. As curing temperature, the aspect ratio and strengths of fibers increase, the characteristic length, which is a measure of ductility of the material, increases significantly.

4 PETROGRAPHIC EXAMINATION

In order to investigate the effects of high temperature curing on the microstructure, polished plane and thin sections of samples were prepared. For the plane sections, composite samples with a cross-sectional area of 10×7.5 cm were cut and impregnated with fluorescence epoxy under vacuum. The fluorescence in the epoxy enters the microcracks and macrocracks, capillary, entrained and entrapped air voids, starting from the surface of the specimen. As the sample is polished and inspected under a UV light it is possible to see the pore structure and microcracks.

The sample was first polished down to a level of 0.25 mm. At this level, under the UV light, it is possible to see the pores and cracks which are connected to the surface. The sample was further polished down to 1 mm and 2 mm levels to investigate if the pores and cracks were still connected to the surface. Figure 4 presents the pictures of matrix (PZ) samples subjected to 200°C curing. The pictures of the cross-section were taken under normal light and for the various polishing levels under UV light.

From the investigation under the normal light and 0.25 mm level under the UV light, it was seen that all the samples presented similar pore volumes and pore sizes distribution. The pictures of 1 mm and 2 mm polished levels under the UV light indicated that there was a map cracking associated with 200°C curing (Figure 4).

As the sample was polished further down, it was observed that the crack system was connected. 20°C and 65°C steam curing levels did not present any cracks, and the volume of surface connected porosity was low. The pore system of 20°C and 65°C were also not connected with each other. For the thin section analysis, composite samples with a cross-sectional area of approximately 3.5×4.5 cm were cut and impregnated with fluorescence epoxy. Then the samples were polished to a thickness of about 30 μm, which allows investigation with a polarising microscope.

With the help of the fluorescence absorbed by the specimen and light filters, it is possible to investigate the crack and pore system. The polarising microscope is also used for mineral identification. Cracks, voids, cement paste and aggregates can be inspected with magnifications of up to 500x.

From Figure 5, the crack system in the samples subjected to 200°C can be observed at 50x magnification. The cracks were further investigated at 200x magnification, crossed polarisation and with a gypsum lamel. With this setting, it was possible to identify calcium hydroxide by its blue-yellow colours. It was observed that the cracks in the sample subjected to 200°C were filled with calcium hydroxide. As seen in thin section of composites cured at 200°C, the interfacial transition zone in the composites with the dense microstructure is characterized by a direct contact between the steel fiber and the matrix (Figure 5d). The CH near the steel fiber reacts with ultrafine silica fume, forming dense C-S-H which fills in the spaces at the interfacial zone, producing increased bond between the matrix and steel fiber. The increased bond allows the efficient transfer of stress between the matrix and steel fibers.

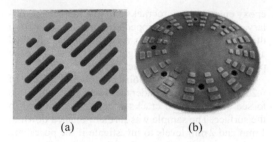

(a)   (b)

Figure 6. Examples produced with high performance hybrid steel fiber reinforced composites (a) gully tops, (b) circular manhole cover.

Evidence of possible thermally-induced damage such as delayed ettringite formation was also checked. No delayed ettringite formation was observed at hot curing conditions during petrographic investigations.

5 RECENT APPLICATIONS IN ISTANBUL

Gully tops and manhole covers produced in Iston which is a company of Istanbul Metropolitan Municipality, can be used in roads and urban environment for rain water drainage. In these structural elements containing classical reinforcing bars, ultra high performance composites with hybrid steel fibers are used as the matrix. The strength of these matrices reaches up to the compressive strength of 350 MPa. The average load carrying capacity of gully tops and manhole covers produced was 460 kN. This means that D400 load carrying ability in EN 124 is valid for these elements. They are available in different sizes and with special designs such as square gully tops and circular manhole covers. Typical examples are shown in Figure 6.

6 CONCLUSION

Based on the tests conducted, it can be concluded that short fibers function as a bridge to eliminate the microcracks, and as a result, the tensile strength of composite increases, and the fibers pull out after macrocracks are formed. Thus, the short fibers have little effect on the post-peak response of load versus displacement at the midspan of the beam. The long fibers have no significant effect on preventing micro cracking, however, there is a substantial effect of long fibers on the post-peak response part of curve and also on peak load, resulting in high value of fracture energy. The composites with high strength long steel fibers show a behavior of enhanced toughness and ductility when compared to the composites with normal strength steel fibers. Curing regimes enhance the microstructure of the composites, as a result the compressive strength, specific fracture energy, and flexural strength can be greatly improved. On the other hand, the hot curing condition significantly reduced flexural strength of plain composite probably because the higher curing condition caused the micro cracks in the matrix phase. Petrographic examination showed that a connected crack system forms at 200°C curing condition which is filled with calcium hydroxide. The calcium hydroxide near the steel fiber reacts with ultrafine silica fume, forming dense C-S-H which fills in the spaces at the interfacial zone, producing increased bond strength between the matrix and steel fiber. The gully tops and manhole covers produced to be used in the city infrastructures can be considered as selected applications.

ACKNOWLEDGMENTS

This research was carried out in the Faculty of Civil Engineering at Istanbul Technical University (ITU). The authors wish to acknowledge the financial support of TUBITAK (The Scientific & Technical Research Council of Turkey): Project:106G122, 1007-Kamu.

REFERENCES

Alaee, F.J. 2002. Retrofitting of Concrete Structures Using High Performance Fiber Reinforced Cementitious Composites (HPFRCC), *PhD Thesis*, University of Wales, Cardiff, 220 p.

Alexander, M.G. 1993. From nanometers to gigapascals cementing future, A University of Cape Town Publication, Inaugural Lecture, 23 p.

Balaguru, P., Narahai, R., and Patel, M. 1992. Flexural toughness of steel fibre reinforced concrete. *ACI Materials Journal* 89, 541–546.

Bonneau, O., Lachemi, M., Dallaire, E., Dugat, J. and Aitcin, P-C. 1997. Mechanical properties and durability of two industrial reactive powder concretes. *ACI Materials Journal* 94(4), 286–290.

Dugat, J., Roux, N. and Bernier, G. 1996. Mechanical properties of reactive powder concrete, *Materials and Structures* 29, 233–240.

Feylessoufi, A., Villieras, F., Michot, L.J., De Donato, P., Cases, J.M. and Richard, P. 1996. Water environment and nanostructural network in a reactive powder concrete, *Cement and Concrete Composites* 18, 23–29.

Fritz, C. 1991. Tensile testing of SIFCON in Eds. Reinhardt and Naaman, *Proceedings of 1st International Workshop on HPFRCCs, June 23–26, Mainz*, RILEM, pp. 518–528.

Grünewald, S. & Walraven, J.C., 2002. High strength self-compacting fibre-reinforced concrete: behaviour in the fresh and hardened state, Eds. König, G. et al. *in Proceedings of 6th International Symposium on HSC/HPC*, l., Leipzig, pp. 977–989.

Hillerborg, A., Modeer, M., and Peterson, P.E. 1976. Analysis of crack formation and crack growths in concrete by means of fracture mechanics and finite elements, *Cement and Concrete Research* 6, 773–782.

Ilki, A., Yilmaz, E., Demir, C. And Kumbasar, N. 2004. Prefabricated SFRC jackets for seismic retrofit of non-ductile reinforced concrete columns, *in Proceedings of 13th World Conference on Earthquake Engineering, Vancouver*.

Lange-Kornbak, D. & Karihaloo, B.L. 1998. Design of fibre-reinforced DSP mixes for minimum brittleness. *Adv. Cement Based Materials* 7, 89–101.

Masazza, F. & Costa, U. 1986. Bond: paste-aggregate, paste-reinforcement, and paste-fibers, *in Proceedings of 8th International Congress on the Chemistry of Cement, Rio de Janeiro, Brasil*, pp. 158–180.

Matte, V. & Moranville, M. 1999. Durability of reactive powder composites: Influence of silica fume on the leaching properties of very low water/binder pastes. *Cement and Concrete Composites* 21, 1–9.

Richard, P. & Cheyrezy, M. 1995. Composition of reactive powder concrete, *Cement and Concrete Research* 25, 1501–1511.

Tasdemir, C., Tasdemir, M.A., Mills, N., Barr, B.I.G. and Lydon, F.D. 1999. Combined effects of silica fume, aggregate type, and size on post peak response of concrete in bending, *ACI Materials Journal* 96, 74–83.

Tasdemir, M.A., Tasdemir, C., Akyuz, S., Jefferson, A.D., Lydon, F.D. and Barr, B.I.G. 1998. Evaluation of strains at peak stresses in concrete: A three phase composite model approach, *Cement and Concrete Composites* 20, 301–318.

Vandewalle, L. 1996. Influence of the yield strength of steel fibers on the toughness of fibre reinforced high strength concrete, *in Proceedings of CCMS Symposium, worldwide advances in Structural Concrete and Masonry, Chicago*, pp. 496–495.

Walraven, J. 1999. The evolution of concrete, *Structural concrete, Journal of fib* 1(1), 3–11.

Mixed-mode fracture tests for application of high-strength fiber-reinforced concrete for seismic isolation

S. Fukuoka & Y. Kitsutaka

Tokyo Metropolitan University, Japan

ABSTRACT: The mixed-mode fracture behavior of a mortar reinforced with vinylon fibers is investigated in this paper, in order to ascertain to what extent this material is suitable to be used in seismic-isolation devices. Since most of the studies on fracture mechanics of cement-based materials have regarded so far Mode I fracture, standard tests are limited to this type of behavior. However, an increasing number of researchers are investigating whether Mode II (shear fracture) exists and how its parameters can be measured. Starting from the assumption that Mode II exists (even if it is generally combined with Mode I), this study aims to investigate the fracture properties of the above-mentioned fiber-reinforced mortar, in order to assess its energy-absorption capability, in view of possible application to seismic devices. In this project, 4-point shear tests on notched prisms were carried out, and both the crack-mouth opening displacement (CMOD) and the crack sliding deformation (CSD) were measured by means of 2-axis LVDTs. Since the specimens were longitudinally confined by tightening two external steel bars, the favorable effect that the confinement has on mortar toughness was investigated as well. The evolution of the load versus CMOD and CSD was also monitored, to have information on mixed-mode fracture, as a step towards the still elusive formulation of appropriate constitutive laws for cementitious materials.

1 INTRODUCTION

Numerous research projects have been so far devoted to fracture in Mode I, in order to determinate the fracture parameters of various types of concrete and mortar, and to make proposals for standard tests (see for instance Bazant & Estenssoro, 1979; Arrea & Ingraffea, 1981; Bazant et al., 1984; Ingraffea & Panthaki, 1985; Rots & de Borst, 1987; Bazant & Pfeiffer, 1987; Swartz et al., 1988; Kaneko & Li, 1993). However several researchers are currently focusing their attention on the other fracture mode characterized by a slip – or slide – between the crack faces (Mode II), that brings in two interrelated problems: (1) whether the fracture process can evolve according to Mode II (without being accompanied by Mode I); and (2) how to evaluate the parameters of Mode II fracture. Both problems require experimental and analytical studies. In order to solve the second problem, achieving a state of pure shear is necessary, as well as forcing the propagation of a crack in such a way that K_{II} prevails over K_I. A specific test should be developed. The specimen should contain a flaw in the pure-shear region, in order to have a slide along the crack, and no – or very limited – opening. Unfortunately, concrete cracking in shear is generally characterized by mixed-mode fracture, because of the mixed boundary conditions. In

this research project, the fracture behavior of a mortar reinforced with vinylon fibers is investigated, by testing a number of beams loaded in 4-point shear. The geometry of this test was first proposed by Iosipescu with a similar objective (shear strength of steel).

In 4-point shear tests, the specimens are designed in such a way that large shear stresses are accompanied by small bending stresses, and the two halves of each specimen do not undergo any mutual rotation. In Fig. 2 the geometry adopted by Arrea & Ingraffea (1981) is shown. The crack propagates from the tip of the notch along a curved path, with an initial angle $\alpha = 68°$ with respect to the horizontal axis.

Bazant & Pfeiffer (1987) used in their tests double-notched specimens (Fig. 1) characterized by a smaller horizontal distance between the point loads (0.167d) than in Arrea and Ingraffea's specimens (0.4d). In the former tests (Fig. 1), a macroscopic vertical fracture connecting the tips of the notches was observed. Such a fracture process was considered by Bazant & Pfeiffer as a demonstration that Mode II fracture does exist. However, the authors concluded the macro-fracture resulted from the cohalescence of many Mode-I micro-cracks at 45° to the horizontal axis, as shown in Fig. 1.

Within this context, the goals of this study are: (1) the presentation of a research program on a new testing procedure for mixed-mode fracture and on

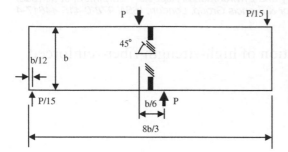

Figure 1. Geometry of Bazant and Pfeiffer's specimens.

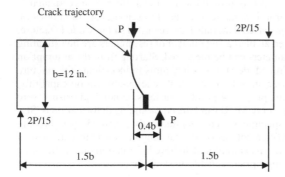

Figure 2. Geometry of Arrea and Ingraffea's specimens.

Table 1. Mix design.

Water/(Cement + Silica fume)	0.17
Fiber/(Cement + Silica fume)	1%
Water	187 kg/m³
Cement	1029 kg/m³
Silica fume	257 kg/m³
Fine Aggregate	859 kg/m³
Vinylon fiber: content/volume fraction	19 kg/m³/1.5%
Superplasticizer	32.15 kg/m³
Curing time	14 days

Table 2. Physical properties of vinylon fibers.

Fiber type	vinylon, VF1500
Length m	30 m
Diameter	240 × 720 µm
Tensile strength	882 MPa
Young's modulus	29 GPa
Mass per unit volume	1.30

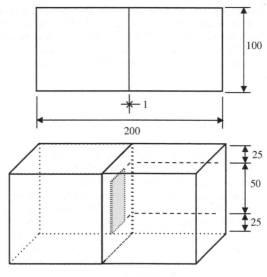

Figure 3. Geometry of the specimens.

the measurement of mixed-mode fracture parameters; and (2) the investigation on the mixed-mode fracture behavior of the afore-mentioned fiber-reinforced mortar, with/without confinement, under both monotonic and cyclic loading.

2 EXPERIMENTAL PROGRAM

2.1 Geometry of the specimens

The mix design of the mortar and the physical properties of the fibers are given in Tables 1 and 2. The cylindrical compressive strength is close to 200 MPa and the maximum aggregate size is 3.5 mm. Prior to testing, the specimens (beams) were cured for 14 days at 20°C and 80% of R.H. The dimensions of the 8 beams (Fig. 3) were: w/d/l = 100/100/200 mm; the notch depth/section depth ratio was $a_0/w = 0.25$. A typical specimen is shown in Fig. 4 (mass $m_g = 4.6$ kg).

2.2 Test set-up

A sketch of the test set-up, with the specimen, the point loads and the LVDTs is shown in Fig. 5. Between the rollers and the specimen thin metal strips are placed in order to avoid any friction. The distance between each couple of point loads is s = 100 mm. Two LVDTs are attached to either side of the specimen, at mid-height, and are connected to a data logger. The details of the two-axis LVDTs are given in Fig. 6.

All tests were displacement-controlled, by controlling the displacement rate of the crosshead of the press (0.1 mm/minute). The force exerted by the MTS press and the stroke were recorded every second.

3 TEST SERIES

The fracture behavior of the mortar reinforced with vynilon fibers was investigated by performing 4 series

Figure 4.   View of an actual specimen.

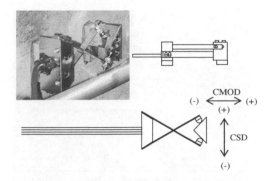

Figure 6.   Details of the 2-axis LVDTs.

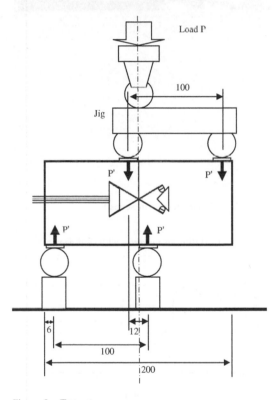

Figure 5.   Test set-up.

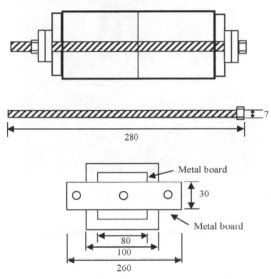

Figure 7.   Lateral view of the prismatic specimens (beams) provided with 2 confining bars (Ø = 7 mm).

was monotonically loaded up to the full cracking of the notched section, and then was unloaded. Subsequently, the two specimens were loaded and reloaded in two slightly-different ways. In the first specimen, the second and third cycles induced a ± 1 mm slide between the crack faces, while the fourth and fifth cycles induced a ± 2 mm slide. In the second specimen, the second and third cycles induced a ± 0.5 mm slide between the crack faces, while the fourth and fifth cycles induced a ±1 mm slide.
Series 4: same as in Series 3, but with confined specimens.

4   RESULTS

The load-displacement curves of the 8 specimens tested in this project are shown in Figs 9 and 10, 12

of tests concerning notched prisms ("beams") loaded in 4-point shear. Each series consisted of 2 specimens:
Series 1: 2 unconfined specimens were tested under monotonic loading.
Series 2: same as in Series 1, but with confined specimens (see Figs. 7 and 8).
Series 3: 2 unconfined specimens were tested under cyclic loading, to assess the bridging properties of vynilon fibers. During the first cycle each specimen

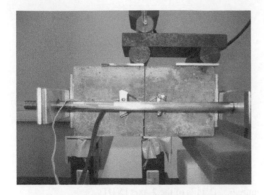

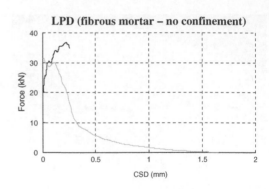

Figure 10. Load-CSD curves in Series 1.

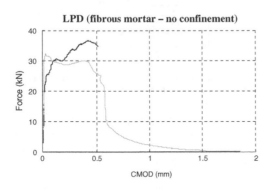

Figure 8. Test set-up in Series 2 and 4.

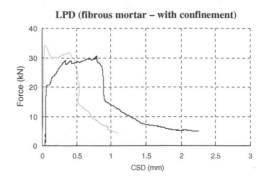

Figure 11. Typical fracture in Series 1.

Figure 9. Load-CMOD curves in Series 1.

Figure 12. Load-CMOD curves in Series 2.

and 13, 15 and 16, 18 and 19. (LPD means Load-Point Displacement). As for the displacements, reference is made to the average CMOD and to the average CSD (average = mean of the two values measured on the front and back faces).

Test Series 1- Monotonic Loading – No Confinement
The force-displacement curves are plotted in Figs.9 and 10. The cracked surface of the 1st specimen is shown in Fig. 11. In both specimens the cracked surface was smooth. Looking at Fig. 9 one can observe that up to the peak load the force-CMOD curves are more extended than the force-CSD curves (Fig. 10), which means that fracture is caused by mixed-mode cracking, with small diagonal micro-cracks joining in a single vertical crack.

1390

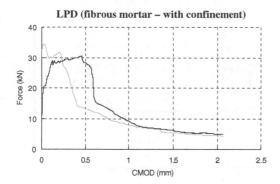

Figure 13. Load-CSD curves in Series 2.

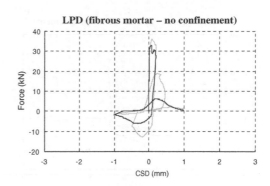

Figure 16. Load-CSD curves in Series 3.

Figure 14. Typical fracture in Series 2.

Figure 17. Typical fracture in Series 3.

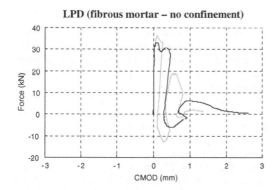

Figure 15. Load-CMOD curves in Series 3.

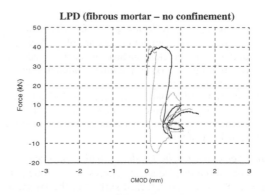

Figure 18. Load-CMOD curves in Series 4.

Test series 2- Monotonic Loading – With Confinement

The force-displacement curves are plotted in Figs 12 and 13. The cracked surface of the 1st specimen is shown in Fig. 14. In both specimens the fracture was caused by mixed-mode cracking, as in the previous case. However, because of the confinement the toughness of the material increased (more extended curves prior to the peak), the local angle α to the horizontal axis was smaller, and the cracked surface was not as smooth as in the previous case.

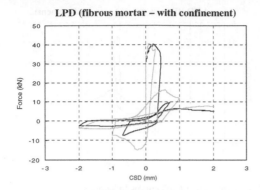

LPD (fibrous mortar – with confinement)

Figure 19.   Load-CSD curves in Series 4.

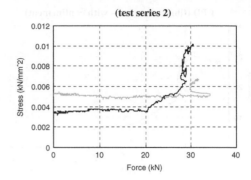

(test series 2)

Figure 21.   Plot of the stress in the confining bars, as a function of the applied load (Test Series 2).

Figure 20.   Typical fracture in Series 4.

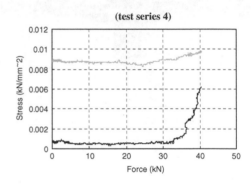

(test series 4)

Figure 22.   Plot of the stress in the confining bars, as a function of the applied load (Test Series 4).

Test Series 3 – Cyclic Loading – No Confinement

The force-displacement curves are plotted in Figs 15 and 16. The cracked surface of the 1st specimen is shown in Fig. 17. Both specimens failed during the 3rd cycle, the former prior to reaching the maximum slip of 1 mm and the latter after reaching the same slip. Contrary to the specimens loaded monotonically, the local angle to the horizontal axis was partly positive and partly negative.

Test Series 4 – Cyclic Loading – With Confinement

The force-displacement curves are plotted in Figs. 18 and 19. The cracked surface of the 1st specimen is shown in Fig. 20. Local cracking was similar to that occurred in the previous tests. However, because of the confinement the toughness of the material increased, the local angle α to the horizontal axis was smaller, and the cracked surface was not as smooth as in the previous case.

As for the confining action, the relationship between the stress in each restraining bar and the applied load is plotted in Figs 21 and 22. Note that

in both Series 2 and Series 4 the initial confinement ("active" confinement) was very small, since the objective of the tests was to investigate crack cohesion due to the fibers, rather than the direct effects of the confinement.

In Series 2 (Fig. 21), the initial stresses in the confining bars were similar (mean value 4.3 MPa) and the total initial confining force was close to 330 N (compressive stress acting on the notched section = 0.13 MPa).

In Series 4 (Fig. 22), the initial stresses in the confining bars were different (0.8 MPa in the 1st test and 8.8 MPa in the 2nd test) and the total initial confining forces were 62 N and 677 N respectively.

Figs 21 and 22 show also that – because of shear-induced crack dilatancy – the stresses in the confining bars started increasing above 20–30 kN in the tests of Series 2 and above 25–35 kN in the tests of Series 4 ("passive" confinement). However, the larger the confinement, the smaller the effects of crack dilatancy (2nd test in both Series 2 and 4).

5 CONCLUSIONS

1. In all fiber-reinforced specimens – and particularly in the unconfined specimens – the shear-induced crack results from the cohalescence of local diagonal micro-cracking.
2. In the confined specimens, both the toughness and the strength markedly increase.
3. Introducing vinylon fibers in the mix design markedly softens the post-peak behavior.
4. In confined specimens, the interface of the shear-induced crack is rather rough, since combining the shear force and the longitudinal confinement tends to diminish the local angle of the micro-cracks to the horizontal axis.
5. The load-CMOD curves exhibit a more extended plateau close to the peak load, than the load-CSD curves, which means that Mode I tends to prevail over Mode II; however, the slide at the crack interface is not at all negligible compared to crack opening; the fracture is characterized by mixed mode, with the final crack resulting from the cohalescence of many diagonal micro-cracks, as mentioned in point 1.
6. The test procedure proposed in this paper may be useful to evaluate the fracture properties in mixed-mode fracture in cementitious composites.
7. The results concerning both CMOD and CSD – under sustained and cyclic loading – give information useful to the formulation of the constitutive laws for mixed-mode fracture.

NOTATION

a_0 = notch depth;

w, d, l, m_g = width, depth, length and mass of each specimen (beam);

s = distance between the two point loads acting at the bottom and the top of each specimen;

CSD = crack sliding displacement;

CMOD = crack-mouth opening displacement;

K_I = stress intensity factor in Mode I;

K_{II} = stress intensity factor in Mode II;

LPD = load-point displacement;

α = angle of the crack to the horizontal axis.

ACKNOWLEDGEMENTS

This research project was financially supported by Tokyo Metropolitan University in 2006, within the Grant-in-Aid Program for Scientific Research-launched by the Japanese Ministry of Education, Science, Sports and Culture, and spanning over the period 2005–07. The cooperation by Dr. Tamura Masaki – Research Associate at Tokyo Metropolitan University – is also gratefully acknowledged.

REFERENCES

Swartz S.E., Lu L.W. & Tang L.D. 1988. Mixed-mode fracture toughness testing of concrete beams in three-point bending. Materials and Structures 1998, 21, 33–40.

Bazant Z.P. & Pfeiffer P.A. 1987. Shear Fracture Tests of Concrete. Materiaux et Constructions 19, 111–121.

Kaneko Y. & Li V.C. 1993. Fracture Behavior of Shear Key Structures with a Softening Process Zone. International Journal of Fracture 59, 345–360.

Rots J.G. & de Borst R. 1987. Analysis of Mixed-Mode Fracture in Concrete. Journal of Engineering Mechanics, Vol. 113, No. 11.

Arrea, M. & Ingraffea, A.R., 1981. Mixed-Mode Crack Propagation in Mortar and Concrete. Rep. 81–13, Dept. of Structure. Eng., Cornell University (N.Y., USA).

Ingraffea, A.R. & Panthaki, M.J., 1985. Analysis of Shear Fracture Tests of Concrete Beams. Proc. U.S. – Japan Seminar on Finite Element Analysis of Reinforced-Concrete Structures, Tokyo, May 21–24, 71–91.

Bazant Z.P. & Estenssoro L.F. 1979. Surface Singularity and Crack Propagation. Int. J. of Solids and Structures, Vol. 15, 405–426, Addendum Vol. 16, 479–481.

Bazant Z.P., Kim J.K. & Pfeiffer P.A. 1984. Determination of Nonlinear Fracture Parameters from Size Effect Tests. NATO Advanced Research Workshop on Application of Fracture Mechanics to Cementitious Composites, ed. by S.P.Shah, held at Northwestern University (Evanston, Illinois, USA), Sept. 4–7, 143–169.

Fracture Mechanics of Concrete and Concrete Structures – High-Performance Concrete,
Brick-Masonry and Environmental Aspects – Carpinteri, et al. (eds)
© 2007 Taylor & Francis Group, London, ISBN 978-0-415-44617-4

New simple method for quality control of Strain Hardening Cementitious Composites (SHCCs)

S. Qian & V.C. Li
ACE-MRL, University of Michigan, Ann Arbor, Michigan, USA.

ABSTRACT: As emerging advanced construction materials, strain hardening cementitious composites (SHCCs) have seen increasing field applications recently to take advantage of its unique tensile strain hardening behavior, yet existing uniaxial tensile tests are relatively complicated and sometime difficult to implement, particularly for quality control purpose in field applications. This paper presents a new simple inverse method for quality control of tensile strain capacity by conducting beam bending test. It is shown through a theoretical model that the beam deflection from a flexural test can be linearly related to tensile strain capacity. A master curve relating this easily measured structural element property to material tensile strain capacity is constructed from parametric studies of a wide range of material tensile and compressive properties. This proposed method (UM method) has been validated with uniaxial tensile test results with reasonable agreement. In addition, this proposed method is also compared with the Japan Concrete Institute (JCI) method. Comparable accuracy is found, yet the present method is characterized with much simpler experiment setup requirement and data interpretation procedure. Therefore, it is expected that this proposed method can greatly simplify the quality control of SHCCs both in execution and interpretation phases, contributing to the wider acceptance of this type of new material in field applications.

1 INRODUCTION

In the past decade, great strides have been made in developing strain hardening cementitious composite (SHCC), characterized by its unique macroscopic pseudo strain hardening behavior after first cracking when it is loaded under uniaxial tension. SHCCs, also referred to as high performance fiber reinforced cementitious composites (HPFRCCs, Naaman and Reinhardt 1996), develop multiple cracks under tensile load in contrast to single crack and tension softening behavior of concrete and conventional fiber reinforced concrete. Multiple cracking provides a means of energy dissipation at the material level and prevent catastrophic fracture failure at the structural level, thus contributing to structural safety. Meanwhile, material tensile strain hardening (ductility) has been gradually recognized as having a close connection with structural durability (Li 2004) by suppressing localized cracks with large width. Many deterioration and premature failure of infrastructure can be traced back to the brittle nature of concrete. Therefore, SHCCs are considered a promising material solution to the global infrastructure deterioration problem and tensile ductility is the most important property of this type of material.

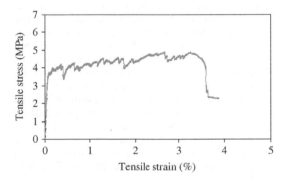

Figure 1. Typical tensile stress-strain curve of ECC.

Engineered Cementitious Composites (ECC, Li 1993) is a unique representative of SHCCs, featuring superior ductility (typically >3%, 300 times that of normal concrete or FRC) (Li and Kanda 1998; Li et al 2001), tight crack width (less than 80μm, Li 2003), and relatively low fiber content (2% or less of short randomly oriented fibers). A typical tensile stress-strain curve of ECC is shown in Figure 1. It attains high ductility with relatively low fiber content via systematic tailoring of the fiber, matrix and interface properties, guided by micromechanics principles. Enhanced with such high tensile ductility and/or tight crack width, ECC has demonstrated superior energy dissipation capacity, high damage tolerance, large deformation capacity, and exceptional durability in many recent experimental investigations (Li 2005). As a result, ECC is now emerging in the field and has seen increasing infrastructure applications, such as dam repair, bridge deck overlay and link slab, coupling beam in high-rise building, and other structural elements and systems (Li 2004).

As aforementioned, tensile ductility is the most important material property of SHCC, yet relatively large variation of tensile ductility was observed in the literature (Kanda et al 2002, 2006; Wang and Li 2004). To address such concern, Wang and Li (2004) have proposed using artificial flaws with prescribed size distribution as defect site initiator to create more saturated multiple cracks, resulting in more consistent tensile strain capacity among different specimens from the same batch. The overall tensile strain capacity shows much more consistent results after implantation of artificial flaws, however, the variation of tensile strain capacity is still relatively large when compared with that of other properties, e.g., first cracking strength. Therefore, test method for quality control of SHCCs onsite should logically focus on tensile strain capacity due to its importance in governing structural response and potentially large variability.

While most characterization of the tensile behavior of SHCCs was carried out using uniaxial tensile test (UTT) in academia, this method is generally considered to be complicated, time-consuming and require advanced equipment and delicate experimental skills. Therefore, it is not suitable for onsite quality control purpose (Stang and Li 2004, Ostergaard et al 2005, Kanakubo 2006). First, special fixtures and/or treatments for the ends of specimens are usually needed in order to transfer tensile loads. Furthermore, the specimen is sensitive to stress concentration induced by misalignment and can fail near the end prematurely. Last but not least, realistic dimensions for specimens large enough to have 3-dimensional random fiber orientation make the UTT even more difficult to conduct.

As a simpler alternative to the UTT, four point bending test (FPBT) was proposed by Stang and Li (2004) for quality control on construction sites, provided that an appropriate interpretation procedure for the test result is available. FPBT, in which the mid-span of the specimen undergoes constant bending moment, may be carried out to determine the moment-curvature or moment-deflection curves. This type of test is much easier to set up and conduct in comparison to UTT, and a large amount of experience in bending test has been accumulated in the user community of cementitious materials. The ultimate goal of this test is to use the moment-curvature or moment-deflection curves so determined to invert for the uniaxial tensile properties. It should be noted, however, that the bending test is not meant to determine whether the material has tensile strain-hardening behavior or tension-softening behavior, but rather to constrain the tensile material parameters, e.g. the tensile strain capacity, as part of the quality control process in the field.

Inverse analyses for FPBT have recently been attempted by Technical University of Denmark (DTU) and Japan Concrete Institute (Ostergaard et al 2005; Kanakubo 2006) with certain success. By adopting a simplified elastic-perfectly plastic tensile model, JCI method generally can predict plateau tensile strength and tensile strain capacity from the FPBT results via a sectional analysis similar to that developed by Maalej and Li (1994). On the other hand, hinge model, including both tensile strain hardening and tension softening effect, was employed in the DTU inverse method along with least square method to invert for tensile material properties from their bending response. The model can predict experimental load–deflection curve fairly well and tensile properties derived based on this method agree well with that from FEM analysis, yet no direct comparison with UTT results has been made so far.

Despite the successes mentioned above, further simplification and/or validation are necessary to make the FPBT widely accepted for quality control of SHCCs. In case of JCI method, significant improvement is needed to simplify the experimental execution and data interpretation procedure. For instance, LVDTs are required in JCI method to measure the beam curvature. This is somewhat burdensome in field conditions, considering quality control may involve a large number of specimens. Furthermore, the inverse process is not user friendly, which require relatively complicated calculation (solving cubic equation). As for the DTU method, firstly it needs complementary UTT results to truly validate the model. Secondly, the uniqueness of solution from such inverse analysis is questionable at times. Finally, the method will need to be packaged into sophisticated software, which may incur additional user cost. A simple engineering chart with reasonable accuracy may be more preferable.

Keeping these considerations in mind, this paper looks to develop a greatly simplified yet reasonably accurate inverse method for determining tensile strain capacity of SHCCs. In the following sections, parametric study to obtain the master curves for inverse analysis will be presented first. Thereafter, the experimental program consisting of both FPBT and complementary UTT will be revealed in detail. The results from FPBT will then be converted to tensile strain capacity and validated with independent UTT test results. Finally, the proposed method will be compared with JCI inverse method, followed by overall conclusions.

Table 1. Range of material parameters used in parametric studies to construct the tensile strain capacity – deflection capacity (curvature) relation.

Material parameters	Tensile properties				Compressive properties	
	σ_{tc} (MPa)	σ_{tu} (MPa)	ε_{tu} (%)	E (GPa)	f'_c (MPa)	ε_{cp} (%)
Range	2.5~13.0	4.0~16.0	0~5	12~53	31~200	0.5~1*

Note: σ_{tc} = tensile first cracking strength; σ_{tu} = ultimate tensile strength; ε_{tu} = tensile strain capacity; E = modulus of elasticity; f'_c = compressive strength; and ε_{cp} = compressive strain capacity; Parameters are in the normal range of test results of SHCC specimens at UM and JCI; Tensile and compressive modulus of elasticity are assumed to be equal; Beam dimensions are 51 × 76 × 356 mm or 100 × 100 × 400 mm with span length of 305 mm or 300 mm for UM and JCI specimens, respectively; *: Estimated range.

2 PARAMETRIC STUDY AND MASTER CURVES

2.1 Flexural behavior model

The flexural behavior model used in this investigation is based on the work of Maalej and Li (1994). Compared with other models, the major distinction of this model is that the contribution of tensile strain hardening property of SHCC's was included. The actual SHCC considered in the model is Polyethylene ECC (PE-ECC) material. To simplify the analysis, the stress – strain behavior of the ECC was assumed as bilinear curves in both tension and compressive. Based on a linear strain profile and equilibrium of forces and moment in a section, the relation between flexural stress and tensile strain at the extreme tension fiber (Simplified as critical tensile strain hereafter) can be determined as a function of basic material properties. Overall, the model predicts experimentally measured flexural response quite well. For more detail, the readers are referred to Maalej and Li (1994).

Based on geometrical considerations, the beam curvature can be computed as the ratio of critical tensile strain to the distance from the extreme tension fiber to the neutral axis. This can be expressed in following equation:

$$\phi = \frac{\varepsilon_t}{c} \tag{1}$$

where ϕ, ε_t, and c are beam curvature, critical tensile strain, and the distance from the extreme tension fiber to the neutral axis.

In a FPBT of SHCC material, if we assume that the curvature is approximately constant along the span length of the beam and equal to the curvature in the middle span, we can obtain a simple equation to relate the deflection of the beam to its curvature and therefore critical tensile strain. For a constant curvature, the load point deflection u for a beam having a span L is given by:

$$u = \frac{L^2 \cdot \phi}{9} = \frac{L^2 \cdot \varepsilon_t}{9 \cdot c} \tag{2}$$

Since the relation between flexural stress and ε_t is already established, we can predict the flexural stress and load point deflection relation based on Equation (2). The peak flexural stress (MOR) and corresponding deflection (deflection capacity) are reached once the strain capacity of the SHCCs is exhausted either at the extreme tensile fiber or at the extreme compression fiber, which is the assumed failure criterion in this model.

2.2 Construction of master curves

Parametric study was conducted to investigate the influence of material uniaxial tensile and compressive properties (parametric values) on the flexural response of SHCCs based on the aforementioned flexural model. The correlation between tensile strain capacity and load point deflection was established and constructed as master curve. All tensile and compressive properties were varied within a wide range of parametric values (Table 1), covering the normal range of test results of SHCC specimens at UM and JCI (Kanakubo 2006). It is expected that the master curves based on this wide range of parametric study can be directly utilized for quality control purpose in field.

Five cases of parametric study were plotted in Figure 2 as examples, showing the flexural stress, load point deflection and corresponding critical tensile strain relation. Beam dimensions are 51 × 76 × 356 mm with span length of 305 mm. From the Figure, load point deflections were observed to correlate very well with critical tensile strains, regardless of the actual parametric material properties (shown in Table 2). Once the critical tensile strain reaches the

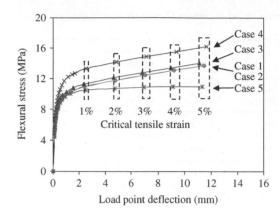

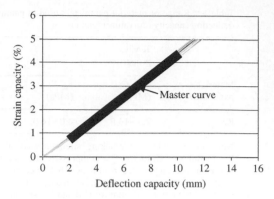

Figure 2. Parametric study for SHCCs with different material parameters (Dashed line boxes include markers corresponding to same critical tensile strain (tensile strain at the extreme tension fiber); markers are plotted on 1% strain interval after 1% strain for all cases).

Table 2. Assumed material properties for different cases of SHCCs.

	Tensile properties				Compressive properties	
	σ_{tc} (MPa)	σ_{tu} (MPa)	ε_{tu} (%)	E (GPa)	f_c' (MPa)	ε_{cp} (%)
Case 1	4.0	5.6	5	18	50	0.005
Case 2	4.0	5.6	5	20	50	0.005
Case 3	4.0	5.6	5	18	75	0.005
Case 4	5.0	6.6	5	18	50	0.005
Case 5	4.0	4.0	5	18	50	0.005

Note: σ_{tc} = tensile first cracking strength; σ_{tu} = ultimate tensile strength; ε_{tu} = tensile strain capacity; E = modulus of elasticity; f_c' = compressive strength; and ε_{cp} = compressive strain capacity.

tensile strain capacity, the beam reaches peak load and the corresponding load point deflection is the deflection capacity. Therefore, it appears that the deflection capacity and tensile strain capacity can be linearly correlated from above cases with variation in all major material properties.

The overall results from the parametric study indeed show a linear relation between tensile strain capacity and deflection capacity, as revealed in Figure 3. Totally 20 cases were investigated in the parametric study, with the range of material parameters shown in Table 1. All linear curves lie in a narrow band regardless of actual material properties, which suggests that the beam deflection capacity is most sensitive to tensile strain capacity for a fixed geometry. For ease of quality control on site, master curve was constructed as a line with uniform thickness to cover all parametric case studies, as shown in Figure 3. The top edge of

the master curve is made to coincide with the upper boundary of all curves for conservativeness.

Additionally, another master curve correlating tensile strain capacity with curvature was constructed by parametric study in order to compare the proposed UM method with JCI method, in which ultimate bending moment and curvature was utilized to derive tensile strain capacity. The range of parametric values is the same as the aforementioned parametric study, as shown in Table 1. The dimension of specimen used in this parametric study is $100 \times 100 \times 400$ mm, with a span length of 300 mm (JCI-S-003-2005).

Similarly, this set of master curve also characterizes a linear relation within a very narrow band regardless of actual material properties. Since curvature may be linearly correlated with deflection using Equation (2), this master curve can be easily transformed into tensile strain capacity to deflection capacity relation, even though the slope should be different from Figure 3 due to different dimensions used in the two parametric studies. In the case when specimens with different dimensions have to be used for quality control, e.g. due to different fiber length, a different set of master curve should be constructed.

2.3 The use method of master curves

Based on the master curves obtained from parametric study, the deflection capacity from simple beam bending test can be easily converted to material tensile strain capacity. A set of equations has been developed to simplify the conversion procedure, as shown below, where Equations (3) and (2) can be used to calculate the average tensile strain capacity and its deviation, respectively.

$$\varepsilon_{tu}' = 0.46 \cdot \delta_u - 0.26 \tag{3}$$

$$PD = 0.46 \cdot SD + 0.17 \tag{4}$$

Figure 3. Tensile strain capacity – deflection capacity relation obtained from parametric study (20 cases) and simplified master curve (with uniform thickness).

Table 3. Mix proportion for different SHCCs.

	Cement	Sand	FA	W/C	SP	Fiber
PVA-ECC 1	1	0.8	1.2	0.27	0.013	0.02
PVA-ECC 2	1	1.1	2	0.26	0.014	0.02
PVA-ECC 3	1	1.4	2.8	0.26	0.016	0.02
PVA-ECC 4*	–	–	–	0.46	–	0.019
Ductal*	–	–	–	0.22	-	0.02

Note: FA = fly ash; W/C = water/cementitious materials (including cement and fly ash); SP = superplasticizer; *: Data from JCI round robin test (Kanakubo, 2006)

where ε'_{tu} is the predicted average tensile strain capacity (%); δ_u is the average deflection capacity obtained from FPBT (mm); PD is the predicted deviation for tensile strain capacity (%) and SD is the standard deviation of the deflection capacity (mm).

It should be noted that this equation can only apply to specimen with the same geometry and same loading conditions as that used by the authors (see Section 3). Should any of these geometry and/or loading conditions change, another set of master curves and corresponding conversion equations should be developed for that purpose. Once the proposed method (or its modified version) is standardized and widely accepted, there should be no need for change in geometry and loading conditions.

Similarly, another set of equation has also been developed to simply the conversion procedure for specimen tested according to JCI method. Equations (5) and (6) can be used to calculate the average tensile strain capacity and its deviation, respectively.

$$\varepsilon_{tu,c} = 0.0094 \cdot \phi_{u,c} - 0.26 \tag{5}$$

$$PD_c = 0.0094 \cdot SD_c + 0.16 \tag{6}$$

where $\varepsilon_{tu,c}$ is the predicted average tensile strain capacity (%); $\phi_{u,c}$ is the average curvature capacity obtained from FPBT (μ/mm); PD_c is the predicted deviation for tensile strain capacity (%) and SD_c is the standard deviation of the curvature capacity (μ/mm). The same limitation as mentioned above for Equations (3) and (4) also applies to Equations (5) and (6), except that the specimen geometry and loading profile should follow those in the JCI method.

3 EXPERIMENTAL PROGRAM

3.1 Materials, specimen preparation and testing

The mix proportion of SHCC materials investigated in this study is shown in Table 3, including PVA-ECC 1, 2 and 3. These SHCC materials feature high amount of fly ash in the mix proportion, with fly ash to cement

Table 4. Material tensile and compressive properties from experiment for different SHCCs.

	σ_{tc} (MPa)	σ_{tu} (MPa)	ε_{tu} (%)	f'_c (MPa)
PVA-ECC 1	4.6 ± 0.3	5.3 ± 0.6	2.1 ± 1.1	54.6 ± 6.5
	(7%)	(11%)	(52%)	(12%)
PVA-ECC 2	3.9 ± 0.5	4.6 ± 0.2	3.5 ± 0.3	46.0 ± 3.8
	(13%)	(4%)	(9%)	(8%)
PVA-ECC 3	4.0 ± 0.2	4.9 ± 0.1	3.7 ± 0.4	37.5 ± 1.7
	(5%)	(2%)	(11%)	(5%)
PVA-ECC 4*	3.7 ± 0.8	5.0 ± 0.5	2.7 ± 0.7	31.3 ± 0.8
	(21%)	(10%)	(26%)	(3%)
Ductal*	13.7 ± 0.9	15.3 ± 1.0	0.5 ± 0.3	198.0 ± 3.7
	(7%)	(7%)	(60%)	(2%)

Note: σ_{tc} = tensile first cracking strength; σ_{tu} = ultimate tensile strength; ε_{tu} = tensile strain capacity; f'_c = compressive strength; *: Experimental data from JCI round robin test (Kanakubo, 2006); Number in parenthesis is coefficient of variation (COV).

ratios of 1.2, 2.0, and 2.8, respectively. Additionally, PVA-ECC 4 and Ductal from JCI round robin test (Kanakubo 2006) are also listed in Table 3, which will be used for comparison between UM method and JCI method.

A Hobart mixer was used in this investigation, with a full capacity of 12 liters. All beam, uniaxial tensile and compressive specimens were cast from the same batch. At least 3 specimens were prepared for each test. After demolding, all specimens were cured in a sealed container with about 99% humidity under room temperature for 28 days before testing. Four point bending test was conducted with a MTS 810 machine. The beam specimen has a dimension of 356 mm long, 50 mm high, and 76 mm deep, all dimensions are at least 4 times that of the PVA fiber length (12 mm), which is the largest length scale among the ingredients of PVA-ECC. The loading span between two supports is 305 mm with a constant moment span length of 102 mm. The beam was tested under displacement control at a loading rate of 0.02 mm/second. The flexural stress was derived based on simple elastic beam theory and the beam deflection at the loading points was measured from machine displacement directly. The test setup is shown in Figure 4 (a) in comparison with the JCI method (Figure 4 (b)).

3.2 Experimental results

The material tensile and compressive properties for different SHCCs can be found in Table 4. With increasing amount of fly ash in PVA-ECC 1–3, the compressive strength continues to decrease as expected, yet PVA-ECC 3 still has a compressive strength of about 38 MPa. For all SHCCs the typical coefficient of variations (COV) of first cracking strength and ultimate

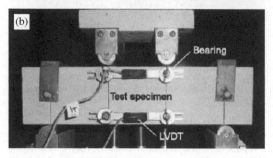

Figure 4. Comparison of test setup for the (a) UM method and (b) JCI method.

tensile strength are less than 15%, similar to that of compressive strength. Conversely, the COV of tensile strain capacity are in the range of 26%–60% except for PVA-ECC 2 and 3, where the robustness of tensile ductility increased (in the form of reduced COV) due to the usage of high volume fly ash (Wang 2005). This general trend – relatively low COV for tensile strength and high COV for tensile strain capacity can also be found in Kanakubo (2006). This further confirmed the rationale of quality control for the tensile strain capacity instead of tensile strength.

PVA-ECC 1–3 show typical deflection hardening behavior under FPBT. More and more saturated microcrack is revealed from PVA-ECC 1 to 3, associated with gradual increase of deflection capacity (Table 5). The modulus of rupture for PVA-ECC 1–3 ranges from 10–12 MPa, about 2.4–3.0 times that of their first cracking strength. This is consistent with the finding of Maalej and Li (1994) that this ratio should be about 2.7 for elastic-perfectly plastic material (for tensile portion of beam), such as the PVA-ECCs investigated in this study.

4 VALIDATION AND VERIFICATION OF THE PROPOSED METHOD

To validate the proposed inverse method, the deflection capacity obtained from FPBT is converted to tensile strain using Equations (3) and (4) (derived for the same beam size as used in the FPBT experiments) and

Table 5. Comparison between predicted tensile strain capacity from FPBT and tensile strain capacity from UTT.

	ε_{tu} from UTT (%)	Deflection capacity from FPBT (mm)	Predicted ε_{tu} (%)	Difference between prediction and test result (%)
PVA-ECC 1	2.1 ± 1.1	5.9 ± 1.6	2.4 ± 0.9	14
PVA-ECC 2	3.5 ± 0.3	7.2 ± 1.3	3.1 ± 0.8	-11
PVA-ECC 3	3.7 ± 0.4	9.4 ± 0.9	4.1 ± 0.6	11

(Note: ε_{tu} = tensile strain capacity.)

then compared with tensile strain capacity obtained directly from uniaxial tensile test for PVA-ECC 1–3. As revealed in Table 5 and Figure 5, the tensile strain capacity derived from FPBT predicts the uniaxial tensile test results with reasonable accuracy, with a difference of less than 15%. This agreement demonstrates the validity of the proposed inverse method.

To further verify the proposed UM method, comparison between UM method and JCI method was conducted based on JCI round robin test data (Kanakubo 2006). As mentioned previously, bending test results from JCI round robin test are presented in the form of moment –curvature relation. To facilitate the comparison, the curvature capacity can be converted to tensile strain capacity using Equations (5) and (6) in UM method. Within the JCI method, the tensile strain capacity is obtained by solving following equations (JCI-S-003-2005):

$$\varepsilon_{tu,b} = \phi_u \cdot D \cdot (1 - x_{nl}) \tag{7}$$

$$x_{nl}^3 + 3x_{nl}^2 - 12m^* = 0 \tag{8}$$

$$m^* = \frac{M_{max}}{E \cdot \phi_u \cdot B \cdot D^3} \tag{9}$$

where $\varepsilon_{tu,b}$ is the predicted tensile strain capacity (%); ϕ_u is the curvature capacity (1/mm), which can be calculated from two LVDTs measurements (Fig. 4(b)); D is depth of the test specimen ($=100$ mm); x_{nl} is the ratio of the distance from compressive edge (extreme compression fiber) to neutral axis over depth of test specimen, which needs to be solved from Equation (8); M_{max} is maximum moment ($N \cdot mm$); E is the static modulus of elasticity (N/mm^2); B is the width of test specimen (100 mm). For more details, readers are referred to the Appendix to JCI-S-003-2005.

As shown in Table 6 and Figure 6, predictions based on both the UM method and the JCI method reveal comparable results with those from uniaxial tensile tests. Furthermore, the UM method shows smaller discrepancy with the uniaxial tensile test result (Table 6)

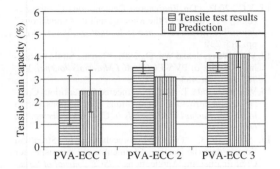

Figure 5. Comparison of tensile strain capacity from UTT test with prediction from proposed UM method for different PVA-ECCs.

Table 6. Comparison between uniaxial tensile test results with predictions based on the JCI method and the UM method

	ε_{tu} from UTT (%)	Curvature capacity from FPBT (μ/mm)	Predicted ε_{tu} (JCI) (method) (%)	Predicted ε_{tu} (UM) (method) (%)
PVA -ECC 4	2.7 ± 0.7	349.2 ± 96.3	3.1 ± 0.9 (15)	3.0 ± 1.1 (11)
Ductal	0.5 ± 0.3	85.0^*	0.6^* (20)	0.5^* (0)

(Note: *: Only two bending specimens were reported. The number in parenthesis is the difference (in percentage) between the predictions and test results from uniaxial tensile test.)

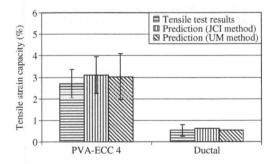

Figure 6. Comparison of tensile strain capacity from UTT with predictions from JCI method and proposed UM method for different SHCCs. (Experimental data for both UTT and FPBT are from JCI round robin test and only two FPBT specimens were reported for Ductal.)

based on limited data. The consistency between the UM method and the JCI method and verification by independent JCI round robin test data further demonstrate the validity of the proposed UM method.

The advantage of the UM method over the JCI method lies in its simplicity, both in experiment and data interpretation phases. In the experiment phase, the UM method only requires machine displacement to be measured. This is not the case for the JCI method, where complicated setup such as LVDTs is needed to measure curvature, as revealed in Figure 4 (a) and (b). In the data interpretation phase, the UM method only needs a simple master curve or linear equation to convert deflection capacity directly into tensile strain capacity, while JCI method requires relatively complicated procedures (solving cubic equation) to obtain tensile strain capacity. Considering the large amount of specimens needed to be tested during construction, the UM method seems to be more suitable for quality control purpose due to its simplicity, efficiency and reasonable accuracy.

5 CONCLUSIONS

To facilitate the quality control of the strain hardening cementitious composites on site, a simplified inverse method is proposed to covert the deflection capacity from simple beam bending test to tensile strain capacity through linear transformation. The linear transformation (in the form of master curves) is derived from parametric study with a wide range of parametric values of material tensile and compressive properties based on a theoretical model. This proposed method has been experimentally validated with uniaxial tensile test results with reasonable agreement. In addition, this proposed method compares favorably with the JCI method in accuracy, but without the associated complexity.

The following specific conclusions can be drawn from this study:

1. A simple inverse method has been successfully developed to derive tensile strain capacity of SHCC from beam bending deflection capacity by using a master curve. This method is expected to greatly ease the on-site quality control for SHCC in terms of much simpler experiment setup requirement (compared with both UTT and the JCI inverse method) and data interpretation procedure (compared with the JCI method), yet with reasonable accuracy (within 15%);

2. The master curve features simple linear transformation from deflection capacity to tensile strain capacity. The master curve decouples the dependence of tensile strain capacity on the moment capacity in contrast with the JCI method where tensile strain capacity is dependent on both curvature capacity and moment capacity. Therefore, this method allows simple linear equations (Equation (3) and (4)) to be used for easy data interpretation;

3. A linear relation between the deflection capacity and the tensile strain capacity is observed based on parametric studies. All linear curves relating

tensile strain capacity and deflection capacity lie in a narrow band regardless of actual material properties. This suggests that beam deflection capacity is most sensitive to tensile strain capacity for a given FPBT geometric dimensions, and much less sensitive to other properties such as compressive strength, Young's Modulus, etc.

It should be noted that the following assumptions are made when the proposed UM method is used: (a) The tested material is truly a strain hardening type; (b) The major target for quality control for this material is tensile ductility; and (c) For this method to be most effective, a standardized beam with fixed geometric dimensions should be agreed upon by the user community.

REFERENCES

JCI-S-003-2005, 2005, Method of test for bending moment–curvature curve of fiber reinforced cementitious composites, Japan Concrete Institute Standard, 7p.

Kanakubo, T., 2006, "Tensile Characteristics Evaluation Method for DFRCC", *Journal of Advanced Concrete Technology*, Vol. 4, No. 1, pp3–17.

Kanda, T., Kanakubo, T., Nagai, S. and Maruta, M., 2006, "Technical Consideration in Producing ECC Pre-Cast Structural Element," *Proceedings of Int'l RILEM workshop on HPFRCC in structural applications*, Published by RILEM SARL, pp. 229–242.

Kanda, T., Saito, T., Sakata, N. and Hiraishi, M., 2002, "Fundamental Properties of Direct Sprayed ECC," *Proceedings of the JCI International Workshop on Ductile Fiber Reinforced Cementitious Composites (DFRCC) – Application and Evaluation (DRFCC-2002)*, Takayama, Japan, pp. 133–142.

Li, V. C., 2005, "Engineered Cementitious Composites", *Proceedings of ConMat'05*, Vancouver, Canada, August 22–24, CD-documents/1-05/SS-GF-01_FP.pdf.

Li, V. C., 2004, "Strategies for High Performance Fiber Reinforced Cementitious Composites Development", *Proceedings of International Workshop on Advances in Fiber Reinforced Concrete*, Bergamo, Italy, pp. 93–98.

Li, V.C., 2003, "On Engineered Cementitious Composites (ECC) – A Review of the Material and its Applications," *Journal of Advanced Concrete Technology*, Vol. 1, No. 3, pp. 215–230.

Li, V. C., Wang, S. and Wu, C., 2001, "Tensile Strain hardening Behavior of PVA-ECC", *ACI Materials Journal*, Vol. 98, No. 6, pp. 483–492.

Li, V.C. and Kanda, T. 1998, "Engineered Cementitious Composites for Structural Applications", Innovations Forum in ASCE, *Journal of Materials in Civil Engineering*, pp. 66–69.

Li, V.C., 1993, "From Micromechanics to Structural Engineering–the Design of Cementitious Composites for Civil Engineering Applications", *JSCE Journal of Structural mechan ics and Earthquake Engineering*, Vol. 10, No. 2, pp. 37–48.

Maalej, M., and Li, V.C., 1994, "Flexural/Tensile Strength Ratio in Engineered Cementitious Composites", *ASCE Journal of Materials in Civil Engineering*, Vol. 6, No. 4, pp. 513–528.

Naaman, A.E. and Reinhardt, H.W., 1996, "Characterization of High Performance Fiber Reinforced Cement Composites (HPFRCC)", in *High Performance Fiber Reinforced Cementitious Composites*, RILEM Proceedings 31, Eds. A. E. Naaman and H. W. Reinhardt, pp. 1–23.

Ostergaard, L., Walter, R., and Olesen, J. F., 2005, "Method for Determination of Tensile Properties of Engineered Cementitious Composites (ECC)", *Proceedings of ConMat'05*, Vancouver, Canada.

Stang, H. and Li, V.C., 2004, "Classification of Fiber Reinforced Cementitious Materials for Structural Applications", *Proceedings of BEFIB*, Varenna, Lake Como, Italy, pp 197–218.

Wang, S. and Li, V.C., 2004, "Tailoring of Pre-existing Flaws in ECC Matrix for Saturated Strain Hardening," *Proceedings of FRAMCOS-5*, Vail, Colorado, USA, pp. 1005–1012.

Wang, S., 2005, *Micromechanics Based Matrix Design for Engineered Cementitious Composites*, PhD Thesis, Univ. of Michigan, Ann Arbor, MI, 222p.

Fracture Mechanics of Concrete and Concrete Structures – High-Performance Concrete, Brick-Masonry and Environmental Aspects – Carpinteri, et al. (eds)
© 2007 Taylor & Francis Group, London, ISBN 978-0-415-44617-4

Fracture stability and micromechanics of Strain Hardening Cementitious Composites

L.G. Sorelli & F.-J. Ulm
Massachusetts Institute of Technology, Cambridge, USA

F. Toutlemonde
Laboratoire des Ponts et Chaussées, Paris, France

ABSTRACT: Innovative cementitious composites endowed with strain hardening behavior, namely UHPFRC, have been successfully employed in several industrial applications, such as bridge engineering and nuclear waste container prototypes, where the crack impact on the durability is of main concern. However, the material ductility of such composites exhibits a complex dependence on the fiber distribution that has been an important issue in the design recommendations (AFGC, 2002). In this paper, we develop an approximate energy-balance approach to link the overall damage to non-uniform fiber distribution. The thermodynamics principles and the micromechanics based secant stiffness formulation allowed us to derive the non linear constitutive stress-strain relationship of the material from the fiber distribution and properties of the composite phases. Finally, the model implication on the stability of the post-cracking behavior is discussed together with an example of industrial application for nuclear waste container.

1 INTRODUCTION

In the last decade, several structural applications of Ultra High Performance Fiber Reinforced Composite (UHPFRC) have been emerging in bridge engineering, rehabilitation and strengthening techniques of damaged structures, nuclear waste container, etc. The driving force of these new materials is the optimized microstructure that has been tailored to achieve high durability, ultra high compressive strength, and tensile pseudo-hardening behavior (Chanvillard & Rigaud, 2003). However, design codes require a reliable assessment of the post-cracking ductility and fracture energy, especially in relation to the microstructure variability. For example, the first design recommendations for UHPFRC (AFGC, 2002) reduces the characteristic tensile law of bended standard specimens by a factor ($K = 1.25$–1.75) to account for the effective fiber distribution within the structure. Moreover, structural applications have been limited to thin elements where the fiber distribution is favored by wall effects (e.g., the fiber efficiency, defined as the average fiber projection in the direction of the external load, is $2/\pi$ and $1/2$ for 2D and 3D random distribution, respectively).

The importance of the fiber distribution on the mechanical properties has been verified experimentally (Bernier & Behloul, 1998; Bayard & Plé, 2003; Chanvillard & Rigaud, 2003). For example, uniaxial tests on UHPFRC blocks reinforced with five different fiber orientations showed the first crack occurs almost always perpendicular to the loading axis regardless the fiber distribution (Bayard & Plé, 2003). Remarkably, the fracture energy reduces by a factor of about four when the fiber alignment changes from 0° to 90°. As for the flexural behavior, the post-cracking response of beams strongly depends on the fiber orientation up to an extent that, when the fibers are unfavorably aligned (e.g., parallel to the major crack direction), the material does not exhibit strain hardening behavior (Bernier & Behloul, 1998).

At observation scale below macroscopic testing, the cracks initiate on a very small length scale within the cementitious matrix, and they evolve into fiber-matrix interface failure. Remarkably, a weak fiber-matrix interface may favor high toughness by promoting fiber-matrix interface failure (Marshall & Cox, 1985; Leung, 1996; Lin & Li, 1997). Moreover, other micromechanics phenomena may be at stake, such as matrix crumbling at the fiber exit point (Zhang & Li, 2002), local fiber bending and yielding (Leung & Chi, 1995), and fiber rupture.

Chuang *et al.* (2001) showed that, in UHPFRC materials, the fiber aspect ratio (i.e., the length-to-diameter ratio of the fiber) is optimally designed to induce simultaneous fiber debonding and matrix cracking. Contrarily, in normal FRC, the fiber length

is often insufficient to satisfactory prevent early interface failure and the average fiber distance is too large
to lead to multiple cracking (i.e., the fiber is over
designed with respect to the matrix strength).

Energy based approaches have been a powerful
analytical tool to study fiber reinforced composites.
Budiansky et al. (1986) applied a variational energy
approach to study fiber-matrix interface failure showing that, in case of initially bonded fibers, a fairly
small interface debonding toughness suffices to inhibit
crack localization. Energy approaches have been also
employed to identify the necessary conditions for the
strain hardening: (i) the complementary energy has
to exceed the crack tip resistant energy at the steady
state crack propagation (Marshall & Cox, 1985); (ii)
the first cracking strength should be lower than the
maximum composite fiber strength (Leung, 1996). By
comparing the energy to form a new crack with the
energy to open the first formed crack, Tjiptobroto and
Hansen (1993) proposed an analytical expression for
the strain at the end of multiple cracking.

Recently, continuum micromechanics theories have
allowed disclosing the link between the microstructure and the overall properties. Karihaloo et al. (1996)
accounted for the interaction of microcrack distribution by solving the elastic problem of a doubly periodic
array of elastically bridged cracks. However, experimental fracture tests indicate the crack distribution is
random in nature. Dormieux et al. (2006) developed a
fracture-micromechanics approach to model interacting random microcracks in presence of internal fluid
pressure. Notably, they found that crack interaction
may be a reason of strain softening behavior and that
the energetic stability depends on the initial critical
crack size and fracture energy.

In this work, we pursue an energy based approach
that derives the constitutive macro stress-strain relationship of UHPFRC materials from their critical
microstructure features (i.e., the fiber distribution and
the distribution of interacting cracks). However, at this
first approach stage, we will focus only on the strain
hardening phase, while disregarding the subsequent
crack localization.

2 A FRACTURE-MICROMECHANICS APPROACH

Every structural system can be characterized by a
length scale L defining the structural dimension. At
a level below, we consider a material point of elementary volume, which includes sufficient matter to be
representative and it is characterized by a macroscopic
length scale D. Below this scale, the matter is heterogeneous and this material system is characterized
by the scale d of its components at the microscopic
level (for instance, the grain diameter or the aggregate
size). A continuum description of a heterogeneous

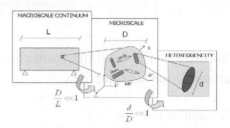

Figure 1. Multi-scale approach for UHPFRC materials.

material requires that characteristic length (D) has to
be much larger than dimension (d) of the components
and much smaller than the structural dimensions (L).
Figure 1 shows the qualitative multi-scale approach
adopted in this work for describing a UHPFRC composite materials made of three phases, namely the
discontinuous steel fibers, penny-shaped cracks, and
the cement-based matrix.

In the following, we assume a micros-scale system made of an elastic-brittle matrix with narrow (or
penny-shaped) cracks, while the fiber is indirectly only
considered in term of the energetic contribution to
delay the crack propagation. However, the fiber energy
is linked to the fiber-matrix interface micromechanics
and the fiber orientation distribution. This approximation disregards the effect of fiber on the elastic
properties of the bulk material, which is expected to
be limited for the considered UHPFRC fiber content
(Chanvillard & Rigaud, 2003).

2.1 Thermodynamics of an elastic cracked medium

The simplest crack system of a single planar crack
embedded in an unbounded linear elastic material is
analyzed within the theoretical framework of classical
thermodynamics, Dormieux et al. (2006). The incremental externally supplied work δW^{ext} provided to the
solid Ω is stored in form of elastic energy density
$\Omega_0 d\psi$ in the solid (where Ω_0 denotes the initial volume of the considered structure and ψ the elastic free
energy density of the solid part). Our starting point is
the Clausius-Duhem inequality which states that the
externally supplied work to the solid (δW^{ext}), which is
not stored as free energy $\Omega_0 d\psi$ (or Helmholtz energy)
in the system, is dissipated into heat.

$$dD = \delta W^{ext} - \Omega_0 d\psi \geq 0 \tag{1}$$

The externally supplied work δW^{ext} is related to the
surface forces T acting on the incremental displace-
ment $d\xi$ over the external surface $\partial\Omega$ (i.e., body force
are neglected) as follows

$$\delta W^{ext} = \int_{\partial\Omega} T \cdot d\xi \, dS \tag{2}$$

1404

We now express the free energy elastic of the solid matrix $\Psi = \Psi(E, \ell)$ as a macro-potential function of two variables, such as the elastic strain E (observable variable) and the crack area ℓ (state variable). In the case of reversible evolutions $dD = 0$, ℓ is constant. By contrast, when the crack propagates, the free energy changes not only with the incremental strain dE, but as well due to the increase of fracture surface ℓ. Introducing (2) into (1), we obtain

$$dD = \left(\int_{\partial\Omega} T \cdot d\xi \, dS - \Omega_0 \frac{\partial\psi}{\partial E} dE \right) - \Omega_0 \frac{\partial\psi}{\partial\ell} d\ell \geq 0 \quad (3)$$

Fracture propagation dissipates energy through creation of additional crack surface $d\ell$ and the dissipation rate is evaluated from Eq. (3).

$$dD = -\Omega_0 \left(\frac{\partial\Psi(E,\ell)}{\partial\ell} \right)_{dE=0} d\ell = G_A(E,\ell) \, d\ell \geq 0 \quad (4)$$

Remarkably, Eq.(4) indicates that the thermodynamic driving force of the crack propagation is the energy release $G_A(E, \ell)$. For example, in the case of a Griffith's crack a fast run crack propagates, in an unstable manner, when the energy release rate $G_A(\ell)$ reaches a critical threshold (R_0), which is called *fracture energy* and is often considered a material property. Owing the linearity of the solid behavior, the free energy is expected to be a quadratic function of E and Σ:

$$\Psi(E,\ell) = \frac{1}{2} E : C^{\text{hom}}(\ell) : E \quad (5)$$

Introducing the above equation in (3) allows us to identify the state equations of the cracked medium:

$$\Sigma = \frac{\partial\psi}{\partial E} = C^{\text{hom}}(\ell) : E \quad (6)$$

Now, Eq. (4) takes the form

$$G_A = \frac{\partial\psi(E,\ell)}{\partial\ell} = -\frac{1}{2} E : \frac{\partial C^{\text{hom}}(\ell)}{\partial\ell} : E \geq 0 \quad (7)$$

The energy release G_A can be estimated from the homogenized secant stiffness in function of the geometric parameter ℓ.

The approach of Dormiuex et al. (2006) can be extended to account the energy contribution of the fiber bridged zone behind the crack tip, we can employ the J-Rice integral, which is an alternative and convenient mathematically form of the energy conservation principle. Thus, the crack energy release can be computed as follows

$$J_B = \oint (\psi \, n_1 + \frac{\partial[\![\xi]\!]}{\partial x_1}) \, dS = \int_\lambda^\ell \sigma([\![\xi]\!])(\partial[\![\xi]\!]/\partial x) \, dx = \\ = -\int_0^{[\![\xi]\!]} \sigma([\![\xi]\!]) \, d[\![\xi]\!] = G_B \quad (8)$$

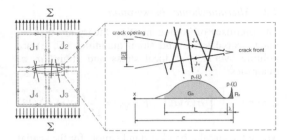

Figure 2. J-Rice integral contours along bridged crack surfaces with enlarged view (right) of the stress-separation function $p_u([\![\xi]\!])$.

where $[\![\xi]\!]$ is the crack opening displacement and σ ($[\![\xi]\!]$) is the fiber bridging stress across the crack surfaces. We emphasize that the fiber energy can be seen as an energy release (G_B) or a toughness term (R_B) depending on the observer's frame. In this case, an observer on the crack enclave will perceive the fiber as applied external forces. Figure 2 qualitatively shows the J-Rice integral contours around the fiber bridged crack.

The presence of two system of loads (i.e., external loads and fibers bridging force) involves additional energy cross-terms, which come from the energy release of the strain energy stored through the indirect work. Although the energy releases are not additive, they are square root additive (Bažant & Planas, 1998), because of the Irwin relationship $G = K^2/E$ between the energy release G and the stress intensity factor K. Hence, the energy balance equation (4) can be written as

$$dD = R_0 = G_A + G_B + 2\sqrt{G_A}\sqrt{G_B} \geq 0 \quad (9)$$

We can define the loading function

$$f(E,\ell) = \Sigma_i G_i(E,\ell) - R_0 \quad (10)$$

which can be used to mathematically describe the loading/unloading conditions:

$$f(\ell) \leq 0, \quad f(\ell) \, d\ell = 0, \quad d\ell \geq 0 \quad (11)$$

If, due to incremental fracture propagation $\ell \to \ell + d\ell$, the energy release decreases from $\Sigma_i G_i(E, \ell) = R(E, \ell)$ to $\Sigma_i G_i(E, \ell + d\ell) < R_0$, the crack propagation will stop. This corresponds to stable fracture propagation if

$$\frac{\partial f}{\partial\ell} = \Sigma_i \left(\frac{\partial G_i(E,\ell)}{\partial\ell} \right)_{dE=0} < 0 \quad (12)$$

or otherwise unstable. In the following, we determine the terms of energy equation (9) from micromechanics theory.

2.2 Micromechanics background

The premises of homogenization theory are that the macroscopic stress and strain quantities (Σ, E) are volume averages of local stress and strain (σ, ε) quantities:

$$\Sigma = \langle \sigma \rangle = \frac{1}{\Omega} \int_\Omega \sigma(x) d\Omega; \quad E = \langle \varepsilon \rangle = \frac{1}{\Omega} \int_\Omega \varepsilon(x) d\Omega \quad (13)$$

where the angular bracket stand simply for the spatial average phase average operator such as

$$\langle b \rangle = \frac{1}{\Omega} \int b \, d\Omega \quad (14)$$

Uniform internal strain is possible only if the material is homogeneous. For uniform boundary condition, the applied strain E on the surface $\partial\Omega$, subjected to a displacement ξ^d, reads

$$\xi^d = E \cdot z \quad \vee z \in \partial\Omega \quad (15)$$

where z = the local coordinate system. Some simplifications in the geometrical description are usually necessary to deal with complex microstructure interactions. First, we assume that each rth phase is linear elastic with isotropic stiffness tensor C_r, and the following constitutive relation holds for each phase.

$$\sigma(z_r) = C_r : \varepsilon(z_r) \quad (16)$$

The strain fields within the rth constituent is approximated by their phase average, and it is linked to the macroscopic strain by a linear strain localization tensors A_r.

$$\overline{\varepsilon}_r = \langle \varepsilon \rangle_r = A_r : \langle \varepsilon \rangle = A_r : E \quad (17)$$

The overall stress Σ can be expressed in terms of the average strain of each rth phase as follows

$$\Sigma = C^{hom} : E = \sum_{r=0}^{N} f_r \langle \sigma_r \rangle_{(r)} = \sum_{r=0}^{N} f_r A_r : E \quad (18)$$

which, after simple mathematical manipulations, leads to the composite stiffness.

$$C^{hom} = \sum_{r=0}^{N} f_r C_r : A_r \quad (19)$$

Note that the summation of the strain localizer or stress tensor over all the phases gives the identity matrix (i.e., compatibility condition):

$$\sum_{r=0}^{N} f_r A_r = I \quad (20)$$

Inserting the Eq. (20) into (19), we obtain the following simplified relation:

$$C^{hom} = C_0 + \sum_{r=1}^{N} f_r (C_r - C_0) : A_r \quad (21)$$

We remark that, although several homogenization models have these previous steps in common, they differ for the localizer tensor A_r that represents the effect of the morphology microstructure. For example, the dilute scheme assumes a single inclusion embedded in an unbounded medium with elastic modulo C_0 of the matrix, i.e., the inclusion interactions are neglected. The strain localization tensor reads

$$A_r^{Esh} = [I + S_r C_0^{-1}(C_r - C_0)]^{-1} \quad (22)$$

where S_r is the Eshelby tensor analytically derived from the problem of a constant strain within an ellipsoidal inclusion body (Eshelby, 1957).

To account for interactions between the inclusions, we will use the Mori-Tanaka (1973) homogenization scheme that considers a single inclusion embedded in a reference material with the average strain of the matrix. The Mori-Tanaka estimate is relevant for morphology of microstructure where the cavity can be regarded as inclusions embedded in a continuum matrix, and its strain concentration tensor reads:

$$A_r^{MT} = [I + S_r C_0^{-1}(C_r - C_0)]^{-1} : \langle I + S_r C_0^{-1}(C_r - C_0) \rangle^{-1} \quad (23)$$

We statistically describe a network of microcracks by the joint probability density distribution $f(a, \vartheta, \varphi)$ of the crack sizes and orientations (where ϑ, φ are two Euler angles describing the crack orientation). Its integral on a sphere of unit radius gives the volume of crack per unit volume N such as

$$N = \frac{1}{4\pi} \int_{a=a_{min}}^{a_{max}} \int_{\varphi=0}^{2\pi} \int_{\vartheta=0}^{\pi} f(a, \vartheta, \varphi) \sin \vartheta \, da \, d\vartheta \, d\varphi \quad (24)$$

Assuming that the crack size distribution is independent of the crack orientation, the density distribution simplifies as $f(a, \vartheta, \varphi) = f_a(a) f_r(\vartheta, \varphi)$ and it follows

$$a = \int_{a=a_{min}}^{a_{max}} f_o(a) da$$

$$N = \frac{1}{4\pi} \int_{\varphi=0}^{2\pi} \int_{\vartheta=0}^{\pi} f(\vartheta, \varphi) \sin \vartheta \, d\vartheta \, d\varphi \quad (25)$$

We define a crack family composed by penny-shaped cracks with the same orientation (n) and crack radius (surface area $\ell = \pi a^2$, crack density N). Moreover, for each crack family, it is convenient to introduce the micromechanics damage parameter $\epsilon = Na^3$, defined by Budiansky et al. (1976). Hence, the crack volume fraction (f_c) of N penny-shaped crack can be expressed in terms of the micromechanical damage density parameter as follows

$$f_c = N \frac{4}{3} \pi a^3 X = \frac{4}{3} \pi \epsilon X \qquad (26)$$

where the crack aspect ratio $X = c/a$ has been used in the ellipsoid volume formula. For example, in the case of a system of microcracks and a solid matrix, relation (26) into (21) yields

$$C^{\text{hom}} = C_0 : (I - f_c A_c) = C_0 : (I - \frac{4}{3} \pi \epsilon X A_c) =$$
$$= C_0 : (I - \frac{4}{3} \pi \epsilon T(\underline{n})) \qquad (27)$$

where the tensor T has the property to have a finite limit when $X \to 0$ such as

$$T = \lim_{x \to 0} X (I - S_{ps}(X, \underline{n}))^{-1} \qquad (28)$$

3 NON LINEAR FRACTURE MECHANICS

In a quasi-static fracture process (without inertia forces), the energy equality $\Sigma G = \Sigma R$ is satisfied whenever the crack is growing. Following the "R-curve" approach (Bažant and Planas, 1998), we analyze a tensile test under mode-I condition by solving the system of the energy balance equation and the (macro) stress-strain relations based on the secant stiffness:

$$f(\epsilon, E) = \sum_i G_i(\epsilon, E) - R(\epsilon) = 0$$
$$\Sigma = C(\epsilon) : E \qquad (29)$$

For numerical implementation it is convenient to cast the Eq. (29) in a rate form and to derive the tangent stiffness between the macro-stress (Σ) and macro-strain (E). When the crack grows ($d\epsilon \neq 0$), the condition $df = 0$ can be re-written as

$$df(\epsilon, E) = \frac{\partial f}{\partial \epsilon} d\epsilon + \frac{\partial f}{\partial E} dE = 0 \to \frac{d\epsilon}{dE} = -\frac{\partial f / \partial E}{\partial f / \partial \epsilon} \qquad (30)$$

Then, the tangent material stiffness at the beginning of each loading increment $d\Sigma$ (or strain increment dE) can be formulated in terms of known quantities in an explicit manner such as

$$\frac{d\Sigma}{dE} = C(\epsilon) + \frac{\partial C(\epsilon)}{\partial \epsilon} \frac{\partial \epsilon}{\partial E} = C(\epsilon) - \frac{\partial C(\epsilon)}{\partial \epsilon} \frac{\partial f / \partial E}{\partial f / \partial \epsilon} \qquad (31)$$

Before solving Eq. (29), we need to define the fracture energy R_0 and the fiber energy G_B in terms of the microstructure crack density parameter ϵ, which is the loading history parameter. Again disregarding fracture

energy interaction, the total dissipation is the sum of the contributions of all the N cracks which are assumed to propagate of the same amount $d\ell$. For simplicity, we assume the crack grows simultaneously with constant number N in order that ℓ and ϵ are equivalent state variables (in accordance with Dormieux et al., 2006):

$$\ell = \pi \left(\frac{\epsilon}{N} \right)^{2/3} ; \quad d\ell = \frac{2\pi}{3N^{2/3}} \epsilon^{-1/3} d\epsilon$$

The elementary contribution of a single crack to the total dissipation is given by

$$R_0 \, d\ell = N \, Gf \, d\ell = R_{0\epsilon} d\epsilon \qquad (32)$$

where

$$R_{0\epsilon} = \frac{2\pi}{3} G_f \left(\frac{N}{\epsilon} \right)^{1/3} = \frac{2\pi}{3} \frac{G_f}{a} \qquad (33)$$

Hence, after this change of variable, R_0 is not anymore an independent variable but is dependent on the crack size, and hence on the loading history. We are now left to find the fiber energy term G_B of equation (29). We will address this point in the following section adopting a statistical fiber distribution.

4 FIBER DEBONDING ENERGY

This section reviews the analytical model for fiber-matrix micromechanics proposed by Lin and Li (1997), which extends the previous work of Marshall & Cox (1985). The constitutive relation between the fiber pull-out force (F) and the crack opening displacement ($[[\xi]]$) holds the key of the underlying role of the fiber micro-mechanisms and several authors (Marshall & Cox, 1985; Li & Leung, 1992; Lin & Li, 1997) have proposed the following form:

$$F_d \left([[\xi]] \right) = k [[\xi]]^{1/2} \quad if \ [[\xi]] \leq [[\xi]]^*$$
$$F_s \left([[\xi]] \right) = k \frac{[[\xi]]^{cr} - [[\xi]]}{[[\xi]]^{cr} - [[\xi]]^*} \quad if \ [[\xi]]^* \leq [[\xi]] \leq [[\xi]]^{cr} \qquad (34)$$

where the load coefficient and the crack opening thresholds are defined as

$$k = \frac{1}{2} \pi \, \phi_f \sqrt{(1+\eta) \tau_u E_f \phi_f} ; \quad \eta = \frac{E_f V_f}{E_m V_m}$$
$$[[\xi]]^* = \frac{4(1+\eta) \tau_u L_f^2}{E_f \phi_f} ; \quad [[\xi]]^{cr} = L_f / 2 \qquad (35)$$

where E_f and E_m are the elastic modulus of fiber and matrix, respectively; L_f = fiber length; ϕ_f = fiber

Table 1. Fiber-matrix interface model parameters.

E_m [GPa]	E_f [GPa]	V_f [%]	L_f [mm]	ϕ_f [mm]
50	210	2.7	13	0.2

diameter; and τ_u = fiber-matrix frictional debonding. The model parameters employed in this work are summarized in Table 1.

Experimental pull-out tests of steel micro-fibers showed that the effect of matrix edge pulley and the fiber local moment tend approximately to compensate each other over the possible fiber inclinations (Shah and Ouyang, 1991). For simplicity, the fiber orientation effect on the pull-out response of a single fiber is here neglected.

Although it is often assumed that the fibers are either unidirectional or randomly dispersed in the cementitious matrix, the actual fiber orientation will never be unique and the state of anisotropy will depend on the fiber geometry, fiber content and fluid flow properties during the casting phase. In this work we assume a statistical π-periodic Gaussian like probability density function, which is axisymmetric in spherical coordinates:

$$W(\theta) = \frac{w_k \cosh(k \cos(\theta - \theta_0))}{\int_{\varphi=0}^{2\pi} \int_{\theta=0}^{\pi} w_k \cosh(k \cos(\theta - \theta_0)) |\sin(\theta)| \, d\theta d\varphi} \quad (36)$$

where $\dfrac{1}{2\pi} \int_{\theta=0}^{\pi} \int_{\varphi=0}^{2\pi} W(\theta,\varphi)|\sin\theta| \, d\theta \, d\varphi = 1$

where $w_k = 2/k \cosh(k)$. Although there is no a physical reason behind the chosen distribution function, one may assume that the axis of symmetry coincide with the average flow direction during the casting phase. The parameters k and θ_0 control, respectively, the degree of uniformity and the angle between the axial symmetric distribution and the crack plane normal direction. Figure 3 shows the distribution function varying the parameter values. Note that the distribution function is uniform when $k = 0$.

Assuming that the probability distribution $p(z)$ of the distance (z) between the fiber centroid and the crack plane is uncorrelated from the fiber orientation and uniformly distributed (i.e., $p(z) = 2/L_f$), the composite fiber stress is evaluated by averaging the pullout force on the crack plane (Lin and Li, 1997), such as

if $[\![\xi]\!] < [\![\xi]\!]^*$

$$\sigma_B([\![\xi]\!]) = \frac{4V_f}{\pi\phi_f} \int_{\vartheta=0}^{\pi/2} \int_{z=0}^{[\![\xi]\!]^0} F_d([\![\xi]\!]) p(z) p(\vartheta) \, dz \, d\vartheta +$$

$$\frac{4V_f}{\pi\phi_f} \int_{\vartheta=0}^{\pi/2} \int_{z=[\![\xi]\!]^0}^{(L_f/2)\cos\vartheta} F_s([\![\xi]\!]) p(z) p(\vartheta) \, dz \, d\vartheta \quad (37)$$

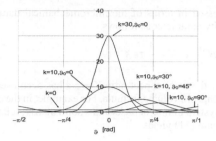

Figure 3. Fiber probability density distribution function for different values of the 2 parameters (k, ϑ_0).

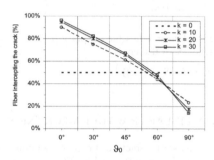

Figure 4. Percentage of fibers intercepting the crack plane varying the parameter (ϑ_0 and k) with respect the loading axis.

if $[\![\xi]\!] > [\![\xi]\!]^*$

$$\sigma_B([\![\xi]\!]) = \frac{4V_f}{\pi\phi_f} \int_{\vartheta=0}^{\pi/2} \int_{z=0}^{(L_f/2)\cos\vartheta} F_s([\![\xi]\!]) p(z) p(\vartheta) \, dz \, d\vartheta \quad (38)$$

where the integration limits are properly defined to account *only* for the fiber intercepting the cracks either for the frictional debonding phase or the slipping phase. In this simplified approach, the fiber percentage intercepting the crack is the key parameter. Figure 4 shows the fiber percentage intercepting the crack plane in function of the fiber distribution parameters (ϑ_0 and k) as estimated from Eq. (36). It is noticed that a non uniform fiber distribution (i.e., high value of the parameter k) is favorable if oriented towards the crack plane. In other words, the higher the value k, the faster the fiber number decreases with the angle ϑ_0.

Finally, the fiber energy is evaluated by substituting, respectively, Eq. (37) and (38) in the J-Rice line integral of Eq. (8). The dependence of the fiber debonding energy on the fiber distribution (defined by parameters ϑ_0 and k) is shown in Figure 5 for a small crack opening displacement equal to 0.01 mm (i.e., $\sim [\![\xi]\!]^*$ of (37)).

To explicit the fiber debonding energy G_B as a function of the crack size, we need the crack profile relationship, this is generally obtained by solving an integral function since the crack opening along

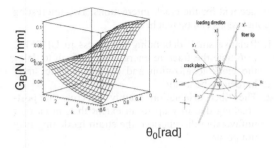

θ_0[rad]

Figure 5. (left) Fiber debonding energy in function of the fiber distribution parameters (ϑ_0 and k) at a crack opening of 0.01 mm; (right) fiber orientation with respect the crack plane.

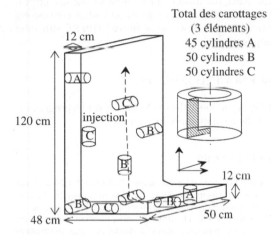

Figure 6. Specimens drilled out from the nuclear waste container with indication of the main direction of casting flow (dashed arrow).

the crack depends on the fiber bridging stress, which itself depends on the crack opening. For simplicity, we assume that the crack profile takes the same elliptical shape as that for a crack with uniform cohesive traction (i.e., unperturbed by fiber bridging) such as

$$[[\xi]] = \frac{8\left(1-v^2\right)}{\pi E} \frac{K_{IC}}{2/\pi f_t \sqrt{\pi a_0}} \sqrt{a^2 - r^2} \qquad (39)$$

where f_t = tensile strength, v = Poisson's ratio, r = radial crack coordinate, K_{IC} = critical stress intensity factor (which can be deduced by the fracture energy R_0 by Irwin's relationship). This assumption, which conveniently decouples the crack opening from the bridging stress σ_B, has been often used in literature (Lawn, 1993; Leung, 1996), and it can be seen as a first order approximation for small crack opening displacement of the penny shape cracks in the strain hardening regime.

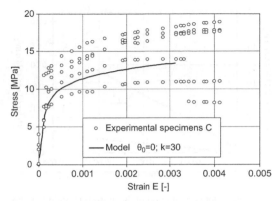

Figure 7. Numerical prediction of the experimental tensile test results for the favorable oriented specimens. "C".

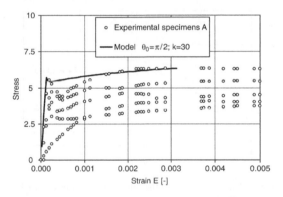

Figure 8. Numerical prediction of the experimental tensile test results for the unfavorable oriented specimens "A".

5 MODEL VERIFICATION

The model is calibrated on the experimental direct tensile response of notched specimens drilled out from an L-shaped prototype made of UHPFRC (Figure 6; Toutlemonde et al., 1999). This application is suitable for validating our model since the fiber orientation was found to be strongly oriented due to the casting process (injection) that is represented by a dashed arrow. More precisely, we have specimens "C" drilled along the favorable fiber orientation induced by the casting process, and specimens "A" in the unfavorable orthogonal direction. The crack opening measurement is converted in strain by assuming a measurement length of 50 mm.

The model parameters reported in Table 1 are identified from Toutlemonde et al. (1999), while further best fitting parameters are $\tau_u = 4$ MPa and $f_t = 6$ MPa. Figures 7 and 8 show the comparison between the experimental results and the model prediction for a uniaxial tensile test in the two orthogonal direction assuming a strongly localized fiber distribution (i.e.,

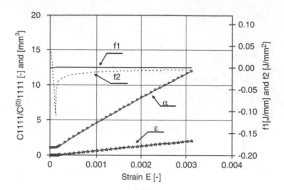

Figure 9. Model prediction of the anisotropy ratio (α), the crack damage parameter, the loading function (f1) of Eq. (10) and the stability condition of Eq. (12) for the specimens "B".

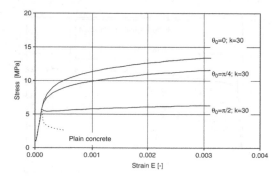

Figure 10. Sensitivity of the constitutive relationship to the fiber orientation.

k = 30). In addition, the model predicts the evolution of the material anisotropy and the micromechanics crack parameter, as shown in Figure 9 for the specimens "C". In the same figure, we report the loading function and the stability condition to show that the cracking process is efficiently stabilized by the fiber debonding.

Finally, Figure 10 shows the model sensitivity to the fiber orientation in terms of the material constitutive relationship prediction.

6 CONCLUDING REMARKS

This work develops a combined fracture-micromechanics model to upscale the UHPFRC material constitutive relation (before crack localization) from the critical microstructure features, such as fiber distribution, fiber-matrix debonding micro mechanism, and initial crack configuration. The approach turns to be a strain formulation of an anisotropic damage model where the homogenized compliance is traced down to the crack density and morphology pattern. The Mori-Tanaka scheme has been used to account for the crack interactions. Two interesting conclusions can be reached:

1. Relative small debonding toughness in UHPFRC (~6 time the fracture energy) are enough to delay the crack localization and allow a ductile strain hardening behavior;
2. The fiber orientation mainly governs the post-cracking ductility up to an extent that, in case of unfavorable distribution, the strain hardening may not occur.

Combined with experimental *in situ* evaluation of the fiber distribution or numerical prediction by flow transport analysis, the model may be used to predict or assess the reliability of UHPFRC post-cracking behavior. Next, the model can be used to identify critical fiber distribution for stability and, after considering mixed fracture mode, to optimize UHPFRC structural applications.

ACKNOWLEDGEMENTS

LGS is grateful to the EU Marie Curie FP6 programme for the research mobility support.

REFERENCES

Association Francaise de Genie Civil (AFGC) 2002. Ultra High Performance Fibre-Reinforced Concrete, Interim Recommendations, J. Resplendino & J. Petitjean eds.

Bayard, O. & Plé, O., 2003. Fracture mechanics of reactive powder concrete: material modelling and experimental investigations, Engineering Fracture Mechanics 70, 839–851.

Bažant, Z.P. & Planas, J. 1998. Fracture and size effect in concrete and other quasi-brittle materials. Boca Raton, Florida, and London: CRC Press, 616.

Bernier G. & Behloul M., 1998. Effet de l'orientation des fibres sur le comportement des BPR, Revue française de génie civil, 1998, vol. 2, no1, pp. 113–122.

Brian Lawn, 1993. Fracture of brittle solids, Cambridge Solid State Science Series, 378.

Budiansky, B., Hutchinson, J.W. & Evans, A.G., 1986. Matrix fracture in fiber-reinforced ceramics. J. Mech. Phys. Solids, 34, 2, 167–189.

Chanvillard, G. & Rigaud, S. (2003) Complete characterisation of tensile properties of Ductal® UHPFRC according to the French Recommendations, HPRFCC-4 symposium, Ann Arbor, Michigan, June 16–18.

Chuang,E., Overland, M. & Ulm, F.-J. 2001. Length scales of fiber reinforced cementitious composites – a Review. Proc. Fracture Mechanics of Concrete Structures, de Borst *et al.* eds., Lisse, 35–42.

Dormieux, L., Kondo, D. & Ulm, F.-J., 2006. A micromechanical analysis of damage propagation in fluid-saturated cracked media. C.R. Mecanique, 334, 440–446.

Eshelby, J. D., 1957, The Determination of the Elastic Field of an Ellipsoidal Inclusion and Related Problems, Proc. R. Soc. London, A241, 376–396.

Karihaloo, B.L., Wang, J. & Grzybowski, M., 1996, Double periodic array of bridged cracks and short fiber-reinforcement cementitious composites, J. Mech. Phys. Solids, 44, 10, 1565–1586.

Leung, C.K.Y. & Chi, J. 1995. Crack-bridging force in random ductile fiber brittle matrix composites, J. of Eng. Mech., December, 1315–1324.

Leung, C.K.Y. 1996. Design criteria for pseudoductile fiber-reinforced composites, J. Eng. Mech., January, 10–18.

Li, V. & Leung, C.K.Y. 1992. Theory of steady state and multiple cracking of random discontinuous fiber reinforced composites, ASCE J. of Eng. Mechanics, 2298.

Lin, Z. & Li, V., 1997. Crack bridging in fiber reinforced cementitious composites with slip-hardening interfaces, J. Mat. Phys. Solids, 45, 5, 763–787.

Marshall, D.B. & Cox, B.N. 1985. The mechanics of matrix cracking in brittle-matrix fiber composites, Acta Metall., 35, 11, 2013–2021.

Mori, T. & Tanaka, K. 1973. Average stress in matrix and average elastic energy of materials with misfitting inclusions, Acta Metall., 21, 571–574.

Rice, J.R. 1968. A path-independent integral and the approximate analysis of strain concentration by notch and cracks, J. Appl. Mech., 35, 379.

Shah, P. & Ouyang, C., 1991. Mechanical Behavior of Fiber-Reinforced Cement-Based Composites, Journal of the American Ceramic Society, 74, 11, 2727–2953.

Tjiptobroto, P. & Hansen, W., 1993, Tensile strain hardening and multiple cracking in high performance cement based composites. ACI Material Journal, 90, 1, Jan.–Feb., 1993.

Toutlemonde, F., Sercombe, J., Torrenti, J.-M. & Adeline, R. 1999. Développement d'un conteneur pour l'entreposage de déchets nucléaires : résistance au choc. Revue Française de Génie Civil, vol. 3, no 7–8, 729–756.

Zhang, J. & Li, V.C., 2002. Effect of inclination angle on fiber rupture load in fiber reinforced cementitious composites, Composites Sciences and Technology, 62, 775–781.

Fracture Mechanics of Concrete and Concrete Structures – High-Performance Concrete,
Brick-Masonry and Environmental Aspects – Carpinteri, et al. (eds)
© 2007 Taylor & Francis Group, London, ISBN 978-0-415-44617-4

Observation of multi-crack formation in Strain Hardening Cement-based Composites

F.H. Wittmann
Aedificat Institute Freiburg, Germany

T. Wilhelm, & P. Grübl
Institute for Concrete Structures, University of Technology (TU), Darmstadt, Germany

F. Beltzung
University of Applied Sciences of North-west Switzerland, Basel, Switzerland

ABSTRACT: Strain hardening cement-based materials can be produced by addition of high performance polymer fibers such as PVA or PE fibers to a fine fresh mortar. It is well known that multi-cracking of the cement-based matrix leads to pseudo ductility. In this contribution records of the surface of a SHCC specimen have been taken by means of a high-resolution digital camera. From the records taken principal strain fields and crack formation can be followed as function of imposed strain. The numerical evaluation method of experimental data is based on pattern recognition. The displacements and deformations of elements of the stochastic pattern on the surface are determined. It turned out that PVA fibers stabilize fictitious cracks in the cement-based matrix. A complex redistribution of stresses and strains can be observed under increasing strain. These mechanisms have to be considered to be at the origin of pseudo ductility of SHCC. Results presented in this contribution may serve as a solid basis for the development of a realistic numerical model, which may help to optimize strain hardening cement-based materials for practical applications.

1 INTRODUCTION

Cement-based materials such as concrete and mortar are most frequently applied materials in industrialized countries. Advanced technologies allow us to produce special types of concrete for many applications. High strength concrete with compressive strength well beyond $120 \, N/mm^2$ has become a usual building material in practice. Service-life of many reinforced concrete structures, however, is not long enough in many cases. One major reason for this is crack formation in the concrete cover. Concrete is a rather brittle material and the combination of hygral, thermal, and mechanical strains imposed under usual conditions cannot be absorbed without crack formation.

With the advent of high performance polymer fibers it became possible to produce cement-based materials with pronounced pseudo strain hardening. In an optimized system ultimate strains of 5% can be achieved. Kanda & Li (1999 and 2006) developed a design theory for pseudo strain hardening cement-based materials based on earlier work by Marshall and Cox (1988). Other authors optimize the composition of cement-based materials using neural networks to evaluate experimental results (Wittmann & Martinola 1993) or just by trial and error.

In any case pseudo strain hardening can be assumed to be due to crack bridging by fibers. Hardened cement paste has a wide pore size distribution. The largest flaw in the porous structure essentially controls strength (Wittmann 1983). It has been shown that very high strength hardened cement paste can be produced by eliminating large flaws (Higgins and Bailey 1976). This material was called MDF: Macro Defect Free cement-based material (Lewis et al. 1994). MDF, however, is a very brittle material.

In the case of materials with pseudo strain hardening micro cracks in the porous structure are not eliminated but bridged by fibers. High ductility is reached when the number of fibers, their size and mechanical properties can avoid unstable crack propagation. This crack arresting mechanism must be activated in pre-existing cracks and in cracks, which occur under applied load (Kanda & Li 2006). Then and only then multi crack formation leads to pseudo ductility. In order to understand these complex mechanisms in concrete better it would be helpful if we could observe crack formation under imposed deformation.

Pseudo ductile deformation is accompanied by continuous crack formation. That means the cement-based material undergoes progressive damage as the deformation increases. With respect to durability of the damaged material it is important to know the width of the produced cracks. As long as the crack width remains below a critical value it may be assumed that durability is not affected (Wittmann 2006 and Lepech & Li 2006). If we could observe the process of crack formation and the widening of cracks under imposed deformation we would be able to discern optimized fiber reinforced cement-based materials from less suitable materials and we could define ultimate strain capacity of strain hardening cement-based materials with respect to durability.

It is not easy to check the validity of models to describe the complex load transfer from cracking material to fibers in fiber reinforced cement-based materials under progressive imposed deformation. One possibility is to observe crack formation on a surface by means of optical methods. Hack et al. (1995) has applied electronic speckle pattern interferometry to observe crack formation in concrete. Rastogi and Denarié (1994) have used the holographic moiré method. Sunderland et al. (1995) observed crack growth in concrete by means of the confocal microscope. Jaquot and Rastogi (1983) have described the optical basis of interferometry in some detail.

We have applied the method of digital image correlation to study strains and crack formation on the surface of cement-based materials. This method is based on pattern recognition. The stochastic pattern on the surface of a sample is deformed by an applied load and from this deformation local strains and crack formation can be deduced by means of suitable software. The method is described in detail by Winter (1993). Choi and Shah (1998) studied the fracture mechanism under compressive load with a similar method. The application of this powerful method to study strain fields and crack formation in composite cement-based materials (2D model concrete) is described in the PhD Thesis of T. Wilhelm (2006).

2 EXPERIMENTAL

Specimens with the following dimensions have been prepared: $80 \times 80 \times 35$ mm. A dry mix provided by Technochem, Italy, together with 2% PVA fibers have been used to prepare a strain hardening cement-based composite (SHCC). The fresh fine mortar has been cast in PVC moulds and cured under sealed conditions for 7 days. The specimens were tested at an age of 3 months.

The specimen has been glued into a testing machine by high strength epoxy. After hardening of the glue an area of 60×60 mm^2 has been photographed by means of a high-resolution digital camera. As a deformation has been imposed by tensile stress digital prints have been taken in regular intervals. At the end of the test more than 200 exposures have been taken and stored in a computer.

The software program ARAMIS was used to evaluate selected exposures. In this way it is possible to determine principal strains, vertical or horizontal displacements.

3 RESULTS AND DISCUSSION

3.1 *Principal strains on the surface*

The testing machine allowed to determine the stress-strain diagram of the SHCC. A typical result is shown in Fig. 1. First a nearly linear relation between stress and strain is observed. After a maxium stress is reached the stress first decreases with further increasing strain until an intermediate minimum stress is reached. Further imposed deformation leads to strain hardening. Fig. 1 indicates that the material under investigation is not optimized. The drop after the maximum stress is reached is too big.

The results obtained with the digital camera can be further evaluated in different ways. In Fig. 2 the principal strains within the observed area are shown as measured at a tensile stress corresponding to 55% of the ultimate tensile load bearing capacity. The direction of the tensile stress is parallel to the left and right borders of the Figure. A characteristic crack pattern is visible. At stresses below this load level, no cracks can be observed with the chosen scale. The width of cracks in Fig. 2 can be estimated. The grey tone corresponds approximately to a strain of $\varepsilon = 0.3\%$. This strain is measured along the edge of one unit, which in this case is $\Delta l = 0.78$ mm. Hence the length change in the damaged zone is

$$\Delta l = \varepsilon \, l_0 = 0.3 \cdot 10^{-2} \cdot 0.78 = 0.0023 \text{ mm} \qquad (1)$$

(In a colour visualisation the crack width can be estimated more precisely.)

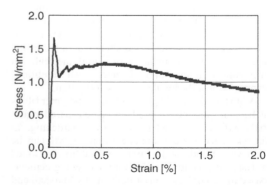

Figure 1. Stress-strain relation of the SHCC under investigation.

In Fig. 3 the crack pattern as observed at the maximum of the stress-strain diagram is shown. When we compare results shown in Figs. 2 and 3 we observe that the crack width has increased and few new cracks have been formed. Near the left lower corner of Fig. 3 a crack normal to the applied tensile stress reaches a width of 0.0055 mm.

All cracks shown in Figs 2 and 3 have a crack width below 10 μm. These fine cracks must be considered to be fictitious cracks in terms of non-linear fracture mechanics. That means that the cement-based matrix contributes to load transfer in cracks shown in Figs. 2 and 3 and that fictitious cracks are stabilized in the presence of PVA fibers. Under tensile load there is obviously intense interaction between the cement-based matrix and the fibers. We must abandon simple models, in which real cracks are assumed to be formed and then bridged by fibers. This may be a suitable model to represent the behavior of brittle ceramic matrices reinforced with fibers. In the case of cement-based matrices the contribution of the load transfer within the fictitious crack without fibers has to be taken into consideration as well.

The patterns shown in Figs 4 and 5 have been observed at macroscopic strains of 0.5% and 1% respectively. It can be observed that the width of some cracks does not increase, while some cracks situated in the center of the observed area widen considerably. Some cracks have reached a width of 0.0078 mm, and the width of the totally white crack is even wider but cannot be determined any more in this way of data processing. We can also observe that the diagonal crack shown in the upper left corner of Fig. 3 practically vanishes at an imposed strain of 0.5%; but it reopens at an imposed strain of about 1%. Quite obviously there is a complex redistribution of stresses and strains in the material before localization of one fatal crack takes place. It is also of interest to note that in going from a macroscopic strain of 0.5% to 1% new and wide cracks are being formed in parallel to the existing cracks.

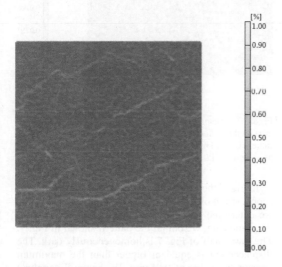

Figure 2. Principal strain field on the surface of SHCC under a tensile stress corresponding to approximately 55% of the ultimate load. The strain of a unit with length 0.78 mm is shown in % as grey tones.

3.2 Displacements in y-direction

In Figures 2 to 5 principal strains are shown. In addition displacements in y-direction have been determined

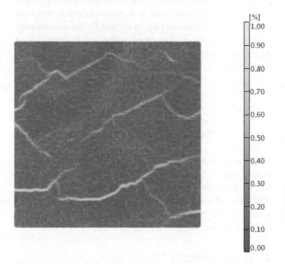

Figure 3. Principal strain field on the surface of SHCC under tensile stress corresponding approximately to the ultimate load capacity.

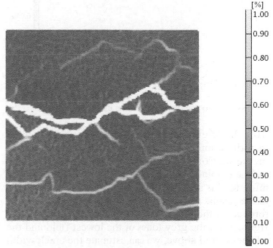

Figure 4. Principal stress field on the surface of SHCC as observed at an imposed macroscopic strain of 0.5%.

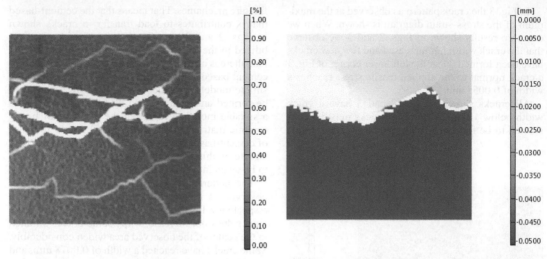

Figure 5. Principal stress field on the surface of SHCC as observed at an imposed macroscopic strain of 1%.

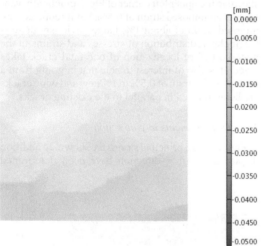

Figure 6. Displacements in y-direction in the observed area on the surface of SHCC at the maximum stress. This figure corresponds to the principal strains as shown in Fig. 3.

from the original data. Tensile stress has been applied in y-direction. In Figures 6 and 7 y-coordinates are measured from the upper edge to the lower edge.

The displacement in y-direction close to the maximum load is shown in Fig. 6. This figure corresponds to the results shown in Fig. 3. If we determine the difference of the displacements in the lower part of the figure from the grey tones of the lowest field and the adjacent field above, we can estimate the crack width to be 0.005 mm. This value agrees very well with the crack width as determined from the principal strains shown in Fig. 3.

Figure 7. Displacements in the y-direction in the observed area on the surface of SHCC at an imposed strain of approximately 0.5%. This figure corresponds to the principal strains as shown in Fig. 4.

In Fig. 4 the principal strains at an imposed macroscopic strain of approximately 0.5% are shown. When we determine the displacements in y-direction from the same data, we obtain the pattern shown in Fig. 7. The lower half of Fig. 7 is homogeneously dark. The displacement is equal or bigger than the maximum measurable value of 0.05 mm. We can still conclude that the big crack running from left to right has an opening bigger than 0.05 mm at the left edge and an opening bigger than 0.025 at the right edge. The left part has been obviously unloaded by the opening of the wide crack. Crack formation imposes a certain rotation of the separated parts. This is certainly the reason why cracks as shown in the upper left corner have closed again at an imposed strain of 0.5%.

The dominating crack, which has been developed at a macroscopic strain of approximately 0.5% and which is clearly visible in Fig. 7 finally became the separating crack of this sample. In Fig. 8 a photograph of the separating crack as observed after failure is shown. It can be seen that failure mechanism essentially is fiber pullout.

Figure 8 can be compared with Fig. 5 in which the crack pattern at an imposed strain of 1% is shown. Most of the fine cracks are closed again after failure and they cannot be observed in Fig. 8. But some of the wider cracks visible in Fig. 5 can still be recognized in Fig 8 as thin lines.

4   CONCLUSIONS

An advanced optical method has been applied to study principal strain fields and crack formation on the

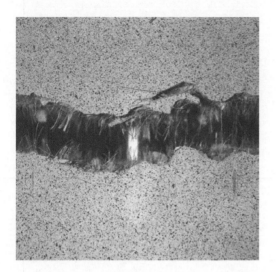

Figure 8. Final macroscopic separating crack in SHCC after failure.

surface of SHCC. Records have been taken by means of a high-resolution digital camera and have been numerically processed.

Multi-crack formation can be followed with this method as function of imposed macroscopic strain.

It turns out that PVA fibers stabilize fictitious cracks in SHCC. In the fictitious crack load is transferred by fibers but in addition by the remaining load bearing capacity of the damaged cement-based matrix. This combination of load transfer must be considered to be the major mechanism for obtaining large deformations, strain hardening, and pseudo ductility of SHCC.

Complex redistribution of strains and fictitious cracks can be observed. Results presented in this contribution shall serve as a basis for further optimizing SHCC for practical applications. The influence of the observed damage on durability has to be further investigated.

ACKNOWLEDGEMENT

The authors gratefully acknowledge supply of the dry mix of SHCC by TECNOCHEM Italiana.

REFERENCES

Choi, S. & Shah, S. P. 1998. Fracture mechanism in cement-based materials subjected to compression. J. Eng. Mechanics, 124 (1): 94–102.

Hack, E., Steiger, T., & Sadouki H. 1995. Application of electronic speckle pattern interferometry (ESPI) to observe the fracture process zone. In Fracture Mechanics of Concrete Structures, Proceedings FRAMCOS-2, Wittmann, F. H. (ed.). Aedificatio Publishers, Freiburg: 229–238.

Higgins D. D. & Bailey J. E. 1976. Fracture measurements on cement paste. Journal of Materials Science 11: 1955–2003.

Jaquot, P. & Rastogi P. K. (1983). Speckle metrology and holographic interferometry applied to the study of cracks in concrete. In Wittmann F. H. (ed.) Fracture Mechanics of Concrete. Elsevier Scientific Publishers, Amsterdam: 113–155.

Kanda, T. & Li, V. C. 1999. New micro-mechanics design theory for pseudo strain hardening cementitious composites. Journal of Engineering Mechnics, ASCE, 124 (4): 373–381.

Kanda, T. & Li, V. C. 2006. Practical criteria for saturated pseudo strain hardening behavior in ECC. Journal of Advanced Concrete Technology 4: 59–72.

Lepech M. D. & Li, V. C. 2006. Long term durability performance of engineered cementitious composites. Int. Journ. Restoration of Buildings and Monuments 12 (2): 119–132.

Lewis J. A., Boyer M. & Bentz, D. P. 1994. Binder distribution in Macro-Defect-Free cements: relation between percolation properties and moisture absorption kinetics. Journal of the American Ceramic Society 77(3): 711–719.

Marshall, T. B. & Cox, B. N. (1988). A J-Integral method for calculating steady-state matrix cracking stresses in composites. Mechanics of Materials 7:127–133.

Rastogi, P. K. & Denarié E. 1994. Measurement of the length of fracture process zone in fiber-reinforced concrete using holographic moiré. Exp. Techn. 18 (4): 11–17.

Sunderland H., Tolou A., Denarié E., L. Job & Huet C. 1995. Use of the confocal microscope to study pre-existing microcracks and crack growth in concrete. In Fracture Mechanics of Concrete Strucutres, Proceedings FRAMCOS-2. Wittmann F. H. (ed.). Aedificatio Publishers: 239–248.

Wilhelm T. 2006. Ein experimentell begründetes mikromechanisches Modell zur Bescheibung von Bruchvorgängen in Beton bei äusserer Krafteinwirkung. Dissertation TU Darmstadt, Germany.

Winter D. 1993. Optische Verschiebungsmessungen nach dem Objektrasterprinzip mit Hilfe eines Flächen orientierten Ansatzes. Internal Report, Institut für Technische Mechanik, Abteilung Experimentelle Mechanik. TU Braunschweig.

Wittmann F. H. 1983. Structure of concrete with respect to crack formation. In Fracture Mechanics of Concrete. Wittmann F. H. (ed.). Elsevier Science Publishers, Amsterdam, pp.43–74.

Wittmann F. H. & Martinola G. 1993. Optimization of concrete properties by neural networks. In Dhir R. K. & Jones M. C. (eds.) Concrete 2000, Economic and Durable Construction through Excellence, E & FN Spon, London, Vol. II: 1889–1893.

Wittmann F. H. 2006. Specific aspects of durability of strain hardening cement-based composites. Int. Journ. Restoration of Buildings and Monuments 12 (2): 109–118.

Fracture Mechanics of Concrete and Concrete Structures – High-Performance Concrete,
Brick-Masonry and Environmental Aspects – Carpinteri, et al. (eds)
© 2007 Taylor & Francis Group, London, ISBN 978-0-415-44617-4

Inverse analyses of bending tests for determining the properties of strain hardening cement-based materials

N. Bretschneider, B. Villmann & V. Slowik
Leipzig University of Applied Sciences, Leipzig, Germany

ABSTRACT: A procedure for determining the mechanical material properties of strain hardening cement-based materials is proposed. It is based on the inverse analysis of bending tests. The experimental results obtained under four-point bending of unnotched specimens allow the determination of the strain hardening curve up to the localization point. A load-crack opening relationship describing the behavior after the localization point is determined on the basis of bending tests at notched specimens. For the numerical simulation of the experiments, a hardening/softening behavior under tension and a non-linear stress-strain curve under compression have been adopted. The optimization required in the inverse analyses is performed by using an evolutionary algorithm which proved be an appropriate as well as efficient numerical tool for the problem to be solved here. The results obtained by inverse analyses of bending tests are compared with those obtained in direct tension tests.

1 INTRODUCTION

Strain hardening materials are characterized by increasing or at least constant stresses after initial cracking. This behavior is accompanied by spatially distributed cracking, i.e. multiple parallel cracks are opening simultaneously. When the first crack starts to exhibit a softening behavior, fracture localization takes place. The further damage is limited to a certain region while outside this region strains are decreasing and cracks are closing. Before fracture localization, the mechanical behavior of the material may be described by a stress-strain relationship, thereafter by a stress-crack opening relationship.

An accepted procedure for determining these material laws are uniaxial tension tests. Such tests are comparably time consuming and expensive. Furthermore, undesirable strain gradients in the ligament cause additional problems as far as the interpretation of the results is concerned. Nonetheless, for the development and optimization of strain hardening materials direct tension tests are inevitable. They allow to directly characterize the uniaxial deformation and fracture behavior which is the standard loading case fundamental studies should be based on.

In practical material testing, the bending test is an alternative way for determining the material parameters of strain hardening materials since such tests are easier to perform and less time consuming. However, the determination of the uniaxial load-deformation relationships is only possible by inverse analyses of

the experiments. These analyses include simulations of the fracture processes in the specimen and an iterative fitting of the numerical results to the experimental ones.

For strain softening cement-based materials, several techniques for the inverse analysis of fracture tests have been proposed and used in the past. As an alternative to manual fitting, Roelfstra & Wittmann (1986) presented the Softfit program. It based on the discrete crack approach and uses a gradient descent optimization. Although the program has been successfully used in numerous investigations (Slowik 1995) it reveals the limitations of the applied optimization method. Experience has shown that this method requires a first guess of the softening parameters which is quite close to the final results. This may be attributed to the risk of getting "caught" by local minima in the error function. The Japan Concrete Institute (2001) recommends a method based on a polylinear softening curve. The individual slopes are determined successively by adjusting corresponding sections of the calculated load-displacement curve to the experimental one. Østergaard (2003) proposed and extensively tested a method for the inverse analysis based on a hinge model. Assuming plane sections in a certain distance from the crack path an analytical solution for the softening curve may be derived. This approach appears to be very attractive because of the comparably short computing time and because of the unique solution of the optimization procedure. However, the assumption of plane section is not always justifiable, especially not in laboratory sized wedge splitting specimens. This shortcoming is taken care of by a calibration of the method in order to adopt a suitable value for the distance between the assumed plane sections. For routine material testing, the inverse analysis based on the cracked hinge model is a very effective tool.

Østergaard et al. (2005) modified the hinge model and the corresponding inverse analysis method for the application to strain hardening materials. In the modified model, the inner region of constant moment in a four-point bending beam is assumed to consist of several hinges which are separated by parallel plane sections. After fracture localization, only one of these hinges continues to open while the other ones are closing. The width of the localizing hinge has to be predefined prior to the analysis. It may be adjusted by comparing the results to those of Finite Element simulations. In addition to the advantages of the hinge model already mentioned above, it has to be pointed out that this method allows to determine the complete fracture behavior including hardening and softening by a single inverse analysis which is based on only one type of experiment.

The authors are following a different approach for the inverse analysis of experiments (Villmann B. et al. 2004 & 2006, Slowik et al. 2006). The main objectives of this work are, firstly, to adopt a mechanical model which requires prior to the analysis only inevitable assumptions concerning material behavior and specimen deformation and, secondly, to use a suitable optimization method for this particular problem. As optimization method an evolutionary algorithm including local neighborhood attraction for convergence improvement is proposed. Its advantages are explained in section 2.2. The method has been utilized by the authors for determining different material parameters of cement-based materials (Villmann B. et al. 2006). Softening curves may be determined on the basis of the discrete crack model. Furthermore, inverse analyses of drying experiments allow the determination of the moisture dependent diffusion coefficient. The evolutionary algorithm provides very good fits which could not be achieved by manual fitting. In its current version, the associated software tool allows to optimize models with up to 12 parameters. This corresponds to seven segments of a multilinear hardening curve. The optimization problem described in the present paper requires this comparably high level of variability in the formulation of the material law.

For the determination of fracture properties of strain hardening materials a two-step method is proposed. Hardening and softening curves are determined in separate inverse analyses of different types of experiments. For the hardening curve, inverse analyses of four-point-bending tests of unnotched beams are required in order to activate the high strain capacity of these materials. The softening properties are determined on the basis of bending tests of notched beams. Although this two-step procedure appears to be expensive it has some advantages. In addition, it has to be taken into account that in many cases only the hardening behavior is of technical interest for the application of these materials. The major advantage of the two-step procedure is that the width of the fracture process zone in the softening mode is not required as a predefined input parameter for the inverse analysis. This width may even be extracted out of the simulation results. The latter depend on the material parameters and on the specimen geometry only. Furthermore, the mechanical model does not require a plane section assumption. It has to be stated, however, that the fracture process zone width in the notched beams is assumed to be constant along the crack path.

The mechanical model and the optimization method have been used for determining the material parameters of two types of strain hardening cement-based composites (SHCC). For verification of the results, different specimen geometries were included in the investigation.

2 PROPOSED METHOD

2.1 Numerical simulation of the experiment

2.1.1 Simulation up to the localization point

The hardening range of strain hardening cement-based materials is characterized by distributed, i.e. multiple cracking. In order to activate it in the experiment, four-point bending tests of unnotched specimens should be performed rather than three-point bending tests or tests of notched specimens. In this way, a wide region of constant moment allows for a spatially distributed damage process. The mechanical model used for the inverse analysis of the experiment is based on non-linear axial members in the inner region of maximum moment and a linear-elastic Finite Element model for the side regions. Taking advantage of symmetry, only half of the beam is considered in the model, see Fig. 1. Cracking outside the inner region of maximum moment is neglected.

The mechanical behavior of the non-linear axial members incorporates a multi-linear stress-strain curve with strain hardening, see Fig. 2. It appeared to be necessary to consider a non-linear stress-strain curve also in the compressive range in order to obtain realistic simulation results. In the analyses presented here, a bi-linear curve was assumed, i.e. a linear-elastic branch and a constant post-peak stress value.

The inverse analysis with the model shown in Fig. 1 and the material law shown in Fig. 2 is performed up to the localization point. After that point, further damage is taking place in a single fracture process zone only and may be described by a discrete crack model as done

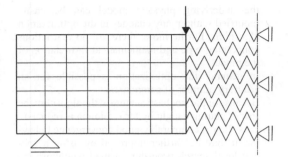

Figure 1. Model for the numerical simulation up to the localization point.

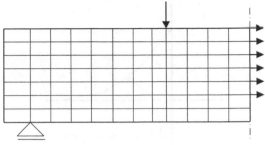

Figure 3. Model for the numerical simulation after the localization point.

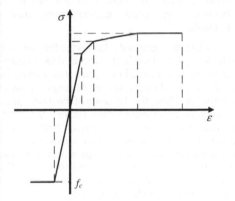

Figure 2. Stress-strain curve assumed for the non-linear axial members.

for softening materials, see section 2.1.2. A continuing simulation with the model shown in Fig. 1 would not be physically correct. Using an effective stress-strain curve which takes into account localization and unloading is not considered an appropriate way of modeling because of the unknown fracture process zone width, because of the unknown unloading behavior and because of required additional compatibility conditions.

In order to improve the computational efficiency, the linear-elastic part is pre-analyzed by Finite Elements and replaced by a linear-elastic macro-element. In this way, the number of degrees of freedom to be considered in the inverse analyses is reduced significantly.

In the numerical model, the specimen self-weight is taken into account. Parametric studies have shown, however, that the influence is negligible for practical specimen sizes.

2.1.2 Simulation after the localization point

In order to enforce fracture localization and to suppress distributed cracking, bending experiments

with notched specimens are undertaken for determining the stress-crack opening behavior of a localized crack. Inverse analyses of such tests are performed by using the same algorithm as for softening materials (Villmann B. et al. 2004). The underlying mechanical model is shown in Fig. 3. Fracture is taking place in the axis of symmetry only where a discrete crack is assumed. The planar part of the specimen exhibits a linear-elastic behavior and is replaced by a pre-analyzed macro-element for the inverse analyses.

For characterizing strain hardening materials, it was necessary to increase the number of the curve parameters to be optimized. A multilinear curve with arbitrary shape and up to six segments has been adopted, as stated before. In order to adequately approximate the long descending branch of the load-deflection curves characteristic for strain hardening materials it was necessary to refine the mesh at the end of the ligament. This led to an irregular discretization.

The major shortcoming of the method is the required assumption of a constant width of the fracture process zone along the crack path. If this simplification is accepted, the determination of fracture properties for a localized crack in strain hardening materials is possible. As a result, a stress-crack opening curve is obtained for an unknown constant fracture process zone width. If the strain at the localization point is known, for example as a result of direct tension tests or of inverse analyses according to section 2.1.1, it is possible to extract the fracture process zone width out of the stress-crack opening curve. It has to be taken into account, however, that the fracture process zone width is dependent on specimen size and geometry.

In order to obtain realistic stress distributions in the ligament the material model includes a limitation of the occurring compressive stresses. This is taken care of by predefining a compressive strength as maximum value. In some cases, the limitation of compressive stress along with the characteristic behavior of the strain hardening materials results in situations of local and momentary crack closing. For that reason,

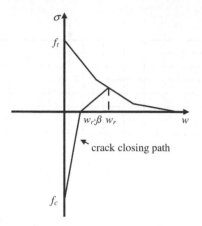

Figure 4. Stress-crack opening curve for the localized crack.

the softening law had to be supplemented by a crack closing path, see Fig. 4.

In parametric studies, the influence of the self-weight on the results of the inverse analyses was found to be negligible for practical specimen sizes and strain hardening materials. For quasi-brittle softening materials this is not always the case.

2.2 Optimization

Optimization in the inverse analyses presented here is a nontrivial problem. For the adequate approximation of the stress-strain curves a comparably large number of parameters is required. For that reason, a manual optimization is not only time consuming but nearly impossible and expected to be not objective. Another problem is that large variations in the hardening function cause only comparably small differences of the numerical stress-strain curve (Slowik et al. 2006). Hence, only quite exact approximations provide reliable results of the inverse analysis, i.e. sufficiently accurate hardening functions. That means the modification of the hardening parameters requires an automatic routine. Furthermore, the n-dimensional error function (n is the number of model parameters to be optimized) of the approximation is quite jagged and there is a risk of getting caught by non-attractive local minima.

A general alternative for such optimization problems are Evolutionary Algorithms (EAs) (Schwefel 1981). EAs are biologically motivated iterative stochastic optimization methods, the roots of which point to biological genetics as described by Mendel's laws of heredity and Darwin's evolution model by natural selection. They have several advantages when compared to classical optimization methods:

– The variation of the model (parameters) to be optimized is separated from its evaluation. Therefore,

the underlying physical model can be easily modified without any changes in the optimization algorithm. Furthermore, there are no restrictions concerning type and mathematical formulation of the model.
– In general, the optimization is possible for an arbitrary number of parameters.
– Because of the stochastic nature of the algorithm the risk of getting caught by local minima is reduced.
– EAs show a comparably good convergence behavior. It may be further improved by using a special local search procedure named neighborhood attraction (Villmann T. 2001).
– The required computing time is moderate.
– After a sufficient number of generations (see below), very good approximations may be obtained.

The basic principle of the EAs is the following: The model to be optimized is characterized by a vector of model parameters called individual. For a given individual a so-called fitness may be calculated, inverse to the error describing the deviation of the model prediction from the experimental data. The task is to find an individual which minimizes the error. Initially, a generation of a given number of individuals is randomly chosen and the fitness of each individual is calculated. In each time step, a new generation (offspring) of new individuals is created by certain genetic operations. Such operations may be (Bäck 1996):

– random variations of vector element values (mutations) or
– randomly cutting two strings and gluing them together in a crossed way (combination).

After the genetic operations, all offspring individuals are evaluated, i.e. their fitness is determined. Then, a fitness based selection procedure follows in order to create a new generation. This selection may comprise either the "best", i.e. fittest, individuals of the offspring only or of both the offspring and the adult generation. In the investigations presented here, a combination of these two selection methods is used (Bäck 1996, Villmann T. 2002). The fittest individual of the actual generation is the currently found solution of the EA. For a more detailed explanation see Bäck (1996), Villmann T. (2001) and Villmann T. et al. (2004).

3 APPLICATION OF THE METHOD

3.1 Material composition and uniaxial tension tests

For testing the proposed method inverse analyses of different experiments were carried out. The development of this method was motivated by the advances in the field of strain hardening cement-based composites (SHCC) with polymeric fiber reinforcement.

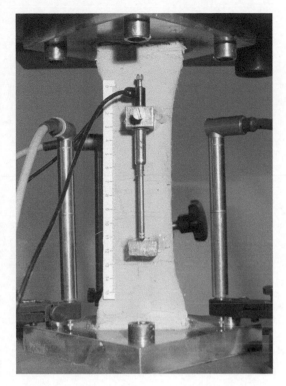

Figure 5. Direct tension test of a strain hardening cement-based composite.

These materials were formerly referred to as Engineered Cement-based Composites (ECC) (Li 1998). Their very high strain capacity is achieved mainly by the addition of high modulus polymeric fibers of short length, in most of the cases polyvinyl alcohol (PVA) fibers.

The materials used in the investigation presented here were a result of an optimization process aimed to a high tensile strain capacity. Their composition is characterized by fine aggregates and moderate water to binder ratios. The uniaxial tensile strength is comparably low in order to provide a smooth transition from initial cracking to the hardening range. Table 1 shows the material composition of the two materials included in this investigation. Both materials were reinforced with 8 mm long PVA fibers (Kuraray REC15). The fiber content was 2.2% by volume.

After casting, the specimens were left in their metal moulds and covered for three days. After that, they were stored at 65% relative air humidity and 20°C. The compressive strength after 14 days amounted to 24.8 N/mm² for material I and to 18.8 N/mm² for material II.

With both materials, direct tension tests have been carried out. Fig. 5 shows the experimental setup. Dog bone shaped specimens with a cross-section of 40 mm

Table 1. Composition of the tested materials.

	Material I	Material II
Cement [kg]	1.00	1.00
Water [l]	1.10	1.20
Fly ash [kg]	2.30	2.30
Silica flour [kg]	1.00	1.00
Silica sand [kg]	0.40	0.00
Plasticizer [kg]	0.040	0.040
Stabilizer [kg]	0.015	0.015
Fiber content [vol.-%]	2.20	2.20

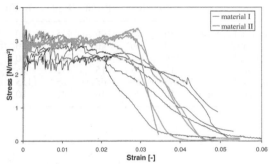

Figure 6. Stress-strain curves obtained in direct tension tests.

by 40 mm were tested under displacement control. The inner part of the specimens with constant cross-section had a length of 100 mm which was also the base length for the displacement measurement.

Fig. 6 shows measured stress-strain curves. The latter are characterized by nearly constant stresses up to the localization point. For material I, the strain at the localization point amounted to about 2%; for material II about 2.5% were reached.

3.2 Determination of hardening curves by inverse analyses of bending tests with unnotched beams

As stated before, the hardening behavior may be determined by inverse analyses of four-point bending tests with unnotched beams. For a verification of the proposed method, different specimen geometries were included in the investigations. Fig. 7 schematically shows the specimen geometry; the corresponding dimensions are given in Table 2.

For material I, some of the load-deflection curves obtained with the beam geometries B 75 and B 150, respectively, are shown in Fig. 8. Whereas approximately the same ultimate load levels were measured for the individual samples having the same geometry, the deflections at the localization point show a comparably large scatter. The best fitting numerical results obtained by optimization are also shown in Fig. 8.

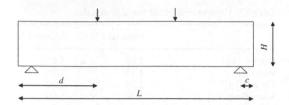

Figure 7. Geometry of the tested beams (dimensions in Table 2).

Table 2. Specimen dimensions.

	B 50	B 75	B 100	B 125	B 150
Midspan [mm]	50	75	100	125	150
H [mm]	40	40	40	40	80
L [mm]	180	205	230	255	400
c [mm]	15	15	15	15	25
d [mm]	65	65	65	65	125
Thickness [mm]	40	40	40	40	80

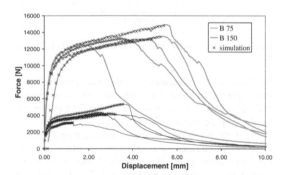

Figure 8. Comparison of measured and simulated load-deflection curves for material I.

It can be seen that the evolutionary algorithm yields excellent approximations of the experimental results.

The obtained hardening curves for the samples used for Fig. 8 are displayed in Fig. 9. As stated before, these curves are physically correct only up to the localization point. For that reason, they were truncated at the point of maximum stress although the numerical simulations were continued slightly beyond this point. Usually, the point of local maximum stress, resulting in localization, is reached before the point of ultimate external load. In the case of bending, fracture localization does not necessarily lead to an immediate decrease of the external load.

The curves shown in Fig. 9 form a band with a clear trend. For the different geometries roughly the same hardening curves were determined although there is a tendency of higher stresses in the smaller beams. The ultimate hardening strain values show a comparably

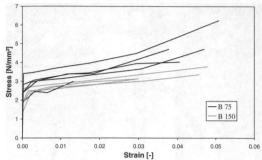

Figure 9. Obtained hardening curves for material I.

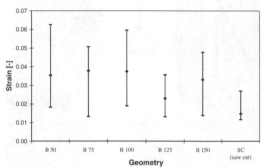

Figure 10. Ultimate hardening strain values obtained in inverse analyses for material I and different geometries.

large variation. Considering the differences between the curves in Fig. 8 it may be expected that the corresponding hardening curves differ as well. However, all the hardening curves show about the same stress level in the hardening range.

In Fig. 10, the variation of the ultimate hardening strain for different geometries is shown. No clear trend may be identified. That means, neglecting the cracks outside of the inner region of the beams with constant moment does not result in significantly size dependent results. For the saw-cut specimens lower hardening strain values were reached in comparison to all the other specimens which had molded surfaces. The random fiber orientation at saw-cut surfaces leads to lower ultimate strains. During the experiments, it was observed that the cracks at saw-cut surfaces were not as equally distributed as those at molded surfaces.

Fig. 11 shows the load-deflection curves for material II. The corresponding results of the inverse analyses are presented in Fig. 12. As in the case of material I, for different geometries roughly the same hardening curves were obtained by the inverse analyses.

Although in general the variation of these results is regarded as being high, it may be noticed that the scatter of the hardening curves obtained for the smaller beams (B 75) is lower.

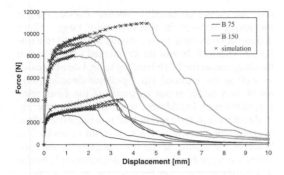

Figure 11. Comparison of measured and simulated load-deflection curves for material II.

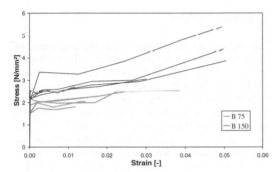

Figure 12. Obtained hardening curves for material II.

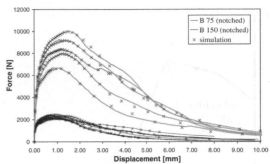

Figure 13. Measured and simulated load-deflection curves for material I (notched beams).

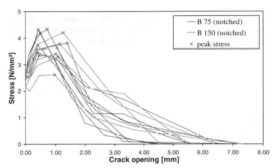

Figure 14. Obtained stress-crack opening curves for material I.

3.3 Determination of softening curves by inverse analyses of bending tests with notched beams

The discrete crack approach is used for simulating the experiments. Thereby, a constant fracture process zone width is assumed. All strains and dislocations within the fracture process zone are considered to be included in the discrete crack opening value.

Two different specimen geometries were included in the investigation, B 75 with a notch length notch 10 mm and B 150 with a notch length of 20 mm.

Fig. 13 shows the experimental and numerical results. The scatter of the maximum load appears to be larger than in the case of unnotched beams. This may be attributed to the spatially delimited fracture process zone controlling the global loading capacity of these inhomogeneous specimens.

The results of the inverse analyses are presented in Fig. 14. As in the case of the hardening curves, see section 3.2, the stress-crack opening curves form a band and for the different specimen geometries the inverse analysis yields nearly the same results. However, the crack opening at maximum stress is clearly higher for the larger specimens. This may be explained by the size dependence of the fracture process zone.

The maximum stress values in the stress-crack opening curves should correspond to the stress at the localization point in the hardening curves. This is confirmed by the obtained results of the inverse analyses.

3.4 Discussion of the results

On the basis of both the crack opening value at maximum stress and the ultimate strain in the hardening curve it is possible to determine the width of the fracture process zone for the particular geometry of the notched beams. For B 75 this width was determined to 20.2 mm and for B 150 to 32.9 mm.

In the pre-peak region of the stress-crack opening curves, see Fig. 14, hardening takes place in the fracture process zone. This phenomenon could also be visually observed during the experiments. Parallel cracks were formed in a zone with a width in same the order of magnitude like the values given above.

The results of the inverse analyses may be used for assembling a complete stress-strain curve for the strain hardening material including both hardening and softening. Up to the localization point, the obtained hardening curve, see section 3.2, is used. Starting with the average ultimate hardening strain, a softening branch is added to the curve. For that, the softening curve needs to be extracted from the stress-crack

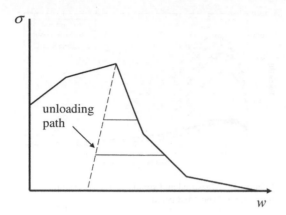

Figure 15. Extraction of the softening curve from the load-crack opening curve obtained by inverse analysis of bending experiments with notched beams.

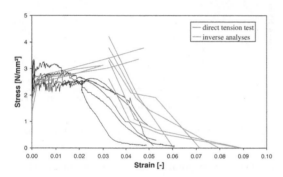

Figure 16. Results obtained by inverse analyses in comparison to those measured in direct tension tests (material I).

opening curve. This is done by subdividing the crack opening displacement obtained by inverse analysis into a portion resulting from the strain in the hardening fracture process zone and into a portion coming from the opening of the final crack, see Fig. 15. Here, a remaining strain of 75% of the maximum has been assumed. This is in good accordance with experimental observations during uniaxial unloading and reloading.

If the discrete crack opening in the softening crack is known, it may be transformed into an equivalent strain value for a given gage length in the direct tension test. Then, the uniaxial stress-strain curve may be completed by adding the calculated softening strain values to the unloading branch starting in the localization point. Fig. 16 shows such curves calculated for several specimens in comparison to results of direct tension tests.

Whereas the hardening stress levels found in direct tension tests and by inverse analyses, respectively, are in good accordance, there are significant deviations in the ultimate hardening strain values. Furthermore,

only the curves obtained by inverse analysis exhibit increasing stresses in the hardening range. These differences may be at least partially attributed to a different material behavior under uniaxial tension and in bending, respectively. In bending tests, lateral stresses as well as the inclined fiber pull-out direction may result in a larger strain capacity of the fiber reinforced material. In addition, strain gradients occurring in direct tension tests might lead to a reduction of the average ultimate strain. Up to a strain of about 2%, however, the results of the inverse analyses are considered to resemble the uniaxial deformation behavior in a satisfactory manner for technical applications.

4 CONCLUSIONS

The method for determining fracture properties of strain hardening cement-based materials by inverse analyses of bending tests yields physically sound results. For different geometries, roughly the same material parameters were obtained. The optimization by using an evolutionary algorithm allows excellent approximations of the experimental results. For strain values above 2%, significant deviations between the results of direct tension tests and those obtained by inverse analyses of bending tests were found. These differences may partially be explained by specific effects occurring in the different experiments and will be subject of further investigations. Furthermore, the model assumptions concerning the compressive failure need to be refined.

REFERENCES

Bäck, T. 1996. *Evolutionary Algorithms in Theory and Practice*. New York, Oxford: Oxford University Press.
Japan Concrete Institute 2001. Determination of tension softening diagram of concrete. JCI-TC992 Test Method for Fracture Property of Concrete.
Li, V.C. 1998. Engineered Cementitious Composites – Tailored Composites through Micromechanical Modeling. In N. Banthania, A. Bentur, and A. Mufti (eds.), *Fiber Reinforced Concrete. Present and the Future*: 64–97. Montreal: Canadian Society for Civil Engineering.
Østergaard, L. 2003. *Early-age fracture mechanics and cracking of concrete*. Thesis. Technical University of Denmark.
Østergaard, L., Walter, R. & Olesen, J.F. 2005. Method for Determination of Tensile Properties of ECC. *International Workshop on High Performance Fiber Reinforced Cementitious Composites in Structural Applications, Honolulu, Hawaii, USA, 23–26 May 2005*.
Roelfstra, P.e. & Wittmann, F.H. 1986. Numerical method to link strain softening with failure of concrete. In F.H. Wittmann (ed.), *Fracture Toughness and Fracture Energy of Concrete*: 163–175. Amsterdamm: Elsevier.
Schwefel, H.-P. 1981. *Numerical Optimization of Computer Models*. Chichester: Wiley and Sons.

Slowik, V. 1995. *Beiträge zur experimentellen Bestimmung bruchmechanischer Materialparameter von Betonen*. Postdoktorandenbericht, ETH Zürich. Freiburg: AEDIFICATIO Publishers.

Slowik, V., Villmann, B. & Bretschneider, N. 2006. Computational aspects of inverse analyses for determining softening curves of concrete. *Computer Methods in Applied Mechanics and Engineering* 195 (52): 7223–7236.

Villmann, B., Bretschneider, N., Slowik, V. & Michel, A. 2006. Bestimmung von Materialeigenschaften zementgebundener Werkstoffe mittles inverser Analyse. *Bautechnik* 83 (11): 747–753.

Villmann, B., Villmann, T. & Slowik, V. 2004. Determination of softening curves by backward analyses of experiments and optimization using evolutionary algorithm. In V.C. Li, C.K.Y. Leung, K.J. William & S.L. Billington (eds.), *Fracture Mechanics of Concrete Structures*: 439–445. IA-FraMCoS.

Villmann, T. 2001. Evolutionary Algorithms and Neural Networks in Hybrid Systems. *Proceeding of European Symposium on Artificial Neural Networks ESANN*: 127–152. Bruges: de facto publications.

Villmann, T. 2002. Evolutionary Algorithms with Subpopulations using Neural Network like Migration Scheme and its Application to Real World Problems. *Integrated Computer-Aided Engineering* 9: 25–35.

Villmann, T., Villmann, B. & Slowik, V. 2004. Evolutionary algorithms with neighborhood cooperativeness according neural maps. *Neurocomputing* 57: 151–169.

Fracture Mechanics of Concrete and Concrete Structures – High-Performance Concrete,
Brick-Masonry and Environmental Aspects – Carpinteri, et al. (eds)
© 2007 Taylor & Francis Group, London, ISBN 978-0-415-44617-4

Direct tensile behavior and size effect of Strain-Hardening fiber-reinforced Cement-based Composites (SHCC)

K. Rokugo & Y. Uchida
Department of Civil Engineering, Gifu University, Japan

M. Moriyama
Central Nippon Expressway Co., Ltd., Japan

S.-C. Lim
Deros Co., Ltd., Japan

ABSTRACT: A simple tension test method on dumbbell specimens for strain-hardening fiber-reinforced cement-based composites (SHCC) was proposed. Tension tests were reliably accomplished using the test apparatus proposed in this study, with all specimens broken in their test zones. Though the tensile strength and the tensile yield strength scarcely varied between the specimen thicknesses, the ultimate tensile strain scattered more widely as the specimen thickness increased. Regardless of the thickness, the crack widths of all specimens increased up to a strain of approximately 1%. When the strain exceeded 1%, the crack width of specimens with an ultimate tensile strain of 4% or more scarcely increased as the strain increased, with the maximum crack width being not more than 0.2 mm.

1 INTRODUCTION

Technology in the concrete engineering field is for the most part intended to control cracking in concrete, to the extent that "concrete engineering" could be renamed as "crack control engineering." Crack width can be controlled to a certain extent by making the structure reinforced with steel, and cracking can be prevented by employing prestressed concrete technique. Various measures are also taken for materials and placing to suppress hydration heat and inhibit shrinkage, which causes cracking in concrete.

Conventionally, the purpose of including short fibers in concrete has been primarily to increase its tensile strength, as in steel fiber-reinforced concrete. Fibers added for the purpose of controlling the crack width have recently attracting attention. A composite material referred to as High Performance Fiber-Reinforced Cement Composite (HPFRCC) has been developed. This material is called Strain-Hardening fiber-reinforced Cement-based Composites (SHCC).

This material allows substantial tensile deformation as narrow cracks occur one after another under increasing tensile forces. The crack width is controlled by short fibers to a level of 0.1 mm and not more than 0.2 mm. This material is characterized by the increase in the number of cracks, instead of the crack width, associated with the increase in the tensile deformation.

These characteristics that distinguish this material from conventional mortars and concretes should be effectively suited to the purposes and conditions of the intended use. To this end, it is vital to elucidate the ultimate tensile strain (peak tensile strain), tensile stress-strain curve, cracking performance (width, number, and spacing), and size effect on these tensile performances, as well as to establish test methods for evaluating these tensile performances(Li 1993, Li 2002, Kunieda & Rokugo 2006).

This study aims to propose a simple tension test method on dumbbell specimens for SHCC showing strain-hardening behavior. Uniaxial direct tension tests were also conducted using the proposed method to investigate the tensile performance of SHCC and the effect of specimen thickness on such performance.

2 PROPOSAL FOR A UNIAXIAL TENSION TEST METHOD AND TEST APPARATUS

2.1 *Performance required for tension test methods and problems*

The key tensile performance items for evaluating SHCC include the following: crack width, number of

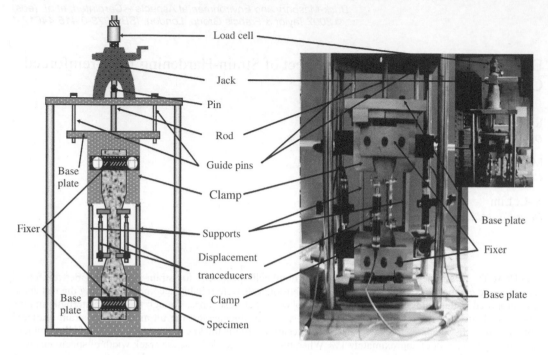

Figure 1. Tension test apparatus.

cracks, crack spacing, tensile strength, tensile yield strength, and ultimate tensile strength. For stable measurement of these items by tension testing, measures should be taken for the following three points: (1) to transfer the tensile forces from the testing machine to the specimen without causing failure of the enlarged ends; (2) to suppress the L-shaped (non-uniform) deformation of the specimen; and (3) to stably control the behavior in the strain-hardening region after the first crack. In regard to point (2) above, L-shaped deformation is essentially unavoidable, as a crack will develop from a point of a specimen to the entire cross-section, but could be reduced to a certain extent. It should be noted that L-shaped deformation of SHCC is not as large as that of normal concrete, as each crack width of SHCC tends to remain small while the crack develops.

In this study, a simple tension testing machine was developed with the aim of solving the above-mentioned problems of tension testing and permitting easy observation of multiple cracks occurring in the specimen, with due consideration given to the operability and safety.

2.2 Tension test apparatus and operation

The tension test apparatus developed in this study is shown in Fig. 1. The shoulders of a dumbbell specimen are engaged in the upper and lower locks in the steel framing with a mass of approximately 30 kg and external size of 250 by 200 by 500 mm. Though frictional clamping of specimens is generally employed, shoulder locking was adopted in this study to transfer tensile forces to specimens.

To facilitate engaging and disengaging of specimens into and from the apparatus, the clamp on one side of each lock (the left side clamps of the upper and lower locks shown in Fig. 1) is fixed to the base plate. The clamp on the other side of each lock (the right side clamps) is designed to slide sideways with a travel of 5 mm. After engaging a specimen, the clamps on the right side are set in place and bolted to the base plates. Both clamps of the upper and lower locks are fastened together using lock fasteners. Tie rods are employed to align the upper and lower locks.

A mechanical jack is mounted on the steel frame to apply tensile forces to the specimen through a force-applying rod. The upper lock is pulled up by the applied force along the two guide pins. In regard to the boundary conditions of tension testing, the lower lock is fixed to the base plate of the steel frame (fixed support), whereas the upper lock is pin-supported by the force-applying rod via a hinge set in the bar.

The load-displacement relationship during tension testing was measured using displacement transducers directly set on the specimen and a load cell placed on top of the loading element.

Table 1. Mix proportions of SHCC.

Units (kg/m³)					Slump flow (mm)	Air content (%)
Mortar	Aggregate	Polymer	Water	Fiber		
1225	323	71	257	26.0	420	6.8

Table 2. Compression test results.

Name	Thickness (mm)	Compressive strength (MPa)	Elastic modulus (GPa)
PVA-13	13		
PVA-30	30	34.4	16.3
PVA-50	50		

3 TENSILE PERFORMANCE AND SIZE EFFECT

3.1 Experiment overview

3.1.1 Materials
Table 1 gives the mix proportions of SHCC and the physical properties of short fibers. Premixed polymer cement mortar containing 323 kg/m³ of coarse aggregate with a maximum size of 10 mm was used as the matrix of SHCC. Polyvinyl alcohol (PVA) fibers 12 mm in length and 40 μm in diameter were blended with the matrix at a ratio of 2% by volume. The unit water content was 257 kg/m³. The slump flow and air content were 420 mm and 6.8%, respectively. All the materials for SHCC were placed in a mixer and mixed for 5 min. When placing SHCC, the molds were tapped with a mallet instead of using a tamping rod or internal vibrator. The specimens were demolded two days after placing and then air-cured in a laboratory with a room temperature of 10 to 20 degrees Celsius until testing at 28 to 36 days. Specimens for compression and bending tests were also fabricated by the same method as tension test specimens.

3.1.2 Specimens
Table 2 and Fig. 2 show the outline of tension specimens and the geometry of dumbbell specimens, respectively. The shape and size of the plan of tension specimens were kept constant, while the thickness was varied in three levels: 13, 30, and 50 mm. Five specimens each were produced for each thickness.

3.1.3 Tensile loading testing
Figure 1 show the outline of crack observation and the state of tensile loading testing, respectively. The displacement of a gage length of 80 mm in the center of each specimen was measured with sensitive displacement transducers with an accuracy of 1/1,000 mm set on both sides. The loading speed was around 0.5 mm/min.

Cracking in three specimens for each size was observed with a microscope (VH-5000 manufactured by Keyence Corporation) while stopping the loading at tensile strain levels of 0.25%, 0.63%, 0.95%, 1.25%, 1.88%, 2.5%, 3.13%, 3.75%, and 5.0%. The observation area was the area 80 mm in length and 10 mm in width on the bottom (molded surface) near the side surface of each specimen where cracking appears to be widest and easy to observe. The microscope with a magnification of 25 was moved along this area to measure the cracks.

3.2 Compression and bending test results

Table 2 and Fig. 3 show the compression and bending test results. Bending tests were conducted by third-point loading. The compressive strength and elastic modulus were 34.4 MPa and 16.3 GPa, respectively. The flexural load-displacement relationship clearly shows increases in the load as the displacement (strain) increases after cracking, the so-called "deflection hardening characteristics." Also, multiple fine cracks were visually observed after bending testing.

3.3 Tension test results and discussion

Tension tests were reliably accomplished using the test apparatus proposed in this study, with all specimens being broken in their test zones.

3.3.1 Size effect on stress-strain relationship
Figure 4 shows the tensile stress-strain relationship of each thickness of tension specimens. Strain hardening characteristics, in which the tensile stress increases as the strain increases without brittle failure after the first cracking, are observed in specimens of all thicknesses. The ultimate tensile strains widely scatter.

The average tensile strength is approximately 4 MPa with no appreciable difference between thicknesses.

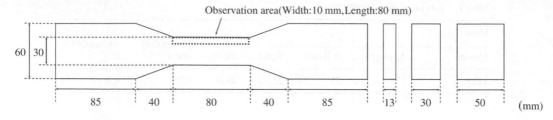

Figure 2.   Specimen geometry.

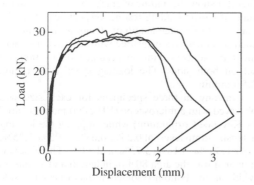

Figure 3.   Flexural load-displacement relationship.

However, the tensile strength of 13 mm-thick specimens widely scatters when compared with other thicknesses.

The tensile yield strength showed no appreciable difference between thicknesses. However, the ultimate tensile strain of 13 mm- and 30 mm-thick specimens concentrates around 3% to 4%, whereas that of 50 mm-thick specimens widely scatters from 2% to 6%.

3.3.2   Cracking tendencies and size effect

Figure 5 shows the crack patterns of specimens at tensile strain stages of 0.63% and 1.88%, as well as the final strain stage observed by a microscope. Cracks expressed as bold lines in the figure represent particularly wide cracks.

Multiple cracks are found in all specimens, with their number increasing as the strain increases. A higher ultimate strain leads to a larger number and area of cracks in specimens of all thicknesses.

In the case of specimen No. 3 with a thickness of 13 mm, however, its ultimate tensile strain is 5% or more, but the number of cracks is smaller than other specimens with an ultimate tensile strain of around 3%. This can be attributed to the small area of microscopic crack observation rather than the effect of different thicknesses.

The cracks of 50 mm-thick specimens tend to be long, whereas those of 13 mm-thick specimens

tend to be very short, and their numbers and length increase as the strain increases, with new short cracks developing. The crack patterns of 30 mm-thick specimens are intermediate between the other two thicknesses.

3.3.3   Crack width and size effect

Figure 6 shows the crack width-strain relationships. The widths of the cracks measured using a microscope, which are shown with blocks (1 to 6) in Fig. 5, are shown with the same blocks in Fig. 6. Measurement was made for the first three cracks and the three widest cracks in each specimen. In regard to specimens with an ultimate tensile strain of 4% or more (as indicated in each graph), the crack width increases up to a strain of approximately 1% but the increment of crack width becomes very small, with the maximum crack width being not more than 0.2 mm. However, the maximum crack width tends to scatter more widely as the specimen thickness increases. Nevertheless, this requires further investigation over a wider range of crack observation.

In regard to specimens with an ultimate tensile strain of less than 4% (as indicated in each graph), crack width development is classified into two types with a strain of over 1%: the crack width rapidly increases in one type and scarcely increases in the other type. As inferred from Fig. 5, this is presumably due to relative localization of cracking. This tendency is common to all thicknesses.

4   CONCLUSIONS

The results of direct uniaxial tension tests are summarized as follows:

– Tension tests were reliably accomplished using the test apparatus proposed in this study, with all specimens broken in their test zones.
– Strain hardening characteristics were observed in specimens of all thicknesses. Though the tensile strength and the tensile yield strength scarcely varied between the specimen thicknesses, the ultimate

Crack observation : — No.1, - - No.2, - ·· No.3, — No.4, No.5

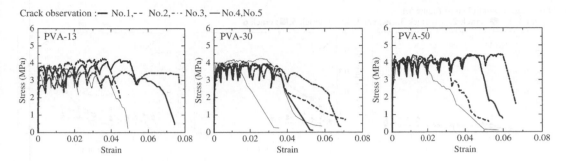

Figure 4.   Tensile stress-strain relationship.

Observation point : 1○, 2●, 3△, 4▲, 5□, 6■

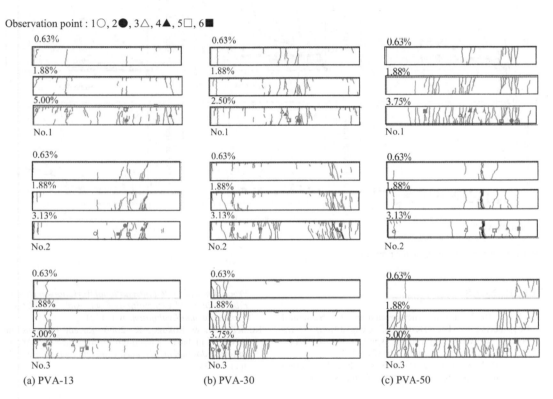

(a) PVA-13   (b) PVA-30   (c) PVA-50

Figure 5.   Crack patterns of specimens at tensile strain stages.

tensile strain scattered more widely as the specimen thickness increased.

– In regard to cracking properties, each crack in 50 mm-thick specimens is long, whereas it is short in 13 mm-thick specimens, and its length increases as the strain increases, while new short cracks begin to appear.

– Regardless of the thickness, the crack widths of all specimens increased up to a strain of approximately 1%. When the strain exceeded 1%, the crack width of specimens with an ultimate tensile strain of 4% or more scarcely increased as the strain increased, with the maximum crack width being not more than 0.2 mm. In specimens with the ultimate tensile strain of less than 4%, however, two increasing patterns in the crack width were observed after the strain exceeded 1%. In one pattern, the crack width significantly increased as the strain increased, whereas it scarcely increased in the other pattern.

Observation point (1~6 of Figure 5.)
○: crack 1, ●: crack 2, △: crack 3, ▲: crack 4, □: crack 5, ■: crack 6

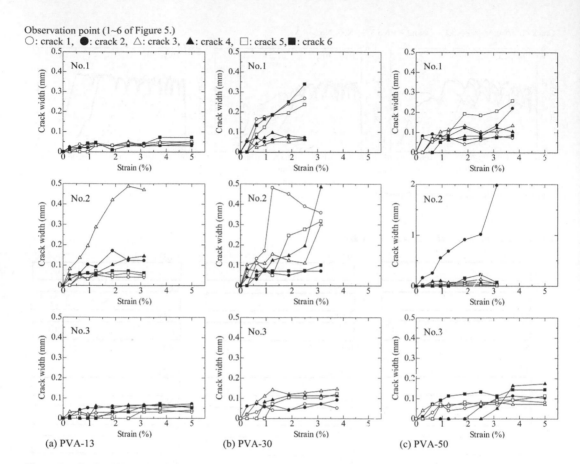

Figure 6.   Crack width-strain relationships.

REFERENCES

Li, V.C. 1993. From Micromechanics to Structural Engineering – The Design of Cementitious Composites for Civil Engineering Applications. *J. Struct. Mech.Earthquake Eng.*, JSCE, 10 (2): 37–48.

Li. V.C. 2002. Reflections on the Research and Development of Engineered Cementitious Composites (ECC), *Proceedings of the JCI International Workshop on Ductile Fiber Reinforced Cementitious Composites (DFRCC) – Application and Evaluation*, JCI: 1–21.

Kunieda, M. & Rokugo, K. 2006. Recent Progress on HPFRCC in Japan – Required Performance and Applications –, *Journal of Advanced Concrete Technology*, JCI, 4(1): 19–33.

Fracture Mechanics of Concrete and Concrete Structures – High-Performance Concrete,
Brick-Masonry and Environmental Aspects – Carpinteri, et al. (eds)
© 2007 Taylor & Francis Group, London, ISBN 978-0-415-44617-4

Tensile/flexural fracture behavior of composite specimens combining SHCC and concrete

T. Mizuta & K. Rokugo
Department of Civil Engineering, Gifu University, Japan

T. Inaguma
Central Japan Railway Consultants Co., Ltd, Japan

ABSTRACT: Tensile and flexural fracture behavior of composite specimens made of SHCC and normal concrete (NC) was investigated. Uniaxial tension tests on these materials were successfully carried out without breaking the enlarged ends of dumbbell specimens by a method in which their shoulders are engaged in locks and subjected to tensile loading. Composite specimens under bending action having NC on the tension side underwent early cracking in NC, resulting in a lower load-bearing capacity of specimens and smaller deformation of SHCC. In composite specimens having SHCC on the outside of NC, cracks in NC under tensile and bending action were rendered fine and multiple ahead in the outer layer of SHCC.

1 INTRODUCTION

Strain-hardening fiber-reinforced cement-based composites (SHCC) (Naaman & Reinhardt 1995, Li 1993, Kunieda & Rokugo 2006) are a highly ductile cementitious material showing multiple-cracking characteristics and pseudo-strain-hardening characteristics under tensile and bending forces. When applied to newly built structural members, SHCC may be used as part of composite members with normal concrete (NC), which shows brittle fracture characteristics. It may also be used for surface repair of concrete structures having cracks to control the surface crack widths.

This study aims to experimentally elucidate the tensile/flexural fracture behavior of composite specimens combining SHCC and NC focusing on the load, deformation, and cracking.

2 OUTLINE OF EXPERIMENT

2.1 Materials

Table 1 gives the mix proportions of SHCC and NC. High-early-strength portland cement was used as the cement for both. SHCC included No. 7 silica sand as fine aggregate and polyethylene fibers 0.012 mm in diameter and 12 mm in length at a fiber ratio of 1.5%

Table 1. Mix proportions of SHCC and NC.

Types of material	Cement (kg/m^3)	Water/Cement ratio (%)	PE fiver (vol%)
SHCC	1264	30	1.5
NH	331	55	–

by volume. Crushed stone with a maximum size of 15 mm was used as coarse aggregate for NC.

2.2 Uniaxial tension testing

2.2.1 Tension specimens

The geometry and types of tension specimens are shown in Fig. 1 (a) and Table 2, respectively. The specimens were of a dumbbell type measuring 400 mm in length, 100 mm in depth, and 100 and 60 mm in width at the enlarged ends and narrowed center, respectively. For composite specimens with NC sandwiched between two SHCC layers ("SNS" specimens), an SHCC layer was placed to a depth of 25 mm, followed by a NC layer of 50 mm and then another SHCC layer of 25 mm. Each layer was placed within 2 hours after placing the previous layer. Composite specimens with SHCC sandwiched between two NC layers ("NSN"

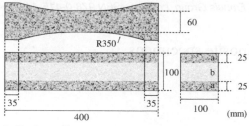

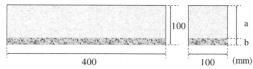

(a) Tension specimens

(b) Flexure specimens

Figure 1. Specimen geometry.

Table 2. Tension specimens.

Series	Layer	Types of material
SNS	a	SHCC
	b	NC
NSN	a	NC
	b	SHCC
SHCC	a	SHCC
	b	
NC	a	NC
	b	

specimens) were also fabricated similarly. Two specimens were prepared for each set of conditions due to limitations of molds.

2.2.2 Tension test procedure

Figure 2 shows the uniaxial tension test setup. The enlarged ends of each specimen were engaged in steel locks. Both the clamps on the right and left sides of each lock were fixed with bolts to prevent them from opening apart during testing. Stiff-mix gypsum that begins to harden in around 10 min was thinly inserted between the clamps and specimens of NSN, SNS, and homogeneous NC to alleviate localized stress concentration.

Loading was applied using a center hole-type hydraulic jack and monitored with a load cell. The displacement was detected using an accurate displacement transducer attached to the central part of the specimen with respect to a gage length of 100 mm. The lower end of each specimen was fixed while the upper end was roller-supported. The test ages ranged from 18 to 25 days.

Figure 2. Uniaxial tension test setup.

2.3 Bending testing

2.3.1 Flexure specimens

The geometry and types of flexure specimens used in this study are shown in Fig. 1 (b) and Table 3, respectively. These beam specimens measured 100 mm in width, 100 mm in depth, and 400 mm in length. For "SC#" composite specimens, NC was placed to a depth of 0, 10, 20, 30 or 40 mm from the tension side, followed by SHCC. For "CS#" composite specimens, SHCC was conversely placed in the tension side ("#" represents the depth of the material on the tension side). Note that the cases of 0 mm are homogeneous specimens. Homogeneous specimens with a depth of 60 to 90 mm were also fabricated in addition. Basically two specimens were fabricated for each set of conditions.

2.3.2 Bending test procedure

Bending tests were conducted by third-point loading at test ages of 18 to 25 days similarly to uniaxial tension tests. The loads were monitored with a load cell, while detecting the displacement using accurate displacement transducers placed at the support points and load points.

2.4 Crack observation

In addition to visual observation, the images of cracks were scanned into a PC using a microscope of 50 magnifications (VH-5000 manufactured by Keyence Corporation).

Table 3. Flexure specimens.

Series	Layer	Types of material	Depth (mm)				
SC#	a	SHCC	100	90	80	70	60
	b	NC	0	10	20	30	40
CS#	a	NC	100	90	80	70	60
	b	SHCC	0	10	20	30	40

#: Represents depth of material on tension side.

Table 4. Peak loads obtained from uniaxial tension tests.

Series	Peak loads (kN)		Average (kN)
SNS	23.0	18.9	20.9
NSN	22.2	20.1	21.1
SHCC	19.8	20.5	20.2
NC	26.5	19.8	23.1

3 TEST RESULTS

3.1 Uniaxial tension testing

3.1.1 Fracture location and peak load
The fracture (rupture of NC and crack widening in SHCC) was localized to the narrowed center (60 mm in width) of each specimen. In other words, uniaxial tension tests on SHCC and NC were successfully carried out without breaking the enlarged ends of dumbbell specimens by engaging their shoulders in the locks and applying tensile forces.

The peak loads obtained from the uniaxial tension tests are given in Table 4. The peak loads of composite specimens made of SHCC and NC were intermediate values between those singly made of SHCC and NC. In other words, the tensile capacities of composite specimens were the sum of those of homogeneous SHCC and NC specimens.

3.1.2 Load-displacement relationship
Figure 3 shows the load-displacement relationships obtained from the uniaxial tension tests. Homogeneous SHCC specimens showed clear pseudo-strain-hardening behavior in which the load increased after the first cracking. In regard to composite specimens made of SHCC and NC at a cross-sectional area ratio of 50:50, cracking occurred in NC at the point of the peak load. The load then rapidly decreased, ending up in brittle failure. This can be attributed to the fact that the tensile force borne by NC cannot be carried only by the load increment of SHCC owing to its post-cracking strain-hardening.

When a combination of NC and SHCC is used for members subjected to tensile action, it is generally expected to include steel reinforcement. In such a case, it is necessary to adequately plan the bar arrangement so that the tensile forces are stably borne by such members after crack onset in NC in a similar manner as the requirement for the minimum reinforcement ratio of reinforced concrete members.

One each of the two specimens of NSN and SNS underwent rapid rupture, with the displacement being out of control after the peak load, but the displacement of the other ones was measurable in the course. The maximum load of composite specimens, which is borne by SHCC after the peak load owing to the bridging of fibers, was found to be 1.2 times the value expected for homogeneous SHCC specimens (half the load of specimens with a depth of 100 mm). This is presumably because the ultimate fracture zone of SHCC is limited to the location of large cracks in NC and because not only SHCC but also crackless portions of NC bear part of the tensile forces.

No marked difference was observed between the deformations of composite specimens after the peak load that were measurable and those of homogeneous SHCC specimens. However, this requires further investigation due to the small numbers of specimens.

3.1.3 Cracking
In the SNS specimen whose displacement after crack onset in NC was under control, the cracking in the outer SHCC layers was consisted of multiple fine cracks even after cracking occurred in the inner NC layer. Cracks in the central SHCC layer in the NSN specimen were similarly fine and multiple.

Figures 4 and 5 show the photographs of cracking on the side surfaces of specimens observed after uniaxial tension testing. Whereas localized wide cracks are found in the NC layer, these are arrested into multiple fine cracks in the SHCC layer. The cracking regions in SHCC were limited by cracking in NC.

3.2 Bending testing

3.2.1 Composite ratio and peak load
Figure 6 shows the relationship between the depth of NC or SHCC from the tension side and the peak load of composite specimens. The results of homogeneous NC and SHCC specimens with a depth of 60 to 90 mm are superimposed as broken lines. The solid line in Fig. 6 (a) represents the expected peak load neglecting NC on the tension side (proportional to the square of the depth of the SHCC layer).

The peak load of composite specimens with NC on the tension side is significantly lower than the level of the broken line and close to the solid line, which neglects NC. This can be attributed to large cracking in NC at an early stage.

On the other hand, the peak load of composite specimens with SHCC on the tension side is similar to the

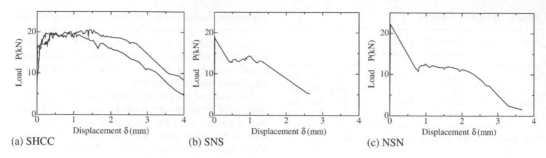

Figure 3. Load-displacement relationships obtained from uniaxial tension tests.

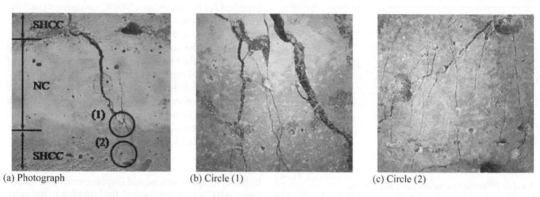

(a) Photograph   (b) Circle (1)   (c) Circle (2)

Figure 4. Photography of cracking on side surfaces of specimens (SNS).

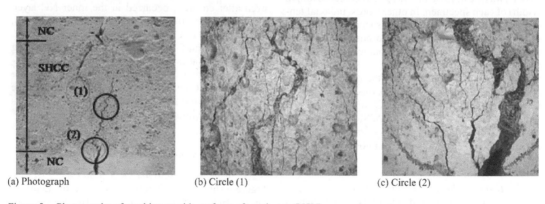

(a) Photograph   (b) Circle (1)   (c) Circle (2)

Figure 5. Photography of cracking on side surfaces of specimens (NSN).

level of the broken line. It is therefore inferred that both SHCC and NC contribute to the resistance to the bending action, thanks to the large deformability of SHCC.

3.2.2 Load-displacement curve

The load-displacement curves obtained from the bending tests are shown in Figs. 7 and 8. The displacement (e.g., those corresponding to the point of the maximum load after the effect of fibers has been developed) of specimens having the NC layer on the tension side is as low as 1/2 to 1/3 of that of homogeneous SHCC

specimens with the same SHCC depth (e.g., SH80). In other words, placing a brittle layer of NC on the outside of highly ductile SHCC restricts the deformability of SHCC. In this light, care should be exercised when using SHCC in combination with normal concrete and asphalt concrete.

On the other hand, SHCC on the tension side bears the load and deforms after cracking occurs in NC, and the displacement increases as the depth of SHCC increases. The load-displacement curve of specimens having SHCC to a depth of 40 mm on the tension

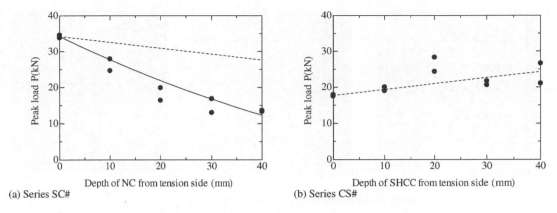

(a) Series SC#   (b) Series CS#

Figure 6.   Relationship between depth of NC of SHCC from tension side and peak load of composite specimens.

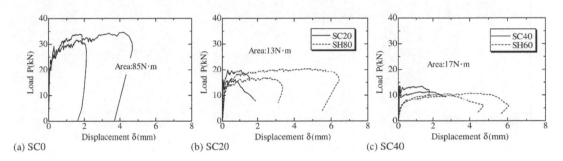

(a) SC0   (b) SC20   (c) SC40

Figure 7.   Load-displacement curves obtained from bending tests (SC#).

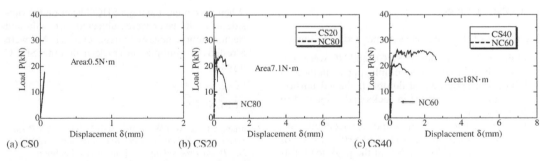

(a) CS0   (b) CS20   (c) CS40

Figure 8.   Load-displacement curves obtained from bending tests (CS#).

side is close to that of homogenous SHCC specimens. Placing SHCC on the tension side of brittle NC subjected to bending action thus improves the load-bearing capacity and deformability of such composite specimens.

3.2.3   Cracking
The state of cracking after bending tests is shown in Figs. 9 and 10. Figures 7 and 8 show the average areas under the load-displacement curves of composite specimens up to the point of the maximum load after the effect of fibers has been developed. Figure 10 clearly shows that cracks that occurred in NC propagate into SHCC, turning into multiple fine cracks. Specimens with a larger area under the load-displacement curve tend to have larger cracking regions on their side surfaces with a larger number of cracks. The authors intend to continue investigation focusing on cracking regions and the number of cracks.

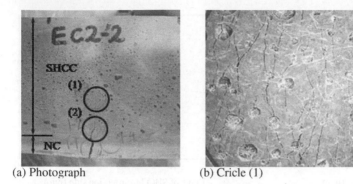
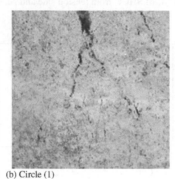
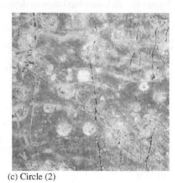

| (a) Photograph | (b) Cricle (1) | (c) Cricle (2) |

Figure 9.   Photography of cracking on side surfaces of specimens (SC20).

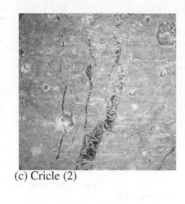
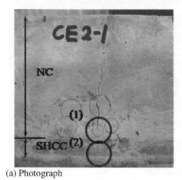

| (a) Photograph | (b) Circle (1) | (c) Circle (2) |

Figure 10.   Photography of cracking on side surfaces of specimens (CS20).

4   CONCLUSIONS

The results of this study are summarized as follows:

– Uniaxial tension tests on SHCC and NC were successfully carried out by employing a method in which the shoulders of dumbbell-shaped tension specimens are engaged in locks and subjected to tensile forces.
– Uniaxial tension tests on composite specimens having a cross-sectional area ratio of SHCC to NC of 50:50 led to cracking in NC at the peak load, followed by rapid reductions in the load, ending up in brittle failure.
– When subjected to bending action, composite specimens having SHCC on the tension side and NC on the inside showed increased load and deformation. However, specimens conversely having NC on the tension side not only showed early cracking in NC, which reduced their load-bearing capacity, but also restricted the deformation of SHCC.

– When specimens having a SHCC layer on the outside of a NC layer were subjected to tensile and flexural action, newly occurring cracks in NC were arrested into multiple fine cracks in the outer SHCC layer.

REFERENCES

Naaman, A.E. & ReH.W inhardt. (eds.) 1995. *High Performance Fiber Reinforced Cement Composites 2 (HPFRCC 2), RILEM Proceedings 31*. London, E&FN Spon.
Li, V.C. 1993. From Micromechanics to Structural Engineering -The Design of Cementitious Composites for Civil Engineering Applications. *J. Struct. Mech.Earthquake Eng.*, JSCE, 10 (2): 37–48.
Kunieda, M. & Rokugo, K. 2006. Recent Progress on HPFRCC in Japan – Required Performance and Applications –, *Journal of Advanced Concrete Technology*, JCI, 4(1): 19–33.

Stress-strain behaviour of Strain-Hardening Cement-based Composites (SHCC) under repeated tensile loading

V. Mechtcherine & P. Jun

Technische Universitaet Dresden, Institute for Building Materials, Dresden, Germany

ABSTRACT: This paper presents results of an experimental investigation on the behaviour of strain hardening cement-based composites (SHCC) subjected to cyclic tensile loading. A series of uniaxial tensile tests on unnotched, dog-bone shaped prisms containing 2.25% by volume of polymeric fibre were performed using both a deformation and load control testing regime, respectively. The experimental program was replenished by a number of deformation controlled monotonic tensile tests as well as tensile creep tests. Two different specimen sizes were used. The effect of the curing conditions before testing was also investigated. The results obtained from the deformation controlled tests revealed no pronounced effect of the cyclic loading on the material performance for the number of loading cycles chosen. However, cracked specimens tested under the load control regime were more prone to failure at lower strain levels. The number of loading cycles prior to failure was considerably larger in comparison to the deformation controlled cyclic tests.

1 INTRODUCTION

This paper addresses the group of fibre reinforced cement-based composites (SHCC) which exhibit strain hardening, quasi-ductile behaviour due to the bridging of fine multiple cracks by short, well distributed fibres. The characteristic behaviour of SHCC in tension under monotonic, quasi-static loading was studied intensively during the last few years; see e.g. Mechtcherine & Schulze (2005), Mechtcherine & Schulze (2006). However, in practice, the majority of concrete structures is exposed to more or less severe cyclic loadings, such as traffic loads, temperature changes, wind gusts and in some cases sea waves, vibrations due to the operation of machinery or, in extreme circumstances, earthquake. Therefore, a profound knowledge of the fatigue behaviour of SHCC is indispensable for a safe and economical design of structural members, as well as building elements for which such materials might be used.

As of yet, only a few investigations on SHCC behaviour under cyclic loading have been performed. Fukuyama et al. (2002) investigated the cyclic tension-compression behaviour of two SHCC materials, which possessed a strain capacity of 0.5% and 1.0%, respectively; only about five cycles were needed until the strain capacity expired, while the cyclic tension response accurately reflected the corresponding curve obtained from a monotonic tension test. In contrast to this result, Douglas & Billington (2006) found that the envelop stress-strain curve from the cyclic tests

laid below the relation as measured in the monotonic regime. The difference was particularly pronounced in the experiments with high strain rates. The investigated SHCC showed a strain capacity of approximately 0.5% when subjected to monotonic, quasi-static loading.

Jun and Mechtcherine (2007) investigated an SHCC with a strain capacity that was clearly above 2% in all tests. A higher number of loading cycles was used compared to earlier studies. Furthermore, two different types of loading regimes were applied: deformation controlled and load controlled tests.

This paper presents results from a subsequent study in which additional test types (tensile creep tests) and test parameters (specimen size, curing conditions) were used. Results obtained will be presented and discussed in concert with previous results, and supplementary evaluation methods will be applied.

2 MATERIAL COMPOSITION

The characteristic behaviour of SHCC under monotonic tensile loading is shown in Figure 1 and can be described as follows. Microscopic defects trigger the formation of matrix cracks at so-called first crack stress (σ_1). As the first crack forms, the fibres bridge the crack transmitting tensile stresses across the crack surfaces. The applied load must be increased in order to enforce a further crack opening. This action leads to the formation of another crack at the second weakest cross-section. The scenario then repeats resulting

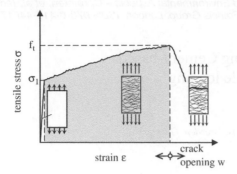

Figure 1. Typical stress-strain response and crack pattern of SHCC specimens under monotonic tensile loading.

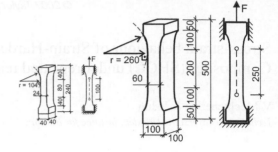

Figure 2. Geometry of the specimen used for the tensile tests.

Table 1. SHCC composition used for the experiments.

Cement [kg/m^3]	Fly ash [kg/m^3]	Silica sand [kg/m^3]	Water [kg/m^3]	SP [kg/m^3]	VA [kg/m^3]	PVA fibres [kg/m^3]
321.0	749.1	535.0	334.5	16.6	3.2	29.3

in a set of almost uniformly distributed cracks. The strain capacity is reached at the maximum load (tensile strength f$_t$), when the localisation of the failure occurs (one main crack develops). Due to a moderate opening of a large number of fine cracks, a strain capacity of several percent can be observed.

SHCC, unlike common fibre reinforced concrete, is a micro-mechanically designed material. The approach for such material design was developed by Li (1993) for a composite which he called ECC (Engineered Cementitious Composite). The material used in this investigation was developed on the basis of this approach in earlier investigations by the authors; see e.g. Mechtcherine & Schulze (2005). However, a more specific term, SHCC, is used in this paper in conjunction with previous research conducted in this field.

Table 1 gives the SHCC composition used for the experiments. A mix containing a combination of Portland cement 42.5 R (30% by mass) and fly ash (70% by mass) was utilized as a binder. The fine aggregate was a uniformly graded silica sand with particle sizes of 0.06 mm to 0.20 mm. Furthermore, PVA fibres, 2.25% by volume with a length of 12 mm, were applied. A superplasticizer (SP) and a viscosity agent (VA) were added to the mix in order to adjust its rheological properties.

The average compressive strength of the SHCC was 33.7 MPa. The compressive strength was derived from 12 displacement controlled tests on cubes with a side length of 100 mm. The displacement rate of the crosshead of the loading machine was 0.01 mm/s. The

findings concerning the stress-strain curves obtained and the observed crack pattern will be published elsewhere.

3 TEST SET-UP, TESTING PROCEDURE AND EXPERIMENTAL PROGRAM

3.1 Specimen geometry, casting, curing, set-up

Based on the findings of previous investigations (Mechtcherine & Schulze, 2005 and Mechtcherine & Schulze, 2006), unnotched, dog-bone shaped prisms were chosen as specimens for this study. Such prisms were produced in two sizes, with identical geometrical shapes. The smaller prisms possessed a cross-section of 24 mm × 40 mm, while the larger ones had a cross-section of 60 mm × 100 mm. The gauge lengths were 100 mm and 250 mm, respectively. Figure 2 gives further geometric data for the specimens.

All specimens were cast horizontally in metal forms. The moulds were stored 2 days in a climate box (T = 25°C, RH = 65%). After demoulding, the specimens were wrapped in a plastic foil (sealed condition) and stored until testing at room temperature. For some test series the specimens were stored after demoulding at the room atmosphere without any protection from desiccation (i.e. unsealed condition) in order to subsequently study on these specimens the effect of the curing conditions on the material performance under cyclic loading. All specimens were tested at a concrete age of 28 to 32 days.

The uniaxial tension tests were performed with non-rotatable boundaries. The deformations were measured by means of two LVDTs fixed to the specimen as displayed in Figure 3. The specimen surfaces were covered with a thin brittle white paint in order to facilitate monitoring of the crack's development.

3.2 Testing procedure

Four types of the experiments were performed with regard to the loading procedure: (1) monotonic

Figure 3. Used test set-up (here for small prisms).

deformation controlled tests; (2) cyclic deformation controlled tests; (3) cyclic load controlled tests, and (4) creep tests.

In the deformation controlled experiments, the deformation rate was always 0.01 mm/s which corresponds to a strain rate of 10^{-4} 1/s for the smaller prisms and $4 \cdot 10^{-5}$ 1/s for the larger specimens. For the deformation controlled cyclic tests the increase of the total deformation within the measuring length was given by the deformation increment $\Delta \delta$ which was chosen to be equal 0.1 mm for the small prisms and 0.25 mm for the large prisms (which corresponds to a strain increment of 0.1%) while being held constant from cycle to cycle. When the preset value $\Delta \delta$ in the following cycle was reached, the specimen was unloaded until the lower reversal point δ_{min} was attained. The lower reversal point δ_{min} was defined as a function of the lower load level $F_{min} = const = 0$ N.

In the load controlled cyclic tests, the specimen was first loaded monotonically until the strain value 0.5% was reached. The specimen was then unloaded and subsequently reloaded cyclically in a load control regime with predefined lower and upper load limits. The preloading in the monotonic regime was needed because of a pronounced scatter in first crack stress σ_1 as had been observed in the monotonic tests. The knowledge of the behaviour of a particular specimen at the beginning of cracking enabled a purposeful choice of the upper stress limit σ_{up} for the cyclic loading. The lower limit was always set to $F_{min} = 0$ N, while the upper limit σ_{up} was chosen under careful consideration of the measured material behaviour during the initial monotonic loading regime.

Since the stress-strain curves under monotonic loading were generally rather unsteady in this investigation (see, for example, the curve shape given in Figure 5b), the choice of the upper limit value σ_{up} was not straightforward. Basically, a stress value lying

between the stress at first cracking σ_1 and the maximum stress σ_{max} measured before changing to the load control was sought. However, in some cases, after the first cracking, a sudden stress drop occurred and the stress level at the strain of 0.5% was below the stress at first cracking σ_1. In such cases, the stress level after the stress drop due to the first cracking and subsequent "recover" was used instead of the σ_1 value.

The load frequency in the load controlled cyclic tests was 0.5 Hz, i.e. each loading cycle took 2 seconds.

In the creep tests the specimens were also loaded monotonically until the strain value of 0.5% was reached. Once reached, the test control mode was switched to the load control regime. In order to derive the creep stress, the same procedure that was used to determine the upper limit value σ_{up} in the load controlled cyclic tests was also used for this procedure.

3.3 Overview of the experimental program

The results of two test series will be presented in the subsequent chapter. These following series, referred to as Series I and II, were performed with a time interval of a few months using the same SHCC composition. Despite the same production procedure, the mechanical performance of the specimens of Series II was somewhat different – in general, a lower tensile strength and a higher strain capacity – then those produced from Series I. A reason for this difference might be the usage of a new charge of the cement and fly-ash for the SHCC production. More comments relevant to this issue will be given in Chapter 4.

In Series I only the small prisms were tested, and all were stored under sealed condition until testing. Five specimens were tested for each loading condition (monotonic, cyclic deformation controlled and cyclic load controlled). Two specimens were subjected to sustained tensile loading for the creep test.

In Series II, four small prisms, stored unsealed after demouding, were tested under the monotonic and deformation controlled regimes, respectively. Additionally, monotonic and cyclic tests (both deformation and load controlled) were performed on the large prisms. Two specimens were tested in each loading regime.

4 EXPERIMENTAL RESULTS

4.1 Behaviour under monotonic loading

The behaviour of the SHCC under monotonic tensile loading was studied in detail in earlier investigations (e.g. Mechtcherine & Schulze 2006). Therefore, the results from the monotonic tests obtained in this study will be presented and discussed solely as a reference to the corresponding results from the cyclic tests. However, one general remark should be made with regard to the shape of the measured stress-strain curves. These curves display some sudden "jumps" and "drops" which were not experienced by the authors in the previous investigations using the same material composition (Mechtcherine & Schulze 2006). One reason for this peculiarity might be the change of the testing machine and the test control regime (the *displacement* of the machine cross-head was used in the earlier studies for the test control). The "new" machine – which was more appropriate for the cyclic tests – working in the deformation controlled regime evidently tended to "overcontrol" the tests in the instance of sudden stiffness changes due to the formation of new cracks. The unsteadiness of the stress-strain curves hampered the evaluation of the results to some degree, but, otherwise, did not seem to affect the information obtained from the experiments.

4.2 Results of deformation controlled cyclic tests

Figure 4 shows representative results from deformation controlled cyclic tests in comparison to the curves obtained using the monotonic loading regime. Nearly no effect of the given, relatively moderate number of loading cycles on the shape of the stress-strain diagram can be observed when considering only the envelop curves. This holds true for the experiments both with small and large prisms, as well as with sealed and unsealed specimens. The lack of a pronounced distinction in the envelope curve course naturally results in only minor differences in the average values of the stress at first cracking σ_1, tensile strength f_t, as well as the strain capacity ε_{tu}, for these two different loading regimes; see Table 2.

The curves obtained from the cyclic tests show characteristic hystereses from which it can be clearly recognised that the high strains are to a great extent due to non-elastic deformations (see strains at zero stress). Furthermore, the stiffness of the composite gradually decreases (the inclination of the hysteresis curves with regard to the strain axis declines).

Table 3 gives values of the respective secant modulus of elasticity for several chosen strain levels. Tests with all three parameter combinations show a very pronounced decrease of this characteristic by a factor of approximately four, when the strain gradually increases from 0.5% to 3.0%.

A comparison of the stress-strain curves obtained for the small and large prisms, as well as the analysis of the derived characteristic values, show some discrepancy. The large prisms seem to provide a little smaller values of the first cracking stress level and the tensile strength, but they display a more apparent hardening and a higher strain capacity. This is in contrast to the results of an earlier investigation (Mechtcherine & Schulze 2005) in which an opposite tendency had been observed. However, unlike the cited investigation, in this study the large and small prisms were produced

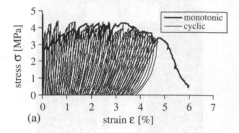

(a)

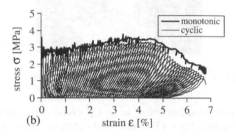

(b)

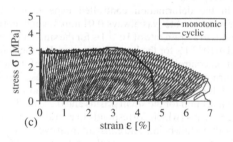

(c)

Figure 4. Representative results of the deformation controlled cyclic tests in comparison to the curves obtained using monotonic loading for: (a) small sealed prisms, (b) large sealed prisms, and (c) small unsealed prisms.

from different batches and with a time interval of several months. Therefore, the observed difference is evidently due to the minor changes of the raw materials used (see also Section 3.3).

A similar note should be made with regard to the apparent effect of the curing conditions in the tests on the small prisms. In previous studies, slightly lower values of the stress at the first cracking, the tensile strength and the strain capacity had been observed in the case of unsealed specimens. In the present study, the difference in the first two characteristic values was more pronounced, and the strain capacity was higher in the tests on the unsealed prisms. Minor changes of raw material might be the reason for such a discrepancy.

Remarkably, the curves obtained from the monotonic and cyclic tests on the specimens subjected to desiccation (i.e. unsealed ones) showed a much smoother shape of the stress-strain diagrams in comparison to those measured on the sealed specimens

Table 2. Statistical evaluation of the mechanical performance of the investigated SHCC for different loading regimes.

Type of loading	Number of cycles $N[-]$	Stress at first cracking σ_1 [MPa]	Tensile strength f_t [MPa]	Upper stress (l. con.) σ_{up} [MPa]	Strain capacity ε_{tu} [%]
		Average value (standard deviation)			
Series I, small prisms, sealed, 5 specimens per loading type					
Monotonic, deform. controlled	1	3.6 (0.7)	4.7 (0.3)	–	2.5 (0.8)
Cyclic, deformation controlled	24 (6)	3.9 (0.7)	4.3 (0.1)	–	2.4 (0.6)
Cyclic, load controlled	1840 (1200)	3.4 (0.4)	3.9 (0.2)*	3.4 (0.3)	2.4 (0.3)
Creep	1	3.4 (-)*	3.9 (-)*	3.7 (-)	3.0 (-)
Series II, large prisms, sealed, 2 specimens per loading type					
Mon., def. control.	1	3.2 (-)	3.8 (-)	–	5.0 (-)
Cyc., def. control.	38 (-)	3.2 (-)	3.4 (-)	–	3.8 (-)
Cyc., load control.	2100 (-)	2.9 (-) *	3.1 (-) *	2.8 (-)	2.0 (-)
Series II, small prisms, unsealed, 4 specimens per load type					
Monotonic, def. control.	1	3.0 (0.4)	3.3 (0.5)	–	3.3 (0.4)
Cyclic, def control.	42 (8)	3.1 (0.3)	2.9 (0.1)	–	4.2 (0.8)

* Values belong to the monotonic curves before changing to the load controlled regime.
(-) Only two measurements were performed.

(compare Figures 4a and 4c). This can probably be traced back to the effect of the SHCC drying on the crack propagation in the specimen. Tensile eigenstresses develop at the surface vicinity due to a pronounced moisture gradient, causing formation of micro-cracks even before any mechanical loading is applied. As a result, when such a specimen is subjected to a mechanical loading, the distribution of stresses over the specimen cross-section is highly non-uniform (i.e. at the surfaces much higher then in the middle). The cracks are prone to develop steadily from the surface into the interior of the prisms inducing no sudden stiffness changes unlike the case of the sealed specimens which has a more uniform stress field and, accordingly, a more spontaneous development of individual cracks.

These suggestions can be confirmed by the analysis of the crack pattern; the cracks were mostly planar and usually propagated throughout the specimen in the

Table 3. Change of the SHCC stiffness with increasing induced strain as observed in deformation controlled cyclic tests.

	Larg prisms	Small prisms	
	sealed	unsealed	
	Secant modulus E [GPa]		
Strain under load	Average value (standard deviation)		
0.5%	11.0 (-)	20.4 (3.1)	13.4 (2.8)
1.0%	8.1 (-)	14.3 (2.7)	6.0 (1.3)
1.5%	5.3 (-)	10.8 (2.2)	5.3 (0.6)
2.0%	5.1 (-)	8.0 (1.5)	4.1 (0.7)
3.0%	3.9 (-)	5.1 (1.4)	3.1 (0.3)

(-) Only two measurements were performed.

case of the sealed prisms. The fracture surface after specimen failure was usually rather smooth, as well. On the contrary, the cracks observed at the surfaces of the unsealed prisms were usually not planar but rather had sophisticated patterns that did not continue throughout the entire specimen.

4.3 Results from load controlled cyclic tests

Figure 5 presents a typical stress-strain curve obtained from the load controlled cyclic tests. After switching to the load controlled regime at the strain of 0.5%, the individual hysteresis lie dense to each other and can hardly be recognized on the scale used. The shape of individual hysteresis for a few chosen strain levels will be presented and discussed in Section 4.5. On average, 1840 load cycles were needed to bring the specimen to failure in the case of the small specimens and 2100 cycles in the case of the large prisms.

In the tests on small prisms, which were performed in the frame of Series I, the strain capacity was practically the same as in the monotonic tests or the deformation controlled cyclic tests (see also Table 2). However, in Series II, the strain capacity obtained from the load controlled cyclic tests was clearly below the corresponding values for the deformation controlled monotonic or cyclic loading. This can be, at least partially, explained by the fact that the material could not relax after the formation of new cracks during the load controlled tests in contrast to the deformation controlled tests where the recurrent unloading took out a part of energy resealed due to cracking. Why this phenomenon was not observed in the tests on small prisms still need to be clarified.

Table 4 gives values of the secant modulus of elasticity for several chosen strain levels. A pronounced decrease of the material stiffness can be stated in all tests. The degree of the stiffness reduction is similar to that obtained in the deformation controlled cyclic tests

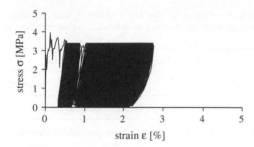

Figure 5. Representative stress-strain curves from the load controlled cyclic tests.

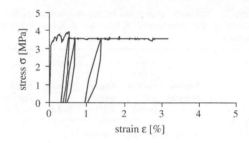

Figure 6. Representative stress-strain curve obtained from creep tests on small sealed prisms.

Table 4. Statistical evaluation of the load controlled tests, the secant modulus of elasticity.

	Large prisms, sealed	Small prisms, unsealed	Small prisms sealed
	Cyclic-load controlled		Creep
	Secant modulus E [GPa]		
Stage	Average value (standard deviation)		
0.5% loaded	18.4**	17.1 (5.7)	16.7**
1.0% loaded	11.8**	13.6 (1.3)	10.5*
1.5% loaded	9.0**	9.9 (1.0)	–
2.0% loaded	7.4*	7.5 (0.8)	6.2*

* Only one measurement was performed
** Only two measurements were performed.

(cf. Table 3); however, on average the apparent stiffness seems to be higher in the load controlled tests. This can likely be traced back to a higher strain rate in these tests in comparison to the deformation controlled tests.

4.4 Results from creep tests

The tensile creep tests were only performed with small prisms that were sealed in foil until testing. Figure 6 shows a stress-strain relation obtained from one of the experiments performed which can be considered representative for both tests. The specimens were first loaded in the monotonic deformation control regime since the creep behaviour of cracked SHCC was the subject of interest. The portion of the curve recorded prior to the change to the load control regime at the strain of 0.5% is comparable to the curves obtained from monotonic tests.

In the creep regime, the load was keep constant, except for two or three intermediate unloading and reloading cycles which were performed in order to monitor the development of the material stiffness as a result of sustained tensile loading. This issue

will be discussed in Section 4.5 together with the corresponding results from other tests.

The strain capacity was found to be slightly higher by comparison to the corresponding deformation controlled tests (cf. Table 2). The time until failure was approximately 5 hours in the first test and approximately 16 hours in the second experiment. In two further tests (not evaluated here in detail), the change to the load control regime occurred at the strain of 1% and 2%, respectively. Higher strain capacity and considerably shorter time to failure were measured in these experiments.

4.5 Shape of the hysteresis in the cyclic tests

In previous sections, the secant modulus of elasticity was presented and discussed as an appropriate measure for the change of the SHCC stiffness resulting from repeated loading. The shape of the individual hysteresis of the stress-strain curves is another feature characterising the material response. This feature will be considered here for all loading regimes used.

Figure 7 shows representative shapes of the chosen individual cycles obtained from deformation controlled and load controlled tests on small sealed prisms. In both types of tests, material stiffness gradually decreases while the hysteresis loops become wider and rounder. The change of the loop shape is more pronounced for deformation controlled tests due to the fact that these contain a considerable portion of in-elastic deformation which increases with the increasing number of loading cycles. In contrast to this, only a very small portion of in-elastic deformation was recorded for the individual hysteresis loops in the load controlled cyclic tests. The SHCC behaviour in the individual load cycles can be described as nearly non-linear elastic. However, one can also observe a clear tendency of wider loops associated with higher strain levels.

It is worth noting that the shape of the un- and reloading loops from the creep tests is similar to those obtained from the load controlled cyclic tests at the same strain levels. It is not surprising since the load

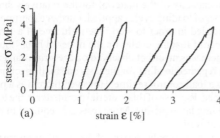

(a)

strain ε [%]

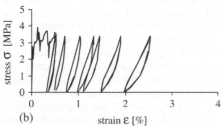

(b)

strain ε [%]

Figure 7. Representative shapes of the chosen individual cycles obtained from (a) deformation controlled and (b) load controlled tests on small sealed prisms.

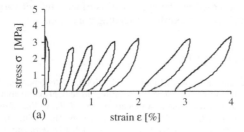

(a)

strain ε [%]

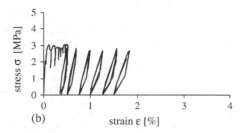

(b)

strain ε [%]

Figure 8. Representative shapes of the chosen individual cycles obtained from (a) deformation controlled and (b) load controlled tests on big sealed prisms.

control regime was applied for the creep tests, as well. The fact that the loops from the creep tests are slightly wider can be traced back to a lower un- and reloading rate in these tests compared to the cyclic load controlled tests.

The results from the cyclic tests on large prisms do not differ in essence from the results obtained for the small prisms (compare Figures 7 and 8). In the case of the deformation controlled tests, the loops of the

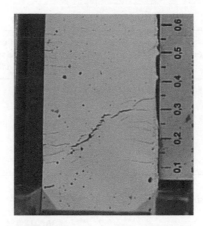

Figure 9. Typical failure pattern from the tests on small unsealed specimens.

stress-strain relations measured on the large prisms are more inclined and, at least for lower strain levels, wider than the corresponding hysteresis loops observed in the tests on the small prisms. A higher number of cracks per unit length of the specimen (cf. Table 5) provides a possible explanation to this phenomenon.

4.6 Comparison of the crack system

Development of the cracks on the specimen's surfaces was monitored during the tests; a number of high resolution digital photographs were taken at given strain levels and visually evaluated afterwards. Only cracks that propagated throughout the specimen were taken into account; no crack branches or one-sided cracks were taken into consideration. This restriction, however, strongly hampered the evaluation of the cracks observed on the small unsealed prisms. Due to pronounced micro-cracking on the specimen surfaces caused by drying, the cracks developed under loading were usually not planar but had complicated shapes and did not propagate throughout the entire specimen. As a result, fracture surfaces (final crack) were also very rough and not perpendicular to the load axis (cf. Figure 9). A more profound evaluation will be needed in order to appropriately describe the crack pattern and geometry in the unsealed prisms. For this reason, the results of these experiments under such curing conditions are not included in this paper.

Table 5 gives a statistical evaluation of the crack numbers observed in the different types of tests at the strain levels of 0.5%, 1.0% and 2.0%, respectively. For Series I of the tests, an evaluation of the crack widths is given in Jun & Mechtcherine (2007). Series II results are still being completed.

Basically, the number of cracks increases with increasing strain level while the average and maximum crack widths become slightly larger (see Jun &

Table 5. Crack number comparison.

Specimen type	Type of loading	Strain ε [%]	Number of cracks n [–] average values (stand. deviation)
Small prisms, sealed	Monotonic	0.5	–
		1	8 (2)
		2	12 (3)
	Cyclic, deformation controlled	0.5	4 (0)
		1	6 (1)
		2	9 (1)
	Cyclic, load controlled	0.5	5 (1)
		1	8 (2)
		2	12 (2)
Big prisms, sealed	Monotonic	0.5	14*/ 5.6**
		1	23*/ 9.2**
		2	36*/ 14.4**
	Cyclic, deformation controlled	0.5	13*/ 5.2**
		1	23*/ 9.2**
		2	34*/ 13.6**

* only one prism was considered, gauge length of 250 mm
** number of crack related to the reference length of 100 mm (i.e. the same gauge length as used in the tests on small prisms)

Mechtcherine, 2007). Generally, the values do not differ much with regard to the loading regime. The deformation controlled cyclic tests provided on average fewer cracks with slightly larger crack openings in comparison to the other two types of tests conducted. More testing is needed in order to prove if this difference is statistically significant.

Table 5 provides the entire number of cracks counted, for the large prisms, over the gauge length used of 250 mm; this number is then divided by 2.5, in order to obtain a direct comparison with the results obtained from the small prisms (gauge length of 100 mm). The large and small prisms revealed similar crack development with increasing strain for both monotonic and cyclic loading; however, the crack density was higher in the case of large prisms. This higher density corresponds well with the higher strain capacity of the specimens tested in Series II (large sealed prisms) in comparison to Series I (small sealed prisms). A possible reason for this difference was discussed in Section 3.3.

5 CONCLUSIONS

The following conclusions can be drawn from the results obtained relative to the effect of the loading regime on the performance of the SHCC tested.

In the deformation controlled regime, the repeated loading caused a decrease in the tensile strength of SHCC compared with the results from the monotonic tests. However, there was no pronounced effect on the

strain capacity of the material for the relatively small number of loading cycles applied. Further experiments are needed in order to study the influence of a larger number of loading cycles on the material behaviour.

There was no effect observed on the strain capacity from repeated loading for the tests on small specimens using a load control regime. However, there was a pronounced decrease of this material parameter in the tests involving large prisms. More tests are required in this area, as well.

The analysis of hysterises of the stress-strain curves showed a pronounced decrease of the material stiffness with an increasing number of loading cycles; the hysteresis loops became wider, as well. The hysterises obtained from the deformation controlled cyclic tests revealed a considerable inelastic deformation portion in every loop. The load controlled cyclic tests provided loop shapes which contained only minimal inelastic portion; however, due to a large number of load cycles in these tests, the accumulated inelastic deformations were comparable to those obtained from the deformation controlled tests for the same strain levels.

The number of cracks, as well as the crack widths, as observed on the specimen's surfaces did not vary much given the different loading conditions.

REFERENCES

Douglas K.S. & Billington S.L. 2006. Rate-dependence in high-performance fiber-reinforced cement-based composites for seismic applications. *Int. RILEM Workshop on HPFRCC in Structural Applications*, Honolulu, May 2005, G. Fischer & V. C. Li (eds.), RILEM Publications S.A.R.L., PRO 49: 17–26.

Fukuyama, H. & Haruhiko, S., & Yang, I. 2002. HPFRCC Damper for Structural Control. *Proceedings of the JCI International Workshop on Ductile Fiber Reinforced Cementitious Composites (DFRCC)*, Takayama, Japan, Japan Concrete Institute: 219–228.

Jun, P. & Mechtcherine, V. 2007. Behaviour of strain-hardening cement-based composites (SHCC) under cyclic tensile loading. *International Conference CONSEC'07*, Tours, France, accepted for publication.

Li, V. C. 1993. From micromechanics to structural engineering – The design of cementitious composites for civil engineering applications. *JSCE J. of Struc. Mechanics and Earthquake Engineering*. 10 (2): 37–48.

Mechtcherine, V. & Schulze, J. 2005. Ultra-ductile concrete – Material design concept and testing. *Ultra-ductile concrete with short fibres – Development, Testing, Applications*, V. Mechtcherine (ed.), ibidem-Verlag, Stuttgart: 11–36.

Mechtcherine, V. & Schulze, J. 2006. Testing the behaviour of strain hardening cementitious composites in tension. *Int. RILEM Workshop on HPFRCC in Structural Applications*, Honolulu, May 2005, G. Fischer & V. C. Li (eds.), RILEM Publications S.A.R.L., PRO 49: 37–46.

Fracture Mechanics of Concrete and Concrete Structures – High-Performance Concrete, Brick-Masonry and Environmental Aspects – Carpinteri, et al. (eds)
© *2007 Taylor & Francis Group, London, ISBN 978-0-415-44617-4*

A plastic damage mechanics model for Engineered Cementitious Composites

L. Dick-Nielsen, H. Stang & P.N. Poulsen
Department of Civil Engineering, The Technical University of Denmark, Lyngby, Denmark

P. Kabele
Czech Technical University in Prague, Czech Republic

ABSTRACT: This paper discusses the establishment of a plasticity-based damage mechanics model for Engineered Cementitious Composites (ECC). The present model differs from existing models by combining a matrix and fiber description in order to describe the behavior of the ECC material. The model provides information about crack opening and spacing, which makes it possible to assess the condition of a structure in the serviceability state. A simulation of a four point bending beam is performed to demonstrate the capability of the model.

1 INTRODUCTION

Engineered Cementitious Composites (ECC) is a strain-hardening fiber reinforced cementitious composite (Li and Leung 1992). In contrast to conventional Fiber Reinforced Concrete (FRC), ECC is characterized by its ability to undergo multiple cracking in tension. Conceptually, the cementitious matrix in ECC is assumed to contain initial flaws, which are randomly distributed throughout the composite material. When the material is loaded in tension, micro-cracks are initiated from the initial flaws due to stress concentrations at the tip of the flaws. The formation and propagation of micro-cracks is taking place under increasing load and gives rise to multiple cracking and strain-hardening. To assess the state of the material in the serviceability state, information about the crack opening and spacing is required and a plasticity-based damage mechanics model providing such information is presented.

Due to the strain hardening behavior of the ECC material, smeared crack models available in commercial Finite Element Method (FEM) programs are often used to simulate the behavior of ECC structures (Walter, Olesen, Stang, and Vejrum 2006) and (Dick-Nielsen, Stang, and Poulsen 2006b). An overview of smeared models is given by (Jirasek 2004). An example of a model specifically developed for ductile fiber-reinforced cement-based composite like ECC is (Han, Feenstra, and Billington 2003). This model is a total strain, rotating smeared crack model. The model is characterized by its detailed description of the unloading phase, which makes it suitable for cyclic loading simulations. In the ECC model proposed by (Kabele 2002) the cracks are fixed when initiated. The model is characterized by the scheme used to describe the stiffness of the fibers in the direction parallel with the crack surface, where the fibers are described as randomly oriented elastic Timoshenko beams. In the latter model the behavior of the cracks after initiation is described solely through the fibers.

A series of finite element simulations on the micro and meso scale has been carried out in previous investigations by the authors. These simulations were performed to get a better understanding of the strain-hardening process in ECC. On the micro scale, the mechanism during micro-crack propagation and subsequently fiber debonding and pull-out were investigated (Dick-Nielsen, Stang, and Poulsen 2005). On the meso scale, investigations on the propagation of single and multiple cracks were performed (Dick-Nielsen, Stang, and Poulsen 2006a) and (Stang, Olesen, Poulsen, and Dick-Nielsen 2006). The present paper describes the establishment of a plasticity-based damage mechanics model on the macro scale.

The present model is based on the smeared fixed, multiple cracking approach. The model differs from existing models by combining a matrix and fiber description in order to describe the behavior of the ECC material. This model is meant for use in the serviceability state. In this state the crack openings would be in the orders of 40–50 μm. A realistic description of the ECC material for these crack openings should therefore include the matrix behavior as well as the fiber behavior as shown in (Dick-Nielsen, Stang, and Poulsen 2005) and (Dick-Nielsen, Stang, and Poulsen 2006a).

The matrix is described by employing an elasto-plastic material model for crack initiation and propagation in plain concrete. The model is a mixed mode cohesive crack model combined with a modified Mohr-Coulomb yield-surface. Even though the fracture energy of the ECC matrix is low, it is shown in (Stang, Olesen, Poulsen, and Dick-Nielsen 2006) that it is best described by a cohesive approach. The matrix model employed in this paper is a modified version of the model originally developed by (Carol, Prat, and López 1997). During sliding of a crack the model is able to capture the dilatation in the normal direction.

The stiffness of the fibers in the normal direction to the crack surface is described through a multi-linear strain-stress curve, which can be found from a uniaxial tensile test. While the stiffness of the fibers parallel to the crack surface is described as randomly orientated Timoshenko beams bridging the crack (Kabele 2002).

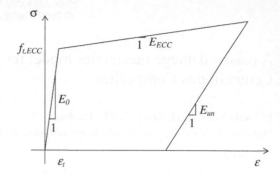

Figure 1. Constitutive relationship of ECC.

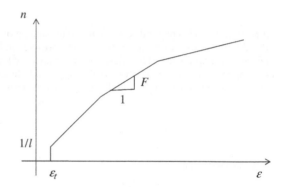

Figure 2. Damage law: strain vs. crack number per length, n.

2 PLASTICITY DAMAGE MODEL

A 2D representative volume element (RVE) in plane stress with the side length, l is considered. It is assumed that this RVE can be considered as a material point, where the constitutive equations are solved. The RVE is subjected to average total stress, σ and strain, ε, and can contain series of parallel multiple cracks, where each series have different orientation. In this section the constitutive equations for a RVE containing multiple cracks in multiple crack directions will be derived for the plastic and elastic state.

2.1 Model input

As input to the model, information concerning the ECC-level, the fibers and the matrix are required.

On the ECC-level the following data are needed: the initial E-modulus, E_0, Poisson's ratio, v, the tensile strength, $f_{t,ECC}$, a constant, b that determines the crack opening at which the crack becomes stress free during unloading and the threshold angle between two crack directions in one iteration point, ϕ. A small angle between two crack directions would be a numerical and not physical phenomenon, why the threshold angle should not be too small. The relationship between total strain, ε and stress, σ (fig. 1) is required. This relationship can be found from a uniaxial tensile test or through a bending test and an inverse analysis (Øestergaard 2003). The uniaxial tensile curve obtained from test is usually jagged (Wang and Li 2004). To avoid numerical problems during computations an idealized smooth curve is employed as shown in the figure. Finally the

relationship between the total normal strain and the numbers of parallel cracks, N in the RVE (fig. 2, $n = N/l$) is needed. Since a smooth overall stress strain response is aimed for, the latter curve has to be continuous. This means that the number of cracks per length, n increases linearly as a function of the strain in the crack direction. It is assumed that all cracks in one direction associated with one integration point has identical crack opening, δ.

For the fibers the following input is needed: a shear stiffness constant, k, which gives the relationship between crack opening and bridging stress. This constant depend on the fiber volume fraction, V_f, the shear modulus for the fibers, G_f and the shape of the fibers (Kabele 2002). The shear stiffness constant will be calibrated through experiments.

The material parameters required for the matrix are the tensile strength, f_t, the cohesion, c, two friction coefficients for the yield surface, μ_f and μ_0, a friction coefficient for the plastic potential, μ_g and the mode I and II fracture energy, $G_{F,I}$ and $G_{F,II}$.

2.2 The constitutive equations

The first crack is initiated when the stress state in the mortar reaches the yield-surface. The normal to the

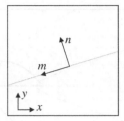

Figure 3.   Local coordinate system in crack.

crack surface is parallel with the normal stress that initiated the crack and the crack direction remains fixed. Due to equilibrium the stress, σ, is equal to the stress in the crack, σ_{cr} and the stress in the uncracked elastic part of the material, σ_e:

$$\sigma = \sigma_{cr} = \sigma_e \qquad (1)$$

The strain, ε can be split into two parts, one related to the uncracked elastic material, ε_e and one related to the additional deformations due to opening of cracks, ε_{cr}:

$$\varepsilon = \varepsilon_e + \varepsilon_{cr} \qquad (2)$$

In fig. 3 a local coordinate system is shown in a crack. The relationship between the global strain in a crack, ε_{cr} and the local strain increment in a crack, e_{cr} can be written as:

$$\varepsilon_{cr} = T e_{cr} \qquad (3)$$

$$\begin{bmatrix} \varepsilon_x^{cr} \\ \varepsilon_y^{cr} \\ \gamma_{xy}^{cr} \end{bmatrix} = \begin{bmatrix} n_\perp^2 & n_x m_x \\ n_y^2 & n_y m_y \\ 2 n_x n_y & n_x m_y + n_y m_x \end{bmatrix} \begin{bmatrix} \varepsilon_{nn}^{cr} \\ \gamma_{nm}^{cr} \end{bmatrix} \qquad (4)$$

where T is the transformation matrix. A similar relationship can be found between the global stress, σ and the traction, s in the crack:

$$s = T^T \sigma \qquad (5)$$

The relationship between the local strain in the crack, e_{cr}, the crack opening for a single crack, δ, the length of the RVE, l and the number of parallel cracks in the RVE, N can be written as:

$$e_{cr} = N/l \delta = n \delta \qquad (6)$$

where n is the number of parallel cracks per length. Thus the total strain formulation (eq. 2) can then be rewritten as:

$$\varepsilon = \varepsilon_e + T n \delta \qquad (7)$$

In incremental form the split of strain gives:

$$d\varepsilon = d\varepsilon_e + T dn \delta + T n d\delta \qquad (8)$$

The relationship between the elastic strain increment, $d\varepsilon_e$ and stress increment, $d\sigma$ is:

$$d\sigma = D_e d\varepsilon_e = D_e (d\varepsilon - T dn \delta - T n d\delta) \qquad (9)$$

where D_e is the elastic stiffness matrix. This matrix refer to the intact material between the cracks and is therefore constant through out the entire analysis.

$$D_e = [C_e]^{-1} = \frac{E_0}{1 - \nu^2} \begin{bmatrix} 1 & \nu & 0 \\ \nu & 1 & 0 \\ 0 & 0 & \frac{1-\nu}{2} \end{bmatrix} \qquad (10)$$

The relationship between crack opening for a single crack, δ and traction in the crack, s in incremental form can be written as:

$$d\delta = C_{cr} ds = C_{cr} T^T d\sigma \qquad (11)$$

where the traction, ds is substituted by use of eq. 5 and C_{cr} is the tangent compliance matrix for a single crack. In order to solve the differential equation (eq. 9) we need to introduce the damage law in incremental form:

$$dn = F'(\varepsilon_{nn}) d\varepsilon_{nn} \qquad (12)$$

where $d\varepsilon_{nn}$ is the normal component of total strain in the crack direction ($d\varepsilon_{nn} = [1\ 0] T^T\ d\varepsilon$) and F is the slope of the damage law (see fig. 2). Inserting eq. 11 and 12 in eq. 9 the differential equation can now be written as:

$$d\sigma = D_e (d\varepsilon - T F d\varepsilon_{nn} \delta - T n C_{cr} T^T d\sigma) \qquad (13)$$

In order to obtain a relationship between total strain increment, $d\varepsilon$ and the stress increment, $d\sigma$ a rearranging of eq. 13 is performed:

$$(C_e + T n C_{cr} T^T) d\sigma = (d\varepsilon - T F \delta d\varepsilon_{nn})$$

$$(C_e + T n C_{cr} T^T) d\sigma = (I - T F \delta [1\ 0] T^T) d\varepsilon$$

$$d\sigma = (C_e + T n C_{cr} T^T)^{-1} (I - T F \delta' T^T) d\varepsilon \qquad (14)$$

where I is a 3 by 3 unit matrix and δ' is a 2 by 2 matrix containing the displacement components:

$$\delta' = \begin{bmatrix} \delta_{nn} & 0 \\ \delta_{mn} & 0 \end{bmatrix} \qquad (15)$$

1451

The tangent stiffness matrix, D_{ep}, is readily identified from eq. 14. For multiple crack directions T, nC_{cr} and $F\delta'$ can be written as:

$$T = [\; T_1 \quad T_2 \quad \cdots \quad T_j \;] \qquad (16)$$

$$nC^{cr} = \begin{bmatrix} n_1 C_1^{cr} & 0 & \cdots & 0 \\ 0 & n_2 C_2^{cr} & & 0 \\ \vdots & & & \vdots \\ 0 & 0 & \cdots & n_j C_j^{cr} \end{bmatrix} \qquad (17)$$

$$F\delta' = \begin{bmatrix} F_1\delta'_1 & 0 & \cdots & 0 \\ 0 & F_2\delta'_2 & & 0 \\ \vdots & & & \vdots \\ 0 & 0 & \cdots & F_j\delta'_j \end{bmatrix} \qquad (18)$$

where j refer to the current crack direction.

2.3 Matrix model

The matrix is modelled employing an elasto-plastic material model for crack initiation and propagation in plain concrete. The employed model is a modified version of the model originally developed by (Carol, Prat, and López 1997). The model is a mixed mode cohesive crack model combined with a modified Mohr-Coulomb yield-surface, f:

$$f = (s_{mn}^m)^2 - (c - s_{nn}^m \mu)^2 + (c - f_t\mu)^2, \qquad (19)$$

where s^m is the traction in the matrix, c is the cohesion, μ is the friction coefficient and f_t is the tensile strength of the matrix. During sliding of a crack the model is able to capture the dilatation in the normal direction. The dilatation phenomenon is essential when modelling crack propagation in cementitious materials. If a crack opening is confined in the normal direction during sliding, large compression normal forces can be build up in the structural member. If the model does not capture this phenomenon the carrying capacity of the structural member can be underestimated. The model includes damage parameters and as the material softens the shape of the yield-surface will gradually tend towards the Coulomb yield-surface. The Coulomb yield-surface eventually models friction between two separate surfaces due to a non-associated flow rule (eq. 20), which represents the plastic potential:

$$g = (s_{mn}^m)^2 - (c - s_{nn}^m \mu_g)^2 + (c - f_t\mu_g)^2, \qquad (20)$$

Figure 4 illustrates the traction evolution in a matrix crack for a material point. The crack is initiated under a pure mode I loading condition. After the crack is

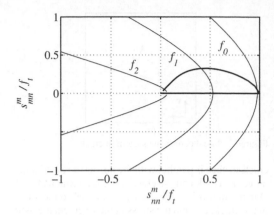

Figure 4. Evolution of traction and yield-surface in the matrix under mixed mode crack opening, where s is matrix traction and f_0 and f_2 are the initial and final yield-surface respectively.

initiated the crack is opened in a mixed mode with at tangential opening that is proportional and twice as large as the normal opening. The thick line is the traction path and the thin curves illustrate the evolution of the yield surface from f_0 to f_2.

2.4 Fiber model

The stiffness of the fibers in the direction normal to the crack surface, E_b is found through information of the global normal stiffness of the ECC, E_{ECC}, the initial E-modulus for plane stress, $E = E_0/(1 - \nu^2)$, the number of parallel cracks per length, n and the stiffness of the matrix for pure mode I opening, $E_{nn,I}^m$. By considering only the stiffness in the direction normal to a crack eq. 14 gives us:

$$(1/E + nC_{11}^{cr})^{-1}(1 - F\delta_{nn}) = E_{ECC} \qquad (21)$$

By substituting the crack compliance, C_{11}^{cr} with the sum of the mode I matrix and fiber stiffness, $C_{11}^{cr} = 1/(E_b + E_{nn,I}^m)$, the mode I fiber stiffness, E_b can be found:

$$E_b = \frac{nE_{ECC}E + E_{nn,I}^m(E_{ECC} - E + E\delta_{nn}F)}{E - E_{ECC} - E\delta_{nn}F} \qquad (22)$$

where the pure mode I stiffness of the matrix, $E_{nn,I}^m$ is found as a function of the current normal opening of a single crack. Due to the present formulation the global tangent stiffness computed agrees with the global tangent stiffness, E_{ECC} found from an idealization of an uniaxial tensile test, when the cracks open in pure mode I. If the cracks open in mixed mode the actual normal stiffness for the matrix, E_{nn}^m will be lower than the pure mode I stiffness, $E_{nn,I}^m$ due to the mixed mode crack formulation. It is assumed that the fiber normal stiffness, E_b is unaffected by mixed mode crack opening.

During sliding of a crack the fibers are modelled as randomly orientated Timoshenko beam and the relationship between crack deformations and shear stresses are found by solving a boundary value problem (Kabele 2002):

$$s_{nm}^b = k \frac{\delta_{nm}}{\delta_{nn}} \qquad (23)$$

where k is a constant calibrated by test, and δ_{nn} and δ_{nm} are the mode I and II opening of the crack. The tangent stiffness matrix for the fibers can then be written as:

$$D_{cr}^b = \begin{bmatrix} E_b & 0 \\ -\frac{k\delta_{nm}}{\delta_{nn}^2} & \frac{k}{\delta_{nn}} \end{bmatrix} \qquad (24)$$

The stiffness of the fibers, D_{cr}^b is a smeared stiffness over the crack length.

2.5 The total compliance crack matrix

Due to the fact that the matrix and fiber bridging in the crack works in parallel, the total stiffness of the crack can be written as:

$$D_{cr} = D_{cr}^b + D_{cr}^m \qquad (25)$$

The superposition of stiffness only holds because the fiber volume concentration, V_f is small and because the fiber bridging stiffness is a smeared stiffness over the crack length. The superposition of stiffness in the crack remains a hypothesis until the model has been validated by experimental results.

2.6 Un- and reloading

During un- and reloading three different elements need to be considered: matrix, fibers and ECC.

Un- and reloading of the matrix is controlled by the yield surface. As observed in experiments (Kesner and Billington 1998) the elastic E-modulus tends to degrade as a function of the largest crack opening obtained. During un- and reloading of the matrix a simple scheme taken this degrading of normal stiffness into consideration is employed (see fig. 5 and eq. 26).

$$E_{nn,unload}^m = \begin{cases} s_{nn}^{m,pl}/((1-b)\delta_{max}) & \delta_{nn} > b\delta_{max} \\ 0 & b\delta_{max} > \delta_{nn} > 0 \\ \infty & \delta = 0 \end{cases} \qquad (26)$$

where b, is a constant calibrated by experiments, $s_{nn}^{m,pl}$ is the normal traction before unloading and δ_{max} is the

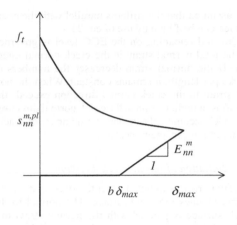

Figure 5. Un- and reloading of matrix.

maximal normal crack opening before unloading. The normal crack opening, δ_{nn} can not assume negative values, because this means that the crack surfaces would overlap. Giving the crack infinitely normal stiffness causes the stiffness of the ECC material in the crack normal direction to be equal to the initial stiffness in compression. Eq. 26 is only valid when unloading occur while the normal traction is positive. If the normal traction is negative before unloading (this can occur during sliding) then the normal-stiffness will be equal to infinity because the matrix is then under compression. If the normal traction becomes positive eq. 26 will again be valid. In order for the traction point to be able to move quickly from one side of the yield surface to the other $((s_{nn}^m, s_{nm}^m) \rightarrow (s_{nn}^m, -s_{nm}^m))$ when the tangential displacement increment change direction, the elastic shear stiffness for the matrix is set to $f_t/(1 \, \mu m)$. The size of the elastic shear stiffness has influence on the distribution of shear stresses between matrix and fibers in the elastic state. An experimental investigation of this phenomenon can decide the real size.

Un- and reloading of the fibers are controlled by the crack opening. The scheme chosen to determine the relationship between the crack opening and the bridging stiffness normal to the crack surface resemble the one chosen for the matrix (fig. 5). The fibers become elastic when the normal crack opening decreases:

$$E_{nn,unload}^b = \begin{cases} s_{nn}^{b,pl}/((1-b)\delta_{max}) & \delta_{nn} > b\delta_{max} \\ 0 & b\delta_{max} > \delta_{nn} > 0 \end{cases} \qquad (27)$$

When the crack is closed the normal traction is transferred entirely through the matrix. It is assumed for simplicity that the fiber bridging stress can not become negative. When the crack normal opening reaches the previous maximal opening, δ_{max}, the bridging stiffness normal to the crack surface is given by eq. 22.

1453

It is assumed that the stiffness parallel with the crack surface can be found by use of eq. 23.

Un- and reloading on the ECC level is governed by the total normal strain in the crack normal direction. If the normal strain decreases the numbers of cracks per length, n remains constant. When the normal strain in the crack normal direction exceeds the previous maximal strain in the crack normal direction, the ECC becomes plastic and the numbers of cracks per length, n can again increase.

2.7 Initiation of second crack direction

The first crack is initiated when the stress state in the matrix reaches the yield-surface. The normal to the crack surface is parallel with the normal stress that initiated the crack and after crack initiation the crack direction remains fixed. An angle threshold around the first crack where new crack directions can not be initiated is introduced. A second crack direction is initiated when the stress state in the matrix outside the angle threshold reaches the yield-surface. The angle threshold ensures that two crack directions in one integration point will not be initiated with too small an angle separating them, which would be a purely numerical phenomenon.

3 FPB SIMULATION

The model is implemented in a user supplied routine in the commercial FEM package 'DIANA'. A simulation of a four point bending (FPB) beam is performed as a test of the present model (see fig. 6). Corresponding experimental results can be found in (Øestergaard, Water, and Olesen 2005). For the simulation a 70 by 17 element mesh is employed. The elements employed are, 8 node, quadrilateral isoparametric plane stress elements. The elements are based on quadratic interpolation and Gauss integration. The dimensions of the beam are: length 500 mm, height 60 mm and width 100 mm. The beam is simply supported and loaded as shown in the figure. Point A and B are used for measuring of vertical displacement, u and point C will be used to evaluate the state of the material.

3.1 Model input

The material data is found from the FPB experiments and an inverse analysis (Øestergaard, Walter, and Olesen 2005): the tensile strength, $f_{t,ECC} = 2.6$ MPa, the initial E-modulus, $E_0 = 33$ MPa, the strain-hardening E-modulus, $E_{ECC} = 0.24$ MPa and the ultimate strain before softening, $\varepsilon_u = 0.007$. After the ultimate strain is reached in the crack normal direction, the normal traction is assumed to decrease linearly until a crack opening of 12 mm is reached. Poisson's ratio, ν is assumed to be 0.2.

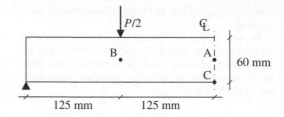

Figure 6. Four point bending beam.

In order to obtain information about crack opening and spacing from the simulation, information about numbers of cracks per length, n as a function of the total normal strain in the crack normal direction, ε_{nn} are required (see fig. 2). These data was not measured in the experiments, therefore some reasonable values are given as input. Because the cracks are mainly opened in mode I these additional input data will only have little influence on global results like global stresses and deflection of the beam. The relation between the numbers of cracks per length, n and strain, ε_{nn} are chosen as: (ε_{nn}; n [mm^{-1}]), $(7.9 \cdot 10^{-5};\ 0.1)$, $(1 \cdot 10^{-4};\ 0.2)$, $(1 \cdot 10^{-3};\ 0.3)$, $(3 \cdot 10^{-3};\ 0.4)$ and $(6 \cdot 10^{-3};\ 0.5)$. Finally the matrix properties related to the elasto-plastic matrix model are chosen as: the matrix tensile strength, $f_t = 2.0$ MPa, the friction coefficient, $c = 2.6$ MPa, the mode I fracture energy, $G_{f,I} = 30$ N/m, the mode II fracture energy, $G_{F,II} = 30$ N/m, the friction coefficients $\mu_f = 0.4$, $\mu_0 = 0.75$ and $\mu_g = 0.375$ and the unloading constant $b = 0.5$.

3.2 Simulation results

In fig. 7 the load-deflection curve from the simulation is plotted together with the upper and lower bound from the experiments. In contrast to the experiments a partial unloading is performed in the simulation to demonstrate the capability of the model. The load is applied in three steps: first the load is increased until a deflection of approximately 0.7 mm is reached (load point a), then a partial unloading is performed (load point b) and finally the load is increased in the remaining simulation. As shown in the figure the model is able to reproduce the experimental results very well.

In fig. 8 the relationship between the traction, s in the normal crack direction at point C vs. the relative deflection are shown. The total traction, s reaches a peak at a relative deflection of approximately 1 mm. After the peak point is reached the ECC material begins to soften. At crack initiation there is a difference between the traction in the ECC material, s and the traction in the matrix, s_m. This is in good agreements with observation made in simulations (Dick-Nielsen, Stang, and Poulsen 2005), where a crack with an opening of only a few nano meter runs through the matrix

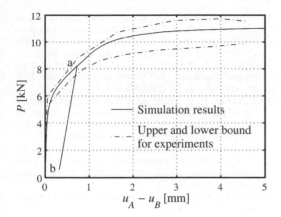

Figure 7. Load deflection curve.

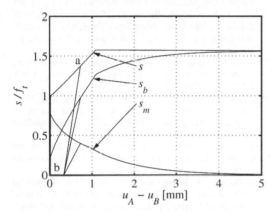

Figure 8. Total traction, s, fiber bridging traction, s_B and matrix traction, s_m in the normal crack direction at point C vs. deflection.

before debonding of the fibers take place. Similar experimental observations has been made by (Wang and Li 2004). The ECC mix 3 in these experiments had a first crack strength of 4 MPa, while experiments performed by Wang at The Technical University of Denmark, showed that the matrix in mix 3 had a tensile strength of 2.83 MPa. As the crack at point C opens the matrix traction begins to soften while the average fiber traction begins to increase. At a relative deflection of approximately 4 mm the crack is only bridged by the fibers. The unloading scheme works as intended, leaving a permanent plastic deformation after unloading.

Figure 9 shows the crack pattern at a deflection of approximately 0.9 mm before localization take place in the bottom of the beam. The line thickness corresponds to the crack opening. Cracks along the entire bottom in the middle section is about to localize, due to the constant moment in this section.

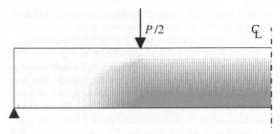

Figure 9. Crack pattern in the beam at a deflection of 0.9 mm. The line thickness corresponds to the crack opening.

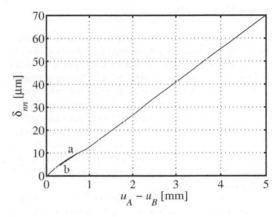

Figure 10. Average crack opening at point C vs. deflection.

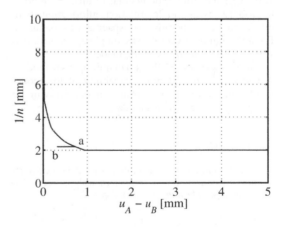

Figure 11. Average crack spacing at point C vs. deflection.

Figure 10 shows the relationship between the relative deflection and the average normal crack opening, δ_{nn} at point C. The slope of the curve changes after a deflection of 1 mm, which is the point at which softening begins to take place in the bottom of the beam. At a relative deflection of 5 mm the average crack opening at point C is 70 µm. During unloading from loading

1455

point a to b the average crack opening associated with point C decreases linearly towards zero.

The average crack spacing at point C is plotted as a function of the relative deflection in fig. 11. After the first crack is initiated the average crack spacing is 10 mm. The spacing decreases until load point a is reached. During unloading from load point a to b the deflection decreases, but the crack spacing remains constant. After reloading to load point b the crack spacing decreases until a spacing of 2 mm is reached after which the spacing remains constant.

4 CONCLUSIONS

In the present paper a plasticity-based damage mechanics model for Engineered Cementitious Composites (ECC) has been introduced. The present model differs from existing models by combining a matrix and a fiber model in order to describe the behavior of the ECC material. Apart from information about the stress and deformation state the model provides information about crack orientation, opening and spacing. The information provided by the model makes it possible to assess the state of an ECC structure in the serviceability state and to determine the serviceability state limit.

A demonstration of the model has been performed by simulating the behavior of a FPB beam made of ECC. The results obtained from the simulation agreed very well with the experimental results. In addition to global stresses and deformations information about crack traction, opening and spacing was obtained. Even though the example was simple it demonstrated very well the capability of the model.

The cracks in the FPB beam mainly opens in mode I. A test of the mixed mode capability of the model is planed, where results from simulations will be compared with experimental data.

REFERENCES

Carol, I., P. Prat, and C. López (1997). Normal/ hear cracking model: Application to discrete crack analysis. *Journal of Engineering Mechanics 123*(8), 765–773.

Dick-Nielsen, L., H. Stang, and P. Poulsen (2005). Micromechanical analysis of fiber reinforced cementitious composites using cohesive crack modeling. *Knud Hojgaard Conference on Advanced Cement-Based Materials.*

Dick-Nielsen, L., H. Stang, and P. Poulsen (2006a). Condition for strain-hardening in ECC uniaxial test specimen. *EFC 16. Alexandroupolis, Grækenland 3–7 juli, 2006.*

Dick-Nielsen, L., H. Stang, and P. Poulsen (2006b). Simulation of strain-hardening in ecc uniaxial test specimen by use of a damage mechanics formulation. *EURO-C 2006 Computational Modelling of Concrete Structures.*

Han, T., P. H. Feenstra, and S. L. Billington (2003). Simulation of highly ductile fiber-rienforced cement-based composite components under cyclic loading. *ACI Structural Journal 100*(6), 749–757.

Jirasek, M. (2004). *Modeling of localized Inelastic Deformation.*

Kabele, P. (2002). Equivalent continuum model of multiple cracking. *Engineering Mechanics (Association for Engineering Mechanics, Czech Republic) 9 (1/2)*, 75–90.

Kesner, K. E. and S. L. Billington (1998). Investigation of ductile cement based composites for seismic strengthening and retrofit. *Concrete Research and Technology*, 19–33.

Li, V. and C. Leung (1992). Steady state and multiple cracking of short random fiber composites. *ASCE J. Eng. Mech. 188*, 2246–2264.

Stang, H., J. Olesen, P. Poulsen, and L. Dick-Nielsen (2006). On the application of cohesive crack modeling in cementitious materials. *EURO-C 2006 Computational Modelling of Concrete Structures.*

Østergaard, L. (2003). Early-age fracture mechanics and cracking of concrete. *Ph.d.-thesis – Technical University of Denmark.*

Østergaard, L., R. Walter, and J. Olesen (2005). Method for determination of tensile properties for ECC II: Inverse analysis and experimental results. *International Workshop on High Performance Fiber Reinforced Cementitious Composites in Structural Applications.*

Walter, R., J. Olesen, H. Stang, and T. Vejrum (2006). Analysis of steel bridge deck stiffened with cement-based overlay. *submitted to: ASCE – Journal of Bridge Engineering.*

Wang, S. and V. Li (2004). Tailoring of preexisting flaws in ECC matrix for saturated strain hardening. *Ia-FraMCos.*

Fracture Mechanics of Concrete and Concrete Structures – High-Performance Concrete,
Brick-Masonry and Environmental Aspects – Carpinteri, et al. (eds)
© 2007 Taylor & Francis Group, London, ISBN 978-0-415-44617-4

Radiographic imaging for the observation of Modes I and II fracture in fibre reinforced concrete

S.J. Foster, G.G. Lee & T.N.S. Htut
School of Civil and Environmental Engineering, The University of New South Wales, Sydney, Australia

ABSTRACT: In this study, radiographic imaging is used to investigate the mechanisms of fracture in fibre reinforced concrete. The investigation looks at the performance of discrete end-hooked and straight fibres crossing a cracking plane at various angles and loaded normal or parallel to the plane. A main finding of the study is that in Mode II fracture, snubbing of fibres in the concrete adjacent to the crack interface provides significant mechanical anchorage. However, the strength benefits of this anchorage come only after significant slip displacement. Further, for acute angled fibres, due to the high anchorage provided by the snubbing, fibre fracture often governs over fibre pullout suggesting an increased brittle response compared to that of fibres subjected to crack opening (Mode I) fracture. Lastly, the tests clearly show that a significant proportion of fibres in Mode II failure pullout from the longer embedded side.

1 INTRODUCTION

Since Romualdi and Batson (1963) first performed tests on modern day fibre reinforced concrete (FRC), a significant body of research has been undertaken and reported in the literature. Studies such as those by Peterson (1980) have been undertaken on fibre reinforced mortar, with various percentages of fibres in the matrix, providing data on the tensile stress versus crack opening displacement (COD) response. Other studies, such as for example Banthia and Trottier (1994), have concentrated on the anchorage and pullout mechanisms of individual fibres crossing a join in a tensile specimen. Space in this paper prohibits an extensive review of the literature on tensile fracture of fibre reinforced cementitious composites For a more comprehensive review we refer the reader to Voo and Foster (2003).

While a significant body of research has been undertaken on Mode I (crack opening) fracture of FRC over a period of 40 years and more, less research has been undertaken on Mode II fracture (also referred to as crack sliding). The first study on direct shear specimens was by Van de Loock (1987). In the time since studies have been conducted by Valle and Büyüköztürk (1993), Balaguru and Dipsia (1993), Khaloo and Kim (1997) and Mirsayah and Banthia (2002). However, these studies were generally limited and have not lead to a general understanding of the behaviour of fibre reinforced concrete subjected to longitudinal shear. A review of each of these test series, highlighting strengths and weaknesses, is given in Lee and Foster (2006a).

Lee and Foster (2006a, 2006b) reported results of 39 discrete fibre test specimens in direct shear (Mode II), in addition to a series of fibre volume tests for straight and end-hooked fibres. The conclusions drawn from the studies were: (i) the load versus crack sliding behaviour is dependent on the fibre orientation across the shear plane; (ii) the critical length of a fibre for fibre fracture is significantly influenced by the angle that the fibre crosses the cracking plane; (iii) the combined effect of bending and axial force has a more pronounced effect on the fracture of acute angle fibres; (iv) due to the snubbing effect for the acute angle fibres, considerable displacement occurs before fibres become effectively engaged; (v) secondary anchor effects due to snubbing increase the load and displacement capacity; (vi) the common modelling assumption of pullout always from the shorter of side of fibre embedment is questioned and its influence requires further investigation; and (vii) the common assumption of uniform bond stress along the length of embedment is questioned and requires more study. In this study radiographic imaging is used to answer some of the questions posed by Foster and Lee (2006b) and to explore fundamental differences in behaviour between Mode I fracture and Mode II fracture of FRC.

2 EXPERIMENTAL PROGRAMME

2.1 Introduction

In this section, details of the uniaxial tension (UT) tests and direct shear (DS) tests are described. For the UT

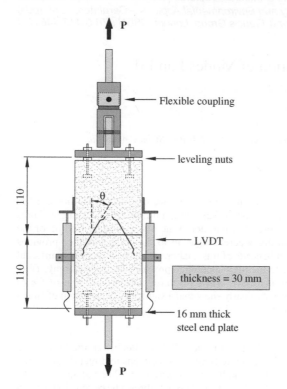

Figure 1. Testing arrangements for tensile specimen.

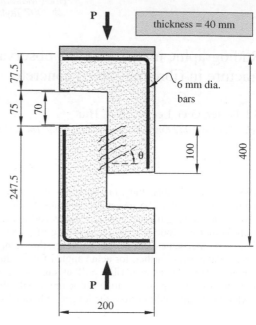

Figure 2. Push-off specimen dimensions and details.

(Mode I) tests, X-ray imaging was performed on end-hooked fibres placed at angles of 0° and 30° to the loading direction. The embedment length ratio (length of fibre each side of the crack relative to the fibre length) was 0.5:0.5.

In the DS (Mode II) tests, gamma ray imaging was used with the fibres oriented at angles of 0°, ±30° and ±60° with respect to a plane perpendicular to the loading direction. Six tests were conducted, three with end-hooked fibres and three with straight fibres, with an embedment length ratio of 0.33:0.67. A further two tests were undertaken with end-hooked fibres at an embedment length ratio of 0.25:0.75. The details of the experimental program are outlined below.

2.2 Materials and fabrication of test specimens

Figure 1 shows the specimen dimensions and reinforcing arrangements for the Mode I fracture specimens and Figure 2 for the Mode II fracture specimens. The specimens were fabricated in two separate concrete pours. The fibre angle (θ) is measured from a horizontal line drawn normal to the interface of the two halves of the specimen. In the L-shaped shear specimens, a clockwise direction is considered as positive in the "2" configuration whereas an anticlockwise direction is

negative. For example, the angle of orientation shown in Figure 2 is negative. The 2 or "2" configuration is taken as the reference or front face direction and the 5 or "5" configuration as the back face.

The thickness of the double L-shaped specimens was 40 mm. The L-shaped bars were 6 mm in diameter with a minimum cover of 5 mm and the yield strength of the bars was nominally 500 MPa. A single row of four fibres were cast into the matrix. The fibres were clamped into position for the first pour using timber packing clamps placed on the side of the specimen selected for the second pour. One day later, the timber clamps were released and the second side of the specimen was cast.

The tensile specimens were constructed in a similar manner to the shear specimens with two fibres were clamped into position on one side of the specimen while the mortar was cast on the other. The thickness of the specimens was 30 mm.

The mortar mix used in the specimens was composed of kiln dried Sydney sand and general purpose Portland cement mixed in a ratio of 3 : 1 (sand : cement) and water at a water : cement ratio of 0.4. No other additives used in the mix design.

The deformed steel fibres used in the tests were high strength, end-hooked Dramix cold drawn fibres produced by Bekaert (Belgium). The fibre dimensions and material properties are given in Table 1. The straight fibres used in the tests were made from end-hooked fibres with the hook portions cut-off.

Table 1. Properties of steel fibres.

Fibre type	Diameter, d (mm)	Length, l (mm)	Aspect ratio l/d	Tensile strength (MPa)
Hooked-ended	0.9	60	67	1085
Straight	0.9	48	53	1085

Table 2. Specimen properties.

Specimen	Test	Fibre type	θ (deg.)	Cylinder strength (MPa)
NDSI − 60H1:3	Shear	hooked	−60	50
NDSI + 0H1:3	Shear	hooked	0	50
NDSI + 60H1:3	Shear	hooked	+60	50
NDSI − 60S1:3	Shear	straight	−60	43
NDSI + 0S1:3	Shear	straight	0	43
NDSI + 60S1:3	Shear	straight	+60	43
NDSI − 30H1:2	Shear	hooked	−30	32
NDSI + 30H1:2	Shear	hooked	+30	32
NUTI ± 0H1:1	Tension	hooked	0	36
NUTI ± 30H1:1	Tension	hooked	30	36

Details of the test specimens for the radiographic imaging are given in Table 2. The specimens are designated as NXXI±θYa : b where N = "non-destructive test", XX = DS for the "direct shear" and XX = UT for the "uniaxial tension" tests, ±θ is the angle of the fibre taken from a line normal to the crack interface, Y is the fibre type (H = end-hooked fibres and S = straight fibres) and a : b is the ratio of the fibre embedded on the short side to that of the long side. For example, specimen NDSI +60H1:3 is a non-destructive (radiographic imaging) test in direct shear with an end-hooked fibre at an angle of +60° and a fibre embedment length (l_e) of $l_e = 0.25 l_f$ on the short side and $l_e = 0.75 l_f$ on the long side (ie. short side embedment to long side embedment ratio is 1:3) where l_f is the total length of the fibre.

The specimens were cast horizontally in two stages into the moulds. The first stage involved casting a section on one side of the cracking plane, with the other half of the specimen blocked out and the fibres protected between two timber sandwich blocks. The specimens were compacted on a vibrating table and then left to cure for 24 hours. After setting of the first half of the specimen, the timber blocks were removed from the second half of the specimen and the casting completed. The completed specimens were then air cured for 24 hours to allow setting then demoulded. Six 200 mm high by 100 mm diameter cylinders were cast for each half specimen for quality control.

Figure 3. Testing arrangements for UT specimens.

After setting of the second half, the specimens and cylinders were stripped and placed in a water bath at 90°C for 3 days to accelerate the curing process. After removal from the hot water bath all the specimens were placed in a constant dry temperature room for 3 days to remove the excess water. Following the drying process all the specimens were removed from the dry temperature room and stored in a laboratory environment until testing. The mean cylinder compressive strength at the time of testing is given in Table 2.

2.3 Testing arrangements

The testing arrangements are shown in Figures 3 and 4 for the UT and DS tests, respectively. The displacements in the direction of movement of the loading jacks were measured using linear variable differential transducers (LVDTs). In the UT tests, two LVDTs were used, one placed on each side of the specimen and the displacement measurements were taken as the average of the LVDT readings. For the DS tests, one LVDT was used to measure the crack sliding displacements.

3 TEST RESULTS

3.1 Introduction

The peak loads, displacements at the peak loads and failure mode are given in Table 3. The results of the radiographic imaging are presented in Sections 3.2 for the UT tests and 3.3 for the DS tests.

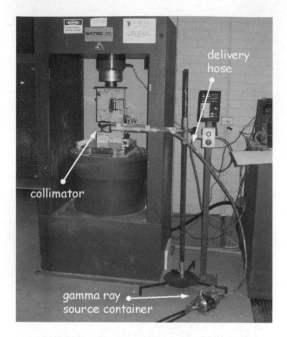

Figure 4. Testing arrangements for DS specimens.

Table 3. Peak loads per fibre, displacements and failure modes.

Specimen	Average peak load per fibre (N)	Slip at peak load (mm)	Failure mode
NDSI − 60H1:3	390	28.6	FF
NDSI + 0H1:3	680	3.3	PO
NDSI + 60H1:3	460	0.6	PO
NDSI − 60S1:3	230	16.9	FF/PO
NDSI + 0S1:3	140	10.5	PO
NDSI + 60S1:3	200	0.1	PO
NDSI − 30H1:2	590	9.9	FF
NDSI + 30H1:2	420	3.5	PO
NUTI ± 0H1:1	460	3.5	PO
NUTI ± 30H1:1	380	4.0	PO

Note: *FF = fibre fracture; PO = fibre pullout

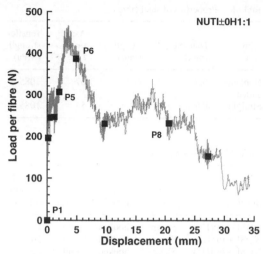

Figure 5. Load-displacement for specimen NUTI ± 0H1:1.

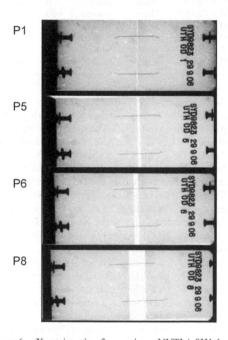

Figure 6. X-ray imaging for specimen NUTI ± 0H1:1.

3.2 Uniaxial Tension (UT) Tests

The UT tests were undertaken in an Instron universal testing machine under displacement control. The resulting load versus crack opening displacement (COD) diagrams are given in Figure 5 for the 0° fibre angle test and Figure 7 for the 30° angle fibre tests and the corresponding X-ray images are presented in Figures 6 and 8, respectively. The photo plate numbers are given as "Px" where x is the image number. The location of each X-ray image taken in the load-displacement space is indicated on the load-displacement figures.

The X-ray negative images were converted into positives using digital imaging and the fibres within the mortar matrix are untouched from those obtained from the X-ray photography. No image is obtained in the opened portion of the specimen (as the X-ray exposure time is set for the density and thickness of the specimen leaving the portion at the opening over exposed).

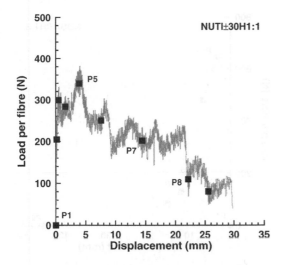

Figure 7. Load-displacement for specimen NUTI ± 30H1:1.

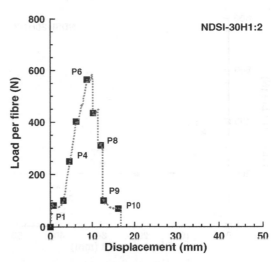

Figure 9. Load-displacement for specimen NDSI + 30H1:2.

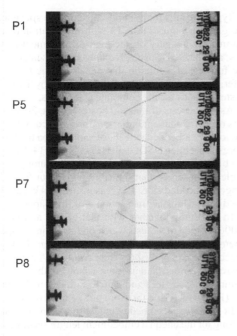

Figure 8. X-ray imaging for specimen NUTI ± 30H1:1.

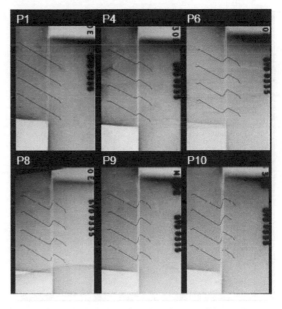

Figure 10. Gamma ray imaging for specimen NDSI − 30H1:2.

In this region the fibres are interpolated based on visual observations.

3.3 Direct Shear (DS) Tests

The DS tests were conducted using a Satec testing machine under displacement control. Before testing, the weak bond on the shear plane of the mortar-mortar interface was broken and a layer of grease was applied along the 70 mm long vertical surfaces to reduce the effect of friction.

With the exception of the points of radiographic imaging, the displacement rate before the displacement corresponding to the peak load was 0.2 mm per minute. After peak, the rate was increased. Each specimen was loaded until full fracture had occurred.

As movement of the specimen blurs the radiographic image, the displacement rate was gradually reduced to zero before the film was exposed and held

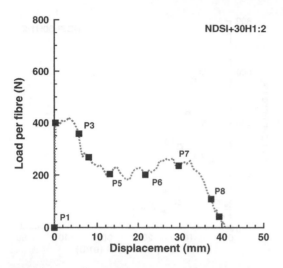

Figure 11. Load-displacement for specimen NDSI + 30H1:2.

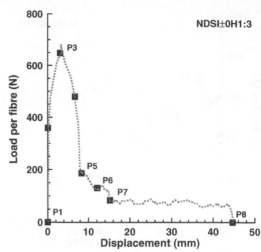

Figure 13. Load-displacement for specimen NDSI + 0H1:3.

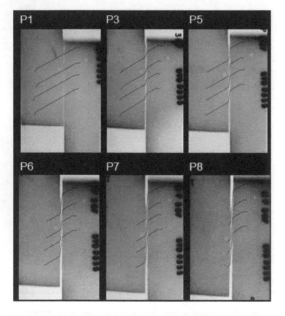

Figure 12. Gamma ray imaging for specimen NDSI + 30H1:2.

constant for the duration of the exposure. After the completion of the imaging the displacement rate was again set to that stated above.

In the DS tests, gamma rays were used for the radiographic imaging. For clear reproduction, the fibres images have been digitally enhanced. As for the UT tests, the fibre images within any gap in the specimen (due to horizontal displacements) are interpolated from visual observations.

Typical results for the end-hooked fibre specimens are given in Figures 9 to 14 with the location of the gamma ray imagery on the load-displacement histories, as indicated. Space precludes the presentation of the full sets of results in this paper and the reader is referred to the laboratory testing report by Lee and Foster (2006c) for the complete test results.

As for the tests of Lee and Foster (2006a, 2006b), it was observed that a number of fibres pulled out from the longer embedded side. For example, see the lower fibre of specimen NSDI + 0H1:3 in Figure 14. Despite the bias by a factor of 3 in favour of the short side it is clearly observed that the fibre pulled out from the longer embedded side. The reasons for this behaviour are discussed further in Section 4 of this paper.

In this study together with that of Lee and Foster (2006a) a total of 47 tests were conducted on DS specimens with 4, 6 or 8 fibres per specimen. In total, observations on 434 fibres were recorded A survey of the results is presented in Table 4 showing the mode of failure for each fibre and, where failure was by pull-out (PO), which side the fibre pulled out from (left or right for the equal embedment length specimens; short or long for the unequal embedment length specimens).

4 ANALYSIS OF RESULTS

The fibre reinforced specimens failed in one of three modes: fibre pullout, fibre fracture or in a mixed mode with some fibres in the specimen pulling out from the matrix while others fractured.

For straight fibres, the initial stage for fibre pullout is an elastic response which is typically linear. In this stage the fibre remain fully bonded to the surrounding matrix. The next stage is partial debonding followed

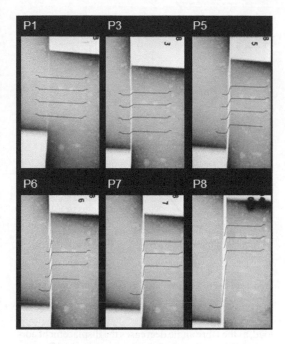

Figure 14. Gamma ray imaging for specimen NDSI + 0H1:3.

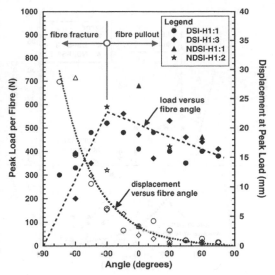

Figure 15. Hooked-end fibre tests: peak load and displacement at peak load versus fibre angle (solid data markers peak load versus fibre angle; hollow data markers displacement at peak load versus fibre angle).

Table 4. Survey of discrete fibre failures.

Specimen Group	Fracture	Total PO	PO$_{Left}$	PO$_{Right}$	PO$_{Long}$	PO$_{Short}$
DSI±θH1:1	26	58	23	35	–	–
DSI±θH1:3	27	51	–	–	8	43
DSI±θS1:1	7	65	36	29		
DSI±θS1:3	12	60	–	–	33	27
NDSI±θH1:3	33	31	–	–	7	24
NDSI±θS1:3	14	30	–	–	1	29
NDSI±θH1:2	8	12	–	–	7	5

by slippage or frictional pullout until the fibre is completely pulled through the fibre tunnel where the fibre tunnel is the space that was previously occupied by the fibre. If end appendages such as hooks are present, then a mechanical clamping stage is introduced before significant slippage takes place (Alwan et al. 1999).

The initial elastic response for the hooked-ended and straight fibres with positive angles of orientation was reasonably linear. However, for the shear tests, as the angle became increasingly more negative, considerable vertical displacement occurred prior to the mechanical engagement of the fibre. A similar observation was made by Voo and Foster (2003, 2004) for fibres subjected to uniaxial tension but this effect is exaggerated in the case of the fibres subjected to crack sliding. The effect of the local damage to the fibre

and the increased snubbing effect for crack sliding are shown clearly when comparing the images in Figures 10 and 12 for the DS tests compared to the 30° UT specimen shown in Figure 8. In fact, the effect of snubbing is shown to dominate the failure of the zero and negative angled fibre specimens.

Examining closely the images of specimen NDSI-30H1:2 in Figure 10, it is seen that the effect of increased bond of the fibres due to snubbing cause the fibres to fracture. In P6 the specimen is observed to be carrying a load of 590 N with all fibres intact. Shortly afterwards, however, it is seen that the bottom fibre has fractured (P8) and in P10 three of the four fibres have fractured. Because of the fibre fractures, the load versus displacement graph (Figure 9) shows the specimen to have a somewhat brittle response. On the other hand, NDSI + 30H1:2 is more ductile albeit with a lower peak load of 420 N. In this case all fibres pullout from the specimen.

In Figure 15 the peak loads are plotted (on the left axis) against the fibre angle for the end hooked (EH) specimens reported here and in Lee and Foster (2006a, 2006b). It is seen that in the EH tests that fibres angled as high as +75° have a significant capacity. The strength increases, with the increased anchorage due to snubbing, until the point where the anchorage is sufficiently high to cause fracture of the fibres. In the tests undertaken in this study, this is at approximately −30°.

A second point of interest is that the relative length of embedment each side of the crack interface appears

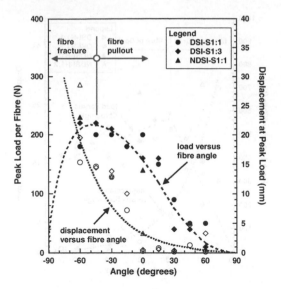

Figure 16. Straight fibre tests: peak load and displacement at peak load versus fibre angle (solid data markers peak load versus fibre angle; hollow data markers displacement at peak load versus fibre angle).

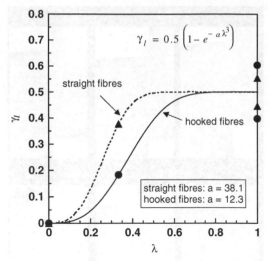

Figure 17. Probability that a fibre will pullout from the longer embedded side.

to have little influence on the peak load. This is because the greater part of the fibre force is developed at the hook and in the snubbing zone and the straight portion of fibre between these points has a significantly smaller influence on the failure load. Note that the difference between the shorter and longer embedded parts of the fibre is the length of this straight portion. Variations in bond in the snubbing zone is sufficient to ensure a number of fibres pullout from the longer embedded side even for specimens with embedment ratios as high as 1:3.

The effect of the high anchorage in the snubbing zone on each side of the crack plane is evidenced by the large number of fibres that were observed to pull out from the longer embedded side. For example, for the DSI $\pm \theta$H1:3 and NDSI $\pm \theta$H1:3 series of tests, of the 82 fibres that pulled out from the specimens, 15 pulled out from the longer embedded side. For the DSI $\pm \theta$S1:3 and NDSI $\pm \theta$S1:3 tests, 34 of 90 fibre pullout failures were from the longer embedded side. The effect of this is that the tail of the curve extends to a displacement approximately equal to the longer embedment length. For example, in specimen NDSI $+$ 0H1:3 (Figure 13) P5 and P6 show that three of the four fibres pulled out from the short side. By P7 (15.2 mm displacement) these fibres had completely pulled out from the matrix. Complete fracture did not occur, however, until 44.6 mm displacement (P8) when the fourth fibre pulled out from the long side.

While in Figure 15 it is seen that fibres oriented at angles between $-30°$ and $+75°$ sustain a significant

load, as the fibre angle decreases the displacement corresponding to this peak load increases significantly. Thus, in a fibre composite only those fibres at high positive angles are likely to be effective in carrying load over the engineering range of a few millimetres. That is, half or more of the fibres do not carry load efficiently. For the straight fibre tests (Figure 16) the fibres at the high positive angles do not carry significant load and those at lower angles do not engage until significant movement has occurred and are prone to fracture.

Comparing Figures 15 and 16 it is demonstrated that straight fibres are inefficient in carrying longitudinal shear forces where concrete blocks are separated by smooth surfaces. Such might be the case, for example, in reactive powder concrete where large aggregate particles are eliminated from the mix. Further research is needed, however, for conventional concrete where sliding maybe combined with crack opening and tension across a crack face forced by aggregate particles sliding over one another.

The survey data in Table 4 can be used to determine the probability that a fibre pulls out from the longer embedded side. In the development of material models for design this observation needs consideration. For the data set that includes only pullout failures, a probability function is developed representing the statistical likelihood that any discrete fibre will pullout from the longer embedded side. The boundary conditions for such a function are: (i) for a fibre with a short to long side embedment ratio of 0:1, the probability of failure from the longer embedded side is zero; (ii) for a fibre with a short to long side embedment ratio of 1:1, the probability of failure from the shorter embedded side is 0.5; (iii) for a fibre with a short to long side embedment ratio of 0:1, the slope of the probability function is zero; and (iv) for a fibre with a short to long side embedment ratio of 1:1, the slope of the probability function approaches zero. A simple function that meets the boundary conditions is

$$\gamma_l = 0.5 \left(1 - e^{-a \lambda^3} \right) \tag{1}$$

where γ_l is the probability that an individual fibre will pullout from the longer embedded side, λ is the ratio of shorter to longer fibre embedment lengths ($l_{e.short}/l_{e.long}$) and a is a fitting constant. The resulting function is shown in Figure 17 with $a = 12.3$ and $a = 38.1$ for the EH and straight fibres, respectively.

5 CONCLUSIONS

Radiographic imaging has proved to be a valuable tool in understanding the behaviour of discrete fibres in a cementitious matrix. Two types of radiographic imaging were used: gamma ray and X-ray. Of the two methods, the more powerful X-rays proved superior with the fibres being clearly identified at all stages of testing.

The imaging has shown the importance of the snubbing effect on the behaviour of fibre reinforced mortar subjected to Mode II fracture. In particular, it is shown that fibres subjected to shear across a crack face behave very differently to that of crack opening. The snubbing effect dominates the behaviour and the angle of a fibre crossing a crack is an important parameter in determining behaviour and failure mode.

Lastly, it has been shown that there is a statistical chance that fibres pullout from the longer embedded side and a probability function is proposed to predict the likelihood of this event.

ACKNOWLEDGMENTS

This study is funded via Australian Research Council (ARC) discovery grant DP0559742. The support of the ARC is acknowledged with thanks.

REFERENCES

Alwan, J.M., Naaman, A.E. & Guerrero, P. 1999. Effect of mechanical clamping on the pull-out response of hooked steel fibers embedded in cementitious matrices. *Concrete Science and Engineering*, 1(3): 15-25.

Balaguru, P. & Dipsia, M.G. 1993. Properties of fiber reinforced high-strength semilightweight concrete. *ACI Materials Journal*, 90(5): 399-405.

Banthia, N. & Trottier, J.F. 1994. Concrete reinforced with deformed steel fibres, part 1: bond-slip mechanisms, *ACI Materials Journal*, 91(5): 435-446.

Khaloo, A.R. & Kim, N. 1997. Influence of concrete and fiber characteristics on behaviour of steel fibre reinforced concrete under direct shear. *ACI Materials Journal*, 94(6): 592-601.

Lee, G.G. & Foster, S.J. 2006a. Behaviour of steel fibre reinforced mortar in shear I: Direct shear tests. *UNICIV Report No R-444, The University of New South Wales*, Sydney: 1185.

Lee G.G. & Foster, S.J. 2006b. Direct shear tests with gamma ray imaging on steel fibres in cementitious materials, *Proceedings, 19th Australasian Conference on the Mechanics of Structures and Materials*, ACMSM 19, Nov 29 – Dec 1, Christchurch, New Zealand.

Lee, G.G. & Foster, S.J. 2006c. Behaviour of steel fibre reinforced mortar in shear II: Gamma ray imaging. *UNICIV Report No R-445, The University of New South Wales*, Sydney: 189.

Mirsayah, A.A. & Banthia, N. 2002. Shear strength of steel fiber-reinforced concrete. *ACI Materials Journal*, 99(5): 473-479.

Peterson, S.R., 1980. Fracture mechanical calculations and tests for fibre-reinforced cementitious materials. *Proceedings, Advanced in Cement Matrix Composites*, Mat. Res. Soc., Annual Meeting, Boston: 95-106.

Romualdi, J.P. & Batson, G.B. 1963. Behaviour of Reinforced Concrete Beams with Closely Spaced Reinforcement. *Journal of the American Concrete Institute*, 60(6): 775-790.

Valle, M. & Büyüköztürk, O. 1993. Behaviour of fiber reinforced high strength concrete under direct shear. *ACI Materials Journal*, 90(2): 122-133.

Van de Loock, L. 1987. Influence of steel fibres on the shear transfer in cracks. *Proceedings of the international symposium on fibre reinforced concrete, Madras, India, 1619 December*. Rotterdam: Balkema.

Voo, J.Y.L. & Foster, S.J. 2003. Variable engagement model for fibre reinforced concrete in tension', UNICIV Report R420, School of Civil and Environmental Engineering, The University of New South Wales, Australia: 86pp.

Voo, J.Y.L. & Foster, S.J. 2004. Tensile fracture of fibre reinforced concrete: Variable engagement model. *Sixth Rilem symposium on fibre reinforced concrete (FRC), Varenna, Italy, 2022 September*. Bagneux: RILEM.

Fracture Mechanics of Concrete and Concrete Structures – High-Performance Concrete,
Brick-Masonry and Environmental Aspects – Carpinteri, et al. (eds)
© 2007 Taylor & Francis Group, London, ISBN 978-0-415-44617-4

Dynamic increase factors for high-performance concrete in compression using split Hopkinson pressure bar

B. Riisgaard
NIRAS Consulting Engineers and Planners, Denmark
Centre for Protective Structures and Materials, Technical University of Denmark

T. Ngo & P. Mendis
Department of Civil and Environmental Engineering, The University of Melbourne

C.T. Georgakis & H. Stang
Department of Civil Engineering, Technical University of Denmark

ABSTRACT: This paper provides dynamic increase factors (DIF) in compression for two different High Performance Concretes (HPC), 100 MPa and 160 MPa, respectively. In the experimental investigation 2 different Split Hopkinson Pressure Bars are used in order to test over a wide range of strain rates, 100 sec[1] to 700 sec^{-1}. The results are compared with the CEB Model Code and the Spilt Hopkinson Pressure Bar technique is briefly described.

1 INTRODUCTION

1.1 Strain rate dependency

It is today well-known and accepted that the dynamic behavior of concrete and concrete like materials are strain-rate dependent. (Grote and Park, 2001).

Compared with statically behavior, increases in strength, strain capacity and fracture energy are observed when such materials are exposed to impact loads (Lok and Zhao, 2004). The term DIF (Dynamic Increase Factor) is used to describe the relative strength enhancement.

1.2 Split Hopkinson Pressure Bar

The dynamic strength enhancement for concrete was first observed by Abrams in 1917 (Bischoff and Perry, 1991) and it has been generally accepted that concrete and concrete like materials are strain rate sensitive

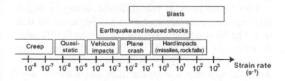

Figure 1. Various strain rate regions.

and the constitutive model of such materials under dynamic loading should include strain-rate effects.

Split Hopkinson Pressure Bar (SHPB) technique has been widely used to measure the dynamic strength enhancement at high strain-rates in the range of 10^1 sec^{-1} to 10^3 sec^{-1}.

1.3 CEB Model Code

The most comprehensive model for predicting the strain rate enhancement of concrete is presented by the CEB Model Code. (Comite Euro-International du Beton – Federation Internationale de la Precontrainte, 1990)

$$DIF = f_{cd}/f_{cs} = \left(\frac{\dot{\varepsilon}}{\varepsilon_s}\right)^{1.026\alpha} \quad \text{for } \dot{\varepsilon} \leq 30 \text{ s}^{-1} \quad (1)$$

$$DIF = f_{cd}/f_{cs} = \gamma\left(\frac{\dot{\varepsilon}}{\varepsilon_s}\right)^{1/3} \quad \text{ffor } \dot{\varepsilon} \geq 30 \text{ s}^{-1} \quad (2)$$

The CEB DIF formulation for concrete has been accepted by most researchers as an accurate representation of the strength enhancement. The formulation takes a bilinear relation between DIF and log($\dot{\varepsilon}$) with a change in slope at strain-rate of 30 s^{-1}. (Malvar and Ross, 1998).

2 SPLIT HOPKINSON PRESSURE BAR

2.1 *How the SHPB works*

The principle of a SHPB device is shown in Figure 2. The axial compression impact is caused by the striker bar impinging the incident bar. When this occurs, an incident stress pulse is developed. The pulse propagates along the incident bar to the interface between the bar and the specimen. At this point, the pulse is both reflected and transmitted. The reflected wave propagates back along the incident bar and the transmitted wave attenuates in the specimen and into the transmitter bar. Both the incident and the reflected waves are measured by a strain gauge mounted on the surface at mid-length of the incident bar. Similarly, the transmitted wave is measured by a strain gauge on the surface at mid-length of the transmitter bar. (Li and Meng, 2003).

The circular specimens are placed between the two long horizontally aligned pressure bars which serve as the medium for the propagation of elastic pulses as well as for measuring the stress-time history. Figure 3 and Figure 4.

2.2 *How to read the data*

A typical output from a SHPB test is shown in Figure 5. All three waves, $\varepsilon_i(t)$, $\varepsilon_r(t)$ and $\varepsilon_t(t)$, are measured at the gauge locations, situated at some distance away from the interface. Therefore, an appropriate time-shifting procedure must be undertaken to transfer the strain histories from the gauge locations to the interfaces Figure 6.

To calculate the specimen stress and the dynamic increase factor, Hooke's law is used to determine the stress of the pressure bars from the measured strain values. Based on (Linholm and Bunshah, 1971) summery of SHPB technique following strain rate, strain and stress history with respect to time, can be calculated, respectively, as

$$\dot{\varepsilon}(t) = \frac{c_0}{L}\left[\varepsilon_i(t) - \varepsilon_r(t) - \varepsilon_t(t)\right] \quad (3)$$

$$\varepsilon(t) = \frac{c_0}{L}\int_0^t\left[\varepsilon_i(t) - \varepsilon_r(t) - \varepsilon_t(t)\right]dt \quad (4)$$

$$\sigma(t) = \frac{AE}{2A_s}\left[\varepsilon_i(t) + \varepsilon_r(t) + \varepsilon_t(t)\right] \quad (5)$$

2.3 *Experimental program*

For obtaining dynamic increase factors (DIF) for HPC in a wide range of strain rates, two different Split Hopkinson Pressure Bars has been used in this test. A 50 mm SHPB for strain rates in the range of 100 sec^{-1}

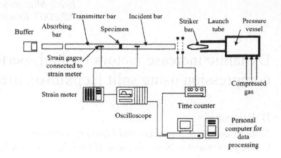

Figure 2. Typical setup for a SHPB device.

Figure 3. Interfaces between pressure bars and specimen.

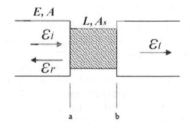

Figure 4. Strain gauge measured wave initiated strains in the SHPB setup. Interfaces at location a and b. Incident, reflected and transmitted, respectively.

to 300 sec^{-1} and a 22 mm SHPB for strain rates in the range of 600 sec^{-1} to 700 sec^{-1}. Dimensions of specimen are shown in Table 1.

HPC specimens of two different strengths and 4 different mix proportions have been prepared for this test. The 50 mm specimens for the 50 mm SHPB were cored out of a 500 × 500 × 150 mm slab. The 15 mm specimen for the 22 mm SHPB were cast in cylinders. All specimens were cured in water for 28 days at a temperature of 20°C. The mix proportions for all tested HPC are shown in Table 2 and Table 3.

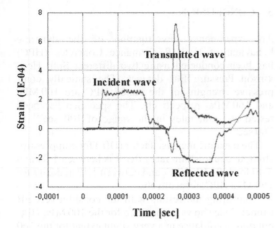

Figure 5.   Output from SHPB test.

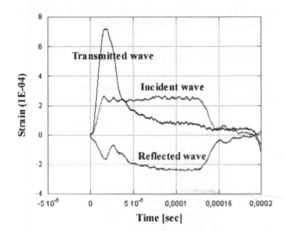

Figure 6.   Time-shifted output from SHPB test.

Table 1.   Specimen dimension.

Test ID #	Specimen diameter, D [mm]	Specimen length, L [mm]	L/D	Pressure bar diameter [mm]
22	15	10	0.67	22
50	50	50	1.00	50

Table 2.   Mix proportions for 15 mm specimen (kg/m^3).

f_{cs} [MPa]	Binder*	Water	Bauxite [1–3 mm]	Sand [1–3 mm]
100	1139	201	433	867
160	1163	193	1300	0

* Densit Binder (ready mix).

Table 3.   Mix proportions for 50mm specimen (kg/m^3).

f_{cs} [MPa]	Cement	Silica Fume	Silica Flower	Aggregate [8 mm]
100	500	50	0	1270
160	657	202	202	0

f_{cs} [MPa]	Sand [1–3 mm]	Water	Super plasticiser
100	630	135	20
160	965	130	43

Table 4.   Experimental results.

Test ID	f_{cs} [MPa]	Strain rate [1/sec]	f_{cd} [MPa]	DIF
01–22	100	710*	336	3.36*
02–22	100	777*	322	3.22*
03–22	100	856*	335	3.35*
04–22	100	722*	340	3.40*
05–22	100	749*	340	3.40*
06–22	160	679*	354	2.36*
07–22	160	674*	338	2.25*
08–22	160	544*	408	2.33*
09–22	160	611*	409	2.34*
10–22	160	516*	433	2.47*
01–50	100	102	184	1.84
02–50	100	145	188	1.88
03–50	100	197	211	2.11
04–50	100	298	240	2.40
05–50	160	81	187	1.17
06–50	160	187	226	1.41
07–50	160	267	241	1.50

* Indicates an average has been used in the plot in Figure 8.

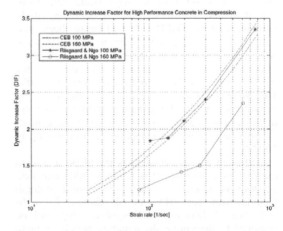

Figure 7.   Plot of experimental results of the dynamic increase factors (DIF) compared with dynamic increase factors (DIF) derived from the CEB concrete model.

3 RESULTS

Results from the experimental investigation are presented in Table 4 and Figure 7. DIF indicates the dynamic increase factors in compression for the tested HPC.

It can be seen that the compressive strength for the 100 MPa HPC increases to 340 MPa at a strain rate of approximately 700 sec^{-1} and the compressive strength for 160 MPa HPC increases to approximately 400 MPa at a strain rate of approximately 600 sec^{-1}.

Also in Figure 7 the dynamic increase factors from the constitutive CEB Model Code are presented for 100 MPa and 160 MPa, respectively. It can be observed that the CEB Model Code give matching results for the 100 MPa HPC but overestimates the dynamic strength enhancement for the 160 MPa HPC.

4 DISCUSSION

The physical mechanisms about the strength enhancement for concrete have yet not been fully understood. At least two factors, the viscoelastic character of the hardened concrete and the time-dependent microcrack growth, may contribute to macroscopic strain rate dependent strength enhancement.

The fact that concrete and concrete like materials are hydrostatic dependent could also cause the dynamic strength enhancement seen using the SHPB device. In the SHPB test lateral confinement could wrongly be initiated from both the contact surface and the lateral inertia during the impact and because of hydrostatic dependency lead to incorrect dynamic increase factors.

As expected the dynamic strength increases with increase in strain rate. Also, as expected, the 160 MPa HPC is less sensitive to high strain rate loading than the 100 MPa HPC. A literature study revealed that no SHPB tests had earlier been conducted on concrete with strength over 100 MPa. Also revealed by that study, is that no constitutive models had been proposed for deriving the dynamic increase factor for concrete over 100 MPa. The presented results was compared with an existing constitutive model for dynamic strength enhancement, and indicates that existing constitutive models only are usable for compressive strengths up to 100 MPa.

Although this programme is based on a limited number of tests only, it is recommended that new constitutive models for deriving dynamic strength enhancement for HPC are developed. In addition extra SHPB tests and special setup will be required to determine the dynamic tensile strength of HPC.

5 CONCLUSION

An experimental investigation of the dynamic behaviour of High Performance Concretes (HPC) has been conducted using two different Split Hopkinson Pressure Bar devices. The statically compressive strengths of the tested HPC are 100 MPa and 160 MPa, respectively. The specimen has been tested at strain rates in the range of 100 sec^{-1} to 700 sec^{-1}.

The dynamic increase factor (DIF) for compressive strength due to strain rate effects is between 1.84 and 3.40 for 100 MPa HPC, and between 1.17 and 2.47 for 160 MPa HPC, respectively.

Comparing the results with the constitutive CEB Model Code shows accordance for the 100 MPa HPC but only accordance in a very slight extent for the 160 MPa HPC.

It is recommended that new and more accurate constitutive models for deriving dynamic strength enhancement for HPC are developed.

6 NOTATION AND REFERENCES

The following symbols and equations are used in this paper:

A	= cross-section area of pressure bar
A	= cross-section area of specimen
D	= specimen diameter
E	= Young's modulus of pressure bar
L	= length of specimen
c_0	= wave velocity in pressure bar
f_{cd}	= dynamic compressive strength
f_{cs}	= static compressive strength
f_{ts}	= static tensile strength
f_{cd}/f_{cs}	= compressive DIF
f_{td}/f_{ts}	= tensile DIF
f'_{co}	= 10 MPa
t	= time
ε	= strain
$\dot{\varepsilon}$	= strain rate (dynamic)
$\dot{\varepsilon}_s$	= strain rate (static)
$\log \gamma$	$= 6.156\,\alpha - 2$
α	$= 1/(5 + 9\,f_{cs}/f'_{co})$

REFERENCES

Bischoff, P. H. and Perry, S. H. Compression behaviour of concrete at high strain-rates. Materials and Structures [24], 425–450. 1991.

Comite Euro-International du Beton – Federation Internationale de la Precontrainte. CEB-FIP Model Code 90 Redwood Books, Trowbridge, Wiltshire, Great Britain. 1990.

Grote, D. L., Park, S. W. and Zhou, M. Dynamic behavior of concrete at high strain rates and pressures: I. experimental characterization. International Journal of Impact Engineering 25[9], 869–886. 2001.

Li, Q. M. and Meng, H. About the dynamic strength enhancement of concrete-like materials in a split Hopkinson pressure bar test. International Journal of Solids and Structures 40[2], 343–360. 2003.

Linholm, U. S. and Bunshah, R. F. High strain rate tests – Measurement of mechanical properties. Interscience, New York 5, 199–216. 1971.

Lok, T. S. and Zhao, P. J. 2004. Impact response of steel fiber-reinforced concrete using a split Hopkinson pressure bar: Journal of Materials in Civil Engineering, v. 16, p. 54–59.

Malvar, L. J. and Ross, C. A., 1998, Review of strain rate effects for concrete in tension: Aci Materials Journal, v. 95, p. 735–739.

Fracture Mechanics of Concrete and Concrete Structures – High-Performance Concrete,
Brick-Masonry and Environmental Aspects – Carpinteri, et al. (eds)
© 2007 Taylor & Francis Group, London, ISBN 978-0-415-44617-4

Experimental tests vs. theoretical modeling for FRC in compression

F. Bencardino, L. Rizzuti & G. Spadea

Department of Structural Engineering, University of Calabria, Rende (CS), Italy

ABSTRACT: Several theoretical models and experimental data for the compressive behavior of Fiber Reinforced Concrete (FRC) are available in literature. In this paper the results of experimental compression tests on Steel Fiber Reinforced Concrete (SFRC) specimens carried out according to standard procedures are shown. The complete stress–strain relationship up to failure for each specimen was monitored and plotted. The different post peak curves of the tested SFRC specimens with fiber content of 1%, 1.6%, 3% were compared and the difference in their behavior was highlighted. Aiming to evaluate the reliability of the models available in literature a comparative study between experimental and theoretical stress–strain relationship was developed. For this purpose theoretical models and experimental data published from 1989 to 2006 were collected in a database and critically analyzed. Many of the theoretical models studied agree more with the experimental curves obtained from the same author rather than with the experimental data obtained from other authors. The validity ranges of the examined stress–strain relationship were critically overviewed.

1 INTRODUCTION

The addition of discontinuous fibers plays an important role in the improvement of mechanical properties of concrete: it increases elastic modulus, decreases brittleness, controls crack growth and crack propagation. Debonding and pull out of fibers require more energy absorption, with a substantial increase in toughness and resistance to cyclic and dynamic loads. In particular, the good performance of Steel Fiber Reinforced Concrete (SFRC) suggests the use of such material in many structural applications, with and without traditional internal reinforcement. Therefore the use of SFRC is particularly suitable for structures loaded over the serviceability limit state in bending, shear, impact forces, as it occurs under seismic or cyclic action. However there is still little knowledge on design/analysis of Fiber Reinforced Concrete (FRC) members. The analysis of structural sections requires, as a basic prerequisite, the definition of a suitable stress–strain relationship for each material to relate its behavior to the structural response. Many stress–strain relationships, in tension and in compression, for FRC materials have been proposed in literature by different authors. With reference to the compressive behaviour, experimental data and analytical models, published from 1989 to 2006, were collected in a database in the following sections. The behaviour of a composite material is influenced by the characteristics of each component and by their proportion in the mixture. In particular, when fibers are added in a concrete mix,

their type, shape, aspect ratio (L_f/D_f) and volume content (V_f) play an important role. It should be mentioned that variations in specimen geometry, loading versus casting direction, loading rate, maximum aggregate size and fibers characteristics modify the compressive behavior of concrete. Extensive experimental data on standard tests are needed to refine a model to consider the effects of different factors. For this purpose only the experimental data carried out according to standard procedures were collected in the database, which, in addition, has been widened with the results obtained from a set of compressive strength tests on cube and cylinder specimens carried out at the University of Calabria.

2 ANALYTICAL MODELS

The addition of steel fibers in the concrete mix significantly affects the tensile and compressive behavior. Fibers contribute to resist crack growth after first crack and allow larger strains, up to concrete crushing. The considerable increase in toughness and ductility of fibrous composites implies that fibers perform an important confining action inside a loaded structural member. Reliable stress–strain relationships are available for plain concrete (Hognestad 1951, Sargin 1971, CEB-FIP 1993) while for FRC a lot of models have been proposed. For a design at ultimate limit states some guidelines (RILEM TC 162-TDF 2003, CNR-DT 204 2006) proposed the same shape of stress–strain relationship in compression used for ordinary concrete (parabolic-rectangular) with an ultimate compressive strain of 0.0035 and a strain at the peak stress of 0.002. The compressive strength of SFRC should be determined by means of standard tests, either on concrete cylinders or concrete cubes. The design principles are based on the characteristic compressive strength at 28 days. Aiming to calculate the actual flexural strength of tested SFRC elements other authors (Lok & Pei 1998, Lok & Xiao 1999) have proposed a similar stress–strain curve used for plain concrete. In evaluating their experimental data some authors (Soroushian & Lee 1989, Ezeldin & Balaguru 1992, Barros & Figueiras 1999, Mansur et al. 1999, Nataraja et al. 1999), have proposed different analytical equations to reproduce the actual behaviour of FRC material in compression. These analytical models are showed below.

2.1 Soroushian & Lee (1989)

The model proposed by Soroushian & Lee (1989) consists of a curvilinear ascending portion followed by a bi-linear descending branch (Eq. 1.a – 1.b).

$$\sigma = -f_{cf}\left(\frac{\varepsilon}{\varepsilon_{pf}}\right)^2 + 2f_{cf}\left(\frac{\varepsilon}{\varepsilon_{pf}}\right) \quad \text{for } \varepsilon \le \varepsilon_{pf} \quad (1.a)$$

$$\sigma = z(\varepsilon - \varepsilon_{pf}) + f_{cf} \ge f_o \quad \text{for } \varepsilon > \varepsilon_{pf} \quad (1.b)$$

In this model the stress and the strain at the peak (f_{cf}, ε_{pf}), the residual stress (f_o) and the slope of the descending branch (z) were evaluated empirically as functions of the matrix compressive strength and fiber reinforcement index ($I_f = V_f(L_f/D_f)$). The compressive strength of fiber concrete (f_{cf}), the strain corresponding to the peak stress (ε_{pf}), are evaluated by adding an additional factor to the matrix strength (f_c), and to the fixed peak strain, $\varepsilon_{co} = 0.0021$. This additional factor is obtained by the fiber reinforcement ratio amplified by another constant value. An empirical equation for the different variables of this proposed model, defined using the least square curve fitting experimental results obtained by the same authors, is shown below:

$$f_{cf} = f_c + 3.6I_f \quad (2)$$

$$f_o = 0.12f_{cf} + 14.8I_f \quad (3)$$

$$z = -343f_c(1 - 0.66\sqrt{I_f}) \le 0 \quad (4)$$

$$\varepsilon_{pf} = 0.0007I_f + 0.0021 \quad (5)$$

In this model the residual ultimate strain of SFRC was not fixed.

2.2 Ezeldin & Balaguru (1992)

Ezeldin & Balaguru (1992) proposed an analytical equation (Eq. 6) to generate the stress–strain curve for normal strength SFRC based on the equation proposed by Carreira & Chu (1985) for uniaxial compression of plain concrete. This equation involves a material parameter β, which is the slope of the inflection point at the descending segment.

$$\frac{\sigma}{f_{cf}} = \frac{\beta\left(\dfrac{\varepsilon}{\varepsilon_{pf}}\right)}{\beta - 1 + \left(\dfrac{\varepsilon}{\varepsilon_{pf}}\right)^\beta} \quad (6)$$

In order to quantify the effect of fibers on the compressive behavior of FRC, a least square fitting analysis was performed to establish a connection between the reinforcement index by weight of hooked end fibers ($RI = W_f(L_f/D_f)$) and the main parameters of the stress–strain curve, namely; the compressive strength (f_{cf}) and the corresponding peak strain (ε_{pf}). Also in this equation f_{cf} and ε_{pf} are calculated by adding an additional factor, linked to fiber property, to the strength and the strain of plain concrete, f_c and ε_{co}, respectively. The following equations were obtained by using the regression analysis performed using experimental data of the same authors.

$$f_{cf} = f_c + 3.51(RI) \quad (7)$$

$$\varepsilon_{pf} = \varepsilon_{co} + 446 \times 10^{-6}(RI) \quad (8)$$

For hooked end fibers:

$$\beta = 1.093 + 0.7132(RI)^{-0.926} \quad (9)$$

Using the experimental results of Fanella & Naaman (1985) for mortar reinforced with straight fibers, the following equation was proposed:

$$\beta = 1.093 + 7.4818(ri)^{-1.387} \quad (10)$$

Where (ri) is the reinforcement index, by weight, of straight fibers.

Without specific experimental data the authors suggest to use $\varepsilon_{co} = 0.002$, according to the International Recommendations (1970).

The equations proposed to evaluate β can be used for reinforcing index values ranging from 0.75 to 2.5 for hooked end fibers and from 2 to 5 for straight fibers. The ultimate strain is not fixed.

2.3 Barros & Figueiras (1999)

Based on their experimental results and following the procedure proposed by Mebarkia & Vipulanandan (1992), Barros & Figueiras (1999) proposed the following compression stress–strain relationship.

$$\sigma = f_{cf} \frac{\dfrac{\varepsilon}{\varepsilon_{pf}}}{(1-p-q)+q\left(\dfrac{\varepsilon}{\varepsilon_{pf}}\right)+p\left(\dfrac{\varepsilon}{\varepsilon_{pf}}\right)^{(1-q)/p}} \qquad (11)$$

with

$$q = 1 - p - \frac{E_{pf}}{E_c} \qquad (12)$$

$$p+q \in \,]0,1[\,, \qquad\qquad \frac{1-q}{p} > 0 \qquad (13)$$

$$E_{pf} = \frac{f_{cf}}{\varepsilon_{pf}}; \qquad\qquad E_c = 21{,}500\sqrt[3]{\frac{f_{cf}}{10}} \quad (14)$$

Where f_{cf} is the average compression strength, experimentally evaluated. The following equations were obtained by the authors by applying the least square method.

For hooked end fibers ($L_f = 30\,mm$, $D_f = 0.5\,mm$, $L_f/D_f = 60$):

$$\varepsilon_{pf} = \varepsilon_{co} + 0.0002 W_f \qquad (15)$$

$$p = 1 - 0.919 e^{-0.394 W_f} \qquad (16)$$

For hooked end fibers ($L_f = 60\,mm$, $D_f = 0.8\,mm$, $L_f/D_f = 75$):

$$\varepsilon_{pf} = \varepsilon_{co} + 0.00026 W_f \qquad (17)$$

$$p = 1 - 0.722 e^{-0.144 W_f} \qquad (18)$$

Where, ε_{co} is the strain at peak for plain concrete. Without specific experimental data the authors suggest to use $\varepsilon_{co} = 0.0022$, according to CEB-FIP (1993). W_f is the fiber weight percentage in the mixture.

The values of ε_{pf} and p can be obtained from the equation proposed, for f_{cf} values ranging from 30 to 60 MPa and for concrete reinforced with similar content of fibers used by the authors.

The value of ultimate strain ε_{fu} is not fixed.

2.4 Nataraja, Dhang, Gupta (1999)

Nataraja et al. (1999) proposed an analytical equation similar to that of Ezeldin & Balaguru (1992) (Eq. 6)

but they used their experimental data by providing other additional factors, related to the fiber reinforcement index, used to determine the strength, the strain corresponding to the peak and the β value.

$$f_{cf} = f_c + 2.1604(RI) \qquad (19)$$

$$\varepsilon_{pf} = \varepsilon_{co} + 0.0006(RI) \qquad (20)$$

$$\beta = 0.5811 + 1.93(RI)^{-0.7406} \qquad (21)$$

According to the authors, without specific experimental data, ε_{co} can assume a value equal to 0.002. The proposed equations can be used for concrete with strength up to 50 MPa, reinforced with crimped fibers with a reinforcing index value ranging from 0.9 to 2.7. The value of ultimate strain ε_{fu} is not fixed.

3 EXPERIMENTAL DATA

Experimental data available in literature were analysed but only the data of the experimental analysis carried out following standard procedure were collected in the present database. For each author the main mechanical parameters and the details of the cylindrical tested specimens (D: diameter; H: height; N.: number of specimens) are given in Table 1.

The authors carried out experimental tests on FRC with maximum aggregate size of: 9 mm (Ezeldin & Balaguru 1992), 10 mm (Dwarakanath & Nagaraj 1991, Wafa & Ashour 1992), 15 mm (Barros & Figueiras 1999), 19 mm (Mansur et al. 1999), 20 mm (Nataraja et al. 1999), 25 mm (Jo et al. 2001).

4 EXPERIMENTAL PROGRAM

The experimental investigation was carried out on 24 specimens, cubes and cylinders, considering plain concrete and SFRC with 1%, 1.6% and 3% of fiber content. Compressive strengths were evaluated and, using a suitable experimental set up, stress–strain curves on cylindrical specimens were also recorded to highlight the role of the fibers in the post peak response.

4.1 Details of materials

The following materials were used: type I Portland cement, crushed coarse aggregate, spherical quartz, water, condensed silica fume and super plasticizer. Maximum size of the coarse aggregate was 15 mm. The steel fibers used in this investigation were hooked end, with a tensile strength of 350–400 MPa, a length of 22 mm, a diameter of 0.5 mm and an aspect ratio of 40.

Table 1. Experimental database.

Authors	Standard adopted	D, H mm	N.	Steel fibers type	V_f (%)	L_f/D_f	ε_{pf}	ε_{fu}	f_{cf} (MPa)	f_{fu} (MPa)
								Specimens		
Soroushian & Lee	–	150, 300	–	Straight	–	–	0.0033	0.0067	41.2	5.2
					2.00	47	0.0033	0.0116	40.7	23.3
					2.00	83	0.0050	0.0116	43.2	31.2
					2.00	100	0.0050	0.0116	44.6	39.5
Dwarakanath & Nagaraj	–	100, 200	–	Straight	–	–	0.0020	0.0044	24.5	17.6
					1.00	72	0.0021	0.0060	26.0	19.4
					2.00	72	0.0025	0.0070	26.7	19.7
					3.00	72	0.0030	0.0072	31.6	26.5
Ezeldin & Balaguru	ASTM C39	100, 200	36	Hooked end	–	–	0.0022	0.0104	35.9	8.2
					0.38	60	0.0025	0.0177	40.7	7.4
					0.57	60	0.0025	0.0200	40.7	9.6
					0.76	60	0.0031	0.0400	37.9	13.5
Wafa & Ashour	ASTM C39, C31, C192	150, 300	126	Hooked end	–	–	0.0020	0.0027	91.5	80.0
					0.5	75	0.0024	0.0107	94.6	18.3
					1.0	75	0.0018	0.0107	95.6	39.5
					1.5	75	0.0018	0.0107	100.0	57.5
Barros & Figueiras	JSCE SF5	150, 300	32	Hooked end	0.76	75	0.0022	0.0400	38.2	1.8
Mansur et al.	–	100, 200	54	Hooked end	–	–	0.0024	0.0045	103.6	27.0
					0.50	60	0.0026	0.0140	104.7	23.5
					1.00	60	0.0027	0.0140	107.0	39.0
					1.50	60	0.0029	0.0140	103.5	40.0
Nataraja et al.	ASTM C39	150, 300	14	Crimped	–	–	0.0027		43.0	
					0.50	55	0.0031	0.0155	45.8	13.8
					0.75	55	0.0033	0.0167	41.6	17.4
					1.00	55	0.0034	0.0170	47.0	21.2
Jo et al.	JCI SF2	150, 300	75	Hooked end	–	–	0.0021	0.0060	41.8	1.0
					0.50	75	0.0026	0.0050	41.8	5.4
					0.75	75	0.0020	0.0100	37.7	11.6
					1.00	75	0.0020	0.0090	34.3	17.7
					1.50	75	0.0023	0.0100	33.1	25.0
					–	–	0.0021	0.0052	61.4	3.8
					0.50	75	0.0021	0.0050	60.0	5.4
					0.75	75	0.0024	0.0043	60.1	30.0
					1.00	75	0.0023	0.0100	53.8	9.2
					1.50	75	0.0021	0.0100	60.9	3.2
					–	–	0.0020	0.0080	64.6	1.3
					0.50	75	0.0022	0.0080	70.0	1.2
					0.75	75	0.0022	0.0100	71.4	0.0
					1.00	75	0.0022	0.0091	71.0	1.1
					1.50	75	0.0027	0.0075	62.8	3.4

4.2 Mix design and casting of specimens

Table 2 shows, the plain concrete (PC) and SFRC (S1%, S1.6% and S3%) mix design for 1 m³ of concrete batch used in the experimental program.

The specimens were mixed, casted and cured in a framework. The concrete was compacted on a vibrating table. All specimens were removed from the moulds within 24hr and cured for 27 more days under water satured sand.

4.3 Test details

The compressive strength tests at 28 days were carried out, according to UNI EN 12390-3, using 150 mm × 150 mm × 150 mm cubes and 150 mm × 300 mm

Table 2. Mix design.

Material		PC	SFRC S1%	S1.6%	S3%
Cement	42.5 R	500	500	500	500
Quartz	0/2 mm	377	377	377	377
	3/6 mm	273	273	273	273
Coarse aggregate	0/5 mm	693	615	567	458
	5/10 mm	290	290	290	290
	10/15 mm	317	317	317	317
Fiber	V_f	0%	1%	1.6%	3%
	(kg)	–	78	126	235
Silica fume	%	6%	6%	6%	6%
	(kg)	30	30	30	30
Super plasticizer	%	1.5%	1.5%	1.5%	1.5%
	(kg)	7.5	7.5	7.5	7.5
Water	w/c	0.35	0.35	0.35	0.35
	(l)	175	175	175	175

Table 3. Cubic compressive strength.

	R_c MPa	R_m MPa	s MPa	δ %
PC_k1	65.5	61.5	3.8	6.2
PC_k2	57.9			
PC_k3	61.1			
S1%_k1	69.8	74.2	4.6	6.2
S1%_k2	79.0			
S1%_k3	73.6			
S1.6%_k1	62.8	60.5	5.4	8.9
S1.6%_k2	54.3			
S1.6%_k3	64.5			
S3%_k1	64.4	62.8	1.6	2.5
S3%_k2	62.8			
S3%_k3	61.2			

Table 4. Cylindrical compressive strength.

	f_c MPa	f_m MPa	s MPa	δ %
PC_c1	64.1	66.7	2.8	4.2
PC_c2	66.3			
PC_c3	69.8			
S1%_c1	67.2	69.6	2.4	3.4
S1%_c2	71.9			
S1%_c3	69.8			
S1.6%_c1	61.1	58.1	2.7	4.6
S1.6%_c2	55.7			
S1.6% c3	57.7			
S3%_c1	59.8	58.5	3.8	6.5
S3%_c2	54.2			
S3%_c3	61.5			

Figure 1. Experimental set-up

cylinders loaded uniaxially. A Zwick/Roell servo hydraulic closed-loop test machine with a 3000 kN capacity was used. Loadings were increased at a rate of 0.05 mm/min.

For cylindrical specimens three HBM LVDTs with a gage length of 20 mm, were mounted at 120-degree intervals along two circumferential ties placed on the specimen at a base length of 100 mm. These aluminum ties were able to support the measuring devices and did not confine specimens. Figure 1 shows the position of the measure instrumentation. The data acquisition and signal control were carried out using a HBM Spider 8 control unit.

4.4 Test results and discussion

All test results are summarized in Tables 3 and 4. In these tables the compressive strength of each specimen and for every set of specimens the mean value of the compressive strength, the standard deviation and the coefficient of variation are given. The graphical representations of the stress–strain curves are given in

Figure 2. The experimental typical curves of the tested plain concrete and SFRC specimens are compared and showed in Figure 3.

Tests results show that the addition of fibers, compared to plain concrete, slightly affects the compressive strength value but more of it influences the post-peak response (Fig. 3). Concrete reinforced with medium content of steel fibers ($V_f = 1\%$) shows a slight improvement in the descending or softening branch compared to plain concrete while concrete reinforced with higher content of fibers ($V_f = 1.6\%$, 3%) shows a more extended softening branch. The presence of fiber modified the failure mode of the concrete cylinder from a brittle to a less brittle failure mode. The post-test aspects of the specimens showed, in the case of plain concrete, either a single shear plane or a cone-type failure. By contrast, SFRC specimens showed a large number of longitudinal cracks near the failure zone, which were oriented in the direction parallel or sub-parallel to the external compressive stresses.

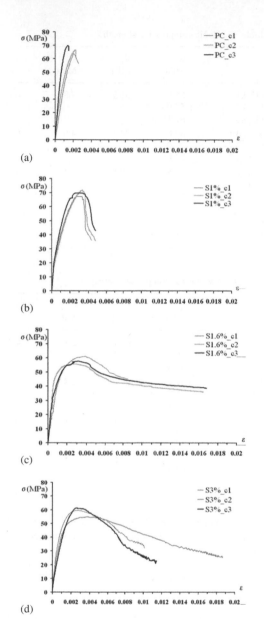

(a)

(b)

(c)

(d)

Figure 2. Stress–strain curves of cylindrical specimens: (a) plain concrete, SFRC (b) $V_f = 1\%$, (c) $V_f = 1.6\%$, d) $V_f = 3\%$.

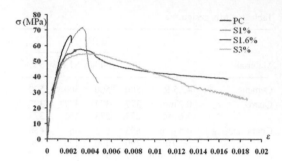

Figure 3. Typical stress–strain curves without and with fibers.

content, the FRC compressive strength and the strain at the peak stress increase.

Example:

Plain concrete: $f_c = 30\,\mathrm{MPa}$, $\varepsilon_{co} = 0.002$.

Steel fibers: $L_f = 30\,\mathrm{mm}$, $D_f = 0.5\,\mathrm{mm}$, $L_f/D_f = 60$.

– *Soroushian & Lee (1989)*

$V_f = 1\%$, $f_{cf} = 32.2\,\mathrm{MPa}$ $(+7.3\%)$, $\varepsilon_{pf} = 0.0025$ $(+25\%)$

$V_f = 2\%$, $f_{cf} = 34.3\,\mathrm{MPa}$ $(+14.3\%)$, $\varepsilon_{pf} = 0.0029$ $(+45\%)$

$V_f = 3\%$, $f_{cf} = 36.5\,\mathrm{MPa}$ $(+21.7\%)$, $\varepsilon_{pf} = 0.0034$ $(+70\%)$

– *Ezeldin & Balaguru (1992)*

$V_f = 1\%$, $f_{cf} = 36.3\,\mathrm{MPa}$ $(+21\%)$, $\varepsilon_{pf} = 0.0028$ $(+40\%)$

$V_f = 2\%$, $f_{cf} = 42.6\,\mathrm{MPa}$ $(+42\%)$, $\varepsilon_{pf} = 0.0036$ $(+80\%)$

$V_f = 3\%$, $f_{cf} = 48.9\,\mathrm{MPa}$ $(+63\%)$, $\varepsilon_{pf} = 0.0044$ $(+120\%)$

– *Nataraja et al. (1999)*

$V_f = 1\%$, $f_{cf} = 34.1\,\mathrm{MPa}$ $(+13.7\%)$, $\varepsilon_{pf} = 0.0032$ $(+60\%)$

$V_f = 2\%$, $f_{cf} = 38.3\,\mathrm{MPa}$ $(+27.7\%)$, $\varepsilon_{pf} = 0.0043$ $(+115\%)$

$V_f = 3\%$, $f_{cf} = 42.4\,\mathrm{MPa}$ $(+41.3\%)$, $\varepsilon_{pf} = 0.0055$ $(+175\%)$

The experimental data collected show that the FRC compressive strength does not increase significantly compared to the strength of plain concrete.

5 EXPERIMENTAL TESTS V/S THEORETICAL MODELING

5.1 Theoretical models: critical review

With reference to fixed mechanical parameters, according to the studied models by increasing fiber

5.2 Cross comparison

A cross comparison between the experimental curve and those obtained by using the theoretical models analysed was carried out. This comparison highlights that every model agrees well with their respective experimental data (Fig. 4) and less well with the experimental data obtained by other authors, because each proposed equation was obtained by using a regression analysis to interpolate their own experimental data. The behavior of a concrete reinforced with a medium content of fibers ($V_f = 1$–2%) is better performed by

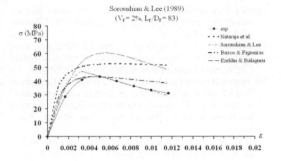

Soroushian & Lee (1989)
($V_f = 2\%$, $L_f/D_f = 83$)

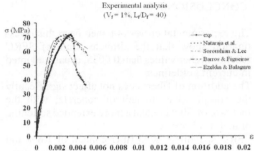

Experimental analysis
($V_f = 1\%$, $L_f/D_f = 40$)

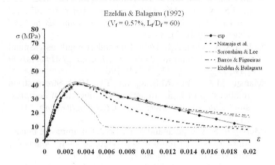

Ezeldin & Balaguru (1992)
($V_f = 0.57\%$, $L_f/D_f = 60$)

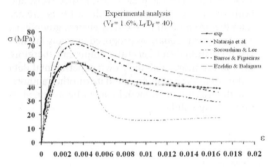

Experimental analysis
($V_f = 1.6\%$, $L_f/D_f = 40$)

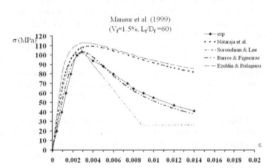

Mansur et al. (1999)
($V_f = 1.5\%$, $L_f/D_f = 60$)

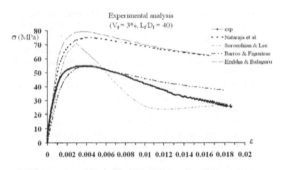

Experimental analysis
($V_f = 3\%$, $L_f/D_f = 40$)

Figure 4. Comparisons: experimental v/s theoretical curves.

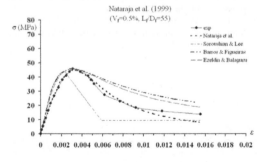

Nataraja et al. (1999)
($V_f = 0.5\%$, $L_f/D_f = 55$)

the model proposed by Soroushian & Lee (1989). According to this model in the case of low content of fibers added into the concrete matrix ($V_f < 1\%$), the post-peak response is underestimated. With reference to concrete reinforced with medium content of fiber ($V_f > 1\%$) the models proposed by Ezeldin

& Balaguru (1992) and by Nataraja et al. (1999), predict a stiffened ascending branch and overestimate the descending branch. These authors worked out an evaluation for the experimental data of SFRC with low content of fibers. The descending branch of these stress–strain relationships is affected by the β value. As the value of β decreases the experimental softening effect is not modelled. The stress–strain relationship proposed by Barros & Figueiras (1999) agrees with the experimental data of Ezeldin & Balaguru (1992) and of Mansur et al. (1999) because the authors also used these experimental data to define their model. In this proposed model the average compression strength of FRC was used while the other models calculated the compressive strength of SFRC and the strain corresponding to the peak stress adding an empirical factor to the values of matrix stress and strain.

6 CONCLUSIONS

- The experimental curves obtained from this investigation show that the ultimate strain of SFRC reaches higher values than 0.0035, usually adopted in current guidelines.
- The addition of fibers does not affect significantly the compressive strength of concrete; with the increase of fiber content a more extended softening branch is observed.
- SFRC specimens with fiber content of 1.6% and 3% show, at 0.01 strain, a residual stress of about 70% and 50% of the peak stress, respectively, and both reach an ultimate strain three times greater than the ultimate strain fixed by current guidelines.
- The analytical equations proposed by many authors are not rational but generally empirical; therefore they agree more with their own experimental data rather than the experimental data of other authors. For this reason each model cannot be assumed as universally valid.

REFERENCES

Barros, J.A.O. & Figueiras, J.A. 1999. Flexural behavior of SFRC: testing and modeling. *Journal of Materials in Civil Engineering* 11(4): 331–339.

Carriera, D.J. & Chu, K.M. 1985. Stress–strain relationship for plain concrete in compression. *ACI Journal* 82(6): 797–804.

CEB-FIP Model Code 1990. 1993. Bullettin d'Information No. 213/214. Comite Euro-International du Beton, Thomas Telford (ed.). London.

CNR-DT 204. 2006. Istruzioni per la progettazione, l'esecuzione ed il controllo di strutture di calcestruzzo fibrorinforzato. CNR, Rome, Italy.

Dwarakanath, H.V. & Nagaraj, T.S. 1991. Comparative study of predictions of flexural strength of steel fiber concrete. *ACI Structural Journal* 88(6): 714–720.

Ezeldin, A.S. & Balaguru, P.N. 1992. Normal and high-strength fiber reinforced concrete under compression. *Journal of Materials in Civil Engineering* 4(4): 415–427.

Fanella, D.A. & Naaman, A.E. (1985). Stress–strain properties of fiber reinforced mortar in compression. *ACI Journal* 82(4): 475–483.

Hognestad, E. 1951. A study of combined bending and axial load in reinforced concrete members. University of Illinois Engineering Experimental Station Bulletin 399: 128.

International recommendations for the design and construction of concrete structures: principles and recommendations. 1970. Commite, Europeen du Beton/Federation International de la Precontrainte, 2nd Edition, Paris, France.

Jo, B.W., Shon, Y.H. & Kim, Y.J. 2001. The evaluation of elastic modulus for steel fiber reinforced concrete. *Russian Journal of Nondestructive Testing* 37(2): 152–161.

Lok, T.S. & Pei, J.S. 1998. Flexural behavior of steel fiber reinforced concrete. *Journal of Materials in Civil Engineering* 10(2): 86–97.

Lok, T.S. & Xiao, J.R. 1999. Flexural strength assessment of steel fiber reinforced concrete. *Journal of Materials in Civil Engineering* 11(3): 188–196.

Mansur, M.A., Chin, M.S. & Wee, T.H. 1999. Stress–strain relationship of high-strength fiber concrete in compression. *Journal of Materials in Civil Engineering* 11(1): 21–29.

Mebarkia, S. & Vipulanandan, C. 1992. Compressive behavior of glass-fiber-reinforced polymer concrete. *Journal of Materials in Civil Engineering* 4(1): 91–105.

Nataraja, M.C., Dhang, N. & Gupta, A.P. 1999. Stress strain curve for steel-fiber reinforced concrete under compression. *Cement and Concrete Composites* 21: 383–390.

RILEM TC 162-TDF. 2003. Test and design method for steel fibre reinforced concrete – σ-ε design method. *Materials and Structures* 36: 560–567.

Sargin, M. 1971. Stress–strain relationship for concrete and the analysis of structural concrete sections. Solid Mechanics Division University of Waterloo, Waterloo, Ontario, Canada. Study No.4.

Soroushian, P. & Lee, C.D. 1989. Constitutive modeling of steel fiber reinforced concrete under direct tension and compression. In R.N. Swamy & B. Barr (eds) *Fibre Reinforced Cements and Concretes, Recent Developments*: 363–375.

UNI EN 12390-3. 2003. Testing hardened concrete – Compressive strength of test specimens.

Wafa, F.F. & Ashour, S.A. 1992. Mechanical properties of high-strength fiber reinforced concrete. *ACI Materials Journal* 89(5): 449–455.

Local bending tests and punching failure of a ribbed UHPFRC bridge deck

F. Toutlemonde, J.-C. Renaud & L. Lauvin
LCPC, Structures Laboratory, Paris, France

S. Brisard
SETRA, Bridge Department, Bagneux, France

J. Resplendino
CETE de Lyon, Bridge Division, L'Isle d'Abeau, France

ABSTRACT: Local bending tests were carried out on critical locations of a new type of steel-concrete composite bridge deck, as a part of the experimental validation of an innovative ultra-high performance fibre reinforced concrete (UHPFRC) ribbed slab made of segments assembled by post-tensioning. Local failure was identified as one of the possible critical aspects of the design, due to lack of information on the real capacities of UHPFRC materials in bidirectional bending and possible punching shear mechanism. Experimental validation of the design was first derived from the representation of Eurocode wheel models, with a standard 0.4 m × 0.4 m loaded zone. A reduced 0.19 m × 0.26 m had to be used for representing concentrated impacts, and the representation of pavement layers was eliminated, so that the local bearing capacity could be reached. It was finally obtained with a characteristic punching failure mechanism, the ultimate load corresponds to two to three times the nominal wheel load of the Eurocode.

1 INTRODUCTION

1.1 *A UHPFRC ribbed slab applied as part of a road bridge deck*

Extending the economic span range of composite bridge decks, classically applied for road bridges from 50 to 100 m-spans, represents an important challenge. For shorter spans, some prototype applications with thinner slabs made of C80 have led to 0.14 m-thick slabs, Causse & Montens (1992), Chevallier & Petitjean (2001). Attempts for even lighter short span bridge solutions using ultra-high performance fiber–reinforced concrete (UHPFRC) are currently under intense study, Tanis (2006), Bouteille et al. (2006).

For longer spans, the concrete slab appears as too heavy, and thinner slabs made of pre-cast high performance concrete (HPC) segments might tend to be developed. This could provide an alternative to steel orthotropic decks, for which fatigue degradations are a major concern. Pursuing this trend of lightness, durability, and savings of natural resources with the use of materials with optimized performance, Bouteille & Resplendino (2005), a preliminary design of a UHPFRC ribbed slab, connected to twin longitudinal steel beams, has been studied within the frame of MIKTI French R & D national project aiming to favor innovative steel-concrete composite applications, Resplendino & Bouteille (2003). The required frame of application consists in a 3-span 90 + 130 + 90 m-long, 9 m-wide road bridge (two 3.5 m-wide lanes + 1 m-wide side strips) with two 1 m-wide sidewalks. The general design, determination of required pre-stressing tendons and detailing, was carried out applying French Recommendations relative to UHPFRC, AFGC-SETRA (2002) in complement of French Bridge design codes for composite and pre-stressed concrete bridges.

Definition of the slab thickness and of the transverse ribs and pre-tensioning was first determined considering the local and transverse bending. Then longitudinal bending was considered, as well as specific design and connection aspects. Length of successive segments was limited to 2.50 m for possible truck delivery. Regular spacing and similar height of longitudinal ribs was searched, except at the ends for the anchoring of safety barriers. The resulting transverse profile of the deck is represented in Figure 1, and the transverse cross-section of one segment in Figure 2. The current slab thickness is 0.05 m, the total thickness with the ribs is 0.38 m, and the current ribs spacing is 0.6 m from axis

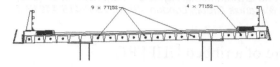

Figure 1. Transverse cross-section including longitudinal post-tensioning (17 strands T15S in sum).

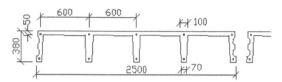

Figure 2. Longitudinal cross-section of precast segments. Longitudinal ribs are 50 mm-wide only at bottom. Lengths in mm.

to axis in both directions. Current longitudinal ribs are 0.05 m-wide at bottom, 0.1 m-wide at the top. The average resulting weight of the slab is about 3.9 kN/m². Transverse pre-stressing is realized by 2 rectilinear T15S tendons along the vertical axis of the ribs. The upper one is anchored from one end to the other, the lower one is sheathed along the corbels and on supports. The resisting bending moment for a current T-shaped 0.6 m-cross-section is around 115 kN m, it is limited by the tendons capacity. Longitudinal post-tensioning ensures a minimum compressive strength equal to 4 MPa for SLS (frequent combination). Low creep of UHPFRC helps keeping the benefit of this post-tensioning very efficiently. An important detailing effort has been carried out for defining anchoring of the safety barrier, connection to the steel main girders, etc. This finally demonstrated the feasibility of realizing a complete bridge design on the basis of AFGC-SETRA UHPFRC Recommendations.

1.2 *Experimental validation program*

However, before this project can be applied, some critical aspects require validation, since the assumptions of beams theory which have been used at preliminary design stage are questionable regarding local bending and possible partial rotation of the ribs around the loaded honeycombs.

Experimental validation was thus undertaken, Toutlemonde et al. (2005), on a 6.1 m-wide model slab made of two ribbed segments, one made of Ductal®-FM and the other of BSI®, at scale 1 for the length and thickness, connected realistically with a UHPFRC cast in place cold joint and longitudinal post-tension (Fig. 3). Transverse span was reduced to 3.98 m (clear span between longitudinal beams used as simple supports) for bending tests aimed at representing the effects of axle loads, and the cantilever side was used for tests of anchoring of the safety barrier.

Figure 3. Model ribbed slab for validation tests. (a) Casting – (b) Cold joint – (c) Honeycombs ; in the center, the connected side ribs of the segments – (d) Longitudinal post-tensioning – (e) Overview of the model under general 'axle' bending test.

1482

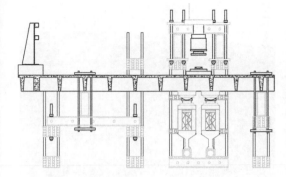

Figure 4. Transverse cross-section of the model in the local bending tests configuration.

Local failure was identified as one of the possible critical aspects of the design, due to lack of information on the real capacities of UHPFRC materials in bidirectional bending and possible punching shear mechanism. Namely, resistance to local bending has a direct implication on the slab thickness (0.05 m) and ribs distance in the project. Experimental validation was first derived from the representation of Eurocode standard wheel models, with a 0.4×0.4 m loaded zone. But reduced wheel surfaces should also be considered. Moreover, the favorable effect of pavement layers should be quantified.

2 LOCAL BENDING TESTS

2.1 Loading setup and objectives

The local bending tests should help quantifying the effective bearing capacity with respect to possibly concentrated loads over the bridge deck. The loads should thus be concentrated at the centre of one particular honeycomb, where the deck is at the thinnest. This location corresponds to the conventional design situation. Assuming 45° diffusion through 9 cm-thick bituminous concrete pavement layers and the deck slab, this design case with a 150-kN service live load according to Eurocode 1 appears as critical with respect to the tensile strength of UHPFRC at the centre of the honeycomb lower sides.

For the experimental validation, since this phase should be at least locally destructive, it was carried out after the fatigue resistance verification described in Toutlemonde et al. (2007). The 3.98 m-span transverse support system was kept, however intermediate supports at the four corners of the loaded honeycomb were provided, so that the load applied by the actuator could be significantly taken by the reactions of the 4 stays (Figs. 4–5). Free rotation was permitted on each support towards the centre of the honeycomb using an intermediate roll, and the reaction on each stay was

Figure 5. Supporting stays at corners of the loaded honeycomb.

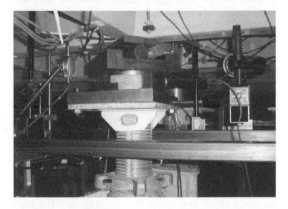

Figure 6. Support detail and reaction load measurement.

measured using a load cell (Fig. 6). The test was load-controlled during a first linear phase with a loading rate of 1 kN/s, then the actuator displacement was used as the load-control signal with a reference 10 μm/s rate. The testing capacity was limited by the stays (200 kN maximum on each) or the actuator (1000 kN).

2.2 Tested zones

The tested honeycombs were chosen so that the results could integrate the influence of the UHPFRC material

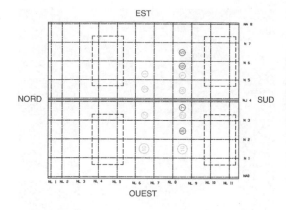

Figure 7. Location of local bending tests (dark circles).

(Ductal®-FM or BSI®), the influence of the twin transverse key rib with cold joint, which can be one side of the loaded honeycomb, and the possible influence of previous fatigue cyclic program, even though it has been shown as non-damaging for the structure, Toutlemonde et al. 2007.

Four honeycombs were thus chosen for the tests (Fig. 7). They are aligned between longitudinal ribs NL8 and NL9 so that the actuator can be easily moved over the zones to be tested, due to its fixation on longitudinal steel rectangular girders 6 m-long over the model (Fig. 8). This choice was also necessary for adapting the supporting system of stays under the slab (Fig. 5). Ends of the slab were avoided, thus the tested zones were limited in the longitudinal direction by transverse ribs referred as N2 and N3, N3 and N4, N5 and N6, and N6 and N7. It was thus expected to get some statistical relevance of the ultimate loading capacity experimentally identified.

2.3 Load diffusion configurations

In the reference case corresponding to the bridge project, referred to as 'configuration 1', application of the 'wheel' load included a 0.09 m-thick polymer material (PA6 commercially available as ®Ertalon) 0.6 m-square plate, representative of the diffusion induced by the bituminous concrete paving layers. Young's modulus of this material is about 2.5 to 3 GPa, close to an average value of bituminous concrete (which highly depends on temperature). For correct load distribution between the rough UHPFRC slab and the ®Ertalon plate confined sand is provided. Over this plate, the load was distributed over a surface deriving from Eurocode 1 wheel model. The reference surface is 0.4 × 0.4 m corresponding to fatigue load models 1 and 3. It is represented in the tests below a point ball hinge under the jack end by a 0.4 × 0.4 × 0.04 m steel plate (Fig. 9).

Figure 8. Fixation of the actuator and superstructures.

However, reduced surfaces are also considered in Eurocode 1 fatigue load models #2 and 4, with wheel types A and C (0.22 × 0.32 m or 0.27 × 0.32 m). Similarly to the fatigue testing program, Toutlemonde et al. (2007), a reduced steel plate of dimensions 0.19 × 0.26 × 0.04 m with corners cut at 2 cm was used instead of the 0.4 m-square one in 'configuration 2' (Fig. 10). The longer side is along the longitudinal direction.

Due to the very low porosity of UHPFRC materials, which might permit to consider bridge decks without watertight and paving overlays, and for a safe evaluation of the slab capacity, tests were also carried out without the 0.09 m-thick polymer layer, possibly representing transient phases where the bridge deck has not got any pavement layer. 'Configurations 3 and 4' correspond to such situations with the standard or reduced wheel surfaces, respectively (Figs. 11–12). Under the steel plate, confined sand under a 3 mm-thin cardboard is still provided.

2.4 Synthesis of tests carried out

A summary of the test configurations and locations is given in Table 1. The first two zones (N5N6 and N3N4) correspond to honeycombs previously submitted to fatigue loading as described in Toutlemonde et al. (2007). Especially, 100,000 cycles between 5 and 155 kN had been applied on each of these zones under

Figure 9. Load distribution: Configuration 1.

Figure 10. Load distribution: Configuration 2.

Figure 11. Load distribution: Configuration 3.

Figure 12. Load distribution: Configuration 4.

Table 1. Test configurations and locations (N4 is the joint).

Date	Zone	Material	Config	Result
Jun 14	N5N6	Ductal®-FM	1	limit of stays capacity
Jun 15	N5N6	Ductal®-FM	2	limit of stays capacity
Jun 26	N5N6	Ductal®-FM	3	limit of stays capacity
Jun 26	N5N6	Ductal®-FM	4	punching shear failure
Jul 12	N3N4	BSI®	4	punching shear failure
Aug 22	N2N3	BSI®	3	limit of stays capacity
Jul 12	N2N3	BSI®	4	punching shear failure
Sept 15	N6N7	Ductal®-FM	4	punching shear failure

'configuration 2'. The local bending tests were carried out from June 2006 to September 2006. The ribbed slab was about 2 years old. Eighty measurement channels were used in order to characterize the global and local bending of the ribbed slab, local strains, deflection at the centre of the honeycombs, vertical displacement at the corners and at mid-span of the loaded honeycomb side ribs. In the following focus is given on the major results in terms of load, deflection and failure modes.

The materials of which the model was made had already been used in real structures, St-Pierre-la-Cour

bridge or the roof of Millau bridge tollgate, Bouteille & Resplendino (2005). One segment is made of Ductal®-FM, having a water to cement ratio equal to 0.21 and a volumetric fiber content equal to 2.15%. Thermal treatment was applied at an age of 48 h, consisting

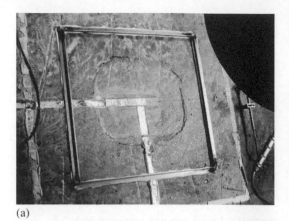

(a)

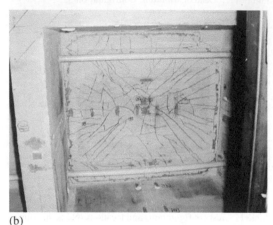

(b)

Figure 13. Punching shear failure. Zone N5N6. (a) Top side. (b) Bottom side.

Figure 14. Punching shear failure. Zone N3N4. Top side.

of linearity 'ftj' equal to 9.3 MPa (mean value). The conventional 'ftu' value which helps accounting for the maximum bending moment of the plates is 8.3 MPa (mean value).

3 GLOBAL RESULTS

3.1 Progress of the experimental program

Before each test, stays were adjusted so that about 90% of the slab dead weight (4×30 kN) is taken by the stays, and contact of the actuator was ensured with 10 kN-pre-load. Then, due to stays and model compliances, about 60% of the actuator load is taken by reaction of the stays, while the remaining load leads mostly to general transverse bending of the slab. In these tests no irreversible degradation was observed due to this residual transverse bending, on the contrary it was searched, in using the different configurations, to focus on local damage. Therefore only the total reaction load (sum of the loads taken by the stays) is considered in the following as significant for the local honeycomb behavior.

Local failure which was searched could be obtained neither with configurations 1, 2 nor 3, even though the total reaction reached about 700 kN, both on Ductal®-FM and BSI® zones. For zone N5N6 significant non-linear extension was observed with configuration 1 beyond 300 kN, but whatever configurations 1 to 3, bending cracks only 0.2 mm-wide were opened with a 700 kN total local load. For zone N2N3 the bending damage signs under 700 kN were not so diffuse: diagonal 'yield lines' had been initiated beyond ca. 400 kN from the center to corners of the loaded honeycomb. Residual deflection however did not exceed 0.4 mm. Finally only with configuration 4 the maximum local bearing capacity could be reached over all tested zones. Next sections consider the observed failure mode and loads.

in 48 hours exposure at 90°C and 95% RH. Average compressive strength measured on cylinders, 70 mm in diameter, is 190 MPa. Average specific gravity determined on the same specimens is 2.53 kg/m³, Young's modulus 55 GPa and Poisson's ratio 0.17. Ductal®-FM characteristics in tension were identified on six $0.05 \times 0.20 \times 0.6$ m plates tested under 4-point bending with a 0.42 m clear span according to 'thin plates' procedure of UHPFRC Recommendations, AFGC-SETRA (2002). Average limit of linearity 'ftj' identified on these specimens corresponds to 9.8 MPa (mean value). The conventional 'ftu' value which helps accounting for the maximum bending moment of these plates is 9.5 MPa (mean value).

The other segment is made of BSI®. Average compressive strength measured after 2 years on cylinders, 110 mm in diameter, is 219 MPa. Young's modulus is 68. Using 'thin plates' bending tests for determining the design constitutive tensile behavior leads to a limit

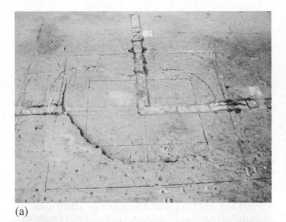

(a)

(b)

Figure 15.   Punching shear failure. Zone N2N3. (a) Top side. (b) Bottom side.

(a)

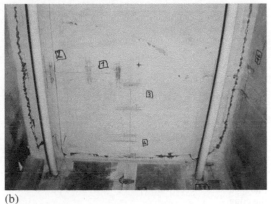

(b)

Figure 16.   Punching shear failure. Zone N6N7. (a) Top side. (b) Bottom side.

3.2   *Tests having reached failure: failure mode*

Punching shear failure was observed in all cases as shown in Figures 13 to 16. On top of the slab, the critical crack reaches roughly the limits of the loading plate (Fig. 14), even after previous other loading configurations (Fig. 13a, Fig. 17). From lower side, failure is located at the inner edge between the ribs and the top slab. Cracks due to bending are visible especially for zone N5N6 (Fig. 13b), where they are very fine and diffuse due to the progressive configuration variation, and zone N2N3 (Fig. 15b) where they are concentrated in a yield line pattern. Only for zone N6N7 rotation was observed on one smaller side (Fig 16a). Otherwise, the failure pattern was fully symmetrical. For zone N3N4 it was even possible to isolate the failed punched element (Fig. 18).

Referring to classical schemes of shear mechanisms, the shape of the failed element corresponds to very inclined struts. This may be induced by the rigid frame constituted by the ribs. When the top slab

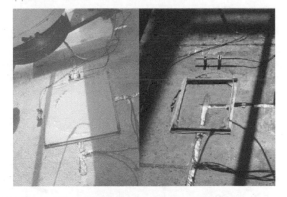

Figure 17.   (Left) Zone N2N3 after failure. Failure on top is concentrated around the reduced loading plate even with sand cushion over the 0.4 m-square surface (configuration 3). (Right) Zone N3N4 after failure (configuration 4).

is loaded, membrane tensile stresses due to this frame may be superimposed to bending and shear stresses. Moreover, the fiber distribution and orientation are probably disturbed at the inner edges between ribs and

(a)

(b)

Figure 18. Punching shear failure. Zone N3N4. (a) The punched slab piece from top side. (b) Fibers pulled out at the edge.

the top slab, which may tend to focus the basis of shear inclined cracks at this location.

3.3 *Failure loads*

Failure loads (in terms of total reaction on the stays) range from 352 to 417 kN (Fig. 19). Previous loading configurations and fatigue cycles pre-loading did not induce any bearing capacity reduction, as induced from comparisons of zone N5N6 (Fail_test1) vs. zone N6N7 (Fail_test4), and zone N3N4 (Fail_test2) vs. zone N2N3 (Fail_test3), respectively.

The failure loads are on average 12% higher for the zones of the Ductal®-FM segment (417 and 391 kN vs. 365 and 352 kN), which may be consistent with the slightly higher tensile characteristics of this UHPFRC material. When referring this punching load to the vertical cross-section along the loading surface (0.05 m thickness × 0.85 m perimeter of the plate), a shear stress value of 8.3 to 9.8 MPa is obtained, which is close to the direct tensile strength of the UHPFRC materials. This indication may be useful for further design rules, due to the lack of other experimental results under such failure modes.

From the load-central deflection curves (Fig. 19) the non-linear behavior of loaded zones of the BSI® segment turns out visible for loads exceeding 180 kN, and the stiffness reduction is more pronounced. This is probably consistent with a more ductile tensile post-peak behavior of Ductal®-FM as identified on thin plates under bending, and could be

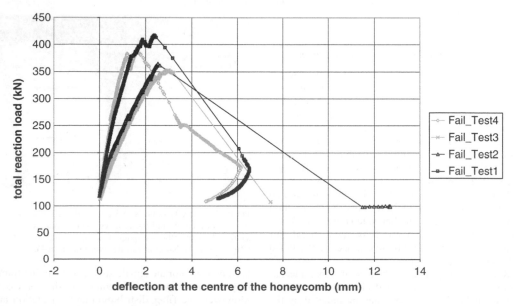

Figure 19. Global load/deflection behavior during tests having reached punching shear failure.

1488

related to its higher number of thinner fibers. However in all cases the punching shear failure takes place very suddenly and leads to a dramatic load decrease and deflection increase.

From the designer's point of view, the safety factor on the maximum concentrated load (150 kN for LM1 in Eurocode 1) ranges from 2.35 to 2.78. If a maximum local pressure of 1125 kN/m^2 is considered (parking loads), the safety factor reaches 6. Concerning the maximum shear force around the concentrated load, maximum design value of Eurocode 1 (LM2) reaches 105 kN/m, the safety factor is thus about 4.6.

4 CONCLUSION

Experimental validation of an innovative design of UHPFRC ribbed slab was carried out, regarding local capacity over 0.05 m-thick, 0.6 m-wide honeycombs. When diffusion was ensured efficiently enough towards the ribs, no local failure was obtained even with loads up to 700 kN. When the surface of load application was reduced enough, punching shear failure was obtained for loads reaching about 2.5 times the Eurocode design load, with a mean shear stress along the load surface close to the tensile strength of the UHPFRC material.

ACKNOWLEDGMENT

This experimental program has been carried out within the R&D 'National Project' MIKTI, funded by the Ministry for Public Works (DRAST/RGCU) and managed by IREX. It has been supervised by a committee chaired by J. Resplendino (CETE de Lyon), also chairman of the *fib* TG 8.6 mirror group. Eiffage Construction (A. Simon) and Lafarge (M. Behloul) are gratefully acknowledged for their contribution in the specimen preparation. The authors are pleased to thank M. Estivin, J. Billo and F.-X. Barin from LCPC Structures Laboratory for their help in the experimental realizations.

REFERENCES

Eurocode 1 – Partie 2: Actions sur les ponts, dues au trafic, NF EN 1991–2:2004.

AFGC-SETRA 2002. *Ultra High Performance Fibre-Reinforced Concretes. Interim Recommendations*. Bagneux: SETRA.

Bouteille S. & Resplendino J. 2005. Derniers développements dans l'utilisation des bétons fibrés ultra-performants en France. In *Performance, Durabilité, esthétique, Proc. GC'2005, Paris, 5–6 October 2005*. Paris: AFGC.

Bouteille S., Le François M. & Resplendino J. 2006. Etude de solutions composites BFUP – matériau composite – a*cier. European project NR2C. WP3 report*. CETE de Lyon.

Causse G. & Montens S. 1992. The Roize Bridge, *High Performance Concrete. From material to structure*. E & FN SPON, 525–536.

Chevallier F. & Petitjean J. 2001. A85. Le PS 13. Les ouvrages mixtes à dalle préfabriquée en BHP. *Travaux. 771*.

Resplendino J. & Bouteille S. 2003. PN MIKTI. Etude d'un pont mixte à dalle BFUP nervurée. *CETE de Lyon. technical report*, 46 p.

Tanis J.-M. 2006. Quelles structures pour demain? In *Proc. Coll. Interoute, Session A2 «NR2C», Rennes (France), 24–26 October 2006*.

Toutlemonde F. et al. 2005. Innovative design of ultra-high performance fiber-reinforced concrete ribbed slab : experimental validation and preliminary detailed analyses. In Henry Russel (ed.), *Proc. 7th Int. Symp. On Utilization of High Strength/High Performance Concrete, Washington D.C. (USA), 20–22 June 2005*. ACI-SP 228, 1187–1206.

Toutlemonde F. et al. 2007. Fatigue performance of UHPFRC ribbed slab applied as a road bridge deck verified according to Eurocodes, In *Proc. 5th Int. Conf. On Concrete under Severe Conditions CONSEC'07, Tours (France), 4–6 June 2007*.

Fracture Mechanics of Concrete and Concrete Structures – High-Performance Concrete,
Brick-Masonry and Environmental Aspects – Carpinteri, et al. (eds)
© 2007 Taylor & Francis Group, London, ISBN 978-0-415-44617-4

A fracture mechanics approach to material testing and structural analysis of FRC beams

A. Jansson & K. Gylltoft
Civil and Environmental Engineering, Concrete Structures, Chalmers University of Technology,
Göteborg, Sweden

I. Löfgren
Thomas Concrete Group AB, Göteborg, Sweden

ABSTRACT: The presented work has been focused on strain-softening FRC and the interrelationship between material properties and structural behaviour. The main purpose was to establish a procedure for structural analysis of flexural members with a combination of conventional reinforcement and steel fibres. A systematic approach for material testing and structural analysis, based on fracture mechanics, has been used and covers: (1) material testing; (2) inverse analysis; (3) adjustment of the σ-w relationship for fibre efficiency; and (4) cross-sectional and structural analysis. The results suggest that the approach used for the material testing provides the necessary properties to perform analyses based on non-linear fracture mechanics. The structural behaviour could be predicted with good agreement, using both FEM and an analytical model, and when comparing the peak loads obtained in the experiments with the results from the analyses, the agreement was good, with a high correlation. This demonstrates the strength of the fracture-mechanics approach for material testing and structural analysis.

1 INTRODUCTION

The number of practical applications of fibre-reinforced concrete (FRC) is increasing, as FRC offers a possibility to greatly simplify in-situ cast concrete construction, e.g. by enabling a decrease of the amount of ordinary reinforcement used for crack width reduction. However, if FRC is to be a more widely used material, general design guidelines which take into account the material properties characteristic of FRC are needed.

A problem with some of the existing test and design methods is that they have not always been consistent in treatment; for example, the tensile behaviour has been characterised by dimensionless toughness indices or by flexural strength parameters, thus failing to distinguish clearly between what is relevant to the behaviour of the material as such and what concerns the structural behaviour of the test specimen. As a consequence, determined parameters (toughness indices or flexural strength parameters) have been found to be size-dependent even though no clear explanation for the size effect has been provided (see e.g. RILEM TC 162-TDF, 2003a & 2003b). The literature gives no clear evidence for a size effect; e.g. Kooiman (2000) investigated the energy absorption for different beam sizes (ligament lengths 125, 250, and 375 mm) but found no evidence for any size effect. Furthermore, di Prisco et al. (2004) investigated size effects in thin plates and found that there was a negligible size effect for the residual strength (i.e. post-cracking) in bending, and suggested that the large scatter may instead support the Weibull theory of statistical defects. Moreover, it is not unlikely that the supposed size effect may partly be explained by the fibre distribution and the fibre efficiency factor; see Löfgren (2005).

One of the problems with the σ-ε approach of RILEM TC 162-TDF is that the basis for the approach was that it should be simple and it should be compatible with the present design regulations for reinforced and prestressed concrete, while still making optimum use of the post-cracking behaviour of fibre-reinforced concrete. Hence, the σ-ε approach of RILEM TC 162-TDF has been shown to give unreliable results and a strong size dependence (which is counteracted by introducing a size effect factor). More disturbing is that the σ-ε relationship, determined from a test, cannot be used to accurately simulate the behaviour of the test beams from which the σ-ε relationship was determined; see e.g. Barros et al. (2005) and Neocleous (2006). In addition, the approach with the σ-ε relationship is not suited for non-linear finite element analyses and, as a consequence, several modifications have been suggested; see e.g. Tlemat (2006).

On the other hand, a consistent framework for material testing and structural analysis is non-linear fracture mechanics. With non-linear fracture mechanics it is possible to accurately predict and simulate the fracture process, and this is necessary for materials like fibre-reinforced concrete – which has a significantly different cracking behaviour compared to plain concrete – and/or when design requirements for the service state are governing. Such an approach has also been suggested by RILEM TC 162-TDF (2002), i.e. the σ-w method.

The purpose of the present study was to investigate, by means of experiments and non-linear fracture mechanics analyses, the flexural behaviour of reinforced FRC members made of self-compacting fibre-reinforced concrete. In order to show the applicability of the fracture mechanics approach to different structural conditions, two separate studies are presented. The tests were carried out on beams reinforced with a combination of steel fibres and either conventional bar reinforcement or a welded mesh. The reinforcement provided a longitudinal geometric reinforcement ratio of $0.25\% < \rho < 0.45\%$, and the fibre volume fractions, V_f, used in these investigations varied from 0.25% to 0.75% (20 to 59 kg/m^3). The post-cracking behaviour of the steel fibre-reinforced concrete was determined through inverse analysis on results from wedge-splitting tests (WST).

2 EXPERIMENTAL PROGRAMME

2.1 Materials

The concrete used was self-compacting (with a slump flow spread of 500 to 650 mm) and had a w/b ratio of 0.55. The fibre content varied from 0.25 vol-% (19.6 kg/m^3) to 0.75 vol-% (58.9 kg/m^3). The mix compositions, as well as the compressive strengths for each mix, for the two investigations can be found in Gustafsson & Karlsson (2006) and in Löfgren (2005).

2.2 Materials testing

The tensile fracture behaviour of the fibre-reinforced concretes was determined by conducting wedge-splitting tests (WST); see Figure 1. The stress-crack opening (σ-w) relationships were obtained, for each mix, by conducting inverse analysis following a procedure presented by Löfgren et al. (2005). This approach, in previous studies by Löfgren and other researchers, has been shown to yield reliable results; see e.g. Meda et al. (2001), Löfgren (2005), and Löfgren et al. (2005). However, the fibre bridging stress is influenced by the number of fibres crossing the fracture plane and, when the stress-crack opening relationship is determined from a material test specimen,

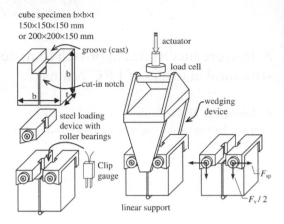

Figure 1. Principle of the wedge-splitting test method.

it may be necessary to consider any difference in fibre orientation between this specimen and the full-scale specimen. Thus the number of fibres was counted in all the WST specimens and an average experimental fibre efficiency factor was determined for each mix.

The experimental fibre efficiency factor, $\eta_{b.WST}$, was calculated as:

$$\eta_{b.WST} = \frac{N_{f.WST}}{V_f/A_f} \qquad (1)$$

where $N_{f.WST}$ is the number of fibres per unit area, V_f is the fibre volume fraction, and A_f is the cross-sectional area of a fibre. The experimental fibre efficiency factor, $\eta_{b.WST}$, should be compared to the fibre efficiency factor, $\eta_{b.beam}$, for the tested beams, which depends on whether the fibres have a free (random) or biased orientation. For the tested beams, the fibre efficiency factor was determined theoretically using an approach suggested by Dupont & Vandewalle (2005).

To account for the differences in fibre efficiency factor between the WST specimens and the tested beams, the stress-crack opening relationship obtained from the inverse analyses was reduced with the ratio between the two fibre efficiency factors, according to:

$$\sigma_{b.beam}(w) = \sigma_{b.WST}(w) \cdot \frac{\eta_{b.beam}}{\eta_{b.WST}} \qquad (2)$$

2.3 Tests of beams with conventional bar reinforcement

The tested beams had a geometry and test set-up according to Figure 2. The experimental programme consists of five series with three beams for each series, resulting in a total of fifteen beams. The programme is listed in Table 1. The beams had a geometric reinforcement ratio of $0.25\% < \rho < 0.45\%$.

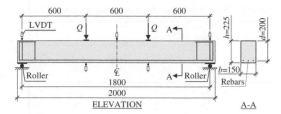

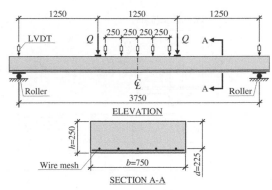

Figure 2. Test set-up for the structural beam tests.

Table 1. Experimental programme.

Series	Reinforcement	Fibre dosage [vol-%] ([kg/m^3])	Type*
V_f 0-ϕ8	3ϕ8 ($\rho=0.45\%$)	0% (0)	–
V_f 05-ϕ8	3ϕ8 ($\rho=0.45\%$)	0.5% (39.3)	RC 65/35
V_f 025-ϕ6	3ϕ6 ($\rho=0.25\%$)	0.25% (19.6)	RC 65/35
V_f 05-ϕ6	3ϕ6 ($\rho=0.25\%$)	0.5% (39.3)	RC 65/35
V_f 075-ϕ6	3ϕ6 ($\rho=0.25\%$)	0.75% (58.9)	RC 65/35

* Dramix® from Bekaert.

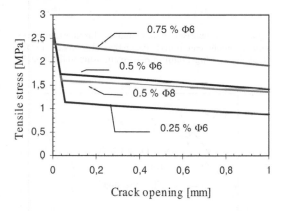

Figure 3. σ-w relationship for the full-scale elements adjusted to account for differences in the fibre efficiency factor.

The compressive strength (determined at 28 days on water-cured cubes 150 mm^3) varied with the fibre content: 47 MPa for $V_f = 0\%$, 39 MPa for $V_f = 0.25\%$, ≈38 MPa for $V_f = 0.5\%$, and 37 MPa for $V_f = 0.75\%$; see Gustafsson & Karlsson (2006).

The fibres used are Dramix® RC 65/35, i.e. with aspect ratio (fibre length to fibre diameter) of 65 and with fibre length = 35 mm. The σ-w relationships for the mixes, obtained with inverse analysis as discussed in Section 2.2, are presented in Figure 3.

2.4 Tests on beams with welded wire mesh reinforcement

The tested beams had a geometry and test set-up according to Figure 4. The experimental programme

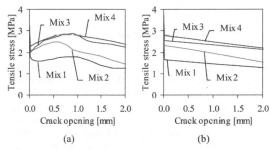

Figure 4. Test set-up for the structural beam tests.

Figure 5. σ-w relationship for the full-scale elements adjusted to account for differences in the fibre efficiency factor: (a) for polylinear and (b) for bilinear σ-w relationship.

consists of four series (for different concrete mixes) with in total twelve beams. The beams had a geometric reinforcement ratio of $0.075\% < \rho < 0.121\%$. Different types of welded wire mesh were used, i.e. bar diameter, bar spacing, and yield strength; see Löfgren (2005) for a full description. The compressive strength of the concrete (determined at 28 days on water-cured cubes 150 mm^3) was 52 to 55 MPa for the four mixes; see Löfgren (2005). The σ-w relationships for the four mixes, obtained with inverse analysis, are presented in Figure 5.

3 ANALYSES

To achieve a deeper understanding of the structural and fracture behaviour, non-linear fracture mechanics was applied, using the finite element method. The general finite element program Diana was used in all analyses; see TNO (2005). The concrete was modelled with four-node quadrilateral isoparametric plane stress elements. For the reinforcement, two different approaches were investigated: with truss elements, where the interaction between the reinforcement and the concrete was modelled by using special interface elements describing the

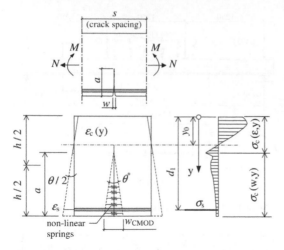

Figure 6. Non-linear hinge model.

Table 2. Maximum loads from experiments and analyses.

| | Q_{max} [kN] | | | | |
	V_f0 $\phi8$	V_f025 $\phi6$	V_f05 $\phi6$	V_f05 $\phi8$	V_f075 $\phi6$
Exp	28.4	18.7	19.8	31.8	21.5
Analytic	29.8	21.2	23.4	33.4	26.1
FE embed	29.1	19	21.5	31.8	23.3
FE bond	29.1	19	20.3	32	22.9

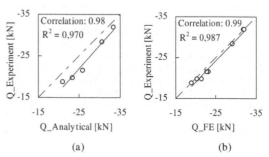

Figure 7. Comparison of maximum load from experiments with maximum load from (a) the analytical approach, and (b) FE analysis, using a bilinear σ-w relationship in the models.

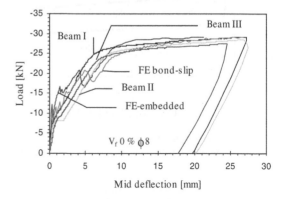

Figure 8. Comparison of load-deflection relationships for full scale elements and models; for the FE model using a bilinear σ-w relationship. No fibres added, rebar diameter 8 mm.

bond-slip relation; and with the concept of 'embedded' reinforcement, where the reinforcement is modelled with perfect bond to the surrounding concrete with no degrees of freedom of its own – see TNO (2005). For the case where bond-slip was considered, its relationship was chosen according to CEB-FIP MC90 (1993), and confined concrete with good bond conditions was assumed.

In addition, an analytical approach was used; see Figure 6. The analytical model is based on the non-linear hinge model, as proposed by Olesen (2001a & 2001b) and described by RILEM TC 162-TDF (2002).

The non-linear hinge model is based on non-linear fracture mechanics and the fictitious crack model, originally proposed by Hillerborg et al. (1976). To be able to model the behaviour of reinforced FRC members, the model was further developed by Löfgren (2003 & 2005) to consider: (1) the non-linear stress-strain behaviour in compression; (2) a multi-linear stress-crack opening relationship; and (3) a multi-linear strain hardening relationship for the reinforcement.

4 COMPARISON OF RESULTS FROM EXPERIMENTS AND ANALYSES

4.1 Beams with conventional bar reinforcement

In general, the peak load was predicted with good agreement for the FE analyses, while it was overestimated by the analytical approach; see Table 2 and Figure 7. Even though the analytical approach consequently overestimates the load-bearing capacity, the correlation is good: 0.99 for the FE analyses and 0.98 for the analytical approach. In addition, the load-deflection curves were predicted with good agreement,

although the models predicted a somewhat stiffer pre-crack and softer post-crack response, especially for the beams with $V_f = 0.75\%$; see Figures 8–12. This could possibly be explained by not having obtained the optimum parameters for the σ-w relationship, and could also be related to difficulties in determining the yield value for the reinforcement bars. Additionally, it can be said that numerical problems were encountered in the analyses of the beams with 0% and 0.75% fibre content.

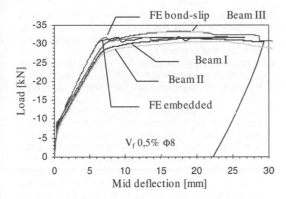

Figure 9. Comparison of load-deflection relationships for full scale elements and models; for the FE analyses using a bilinear σ-w relationship. $V_f - 0.5\%$ and rebar diameter 8 mm.

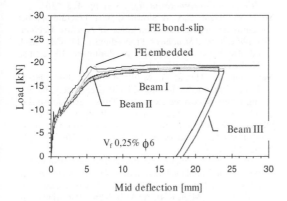

Figure 10. Comparison of load-deflection relationships for full scale elements and models; for the FE analyses using a bilinear σ-w relationship. $V_f = 0.25\%$ and rebar diameter 6 mm.

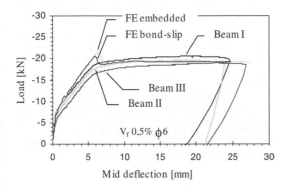

Figure 11. Comparison of load-deflection relationships for full scale elements and models; for the FE analyses using a bilinear σ-w relationship. $V_f = 0.5\%$ and rebar diameter 6 mm.

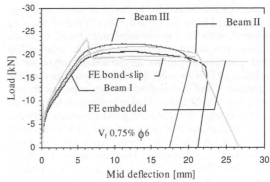

Figure 12. Comparison of load-deflection relationships for full scale elements and models; for the FE model using a bilinear σ-w relationship. Fibre volume, $V_f = 0.75\%$ and rebar diameter 6 mm.

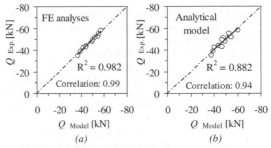

Figure 13. Comparison of maximum load from experiments and models: (a) for the FE analyses using a polylinear σ-w relationship; (b) for the analytical approach using a bilinear σ-w relationship.

4.2 Beams with welded wire mesh reinforcement

Again, the peak load was predicted with good agreement; see Figure 13. The average ratio between analysis and experiment is close to unity (1.01 for the FE analyses and 1.00 for the analytical); moreover, the correlation is good, 0.99 for the FE analyses and 0.94 for the analytical approach. In addition, both the crack width and the load-deflection curves were predicted with good agreement (although the models predicted a somewhat stiffer response); see Löfgren (2005).

5 CONCLUSIONS

To study the flexural behaviour of reinforced concrete members made of self-compacting fibre-reinforced concrete, two series of tests were carried out and non-linear fracture mechanics were used to simulate the response. The beams had a longitudinal geometric reinforcement ratio of $0.075\% < \rho < 0.45\%$ and the fibre volume fractions, V_f, used in these investigations varied from 0.25% to 0.75% (20 to 59 kg/m³). In general, with the fracture-mechanics-based approach it was possible to determine the σ-w relationship and use this to predict the structural behaviour of reinforced FRC beams; this was done with acceptable agreement and correlation between experiments and analyses. For the finite element analyses on beams with conventional reinforcement, good agreement between experiments and analyses was found for fibre contents 0.0% to 0.5%, while for fibre content 0.75% the agreement was less satisfactory. The finite element analyses on beams with welded mesh reinforcement resulted in overall good agreement. The analytical approach is a fast and simple tool that can be used for cross-sectional analyses and the behaviour of simple structures can be determined (e.g. beams).

Based on the test results and the analyses, the following conclusions can be drawn:

– the WST method provided pertinent information regarding the post-cracking behaviour of the fibre-reinforced concrete;
– by considering the fibre distribution in the material test specimens and the full-scale elements, it was possible to adjust the stress-crack opening relationship obtained from the inverse analysis so that it could be used in the analyses of the full-scale tests with good agreement;
– in the experiments, although multiple cracking was obtained, the peak-load and post-peak behaviours were determined by a single crack.

REFERENCES

Barros, J.A.O., Cunha, V.M.C.F., Ribeiro, A.F. & Antunes, J.A.B. 2005. Post-cracking behaviour of steel fibre reinforced concrete. *Materials and Structures* 38 (Jan–Feb 2005), pp. 47–56.

CEB 1993. CEB-FIP *Model Code 1990*. Bulletin d'Information 213/214, Lausanne, Switzerland.

di Prisco, M., Felicetti, R., Lamperti, M.G.L, & Menotti, G. 2004. On size effect in tension of SFRC thin plates. In *Fracture Mechanics of Concrete Structures*, Vol. 2, Li et al. (eds.), Proceedings of FRAMCOS-5, Vail, Colorado, USA, April 2004, pp. 1075–1082.

Dupont, D. & Vandewalle, L. 2005. Distribution of steel fibres in rectangular sections. *Cement and Concrete Composites*, 27, 2005, pp. 391–398.

Gustafsson, M. & Karlsson, S. 2006. Master's Thesis 2006:105. *Fiberarmerade betongkonstruktioner –Analys av sprickavstånd och sprickbredd* Dept. of Civil and Environmental Engineering, Chalmers University of Technology, Göteborg, Sweden, 2006 (in Swedish). 99p.

Hillerborg, A., Modeer, M., & Petersson, P.E. 1976. Analysis of Crack Formation and Crack Growth in Concrete by Means of Fracture Mechanics and Finite Elements. *Cem. and Concrete Research* 6, 1976, pp. 773–782.

Kooiman, A.G. 2000. *Modelling Steel Fibre Reinforced Concrete for Structural Design*. Ph.D. Thesis, Dep. of Structural and Building Eng., Delft University of Technology. 184p.

Löfgren I. 2003. Analysis of Flexural Behaviour and Crack Propagation of Reinforced FRC Members. In *Design Rules for Steel Fibre Reinforced Concrete Structures* (ed. T. Kanstad), Proc. Nordic Miniseminar, Oslo, Oct 2003. The Norwegian Concrete Association, Oslo, pp. 25–34.

Löfgren, I. 2005. *Fibre-reinforced Concrete for Industrial Construction – a fracture mechanics approach to material testing and structural analysis*. Ph.D Thesis, Dept. of Civil and Env. Eng., Div of Structural Eng, Chalmers University of Technology, Göteborg, 268 p.

Löfgren, I., Stang, H. & Olesen, J.F. 2005. Fracture properties of FRC determined through inverse analysis of wedge splitting and three-point bending tests. *Journal of Advanced Concrete Technology*, 3(3), pp. 425–436.

Meda A., Plizzari G.A., & Slowik V. 2001. Fracture of fiber reinforced concrete slabs on grade. In *Fracture Mechanics of Concrete Structures*, Proceedings of FRAMCOS-4, ed. De Borst et al., Swets & Zeitlinger, Lisse, The Netherlands, pp. 1013–1020.

Neocleous, K., Tlemat, H., & Pilakoutas, K. 2006. Design Issues for Concrete Reinforced with Steel Fibers, Including Fibers Recovered from Used Tires. *ASCE Journal of Materials in Civil Eng.*, Sep/Oct 2006, pp. 677–685.

Olesen, J. F. 2001a. Fictitious crack propagation in fibre-reinforced concrete beams. *ASCE, J. of Eng. Mech.* 127(3), pp. 272–280.

Olesen, J.F. 2001b. Cracks in reinforced FRC beams subject to bending and axial load. In *Fracture Mechanics of Concrete Structures* (R. de Borst, J. Mazars, G. Pijaudier-Cabot and J. G. M. van Mier, eds), A. A. Balkema Publishers, pp. 1027–1033.

Tlemat, H., Pilakoutas, K., & Neocleous, K. 2006. Modelling of SFRC using inverse finite element analysis. *Materials and Structures* 39(2), pp. 197–207.

RILEM TC 162 – TDF. 2002. Design of steel fibre reinforced concrete using the σ-w method – principles and applications, *Materials and Structures*, Vol. 35, June 2002, pp. 262–278.

RILEM TC 162 – TDF. 2003a. Test and design methods for steel fibre reinforced concrete: σ-ε Design Method Final Recommendation, *Materials and Structures*, Vol. 36, Oct 2003, pp. 560–567.

RILEM TC 162 – TDF. 2003b. *Test and design methods for steel fibre reinforced concrete: Background and Experiences*. Proceedings of the RILEM TC 162-TDF Workshop, eds. B. Schnütgen and L. Vandewalle. PRO 31, RILEM Publications S.A.R.L., Bagneaux, 2003.

TNO. 2005. *DIANA – Finite Element Analysis*, User's Manual release 9. Delft, The Netherlands, 2005.

Fracture Mechanics of Concrete and Concrete Structures – High-Performance Concrete,
Brick-Masonry and Environmental Aspects – Carpinteri, et al. (eds)
© 2007 Taylor & Francis Group, London, ISBN 978-0-415-44617-4

Development of bottom ash and PP fiber added fire protection coating mortar for NATM tunnel concrete lining

J.H.J. Kim
Department of Civil & Environmental Engineering, Sejong University, Seoul, Korea

H.G. Park & M.S. Lee
Civil Division, Samsung Engineering & Construction, Gyonggi-do, Korea

J.P. Won
Department of Civil & Environmental System Engineering, Konkuk University, Seoul, Korea

Y.M. Lim
Department of Civil & Environmental Engineering, Yonsei University, Seoul, Korea

ABSTRACT: Underground spaces, like tunnels, have confined conditions. So tunnel fires can result in severe human casualties and structural damages. In recent years, in order to prevent these problems, clients are requesting that tunnel linings should be fire-resistant. Fire protection coating on the surface of concrete can give sufficient fire-resistant capacity to tunnel linings, reducing the risk of lining and ground collapse during or after a fire. Also the cost is very sensitive issue in popularizing the material. The newly developed cementitious fire protection coating material contains bottom ash and PP (polypropylene) fibers, which play important role in fire-resisting ability. The newly developed cementitious fire protection coating material has much better strength than the other fire protection materials currently available in the market. This paper describes the results of fire tests of the newly developed cementitious fire protection coating material applied concrete tunnel lining specimens for the application in tunnels.

1 INTRODUCTION

In March 24, 1999, a fire took 39 people's lives in Mont Blanc Tunnel, which connects France and Italy. Underground spaces, like tunnels, have confined conditions, so tunnel fires can result in severe human casualties and structural damages. Therefore, tunnels are enforced with very stringent fire safety requirements and preventive measures, which are gaining great importance in tunnel design. In recent years, in order to prevent these problems, clients are requesting that tunnel linings should be fire-resistant and new and old tunnels are applied with fire protection coating on the surface of concrete tunnel lining.

1.1 Damage mechanisms

When a fire occurs in an unprotected tunnel, the main damage occurs by concrete spalling due to the rapidly increasing temperature. Moreover, in case of burning vehicles, there is a very long duration of the fire with extremely high temperatures of up to 1200°C.

Concrete spalling develops due to various damage mechanisms, resulting from interactions between working loads and thermally induced additional strains.

In the course of the thermo-hydraulic process, the temperature increase causes a transport of mass in the form of water, steam, and air through the pore structure of the concrete. Starting from the fire-exposed concrete surface, the concrete develops multiple zones of a dried-up, a drying, and a quasi water saturated zone. The formation of various zones is due to the steam that not only escapes through the fire-exposed surface, but also makes its way into the concrete. The steam condenses in the cooler zone, which increases the local water content up to the point of water saturation and, thus, brings about an extreme reduction in vapor permeability. Subsequently, the steam pressures generated in front of this zone become extremely high, up to the point where the tensile strength of concrete is exceeded, giving rise to sudden spalling of sections of the concrete.

The thermo-mechanical process describes the impact of thermally generated strains of the stresses and strength of concrete. On the one hand, inherent stresses are caused by a non-linear temperature distribution in the cross section. On the other hand, the different and temperature-dependent properties of the reinforcement and the concrete components give rise to different temperature expansion behavior. The aggregate will expand depending on the type of rock, while the cement matrix is subjected to shrinkage processes caused by drying-out and chemical changes. Further aggregate changes may be triggered by chemical reactions and mineralogical transformations as well as disintegration process.

The first step towards developing a fire-resistant concrete consists in preventing the thermo-hydraulic processes, i.e. the formation of a saturation zone and high steam pressures linked with it. This may be achieved by adding synthetic fibers to the concrete, as these fibers melt at temperatures of 160°C, depending on the type of fiber used. The capillary pores and micro cracks induced this way form the preferred steam transportation path. Theoretical studies corroborate the experimental finding that water vapor transportation and, thus, steam pressure equalization through fine cracks is much more effective than through larger individual cracks.

1.2 Methods of fire-resistant

Nowadays, to protect tunnel from the fire, several fire-resisting methods (i.e., spray type, cladding board type and blanket type and so on) are used.

Spray type fire-resistant is spraying the wet fire-resistant materials onto the internal surfaces of the tunnel. Before spraying the materials, the surfaces have to be thoroughly cleaned and steel mesh attached to prevent the sprayed lining from breaking off in pieces. Spray type coatings are very well suited to conforming with irregular shaped linings.

Cladding board type is attaching fire-resistant boards onto the internal surfaces of the tunnel. The boards can be attached either directly against the surface of with a gap, as controlled by the arrangement of attachment bolts.

Fire-resistant blanket types are made of materials such as ceramic fiber and organic fiber. The product is made by electrically melting the raw materials and then spinning them into laminated fibers which are formed into a blanket. The blanket can be applied to the internal tunnel surfaces using some type of framework such as punched metal. The blanket method is well suited for rounded surfaces such as hollows and depressions.

Spray type fire protection coating material is the one of the most efficient way to protect tunnels from the fire. So the fire protection coating material in this paper is also developed for spray type. But fire protection material is coated on concrete linings by hand and tested to confirm the performance of the material.

2 TESTING PROCEDURE

2.1 Distinctive features of newly developed fire protection material

Newly developed cementitious fire protection coating material is produced to meet safety requirement of tunnel structure for the fire damage and to facilitate casting and coating on construction site. Especially, this fire protection coating material is developed focusing on improving compressive strength to resist spalling or exfoliation due to severe vibration induced by train and traffic or by wind pressure.

This new coating material is primarily composed of cement, fire proof aggregate, and PP (polypropylene) fibers, which play vital role in fire-resisting ability. Fire protection performance is more strongly dependent on aggregate type than cement type, so type I ordinary Portland cement is used to produce the coating material. For aggregate, bottom ash from coal generated electric power plant is used as light weight fire proof aggregate over shell sand, which is used broadly in Europe and Japan. Bottom ash is 20% of burnt coal ash, which remains at the bottom of coal burner and has porous micro structure. Because of porous micro structure, bottom ash has superior heat insulating characteristic. Also, utilization of bottom ash is environmentally beneficial since it is a waste material which needs to be disposed. Because density of bottom ash is higher than shell sand, it enhances the strength of coating material. In order to minimize the reduction of bond strength during shotcreting due to the increase in density, accelerated setting agent is used. When concrete lining is applied with fire loading, PP fiber melts at temperatures of 160°C ~ 180°C forming a transportation path for free water steam to be released and prevent catastrophic spalling.

In order to form effective passageway for release of steam, fiber length and thicknesses of 18 mm and 2.1 denier, respectively, are used. This fireproof material is pre-mixing type where shotcreting and on site casting are possible. Material mixture design of cement to aggregate ratio of 1 to 1.5 and the fiber volume ratio of 0.25% are used. Also, acceleration setting agent of 1 volume percent is used to improve bond strength of the material for shotcreting. The water to coating material ratio of 0.395 is used for the mixture design. Since the material emphasizes strength as the main material property enhancement, compressive, flexural, and bond strength experiments are performed. The test datum are compared with the properties of the popular European fire proof coating material for shotcreting and on site casting. Experiment results are shown in Figures 1, 2 and 3.

As shown in Figure 1, 28 day compressive strengths of the newly developed material and the commercially available material are 20.89 MPa and 7.44 MPa, respectively. Also, as shown in Figure 2, the 28 day

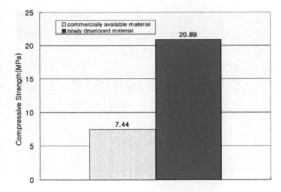

Figure 1. Comparison of compressive strength of commercially available material and new material.

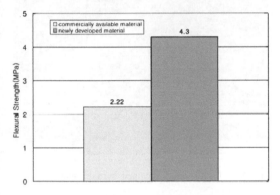

Figure 2. Comparison of flexural strength of commercially available material and new material.

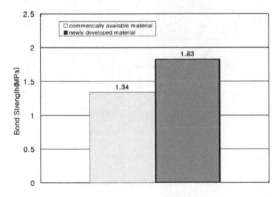

Figure 3. Comparison of bond strength of commercially available material and new material.

flexural strengths of the newly developed material and the commercially available material are 4.30 MPa and 2.22 MPa, respectively. Finally, as shown in Figure 3, the bond strength of the newly developed material and the commercially available material are

Table 1. Concrete tunnel lining mix design.

Material	Unit (kg/m³)
Cement	334
Water	167
Fine aggregate	728.33
Coarse aggregate(G)	1025.13
	Unit (%)
Air	4.5
Plasticizer	0.2
AE agent	0.03
	Unit (mm)
G_{max}	25
slump	140

1.83 MPa and 1.34 MPa, respectively. The improvement in bond strength is due to the usage of bottom ash and accelerated curing agent.

The experimental results confirm the superior strength performance of the newly developed cementitious fireproof coating material.

2.2 Production of specimens

In order to evaluate the newly developed fire protection material, the general NATM tunnel concrete lining with the coating material is fire tested. Concrete lining specimens are based on the original RC tunnel lining. The mix design of concrete of Korean Highway Corporation is used to cast original RC tunnel lining. Table 1 shows mixture design contents of concrete of RC tunnel lining.

Specimen size and reinforcement type was selected based on typical NATM tunnel lining which is recently constructed. Specimen panel size is 1400 mm × 1000 mm × 400 mm. D16 and D13 steel bars are used as main reinforcement and hoop reinforcement, respectively.

K-type sheathed thermocouples were embedded at specified locations in the tunnel lining specimen to obtain the temperature data. Five thermo couples for fire protection coated specimens and four thermo-couples for uncoated concrete tunnel lining specimen are placed at specific locations of specimen to measure temperatures. Thermo-couple locations are ①interface between concrete and fire protection coating, ②mid-depth of concrete cover thickness(37.5 mm), ③surface of bottom steel reinforcing bar(75 mm), ④mid-depth of specimen(207.5 mm), and ⑤mid-depth of back concrete lining(370 mm). Figures 4 and 5 show schematic drawing of reinforcement arrangement of the specimen and sheathed

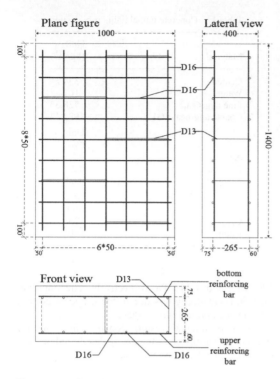

Figure 4. Schematic drawing of reinforcement arranged in the specimen.

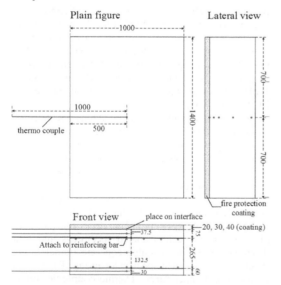

Figure 5. Schematic drawing of sheathed thermocouples locations.

thermocouple locations, respectively. Figure 6 shows formwork and reinforcement layout. Figure 7 shows installation setup photo of sheathed thermocouples. Figures 8 show photos of concrete casted specimens.

Figure 6. Formwork and reinforcing.

Figure 7. Installation of sheathed thermocouples.

Figure 8. Concrete casted specimens.

2.3 Coating of fire protection material

The developed fire protection coating material is applied on the concrete tunnel lining surface. Coating thicknesses of 20 mm, 30 mm and 40 mm were selected to verify the fire protection performance according to the thickness of fire protection coating.

First, 1400 mm × 1000 mm size of steel wire mesh was fixed to enhance interface bonding between specimen and fire protection coating. And after 28 days, dry cured specimens were coated with fire protection

Figure 9.   Steel wire mesh.

Figure 10.   Coating the fire protection material by hand.

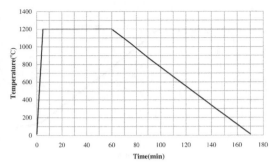

Figure 11.   Time and temperature curve of RABT fire loading.

Figure 12.   Concrete lining specimen without fire protection coating.

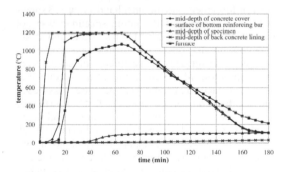

Figure 13.   Test result of concrete lining specimen without fire protection coating.

material by hand. Single specimen was produced for each thickness and single non-coated specimen was tested as a control specimen. Figures 9 and 10 are the photos of steel wire mesh to strengthen interface and application of the newly developed fire protection material on the specimen, respectively.

2.4   Fire loading curve and fire test

Fire test was carried out in KICT (Korea Institute of Construction Technology) after 28 days of dry curing of fire protection material. Fire was applied to the coated face of the specimen using LPG furnace. RABT fire curve was used as control temperature and temperatures obtained from thermo couples were recorded. Figure 11 shows time-temperature curve of RABT fire loading methods.

3   TEST RESULTS

3.1   Concrete lining specimen without fire protection coating (control specimen)

Figures 12 and 13 are photos of uncoated RC lining specimen setup on the furnace and temperature versus time test result, respectively. Figures 14 and 15 are individual temperature-time curves at mid depth of concrete cover and surface of bottom reinforcing bar, respectively. In all of the test result figures, the furnace temperature-time curve is shown for comparison purpose.

As shown on Figure 14, when the furnace temperature reaches 1200°, temperature at mid depth of concrete cover abruptly increased due to the cover thickness spalling caused by high temperature. About 20 minutes after the start of the test, concrete surrounding the reinforcements also spalled off

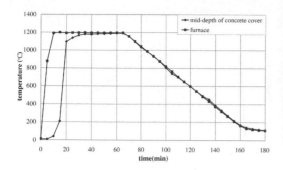

Figure 14. Time-temperature curve at mid-depth of concrete cover.

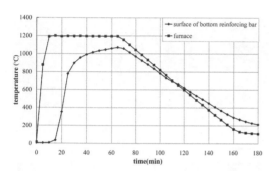

Figure 15. Time-temperature curve at surface of upper reinforcing bar.

Figure 16. Surface of the specimen inside of the furnace after spalling.

Figure 17. Vapor eruption from cracks of the specimen.

Figure 18. Bottom surface of the specimen after test.

followed by rapid increase in temperature of the bottom reinforcement. On the other hand, Figure 13 shows that the mid-depth of RC lining specimen was not affected by the fire. During the test, water and steam were continuously released from the cracks progressed from the side surfaces of specimen. The cracks were propagated toward the upper part of the specimen at the side surfaces. The highest temperature measured at mid-depth of concrete cover was 1197°C, at surface of bottom reinforcing bar was 1075°C, at the mid-depth of specimen was 111°C, and at the mid-depth of back concrete lining was 33°C. Figure 16 is a photo of concrete lining specimen's bottom surface where the fire is applied. After the concrete lining cover was spalled off, exposed surface of concrete and aggregates could be identified. Figure 17 is a photo of cracks and eruption of steam from the cracks. Figure 18 shows surface of concrete lining cover after the test. Concrete lining cover was severely damaged by spalling and main reinforcement bar was exposed.

3.2 Concrete lining specimen with thickness of 20 mm newly developed fire protection coating

Figures 19 and 20 are photos of concrete lining specimen with newly developed fire protection coating

material of 20 mm setup on the furnace and obtained temperature-time test results, respectively. Figures 21, 22 and 23 are individual time-temperature curve at interface of concrete lining cover and newly developed fire protection coating, mid-depth of concrete cover, and surface of bottom reinforcing bar, respectively.

In the test of concrete lining specimen with coating thickness of 20 mm, the fire protection coating was exploded abruptly by spalling after 50 minutes from

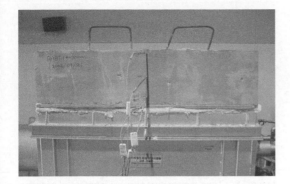

Figure 19. Specimen with 20 mm thickness of fire protection coating.

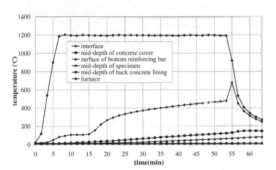

Figure 20. Test result of concrete lining specimen with thickness of 20 mm newly developed fire protection coating.

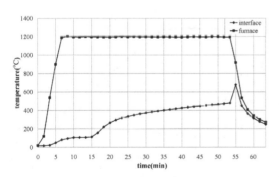

Figure 21. Time-temperature curve at interface of concrete lining cover and newly developed fire protection coating.

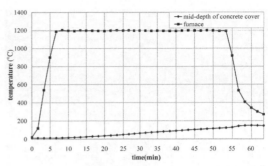

Figure 22. Time-temperature curve at mid-depth of concrete cover.

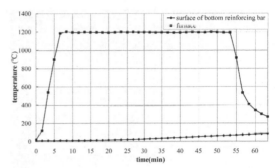

Figure 23. Time-temperature curve at surface of bottom reinforcing bar.

Figure 24. Fire protection coating surface in the furnace before explosion.

the start of the test. Most of the fire protection coating was spalled off and test was stopped for a safety reason. When the test was stopped, temperatures at all positions were constantly increasing. From Figures 21, 22 and 23, the highest temperature at interface, mid-depth of concrete cover, surface of bottom reinforcing bar, mid-depth of specimen and mid-depth of back concrete lining was 839°C, 151°C, 85°C, 11°C and 12°C, respectively.

Figure 24 is photo of fire protection coating surface in the furnace before explosion of fire protection coating. Figure 25 is bottom surface of the specimen after the test. The figure shows that concrete lining cover and fire protection coating was severely damaged by impact of explosion. And steel wire mesh for strengthening the interface was separated from the specimen.

1503

Figure 25. Separated fire protection coating.

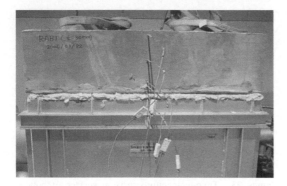

Figure 26. Specimen with 30 mm thickness of fire protection coating.

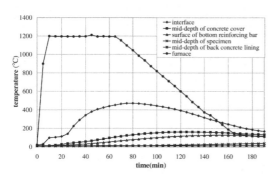

Figure 27. Test result of concrete lining specimen with thickness of 30 mm newly developed fire protection coating.

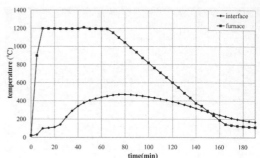

Figure 28. Time-temperature curve at interface of concrete lining cover and newly developed fire protection coating.

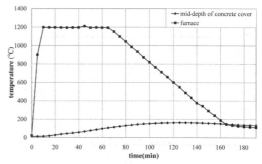

Figure 29. Time-temperature curve at mid-depth of concrete cover.

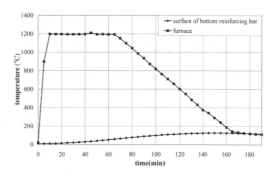

Figure 30. Time-temperature curve at surface of bottom reinforcing bar.

The spalling of the coated material shows that 20 mm thickness is insufficient for resisting RABT fire curve's maximum temperature of 1600°C.

3.3 Concrete lining specimen with thickness of 30 mm newly developed fire protection coating

Figures 26 and 27 are photos of 30 mm coated concrete lining specimen setup on the furnace and obtained temperature-time test results, respectively. Figures 28, 29 and 30 are individual time-temperature curve at interface of concrete lining cover and newly developed fire protection coating, mid-depth of concrete cover, and surface of bottom reinforcing bar, respectively.

The behavior of concrete lining specimen with thickness of 30 mm fire protection coating which was different than that of the specimen with thickness of 20 mm fire protection coating, was stable during the

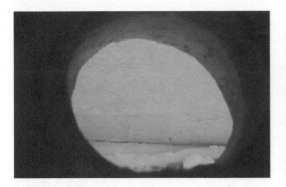

Figure 31. Surface of fire protection coating in the furnace.

Figure 33. Surface of fire protection coating after test.

Figure 32. Water drops formed on the fire protection coating.

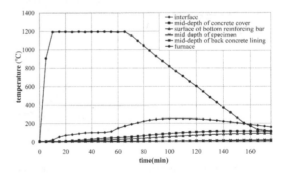

Figure 34. Specimen with 40 mm thickness of fire protection coating.

test. 20 minutes later after the start of the test, temperature of interface between concrete lining cover and fire protection coating increased significantly. Significant amount of steam is released from the gaps between thermo-couples and specimen and also fire protection coating during the test, but the amount of steam is less than that shown from the specimen with thickness of 20 mm coating. There were no cracks found on the specimen. The highest temperature measured at interface were 473°C, at mid-depth of concrete cover was 163°C, at surface of lower reinforcing bar was 129°C, at the mid-depth of specimen was 40°C, and at the mid-depth of back concrete lining was 16°C. Figure 31 is a photo of surface of the fire protection coating in the furnace from window. There were no changes of the surface during the test. Figure 32 shows vapors formed from released steam at the fire protection coating. Figure 33 is fire applied surface of the fire protection coating after the test and showed no significant damages except slight color change. The results show that the newly developed coating material with 30 mm thickness is sufficient in resisting the temperature up to 1600°C.

Figure 35. Test result of concrete lining specimen with thickness of 30 mm newly developed fire protection coating.

3.4 Concrete lining specimen with thickness of 40 mm newly developed fire protection coating

Figures 34 and 35 are photos of 40 mm coated concrete lining specimen setup on the furnace and obtained temperature-time test results, respectively.

Figures 36, 37 and 38 are individual time-temperature curve at interface of concrete lining cover and newly developed fire protection coating,

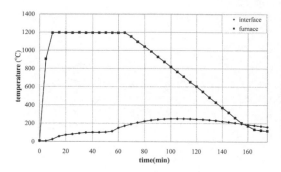

Figure 36. Time-temperature curve at interface of concrete lining cover and newly developed fire protection coating.

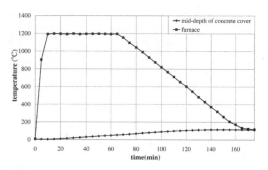

Figure 37. Time-temperature curve at mid-depth of concrete cover.

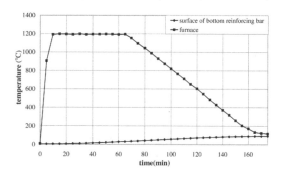

Figure 38. Time-temperature curve at surface of bottom reinforcing bar.

mid-depth of concrete cover, and surface of bottom reinforcing bar, respectively.

Test result of the concrete lining specimen with thickness of 40 mm newly developed fire protection coating shows very similar behavior to that of the specimen with thickness of 30 mm fire protection coating. However, the temperatures measured from the thermo couples were lower than that of 30 mm coated specimen. Also, lesser amount of steam release was observed from the specimen and fire protection coating. The highest temperature measured at

Figure 39. Surface of fire protection coating in the furnace.

Figure 40. Water steam stain at the interface.

Figure 41. Surface of fire protection coating after test.

interface were 252°C, at mid-depth of concrete cover was 117°C, at surface of lower reinforcing bar was 93°C, at the mid-depth of specimen was 24°C, and at the mid-depth of back concrete lining was 12°C. Figure 39 is surface of the fire protection coating in the furnace during testing and no special surface changes were not found. Figure 40 is a photo of specimen with moisture stains, which is a evidence of escaped steam from the fire protection coating. Final Figure 41 shows surface of fire protection coating after the test and no significant damages were found. The results show

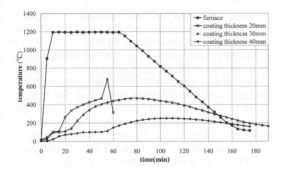

Figure 42. Test result of temperature at interface according to thickness of fire protection coating.

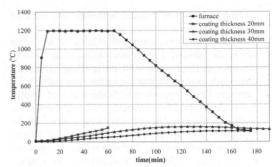

Figure 43. Test result of temperature at mid-depth of concrete cover(37.5 mm) according to thickness of fire protection coating.

that the newly developed coating material with 40 mm thickness is most effective in resisting the temperature up to 1600°C.

4 COMPARISON OF TEST RESULTS

(comparing the temperature at each location according to thickness of fire protection coating)

4.1 Temperature at lining and coating interface

Figure 42 is the test result comparison of temperature – time curve measured at the coating and lining interface.

For 20 mm thickness coated specimen, the coated layer abruptly spalled off at approximately 20 minutes after the start of the test. For safety reasons the test was stopped for this specimen. For 30 mm and 40 mm specimens, even though the coating thickness increased by 1 cm, the temperature difference was two times larger in 30 mm specimen than 40 mm specimen. This proves that there is optimum coating thickness for fire protection of this newly developed material. Figure 42 shows that the maximum temperatures at the interface are reached at few minutes. After the furnace temperature descended from 1200°C, this trend means that the material is able to absorb the heat there by proving the effectiveness of the porous bottom ash usage in the material.

4.2 Temperature at mid-depth of concrete cover

Figure 43 is the test result comparison of temperature – time curve measured at mid-depth of concrete cover.

Obviously, the maximum temperatures at the mid-depth of concrete cover are lower than the maximum temperatures at the interface. This means that the applied fire was effectively resisted by the coating material since the maximum temperature was less than 200°C for 40 mm coated specimen. Between 30 mm and 40 mm coating thickness, the maximum temperature for 30 mm thickness is approximately 139% than

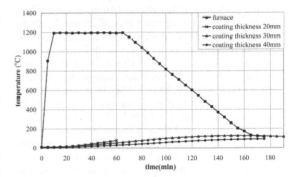

Figure 44. Test result of temperature at surface of bottom reinforcing bar(75 mm) according to thickness of fire protection coating.

that of 40 mm thickness. In any case, the damage from RABT fire curve is prevented using coating material.

4.3 Temperature of surface of bottom reinforcing bar

Figure 44 is the test result comparison of temperature – time curve measured at surface of bottom reinforcing bar.

The maximum temperatures measured at this location were slightly less than the temperature measured at mid-depth of concrete cover, once again showing the effectiveness of the coating material in resisting fire damage. Since the maximum temperature at this position is approximately 129°C, the fire will not affect the reinforcing bars performance. Between 30 mm and 40 mm thickness coated specimen, the differences are very small and negligible.

5 CONCLUSION AND SUMMARY

• During the test, concrete lining specimen with thickness of 20 mm newly developed fire protection coating is destroyed because of the spalling induced high temperature.

- concrete lining specimen with thickness of 30 mm newly developed fire protection coating showed ordinary performance with no spalling effect but little high inner-temperature of the specimen.
- concrete lining specimen with thickness of 40 mm newly developed fire protection coating showed no spalling effect and lower inner-temperature of the specimen than specimen with thickness of 30 mm fire protection coating. Therefore, this thickness of 40 mm coating specimen showed best performance among all three specimens.

There were not any special damages in concrete lining specimens with newly developed fire protection coating thickness of 30 mm and 40 mm except spalled specimen with coating thickness of 20 mm by high temperature. And also, there were few damages in coating. But unfortunately, specimens with fire protection coating thickness of 30 mm showed very high temperature at the interface between coating and concrete lining which is nearly 500°C. When inner-temperature of concrete lining increases, concrete lining itself will have much probability to be failed by spalling caused high temperature and tension of reinforcement. So, in substance fire protection coating thickness of 30 mm is insufficient for protecting concrete lining from the fire. On the other hand, fire protection coating thickness of 40 mm has best performance. Coating thickness between 30 mm and 40 mm will be suitable choice considering economical aspects.

Therefore, this test proved that the newly developed fire protection coating material has enhanced fire protection ability and the thicker coating has more performance to protect concrete lining from fire events. Using bottom ash which has many minute holes in it, when produce fire protection material, makes fire protection ability higher. Moreover, industrial wastes from coal generated electric power plant are decreased and this is very beneficial effect for environment.

ACKNOWLEDGEMENT

This research was supported by Civil Division of Samsung Engineering & Construction Company research program, "Fire Protection of RC Tunnel Lining System Development". And this financial support is gratefully acknowledged.

REFERENCES

Both, C. et al. 2003. Spalling of concrete tunnel linings in fire. In J. Saveur(ed.), *(Re)Claiming the Underground Space*: 227–231. Swets & Zeitlinger: Lisse.
Khoury G.A. 2002. Passive protection against fire. *Tunnels and Tunneling International*: 40–42.

Fracture Mechanics of Concrete and Concrete Structures – High-Performance Concrete,
Brick-Masonry and Environmental Aspects – Carpinteri, et al. (eds)
© 2007 Taylor & Francis Group, London, ISBN 978-0-415-44617-4

Development of a zero-span tensile test for HPFRCC used for structural repairing

A. Kamal, M. Kunieda, N. Ueda & H. Nakamura
Department of Civil Engineering, Nagoya University, Nagoya, Japan

ABSTRACT: The most important characteristic of High Performance Fiber Reinforced Cement Composites (HPFRCC) is the formation of multiple fine cracks in tension, which provides novel mechanical properties, such as strain hardening behavior. For these reasons, HPFRCC can be widely used in retrofitting and repairing of existing concrete structures. In a repair application, the crack in the substrate induces localized fracture within HPFRCC. Elongation performance of HPFRCC on an existing crack, which represents resistance against the localized fracture, is required in repair applications. This research proposes the zero-span tensile test to evaluate the crack elongation performance for HPFRCC, and examines the effects of test conditions (i.e. artificial crack width, steel plate thickness, and specimen thickness) on the zero-span tensile test results. This paper also reveals that the crack elongation obtained from the test is lower than the deformation capacity evaluated by ordinary uni-axial tensile tests.

1 INTRODUCTION

In the past decade, fiber reinforced cementitious composites with higher ductility such as High Performance Fiber Reinforced Cement Composites (HPFRCC) have been developed. This progress is due to the development in fiber, matrix, and process technology, as well as better understanding of the fundamental micromechanics governing composite's behavior (Li et al. 2004).

Engineered Cementitious Composite (ECC) is one type of HPFRCC, and is designed based on fracture mechanics and micromechanics. It exhibits the ultimate strength higher than its initial cracking strength and provides multiple cracking during the inelastic deformation process, as shown in Figure 1. In conventional Fiber Reinforced Concrete (FRC), apparent softening behavior can be observed with localization after cracking. However, the deformation of HPFRCC is approximately uniform in a macro scale, and is considered as pseudo-strain hardening materials (Fisher & Li 2004, Li 1998a).

HPFRCC promises to be used in a wide variety of civil engineering applications, as summarized in JCI (2002). In Japan, there are many applications using HPFRCC, such as seismic dampers of multi storey building, overlay repair for steel deck plate, surface repair of retaining wall deteriorated by Alkali-Silica Reaction (ASR), surface repair of irrigation channel etc. (Kunieda & Rokugo 2006). Especially, multiple fine cracks in HPFRCC impart durability to concrete structures, because fine cracks reduce the penetration

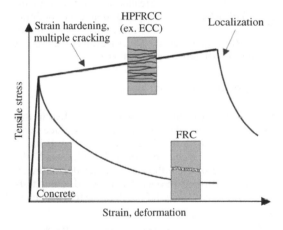

Figure 1. Tensile stress-strain behaviour of cementitious composites.

of substance (i.e. water, oxygen, chloride ion etc.) through the cracks.

Several researches subjected to the advantages of structures repaired by HPFRCC have been carried out. Lim & Li (1997) found the mechanical advantages on the interface crack trapping mechanism within HPFRCC/RC composites. Horii et al. (1998), Li (1998b), and Li et al. (2000) tried to apply HPFRCC to repair or retrofit of concrete structures, and confirmed the effect of the material ductility of repair materials on the structural performance. Kamada & Li (2000) discussed the effects of surface preparation on the fracture behavior of HPFRCC/RC composites. Kesner & Billington (2000) investigated an infill wall system using HPFRCC for seismic retrofit applications. Li (2004) addressed the required properties for repair materials to obtain durable repaired concrete structures.

The previous researches have been taken into account the advantages of high ductility of HPFRCC, which can be evaluated by uni-axial tensile tests or flexural tests. However, these tests have boundary conditions different from the existing ones in repair or retrofit applications. Kunieda et al. (2004) and Rokugo et al. (2005) showed the localized fracture of HPFRCC in patch repair systems, experimentally and analytically. One of the advantages of HPFRCC is to provide the widely distributed cracks having fine crack width. However, in repair applications having an existing crack within the substrate, crack distribution of HPFRCC was limited adjacent to the existing crack, as shown in Figure 2. In repair applications using HPFRCC, it is important to recognize the resistance against the localized fracture (crack elongation performance) in a specific boundary condition, and to evaluate the crack elongation performance of HPFRCC itself.

This paper presents the reduction in apparent ductility of HPFRCC under the special boundary condition that represent repair applications. This paper also proposes Zero-Span Tensile Test to evaluate the crack elongation performance of HPFRCC.

2 ZERO-SPAN TENSILE TEST

The basic idea of the Zero-Span Tensile Test in this study came from the test method for surface coating materials with thin layer such as epoxy painting etc., which was specified in JSCE-K-532. Usually, the performance of the HPFRCC or other ductile materials is confirmed through tensile or bending tests with no restraint along the length of the test specimen.

In surface coating repair materials evaluation should not be performed only with free film elongation tests, as shown in Figure 3 (Kunieda et al. 2004). In a deteriorated concrete structure having cracks, the deformation of the members increases as cracks open wider. It means that localized fracture occurs on the repair material near the cracks in substrate concrete. So this led to use the idea of the Zero-Span Tensile Test. The test method developed and specified to compare the material property of surface coating materials with flexibility. Because cracking of the coating material itself occurred adjacent to an existing crack, the specific boundary condition has been modeled. In the specified test method in JSCE-K-532, mortar specimens of size $120 \times 40 \times 10$ mm are prepared. Then, a crack is induced in the mortar specimens, and each specimen half is carefully positioned, so that the artificial crack width is about 0 mm. The specimen with the induced crack is repaired by a surface coating material, and uni-axial tensile tests for the specimen is conducted after the curing of the surface coating material, as shown in Figure 4. The measured displacement at

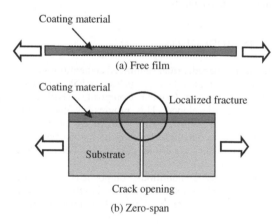

Figure 3. Elongation tests for surface coating materials (Kunieda et al. 2004).

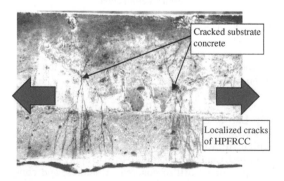

Figure 2. Localization of cracks distribution of HPFRCC adjacent to existing cracks of substrate concrete.

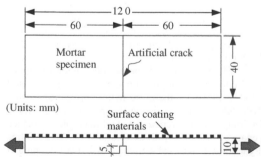

Figure 4. The test of concrete surface coating material over concrete crack (JSCE-K-532).

the peak load is defined as the crack elongation performance of the surface coating material. As described in the previous section, the boundary condition in the repair application with HPFRCC is similar to that of this Zero-Span Tensile Test.

3 EXPERIMENTAL PROCEDURES

3.1 Materials

HPFRCC with Polyvinyl Alcohol fibers (HPFRCC-PVA) was used in this test. The volume fraction of

fiber was 2%. The mix proportions of the HPFRCC-PVA, which is developed by Kanda et al. (2005), are shown in Table 1. The compressive strength and Young's modulus measured by cylindrical specimens of ø100 × 200 mm were 49.9 MPa and 18.6 GPa, respectively. By conducting uni-axial tensile tests on dumbbell-shaped specimens (tested cross section: 30 × 30 mm), the tensile strength of the HPFRCC was 3.5 MPa and the maximum strain was approximately 0.8–1% in the gauge length of 100 mm, as shown in Figure 5.

Table 1. Mix proportions (Kanda et al. 2005).

Water to binder ratio	Unit water (kg/m³)	Sand by binder ratio	Shrinkage reducing agent (kg/m³)	Fiber volume fraction (%)	Air content (%)
0.46	364	0.64	15	2	10

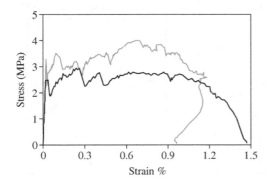

Figure 5. Stress-strain relationship of tensile test of HPFRCC-PVA with 2% PVA fiber.

3.2 Specimens and test setup

This research proposed the zero-span tensile test using steel plates to evaluate the crack elongation performance of HPFRCC, as shown in Figure 6. The proposed test method used the steel plates not mortar as a substrate, and the steel plates and HPFRCC (repair material) was glued each other by epoxy adhesive. The other substrates such as mortar or concrete that represent similar substrate in concrete structures involve a surface preparation to obtain good bonding between HPFRCC and the substrate. However, the procedures on surface preparation impart much work to the test method. In addition, some kinds of substrate induce delamination at interface between the substrate and HPFRCC, which gives better result due to distributed cracked area widely. The influence of bond property on the crack elongation performance of HPFRCC should be removed. However, the appropriate geometry of the test specimen was neither discussed nor proposed before. In this research, three variables were examined to know their effects on the zero-span tensile test results: (1) artificial crack width, (2) steel plate thickness, (3) specimen thickness, as described in following section.

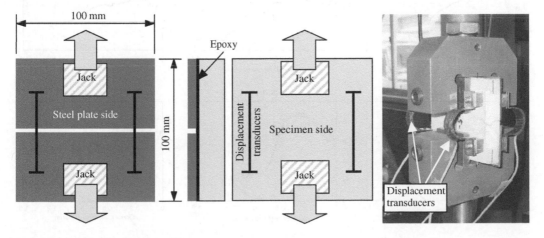

Figure 6. Test setup on zero-span tensile tests.

The size of the HPFRCC specimen was 100×100 mm. Combinations among the specimen thickness, steel plate thickness, and artificial crack width were adopted in this test, as shown in Table 2. Three specimens were tested in each case. In the tests, four displacement transducers were glued on both surfaces

Table 2. Tested specimens and its variables.

Specimen thickness (mm)	Steel plate thickness (mm)	Artificial crack width (mm)
10	1	0, 1, 5
	3	0, 1, 5
	5	0, 1, 5
15	1	0, 1, 5
	3	0, 1, 5
	5	0, 1, 5
20	1	0, 1, 5
	3	0, 1, 5
	5	0, 1, 5

to measure the opening displacement at artificial crack (i.e. two transducers were fixed on the specimen side and the two other transducers were fixed on the steel plate side), as shown in Figure 6. The measurement length of the transducers was 50 mm and the sensitivity of each transducer was about 1/2000 mm. The loading rate was about 0.2 mm/min, and the loading was terminated when the displacement was over 2 mm, which was equal to the capacity of the transducers. The load was measured by load-cell having the capacity of 50 kN. All tests were carried out during the age of 28–32 days.

4 EXPERIMENTAL RESULTS

4.1 Load-opening displacement relations

Figure 7 shows examples of the measured load-displacement curves through the zero-span tensile

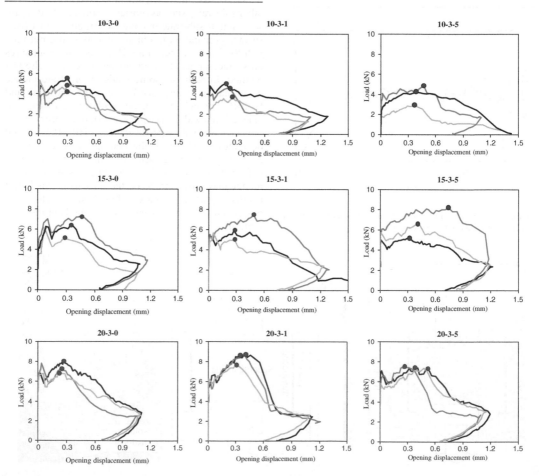

Figure 7. Examples of measured load-displacement curves from zero-span tensile tests.

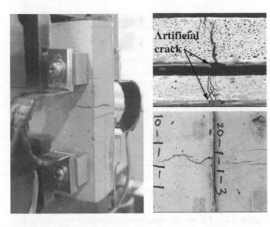

Figure 8. Examples of crack patterns on HPFRCC-PVA specimens under zero-span tensile tests.

tests, which are the averaged curves of the four displacement transducers. The legends of each graph, such as (10-3-1), represent specimen thickness (mm), steel plate thickness (mm), and artificial crack width (mm), respectively. In this study, the crack elongation is defined as displacement value at the peak load in the averaged curves of the four displacement transducers, avoiding the point of first crack load to cancel the influence of bond property on the crack elongation, as shown in Figure 7.

In these figures, slight strain hardening after the initial cracking can be observed, and HPFRCC-PVA provided crack elongation in ranging from 0.1 to 0.6 mm. In the uni-axial tensile response, HPFRCC-PVA exhibited strain of about 1% at peak load, as shown in Figure 5. Each strain value obtained from the uni-axial tensile tests can be converted to the displacement values by means of the measurement length (i.e. 100 mm). According to that, displacement at peak load in uni-axial tensile tests was 1 mm.

In the results of the zero-span tensile test, most cases gave smaller displacement values than 1 mm. Figure 8 shows examples of crack patterns on 10 and 20 mm thickness specimens with artificial crack width of 1 mm and steel plate thickness of 1 mm. As shown in these figures, localized fracture adjacent to the artificial crack can be observed. Normally, HPFRCC exhibits distributed fine cracks, which imparts ductility to the member. That means HPFRCC requires appropriate specimen shape and boundary condition to obtain the distributed cracking.

4.2 Effects of artificial crack width

Figure 9 shows examples of crack patterns obtained in the cases of different artificial crack width. As shown in Figure 9, the number of cracks was decreased with decreasing the artificial crack width.

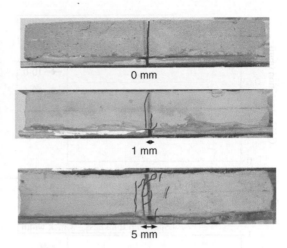

Figure 9. Effect of artificial crack width on the number of allowed cracks.

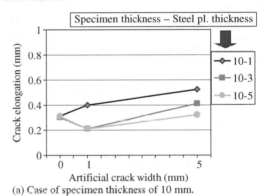

(a) Case of specimen thickness of 10 mm.

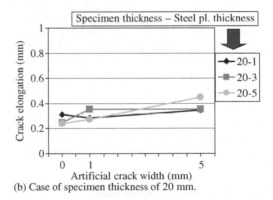

(b) Case of specimen thickness of 20 mm.

Figure 10. Effect of artificial crack width on the crack elongation through zero-span tensile test.

Figure 10 shows the relationship between crack elongation and artificial crack width, in the case of the specimen thickness of 10 mm and 20 mm. The legend of each figure shows the specimen thickness and

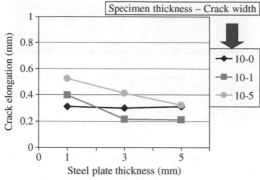

(a) Case of specimen thickness of 10 mm.

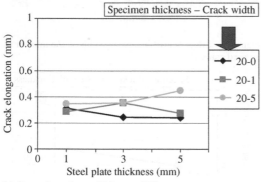

(b) Case of specimen thickness of 20 mm.

Figure 11. Effect of steel plate thickness on the crack elongation through zero-span tensile test.

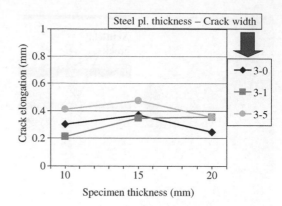

Figure 12. Effect of specimen thickness on the crack elongation through zero-span tensile test.

steel plate thickness, respectively. The obtained crack elongation became smaller with decreasing the artificial crack width. That is due to the number of allowed cracks increases around the artificial crack region by increasing the artificial crack width. The crack elongations of the specimen with artificial crack width of 0 mm were in ranging from 0.2 mm to 0.5 mm. However, artificial crack widths of 1 mm and 5 mm do not reflect the existing crack width in concrete structures for repair. As proposed in ordinary standard test method (JSCE-K-532), artificial crack width of 0 mm might be useful for crack elongation of HPFRCC.

4.3 *Effects of steel plate thickness*

Figure 11 shows the relationship between crack elongation and steel plate thickness, in the case of the specimen thickness of 10 mm and 20 mm. The legend shows the specimen thickness and artificial crack width, respectively. There are two typical behaviors either increasing the crack elongation with increasing the steel plate thickness, or decreasing the crack

elongation with increasing the steel plate thickness. No significant tendency can be observed through the tests. It seems that both the stiffness and thickness of HPFRCC and the steel plate affect the deformation of the composite in tension. However, the results using the steel plate thickness of 3 mm provided similar results in both 10 mm and 20 mm cases.

4.4 *Effects of specimen thickness*

Figure 12 shows the relationship between crack elongation and specimen thickness, in the case of the steel plate thickness of 3 mm. The legend shows the steel plate thickness and artificial crack width, respectively. There is no influence of the specimen thickness on the crack elongation in this test. It was clarified that the specimen size in ranging from 10 mm to 20 mm can be applied to the zero-span tensile test.

In the future, specimen thickness, which reflects the thickness of repaired layer, should be used.

5 CONCLUSIONS

(1) The evaluated crack elongation through Zero-Span Tensile Tests is much smaller than that of deformation capacity in ordinary uni-axial tensile tests. That is because a localized cracking can be observed adjacent to the artificial crack in this test. These results show that evaluating the crack elongation performance of HPFRCC through the zero-span tensile tests is quite important for repairing applications.

(2) The influence of both steel plate thickness and specimen thickness on the crack elongation was not significant. However, the obtained crack elongation became smaller with decreasing the artificial crack width.

REFERENCES

Fischer, G. & Li, V.C. 2004. Effect of Fiber Reinforcement on the Response of Structural Members. Proceedings of FRAMCOS-5, Vail, Colorado, USA, April 2004, pp. 831–838.

Horii, H., Matsuoka, S., Kabele, P., Takeuchi, S., Li, V.C. & Kanda, T. 1998. On the Prediction Method for the Structural Performance of Repaired/Retrofitted Structures. Proceedings FRAMCOS-1, Fracture Mechanics of Concrete Structures, AEDIFICATIO Publishers, D-79104 Freiburg, Germany, pp. 1739–1750.

JCI 2002. Proceedings of the JCI International Workshop on Ductile Fiber Reinforced Cementitious Composites (DFRCC), Japan Concrete Institute, Oct. 2002.

JSCE 2002. Test method for elongation performance of concrete surface coating materials over concrete crack (JSCE-K-532-1999). *Standard specification for concrete structures-2002, Test methods and specification*, pp. 247–250.

Kamada, T. & Li, V.C. 2000. The Effects of Surface Preparation on the Fracture Behavior of ECC/Concrete Repair System. *Journal of Cement and Concrete Composites*, Vol. 22, No. 6, pp. 423–431.

Kanda, T., Kanakubo, T., Nagai, S. & Maruta, M. 2005. Technical Consideration In Producing ECC Pre-Cast Structural Element. Proceedings of International workshop on HPFRCC in structural applications, Honolulu, Hawaii, USA, 23–26 May 2005.

Kesner, K. & Billington, S.L. 2000. Engineered Cementitious Composites for the Retrofit of Critical Facilities. Proceedings of the Second MCEER: Workshop on Mitigation of Earthquake Disaster by Advanced Technologies (MEDAT-2), Las Vegas, Nevada, USA, pp. 77–86.

Kunieda, M., Kamada, T., Rokugo, K. & Bolander, J.E. 2004. Localized Fracture of Repair Material in Patch Repair Systems. Proceedings of FRAMCOS-5, Vail, Colorado, USA, April 2004, pp. 765–772.

Kunieda, M. & Rokugo, K. 2006. Recent Progress of HPFRCC in Japan – Required Performance and Applications. *Journal of Advanced Concrete Technology*, Japan Concrete Institute, Vol. 4, No. 1, pp. 19–33.

Li, V.C. 1998a. Engineered Cementitious Composites – Tailored Composites Through Micromechanical Modeling. Fiber Reinforced Concrete: Present and the Future edited by N. Banthia, A. Bentur & A. Mufti, Canadian Society for Civil Engineering, Montreal, pp. 64–97.

Li, V.C. 1998b. ECC for Repair and Retrofit in Concrete Structures. Fracture Mechanics of Concrete Structures. Proceedings FRAMCOS-3, AEDIFICATIO Publishers, D-79104 Freiburg, Germany, pp. 1715–1726.

Li, V.C. 2004. High Performance Fiber Reinforced Cementitious Composites as Durable Material for Concrete Structure Repair. *International Journal for Restoration of Buildings and Monuments*, Vol. 10, No. 2, pp. 163–180.

Li, V.C., Horii, H., Kabele, P., Kanda, T. & Lim, Y.M. 2000. Repair And Retrofit With Engineered Cementitious Composites. *International Journal of Engineering Fracture Mechanics*, Vol. 65, No. 2–3, pp. 317–334.

Li, V.C., Horikoshi, T., Ogawa, A., Torigoe, S. & Saito, T. 2004. Micromechanics-based Durability Study of Polyvinyl Alcohol-Engineered Cementitious Composite (PVA-ECC). *ACI Materials Journal*, Vol. 101, No. 1, pp. 242–248.

Lim, Y.M. & Li, V.C. 1997. Durable Repair of Aged Infrastructures Using Trapping Mechanism of Engineered Cementitious Composites. *Cement and Concrete Composites Journal*, Vol. 19, No. 4, pp. 171–185.

Rokugo, K., Kunieda, M. & Lim, S.C. 2005. Patching Repair with ECC on Cracked Concrete Surface. Proceedings of ConMat'05, Vancouver, Canada, 22–24 August 2005.

Fracture Mechanics of Concrete and Concrete Structures – High-Performance Concrete,
Brick-Masonry and Environmental Aspects – Carpinteri, et al. (eds)
© 2007 Taylor & Francis Group, London, ISBN 978-0-415-44617-4

Initiation and development of cracking in ECC materials: Experimental observations and modeling

G. Fischer, H. Stang & L. Dick-Nielsen

Technical University Denmark, Kgs. Lyngby, Denmark

ABSTRACT: Applications of fiber reinforced engineered cementitious composites (ECC) and other strain hardening cementitious composites (SHCC) require information on their material properties for structural purposes as well as for durability. The distinctive material properties of ECC and other SHCC materials are the tensile stress-strain relationship and the cracking characteristics including crack width and spacing prior to localization of cracking. The investigations presented in this paper will focus on these distinctive features and aim at developing tools to predict and design for specific composite behavior.

The initiation and development of cracking are principal mechanisms in the tensile deformation behavior of ECC and other SHCC materials. In this paper, the parameters affecting the initiation and propagation of cracking and the fiber bridging stress-crack opening relationship are investigated. Furthermore, the observations will form the basis for calibration of a plasticity based damage mechanics model for ECC.

It is found that the size and distribution of flaws in the cementitious matrix and the range of fiber bridging crack opening characteristics throughout the composite material are the governing parameters for the tensile stress-strain response. Particularly important for the durability of structures containing ECC is the evolution of crack widths under increasing tensile loads up to the tensile strength of the material. The experimental data and analytical tools presented in this paper will provide detailed information on this particular feature of ECC materials.

1 INTRODUCTION

Fiber reinforced concrete can be categorized into conventional Fiber Reinforced Cementitious Composites (FRCC) and Strain Hardening Cementitious Composite (SHCC). The former shows a tension softening behavior after first cracking, whereas the later develops multiple cracking and a strain hardening behavior in tension. Compared to plain concrete and conventional FRCC, SHCC shows a significant improvement in ductility. SHCC characteristics at low volume fraction of discontinuous, randomly oriented fibers have been achieved in Engineered Cementitious Composites (ECC) such that industrial application of this composite material can be economically feasible. For better understanding the properties of this promising material, a significant amount of research effort has been focused on micro mechanics of the composite, and various models have been established to characterize the fiber bridging behavior leading to such composite properties. An analytical model that accounts for the multiple cracking of short randomly distributed fiber composites was introduced by Li & Leung (1992) under the assumption of no fiber rupture in the fiber bridging-crack opening process. Later,

Maalej et al. (1995) extended the model to account for the possibility of fiber rupture, which is referred to as fiber pullout and rupture model (FPRM). However, these analytical models assume a deterministic tensile rupture strength of the fibers bridging a crack. Maalej (2001) proposed a model to predict the bridging stress-crack opening relationship (hereafter abbreviated as $\sigma_B - \delta$ curve) – a fundamental material property necessary for the analysis of steady state cracking in short random fiber composites. The model was derived for fibers having a statistical tensile strength distribution. For ECC to develop multiple cracking in the strain hardening stage, the flaw size distribution of the matrix also plays an important role. Wu & Li (1995) predicted the multiple cracking process of short random fiber reinforced brittle matrix composites through Monte Carlo simulation of the flaw size distribution. Furthermore, Wang & Li (2004) tailored the material including pre-existing flaws in Engineered Cementitious Composites (ECC) by introducing artificial flaws and greatly enhanced the multiple cracking behavior and ductility of the composite in uniaxial tension.

The above mentioned research efforts, which account for the influences of the distribution of fiber tensile strength and matrix flaw size, are invaluable in combining the micro mechanics of fiber bridging with random properties of fiber and matrix. In this paper we examine directly the experimentally obtained $\sigma_B - \delta$ curve, and utilize this information as a parameter to simulate the multiple cracking and strain hardening behavior of ECC in uniaxial tension. In this way, parameters that are difficult to realistically predict, such as randomness in fiber tensile strength, matrix flaw size distribution, fiber matrix interfacial bonding, and other factors are inherently included. Based on the experimentally obtained data of the fiber bridging stress-crack opening relationship, a simulation method is proposed and compared to experimental test results.

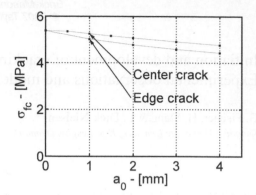

Figure 1a. Relation between first crack strength, σ_{fc} and initial crack length, a_0.

2 CONCEPT

The load-deformation behavior of ECC materials was investigated with respect to the initiation and propagation of cracking as well as the formation of multiple cracking.

The influence of the initial flaw size, a_0 in a typical ECC material has been investigated by Dick-Nielsen et al. (2006) by use of a semi-analytical approach. Here, simulations have been performed for infinite sheets containing initials flaws with different realistic lengths.

The results are shown in Figure 1 for center cracks and edge cracks respectively. The entire length of the center flaw is denoted $2a_0$ while the entire length of the edge flaw is denoted a_0. Due to this definition flaw lengths in the two situations can be directly compared.

Increasing length of the initial stress free flaw resulted in a decrease of the first crack strength (see Figure 1a). For center cracks the decrease in first crack strength is weak while it for edge cracks is more pronounced. For crack lengths, a smaller than 200 mm the maximal crack opening found in the simulations was less than 20 μm independent of crack position and initial flaw size. At no time during these crack propagations was the process zones fully evolved.

The particular SHCC investigated experimentally in this study was an Engineered Cementitious Composite (ECC) reinforced with PVA (Polyvinyl Alcohol) fibers (2.0 Vol-%). The PVA fibers have a nominal tensile strength of 1600 MPa and are 8 mm long with a diameter of 39 μm. The ECC mix proportions are listed in Table 1, with proportions of ingredients given by ratio of weight, except for the fibers, which are quantified by volume fraction.

The fiber reinforcement was added to the cementitious matrix during the mixing procedure and the composite was cast into blocks (3 in × 4 in × 16 in) from which the test specimens were cut into the shape shown in Figure 2 (Yang and Fischer, 2006). The specimens are notched on all sides to facilitate the formation of a single crack. Ideally, a single crack should be

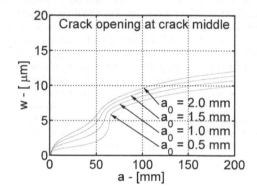

Figure 1b. Center crack opening w and matching total crack length, a.

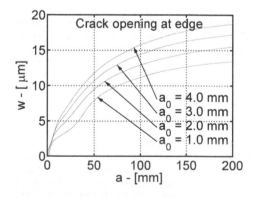

Figure 1c. Edge crack opening w and corresponding total crack length, a.

generated at the notch when a tensile load is applied and the crack opening is measured using a clip gauge.

The stress is calculated by normalizing the tensile load by the cross-sectional area at the notch and is plotted against the measured crack opening to obtain the

Table 1. Mix proportions of PVA-ECC.

Cement	Fly Ash	Sand	Water	SP	PVA Fiber (Vol. %)
1	2	1.4	1	0.023	2

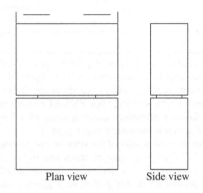

Plan view   Side view

Figure 2. Specimen dimensions.

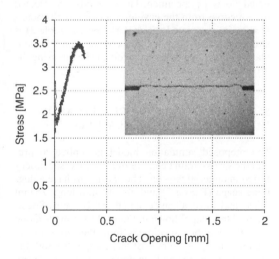

Stress [MPa]

Crack Opening [mm]

Figure 3. $\sigma_B - \delta$ curve of a single crack.

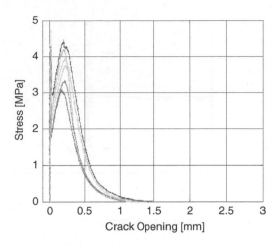

Figure 4. $\sigma_B - \delta$ curves of specimens.

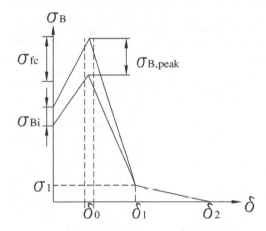

Figure 5. Envelope of the simplified $\sigma_B - \delta$ curves.

$\sigma_B - \delta$ curve. To check whether there is indeed only a single crack at the predefined crack location, the test is terminated immediately after the load has reached the peak level, and the specimen is cut vertically along the center line of its thickness to visually inspect the crack formation. Figure 3 shows typical observations of cut specimens with a single crack and the corresponding $\sigma_B - \delta$ curve.

The tensile tests were carried out on specimens that were cured in air for 28 days. Figure 4 shows the $\sigma_B - \delta$ curves obtained. From these curves it can be observed

that while the ultimate fiber bridging stress has relatively large variability, the crack opening at the peak load remains relatively constant.

Based on the range of fiber bridging stress-crack opening responses obtained from the experimental investigation, a tri-linear simplification as shown in Figure 5 is chosen to represent the experimentally obtained responses. To account for the variability in the $\sigma_B - \delta$ curve, a range in characteristic composite parameters is assumed in the model based on the experimentally obtained data. The parameters that affect the behavior of the composite are the first cracking strength σ_{fc}, the peak bridging stress $\sigma_{B,peak}$, the crack opening δ_0 at peak bridging stress, and the fiber bridging stiffness K_f. These parameters may vary randomly within the experimentally defined range and will inherently account for the variability in fiber

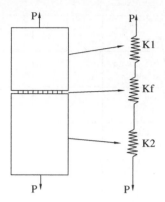

Figure 6. Equivalent spring system.

Table 2. Parameters of $\sigma_B - \delta$ envelope.

Parameter name	Value
σ_{fc}	3.2–4.8 MPa
σ_i	1.82–2.97 MPa
$\sigma_{B,peak}$	2.9–4.6 MPa
δ_0	0.17–0.22 mm
σ_1	1.0 MPa
δ_1	0.60 mm
δ_2	2.0 mm

tensile strength, orientation, interfacial bond characteristics, and matrix flaw size distribution. The simulation model for the composite stress-strain behavior developed in this study will include potential multiple cracking and strain hardening features of the ECC in direct tensile loading. For general applicability of the model to SHCC materials as well as tension softening FRCC, model parameters can be adjusted to experimental data obtained for the fiber bridging stress-crack opening relationship of other FRCC materials.

During the displacement controlled deformation process of the composite, the tensile specimen initially behaves like a spring with an effective stiffness K_m of the un-cracked composite. At increasing deformations, the tensile load increases until the first crack forms at the largest flaw and lowest first cracking strength in the specimen. After the first tensile crack is formed, the cracked composite is represented by inserting a second spring element with a stiffness corresponding to the fiber bridging stiffness K_{fl}, as shown in Figure 6. The load will subsequently drop to a value where force equilibrium between the fiber bridging section and adjacent uncracked composite section is achieved. When the induced tensile deformations in the composite are further increased, the load will again increase with a modified total stiffness K until the tensile stresses are sufficient to cause the formation of a second tensile crack in the composite. This process is continued and multiple cracking can initiate until the tensile stress reaches the lowest peak bridging strength as defined by the parameter envelope (Figure 5). At this point, localization of cracking occurs and the composite fails in a tension softening manner.

The maximum number of potential cracks in a given composite is governed by the specimen length and the minimum crack spacing x_d, which has been theoretically derived[5] for the case of random short fiber reinforced composites. A crack spacing between x_d and $2xd$ is expected at crack saturation. However, due to the

variation of matrix properties, fiber matrix interfacial bonding, and fiber distribution, the observed minimum crack spacing often exceeds twice the derived minimum crack spacing. For PVA-ECC with preexisting flaws, a minimum crack spacing of 1.7 mm to 2.5 mm at crack saturation was reported[7].

Experimental results of the stress-strain response of PVA-ECC show a minimum crack spacing of 2 mm, which is adopted as a parameter for the simulation model presented in this paper. At a given specimen length and minimum crack spacing, the maximum number of potential cracks and their locations are identified along the specimen. Then, a randomly selected $\sigma_B - \delta$ curve from the envelope (Figure 5) characterized by its associated parameters (see e.g. Table 2) is assigned to each potential crack location and the simulation of the uniaxial tension test can be run using the procedure described above.

3 EXPERIMENTAL TEST RESULTS AND COMPARISON TO SIMULATIONS

The proposed simulation model is applied to predict the stress-strain behavior of PVA-ECC specimens tested in uniaxial tension. The specimens have a dogbone shape (Figure 7) and are 305 mm long, 25 mm thick, and have a gauge length of 100 mm. Specimens of the shape shown in Figure 2 were made along with the dogbone specimens from the same mix to identify the $\sigma_B - \delta$ curves of this composite. Both the notched specimens and the dogbone specimens were tested under uniaxial load at the age of 28 days. The experimentally obtained $\sigma_B - \delta$ curves of the notched specimens are shown in Figure 4. The parameters that describe the envelop of the $\sigma_B - \delta$ curves are summarized in Table 2. This information is taken as input parameters of the simulation model and the predicted stress-strain curves for the dogbone specimens are compared with experimentally obtained stress-strain curves in the uniaxial tension test in Figure 8(a) and (b).

Since a random $\sigma_B - \delta$ curve from within the experimentally obtained envelope is assigned to each potential crack location in each simulation run, the

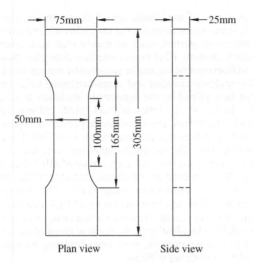

Figure 7. Dimensions of tensile specimen.

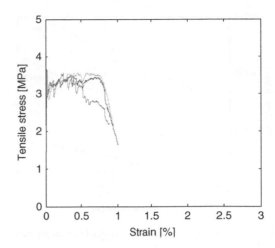

Figure 8b. Experimentally obtained $\sigma - \varepsilon$ curves.

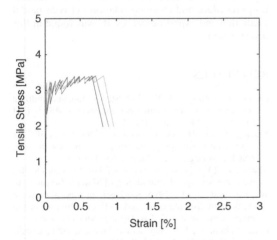

Figure 8a. Simulated $\sigma - \varepsilon$ curves.

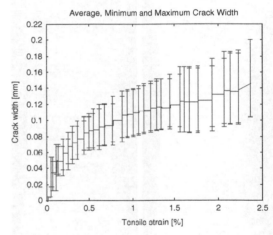

Figure 9a. Range of crack widths obtained from simulation procedure

obtained results for the composite stress-strain curve differ as well in each simulation. Figure 8(a) shows three simulated results plotted in the same figure. The comparison shows that the simulation results agree well with experimental data. In addition, the plotted results demonstrate that the simulation program developed using aforementioned concept is capable of capturing the multiple cracking phenomenon and the strain hardening behavior of ECC under direct tensile loading.

This simulation procedure can offer other related results such as evolution of crack spacing and a quantitative assessment of crack widths as a function of composite tensile strain (Figure 9a).

The predictions of crack widths as a function of composite tensile strain resulting from the simulation

procedure have been compared to experimental data and showed close correlation (Figure 9b). The crack formation (Figure 10) of an ECC specimen under direct tensile loading was continuously monitored during the experiment using an image capturing and analysis system (ARAMIS). This system allows placement of virtual deformation gages after the test has been conducted and provides information on the deformations of the entire specimen surface and at particular user-defined locations.

4 CONCLUSIONS

A method to simulate and predict the multiple cracking and strain hardening behavior of Strain Hardening

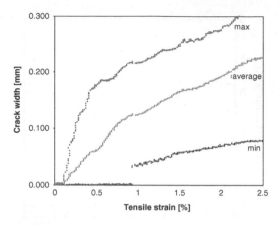

Figure 9b. Range of crack widths obtained from experimental testing.

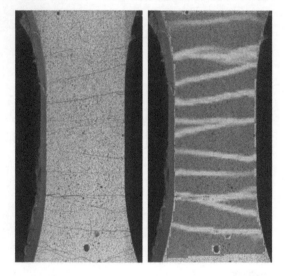

Figure 10. Cracking in ECC tensile specimen and visualization with ARAMIS.

Cementitious Composites (SHCC) under uniaxial tension was developed based on experimental information of the fiber bridging stress-crack opening relationship of the composite. Utilizing this information as input

parameters of the simulation of the composite tensile stress-strain behavior, the variability of composite material properties, such as matrix flaw size, fiber tensile strength, fiber matrix interface characteristics, and fiber orientation, can be realistically incorporated. The multiple cracking and strain-hardening behavior can be captured by the suggested simulation model. In addition, the evolution of crack width and spacing can be quantified. The simulated response and experimentally obtained stress-strain behavior of SHCC are in agreement. The proposed method can serve as a tool in estimating the stress-strain behavior of SHCC based on $\sigma B - \delta$ information, which is essential in the design of structural applications using SHCC. The simulation model can also be used to design SHCC materials with a target tensile stress-strain response and crack width limit by identifying the optimal range of matrix first cracking strength, peak fiber bridging strength and fiber bridging stiffness.

The image capturing and analysis system ARAMIS has been used to confirm the results of the simulation procedure and close correlation between model prediction and experimental results was found in this investigation.

REFERENCES

Li, V.C. and Leung, C.K.Y., 1992, Steady state and multiple cracking of short random fiber composites, ASCE J. of Engineering Mechanics, 188 (11).

Maalej, M., Li, V.C. and Hashida, T., 1995, Effect of fiber rupture on tensile properties of short fiber composites, ASCE J. of Engineering Mechanics, 121 (8).

Maalej, M., 2001, Tensile properties of short fiber composites with fiber strength distribution, J. of Material Science, 36.

Wu, H.C. and Li, V.C., 1995, Stochastic process of multiple cracking in discontinuous random fiber reinforced brittle matrix composite, Int'l J. of Damage Mechanics 4(1).

Dick-Nielsen, L., H. Stang, and P. Poulsen, 2006, Condition for Strain-Hardening in ECC Uniaxial Test Specimen, Proceedings of MMMCP, Alexandroupolis, Greece.

Wang, S. and Li, V.C., 2004, Tailoring of pre-existing flaws in ECC matrix for saturated strain hardening", Proceedings of FRAMCOS-5, Vail, Colorado, USA.

Yang, J. and Fischer, G., 2006, Simulation of the tensile stressstrain behavior of strain hardening cementitious composites, Proceedings of MMMCP, Alexandroupolis, Greece.

Fracture Mechanics of Concrete and Concrete Structures – High-Performance Concrete,
Brick-Masonry and Environmental Aspects – Carpinteri, et al. (eds)
© 2007 Taylor & Francis Group, London, ISBN 978-0-415-44617-4

Influence of steel fibers on full-scale RC beams under shear loading

F. Minelli & G.A. Plizzari
University of Brescia, Italy

F.J. Vecchio
University of Toronto, Canada

ABSTRACT: The use of Fiber Reinforced Concrete (FRC) has gained considerable attention in the last few years, particularly when crack propagation control is of primary importance, such as in slab-on-grade applications or in beams when shear reinforcement is partly or totally absent. Many experiments available in the literature showed that fibers, if provided in sufficient amount, are significantly effective as shear reinforcement. Fibers limit the growth of shear inclined crack, give visible warning prior the structure collapse and also provide a stable and diffused crack pattern within the shear critical area. However, the issue of size effect in members containing steel fibers, has not been deeply investigated and evaluated yet. This paper studies the beneficial influence of fibers on the size effect in members without conventional shear reinforcement. Moreover an extensive parametrical numerical study, performed by means of a FE program based on the Modified Compression Field Theory (MCFT), suitably adapted to FRC materials, is herein presented.

1 INTRODUCTION

During the last few decades many tests investigating shear behavior have been carried out on relatively small beams. It was found that the results of these tests can not be directly extended to full size beams. It was shown by Kani (1967) that there is a very significant size effect on the shear strength of members without transverse reinforcement. The shear strength of these members appears to decrease as the effective depth increases. Shioya et al. (1989) reaffirmed this fact and extended the available data to beams with a depth up to 3000 mm. Figure 1 shows that the average shear stress to cause failure of the largest beam was about one-third the average shear stress to cause failure of the smallest beam.

There is a general agreement that the main reason for this size effect is the larger diagonal crack width in larger beams. On the other hand, there is a disagreement concerning the modeling of this phenomenon, which has prompted a strong debate within the research community (Bentz, 2005).

A different approach was proposed by Bažant & Kim (1984), who stated that the most important consequence of wider cracks is the reduced residual tensile stress and aggregate interlock. Most of current formulas for predicting shear capacity of a member are in fact based on the concept of tensile strength.

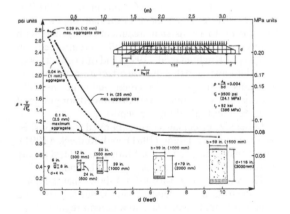

Figure 1. Shear stress at failure; results from Shioya (1989).

However, this concept is theoretically justified only in the case of ductile failures governed by the theory of plasticity. For failures in which the stress decreases after reaching the strength limit, as in the case of tensile cracking, the strength concept is inconsistent when applied in a continuum analysis.

New approaches do not treat fracture like a point phenomenon, but recognizes that in brittle heterogeneous materials such as concrete, the fracture propagates with a relatively large fracture process zone in which progressive microcracking gradually reduces the tensile stress to zero.

Bažant & Kim (1984) investigated also the consequences of this nonlinear fracture mechanics approach for diagonal shear failure and proposed a size reduction factor for shear members (Bažant & Kim, 1984). The model represents a gradual transition from the strength criterion for small structures to the linear elastic fracture mechanics for very large structures.

Bažant & Cao (1986) and Bažant & Kazemi (1991) confirmed the significant agreement between nonlinear fracture mechanics and experimental evidence for prestressed concrete beams and a wide series of tests on reinforced concrete beams, with a size range being of 1:16. They also suggested important adjustments to the current design codes accounting for size effect coefficients.

Other researchers believe that wider cracks reduce the ability to transmit crack interface shear stresses. The crack width is scaled like any other dimension of the member. The crack opening in the middle of the web is controlled by the strain of the longitudinal reinforcement, but this reinforcement is too far away to control the crack spacing. With the larger crack width, the friction of the shear forces is reduced and, consequently, the ultimate capacity is reduced as well.

As the crack spacing used to determine the shear stress at the limiting crack interface is a function of the specimen depth, no special factor is required to account for the size effect (Collins et al., 1996) Perhaps the strongest argument for this latter approach is that it leads to a consistent treatment of members with different arrangements of longitudinal reinforcement. Tests of Kuchma et al. (1997) and of Frosch (2000) have demonstrated that the size effect is not significant in beams with web reinforcement and in beams without stirrups containing well distributed longitudinal reinforcement (all along the depth).

Other size effect laws were calibrated and published by Okamura & Higai (1981) and Reineck (1991), who considered the tests of Shioya et al. (1989) and, therefore, represent reasonable lower bound.

The analysis of Zararis (2001) shows that the diagonal shear failure in slender beams is due to a splitting of concrete that takes place in a certain region of the beam. It was realized that in this section the problem of the size effect on diagonal shear failure can be reduced to the problem of size effect on split-tensile failure. Tests on cylindrical disks of constant thickness done by many researchers (i.e. Sabnis & Mirza, 1979) confirmed the existence of size effect on split-tensile failure, and showed that up to a certain critical diameter the split-cylinder strength decreases as the diameter increases (ASCE-ACI Committee 445, 1998).

Recently, an interesting campaign was carried out at the University of Toronto with the purpose of investigating the safety of the shear design of large, wide

Figure 2. Shear failure of a deep beam tested at the University of Toronto.

beams (Lubell et al., 2004). The beam tested exhibited a brittle shear failure, typical for high-strength concrete beams, with a loud noise as the mid-span load reached 2440 kN. This failure load was only 52% of the failure load predicted by the ACI shear provisions, meaning that the beam would fail under the actual service loads. The maximum crack width measured was only 0.25 mm while the midspan deflection at failure was less than 1/500 of the span.

The Authors concluded that the ACI design procedure is currently unsafe, as it neglects the size effects in shear. On the contrary, the MCFT (Bentz et al., 2006) design procedures predicts that, if the depth of a member is doubled, the shear capacity will be less than double and that this strength ratio will become smaller as the size of the beam increases and as the aggregate size decreases.

The ACI 318-02 shear design expression for members without stirrups can be very unconservative not only because it was based on test results from beams that had rather small depths, but also because those test beams typically had very large amounts of longitudinal reinforcement ($\rho > 2\%$) to avoid any possibility of flexural failures.

Recent advancements in material technologies and productions, as well in standardization, provide new materials to introduce in the design process. A very promising material, easily available in the market, is Fiber Reinforced Concrete (FRC): it is characterized by a significant toughness enhancement, especially under tension. Fiber reinforced concrete was shown to give rather good performance in shear-critical beams (Minelli, 2005) and in other structural elements, as a valuable and effective substitution of secondary reinforcement (Di Prisco & Ferrara, 2001; Meda et al., 2005; Minelli et al., 2005).

In the work of Minelli & Plizzari (2006), a modification of the current EC2 (2003) equation for members not containing stirrups under shear was proposed for FRC, with a good agreement against more than 30 experimental results taken from the database of the University of Brescia. Among the principal assumption of such modeling, fibers were treated as reinforcement spread over the entire depth of a member, giving a shear contribution which is similar to that offered by small diameter bars placed all along the beam, as previously shown (Kuchma et al., 1997).

In other words, fibers can effectively reduce the size effect issue in reinforced concrete members not containing stirrups, if provided in sufficient amount and giving significant toughness and energy dissipation ability to the matrix.

The aim of this paper is to discuss this critical issue which has not been properly treated yet in the scientific literature.

Firstly, a number of experimental results on full scale shear-critical beams will be presented, focusing with special emphasis on recent tests conducted on 1 m deep members not containing stirrups. The effect of fibers and of the minimum amount of transverse reinforcement, as stated by EC2, will be compared against the response of identical plain concrete members. Secondly, a set of numerical analyses, carried out using a finite element program based on the compression field model procedures, will be presented, showing the promising effect of steel fibers in reducing the scale effects.

The aim of the ongoing research at the Universities of Brescia and Toronto is to develop a simple, although rational and reliable, procedure allowing engineers to incorporate SFRC in the design process.

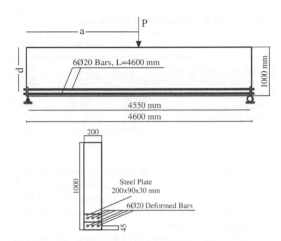

Figure 3. Geometry of Large Size Specimen.

Table 1. Geometry characteristics of specimens.

	H = 500 mm	H = 1000 mm
Total length	2400 mm	4600 mm
Span	2280 mm	4550 mm
Shear span a	1140 mm	2275 mm
Width	200 mm	200 mm
Total depth	500 mm	1000 mm
Gross cover	45 mm	90 mm
Effective depth d	455 mm	910 mm
Reinforcement area	905 mm^2	1884 mm^2
Reinforcement ratio	1.04%	1.03%

2 EXPERIMENTS

2.1 Materials and specimen geometry

Among the 47 shear tests carried out at the University of Brescia since 2001 (most of them can be found in Minelli, 2005), the following discussion will focus on those experiments conducted to assess the influence of the minimum shear reinforcement, represented either by classical transverse reinforcement or steel fibers, and the size effect in shear.

Two series of specimens are presented herein: the first refers to sample beams having a total depth of 500 mm ("Small Size Specimens"), while the second consists of elements 1000 mm deep ("Large Size Specimens").

Concerning the first set of experiments, five shear-critical beams loaded with a three point loading system having a shear span-to-depth ratio of 2.5 (which is recognized to be the most critical in terms of shear strength (Kani, 1967) were tested. All beams had the same geometry aimed at analyzing the effect of the addition of a randomly distributed fibrous reinforcement to concrete.

A beam depth of 500 mm was chosen, with a gross cover of 45 mm. The beam spanned 2280 mm, while the overall length of the specimen was 2400 mm. Two deformed longitudinal bars, having a diameter of 24 mm, were present in each specimen, corresponding to a reinforcement ratio of 1.04%.

Deeper beams (Large Size Specimens) were cast with a total depth of 1000 mm, an effective depth of 910 mm, a width of 200 mm and a span of 4550 mm (Figure 3). A three point loading scheme was chosen resulting in a span-to-depth ratio again of 2.5. The steel reinforcement (6φ20 mm rebars) was located in two identical bottom layers giving a reinforcing ratio of 1.03%.

Table 1 summarizes the geometrical characteristics of the beams. All smaller specimens were cast by using a normal strength concrete (NSC), while two different series of larger beams were tested, the first one cast in the same batch as the smaller beams, and the second using a high strength matrix (HSC). Table 2 reports the composition of the two different concrete batches.

One of the Small Specimens was cast without any transverse reinforcement (PC-50), two with the

Table 2. Mechanical properties of concrete.

	NSC	HSC
Cement content [kg/m³]	345	380
Maximum aggregate size [mm]	20	20
Plasticizer [l/m³]	5.2	3.8
Compressive cubic strength [MPa]	25.7	55
Elastic modulus [GPa]	31	37

minimum amount of transverse reinforcement (MSR-50, with stirrups 2ϕ8@300 mm) as required by EC2 (2003), and two with 20 kg/m³ of steel fibers (FRC-50) having a length of 50 mm and a diameter of 1 mm (aspect ratio l/ϕ = 50).

The fracture properties of FRC were determined according to the Italian Standard (UNI 11039, 2003), which requires bending tests (4PBT) be performed on small beam specimens (150 × 150 × 600 mm).

The equivalent post-cracking strengths related to the SLS and ULS were equal to $f_{eq(0--0.6)}$ = 2.53 MPa and $f_{eq(0.6-3)}$ = 2.50 MPa, respectively.

Three Large Size Specimens were cast for each concrete strength (NSC and HSC): the reference element (PC-100), the sample containing the minimum amount of shear reinforcement (MSR-100, with stirrups 2ϕ8@650 mm) and the latter containing 20 kg/m³ of hooked steel fibers (FRC-100). Note that the notation PC refers to a beam cast with plain concrete, whereas FRC always indicates a beam with fibers as the only shear reinforcement. Finally, MSR refers to minimum shear reinforcement specimen.

One should note that the fiber amount of 20 kg/m³ is quite low; some structural (e.g., CNR DT 204, 2006) do not allow for the usage of a such a low fiber content. However, aim of the present comparison is the evaluation of steel fibers as a minimum transverse shear reinforcement and how one can achieve a good shear response with a relatively low fiber content. The Authors, in other words, decided to use a low fiber content to enforce their study toward an extensive and, at the same time, economical utilization of FRC.

2.2 Experimental results

In the following, the main experimental results (from Figure 4 to Figure 8) are presented and discussed. With regard to the small size specimens (H = 500), Figure 4 plots the load-displacement curve of the five experiments and Figure 5 shows the corresponding shear-crack width vs. load curve.

The specimen made of plain concrete (PC-50) showed the well known brittle shear collapse, characterized by early diagonal cracking and unstable propagation for a low load level. No ductility after cracking nor visible warning were detected.

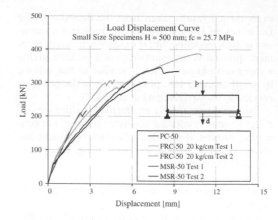

Figure 4. Load-Displacement Curve; series H = 500 mm.

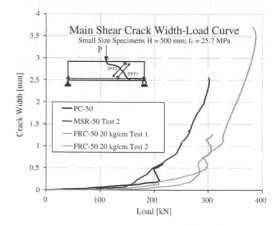

Figure 5. Shear Crack Width-Load Curve; series H = 500 mm.

On the contrary, the two elements made of FRC as well as the two cast with the minimum amount of stirrups showed visible cracking with a stable propagation, accompanied also by a not negligible increase in bearing capacity; in fact, specimen FRC-20-2 doubled the ultimate load and showed a displacement at failure four time bigger than the reference sample (PC-50). Also the plot of the main shear crack (Figure 5) confirms the role played by a minimum amount of transverse reinforcement, constituted either by fibers or stirrups.

As diffusely reported in the literature, the primary role of the minimum shear reinforcement is to limit the growth of inclined cracks, to improve ductility, and to ensure that the concrete contribution to shear resistance is maintained at least until yield of the shear reinforcement. In other words, it is commonly agreed that, before failure, the R/C structure must give a warning by cracking and visible deflection; this represents the requirements for the minimum reinforcement ratio.

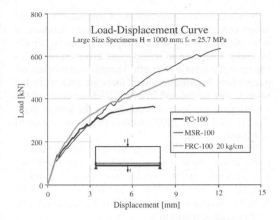

Figure 6. Load-Displacement Curve; series 1 (H = 1000 mm).

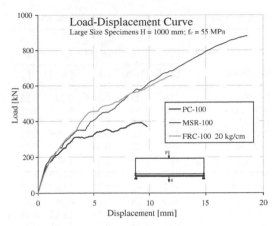

Figure 7. Load-Displacement Curve; series 2 (H = 1000 mm).

Beams that do not contain web reinforcement may fail in a relatively brittle manner immediately after the formation of the first diagonal crack (Figure 5).

Moreover, one should keep in mind that the shear capacity of such members can be substantially reduced by factors such as (1) repeated loading which propagates existing cracks and lowers the apparent tensile strength of concrete; (2) tensile stresses caused by restrained shrinkage strains; (3) thermal strains or creep strains; stress concentrations due to discontinuities such as web openings, (4) termination of flexural reinforcement, or (5) local deviation of tendon profiles (Collins & Mitchell, 1993).

Such reinforcement can be omitted if there is no significant chance of diagonal cracking. It can also be omitted if the member is of minor importance, or if the member is part of a redundant structural system that allows substantial redistribution and, hence, will show adequate ductility.

A low amount of fibers provides all the aforementioned benefits, being a valuable and economical alternative to the traditional stirrups, whose handling and placing can be expensive especially when dealing with precast beams or, in general, with structural elements characterized by non-rectangular or square cross-sections. FRC gives the same structural response of members containing the minimum transverse reinforcement.

Figure 6 and Figure 7 show the load-displacement curves for the large size specimens, respectively for the NSC series and HSC series.

Figure 8 exhibits the width of the main shear crack vs. the load for the HSC series. Note that the main shear crack is the average of 6 measurement performed in both shear spans, as recalled in the plot of Figure 8.

Differently form the small size specimens, the traditional shear reinforcement turned out to be significantly more effective than steel fibers, both in terms

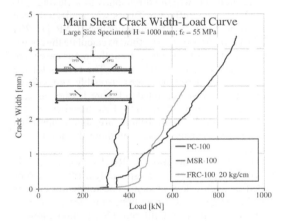

Figure 8. Shear Crack Width-Load Curve; series 2 (H = 1000 mm).

of bearing capacity and ductility. Note that fibers gives better performance, compared to the MSR specimens at the service level, improving the tension stiffening effect and therefore reducing the displacement. Also cracks are fewer and narrower, at service level, then those of MSR specimens, for both NSC and HSC series. However, with increased loads and displacements, and after a shear crack of around 3 mm, fibers are no longer able to resist further load and displacement, bringing the member to the "block mechanism", which characterizes the shear collapse.

Table 3 reports a summary of the main experimental results of the experimental campaign herein discussed. Note that the ultimate shear stress (v_u) as well as the ultimate shear force (V_U), was determined by considering also the self-weight of the members. Moreover, in the calculation of the bearing capacity

Table 3. Summary of the main experimental values.

Specimen		P_u	v_u	$v_u/$ $(f_c)^{1/2}$	δ_u	$V_u/$ $V_{u,FL}$
		[kN]	[MPa]	[-]	[mm]	[-]
PC-50		216	1.22	0.24	2.74	0.52
MSR-50 1		346	1.93	0.38	9.33	0.82
MSR-50 2		302	1.69	0.33	7.03	0.72
FRC-50 1		388	2.16	0.43	10.95	0.92
FRC-50 2		308	1.72	0.34	4.77	0.73
PC-100	NSC	365	1.07	0.21	7.60	0.43
MSR-100		635	1.81	0.36	12.60	0.73
FRC-100		494	1.42	0.28	11.05	0.57
PC-100	HSC	393	1.14	0.15	9.79	0.45
MSR-100		880	2.48	0.33	18.62	0.99
FRC-100		656	1.86	0.25	12.01	0.74

of the member, which corresponds to a flexure failure ($V_{U,FL}$), the effect of fibers was neglected. One should note that the design guidelines CNR DT 204 (2006) provide a simple method to calculate the contribution of fibers to bending strength; in our case there is an improvement of 10% only. The results for the plain concrete members agree reasonably well with the experimental results plotted in Figure 1, whereas very good results can be obtained for FRC members which, especially for the small beams, are able to approach the flexural bearing capacity without the addition of stirrups.

Definitely, the addition of FRC does not completely solve the size effect issue, especially if fibers are provided in such a low amount. The following chapter will investigate this aspect better through a series of numerical analyses on different FRC compositions.

3 NUMERICAL ANALYSES

3.1 Introduction

Numerical analyses were initially performed to validate the FE program VecTor2 (Wong & Vecchio, 2002) which is based on the MCFT (Vecchio & Collins, 1986). The latter represents a well known model for representing the nonlinear behavior of reinforced concrete structures; it is essentially a smeared, rotating crack model for cracked reinforced concrete elements. On the basis of a number of panel tests, constitutive relationships were developed to describe the behavior of cracked reinforced concrete in compression and in tension.

Numerical analyses were performed by assuming perfect steel-to-concrete bond (no slip). The beam was discretised with four-node 2D plane stress elements. Transverse reinforcement was modeled as embedded in concrete elements.

While concrete in compression was easily modeled using constitutive laws available in the literature, it is worthy discussing the assumption regarding tensile behavior. Concrete in tension was assumed to be linear up to the tensile strength. The post-peak behavior, particularly important when fibers are present in the matrix, was calculated from FE back analyses of the experiments performed according to the Italian Standard (UNI 11039, 2003); further details are reported in Minelli (2005).

The cohesive laws were initially determined as stress-crack opening (σ-w) laws and, since VecTor2 is a smeared cracking model, were eventually transformed into stress–strain relationships ($\sigma - \varepsilon$) by dividing the fracture opening (from back analysis) by a "characteristic length".

These relationships were then incorporated in VecTor2, and used to perform the FE analyses of all small and large beams with steel fibers. The characteristic length was assumed as the effective flexural depth of members, which represents an approximation of the crack spacing of beams with little or no shear reinforcement. The maximum size of the aggregate, which is a quite important parameter for shear, was assumed equal to 20 mm, as in the experiments.

Special emphasis was devoted to the crack-width control, which is based on a maximum crack width that can be selected by the user among a series of proposal. This parameter turned out to be quite influential on the overall response of members subject to a shear mode of failure. The maximum crack width was here selected equal to a fifth of the aggregate size for plain concrete members (Vecchio, 2000) whereas it was chosen equal to 3 mm for the FRC members (Minelli & Vecchio, 2006), according to the experimental results. Note that such a limit does not significantly affect the structural response of members with transverse reinforcement (Vecchio, 2000).

Finally, rebars were simulated by using a polylinear stress-strain curve, including strain hardening, according to the experimental data.

Figure 9 reports the results of numerical analyses performed on the Large Size Elements made of NSC. Space restrictions do not allow the authors to show all results, even though the following discussion fully applies to all analyses referred to the experiments presented above.

The MCFT-based procedures already gave a number of examples demonstrating their ability in accurately representing the behavior of R/C members with little or no shear reinforcement (Vecchio, 2000). Fewer cases are reported for FRC elements (Minelli, 2005). Adding the corresponding tension softening law (as determined form back analysis) and suitably improving the crack width control, the compression field model was able to produce reasonably well the experiments, in terms of bearing capacity, stiffness, crack

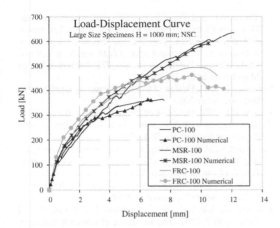

Figure 9. Numerical analyses vs. experiments, Large Size Specimens, NSC.

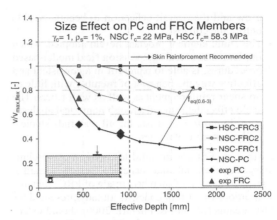

Figure 10. Size effect development for different fiber reinforced concretes; results from numerical analyses.

patterns and ductility. Modeling of PC-100 specimen turned out to be very accurate, whereas a general overestimation of the stiffness of the cracked stage was found for FRC-100 specimen. However, good accuracy in the ultimate bearing capacity and in the crack pattern was seen. As far as the MSR-100 specimen is concerned, a 7% underestimation of the peak load was observed, even though the crack pattern and the overall behavior was well modeled. In conclusion, the numerical study confirmed the ability of fibers in improving the post cracking stiffness and in reducing the shear cracking.

Once the FE program was validated against many experiments (Figure 9), an extensive numerical study was performed to better clarify the influence of fibers in reducing the scale effects. Figure 10 plots the ultimate shear (from numerical analyses) to maximum shear (from flexure failure) ratio vs. the effective depth. The ratio between the two shear stress values has been selected to show both scale effects and modes of failure. Different constitutive tension softening laws were chosen for a series of beams having a depth ranging form 250 to 2000 mm and characterized by the critical a/d ratio of 2.5. The longitudinal reinforcement was selected in all case as equal to 1%. NSC-PC and NSC-FRC1 refer to the concrete mixtures reported in the present paper, while NSC-FRC2 refers to a content of fibers of 40 kg/m³. Finally, HSC-FRC3 concrete composition concerns a high strength concrete with 50 kg/m³ of high-carbon steel fibers, having a considerably high performance and a strain-hardening behavior under flexure (Minelli & Vecchio, 2006). Numerical results are plotted against the experimental values related to Table 3, showing a discrete agreement.

From this plot one can notice that the decay in the ultimate shear strength with increasing dimensions diminishes as the fiber toughness increases, as one would have expected. When fibers are added in large amounts and sufficient toughness, they result, even for significantly deep members, in a collapse failure which moves from shear (brittle) to flexure (generally ductile) without the addition of any stirrup reinforcement. This is of great importance from a design point of view, giving the possibility to reduce or also totally substitute stirrups with fibers and, moreover, to significantly reduce the scale effects. Note that the plot shows a vertical line in correspondence of an effective depth of 1 m. In many cases, in fact, codes require, from that point on, the addition of a distributed skin reinforcement in the two directions, basically for crack control. This requirement, as well as the one of the minimum shear reinforcement, can be met by using a relatively low amount of fibers, according to Figure 10 and to the experimental results herein discussed.

4 CONCLUSIONS

In the present paper, some critical issues related to the behavior of structural R/C members with little or no shear reinforcement or with the addition of fiber reinforced concrete were discussed.

The following main conclusions can be drawn:

- Fibers in reinforced concrete elements under shear significantly reduce the scale effects, even if provided in relatively low amount. To fully avoid such an experimental drawback, a greater fiber content (which means a bigger toughness) is required;
- Fibers can effectively replace the minimum amount of transverse reinforcement and the skin reinforcement required to control the crack propagation and give visible warning before the structural collapse; fibers, in fact, allow the shear cracking to gradually

grow and the overall deflection to become visible before the collapse.

- The finite element program herein adopted, which implements several compression field model-based procedures, is able to represent in a consistent and effective way the behavior of member with little or no shear reinforcement as well as with fibers. The scale effect determined through numerical analyses were found to be similar, for all materials studied, to the experimental ones.

Further studies are intended to assess these preliminary experimental and numerical results toward a consistent design of members made of FRC. Suitable adaptations of current codes, as already presented for FRC beams (Minelli & Plizzari, 2006) should extensively evaluated with regard to the size effect issue.

ACKNOWLEDGMENTS

The Authors would like to give their appreciation to steel factory Alfa Acciai (Brescia) for supplying the reinforcement for the experiments. Moreover, the assistance of engineers Bertozzi Alessandro, Federico Ginesi and Adriano Reggia is gratefully acknowledged.

REFERENCES

ASCE-ACI Committee 445 on Shear and Torsion. 1998. Recent Approaches to Shear Design of Structural Concrete, *Journal of Structural Engineering*, 124(12), 1375–1417.

Bažant, Z. P., and Kim, J,-K. 1984 Size Effect in Shear Failure of longitudinally Reinforced Beams. *ACI Journal,* 81(5), 456–468.

Bažant, Z.P., and Cao, Z. 1986. Size Effect of Shear Failure in Prestressed Concrete Beams. *ACI Journal,* 83(2), 260–268.

Bažant, Z.P., and Kazemi, M.T. 1991. Size Effect on Diagonal Shear Failure of Beams without Stirrups. *ACI Structural Journal,* 88(3), 268–276.

Bentz. E.C. 2005. Empirical Modeling of Reinforced Concrete Shear Strength Size Effect for Members without Stirrups. *ACI Structural Journal,* 102(2), 232–241.

Bentz, E.C., Vecchio, F.J. and Collins, M.P. 2006. Simplified Modified Compression Field Theory for Calculating Shear Strength of Reinforced Concrete Elements. *ACI Structural Journal,* 103(4), 614–624.

CNR DT 204/2006. 2006. "Guidelines for the Design, Construction and Production Control of Fibre Reinforced Concrete Structures", *National Research Council of Italy.*

Collins, M.P., and Mitchell, D. 1997. Prestressed Concrete Structures. *Chapter 7, Response Publication,* Toronto and Montreal, Canada.

Collins, M.P., Mitchell, D., Adebar, P.E. and Vecchio, F.J. 1996. A General Shear Design Method. *ACI Structural Journal,* 93(1), 36–45.

Di Prisco, M. and Ferrara, L. 2001. HPFRC Pre-stressed Thin-web Elements: Some Results on Shear Resistance. *IV Fracture Mechanics of Concrete Structures, R.de Borst, J. Mazars, G. Pijaudier-Cabot, J.G.M. van Mier Editors, (A.A.Balkema Publishers, Lisse (The Netherland)),* 28 May-1 June 2001, 2, 895–902.

Eurocode 2. 2003. Design of Concrete Structures, prEN 1992 Ver. December 2003.

Frosch, R.J. 2000. Behavior of Large-Scale Reinforced Concrete Beams with Minimum Shear Reinforcement. *ACI Structural Journal,* 97, 814–820.

Hasegawa, T., Shioya, T., Okada, T. 1985. Size Effect on Splitting Tensile Strength of Concrete. *Proc., Japan Concrete Inst. 7th Conf.,* 1985, 309–312.

Kani, G.N.J. 1967. How Safe are our Large Reinforced Concrete Beams?. *Journal ACI,* 64, 128–141.

Kuchma, D., Végh, P., Simionopoulos, K., Stanik, B. and Collins, M.P. 1997. The Influence of Concrete Strength, Distribution of Longitudinal Reinforcement, and Member Size, on the Shear Strength of Reinforced Concrete Beams. *"Concrete Tension and Size Effect",* CEB Bulletin 237, Lausanne, 258 pp.

Lubell, A., Sherwood, T., Bentz, E., and Collins. M.P. 2004. Safe Shear Design of Large, Wide Beams. *Concrete International,* 26(1), 67–78.

Meda, A., Minelli, F., Plizzari, G.A. and Riva, P. 2005. Shear Behavior of Steel Fibre Reinforced Concrete Beams. *Materials and Structures,* 38, 343–351.

Minelli, F. 2005. Plain and Fiber Reinforced Concrete Beams under Shear Loading: Structural Behavior and Design Aspects. *Ph.D. Thesis, Department of Civil Engineering, University of Brescia, Starrylink Editor,* pp. 430.

Minelli, F. and Plizzari, G.A. 2006. Steel Fibers as Shear Reinforcement for Beams. *Proceedings of The Second Fib Congress, Naples, Italy, 5–8 June, 2006, abstract on page 282–283, full length paper available on accompanied CD,* pp.12.

Minelli, F. and Vecchio, F.J. 2006. Compression Field Modeling of Fiber Reinforced Concrete Members under Shear Loading. *ACI Structural Journal,* 103(2), 244–252.

Minelli, F., Cominoli, L., Meda, A., Plizzari, G.A., and Riva, P. 2005. Full Scale Tests on HPSFRC Prestressed Roof Elements Subjected to Longitudinal Flexure. *Proceedings of the International Workshop on High Performance Fiber Reinforced Cementitious Composites in Structural Applications, Honolulu, Hawaii, USA, May 23–26, 2005,* 323–332.

Okamura, H., and Higai, T. 1981. Proposed Design Equation for Shear Strength of R.C. Beams without Web Reinforcement. *Proc., Japan Soc. Civ. Engrg.,* 300, 131–141.

Reineck, K.-H. 1991. Ultimate Shear Force of Structural Concrete Members without Transverse Reinforcement derived from a Mechanical Model. *ACI Structural Journal,* 88(5), 592–602.

Sabnis, G.M., Mirza, S.M. 1979. Size Effect in Model Concretes. *Journal Struct. Div., ASCE,* 106(6), 1007–1020.

Shioya, T., Iguro, M., Nojiri, Y., Akiayma, H., Okada, T. 1989. Shear Strength of Large Reinforced Concrete Beams, Fracture Mechanics: Application to Concrete. *SP-118, ACI, Detroit,* 259–276.

UNI 11039. 2003. Steel Fiber Reinforced Concrete – Part I: Definitions, Classification Specification and Conformity – Part II: Test Method for Measuring First

Crack Strength and Ductility Indexes. *Italian Board for Standardization.*

Vecchio, F.J. 2000. Analysis of Shear-Critical Reinforced Concrete Beams. *ACI Structural Journal*, 97(1), 102–110.

Vecchio, F.J., and Collins, M.P. 1986. Modified Compression Field Theory for Reinforced Concrete Elements Subjected to Shear. *ACI Journal, Proceedings*, 83(2), 219–231.

Wong, P.S. and Vecchio, F.J. 2002. VecTor2 and FormWorks User's Manual. *Technical Report of the Department of Civil Engineering, University of Toronto, August 2002, pp.217. (available @ http://www.civ.utoronto.ca/vector/).*

Zararis, P.D., Papadakis, G.Ch. 2001. Diagonal Shear Failure and Size Effect in RC Beams without Web Reinforcement. *Journal of Structural Engineering*, 127(1), 733–742.

Fracture Mechanics of Concrete and Concrete Structures – High-Performance Concrete,
Brick-Masonry and Environmental Aspects – Carpinteri, et al. (eds)
© 2007 Taylor & Francis Group, London, ISBN 978-0-415-44617-4

Use of tension softening diagrams for predicting the post-cracking behaviour of steel fibre reinforced concrete panels

A. Nour & B. Massicotte

École Polytechnique de Montréal, Montréal, Québec, Canada

ABSTRACT: In this paper, a semi-analytical model aimed to investigate the post-cracking behaviour of FRC panels is presented. The model is based on force equilibrium in the critical cracked section and uses an arbitrary tension softening diagram as input. An additional relation that links the crack length parameter to the deflection needs to be determined. Due to the random fibre distribution and others uncertainties involved in concrete mix, deterministic approaches are not suitable for deriving the missing relationship for panels. The load deflection response is predicted using yield line theory based on the crack length parameter normalized-deflection diagram developed from the analysis of beams having the same thickness using Monte Carlo simulation (MCS) technique. The proposed approach is applied to a square panel supported on four corners with a comparison to an existing theory based on predefined parabolic softening law. A good agreement has been found between the model predictions and finite element calculations.

1 INTRODUCTION

The use of fibre reinforced concrete (FRC) as construction material has continuously grown over the last decades due to economical advantages in labour and material costs as compared to conventional reinforced concrete but also due to the enhanced mechanical characteristics of concrete. Randomly distributed fibres in concrete contribute to enhance concrete mechanical performance and durability when well proportioned and good quality FRC mixes used. Fibres significantly prevent shrinkage cracking, reduce brittleness due to impact loading, compression or in unreinforced members subjected to shear or tensile forces, control flexural cracking, improve water tightness, etc. For these reasons, FRC has been successfully used in shotcrete, precast concrete, elevated slabs, bridge decks, pavements, industrial floors, seismic resisting structures, repair, etc.

The post-cracking behaviour of FRC has been experimentally investigated based on a variety of tests. Although the actual post-cracking response of FRC is better defined by the stress-crack width $\sigma(w)$ relationship measured in a direct tension test, the vast majority of reported test results in the literature involve beams due to the ease of performing bending tests. Flexural tests have however two main disadvantages: the $\sigma(w)$ relationship cannot be obtained directly whereas tests on small FRC elements engender important scatters in results that are not representative of actual in situ

conditions. The later has encouraged many investigators to propose tests on square or round panels of large dimension. Despite the better representation of concrete volumes in panel tests, the actual post-cracking behaviour of FRC in tension cannot be determined satisfactory using analytical approaches by any of the proposed tests in literature unless the softening mechanism is well understood, and the level of deformation and strain softening characteristics of the material are known.

Contributions in literature dealing with the post-cracking response determination of FRC panels using simple approaches are very limited. Marti et al. (1999) developed a simple theoretical approach that accounts for the random fibre distribution for the analysis of slabs. Their proposed model is based a priori on a predefined parabolic softening relationship which simplifies considerably the derivation of load-deflection curves from yield line theory. Tran et al. (2005) proposed an interesting formulation for determining the nonlinear load-deflection response of the ASTM C-1550 round panel. Their formulation uses yield line theory based on the flexural capacity of beams of similar composition and thickness.

For the case of FRC beams, Zhang and Stang (1998) proposed an analytical formulation which provides with satisfying accuracy the load-displacement response for an arbitrary inputted tension softening diagram. They use an additional relationship derived from fracture mechanics that links the crack mouth opening displacement (CMOD) to the external moment and a crack length parameter α (Figure 1). To the authors' knowledge, similar analytical formulation is either scarce or does not exist for the case of rectangular of round FRC panels, and such additional relationship is very difficult to derive analytically.

This study is therefore aimed to propose a simple analytical formulation to investigate the post-cracking behaviour of FRC panels of rectangular or circular geometry using an arbitrary tension softening diagram as input. The formulation assumes a symmetrical or axi-symmetrical crack pattern. In this situation, a relation that links the crack length parameter α to the deflection is missing. Due to the stochastic nature of material properties, the random fibre distribution, and others uncertainties involved in concrete mixes, deterministic approaches are not suitable for deriving that additional relationship for panels. The resort to probabilistic techniques enables modeling uncertainties and analyzing their dispersion effect. In this frame, the load deflection response is predicted using yield line theory based on the crack length parameter-normalized deflection diagram developed from the analysis of beams having the same thickness using Monte Carlo simulation (MCS) technique.

The proposed approach is applied in this paper to a square panel on four corners with a comparison to Marti's model. The model performance is also checked using EPM3D constitutive model (Massicotte et al., 2007) merged at Gauss integration point in FE computer program ABAQUS (2004).

2 MODEL DERIVATION FOR FLEXURAL ANALYSIS OF FRC PANELS

There is limited published information on analytical techniques suitable for predicting the complete flexural history of panels in bending made of strain softening FRC. Classical yield line theory (Johansen, 1972) can only predict the maximum load because it assumes that the level of resistance offered by the chosen collapse mechanism remains constant over the range of deformations associated with the introduction of load. If the resistance of a component within an assumed mechanism changes as load is introduced, the work done in resisting the external load is altered and the overall capacity changes. The magnitude of the change in load capacity can be determined only if the level of deformation and strain softening characteristics of the material are known.

The model adopted in the present study is based on the yield approach. In the initial stage the behaviour is assumed elastic until the maximum stresses reach concrete tensile strength. Beyond that point the model assumes that several fictitious cracks can develop depending on the failure pattern of the analyzed panel.

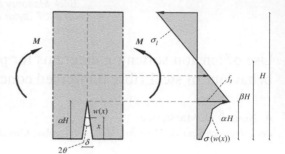

Figure 1. Stress distribution at cracked panel section.

Figure 1 depicts a typical stress distribution at a cracked section. The fictitious crack develops when the tensile stress reaches its ultimate value f_t and spreads along a part of the panel thickness H. After the crack has initiated, the fictitious crack progresses and the material is softened by cohesive forces in the fracture process zone where a nonlinear $\sigma(w)$ relationship is used.

When the crack opening displacement (COD) w reaches a critical value W_C, the stress transfer becomes zero and real crack starts to grow freely. The same cracked section proposed by Zhang and Stang (1998) for beams is adopted in the present study where it is assumed that the crack has a linear profile:

$$w = \delta\left(1 - \frac{x}{\alpha H}\right) \qquad (1)$$

for which:

$$\delta = \frac{\sum_{i=1}^{n_c} \delta_i}{n_c} \qquad (2)$$

δ is the mean CMOD, n_c is the number of cracks depending on the chosen failure pattern. Therefore, w is the mean COD at location x, and αH is the mean crack length for which the crack length parameter $\alpha \in [0, 1]$. With these assumptions, $\sigma(w)$ represents the mean softening diagram of the panel. From equilibrium conditions, we have:

$$\int_0^{\alpha H} \sigma(w(x))dx + \int_{\alpha H}^{H} \sigma_l(x)dx = 0 \qquad (3.a)$$

$$\int_0^{\alpha H} \sigma(w(x))(H-x)dx + \int_{\alpha H}^{H} \sigma_l(x)(H-x)dx = M \qquad (3.b)$$

M is the resisting moment per unit length along the yield line; $\sigma(w(x))$ and $\sigma_l(x)$ are the normal stress functions in the cracked (nonlinear) and the uncracked (linear) parts, respectively. $\sigma(w(x))$ is related to αH

and δ using the σ-w relationship together with Eq. 1. $\sigma_l(x)$ can be related to αH, βH and δ for which βH stands for the total depth of the tensile zone with $\beta \in [0.5, 1]$.

According to the principle of virtual work, one can derive a relationship between the applied load F and the generated bending moment M as follow (Johansen, 1972):

$$\sum F \cdot \Delta = \sum M \cdot \theta \cdot L \qquad (4)$$

Here Δ stands for the deflection, L is a characteristic length of the panel and θ is the corresponding crack angle of rotation (in radian) between the adjoining uncracked parts of the panel (Figure 1). By modeling the crack as a generalized plastic hinge, θ can be estimated by:

$$\theta = \frac{\delta}{\alpha \cdot H} \qquad (5)$$

This approach necessitates an additional relation linking the crack length parameter to the deflection to obtain the complete solution between the external load and the deflection. In the case of beams under three point bending, Zhang and Stang (1998) used for the required additional equation a relationship derived analytically from fracture mechanics in which δ is linked to the external moment M and the crack length parameter α. The additional relationship would be difficult to derive analytically in the case of panels for which an arbitrary σ-w softening law is used.

Marti et al. (1999) developed a theoretical approach based a priori on a well defined parabolic softening law. Such assumption simplifies considerably the derivation of load-deflection curves from the yield line theory. Tran et al. (2005) performed a series of analysis using yield line theory to derive the non-linear load-deflection response of the ASTM C-1550 round panels based on the flexural capacity of beams of similar composition and thickness. They did not explicitly use a σ-w softening relationship as an input in their approach. Instead they adopted the moment crack rotation angle diagrams (M-θ) of beams as input. According to Eq. 4, the post-cracking load-deflection curve can be determined by increasing the displacement at the centre of the panel and, using the resisting moment offered by beams for each corresponding crack rotation angle, find the load at equilibrium for these moments.

This study proposes a simple analytical formulation to investigate the post-cracking behaviour of FRC panels using an arbitrary σ-w diagram as input. The post-cracking behaviour of FRC panels is studied using yield line theory, where the load deflection response is predicted based on the crack length parameter-normalized deflection relationship developed from the

analysis of beams having the same thickness using Monte Carlo simulation (MCS) technique. This issue is described in detail in the next section.

3 CRACK LENGTH PARAMETER-NORMALIZED DEFLECTION CURVE FOR BEAMS USING MCS TECHNIQUE

3.1 Procedure

To derive the additional relation relating the crack length parameter α to the deflection Δ, one follows the idea proposed by Tran et al. (2005) where the missing information is obtained from beams which is then used for analyzing panels. Because in our study any arbitrary σ-w relationship can be inputted in the numerical analysis, the α-Δ diagrams are determined from the analysis of beams. However, since panel maximum deflections are larger than beams, the crack length parameter α is obtained as a function of the normalized displacement $\bar{\Delta}$. Here, $\bar{\Delta} = \Delta / \Delta_{max}$ where Δ_{max} is the maximum analytical beam deflection. In the beam analysis, the following σ-w relationship is chosen for its generality and versatility:

$$\sigma(w) = f_t \cdot \left(1 - \frac{w}{W_C}\right)^N \qquad (6)$$

N is the softening index. For a given value of N ($0 < N < \infty$), equation (6) covers all types of engineering materials. For instance, with $N = 0, \sigma(w) = f_t$ which describes the behaviour of all elastic perfectly plastic solids as it is used in Dugdale model. With $N = \infty, \sigma(w) = 0$, in which case Eq.(6) represents brittle materials without softening region. Expressions obtained for $0 < N < 1$, are typical of ductile metals and polymers in plane stress conditions characterized by strain hardening behaviour (Roger et al., 1986). For the materials like steel fibre reinforced concretes (SFRC), the softening index range is $1 < N < \infty$ (Ballarini and Shah, 1984).

Due to the stochastic nature of the material properties, the random fibre distribution, and others uncertainties involved in concrete mix, deterministic approaches are not suitable for deriving the α-$\bar{\Delta}$ formula for panels. The resort to probabilistic techniques enables modeling uncertainties and analyzing their dispersion effect. For this raison, a stochastic model that accounts for the randomness of the three variables (f_t, W_C and N) defining the adopted σ-w law is used. In this study, α-$\bar{\Delta}$ diagrams for panels are obtained using the analytic formulation proposed by Zhang and Stang (1998) for beams combined with the MCS technique.

3.2 Stochastic model for σ-w diagrams

Tensile strength f_t, critical crack opening displacement W_C and softening index N in Eq. 6 are modeled herein as random fields. For the random simulation of these properties, the chosen random variables are defined by their moments of order 1 and 2, which are respectively the mean, and the variance supposed in accordance with laboratory samples. Let T_p standing for one typical random variable, defined as a function of the deterministic function T_{0p} describing the trend, taken in practice as the mean of measured values, and also function of zero mean, unit variance Gaussian random field ΔT_p. One can write:

$$T_p = \Re\left[T_{0p} + \tau_p \, \Delta T_p\right] \qquad (7)$$

$\Re$ is a transformation taking the Gaussian process ΔT_p, into the distribution appropriate for T_p and τ_p is the standard deviation. Here $p = 1$ corresponds to softening index, $p = 2$ for tensile stress, and $p = 3$ for critical crack opening displacement. The zero mean, unit variance, three-variate Gaussian random field ΔT_p, can be simulated as follow:

$$\Delta T_p = \sqrt{\frac{2}{K_p}} \cdot \sum_{l=1}^{K_p} \cos(\Omega_{l,p}) \qquad (8)$$

$$mean\,(\Delta T_p) = 0; \quad p = 1,3 \qquad (9)$$

Ω_l is a random phase angle distributed uniformly over the interval $[0, 2\pi]$. K is a large enough integer. A stochastic independence between f_t and the other random variables W_C and N is assumed which is preferable than assuming any erroneous correlation. In practice, there is a direct correlation between N and W_C. Small values of N are typical of high fibre dosage leading to large values of W_C, whereas large values of N are more representative of materials with small fibre dosage and therefore more brittle material, leading to small values of W_C. For this reason, the stochastic correlation is considered between N and W_C. The three properties are simulated using the stochastic formulation proposed by Nour et al. (2002). The tensile softening index is simulated using the beta distribution whereas the tensile stress and the crack opening displacement are modeled using the lognormal distribution.

3.3 Numerical analysis for beams

The above described procedure is used to derive the crack length parameter-normalized deflection α-$\bar{\Delta}$ for beams to be used for panel analysis. Monte Carlo simulations are used to generate samples having characteristics close to specimens produced in laboratory.

It is reported in literature (MacGregor et al., 1983) that the variability of f_t is roughly of the same order as for the compression strength, which is around $CV_{f_t} \approx 0.15$ to 0.25. However Bungey and Millard (1996) indicated that CV_{f_t} could be greater than 0.4 for poor concrete and for this reason CV_{f_t} is varied in this study from 0.15 up to 0.5. For the critical crack opening displacement W_C, Bazant et al. (2002) reported that the ratio between the areas under the complete σ-w curve and under the initial tangent of this curve is in the order of 2.5 with a variation coefficient around 40%. Because the complete area of the σ-w curve is controlled by W_C, a variability up to 40% for W_C seems reasonable. For the softening index N, there is practically no available information about its variability. The variation coefficient CV_N is chosen to be equal to $0.5 \cdot CV_{N,cr}$ with $CV_{N,cr} = N_0/\tau_{N,cr}$. (Nour et al., 2002). $CV_{N,cr}$ stands for the critical variation coefficient for N, N_0 for the mean value for N, and $\tau_{N,cr}$ for the critical standard deviation for N. Hence, the following data are used:

- Mean tensile stress: $\mu_{f_t} = 3.5$ MPa;
- Mean crack opening displacement $\mu_{W_C} = 8$ mm;
- $N_{min} = 1$ and $N_{max} = 7$ with the mean value $N_0 \in [N_{min} N_{max}]$:
- The ratio $H/S = 4$ (H stands for the beam height and S for the beam clean span).

The softening index N is in direct correlation with the parameters influencing the shape of the post-cracking load-deflection response such as the type of fibres, the dosage and the mix quality. For this reason, three representative situations were considered. The first one represents specimens dominated by FRC mixes having high percentage of fibres, idealized here by samples of Monte Carlo simulations having N_0 close to N_{min} i.e. $N_0 = (3N_{min} + N_{max})/4 = 2.5$. The second situation considers specimens covering all possible percentage of fibres dosages, idealized here by samples of Monte Carlo simulations having N_0 equal to the central value i.e. $N_0 = (N_{min} + N_{max})/2 = 4$. Finally the third situation is the opposite of the first one with the majority of specimens made with low percentage of fibres, idealized here by samples of Monte Carlo simulations having N_0 close to N_{max} i.e. $N_0 = (N_{min} + 3N_{max})/4 = 5.5$.

Using Eq. 6, 1000 independent realizations of σ-w diagrams were randomly generated and were directly considered in Monte Carlo simulations of beam analysis. The result of this exercise is the required mean α-$\bar{\Delta}$ diagram to be used for panels. Figure 2 illustrates 25 typical realizations of σ-w diagrams for the case of $N_0 = 4$ for which a strong negative correlation between N and W_C was considered, i.e. $R_{W_C N} = -0.75$.

In this study, for a given beam thickness H, the required α-$\bar{\Delta}$ diagrams are determined after achieving 1000 Monte Carlo simulations for each representative situation of N_0 ($N_0 = 2.5$, 4 and 5.5). This allows

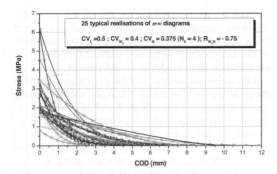

Figure 2. Typical realizations of σ-w diagrams.

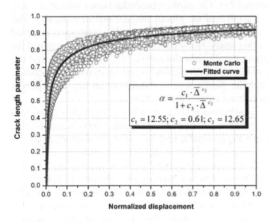

Figure 3. Crack length parameter-normalized displacement diagram for $h = 75$ mm.

covering a maximum range of fibre dosage. In reality 3000 samples of Monte Carlo simulations are superimposed all together in Figure 3. As shown on this figure, the randomness in f_t, W_C and N produces a scatter in α-$\bar{\Delta}$ results with an interesting trend. This permits to easily fit the obtained results with an appropriate function. The function given by Eq. 10 is chosen because it captures with fidelity the full trend observed in MCS results. Constants c_1, c_2 and c_3 are estimated using a nonlinear fitting scheme from the ensemble of realizations, whereas the assumed function satisfactory passed the Chi-square goodness of fit test.

$$\alpha = \frac{c_1 \cdot \bar{\Delta}^{c_2}}{1 + c_3 \cdot \bar{\Delta}^{c_2}} \tag{10}$$

Results reported in Figure 3 corresponds to $H = 75$ mm. The same routine could be easily repeated for other thicknesses.

4 APPLICATION TO A SQUARE PANEL SUPPORTED ON FOUR CORNERS

Equation 10 constitutes the additional relation that links the crack length parameter α to the deflection Δ. It is of worth to note that very limited contributions dealing with the post-cracking response for FRC panels via simple approaches are available in literature. The model proposed by Marti et al. (1999) belongs to this category; it is based on a fixed parabolic softening law and is used herein for comparison purposes. The next validations uses the following data: $H = 75$ mm and $f_t = 3$ MPa.

4.1 Comparison with Marti's model

Khaloo and Afshari (2005) performed several tests to determine the flexural strength, load-deflection curve and energy absorption of small concrete slabs. Figure 4 shows the test principle and typical crack pattern for the adopted square panel supported on four corners under a central point loading. Using yield line theory, Khaloo and Afshari (2005) estimated the resisting moment M according to Marti et al. (1999) model, and developed the equations defining the crack angle rotation θ and the load F function of the central displacement Δ and the moment M, respectively:

$$\theta = \frac{4\Delta}{b - a} \tag{11.a}$$

$$F = \frac{4(b + 2c)}{b - a} M \tag{11.b}$$

For the numerical application, one considers $b = 680$ mm, $a = 80$ and $c = 70$ mm, $N = 2$ (parabolic softening law) and $W_C = 15$ mm. Figure 4.b illustrates the comparison between the proposed approach and the adopted Khaloo and Afshari (2005) theory. In the proposed approach, three regions describe the load–deflection curves. The first region has an ascending slope covers the response form the onset of concrete cracking up to the maximum load. Marti et al. (1999) neglected the contribution of the elastic part of the sound ligament in tension. Therefore their theory predicts only the behaviour of FRC after ultimate. The second region begins at ultimate load and ends at a point where the tensile forces are resisted totally by bond between fibres and concrete. This corresponds to the steepest softening portion of the curve. In the third region, the slope of the curve reduces accompanied by an asymptotic residual load. One sees that both approaches predict practically the same ultimate value, but they exhibit different behaviour in the softening region. The results indicate that the theoretical predictions from Khaloo and Afshari (2005) are not conservative and present higher energy absorption

1537

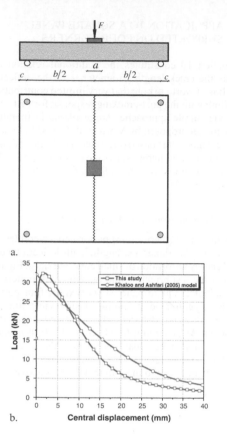

a.

b.

Figure 4. Failure mechanism and load displacement curves for a square panel on four corners.

compared to the proposed approach. Their model predicts also higher energy absorption than obtained experimentally.

4.2 Comparison with finite element method

In this section, the proposed approach is compared to the finite element method. To this end, EPM3D concrete model is used for the analysis. This model was originally developed by Bouzaiene and Massicotte (1997) and was recently merged for standard and explicit computations in ABAQUS (2004) by Ben Ftima and Massicotte (2004). It was also used for modelling concrete structures reinforced with internal and external FRP (Nour et al., 2006). In compression, the concrete model follows a three-dimensional hypoelastic approach which accounts for anisotropy and inelastic volume expansion using a compression scalar damage parameter λ (Bouzaiene and Massicotte, 1997). Degradation of material properties due to cracks propagation is described by means of a scalar parameter which enables coupling the compressive

and tension damage parameters for the tensile residual stress calculation. The model is based on smeared crack approach with cracks spread over the elements, so the analyses are performed without introducing to the finite element model any crack pattern.

Finite element analyses of FRC panels involve modeling severe nonlinearities. Strategy solution using standard computations through ABAQUS leads to serious converging difficulties resulting in a large number of iterations, which complicates exploring the post-cracking behaviour of FRC panels. In this case, the analyses are more efficient using explicit computations. Whereas ABAQUS standard must iterate to determine the solution to a nonlinear problem, ABAQUS explicit determines the solution by explicitly advancing the kinematic state from the previous increment. The explicit procedure does not require any iteration and no global tangent stiffness matrix. For these reasons, the explicit solution strategy is adopted in this study. However, a special caution is required for choosing the time duration of the analysis, because the explicit solution method is a truly dynamic procedure and the inertia forces should not in any case play a dominant role in the solution.

Furthermore, the energy balance check is employed to evaluate whether or not a simulation is in accordance with a quasi-static response. For the numerical analysis, one considers $b = 680$ mm, $a = 0$ and $c = 70$ mm. The analyses are carried out considering the bilinear and the tri-linear softening laws shown in Figure 5. In EPM3D, any polylinear softening diagram defined by a series of points $(\sigma_i, w_i; i = 1, \ldots, N_p)$ can be used for the analysis according to the following expression:

$$\sigma(w) = (1 - 1.25\lambda) \sum_{i=1}^{N_p - 1} \sigma_i + \frac{\sigma_{i+1} - \sigma_i}{w_{i+1} - w_i} (w - w_i) \quad (12.a)$$

$$w = \frac{\varepsilon}{L_{cr}} \quad (12.b)$$

Here N_p is the number of points in the softening diagram, ε is the tensile strain and L_{cr} is the critical length transforming the crack opening in tensile strain, assumed in this study equal to $0.5H$. The comparison results are clearly illustrated in Figure 5.

If one considers finite element solution as reference, one sees that the proposed method captures with fidelity the global trend in the load-deflection curve. With the proposed approach, the predicted maximum load is roughly 3% less than finite element result for the bilinear softening law and is 8.5% less for the tri-linear law; also the observed difference in post-peak is acceptable for both softening laws. The finite element model is shown in Figure 6.a along with the predicted failure mode (Figure 6.b). As shown in Figures 6.b and 6.c, the failure pattern takes the form of a straight line and a flexural hinge divides the slab into two rigid

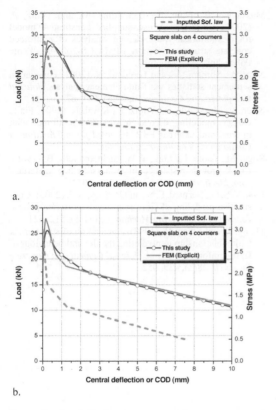

a.

b.

Figure 5. Load displacement curves for a square panel on four corners. a. bilinear softening diagram. b. tri-linear softening diagram.

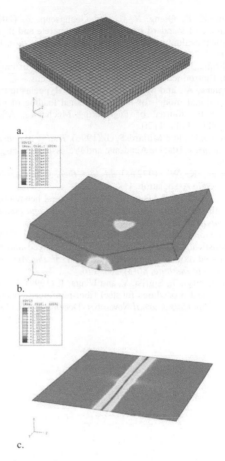

a.

b.

c.

Figure 6. Finite element model and residual tensile stress distribution. a. Finite element mesh. b. Predicted failure mode. c. Tensile stress distribution at the slab bottom surface.

parts. In term of failure mode, EPM3D predictions are in accordance with experimental observations of Khaloo and Afshari (2005). One sees that the tensile stress decreases significantly in the vicinity of the middle straight line of the slab as well as around the applied load.

Despite EPM3D is based on a smeared crack approach, it allows to determine the crack pattern for subsequent analysis of complicated slab geometries using yield line theory, and the complete load-deflection response can be easily obtained via the proposed approach.

5 CONCLUSIONS

In this paper, a semi-analytical model aimed to investigate the post-cracking behaviour of FRC panels has been presented. The load deflection response is predicted using yield line theory based on the crack length parameter-normalized deflection diagram developed from the analysis of beams having the same thickness using Monte Carlo simulation (MCS) technique. The proposed model has been applied to a square panel supported on four corners with a comparison to an existing theory based on predefined parabolic softening law. Also, a good agreement has been obtained between the model predictions and finite element calculations. This study indicates that it is possible to obtain satisfactory predictions of the post-cracking load-deflection response for panels with independently obtained experimental/analytical data for the stress-crack width relationship using this present simple model.

REFERENCES

ABAQUS. (2004). User's Manual Version 6.5. Hibbit, Karlsson & Sorensen, Inc., USA.
Ballarini, R., Shah, S. P., and Keer L. M. (1984). Crack growth in cement-based composites. *Engineering Fracture Mechanics*, 20(3), 433–445.

Bazant, Z. P., Qiang, Y. U., and Goangseup, Z. (2002). Choice of standard fracture test for concrete and its statistical evaluation. *International journal of fracture*, 118, 303–337.

Ben Ftima, M., and Massicotte, B. (2004). Rapport interne sur l'introduction du model de béton a Abaqus.

Bouzaiene, A., and Massicotte, B. 1997. Hypoelastic tridimensional model for non proportional loading of plain concrete. Journal of Engineering Mechanics, ASCE, 123(11): 1111–1120.

Bungey, J. H., and Millard, S. G. (1996). *Testing of concrete structures*. Blackie Academic and Professional, Glasgow, UK.

Johansen, K. W. (1972). *Yield line theory*, Cement and Concrete Association, U.K.

Khaloo, A. R., and Afsari M. (2005). Flexural behaviour of small steel fibre reinforced slabs. *Cement and concrete composites*, 27, 141–149.

MacGregor, J. G., Mirza, S. A., and Ellingwood, B. (1983). Statistical analysis of resistance of reinforced and prestressed concrete members. *Journal of the American Concrete Institute*, 80(3), 167–176.

Marti, P., Pfyl, T., Sigrist, V., and Ulaga, T. (1999). Harmonized test procedures for steel fiber-reinforced concrete. *ACI Materials Journal* November–December, 676–685.

Massicotte, B., Nour, A., Ben Ftima, M. and Yildiz, E. (2007). EPM3D – A user-supplied constitutive model for the nonlinear finite element analysis of reinforced concrete structures. Report EPM/GCS-2007-01, École Polytechnique de Montréal.

Nour, A., Slimani, A., and Laouami, N. (2002) Foundation settlement statistics via finite element analysis. *Journal Computers and Geotechnics*, 29(8), 641–672.

Nour, A., Massicotte, B., Yildiz, E., Koval, K. (2006) Finite element modeling of concrete structures reinforced with internal and external FRP. *Accepted for publication in Canadian civil engineering journal.*

Roger M., Foote, L., Mai., Y. W., and Cotterell., B. (1986). Crack growth resistance in strain-softening materials. *Jour. Mech. Phys. Solids*, 34(6), 593–607.

Tran, V. N. G., Bernard, E. S., and Beasley, A. J. (2005). Constitutive modeling of fibre reinforced shortcrete panels. *Journal of Engineering Mechanics*, 131(5), 512–521.

Zhang, J., and Stang, H. (1998). Application of stress crack width relationship in predicting the flexural behaviour of fibre-reinforced concrete. *Journal Cement and Concrete Research*, 28(3), 439–452.

An application of high performance fiber reinforced cementitious composites for RC beam strengthening

G. Martinola
Concretum Construction Science AG, Switzerland

A. Meda
University of Bergamo, Italy

G.A. Plizzari
University of Brescia, Italy

Z. Rinaldi
University of Roma "Tor Vergata", Italy

ABSTRACT: The possibility of using concrete materials with tensile hardening behavior (High Performance Fiber Reinforced Cementitious Composites, HPFRCC) for strengthening R/C beams is investigated. In order to verify the effectiveness of the proposed solution, full scale tests have been performed on 4.5 m long beams. A beam without any reinforcement and a beam with a low reinforcement ratio, equal to 0.03%, have been strengthened with a jacket in HPFRCC having a thickness of 40 mm. A third beam with the same low reinforcement ratio but without HPFRCC jacket, has been used as reference specimen. The obtained results show the effectiveness of the proposed technique both at ultimate and serviceability limit states.

1 INTRODUCTION

The interest for strengthening and repair applications on existing reinforced concrete structures is increased in the last few years. In addition to the well known problems of the seismic retrofitting regarding R/C structures, the strengthening of constructions can be also required by an increase of vertical loads (e.g. due to a change in the design loads or to problems in the construction phase). Moreover, it should be mentioned the urgent need of repairing structures damaged because of lack of durability. Finally, there are cases of important infrastructures, such as bridges or tunnels, that have to be necessarily repaired due to the difficulty in substituting with new structures.

Besides the traditional strengthening techniques, such as the beton-plaque or the R/C jacketing (Fib Report 1991), new solutions have been recently introduced; among these solutions, a great favor has been encountered by the application of FRP (Fiber Reinforced Polymer, Fib Bulletin 14, 2001). A new technique based on the use of high performance Fiber Reinforced Concrete (FRC) is presented herein.

In the last years the use of concretes reinforced with fibers is increased due to their enhanced properties in cracking stage (Rossi & Chanvillard, 2000, di Prisco et al., 2004). Fiber Reinforced Concrete is nowadays extensively used in applications where the fiber reinforcement is not essential for the structure safe (e.g. industrial pavements or shotcrete in tunnels). Besides these examples there are structures where the fiber reinforcement is used as total substitute of the traditional reinforcement (Falkner et al., 1997, ACI 544.4R, 1988). In particular, several studies demonstrated that fiber can be used to replace transverse reinforcement (Meda et al., 2005, Minelli et al., 2006). It has to be noticed that in all these practical applications, characterized by low fiber contents (<<1% by volume), fiber reinforced concrete exhibits a post peak softening behavior in tension.

Recently, FRC materials having a hardening behavior in tension, usually named High Performance Fiber Reinforced Cementitious Composite (HPFRCC), are available for practical uses (Li, 1993, Rossi, 1997, Rilem-Pro 49, 2006, van Mier, 2004) and allow interesting applications. As a matter of fact, the possibility of having a hardening behavior avoids brittle collapse and, as a consequence, the traditional reinforcement can be removed (Shimoyama & Uzawa, 2002; Vicenzino et al., 2005). It is possible, in this

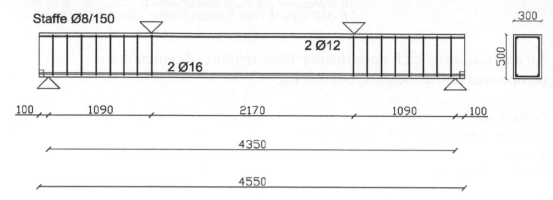

Figure 1. Geometry of the specimens.

way, to design structures with new geometries and shapes that are not any longer bounded to the reinforcement placement limitations. Unfortunately, the cost of these materials is not comparable with the traditional reinforced concrete and this is a limit to common applications. Nevertheless, particular solutions can be conveniently proposed such as suggested herein, where a HPRCC jacket is applied to existing beams.

As an example, the effectiveness of this application is studied by performing experimental flexural tests of beams with a span of 4.35 m, reinforced with a 40 mm thick jacket of a HPFRCC.

The proposed technique allows several advantages due to the easy placing procedures and due to the increased bearing capacity of the reinforced beams.

It has to be noticed that the HPFRCC jacketing is able to increase not only the ultimate bearing capacity since it remarkably enhances the behavior at serviceability limit state, increasing the stiffness of the beams. This effect is not often obtained with other solutions such as the beton-plaque or the FRP applications.

2 BEAMS GEOMETRIES

The effectiveness of the proposed strengthening technique with HPFRCC jacketing is investigated by performing full-scale experimental tests on three 4.55 mm long beams having a depth of 500 mm and a with of 300 mm (Figure 1). One of the beams was cast without any reinforcement while, in the other two beams, a longitudinal reinforcement of 2Ø16 mm bars (with the behavior shown in Figure 2) in the bottom part and 2 Ø 12 mm in the top part was placed. The bars ends were welded to steel plates in order to avoid any slip at the anchorage (Figure 3). Stirrups were placed at the beams ends in order to avoid shear failure.

The beams were cast with a C20/25 concrete grade. Such low resistance, with the low reinforcement percentage (0.3%), was chosen to highlight the strengthening effectiveness.

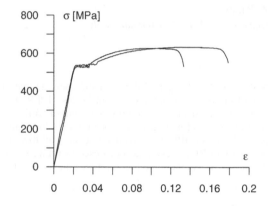

Figure 2. Constitutive relationship of the steel rebars (Ø 16).

Figure 3. Detail of the steel rebars anchorage.

One of the reinforced beams was used as the reference specimen while a 40 mm thick layer of HPFRCC was applied on the other two beams, as shown in Figure 4.

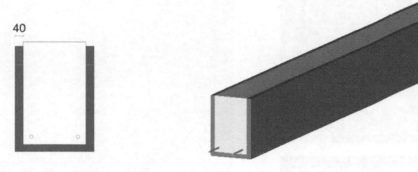

Figure 4. Strengthening scheme.

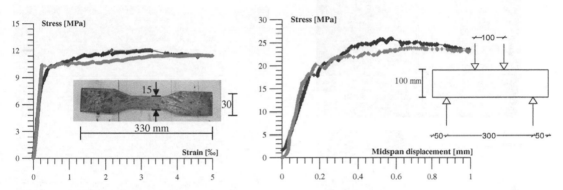

Figure 5. Material characterization: (a) direct tensile test on dog-bone specimen (thickness = 13 mm); (b) flexural test (100 × 100 mm cross-section).

The strengthening material is a concrete reinforced with a volume fraction of 2.5% of straight steel fibers having a length of 12 mm and a diameter of 0.18 mm.

Direct tensile tests on dog-bone specimens and bending tests on small un-notched beams were performed in order to determine the material properties. The results of the tests together with the specimen geometries are reported in Figure 5.

The compressive strength measured on cubes having a side of 100 mm was 176.8 MPa.

3 JACKET APPLICATION

A preliminary investigation was carried out in order to define the procedure for the application of the HPFRCC strengthening layer. Particular attention was devoted to the optimization of the adhesion between the base concrete and the new material.

To this aim, a first series of tests was performed on 150 × 150 × 600 mm specimens made with the same concrete used for the full-scale beams. After a sandblasting of the surface, a layer of 40 mm of HPFRCC material was cast.

The adhesion between HPFRCC and existing concrete was verified by performing four point bending tests on the beam specimens (Figure 6); after the experiments, no slip was noted between the two materials. Afterwards, the technology of the HPFRCC layer application was defined and the strengthening jacket was applied on the full scale beams after a sandblasting that allowed obtaining a roughness of 1–2 mm (Figure 7), considered enough to avoid the use of primer products.

The HPFRCC material was prepared in mixers and placed without any vibration. After placing a plastic layer on the surface to limit the evaporation of water from the specimen, curing at ambient temperature and humidity was carried out.

4 TESTS DESCRIPTION

The full-scale beams were tested under flexure with a four point bending set-up. The beams were placed on a 4.35 m span and loaded in two points located at a distance of 1.09 m from the supports (shear length equal to 2.4), as shown in Figures 1 and 9.

Figure 6. Preliminary tests for the evaluation of the adhesion.

The tests were performed under displacement control by adopting a 1000 kN electromechanical jack. The jack was placed under the beam with a ties system and the applied load was measured with load cells placed on the ties (Figure 10).

Figure 11 shows the potentiometer and LVDT transducers adopted for measuring the vertical displacement and the horizontal deformation (mainly the crack opening).

5 RESULTS

5.1 R/C beam without HPFRCC jacket

Firstly, the test on the R/C beam without the HPFRCC strengthening jacket has been carried out. Figure 12

Figure 7. Sandblasting of the surfaces.

Figure 8. Casting of the HPFRCC layer.

shows the total applied load versus midspan displacement curve (the displacement does not include the support settlement as well as the contribution of the self weight of the loading frame).

Figure 9.   Testing set-up.

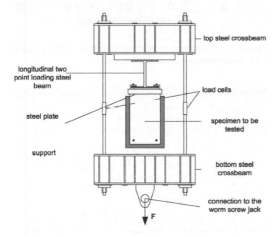

Figure 10.   Loading system.

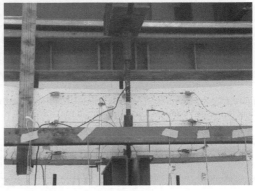

Figure 11.   Transducer positions.

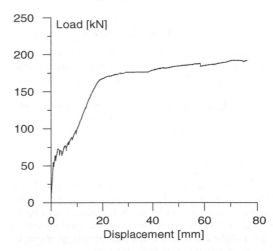

Figure 12.   Reference beam (R/C): curve of the load versus midspan displacement.

When the load reached a value of 50 kN a first crack occurred. Afterwards, other cracks developed in the zone between the two point loads, with a crack spacing ranging between 300 and 400 mm. The crack depth was around 430 mm; as a consequence, the width of the compressive chord was 70 mm.

After the complete crack development the load still increased with a lower stiffness up to yielding of the longitudinal reinforcement; this occurred when the load was equal to 190 kN.

The collapse was characterized by the development of an arch mechanism. The main experimental observations are the following:

– starting from the bottom part of every crack, at reinforcement level, a splitting crack developed and led to a loss of bond between longitudinal bars and concrete (Figure 13a);
– due to the bond failure, the stiffness of the beam remarkably decreased with great deformations;

– at the upper compressive chord, horizontal cracks developed, defining the arch mechanism geometry (Figure 13b).

Figure 14 shows the final crack pattern at collapse.

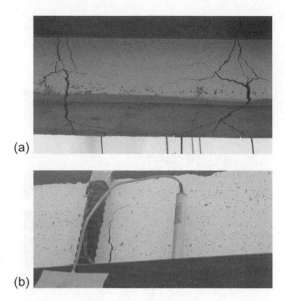

(a)

(b)

Figure 13. Un-reinforced beam: cracks distribution close to the collapse: (a) bottom side; (b) upper side.

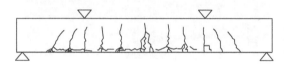

Figure 14. Final crack pattern at collapse of the reference beam (reinforced concrete) without jacket.

5.2 Plain beam with HPFRCC jacket

The load versus midspan displacement curve for the beam without the longitudinal reinforcement, strengthened with the HPFRCC jacket is shown in Figure 15.

The collapse occurred with a brittle behavior when the load was equal to 258 kN. A single main crack developed close to the midspan (Figure 16).

5.3 R/C beam with HPFRCC jacket

The behavior of the R/C beam strengthened with the HPFRCC jacket is shown in Figure 17. The beam exhibited the same behavior of the plain beam up to a load of 250 kN. After this level, the load increased with a slight stiffness due to the development of cracks spaced at a distance of 500–600 mm.

The maximum load was equal to 410 kN; afterwards, the load decreased and stabilized at a level of 260 kN, with the development of a single crack under the point load (Fig. 18).

Beam collapse occurred because of the yielding of the longitudinal reinforcement, with the crack pattern shown in Figure 19.

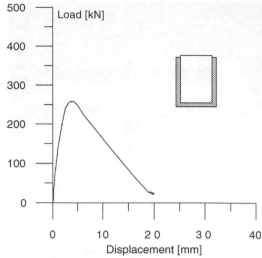

Figure 15. Plain concrete beam with HPFRCC jacket: curve of the load versus midspan displacement.

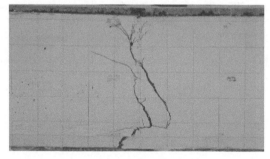

Figure 16. Plain concrete beam with HPFRCC jacket: main crack at failure.

6 DISCUSSION OF THE RESULTS

The comparison between the experimental results obtained from the three full-scale beams is shown in Figure 20.

For the sake of brevity, in the following the discussion will be focalized on two R/C beams with and without strengthening jacket. It can be noticed as the HPFRCC use allows increasing the bearing capacity of the beam (2.15 times), even if the post peak behavior becomes softening. In any case, at the end of the softening branch the load stabilizes with a plastic branch, with a value higher than that obtained in the R/C beam without jacket.

During strengthening design of an existing structures, designers should refer to both Ultimate and Serviceability Limit States. As an example, for beam elements under flexure, an inadequate bearing

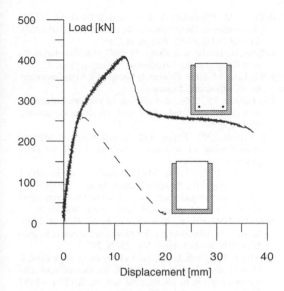

Figure 17. R/C beam with HPFRCC jacket: curve of the load versus midspan displacement.

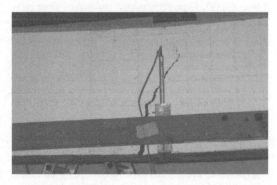

Figure 18. R/C HPFRCC jacketed beam: localization of the main crack.

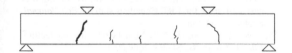

Figure 19. R/C beam with HPFRCC jacket: crack pattern at failure.

capacity leads also to significant displacements and great crack openings.

As far as the service conditions are concerned, the proposed technique allows to remarkably increase the beam stiffness, with a behavior similar to the uncracked stage of the existing (before strengthening) beam (Figure 21). Indeed, the HPFRCC jacket limits the development of macrocracking with evident advantages in terms of stiffness.

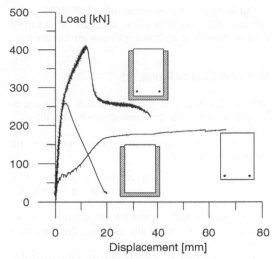

Figure 20. Comparison between the experimental results.

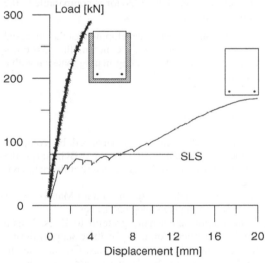

Figure 21. Serviceability limit state behavior of R/C beams with and without HPFRCC jacket.

By assuming a service load of 80 kN, the use of the HPFRCC jacket leads to a decrease of the midspan displacement from 6 mm to 0.5 mm (i.e. about 12 times lower; Fig. 21). This effect is comparable to that obtained with an external prestressing, where the cracking is avoided with a fully compressed section. From the technological point of view, the use of a HPFRCC jacket can be easily proposed as a convenient alternative to the use of external prestressing cables.

It should also be remarked that the use of a FRP strengthening does not limit cracking phenomena and, as a consequence, it cannot increase the beam stiffness.

Furthermore, the adoption of a HPFRCC jacket enhances the durability of the structure, as it can be used as protective layer in aggressive environment.

7 CONCLUDING REMARKS

The possible use of HPFRCC materials for strengthening R/C beams has been investigated with full scale applications.

The following remarks can be drawn from the result obtained in this study:

- a simple sandblasting ensures a perfect bond between the base concrete material and the strengthening HPFRCC layer;
- the application of a layer of HPFRCC having a thickness of 40 mm remarkably increases the maximum load (more than double);
- the strengthening layer has provided a remarkable stiffness increase; as a consequence the midspan displacement at service conditions has been reduced of about 12 times. This behavior is comparable to the application of prestressing.

Finally, an interesting development of the proposed technique, now under study, concerns the possibility of the application of the strengthening material with a spray technique.

ACKNOWLEDGEMENTS

The presented research was financed by company Tecnochem Italiana S.p.a. (Barzana, BG, Italy); the support of Mr. Dario Rosignoli is gratefully acknowledged.

A special thank goes to Eng. Laura Maisto for her assistance on the technological aspects.

The authors are finally grateful to Engs. Fausto Minelli and Luca Cominoli for their support during the experimental tests, and to Engs. Cristina Zanotti and Nicola Rossini for the work carried out during their graduation thesis.

REFERENCES

ACI 544.4R Design consideration for steel fiber reinforced concrete (Reported by ACI Committee 544), *ACI Structural Journal* 85(5), 1988.

di Prisco, M., Plizzari, G. A. & Felicetti. R. 2004. *6th RILEM Symposium on Fibre Reinforced Concretes* (BEFIB 2004), RILEM Publications, Bagneaux France.

Falkner, H., Henke, V. & Hinke, U. 1997. Stahlfaserbeton für tiefe Baugruben im Grundwasser, Bauingenieur, 72.

Fib bulletin 14. 2001. Externally bonded FRP reinforcement for RC structures.

Fib bulletin 35. 2006. Retrofitting of concrete structures by externally bonded FRPs, with emphasis on seismic applications.

Fib Report. 1991. Repair and strengthening of concrete strengthening of concrete structures. Guide to good practise.

Li, V. C. 1993. From Micromechanics to Structural Engineering – the design of cementitious composites for civil engineering applications. *JSCE Journal of Structural Mechanics and Earthquakes Engineering*, Vol. 10, N. 2.

Meda, A., Minelli, F., Plizzari, G. A. & Riva, P. 2005. Shear behaviour of steel fiber reinforced concrete beams. *Materials and Structures*. Vol. 38. N. 277.

Minelli, F., Cominoli, L., Meda, A., Plizzari, G. A. & Riva, P. 2006. Full-scale tests on HPSFR prestressed roof elements subjected to longitudinal flexure. RILEM – PRO 49 *International Rilem Workshop on High performance fiber reinforced cementitious composites (HPFRCC) in structural applications*. Rilem Publications S.A.R.L.

Rossi, P. 1997. High Performance multimodal fiber reinforced fibre reinforced cement composite (HPMFRCC): the LPC experience. *ACI Materials Journal*, Vol. 94, N. 6.

Rossi, P. & Chanvillard, G. 2000. *5th RILEM Symposium on Fibre Reinforced Concretes* (BEFIB 2000), RILEM Publications, Cachan France.

RILEM – PRO 49 2006. *International Rilem Workshop on High performance fiber reinforced cementitious composites (HPFRCC) in structural applications*. Rilem Publications S.A.R.L.

Shimoyama, Y. & Uzawa, M. 2002. Taiheiyo Cement Kenkyu Hokoku. *Journal of the Taiheiyo Cement Corporation*, Japan, no. 142, pp. 55–62.

van Mier, J. G. M. 2004. Cementitious composites with high tensile strength and ductility through hybrid fibres. *6th RILEM Symposium on Fibre-Reinforced Concretes*, BEFIB 2004, 20–22 Settembre, Varenna.

Vicenzino, E., Culhman, G., Perry, V. H.,. Zakariasen, D. & Chow, T. S. 2005. The first use of UHPFRC in thin precast roof shell for LRT Canadian station. *PCI Journal*. September–October.

Walraven, J. C. 2006. The New Model Code Keynote lecture. *Fib 2nd Congress*, Naples, June 2006.

Fracture Mechanics of Concrete and Concrete Structures – High-Performance Concrete,
Brick-Masonry and Environmental Aspects – Carpinteri, et al. (eds)
© 2007 Taylor & Francis Group, London, ISBN 978-0-415-44617-4

On the fracture behavior of thin-walled SFRC roof elements

M. di Prisco & D. Dozio
Politecnico di Milano, Milano, Italy

B. Belletti
University of Parma, Parma, Italy

ABSTRACT: Several experimental investigations aimed at the analysis of the structural behaviour of thin walled steel fibre reinforced elements highlighted the effectiveness of design rules suggested in standards, recently proposed for the prediction of serviceability and ultimate limit states, but also a significant interaction between the longitudinal and the crosswise bending of such elements at failure. The interaction due to second order effects is caused by both the open cross-section shape of the prefabricated elements and the reduced thickness of the profiles. The tests correspond to a large scale tests of a strip plate 1 m wide and 30 m long subjected to an eccentric tension. The shape loss of the cross section drastically anticipated the longitudinal bending collapse. A model based on the equilibrium of a segment assumed perfectly stiff in the cross section plane and a Finite Element approach are compared to explain the onset of the collapse. Very precious indications are pointed out in order to realize what happens in this large scale fracture mechanical test.

1 INTRODUCTION

Fibre reinforced concrete is a promising material when introduced in thin walled structures: a recent research programme carried out by an Italian factory, was aimed at the substitution in prefabricated roof elements of steel welded mesh fabric, made of high bond wires usually adopted as transverse reinforcement, with steel fibres (di Prisco & Felicetti, 1999; di Prisco et al. 2003). The main advantages are related to the thickness reduction connected to the overcoming of cover limitations, the increase of freedom in specimen shape and geometry, that are no longer constrained by reinforcement detailing, and the major industrialization degree reachable in the production process, due to the exclusion of transverse reinforcement handling and placing. The diffused reinforcement is much more well spread and continuously adjustable with respect to steel mesh, limiting the positioning tolerances which can significantly reduce the crosswise bearing of thin webs. The introduction of this material suggests the designer to make the section profile thinner and thinner, thus emphasizing all the problems typical of the high performance materials like instability (di Prisco & Dozio, 2005). The aim of this research is to show how the loads acting on open cross section thin walled elements reproducing snow distribution on their extrados induce an eccentric tension on the horizontal bottom plate. The huge size of the plate (30 m × 1 m) subjected to a

constant axial force and a variable crosswise bending moment due to vertical load distribution and second order effects can be instrumental in the suggestion of a proper proportioning valid for large thin walled plates made of fibre reinforced concrete. The main point is related to the right design ratio between the ultimate and the cracking resistant bending moment in the crosswise direction.

2 STRUCTURAL TESTS

Three simply supported prefabricated roof elements 30 m long, 2.5 m wide and 1 m deep made of high performance concrete were loaded by means of steel bundles. The reference element, made of conventional R/C, was compared with two Steel Fibre Reinforced Concrete (SFRC) elements. Both the SFRC elements were reinforced with 50 kg/m^3 of steel fibres: low and high carbon steel fibres were respectively introduced in the same concrete mix. The total weight of each element was about 24 t; it was characterized by only one closed end. The prestressed reinforcement consisted of 20 × 0.5″ strands.

The prestressed strands were introduced only in the bottom chords (Fig. 1a), where a significant fire protection due to thick covers can be achieved.

Two square welded meshes (2 × 1ø7/150 mm) located in the wings and in the horizontal slab,

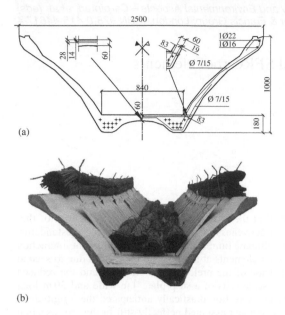

(a)

(b)

Figure 1. (a) Geometry and structural reinforcement of prefabricated roof elements, (b) loaded cross section.

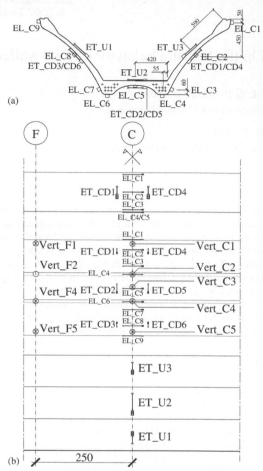

Figure 2. Instrumental equipment in the central segment: (a) cross section view; (b) longitudinal projections.

guaranteed a geometrical reinforcement ratio of about 8.5‰, in the crosswise direction, in the middle of the slab and 6.2‰ at the bottom of the wings. Each element was widely instrumented in the central section: in fact, the midspan zone was free of any steel bundle for about 1 m. In this section five vertical transducers were located to measure the vertical displacements of the bottom and top chords (Vert_C1, C2, C4, C5; Fig. 2a) and one in the middle, at the intrados of the horizontal slab (Vert_C3). Nine resistive transducers measured the longitudinal strains on a gauge length of 500 mm (EL_C1 ÷ C9; Fig. 2b) and six resistive transducers measured the transversal strains in two sections t 350 mm from the middle section, at the intrados, on a gauge length of 200 mm (ET_CD1 ÷ CD6; Fig. 2b), while further three resistive transducers measured the transverse strain in the middle section at the extrados, always on a gauge length of 200 mm (ET_U1-U3; Fig. 2b). The instrumental set-up was completed by the measure of the relative displacement between the top chords by means of a suitable potentiometer.

Four further sections were instrumented to measure the vertical displacements at the support and at the span quarters. Finally, one section was instrumented to measure the vertical displacements of the top chords, in order to control the expected instability of the inclined wings at a distance of 2.5 m from the central section. The instrumental equipment consisted of potentiometer transducers connected via a multiplexer to an Analog/Digital card directly inserted in a PC. The vertical displacement can be affected by

a maximum precision error of about 0.1%. By contrast, the longitudinal strains were measured by means of resistive transducers with a lower precision (linearity error less than 0.3%; gauge stroke = 10 mm), but the small values recorded justify the choice made. Each steel bar bundle had a weight in the range of about 9–23 kN and was characterized by an eccentricity with respect to the barycentre of the cross section. Due to the long span of these elements, two bundles (each 12 m long) were located on the extrados in the same position on the right and on the left side of the middle section, and for each bundle the eccentricity was measured (with an approximation of about 5 cm). More detailed information can be found in the technical report, where some sliding shear experimental data is also shown (di Prisco & Trintinaglia, 2002).

In the experimental investigation the steel bundles were maintained on the top chords by six suitable steel

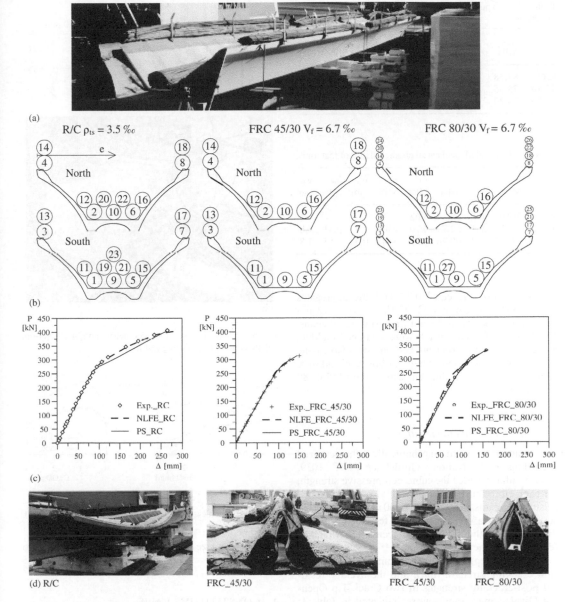

Figure 3. Structural tests: (a) typical loading situation view; (b) steel bundle location, (c) total load vs. midspan deflection measured with bottom LVDTs; (d) collapse views.

forks (Fig. 3a). The symmetric load condition is here considered up to collapse, even if at the serviceability conditions the structural element was also loaded with a weak torsion in the test due to bundle eccentricities. In the last step, the vertical displacement (C5) was removed to prevent the transducer failure.

3 MATERIAL CHARACTERISATION

3.1 *Mix design*

Three materials were analysed. They had similar HPC matrix ($R_{cm} = 73$–82 MPa), characterized by 380 kg/m^3 of cement I 525 and 60 kg/m^3 of fly ash.

Table 1. Experimental mechanical characteristics of materials.

	R_{cm} MPa	$f_{lf,m}$ MPa	$f_{eq0-0.6\,m}$ MPa	$f_{eq0.6-3\,m}$ MPa	f_{yk} MPa	f_{ptk} MPa
R/C	82.58	–	–	–	500	1860
45/30	75.65	5.22	5.44	2.80	–	1860
80/30	73.20	5.22	7.56	8.12	–	1860

Table 2. Computed mechanical characteristics of materials.

	E_c MPa	ν	f_c MPa	f_{ct} MPa	σ_a MPa	w_a mm	σ_b MPa	w_b mm
R/C	39193	0.2	68.54	5.02	–	–	–	–
45/30	38176	0.2	62.79	4.70	2.45	0.3	0.31	1.8
80/30	37801	0.2	60.76	4.70	3.40	0.3	2.55	1.8

The w/b ratio was equal 0.34 and the aggregates (120 kg/m³ of 0–3 mm sand, 970 kg/m³ of 0–12 mm sand and 815 kg/m³ 8–15 mm gravel) were siliceous. Two fibre reinforced concrete were obtained by adding 50 kg/m³ of 30 mm long, hooked-end steel fibres: the fibres were made of low or high-carbon steel and were characterized by two different aspect ratio (45 or 80 respectively).

3.2 Mechanical characteristics

The materials were mechanically characterized according to the National Guidelines CNR 11039. In particular, besides the cubic compressive strength, for each fibre reinforced material, also three notched prismatic beams (150 × 150 × 600 mm) were tested ac cording to a 4 point bending test set up, taking the Crack Mouth Opening Displacement (CMOD) as feedback parameter.

The notch/depth ratio of the specimens was 0.3. The first cracking strength (f_{lfm}), and the average residual post-cracking strengths in two Crack Tip Opening Displacement ranges were indicated in Table 1: the serviceability (CTOD = 0–0.6 mm; $\sigma_{0-0.6}$) and the ultimate average values (CTOD = 0.6–3 mm; $\sigma_{0.6-3}$) are both there shown. Finally, the yielding and the ultimate strengths of steel for welded mesh and pre-stressing reinforcement are indicated. According to Model Code and CNR DT-204, other basic mechanical parameters were computed (Table 2). It is worth to note that the same FRC materials were carefully investigated also by means of an extensive experimental programme aimed at the characterization in uniaxial tension, eccentric tension and simple bending of unnotched prismatic thin specimens, taking into

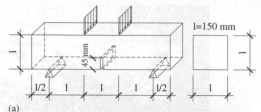

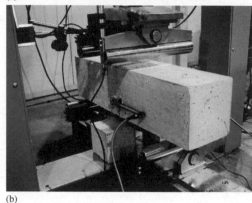

(a)

(b)

Figure 4. UNI test: (a) geometry and test set-up; (b) specimen image during testing.

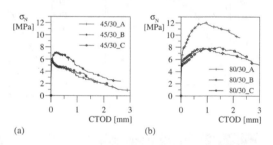

(a)

(b)

Figure 5. Load vs. CTOD for UNI tests: (a) 45/30; (b) 80/30.

account also size effect in the range 1:6.25 (di Prisco et al. 2004).

4 CONSTITUTIVE LAWS

The modelling adopted, that uses both the plane section model and the shell elements in a Finite Element approach, requires only uniaxial constitutive laws. In fact, in the F.E. approach, the biaxial state of stress is described by means of two equivalent uniaxial states. If 1-2 denote the principal directions of strains, the strength values, corresponding to the "equivalent" uniaxial states, were determined through the analytical biaxial strength envelope suggested by Kupfer (Belletti et al. 2001). Depending on the uniaxial stress-strain curve, the secant moduli E_1 and E_2 are determined and the stiffness matrix for the orthotropic material takes the following form:

$$\left[D_{1,2}^c\right]=\frac{1}{1-v^2}\begin{bmatrix} E_1 & v\sqrt{E_1E_2} & 0 \\ v\sqrt{E_1E_2} & E_2 & 0 \\ 0 & 0 & \frac{1}{4}\left(E_1+E_2-2v\sqrt{E_1E_2}\right) \end{bmatrix} \quad (1)$$

Uncracked reinforced concrete stiffness matrix is obtained by adding to concrete stiffness matrix $\left[D_{1,2}^c\right]$ the contribution of i-th steel bars, assumed smeared and inclined by an angle $\varphi_i=\theta_i-\psi$, where θ, ψ indicate the steel axis and the 1 principal direction in relation to x axis, respectively, with respect to 1 axis:

$$\left[D_{1,2}\right]=\left[D_{1,2}^c\right]+\left[D_{1,2}^s\right]=\left[D_{1,2}^c\right]+\sum_{i=1}^{n}\left[T_{(\varphi i)}^t\right]\left[D_{si}\right]\left[T_{(\varphi i)}\right] \quad (2)$$

boging:

$$\left[T_{(\varphi_i)}\right]=\begin{bmatrix} c_i^2 & s_i^2 & c_is_i \\ s_i^2 & c_i^2 & -c_is_i \\ -2c_is_i & 2c_is_i & c_i^2-s_i^2 \end{bmatrix} \qquad \left[D_{si}\right]=\begin{bmatrix} \rho_i E_{si} & 0 & 0 \\ 0 & 0 & 0 \\ 0 & 0 & 0 \end{bmatrix}$$

$$c_i=\cos\varphi_i \quad s_i=\sin\varphi_i$$

When the principal tensile stress, calculated at the integration point, becomes greater than the concrete tensile strength, crack occurs and the procedure adopts the stiffness matrix for cracked reinforced concrete, having orthotropic axes coincident with the direction perpendicular, 1, and parallel, 2, to the fixed crack. The strain field is expressed as a function of three fundamental variables: the crack width w, the crack slip v and the concrete axial strain ε_{c2} in the direction parallel to the crack. The strain vector, expressed in the orthotropic co-ordinate system, takes the following form:

$$\left\{\varepsilon_{1,2}\right\}=\left\{\varepsilon_1 \quad \varepsilon_2 \quad \gamma_{12}\right\}^t=\left\{\frac{w}{a_m} \quad \varepsilon_{c2} \quad \frac{v}{a_m}\right\}^t$$

a_m being the crack spacing, assumed constant and depending on rebar spacing.

The stiffness matrix of cracked reinforced concrete takes into account all the fundamental phenomena occurring after cracking, such as tension stiffening, dowel action, bridging effect, confinement and aggregate interlock (Belletti et al. 2001). The constitutive relationship adopted for steel is represented by the bilateral elastic-hardening curve, while for concrete in compression the softened Thorenfeld curve has been implemented. The stiffness matrix, both for uncracked and cracked concrete, is referred to the 1-2 orthotropic coordinate system, and a fixed crack approach is followed.

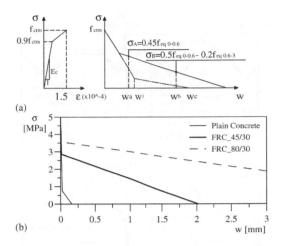

(a)

(b)

Figure 6. Uniaxial tension constitutive relationships: (a) general description; (b) σ-w curves adopted for the materials.

Under loading, if the minor principal moment reaches cracking moment values (i.e. as happens in the wings and in the bottom slab for the analysed precast element), two sets of primary and secondary cracks may occur at the integration point. Following the same approach proposed in (Belletti et al. 2003), in R/C case doubly cracked concrete is assumed to have zero secant Young's modulus in the direction parallel to the first crack, after the peak, while reduced softening is taken into account for FRC cases.

4.1 Constitutive relationships for the material

The uniaxial constitutive laws assumed for plain concrete material were those suggested in Model Code: Sargin equation for uniaxial compression and the trilinear relation for uniaxial tension (Fig. 6a). According to Model Code, the values w_1 and w_c are function of fracture energy that is related to the maximum aggregate size and the cylindrical compressive strength: the computed values are respectively 21 and 161 μm.

FRC materials adopt the same constitutive law in uniaxial compression due to the small fibre content, while in uniaxial tension only the final branch is substituted according to CNR DT-204 (di Prisco et al., 2004a,b; Fig. 6a) in order to suitably take into account pull-out mechanism. The significant increase of dissipated energy due to fiber addition is immediately perceived by observing the softening laws (Fig. 6b). According to the experimental data, there is a significant difference between the two types of steel fibre.

4.2 Generalized constitutive relationships

According to plane section model, that is the smeared version of the multi-layer procedure first introduced by

1553

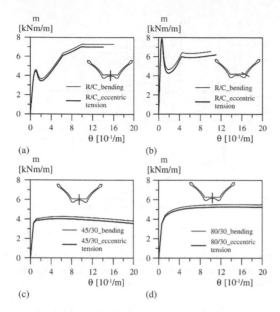

(a)

(b)

(c)

(d)

Figure 7. Bending moment vs. average curvature: current longitudinal cross section (width = 1 m): (a) R/C along the symmetric plane; (b) R/C at the inclined wing – bottom slab connection; (c) SFRC 45/30 along the symmetric plane; (d) SFRC 80/30 along the symmetric plane.

Hordjik (1991; see also di Prisco et al. 2001), the bending moment vs. curvature curve for the softened region in R/C and FRC thin walled plates was computed. The structural characteristic length l_c, representing the size of the softening region, was assumed equal to the crack distance and the plate thickness for R/C and FRC elements respectively.

The results (Fig. 7) highlight as R/C element is characterized by a reinforcement ratio larger than the minimum in the central section, but not at the connection between the wing and the bottom chord. On the contrary, SFRC bending behaviour appears almost elasto-plastic, even though sometimes this is not enough to guarantee a multi crack pattern. The comparison with the results of a previous bending characterization of the same material on unnotched plates (di Prisco et al, 2004a) gives a reliable fitting.

5 STRUCTURAL MODELLING

The anticipated failure of the structural tests presented were predicted by investigating the crosswise bending with a second-order approach. According to this approach (di Prisco et al. 1998, 2002), that regards each longitudinal fibre as a wire, an additive distributed load can be computed by multiplying the longitudinal curvature times the longitudinal stress acting on the cross section (Fig. 8a) Each static contribution is evaluated

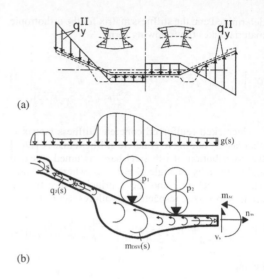

(a)

(b)

Figure 8. Second-order plane-section model [di Prisco et al. 2002]: (a) transverse bending introduced by second-order distributed loads for thin-walled open sections and (b) equilibrium according to Vlasov approach.

on the basis of the equilibrium of a beam longitudinal segment, by assuming that the projection of the cross-section cannot deform itself in its plane. Therefore, besides the external load contribution offered by self-weight and steel bundles (Fig. 8b), second-order effects associated to the main longitudinal curvature are also taken into account.

The superposition is strictly valid up to first cracking. The longitudinal stresses caused by longitudinal bending can be evaluated also for cracked section according to a plane-section model and a perfect compatibility between concrete and steel ($\varepsilon_s = \varepsilon_c$); in this way the main mechanical non-linearity can be taken into account. The shear flows $q_J(s)$ are distributed according to Jourawski's theory; the specific torque $m_{DSV}(s)$, is computed according to circular torsion owing to the large values of the warping factor (Vlasov, 1961), estimated only by taking into account the uncracked portions of the cross-section. The equilibrium allows us also to compute the shear (v_s) and the axial force (n_{ss}) along the wing profile (Fig. 8b).

The total crosswise bending moment (m_{sz}) can be compared with the resistant one. With reference to F.E. approach, due to the symmetry in the geometry, as well as in the loading along the cross-section and support conditions, only half of the cross section was modelled for the entire length of the element (Fig. 9a). The support condition (Fig. 9b) was modelled through spring elements having no tension behaviour (Fig. 9c). Rebar layers, embedded in the "host" shell elements, were used in order to model strands. The prestressing action in the rebars was defined as a given initial condition.

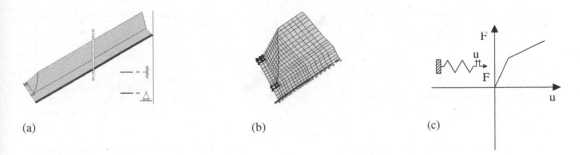

(a)   (b)   (c)

Figure 9.   (a) Mesh adopted for NLFEA; (b) a zoom of the closed end; (c) no-tension spring element adopted as support condition.

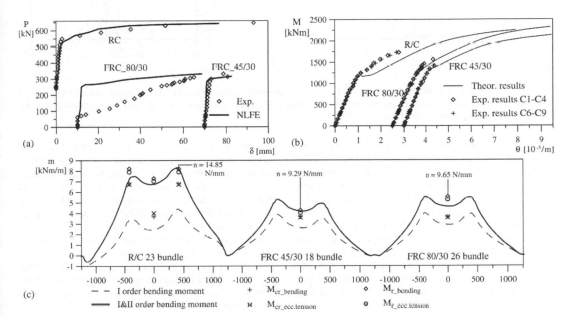

Figure 10.   Longitudinal and transverse behaviour of the three prefabricated structures: (a) load vs. transverse opening of the wings; (b) bending moment vs. longitudinal curvature; (c) transverse bending moment along the curvilinear abscissa.

In F.E. modelling, the related distributed load was concentrated in six loads acting on each top chord, while in the simple model these bundles were regarded as longitudinally distributed load strips according to the eccentricities experimentally measured, and therefore, along the cross-section profile each bundle was regarded as a specific concentrated load.

The F.E. prediction expressed in terms of load vs. crosswise opening of the wings measured in the midspan fits very well the experimental curves (Fig. 10a); only the experimental behaviour of 80/30 FRC element seems be affected by a previous crack pattern that can be expected in such a long element. This defect transforms a predicted abrupt opening of the inclined wing in a progressive phenomenon, without changing the asymptotic load value. Furthermore,

the good fitting of the crosswise cracking in the other two cases was obtained by a reduction of 15% of the average peak strength, that can be associated to the statistical strength reduction due to small defects distributed in the central zone. The plane section model clarifies the significant reduction in terms of longitudinal bending moment, due to the crosswise failure of the section (Fig. 10b), although the lack of tension stiffening effects in the model causes larger curvatures after cracking in the prediction. The crosswise equilibrium analysis of the section at the last experimental load step (Fig. 10c) highlights the reliability of this design approach as well as the key role of second order effects in the crosswise bending.

The deformed shape of R/C element is very well reproduced by F.E. modelling (Figs. 11a,b), but also

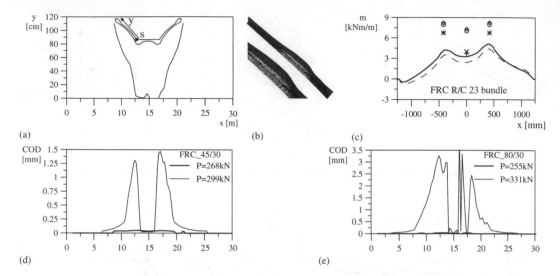

Figure 11. (a) fracture profile in the wing measured along the longitudinal axis; (b) F.E. deformed shape at failure; (c) transverse bending moment in a section at 9 m from the support; F.E. approach: crack opening along the longitudinal axis for (d) 45/30 and (e) 80/30 SFRC elements.

an equilibrium analysis of a cross section performed by means of the plane section model at the border of the fracture profile, at 9 m from the supports, shows the strong reduction of the bending moment crosswisely acting on the wing/bottom chord connection, moving from the midspan towards the supports. Second order effects are also highlighted by the crack opening of the extrados bottom plate measured along the longitudinal axis (Figs. 11d,e). Also in this case, the crack propagation from the centre towards the supports emphasizes second order effects, taking into account also the lack of steel bundles in the midspan for about a 1 m long segment.

both the matrix and the residual strengths. The plane section model gives a reliable instrument to capture second order effects, even it could be weakly unsafe, because it does not take into account the limited deformability in the cross section plane. The strong reduction of the longitudinal bending moment capacity computed with reference to the average strength values (from 23% for R/C element up to 38% for 45/30 FRC element!), highlights the need for such element of higher crosswise bending resistance in the central regions, attainable with larger reinforcement ratios, increased fibre contents or special ties and suitable closed ends.

6 CONCLUSIONS

Finite Element modelling of R/C and FRC structural elements is able to predict the deformation and the failure mechanisms, experimentally observed, up to the last load-step; the only exception concerns 80/30 FRC element, where an existing longitudinal crack anticipated the wing opening without significant changes on the ultimate load, due to tensile residual strength guaranteed by fibre pull-out. Second order effects significantly affect these structural elements: the very brittle failure observed was also emphasized by the lack of one closed end, that prevents an alternative equilibrium condition due to torsional resistance. The good fitting of the structural behaviour for R/C and 45/30 FRC elements confirms the reliability of the constitutive relationships chosen for the materials and suggests quite negligible influence of size effect on

REFERENCES

Belletti, B., Cerioni, R. & Iori, I. 2001. A Physical Approach for Reinforced Concrete (PARC) membrane elements. *J. Struct. Engrg., ASCE*, 127 (12), 1412–1426.

Belletti, B., Cerioni, R. & Iori, I. 2002. Theoretical and Experimental Analyses of Precast Prestressed Concrete Roof Elements for Large Span. *First fib Congress, Osaka*, on CD ROM.

Belletti, B. 2002. Nonlinear FE analysis of precast prestressed concrete roof elements (in italian). *Proc. of 14° Congress CTE, Mantova*, 13–22.

Belletti, B., Bernardi, P., Cerioni, R. & Iori, I. 2003. Nonlinear analysis of reinforced concrete slabs. *Proceedings of ISEC-02 Conference*, 663–668.

Belletti, B., di Prisco, M. & Dozio, D. 2006. Thin-webbed Open Cross-section Roof Elements: Modelling of Second Order Effects up to Failure, *Proc. of the 2nd fib Congress, Napoli (Italy), June 5–8*, ID 3–69 on CD-ROM.

CNR-DT 204, 2006. Guidelines for the design, construction and production control of the fibre reinforced concrete structures.

di Prisco, M., Felicetti, R. & Ferrara, L. 1998. Mono-cellular element for large-span roof: a theoretical and experimental investigation (in Italian), Proc. of 12° CTE Conference, Padova, 525–535.

di Prisco, M. & Felicetti, R. 1999. HSC Thin-Web Roof-Elements: an Experimental Investigation on Steel fibre Benefits. *Proc. 5th Int. Symposium on Utilization of HS/HP Concrete, Sandefjord, Norway*: 546–555.

di Prisco, M., Felicetti, R., Iorio, F. & Gettu, R. 2001. On the identification of SFRC tensile constitutive behavior. In R. de Borst, J.Mazars, G. Pijaudier-Cabot & J.G.M. van Mier (eds.), *Fracture Mechanics of Concrete Structures*: 541–548. A.A. Balkema Publishers.

di Prisco M. & Trintinaglia, C. 2002. Investigation on the structural behaviour of precast prestressed roof elements –TP elements. *Technical report DIS Politecnico di Milano*.

di Prisco, M., Iorio, F., Trintinaglia, & C., Signorini, S. 2002. Thin-web open-section roof elements: geometrical non linearity effects, (in Italian), *Proceedings 14° CTE Conference*, Mantova: 569–578.

di Prisco, M., Iorio, F. & Plizzari, G. 2003. HPSFRC prestressed roof elements. In B. Schnütgen and L. Vandewalle (eds.), *Test and Design Methods for Steel Fibre Reinforced Concrete – Background and Experiences*, PRO31: 161–188. Bochum: Rilem Publications S.A.R.L.

di Prisco, M., Felicetti, R., Lamperti M. & Menotti, G. 2004a. On size effect in tension of SFRC thin plates. Fracture Mechanics of Concrete Structures, (eds.). In V.C. Li C.K.Y.Leung, K.J.Willam, S.L. Billington (eds.) vol.2: 1075–1082. B.L.Schmick & A.D.Pollington Publishers.

di Prisco, M., Ferrara, L., Colombo, & M., Mauri, M. 2004b. On the identification of SFRC costitutive law in uniaxial tension, in Fibre-reinforced concretes. In M. di Prisco, R. Felicetti, G.A. Plizzari (eds.), Fiber reinforced concrete – BEFIB: 827–836. Cachan: Rilem Publications S.A.R.L.

di Prisco, M. & Dozio, D. 2005. Thin-webbed open cross-section roof elements: second order effects. *Proceedings of the International FIB Symposium Keep concrete attractive, Budapest, May 23–25:* 619–624.

Hordijk, D. 1991. Local approach to fatigue of concrete, Ph.D.Thesis, Deft University of Tecnology, 1–207.

UNI 11039 2003. Steel fibre reinforced concrete. Test method for the determination of first crack strength and ductility indexes.

Vlasov, V.Z. 1961. Thin walled elastic beams, Israel Programme for Scientific Translation, Jerusalem (1961).

Synthetic fibre reinforced concrete for precast panels: Material characterization and experimental study

L. Cominoli
Dept. of Eng. Design and Technologies, Univ. of Bergamo, Dalmine (BG), Italy

G.A. Plizzari
Dept. DICATA, University of Brescia, Brescia, Italy

P. Massinari
Chryso Italia s.p.a., Lallio (BG), Italy

ABSTRACT: The present paper describes an experimental study on structural behaviour of precast façade panels made of Self Compacting Concrete reinforced with synthetic fibres (SCFRC). The main research aims concern the material characterization and, at structural level, the possibility of replacing, in the external plates of precast panels, the traditional welded mesh with synthetic fibers. Fracture tests were performed on both normal weight and lightweight concrete where fibres may be particularly efficient in reducing the well-known brittleness of this material. The application of FRC to the production of precast panels is a promising technique since it allows several advantages; in fact the possibility of replacing conventional reinforcement with fibers allows for time and cost-savings and, since the minimum concrete cover is no longer required, may reduce the panel weight and the transportation costs. The experiments were performed on both traditional (RC) and SCFRC full-scale panels; they have been loaded up to failure to observe their response in terms of bending moment versus displacement and crack development.

1 INTRODUCTION

Concrete is a brittle material; in fact, when subjected to tensile stresses, it cracks already under small loads. Concrete may already crack before applying loads because of shrinkage or thermal effects. For this reason, steel reinforcement has been used to overcome the low concrete tensile strength and reinforced concrete (RC) became one of the most important construction techniques (Walraven, 1999). As a composite system, the reinforcing steel is assumed to carry all tensile loads.

Another approach to limit concrete brittleness concerns the use of discontinuous fibers to produce Fibre Reinforced Concrete (Romualdi & Mandel, 1964, Shah et al., 2004). In Fibre Reinforced Concrete, thousands of small fibres are dispersed and randomly distributed in the concrete matrix; therefore, they improve concrete properties in all directions (Balaguru & Shah, 1992).

Fibres enhance the post–cracking strength in tension, the fatigue strength, the resistance to impact loading and reduce temperature and shrinkage cracks (Ziad & Gregory, 1989; di Prisco et al., 2004). Only a few of the possible hundreds of fibre types have been found suitable for commercial applications (ACI 544, 1996).

Fibre Reinforced Concrete (FRC) finds applications in many areas of civil engineering where needs for repairing and durability arise.

Aim of the present study is to explore the mechanical properties of concrete reinforced with polypropylene fibres adopted to enhance its toughness. In addition, the paper reports preliminary results from full-scale tests on precast façade panels (Cominoli et al., 2006) made of ordinary reinforced concrete (as a reference) and SCFRC, where fibres substitute the welded mesh in the two external plates of the panel.

Material tests were performed on both normal and lightweight concrete where fibres may reduce the remarkable brittleness of this material.

The use of fibres in the precast industry is gaining particular interest (Failla et al., 2004). In particular, the use of FRC in the production of precast panels is a promising technique since it brings to several advantages. First of all fibres may replace the conventional transverse reinforcement, bringing to time and cost-savings, contributing to a major

Table 1. Composition of the Normal and Lightweight Self Compacting Fibre Reinforced Concrete.

Composition of Normal SCFRC		
Portland Cement 42.5R II A-LL	320	[kg/m³]
Acrylic plasticizer	3.3	[l/m³]
Water/cement ratio	0.56	[-]
Fresh Concrete Density	2364	[kg/m³]
Slump Flow	700	[mm]
Composition of Lightweight SCFRC		
Portland Cement 42.5R II A-LL	400	[kg/m³]
Acrylic plasticizer	4.3	[l/m³]
Water/cement ratio	0.56	[-]
Fresh Concrete Density	1835	[kg/m³]
Slump Flow	650	[mm]

Table 2. Geometrical and mechanical characteristics of the synthetic fibers adopted in the present research work.

Fibre	S40	S25
Length (L_f)	40 mm	25 mm
Equivalent diameter (ϕ_f)	1.0 mm	1.0 mm
Aspect ratio (L_f/ϕ_f)	40	25
Tensile strength [MPa]	650	650
Density [kg/m³]	920	920
Elastic modulus [GPa]	5.0	5.0

Table 3. Fiber type and content in the different concrete mixes.

Mix		Fibre dosage [kg/m³]	
		S40	S25
Mix-1	N-SCC	–	–
Mix-2	N-SCFRC	3.0 (0.33%)	–
Mix-3	N-SCFRC	4.0 (0.44%)	–
Mix-4	N-SCFRC	5.0 (0.55%)	–
Mix-5	N-SCFRC	–	4.0 (0.44%)
Mix-6	L-SCC	–	–
Mix-7	L-SCFRC	3.0 (0.33%)	–
Mix-8	L-SCFRC	4.0 (0.44%)	–

industrialization degree of the production process and avoiding the areas used for the storage of the welded mesh. Secondly FRC allows the design of wall panels to be dependent only on static requisites and not on cover limitations, and, finally, the use of lightweight concrete, beside the smaller concrete cover, may allow a reduction of the transportation costs.

For these reasons, FRC panels represent a competitive and cost-effective solution.

2 BEHAVIOUR OF SYNTHETIC FIBRES IN A CEMENT MATRIX

The precast panels studied in the present research work are made of Self Compacting Concrete reinforced with polypropylene fibres (*Self Compacting Fibre Reinforce Concrete, SCFRC*). The initial part of the research focused its attention on toughness properties of normal and lightweight concrete reinforced with structural polypropylene fibres.

The real advantage of adding fibres into concrete matrices becomes evident when fibres bridge these cracks and provide residual strength during their pull-out process. The residual strength is significantly influenced by the volume fraction and by the aspect ratio of the fibers, as well as by the orientation in the concrete matrix. The other factors that control the performance of the composite material are the physical properties of the concrete, matrix as well as the bond between fibres and matrix.

Material tests were carried out on specimens made of Normal Self Compacting Fibre Reinforced Concrete (N-SCFRC) and Lightweight Self Compacting Fibre Reinforced Concrete (L-SCFRC), as summarized in Table 1.

Two different types of polypropylene fibres were adopted. The first type (S40) had an equivalent diameter of 1.0 mm and a length of 40 mm (aspect ratio equal to 40), while the second type (S25) had the same

diameter, but a length of 25 mm (aspect ratio equal to 25); their geometrical and mechanical characteristics are reported in Table 2.

Fibres were added to the concrete matrix in different combinations; in particular, fibres S40 were used with three different dosages, equal to 3, 4 and 5 kg/m³ (corresponding to a volume fraction, V_f, of 0.33%, 0.44% and 0.55% respectively), while S25 fibres were used only with a dosage of 4 kg/m³ ($V_f \approx 0.44\%$; Tab. 3).

Table 4 shows the mechanical properties of the different concrete mixes adopted, in terms of compressive strength from cubes ($f_{c,cube}$), elastic modulus (E_c) and tensile strength (f_{ct}).

Fracture properties were determined by using notched beams ($150 \times 150 \times 600$ mm), tested under four-point bending according to the Italian Standard (UNI 11039, Fig. 1a). Fracture tests were carried out with a closed-loop hydraulic testing machine (INSTRON-1274) by using the Crack Mouth Opening Displacement (CMOD) as control parameter. Additional Linear Variable Differential Transformers (LVDT) were used to measure the Crack Tip Opening Displacement (CTOD) and the vertical displacement at midspan and under the load points (Fig. 1b).

The experimental results obtained from tests on SCFRC specimens are summarized, in terms of load versus CTOD curves, in Figure 2 (N-SCFRC) and in Figure 3 (L-SCFRC). It should be observed that

Table 4. Mechanical properties of the different concrete mixes.

Mix	Elastic modulus E_c [GPa]	Tensile strength f_{ct} [MPa]	Compressive strength $f_{c,cube}$ [MPa]
Mix-1	27.7	3.20	42.5
Mix-2	26.5	3.71	39.7
Mix-3	26.2	3.57	39.5
Mix-4	29.5	3.46	43.2
Mix-5	23.6	3.31	44.5
Mix-6	29.8*	2.17*	33.1
Mix-7	28.8*	1.94*	29.5
Mix-8	28.8*	1.96*	29.7

* Values calculated according to EC2 (2003).

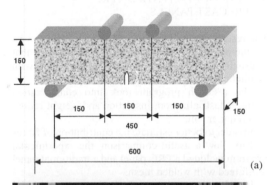

(a)

(b)

Figure 1. Geometry (a) and instrumentation (b) of the notched specimen for the bending tests, according to UNI 11039.

specimens were characterized by different volume fraction of fibres (0.33%÷0.55% for N-SCFRC and 0.33% ÷ 0.44% for L-SCFRC).

For an easier comparison, the figures show only one representative curve for each material which is chosen a the one closest to the average curve. It can be observed that fibre geometry and content remarkably

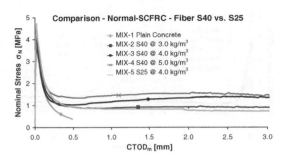

Figure 2. Nominal stress versus CTOD curves (average curves) obtained from bending tests on N-SCFRC specimens.

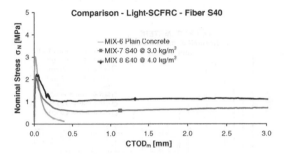

Figure 3. Nominal stress versus CTOD curves (average curves) obtained from bending tests on L-SCFRC specimens.

influence concrete toughness (evidenced by the post-peak strength) while the peak stress is not significantly influenced by the presence of fibres (it mainly depends on the tensile strength of the concrete matrix).

Although polypropylene fibres are characterized by a low elastic modulus, it is quite apparent that the load carrying capacity of FRC under flexural loading is considerably increased (Figures 2 and 3). The residual strength is maintained until large values of the crack opening, greater of 3 mm (normally accepted in the concrete structures).

As far as the comparison between Lightweight and Normal SCFRC is concerned, in Figure 4 it can be noticed how the first one evidences a post-peak strength similar to those of the Normal-SCFRC, although the peak stress is lower (2.5 MPa for L-SCFRC and 4 MPa for N-SCFRC).

This is a significant potentiality of lightweight concrete that could be used in the prefabrication industry, with remarkable advantages from the structural and economical point of view.

In order to better identify the post-cracking response of FRC, three independent parameters are proposed by the Italian Standard (UNI 11039, 2003): the first crack strength (f_{If}) and two equivalent flexural strengths ($f_{eq,(0-0.6)}$ and $f_{eq,(0.6-3)}$). The first flexural strength ($f_{eq,(0-0.6)}$) corresponds to a Crack Tip Opening Displacement range of 0-0,6 mm (significant

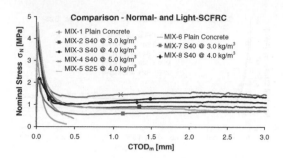

Figure 4. Comparison of the nominal stress versus CTOD curves (average curves) obtained from bending tests on both Normal SCFRC and Light-Weight SCFRC specimens.

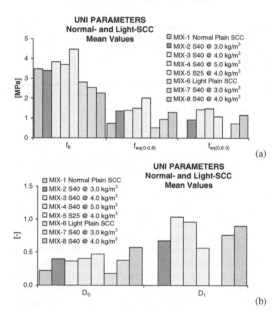

Figure 5. First-crack strength and post-cracking equivalent strengths (a) and ductility indexes (b) according to UNI 11039.

for the Serviceability Limit State) while the second one corresponds to a CTOD range of 0,6–3 mm (significant for the Ultimate Limit State). In order to better identify the reduced FRC brittleness, the Italian Standard also requires two Ductility Indexes that are defined as the ratios between the equivalent strength as follows:

$$D_0 = f_{eq,(0-0.6)}/f_{If} ; \quad D_1 = f_{eq,(0.6-3)}/f_{eq,(0-0.6)}.$$

The equivalent flexural strengths obtained from bending tests on FRC specimens are given in Figure 5a (mean values), while the two ductility indexes D_0 and D_1 are shown in Figure 5b.

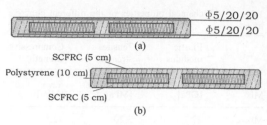

Figure 6. Geometrical characteristics of the transverse section of a traditional SCC panel (a) and a SCFRC panel (b).

Based on the experimental results obtained from the material characterization, a SCFRC panel was realized by using the SCFRC matrix reinforced with 3 kg/m³ of S40 fibres.

3 FULL-SCALE TESTS ON FRC PRECAST PANELS

The research on full-scale façade panels aims to verify the possibility of replacing the welded mesh that is usually placed on the two faces of the panel, with synthetic fibres.

The research program took into consideration static, industrialization, insulation and weight reduction requirements.

In order to better estimate the contribution of fibres and to allow a useful comparison, the experimental program included a FRC panel and a traditional panel reinforced with welded mesh.

In order to reproduce realistic situations, a special experimental set-up was used to reproduce the combined bending effects of the panel weight and of the wind load.

3.1 Panel geometry

The panels are realized with two external plates, connected with stiffening ribs with interposed a layer of insulating material (characterized by low specific weight), with the function of lightening and thermal isolation. The density of this insulating material ($10 \div 30 \, kg/m^3$) and the extension of the ribbings determine the thermal insulation of the panel (Fig. 6).

The preliminary experimental program included two full-scale panels, having a length of 11,2 m, a height of 2,5 m and a thickness of 0,2 m, characterized by different reinforcement and materials, as described in the following:

- a first panel (RC) was reinforced with traditional welded mesh on the two plates and with reinforcing bars in the ribs (Fig. 7a);
- a second panel (FRC) was reinforced with 3.0 kg/m³ of synthetic fibres, in substitution of the mesh fabric (Fig. 7b).

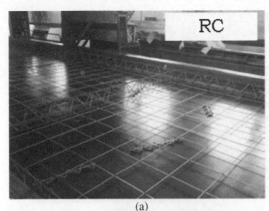

(a)

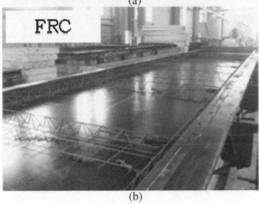

(b)

Figure 7. Traditional reinforcement of the RC panel (a) and FRC panels (b).

The cross section of the panels is composed of three different layers. In particular, from the internal toward the external side of the building, there is (Fig. 6):

- a layer of concrete (5 cm);
- a layer of polystyrene material (10 cm);
- another layer of concrete (5 cm).

Table 5 shows the compressive strength of concrete measured on 150 mm side cubes ($f_{c,cube}$), the reinforcement type, the self-weight of the panels and the age of concrete at the time of the test. One should notice that panels had similar mechanical properties since they were made by using the same concrete mix (Tab.1).

3.2 Test set-up

One of the main issues concerning experimental tests on prefabricated panels is related to the test set-up. The experimental test set-up aims to reproduce the latter configuration (Fig. 8), because this is the most critical working condition. The bilateral restraint on the vertical edges, which is represented, in a real structure, by the columns, was realized by means of two

Table 5. Concrete and reinforcement characteristics.

Panel	$f_{c,cube}$ [MPa]	Curing [days]	Fibre Type	V_f [kg/m^3]	Weight [ton]
RC	40.4	78	Mesh	–	≈9.5
FRC	39.1	80	S40	3.0	≈9.5

Figure 8. Panel placement on the reaction frame.

vertical steel profiles which are also required to avoid overturning of the panel.

Another problem is related with the application of the transverse load to simulate wind effects. This was accomplished by means of an electrical screw-jack, which acts on a loading distribution system, made of three different levels of steel beams, properly connected to each other by means of bilateral pinned restraints, necessary to allow panel unloading.

In order to simulate the wind effects, the front face of the panel was loaded by means of four loading strips, positioned in such a way to obtain the same maximum shear force and bending moment, as for an uniformly distributed pressure (Fig. 9).

During the test, the horizontal displacements were recorded in several points, as shown in Figure 10. Compressive and tensile deformations (or the opening of a flexural cracks) were also measured on the panel faces. The displacements were measured by means of either Linear Variable Differential Transducers (LVDT), or potentiometric transducers with 250 mm stroke.

More details about the test set-up can be found in Cominoli et al., 2006.

3.3 Experimental results and discussion

The ultimate load of the panels was reached after submitting the specimens to different cycles of increasing intensity (fraction and multiple of the service bending

Figure 9.   Global view of the experimental set-up for the full-scale panels.

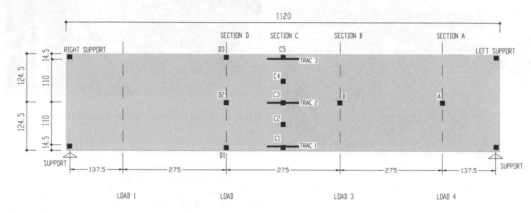

Figure 10.   Instrumentation used during the tests.

moment, M_s). At the end of each step, the panels were unloaded in order to check the residual deformation. Finally the load was increased up to collapse.

The bending moment under service load ($M_s = 25.3$ kNm) was determined by assuming the panels as part of a precast building located in wind Zone I and Category IV, according to the Italian Standard (D.M. 14-09-2005).

The curves of the maximum bending moment versus the horizontal displacement of the midspan section, the final crack patterns and the effects of fiber reinforcement are presented and discussed in the following.

Figure 11 shows the experimental curves of the bending moment as a function of the horizontal deflection (C3 in Fig. 10) for the two tested panels. It can be observed that the traditional panel (RC; Fig. 11a) exhibits an ultimate load and a stiffness, in the cracked stage, different from the FRC panel (Fig. 11b), even if the latter is characterized by better ductility. In both cases, a considerable stiffness reduction is observed when the first crack appears (I stage); as a result of the

formation of this crack, the behaviour is still almost linear, even if there is a remarkable loss of stiffness (II stage).

The transition from the first to the second stage in SCFRC panels occurred for a smaller bending moment than in the traditional panel (approximately 12 kNm versus 20 kNm). However, the stiffness in the initial (uncracked) stage is similar.

The tests continued until reaching the ultimate moment, approximately of 50 kNm (III stage).

One should notice that the horizontal displacement in service conditions for FRC panel is about 1/600 of the span length and that the maximum moment is about 1.4 times the service moment.

It is necessary to emphasize that panels were already cracked before starting the test, due to shrinkage phenomena; when increasing the load, these cracks tends to concentrate in the midspan zone of the element, where the maximum bending moment is present.

The use of fiber reinforcement is also effective in reducing the crack width so that it enhances the durability of the panels.

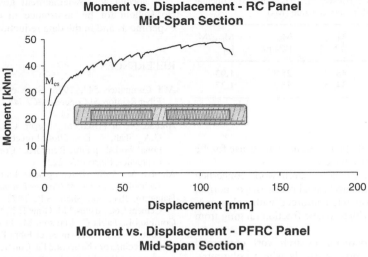

Moment vs. Displacement - RC Panel Mid-Span Section

M_{es}

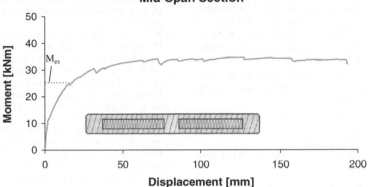

Moment vs. Displacement - PFRC Panel Mid-Span Section

M_{es}

Figure 11. Comparison of the mid-span bending moment – deflection curves for the tested panels.

Figure 12 illustrates the final crack pattern of the panels; in all cases, failure involved longitudinal bending with transversal cracks along the whole height.

It can be noticed that the RC panel is characterized by the formation of a main crack (Fig. 12a), whose width is already important at the moment of its appearance. In the FRC panel, instead, the crack pattern is diffused and the maximum cracks width is reduced (Fig. 12b).

Table 6 shows a comparison between the maximum experimental bending moment (M_{max}) and the service moment (M_s).

Once again, it should be noticed that, in the two preliminary panels, the maximum bending moment was larger than its design value for ultimate limit state.

4 CONCLUDING REMARKS

The present paper deals with the mechanical characterization of a self compacting concrete reinforced

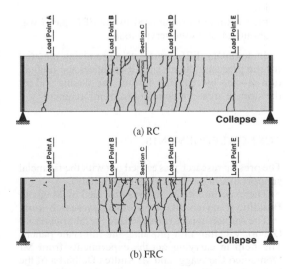

Figure 12. Comparison of the cracks distribution in the tested panels after the test.

Table 6. Comparison between the maximum experimental bending moment and its service value (Ms).

Panel	$f_{c,cube}$ [MPa]	M_{max} [kNm]	M_s [kNm]	M_{max}/M_s [-]
RC	40.4	48.74	25.3	1.93
FRC	39.1	34.65	25.3	1.37

with structural synthetic fibres and with its use for the production of precast panels.

The initial part of the research work concerned the evaluation of the mechanical properties of normal and lightweight concrete reinforced with polypropylene fibers, used with a volume fraction ranging from 0.33% to 0.55%.

The second part of the research work concerned full-scale tests on precast panels where polypropylene fibers replaced the welded mesh in the external plates of the panel. By doing so, fibers allow for time and cost-savings and may reduce the structural weight, decreasing the transportation costs. In fact, the absence of conventional reinforcement may allow for a reduction of the slab thickness, since the minimum concrete cover is no longer required.

In order to better appreciate the benefits provided by fibres, a test on a panel reinforced with the traditional welded mesh was also performed.

From the experimental results, the following conclusions can be drawn:

• the Fibre Reinforced Concrete panel exhibited a behaviour similar to the traditional panel reinforced with mesh fabric, even if the ultimate load was lower;
• the maximum load applied to the FRC panel was about 1.4 times the service load.

In summary, FRC represents a promising material for the precast industry since it enhances the production process by reducing the time for placing and for the storage areas for conventional reinforcement.

ACKNOWLEDGEMENTS

The present research was carried out with the financial support of Chryso Italia (Lallio, BG, Italy).

The experimental set-up was provided by Magnetti Building (Carvico, BG) for a previous research work.

The Authors wish to thank Mozzo Prefabbricati s.r.l. (Zevio, VR, Italy) for providing the full-scale panels. The help in carrying out the experiments from Mr Domenico Caravaggi and Mr Andrea Delbarba of the Laboratory P. Pisa of the University of Brescia, is also gratefully acknowledged.

A special acknowledgement goes to Eng. Valerio Belloni for his assistance in carrying out the experiments and in the data reduction.

REFERENCES

ACI Committee 544, 1996, State of the art Report on Fiber Reinforced Concrete, *ACI 544.1R-96*, ACI, Detroit, Michigan.

Ahmad, A., di Prisco, M., Meyer, C., Plizzari, G.A. and G.A., Shah, S., Eds. 2004, Proceedings of the International Workshop Fibre Reinforced Concrete: From theory to practice, Starrylink: 222.

Ahmed, M., 1982, Polypropylene Fibres – Science and Technology, *Society of Plastics Engineers*, New York.

Balaguru, P.N., and Shah, S.P., 1992, "Fibre Reinforced Cement Composites," McGraw Hill, New York.

Cominoli L., Failla C., Manzoni M., Plizzari G.A., Riva P., 2006, Precast panels in Steel Fibre Reinforced Concrete, Proceedings of The Second Fib Congress, Naples, abstract on page 448–449, full length paper available on accompanied CD.

di Prisco, M., Felicetti, R., and Plizzari, G.A. (eds) 2004. Proceedings of the 6th RILEM Symposium on Fiber Reinforced Concretes (FRC), Varenna (Italy), September 20–22, RILEM PRO 39, Bagneaux (France): 1514.

Eurocode 2. 2003. Design of concrete structures – Part 1-1: General rules and rules for buildings. ENV 1992-1-1.

Failla, C., Magnetti, P. and Pasini, F., 2004, Application of SFRC to precast building, Proc. of the RILEM International Conference BEFIB 2004, Varenna, Italy: 81–100.

Hannant, D.J. 1978, Fibre cements and fibre concretes, John Wiley and Sons, New York: 213.

Italian Ministry for Public Works (D.M. 14-09-2005), Norme tecniche per le costruzioni (Italian Standard for loads on Structures).

Minelli, F., Cominoli, L., Meda, A., Plizzari, G.A., and Riva P., 2005, Full-scale tests on HPSFRC prestressed roof elements subjected to longitudinal flexure, International Workshop on High Performance Fiber Reinforced Cementitious Composites in Structural Applications, Li V.C. e Fisher G. Eds., Honolulu (USA).

Romualdi, J. and Mandel, J. 1964. Tensile strength of concrete affected by uniformly distributed and closely spaced short lengths of wire reinforcement. *Journal of the American Concrete Institute*, Vol. 61, No. 6: 657–671.

Shah, S.P., Kuder, K.G., and Mu, B., 2004, Fiber-reinforced cement-based composites: a forty year odyssey, in BEFIB 2004, di Prisco, M., Felicetti, R. e Plizzari, G.A. Eds., Proceedings of the Sixth RILEM symposium on fibre reinforced concrete (FRC): 3–30.

UNI 11039. 2003, Steel fibre reinforced concrete – Part I: Definitions, classification specification and conformity – Part II: test method for measuring first crack strength and ductility indexes. Italian Board for Standardization.

Walraven, J., 1999, The evolution of Concrete. *Structural concrete*, Vol. 1: 3–11.

Ziad, B., and Gregory, P., 1989, Use of Small Diameter Polypropylene fibres in Cement based materials, in *Fibre Reinforced Cements and Concretes, Recent Developments*, Edited by Swamy, R.N., Barr, B.:200–208.

SFRC bending behaviour at high temperatures: An experimental investigation

M. Colombo, M. di Prisco & R. Felicetti

Politecnico di Milano, Milano, Italy

ABSTRACT: Steel fibre reinforced concrete (SFRC) is increasingly considered as a profitable replacement for diffused reinforcement like welded steel mesh, especially for thin cross sections. In this case fire becomes a very important condition in the design. Previous experimental research showed the benefits in fire resistance of steel fibres when structural elements are bent. A more complete mechanical characterization of the material, when exposed to high temperature, is here presented. The research was instrumental to validate the fire design approach suggested in the recent Italian National Recommendations (CNR-DT 204/06) for design of fibre reinforced concrete structures. Particular attention is given to the reliability of a linear softening constitutive law proposed for uniaxial tension.

1 INTRODUCTION

Cementitious composites are tipically regarded as brittle materials, with low tensile strength and strain capacity. However with the use of fibre reinforcement, the brittleness shown by plain concrete structures can be overcome producing structures with improved load bearing capacity, ductility and durability. In the last forty years the effect of fibres on toughness, fracture behaviour, impact performance and composite hybridization has been investigated as well as the role of the fibre-matrix interface (di Prisco et al. 2004a).

In the last years new High-Performance cementitious composites have been also developed exhibiting an enanched elastic limit as well as a strain-hardening response after cracking in bending or even in uniaxial tension (Naaman & Reinhardt 2003).

The material performance improvement and the new applications of these materials ask the researchers to better investigate some new aspects such as fatigue, impact, fire behaviour, durability and shrinkage.

Fire condition is, as a matter of fact, a very important issue in designing precast concrete structures and design in such a condition is now regulated in Europe by European Standard EN 1992-1-2 *Eurocode 2: Design of concrete structure – Part 1.2: General rules – Structural fire design*.

Researchers are very interested in this matter and some open projects are now focused on fire effects on different kind of structures and in particular on tunnel linings: two large-scale experimental investigations on

tunnel lines have being carried out in Austria and in Germany (Dehn & Werther 2006).

The use of steel fibres coupled with polypropylene fibres (Kalifa et al. 2001) can bear some benefits to a structure. The absence of spalling phenomena, first of all, prevents the hot surface to move deeper inside the structure and to reach the inner layers with the consequent mechanical decay of the reinforcement. Steel fibres give to the material a certain residual bending resistance even when exposed to high temperature; this improves the bearing capacity of the structure itself, but it also ensure the conservation of the initial cross section preventing any detachment of external layers.

A recent experimental investigation has shown the benefits given by steel fibre either to mechanical properties of the material at high temperature (di Prisco et al. 2002, di Prisco et al. 2003) and to fire resistance of bent element (di Prisco et al. 2003). In the same research some tests were carried out in order to demonstrate that the mechanical properties of the material does not mainly depend on the temperature at testing, but on the maximum experienced temperature (Colombo et al. 2004). The thermal diffusivity of the material was shown to be scantly affected by steel fibres up to a volume content of 1%.

This research (Colombo 2006) was instrumental to define and validate the fire design approach of the National Recommendation CNR DT204/06 recently issued about the design of fibre reinforced concrete structures in Italy. A proper experimental

Table 1. Mix design of steel fibre reinforced concrete.

Constituent	Type	Content	Unit
Cement	I 52.5R	450	kg/m³
	0/3	620	kg/m³
Aggregates	0/12	440	kg/m³
	8/15	710	kg/m³
Plasticizer	Acrylic	5.5	kg/m³
Total water		195	l/m³
Filler		30	kg/m³
Fibre		50	kg/m³

investigation is here presented aimed at identifying the material properties after a thermal damage.

2 EXPERIMENTAL PROGRAMME

The experimental programme here presented was planned in order to perform a proper identification of the mechanical properties of a steel fibre reinforced concrete (SFRC) and to investigate the stiffness degradation in bending, uniaxial tension and compression when it is exposed to high temperatures. This material is being used for roof elements production in the precast industry.

The material compressive strength is 75 MPa and the fibre content is 50 kg/m³; fibres are low-carbon, hooked end, 30 mm long and with an aspect ratio (l_f/d_f) equal to 45; all the aggregates are siliceous. The mix design is shown in Table 1.

Twelve prismatic were casted specimens according to National Recommendation UNI 11039.

The thermal treatment of the material was carried out in a oven by performing some thermal cycles up to different maximum temperatures. Three different maximum temperatures (200, 400 and 600°C) were considered. An heating rate equal to 30°C/h was used up to the maximum threshold; after this a 2 hours stabilization phase was imposed in order to guarantee an homogeneous temperature within the specimen. The cooling phase was performed with a rate of 12°C/h down to 200°C, temperature from which the oven was opened in order to faster the natural cooling which tends to slow down approaching the room temperature. In each cycle, three nominally identical specimens were introduced into the oven. In this way all the specimens characterized by the same maximum temperature had the same thermal history. The thermal cycles for the three maximum temperatures considered are summarized in Figure 1.

The experimental programme is organized in three different phases (Fig. 2). The first one refers to the mechanical characterization of the material by means of four point bending tests on notched specimen

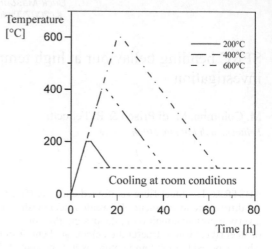

Figure 1. Thermal cycles.

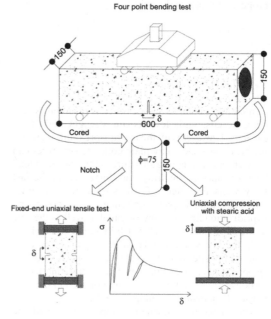

Figure 2. Experimental programme.

according to National Recommendation UNI 11039. Once tested, two cylinders 150 mm high with a 75 mm diameter were cored from each specimen: the first one was tested in uniaxial compression, the second one was notched and tested according to a fixed end uniaxial tension test.

In order to investigate the stiffnes degradation of the material, some unloading-reloading cycles were performed in all the tests.

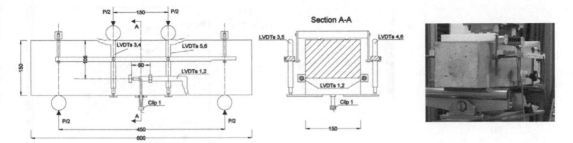

Figure 3. Four point bending test set-up.

2.1 Four point bending tests

Four point bending tests were performed according to National Recommendation UNI 11039 on $150 \times 150 \times 600$ mm specimens notched with a notch ratio equal to 0.3.

Tests were performed by considering the CMOD measurement (Crack Mouth Opening Displacement, Clip 1) as feedback parameter. In order to measure the crack opening at the notch tip (CTOD) two LVDT transducers were used (LVDTs 1,2). Four different LVDTs were instrumental to measure the deflection under the load application points on both the specimen sides (LVDTs 3–6). In order to prevent that deflection measurements could be affected by crushing of the material at the supports, a proper frame was used (Fig. 3).

The experimental results of four point bending tests are shown in Figure 4 in terms of the nominal stress ($\sigma_N = 6M / bh^2$; b = width, h = depth) vs. CTOD curves. The average envelope curves of three nominally identical tests are shown in Figure 5.

The mechanical strengths at increasing temperature are listed in Table 2 and plotted in Figure 6. The meaning of each parameter can be explained as follows:

f_{If} : first cracking strength representing the matrix tensile strength; it's the nominal stress corresponding to a CTOD equal to 25 μm;

$f_{eq0-0.6}$: average nominal strength in the CTOD range between 0.025 mm and 0.625 mm representing the Serviceability Limit State (SLS) residual strength;

$f_{eq0.6-3}$: average nominal strength in the CTOD range between 0.625 mm and 3.025 mm representing the Ultimate Limit State (ULS) residual strength, when material behaviour is governed only by pull-out mechanism.

Looking at the experimental results, a remarkable decay of material properties between 200°C and 400°C is observed; for higher temperatures the material properties seem to remain quite constant; for temperatures lower than 200°C the first cracking nominal strengths do not seem to be affected by thermal treatment.

Table 2. Four point bending tests nominal stresses: first cracking and average nominal stresses in two different CTOD ranges: 0–0.6 mm and 0.6–3 mm.

Temp.	Test	f_{If} [MPa]	$f_{eq0-0.6}$ [MPa]	$f_{eq0.6-3}$ [MPa]
20°C	1	5.73	5.53	4.18
	2	5.84	4.88	3.29
	Average	5.78	5.21	3.73
	Max$|x_i - \bar{x}|$	1.04%	6.33%	12.06%
200°C	1	5.79	4.15	3.10
	2	5.57	4.58	3.52
	3	5.47	4.16	3.20
	Average	5.61	4.30	3.27
	Max$|x_i - \bar{x}|$	3.21%	6.51%	7.65%
400°C	1	2.91	2.98	1.50
	2	2.31	2.66	1.44
	3	2.78	3.07	2.27
	Average	2.67	2.90	1.74
	Max$|x_i - \bar{x}|$	13.48%	8.28%	30.46%
600°C	1	3.01	2.69	1.52
	2	2.80	2.67	2.04
	3	2.63	2.60	1.00
	Average	2.81	2.65	1.52
	Max$|x_i - \bar{x}|$	7.12%	1.89%	34.21%

2.2 Uniaxial compression tests

Uniaxial compression tests were performed on cylinders cored from the beam after bending test. Specimens were instrumented by means of seven LVDT transducers: three disposed at 120° around the specimen in order to measure the displacement between the loading platens of the press, other three LVDTs, disposed at 120° as well, were applied to the central zone of the specimen with a gauge equal to 50 mm; finally one LVDT connected to a chain surrounding the specimen was used to measure the average circumferential relative displacement in the central region. The geometry and the specimen set-up in uniaxial compression are shown in Figure 7.

In order to reduce friction with the press platens some stearic acid was smeared on the specimen ends.

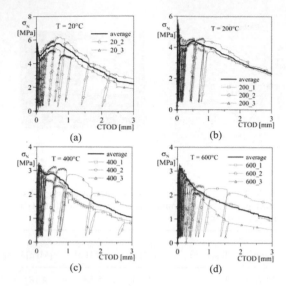

(a)

(b)

(c)

(d)

Figure 4. Four point bending tests results: nominal stress vs. CTOD curves for different temperatures (a) T =20°C, (b) T =200°C, (c) T =400°C, (d) T =600°C.

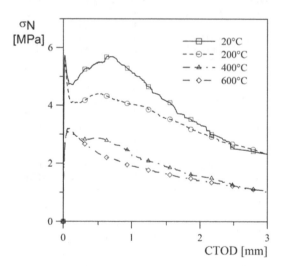

Figure 5. Four pint bending tests: envelope average nominal stress vs. CTOD curves.

All the tests were displacement controlled by using as feedback parameter the total vertical displacement of the specimen measured by one of the full bridge LVDT.

The displacement rate imposed was equal to 0.1 mm/min either in loading and unloading phases.

The experimental results are shown in Figure 8 by means of the nominal stress ($\sigma_N = P/A$) vs. normalized total vertical displacement (δ/l) curves.

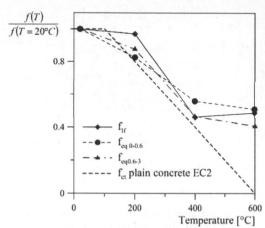

Figure 6. Four point bending tests: nominal stress behaviour with temperature.

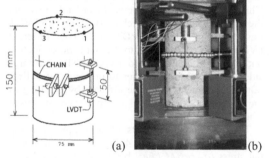

(a)

(b)

Figure 7. Uniaxial compression test set-up.

In testing specimens at room condition some instabilities in controlling the test caused in two cases a sudden failure before reaching the peak load. The variation of peak nominal strength is summarized in Table 3 and Figure 9; in compression tests, differently from what happened in bending, the degradation of the material is quite linear in all the temperature range investigated, even if in the range between 200°C and 400°C the peak nominal strength seems to be less influenced by temperature increasing.

The behaviour experimentally detected is similar to the one proposed by Eurocode 2 for plain concrete (Fig. 9).

The average envelope behaviour of the material, for different maximum temperatures, is shown in Figure 10.

In this figure the nominal stress is respectively plotted with respect on the left side, to the normalized total vertical displacement (δ/l) and on the right side to the normalized circumferential displacement. In the latter case the circumferential displacement (δ_r), read from the transducer placed on the chain clamped on the

1570

Table 3. Uniaxial compression tests: peak nominal strengths.

Temperature	Test	Peak nominal strength [MPa]
20°C	2	79.25
	3	70.70
	Average	74.98
	$\text{Max}|x_i - \bar{x}|$	5.71%
200°C	1	60.26
	2	64.81
	3	60.89
	Average	61.99
	$\text{Max}|x_i - \bar{x}|$	4.55%
400°C	1	59.52
	2	56.46
	3	60.02
	Average	58.67
	$\text{Max}|x_i - \bar{x}|$	3.77%
600°C	1	44.98
	2	57.57
	3	38.20
	Average	46.92
	$\text{Max}|x_i - \bar{x}|$	22.70%

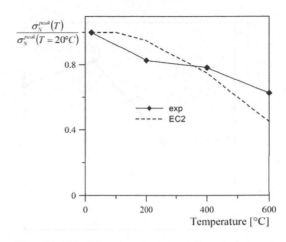

Figure 9. Uniaxial compressive tests: peak nominal strength at different temperatures.

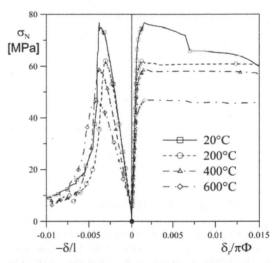

Figure 10. Uniaxial compression tests: envelope average curves.

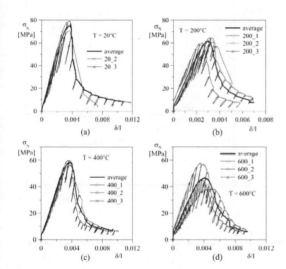

Figure 8. Uniaxial compression tests: nominal stress vs. normalized vertical displacement curves for different temperatures (a) $T = 20°C$, (b) $T = 200°C$, (c) $T = 400°C$, (d) $T = 600°C$.

specimen, is divided by the perimeter of the cylinder cross section ($\pi\phi$).

Comparing the normalized circumferential displacement with the total vertical one the evolution of Poisson's ratio at increasing temperature can be detected by taking into account the ratio between these two quantities in the initial elastic branch of each curve in the range between 5% and 30% of the peak load. The evolution of the Poisson ratio is represented in Figure 11; the main contribution of temperature occurs in the range between 200°C and 400°C.

2.3 Fixed end uniaxial tension tests

Uniaxial tension tests were performed on notched cylinders (Fig. 12) glued to the press platens by means of an epoxy glue made of two components: binder and hardener. The specimen was instrumented by six displacement transducers LVDTs: three were placed astride the notch (gauge length = 50 mm) to measure crack opening displacement (COD) and three were

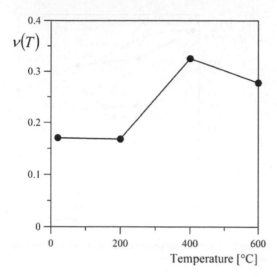

Figure 11. Poisson ratio evolution.

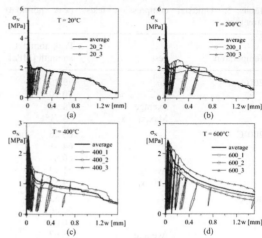

Figure 13. Uniaxial tensile tests: nominal stress vs. crack opening displacement (w) curves for different temperatures (a) $T = 20°C$, (b) $T = 200°C$, (c) $T = 400°C$, (d) $T = 600°C$.

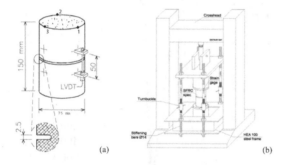

Figure 12. Uniaxial tensile test set-up.

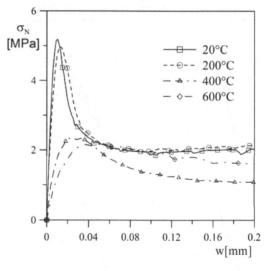

Figure 14. Uniaxial tensile tests: envelope average curves.

used to measure the displacement between the two end platens of the press. One of the latter was used as the feedback parameter in performing tests. The displacement rate imposed during the tests was equal to $0.04 \, \mu\text{m/s}$ either in loading and unloading branch up to crack opening displacement equal to 0.6 mm and was then increased to $0.4 \, \mu\text{m/s}$ for higher COD.

In order to keep the platens parallel during the test an active control was performed (Fig. 12). Four steel 14 mm diameter bars were used to connect the fixed base of the press with the upper plate connected to the specimen and to an articulated joint. The bars were fixed to a steel frame made of HEA100 beams connected to the basement of the press. These bars have an adjustable length by means of a turnbuckle and each bar is instrumented with two strain-gages to measure its elongation. Acting on the turnbuckles during the tests it was ensured that all the crack opening measurements were very close in order to consider the plates as fixed. This test set-up induces in the specimen also a bending moment: the maximum value

reached in both direction (x and y) was equal to 0.15 kNm to which correspond the maximum rotation $\vartheta_x^{\max} = 1.31 \cdot 10^{-5}$ and $\vartheta_y^{\max} = 7.40 \cdot 10^{-6}$.

The experimental results are shown in Figure 13 by means of nominal stress ($\sigma_N = P/A$) vs. crack opening displacement (w) curves.

In performing two tests (one at room condition and one at 200°) some problems occurred in controlling the press: this caused these two tests to be lost. The envelope average curves σ_N vs. w are shown in Figure 14. In order to better investigate the uniaxial tensile

Table 4. Uniaxial tensile tests: nominal stress at peak and average nominal stresses in COD range: 0.9 mm ± 20%.

Temperature	Test	σ_N^p [MPa]	$\sigma_N^{0.9}$ [MPa]
20°C	1	5.21	1.32
	2	5.27	1.32
	Average	5.24	1.32
	Max$|x_i - \bar{x}|$	0.57%	0.16%
200°C	1	5.06	1.30
	2	4.92	1.30
	Average	4.99	1.30
	Max$|x_i - \bar{x}|$	1.4%	0.11%
400°C	1	2.61	–
	2	2.06	0.65
	3	2.47	0.99
	Average	2.38	0.82
	Max$|x_i - x|$	13.45%	20.73%
600°C	1	1.73	0.80
	2	2.46	0.71
	3	2.37	1.09
	Average	2.19	0.87
	Max$|x_i - \bar{x}|$	21.00%	25.29%

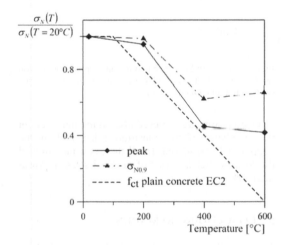

Figure 15. Uniaxial tensile tests: nominal stress at peak and average nominal stresses in COD range: 0.9 mm ± 20%.

behaviour of SFRC when exposed to high temperatures the evolution of two different nominal strength is considered:

σ_N^{peak}: nominal stress at peak that represents the matrix behaviour;

$\sigma_N^{0.9}$: average nominal stress in the crack opening displacement (w) range $0.9 \pm 20\%$ related to pull-out mechanism.

The evolution of these two parameters is summarized in Table 4 and shown in Figure 15. As already

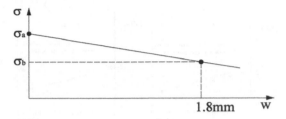

Figure 16. Uniaxial tension linear constitutive law.

seen in the bending behaviour the highest material degradation occurs in the temperature range between 200°C and 400°C; out of this range the nominal stresses considered are quite constant. The pull-out mechanism seems to increase the strengths between 400°C and 600°C; this behaviour is similar to the one observed in a previous experimental investigation (di Prisco et al. 2003b) on a nominally identical material.

3 UNIAXIAL TENSILE CONSTITUTIVE LAW

The knowledge of the uniaxial constitutive law is a very important issue in the design of steel fibre reinforced concrete structures. The new National recommendation CNR DT 204/06 propose for fibre reinforced concrete the linear softening stress (σ) – crack opening relationship shown in Figure 16.

The linear softening law recommends the identification of two stress parameter (σ_a and σ_b) by means of four point bending tests according to UNI 11039. These stress parameters are defined according to the following formulation:

$$\sigma_a = 0.45 \cdot f_{eq0-0.6}$$

$$\sigma_b = 0.5 \cdot f_{eq0.6-3} - 0.2 \cdot f_{eq0-0.6}$$

The constitutive law proposed is aimed to describe the behaviour of the material governed by fibre pull-out mechanism, neglecting in the design procedure the matrix strength and the unstable crack propagation in the concrete matrix.

The National Recommendation proposes the same uniaxial tension constitutive law also in fire condition, providing a proper identification procedure to be performed by means of four point bending tests on specimens damaged by a thermal cycle. This procedure uses tests at room condition, computes the residual equivalent strengths $f_{eq0-0.6}$ and $f_{eq0.6-3}$ in specimen previously subjected at different maximum temperature reached during a thermal cycle.

The reliability of the constitutive law, also at high temperatures, is shown in Figure 17 where the constitutive law identified from bending tests performed at different maximum temperatures ($T_{max} = 20, 200, 400$ and 600°C) are compared with the experimental

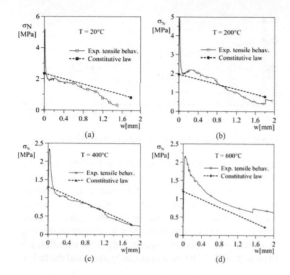

(a)

(b)

(c)

(d)

Figure 17. Uniaxial tensile behaviour: constitutive law validation (a)T = 20°C, (b) T = 200°C, (c) T = 400°C, (d) T = 600°C in term of average curves.

results of the uniaxial tensile tests performed on cylindrical specimens cored from the same beam on which the bending tests was carried out.

Looking at the tensile behaviour governed by fibre pull-out mechanism, all the cases investigated show a good reliability of the uniaxial constitutive law proposed; the only exception is observed for $T_{max} = 600°C$ where the scatter is equal to 20%. It's worth noting that, in this last case, the constitutive law is in favour of safety.

4 DESIGN PREDICTIONS

In this section the bending behaviour of the material previously investigated is taken into account (di Prisco et al. 2003a,b). The case of bent elements in which the element cross section is characterized by a uniform thermal damage (Colombo & di Prisco 2006) is taken into account by considering the design approach suggested by National Recommendation and just explained.

Considering the situation of uniformly distributed thermal damage, the comparison between the experimental tests previously described and the numerical predictions is shown in Figure 18 for the different temperatures considered. The numerical simulations were performed by considering the plane section assumption (di Prisco et al. 2004b) according to the multi-layer procedure proposed by Hordijk (1991). In this way, the notch effect was neglected and the net section of the specimen was assumed as critical section.

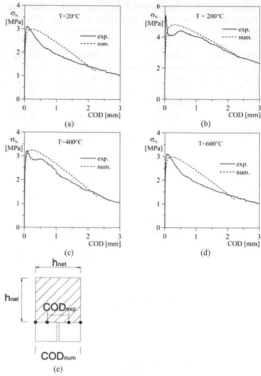

(a)

(b)

(c)

(d)

(e)

Figure 18. Bending behaviour: numerical prediction (a) T = 20°C, (b) T = 200°C, (c) T = 400°C, (d) T = 600°C. (e) Notation in COD definition.

Good results are achieved also taking into account the simplicity of the constitutive law and of the approached considered. In all the cases investigated the local scatter in terms of nominal stress is lower than 20%.

5 CONCLUDING REMARKS

The research here presented is the final step of a long research programme of about three years aimed to evaluate fire resistance of fibre reinforced concrete structure with a fibre content lower than 1%. This research allowed us to conclude that is possible to performe a mechanical identification at room condition after thermal cycles characterized by a maximum temperature that causes an irreversible thermal damage of the material. The thermal diffusivity is not significantly affected by the fibres content considered. In order to design steel fibre reinforced concrete structures in fire conditions is possible to adopt a linear constitutive softening law to describe the post-peak uniaxial tensile behaviour and to use a simply plane section model. This conclusion moved the fire design approach suggested by the recent National Recommendation CNR DT 204/06.

ACKNOWLEDGMENTS

The authors thank Magnetti Larco-Building for the technical support in the cast of the specimens. The research was financially supported by the Italian Ministry for the Education, University and Research (PRIN 2004).

REFERENCES

CNR DT204. 2006. Instruction for Design Execution and Control of Fibre Reinforced Concrete Structures (in Italian).

Colombo, M, Felicetti, R., Manzoni, M. & Bergamini, E. 2004. On the bending behaviour of SFRC exposed to high temperature. In di Prisco, Felicetti & Plizzari (eds.), *Proc. Of 6th Sympoium on Fibre Reinfoced Concrete BEFIB 2004 – PRO39*: 647–658. Bagneux, France: RILEM Publications S.A.R.L.

Colombo, M. 2006. FRC bending behaviour: a damage model for high temperatures. *Ph.D. Thesis, Department of Structural Engineering, Politecnico di Milano, Italy.*

Colombo, M & di Prisco, M. 2006. SFRC: a damage model to investigate the high temperature mechanical behaviour. In Meschke, de Borst, Mang & Bicanic (eds.), *Computational Modelling of Concrete Structures EURO-C*: 309–318. Rotterdam: Balkema.

Dehn, F. & Werther, N. 2006. Brandversuche an Tunnelschalenbetonen fur den M30-Nordtunnel in Madrid (Fire tests on tunnel concretes for the M30-Northtunnel in Madrid). *Beton- und Stahlbetonbau 9*: 729–731 (in German).

di Prisco, M., Felicetti, R. & Gambarova, P. 2003a. On the fire behavior of SFRC and SFRC structures in tension and bending. In A.E. Naaman & H.W. Reinhardt (eds.), *High Performance Fiber Reinforced Cement Composite*: 205–220. Bagneux, France: RILEM Publications S.A.R.L.

di Prisco, M., Felicetti, R. & Colombo, M. 2003b. Fire resistance of SFRC thin plates. In Bicanic, de Borst, Mang & Meschke (eds.), *Computational Modelling of Concrete Structures EURO-C*: 783–792. Rotterdam: Balkema.

di Prisco, M., Felicetti, R. & Plizzari, G. (eds.) 2004a, *Proc. Of 6th Sympoium on Fibre Reinfoced Concrete BEFIB 2004 – PRO39*. Bagneux, France: RILEM Publications S.A.R.L.

di Prisco, M., Ferrara, L., Colombo, M. & Mauri, M. 2004b. On the identification of SFRC constitutive law in uniaxial tension. In di Prisco, Felicetti & Plizzari (eds.), *Proc. Of 6th Sympoium on Fibre Reinfoced Concrete BEFIB 2004 – PRO39*: 827–836. Bagneux, France: RILEM Publications S.A.R.L.

EN 1992-1-2. 2002. Eurocode 2: Design of concrete structure – Part 1.2: General rules – Structural fire design.

Hordijk, D. 1991. Local approach to fatigue of concrete. *Ph.D. Thesis, Department of Concrete Structures, Faculty of Civil Engineering, Delft University of Technology, The Netherlands*:131–134.

Kalifa, P., Chene, G. & Galle, C. 2001. High temperature behaviour of HPC with polypropylene fibres. From spalling to microstructure. *Cement and concrete research* 31: 1487–1499.

Naaman, A.E. & Reinhardt, H.W. (Eds.) 2003. *High Performance Fiber Reinforced Cement Composites (HPFRCC4)* – RILEM PRO 30. Bagneux, France: RILEM Publications S.A.R.L.

UNI 11039. 2003. Concrete reinforced with steel fibres. Part II Test method for the determination of first cracking strength and ductility indexes (in Italian).

Fracture Mechanics of Concrete and Concrete Structures – High-Performance Concrete,
Brick-Masonry and Environmental Aspects – Carpinteri, et al. (eds)
© 2007 Taylor & Francis Group, London, ISBN 978-0-415-44617-4

Structural behavior of SFRC tunnel segments

G.A. Plizzari & G. Tiberti

University of Brescia, Brescia, Italy

ABSTRACT: The paper addresses precast tunnel segments for the Line T02 of Saronno-Malpensa railway (Italy). After the experimental determination of the material properties, the structural behavior of the lining is simulated with FE analyses based on Non Linear Fracture Mechanics (NLFM). The numerical analyses presented herein will be mainly focused on the force exerted by hydraulic jacks on the ring during the excavation process. The numerical analyses were carried out considering different loading configurations and atypical boundary conditions in order to investigate the advantages coming from the use of Steel Fiber Reinforced Concrete (SFRC) in terms of crack control and of bearing capacity. An optimized reinforcement, based on the combination of rebars and steel fibers, is proposed.

1 INTRODUCTION

After more than four decades of research, interest in Fiber Reinforced Concrete (FRC) is continuously growing in many application fields. FRC is already widely used in structures where fiber reinforcement is inessential for integrity and safety, such as in slabs on grade or in first phase linings.

For structural purposes, steel fibers are becoming widely utilized since they enhance concrete toughness under tensile loading. This makes the material able to sustain tensile stresses after cracking and so allows stress redistribution.

Among the structural applications of steel fiber reinforced concrete (SFRC), there is a growing interest in precast tunnel segments to be used with the Tunnel Boring Machines, TBM (de Waal, 1999, Blom, 2002). These segments are generally made of ordinary reinforced concrete; however, the addition of fibers is gaining considerable attention among designers and producers due to the enhanced toughness that allows a partial or even total substitution of the conventional reinforcement.

In fact, the internal forces in a shield lining, in the final state, are usually a combination of a bending moment and a hoop force, which could make possible to replace the traditional reinforcement mesh by steel fibers. Since fiber reinforcement is present in the cover zone also, tunnel segments have a better resistance against impact loads during installation.

In the present paper, structural behavior of the segments for the Line T02 of Saronno-Malpensa railway (Italy) are numerically simulated with nonlinear analyses.

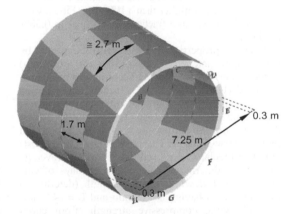

Figure 1. 3D scheme of the tunnel for the Line T02 of the Saronno-Malpensa railway, with evidenced the nine precast segments.

The line consists in two TBM-excavated tunnels with a relative distance of about 7 m. Each tunnel has got an average diameter of about 7.55 m and is located from about 11 to 16 m below the surface. Each tunnel is made of 7 parallelogram segments, a trapezoidal counter-key segment with an average length of about 2.7 m and a smaller key segment. These last segments are respectively evidenced as type "A" and "F" in Figure 1.

Experiments have been performed to determine the materials properties and the constitutive laws of SFRC for use in tunnels.

Numerical analyses were performed by using a finite element model based on Non Linear Fracture

Mechanics (NLFM; Hillerborg et al., 1976) with a smeared crack approach. These analyses allowed to study the structural behavior of the segments with several combinations of reinforcement under different loading conditions.

The numerical model was validated by using the experimental results of full-scale tests performed on SFRC precast tunnel segments without curvature (Hemmy, 2001). During the validation, different concrete crack models have been tested and compared.

An optimized reinforcement, based on a combination of rebars and steel fibers, is proposed.

2 MATERIALS

The experimental characterization of the material properties of SFRC was performed on a concrete matrix having the same strength of the one used for the railway Line T02 between Saronno and Malpensa.

Segments were reinforced with steel fibers 50/0.75, having a length L_f of 50 mm, a diameter ϕ_f of 0.75 mm and an aspect ratio L_f/ϕ_f equal to 67. These fibers are cold drawn, have a hooked shape, a rounded shaft and a tensile strength higher than 1100 MPa. Fibers were used with the volume fractions (V_f) equal to 0.38% (30 kg/m³).

Fracture properties of FRC were determined by using eight notched beams ($150 \times 150 \times 600$ mm) tested under 4-point bending according to the Italian Standard (UNI, 2003; Fig. 2). In addition, 4 beams of plain concrete were used as reference specimens. The slump of the fresh concrete was always greater than 150 mm. Specimens were stored in a fog room (R.H. >95%; T = $20 \pm 2°$C) until 24 hours before testing. The mechanical properties of concrete (average values), as determined after about 60 days of curing are the following: tensile strength (determined from cylinders having $\phi = 80$ mm and L = 240 mm) $f_{ct} = 3.28$ MPa, compressive strength (from cubes with a side of 150 mm) $f_{c,cube} = 53.16$ MPa and Young's modulus (from compression test on cylinders $\phi = 80$ mm, L = 240 mm), $E_c = 39625$ MPa.

Fracture tests were carried out with a closed-loop hydraulic testing machine by using the Crack Mouth Opening Displacement (CMOD) as control parameter, which was measured by means of a clip gauge positioned astride a notch at midspan, having a depth of 45 mm. Additional Linear Variable Differential Transformers (LVDTs) were used to measure the Crack Tip Opening Displacement (CTOD) and the vertical displacement at midspan and under the load points (Fig. 2).

Inverse analyses of the bending tests, based on NLFM (Hillerborg et al., 1976), allowed to determine the best-fitting post-cracking law (σ-w) for the SFRC adopted in the present research (Roelfstra and Wittmann 1986).

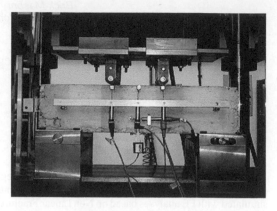

Figure 2. Instrumentation for the notched specimen used for the 4 point bending beam tests.

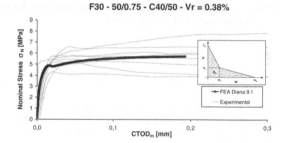

F30 - 50/0.75 - C40/50 - Vr = 0.38%

Figure 3. Experimental and numerical results obtained from SFRC beams and the bilinear law adopted.

The Young's modulus (E_c) was the one experimentally measured from the cylinders while the Poisson ratio (ν) was assumed equal to 0.2. The softening law was approximated as bilinear where the first steeper branch can be associated with the bridge action of concrete between microcracks while the second branch represents the residual stress due to fiber bridging (Fig. 3). The numerical analyses were performed by assuming both a discrete crack and a smeared crack approach with Diana ver. 9.1 (2005); the latter was also used for the numerical simulations of the tunnel segments.

The material parameters identified from the bending tests are the following (see Fig. 3): $f_{ct} = 3.28$ MPa, $\sigma_1 = 2.045$ MPa, $w_1 = 0.020$ mm, $w_{cr} = 3.851$ mm. The best-fitting numerical curves obtained with Diana are compared with the experimental ones in Figure 3.

3 VALIDATION OF THE FE MODEL

Experimental studies on steel fiber reinforced concrete under splitting actions were carried out at the University of Braunschweig by Hemmy, 2001. Experiments included the material characterization and the

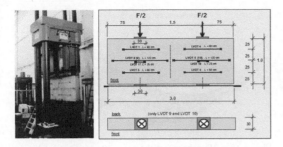

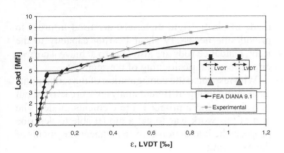

Figure 4. Configuration of the splitting test on full-scale segments without curvature (Hemmy, 2001).

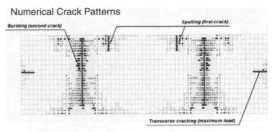

Numerical Crack Patterns

Experimental Crack Patterns

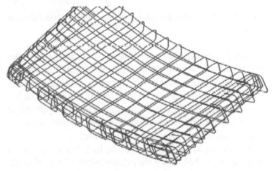

Figure 6. Comparison between numerical and experimental crack patterns from splitting test in a SFRC segment (Hemmy, 2001).

Figure 5. Comparison between numerical and experimental load-LVDTS curves: splitting test in a SFRC segment without curvature.

structural tests under point or line loads to assess the performance of SFRC under bi-axial and tri-axial stress concentrations.

The experimental results were used to validate the FE model used in the present research work. In particular, the simulation of the splitting test on a full-scale SFRC segment without curvature is presented herein.

Figure 4 shows the experimental set-up that consists of a segment on two supports subjected to two concentrated loads. This test aimed to simulate typical actions for precast tunnel segments, represented by the high concentrated forces of the TBM hydraulic jacks.

The concrete used in the test has a strength class C40/50. The specimen has been reinforced with 35 kg/m³ of steel fibers having an aspect ratio of 65. Material properties of concrete with 40 kg/m³ of fibers 50/0.75 were used for the numerical simulation, since they were determined from previous tests. As the materials are not exactly the same, the main aim of this comparison was to capture the general trend of the experimental results.

Figure 4 shows the test configuration with the LVDTs (Linear Variable Differential Transformers) used to measure the crack openings under loading areas. A comparison between the experimental and the numerical load-displacement curve is shown in Figure 5. One should notice the good agreement between the two curves; also the numerical crack patterns are similar to the experimental ones (Fig. 6).

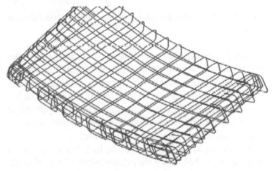

Figure 7. Scheme of conventional reinforcement (rebars) used in precast tunnel segments of the Saronno-Malpensa railway.

4 DESIGN ASPECTS

An open question for the construction companies and the designers concerns the reinforcement for these precast elements. In fact, an heavy conventional reinforcement is quite complex and labor-intensive to construct; it also needs a relatively large storage area before placing it in the segment. Figure 7 shows the conventional reinforcement adopted for the segments of the Saronno-Malpensa railway; it corresponds to an equivalent weight of 82 kg/m³.

Generally, reinforcement is designed according to the design actions on the tunnel segments, resulting from segment transportation, placing process and soil pressure in the final state.

In particular, during construction, the thrust forces and the grout pressure are the most critical factors. Cracks often appears in the tunnel lining in the phase

Figure 8. Cracks that typically appear in segmental tunnel linings during the construction phase.

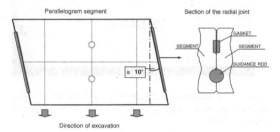

Figure 9. Plan of the parallelogram segment with the section of the radial joint.

in which these forces occur. Some examples of cracks that typically appear in segmental tunnel linings are shown in Figure 8. Possible causes of these cracks could be eccentricity or inclination of the thrust jacks. Also, a number of phenomena due to the trumpet shape as the torsional deformation or a non-smooth support of the ring joint, may cause these cracks.

It is desirable to mitigate or reduce these cracks as much as possible since they determine a loss of quality, leakage and high repair costs. Cracking phenomena can be limited by introducing a change in tunnel design by using, for example, an opportune configurations of the thrust jacks and supports. Alternatively, they can be reduced by using an opportune combination of SFRC and conventional reinforcement localized in proper regions of the precast tunnel segment (Plizzari and Tiberti, 2006). This research work focuses on the last aspect since conventional reinforcement for these precast elements is quite costly because of the placement of many curved rebars. In order to investigate the advantages due to a combination of rebar (for localized stress) and fibers (for diffusive stress), several non-linear analyses were carried out, by considering also very un-favorable boundary conditions.

Based on the material properties described in § 2, the total strain rotating crack model provided by DIANA was adopted for FE simulations.

The precast tunnel segments are universal tapered segments shaped like a parallelogram. Accordingly, the radial joints of the lining are inclined with respect to the longitudinal axis of the tunnel of about 10 degrees (Fig. 9). In these radial joints, guidance rods,

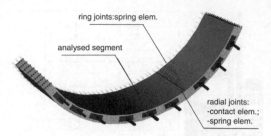

Figure 10. 3D scheme of the segment: nominal load configuration. The scheme evidenced the rear springs in the ring joints, the contact and the spring elements in the radial joints.

made of plastic rigid materials, are used in order to provide the necessary shear bearing capacity to the joint. Guidance rods are also able to achieve a correct circumferential positioning of the segments.

With respect to the configuration of the radial joints used in the lining, different analyses were carried out in order to find a proper finite elements model. The aim of this work was to get a model able to describe properly the interaction between adjacent segments of the same ring.

The final model adopted is presented herein. It consists of the central segment, together with his corresponding adjacent two segments (Fig. 10). The interaction between the analysed segment and the adjacent ones, through the radial joints, has been modelled by contact elements in order to simulate the friction along the lateral surfaces of the segments. The guidance rods between the segments have been simulated as springs, whose stiffness has been obtained from shear tests.

The numerical analyses were carried out by adopting a 3D solid model with 11151 second order hexahedral elements (twenty nodes brick elements CHX60). The constitutive law for concrete under compression was assumed according to Thorenfeldt (1987). Moreover, the increase of compressive strength due to lateral confinement has been implemented according to Selby and Vecchio (1993).

During the excavation process, 29 TBM actuators acted on the tunnel lining. The average service load applied by a single actuator is 2.21 MN. The load configuration used in the Saronno-Malpensa railway is a Japanese jack configuration. This means that the hydraulic jacks are almost uniformly distributed along the ring; in practice, three thrust jacks act on each tunnel segments.

In a first phase of the research, a nominal load configuration with thrust jacks exactly placed in position and with uniform support of the ring joint, has been considered (Fig. 10).

Notice that no bearing pads are positioned in the rear face of the tunnel segments (ring joints). On that face, only a slightly protusion of concrete, about

Figure 11. Rear faces of the tunnel segments without bearing pads: only a little concrete protusion has been provided.

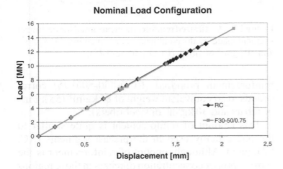

Figure 12. Comparison of the load-displacement curves numerically obtained between F30-50/0.75 and RC specimens: nominal load configuration.

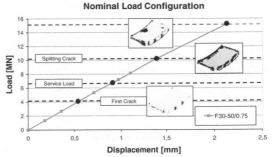

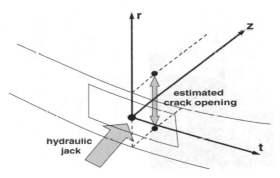

Figure 13. Load-displacement curve obtained from the specimen F30-50/0.75: nominal load configuration.

Figure 14. Scheme of local coordinate system and of the points of measurement used to estimate the splitting cracks.

3 mm thick, has been provided. This little protusion involves about all the rear surfaces of the segments, as shown in Figure 11. Accordingly, ring joints have been simulated by using "no-tension" elastic springs uniformly positioned on the rear surfaces of the segments (Figure 10).

Structural behavior of segments with different type of reinforcement was compared. Numerical results from segments with conventional reinforcement only (RC, foreseen by designers) and with 30 kg/m³ of 50/0.75 fibers (F30-50/0.75) are presented herein.

Figure 12 shows the numerical results in terms of load (from the three actuators of the central segment) versus the longitudinal displacement under the loading surface of the central hydraulic jack. It can be noticed that the specimen with only fibers has a slightly higher ultimate load with respect to a RC segment.

To better understand the structural behavior of the segments with fiber reinforcement only, Figure 13 shows the load-displacement curve of the specimen with 30 kg/m³ of fibers. The most significant crack patterns at different load levels are also shown.

The segment collapsed at a total load of 14.6 MN; since the service load is 6.65 MN, the safety factor is 2.20. The compressive strength of concrete is almost reached in longitudinal stresses under loading surfaces. As shown in Figure 12, the first cracks appear between the loading surfaces, due to spalling, at approximately 5.32 MN (0.9 times the service load).

The spalling cracks develop with an increase of load but not penetrate deeply in the segment.

Splitting cracks appear under the loading surfaces at 10.5 MN (1.5 times the service load), due to splitting stresses in radial direction (σ_r), according to the local coordinate system shown in Figure 14.

Figure 15 shows the development of the splitting crack width during loading. In particular, a comparison between RC and F30-50/0.75 specimen is presented. The graph represents the relative displacements of two nodes: one located inside and the other outside the segment (Figure 14).

As clearly shown in Figure 15, the load-crack opening curves obtained from F30-50/0.75 and RC specimens are approximately the same. This means that locally, under the loading areas, because of the concrete toughness provided by fibers, cracks develop in a stable way. The same behavior appears in the RC specimen due to the presence of stirrups.

With this nominal load configuration, the use of 30 kg/m³ of steel fibers in the precast tunnel segments, ensures the same global and local behavior of the more expensive conventional reinforcement.

However, a number of irregularities can occur in practice. As a consequence, the ring joint may not be plane as considered in the nominal load configuration.

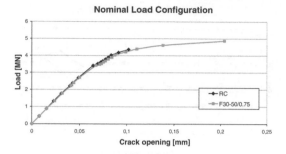

Nominal Load Configuration

Figure 15. Development of splitting crack width under the load: comparison between results obtained from F30-50/0.75 and RC specimens.

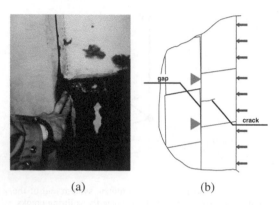

(a)   (b)

Figure 16. Possible gap between rings due to a no-perfect placing process (a); possible non-smooth support configuration (b).

In fact, it is almost impossible to build a tunnel lining with perfect planes at interface between rings (this is generally confirmed during inspection of a tunnel lining); therefore, the tunnel segments are not supported uniformly by the previous tunnel ring, as shown in Figure 16a. Consequently, a bending moment arise in the segment and may cause the cracks shown in Figure 16b. This could be one of the possible explanations to the crack patterns show, for instance, in Figure 8.

In fact, it is clear that steel fiber reinforcement can not compete with the concentrated rebars in a load condition governed by bending, as occurs in a segment with two supports (Figure 16b).

For this reason, the effect of different possible non-smooth support configuration has been investigated by means of the finite element calculation, in order to achieve an optimized reinforcement made of the combination of steel fibers and rebars.

In a previous study it was demonstrated that the conventional rebars could be limited to the chords along the two longer sides (Plizzari & Tiberti, 2006). In the present paper, similar chords combined with steel fiber are adopted (Fig. 17). The optimized

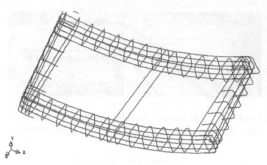

Figure 17. Optimized reinforcement proposed for the tunnel segment made by the combination of two chords (55 kg/m^3) and 30 kg/m^3 of fibers (RCO-F30-50/0.75).

reinforcement proposed (RCO-F30-50/0.75) corresponds to an equivalent weight of 85 kg/m^3 (55 kg/m^3 of rebars and 30 kg/m^3 of steel fibers).

This amount of reinforcement is quite similar to the one actually used in the precast segments (RC, 82 kg/m^3). Although amount of reinforcement is the same, a proper concentration of rebars in those regions allows to improve the bending bearing capacity of the segment even in severe restraint conditions (e.g.: non-smooth support).

Moreover, by adding fibers, it is also possible to reduce significantly the amount of stirrups usually placed under the loading areas, as already demonstrated in the nominal loading condition. Notice that this proposed optimized reinforcement has been verified with respect to the final service condition (lining loaded by the ground pressure) according to the Italian guidelines for design of FRC structures (CNR-DT 204, 2006).

The results of a very severe non-smooth support configuration are presented herein. In particular, the two supports were localized at the ends of the tunnel segment, as shown in Figure 18, acting on a surface of about 300 × 300 mm^2 each ones.

Moreover, this boundary conditions have been maintained during all the thrust phase. This assumption implies that no further supports could arise in the middle of the segment where the deflection is maximum.

According to these very restrictive hypothesis, the rebars in the two chords have been designed in order to guarantee, in the assumed severe condition, a sufficient bending strength under service loads.

The finite element model is the same described in the previous nominal load configuration. The basic hypothesis is that the non-smooth support configuration occurs only in the analyzed segment (the central one).

Figure 19 shows the numerical results in terms of load (from the three actuators of the central segment) compared to the deflection measured in the mid-span

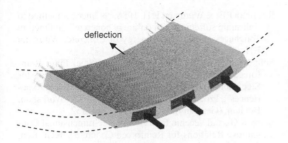

Figure 18. Non-smooth support configuration adopted for the central precast tunnel segment of the invert.

Non-smooth Support Configuration

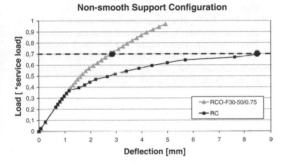

Figure 19. Comparison of the load-deflection curve obtained from RCO-F30-50/0.75 and RC specimens with a two-support configuration.

of the segment. Specimen with optimized reinforcement reaches an ultimate load approximately equal to the service load, as expected from the reinforcement design. On the contrary, the conventional RC specimen is able to carry only 70% of the service load. With this load, the deflection obtained in the RC specimen is about three times the RCO-F30-50/0.75 ones.

To better understand the structural behavior of the segments, Figure 20 shows the comparison between the two specimens in term of crack-opening. The latter has been estimated as the relative displacement between two nodes positioned astride the crack. The initial distance between these nodes is 550 mm. The curves clearly evidenced the advantages coming from the use of steel fibers, in combination with rebars, for reducing the crack-opening. In particular, looking at a load equals to 0.7 times the service load, RCO-F30-50/0.75 specimen presents a crack-opening that is ten time smaller than the one present in the RC segment.

5 CONCLUDING REMARKS

The present paper shows results of a numerical study of precast tunnel segments for the Line T02 of the Saronno – Malpensa railway. The numerical model allowed to determine the bearing capacity and crack development in a segment during TBM operations.

Non-smooth Support Configuration

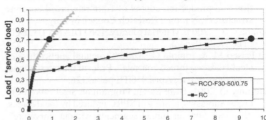

Figure 20. Comparison of the load-crack opening curve obtained from RCO-F30-50/0.75 and RC specimens with a two-support configuration.

It has been demonstrated that in the nominal load configuration, where the support of the ring joint is uniform, by using 30 kg/m^3 of steel fibers it is possible to ensure, globally and locally, the same behavior of a segment reinforced with rebars.

However, in practice, some irregularities usually occurs; consequently, non-smooth ring joints are present. This leads to a non-smooth support of the segments on the previous (placed) ring. This could be one of the possible explanations of longitudinal cracks that are often present in segmental tunnel linings. In fact, with this severe loading condition, a very high bending moment is present in the segment.

The proposed optimized reinforcement, based on a combination of fibers and rebars, represents a competitive solution for the precast tunnel segments. In fact, even with localized supports, it was possible to demonstrate that the reinforcement proposed (RCO-F30-50/0.75) provides a better behavior than the conventional one (RC). This has been clearly evidenced both in terms of bearing capacity and crack patterns. In fact, it is well known that the crack width gets benefits from the presence of fibers.

This results underline the importance of fibers in concrete or r.c. structures for the durability aspects.

ACKNOWLEDGEMENTS

This research project was financed by Officine Maccaferri S.p.A. (Bologna, Italy) whose support is gratefully acknowledged. Moreover, a special thank goes to engineer Chiara Benedetti for the assistance in running the nonlinear FE analyses.

REFERENCES

Blom C.B.M. 2002. "Design philosophy of concrete linings in soft soils", ISBN 90-407-2366-4, Delft.
Bloemhof K.C. 2001. "Geometrical tunnel model; damages on a shield tunnel", Delft.

CNR-DT 204, 2006. "Guidelines for the design, construction and production control of fibre reinforced concrete structures", National Research Council of Italy.

Diana v. 9.1. 2005. Material Library. TNO DIANA BV, Delft (The Netherlands).

di Prisco M. & Toniolo G. 2000. "Structural applications of steel fibre reinforced concrete", Proc. of the international workshop. Milan, April 4, CTE publ., 126.

Hemmy O. 2001. Brite Euram Program on Steel Fibre Concrete, Subtask: Splitting of SFRC induced by local forces, "Investigation on tunnel segments without curvature", University of Braunschweig (Germany).

Kooiman A.G. 1996. "Steel fibre reinforced concrete in shield tunnel linings, a state of art". Stevin report 25.5-96-10. Delft University of Technology.

Plizzari G.A. & Cominoli L. 2005. "Numerical Simulations of SFRC Precast Tunnel Segments." In Proceedings of ITA-AITES 2005, Underground space use: Analysis of the past and lessons for the future, Erdem Y. and Solak T. Eds., Istanbul (Turkey), May 7–12, pp. 1105–1111.

Plizzari G.A. & Tiberti G. 2006. "Steel fibers as a reinforcement for precast tunnel segments", in the World Tunnel Congress ITA AITES 2006, Seoul Corea, April 22–27.

Roelfstra P.E. & Wittmann F. H. 1986. "Numerical method to link strain softening with failure of concrete". In Fracture Toughness and Fracture Energy of Concrete, Wittmann F.H. Ed., Elsevier, Amsterdam, 163–175.

Schnütgen B. 2003. "Design of Precast Steel Fibre Reinforced Tunnel Elements", in "Test and Design Methods for Steel Fibre Reinforced Concrete – Background and Experiences", Proc. of the RILEM TC 162-TDF Workshop, Bochum (Germany), March 20–21, 145–152.

Selby R.G., and Vecchio F.J. 1993. "Three-dimensional Constitutive Relations for Reinforced Concrete". Tech. Rep. 93–02, Univ. Toronto, Civil Eng., Toronto, Canada.

Thorenfeldt E., Tomaszewicz A., and Jensen J.J. 1987. "Mechanical properties of high-strength concrete and applications in design". In Proc. Symp. Utilization of High-Strength Concrete (Stavanger, Norway).

UNI 11039. 2003. Steel fibre reinforced concrete – Part I: Definitions, classification specification and conformity – Part II: Test method for measuring first crack strength and ductility indexes. Italian Board for Standardization (UNI).

Waal R.G.A. de. 1999. "Steel fibre reinforced tunnel segments", Delft, ISBN 90-407-1965-9, Delft.

Part XI
Brick-masonry and other Quasi-brittle materials

Fracture Mechanics of Concrete and Concrete Structures – High-Performance Concrete,
Brick-Masonry and Environmental Aspects – Carpinteri, et al. (eds)
© 2007 Taylor & Francis Group, London, ISBN 978-0-415-44617-4

Failure issues of brick masonry

B. Blackard, B. Kim, C. Citto, K. Willam & S. Mettupalayam
CEAE Department, University of Colorado-Boulder, Boulder, Colorado

ABSTRACT: This paper focuses on the fundamental meso-mechanical failure processes in masonry prisms when composite clay bricks bonded by cement mortar are subjected to axial compression. There are several intriguing issues which complicate the failure of the masonry composite because of the subtle interaction effects between the clay brick units and the weaker cement mortar. In fact, 3-D analysis is required to fully penetrate the failure mode which is governed by biaxial tension-compression of the solid brick rather than triaxial compression of the mortar. Dimensional reduction in the form of 2-D plane stress analysis leads to erroneous mode conversion and to significant underestimation of the compression capacity of the masonry prism which depends primarily on the tensile strength of the brick unit. In other terms, the out-of-plane confinement plays a critical role in the assessment of the overall compression capacity of masonry walls.

1 INTRODUCTION

Due to the composite construction of masonry, progressive failure is a complicated process. Masonry consists of two components, brick units and mortar joints which exhibit very different stiffness and strength properties. Moreover, both constituents have a low tensile strength when compared to their compressive counterparts. Consequently, a masonry wall in compression does not necessarily fail in compression of the weakest component but can fail in tension due to mismatch conditions in the composite. Aside from the large strength variations of the two components it is the bond among the brick unit and the mortar which determines the critical failure path in a masonry wall.

The current computational-experimental study focuses on the critical failure mode and the question whether three dimensional analysis is required to capture in-plane as well as out-of-plane failure modes or whether two dimensional failure simulations do suffice. To this end continuum based finite elements are employed for meso-modeling of the brick and mortar response of a masonry prism subjected to compression. The test article consists of five solid clay brick units which are mortared by four layers of cement mortar. The nonlinear failure simulations resort to the damage-plasticity model by Lee and Fenves [1998] which has been incorporated into Abaqus [Version 6.5]. Appropriate material tests on brick units and cement mortar were carried out in-house in conjunction with the current NEESR-SG research project [Seismic Performance Assessment and Retrofit of Non-Ductile RC Frames with Infill Walls]. Thereby the main objective is to validate the computational results with the experimental data of masonry prisms.

2 MATERIAL MISMATCH

A key issue in the failure mechanism of a masonry prism is the initial mismatch of the elastic properties of brick and mortar. Assuming isotropy the elastic moduli E^m, E^b in concert with the Poisson numbers v^m, v^b govern the cross-effect of axial compression in the form of the in-plane and out-of-plane lateral strain. If the interfaces between the mortar and brick were allowed to slip freely, the two material components would deform laterally by differing amounts except for a special condition which is detailed below. The result of this mismatch of lateral deformation is that friction restrains and hence compresses the mortar joints from free lateral expansion because of the stiffer brick units, which in turn experience lateral tension. This can be readily explained from examining the compliance relationships of isotropic linear elastic materials and their interaction along bimaterial interfaces.

Considering the masonry prism in Figure 2 where the y-axis refers to the direction of axial compression and the x- and z- coordinates the lateral directions. The normal strains in the x-direction of the brick unit and the cement mortar are according to generalized Hooke's law,

$$\varepsilon_x^{\ b} = \frac{1}{E^b}\left[\sigma_x - v\sigma_y - v\sigma_z\right]^b$$

$$\varepsilon_x^{\ m} = \frac{1}{E^m}\left[\sigma_x - v\sigma_y - v\sigma_z\right]^m$$

(1)

Under axial compression equilibrium dictates that $\sigma_y^b = \sigma_y^m = \sigma_y$. Assuming for the time being that the lateral stresses remain zero because of no frictional restraint then the lateral strains are directly proportional to the compliance relations,

$$\varepsilon_x^{\,b} = -\frac{v^b}{E^b}\sigma_y \;;\quad \varepsilon_x^{\,m} = -\frac{v^m}{E^m}\sigma_y \qquad (2)$$

In other terms the lateral compliance is directly proportional to the ratio $c_{lat} = v/E$. In contrast under full bond and adherence of the two components at the bimaterial interface strong compatibility requires that the lateral strains in the brick and mortar in Equation (1) are equal. Assuming $\sigma_x^m = \sigma_z^m$ and $\sigma_x^b = \sigma_z^b$ this leads to the critical condition of no lateral mismatch under axial stress σ_y when,

$$v^b E^m = v^m E^b \quad \text{or} \quad n = \frac{c_{lat}^m}{c_{lat}^b} = 1 \qquad (3)$$

In other terms, if the ratio of the lateral compliances $n = 1$, then there is no mismatch between the lateral deformations of the two materials.

Using equilibrium the lateral stresses in the brick and mortar components may be related to each other by a statement of self-equilibrium, $\sigma_x^m = -r\sigma_x^b$ and $\sigma_z^m = -r\sigma_z^b$ where $r = t^b/t^m$ denotes the thickness ratio of the brick unit over the mortar joint. This condition has been introduced by Berto et al. [2005] in order to develop the explicit expressions below for the lateral stresses in the brick unit and the mortar joint in terms of the axial stress σ_y,

$$\sigma_x^{\,b} = \sigma_z^{\,b} = \frac{v^b E^m - v^m E^b}{E^m + \dfrac{t^b}{t^m}E^b - E^m v^b - \dfrac{t^b}{t^m}E^b v^m}\sigma_y$$

$$\sigma_x^{\,m} = -\frac{t^b}{t^m}\sigma_x^{\,b} \quad \text{and} \quad \sigma_z^{\,m} = -\frac{t^b}{t^m}\sigma_z^{\,b} \qquad (4)$$

The assumptions for this expression are as follows:

1. The vertical stress is uniform throughout both components of the prism
2. There is no slip along the brick/mortar interface
3. There is a state of uniform lateral stress within each material
4. There is no stress transfer through interface shear.

It is the differential equilibrium condition which restricts the validity of the two expressions in Equation (4) to the two axial planes of symmetry in the center of the masonry prism. Clearly at the surface of the prism all lateral tractions must vanish. Therefore large shear stresses must develop near the lateral faces of the prism (a) in order to diminish the mismatch of the lateral

normal stresses and (b) to reduce the magnitude of the surface tractions to zero ($\sigma_x^b = \sigma_x^m = 0$ at the lateral x-faces and $\sigma_z^b = \sigma_z^m = 0$ at the free lateral z-faces).

In Equation (4) σ_y is negative for the case of a prism loaded in axial compression. It can be easily verified from Equation (4) that the lateral normal stresses are zero when the mismatch condition $n = 1$. It is also elementary to recognize that the lateral brick stresses are positive in tension, $\sigma_x^b = \sigma_z^b > 0$, when $n > 1$, while the lateral mortar stresses are in compression, $\sigma_x^m = \sigma_z^m < 0$. Together with the axial compression $\sigma_y < 0$ this leads to a fracture critical state of equibiaxial tension-compression in the brick and to a forgiving state of triaxial compression in the mortar. Note the opposite result occurs for $n < 1$ when the brick unit is laterally more compliant than the mortar, resulting in a fracture critical state of equibiaxial tension-compression in the mortar rather than in the brick.

An extension of these results can be obtained by considering a transversely anisotropic brick model. Assuming the extrusion direction (vertical prism direction) of the brick to be the principal axis of transverse anisotropy, the axial brick stiffness differs from the lateral one. In this case three additional elastic material properties are needed, E_y^b, $v_{yx}^b = v_{yz}^b$, $G_{yx}^b = G_{yz}^b$ beyond the two isotropic moduli in the plane of the brick unit, E_{xz}^b and v_{xz}^b:

E_y^b = brick modulus of elasticity in axial y-direction

E_{xz}^b = brick modulus of elasticity in the lateral x, z-directions (isotropic x-z plane)

$v_{yx}^b = v_{yz}^b$ = brick Poisson ratio relating the axial and any lateral x-, z-direction

v_{xz}^b = brick Poisson ratio relating the two lateral x-z directions (isotropic x-z plane)

$G_{yx}^b = G_{yz}^b$ = brick shear modulus relating axial and lateral directions

E^m = [isotropic] mortar modulus of elasticity

v^m = [isotropic] mortar Poisson ratio

Starting from the transversely anisotropic elastic compliance equations for brick and mortar:

$$\varepsilon_x^{\,b} = \frac{1}{E_{xz}^b}\sigma_x^{\,b} - \frac{v_{yx}^b}{E_y^b}\sigma_y^{\,b} - \frac{v_{xz}^b}{E_{xz}^b}\sigma_z^{\,b}$$

$$\varepsilon_z^{\,b} = -\frac{v_{xz}^b}{E_{xz}^b}\sigma_x^{\,b} - \frac{v_{yx}^b}{E_y^b}\sigma_y^{\,b} + \frac{1}{E_{xz}^b}\sigma_z^{\,b}$$

$$(5)$$

$$\varepsilon_x^{\,m} = \frac{1}{E^m}\sigma_x^{\,m} - \frac{v^m}{E^m}\sigma_y^{\,m} - \frac{v^m}{E^m}\sigma_z^{\,m}$$

$$\varepsilon_z^{\,m} = -\frac{v^m}{E^m}\sigma_x^{\,m} - \frac{v^m}{E^m}\sigma_y^{\,m} + \frac{1}{E^m}\sigma_z^{\,m}$$

and making the same assumptions as with the isotropic model, the lateral normal stresses in the brick units and mortar joints turn out to be:

$$\sigma_z^{\,b} = \sigma_x^{\,b} = \frac{E_{xz}^b}{E_y^b}\frac{A+B}{C^2-D^2}\sigma_y$$

$$\sigma_z^{\,m} = \sigma_x^{\,m} = -r\sigma_x^{\,b}$$

where (6)

$$A = \left(E^m v_{yx}^{\,b} - E_y^{\,b} v^m\right)\left(E^m + E_{xz}^{\,b} r\right)$$

$$B = \left(E^m v_{yx}^{\,b} - E_y^{\,b} v^m\right)\left(E^m v_{xz}^{\,b} + E_{xz}^{\,b} v^m r\right)$$

$$C = E^m + E_{xz}^{\,b} r$$

$$D = E^m v_{xz}^{\,b} + E_{xz}^{\,b} v^m r$$

Although this complicates direct interpretation of the cross effect between axial compression and lateral expansion/stresses the condition for zero mismatch leads to the same condition $n = 1$ in analogy to Equation (3) whereby $c_{lat}^b = v_{yx}^b / E_y^b$. Also note that the shear stiffness properties do not appear in these expressions because the principal stress coordinates are assumed to coincide with the principal axes of transverse anisotropy, and because only normal components are considered in the derivation of Equation (6).

3 EXPERIMENTAL RESULTS

Bricks, mortar, and masonry prisms were tested at the University of Colorado structural materials laboratory. Table 1 displays the average results of these experiments indicating reasonable correlation with COV values varying between 3–15%. So far a total of 9 bricks, 4 mortar cylinders, and 5 prisms were tested in axial compression and tension. The bricks had dimensions $4.5'' \times 3.75'' \times 2.25''$ and the mortar joints were $0.375''$ thick. The mortar mix consisted of a sand:cement:lime ratio of 5:1:1. The normalized axial response behavior of the five prism experiments are shown in Figure 1 together with that of brick and mortar. Note that complete stress-strain data was collected for only two brick specimens.

The brick tests were performed after capping the clay units with a thin layer of gypsum capping compound. The bricks were stored under ambient conditions. The compression tests on the cement mortar were performed on $4'' \times 8''$ cylinders which were cured 28 days in the fog room before capping and testing. The prisms were covered with plastic and cured for 28 days under ambient conditions before testing. The construction of the prisms included wetting each brick for approximately 10 seconds before placement. The test data are summarized in Table 1 and shown in Figure 1.

Table 1. Average material properties from testing.

	Brick	Mortar	Prism
Comp. strength (psi) [COV]	4,840 [15.2%]	732 [8.4%]	2,490 [9.8%]
Tensile splitting strength (psi) [COV]	372 [12.5%]	141 [3.2%]	–
Modulus of rupture (psi) [COV]	640 [8.6%]	–	–

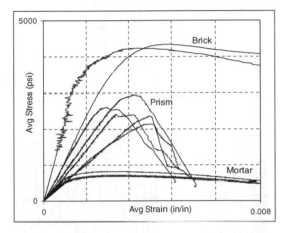

Figure 1. Experimental data from prism, brick and mortar tests.

4 3D NUMERICAL SIMULATIONS

A three dimensional finite element prism model was created with 4653 nodes and 3600 first order hexahedra elements resulting in 13959 DOF. The commercial software Abaqus Version 6.5 was used for the finite element analyses. Figure 2 illustrates the 3D model, and Figure 3 shows the deformed mesh along with the equivalent plastic strain. Figure 3 clearly shows the lateral "bursting" in the middle brick which is the dominant failure mechanism in the masonry prism.

The material properties used for the numerical analyses are listed in Table 2. The damage plasticity concrete formulation by Lee & Fenves [1998] was used for both materials. The parameters K, f_{bo}/f_{co} and ecc determine the initial shape of the failure surface, where K defines the out-of roundness of the deviatoric trace, the strength ratio f_{bo}/f_{co} the increase of equibiaxial compression over uniaxial compression and hence the internal friction, and ecc the rounding factor of the equitriaxial tensile vertex on the hydrostat. The dilation angle Ψ specifies the direction of the (non-associated) plastic flow. The tensile fracture energy describes the initial energy release after the peak stress is reached in tension. Tensile exponential decay was

1589

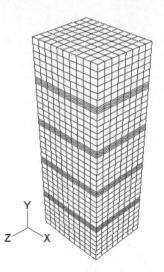

Figure 2. 3D finite element mesh.

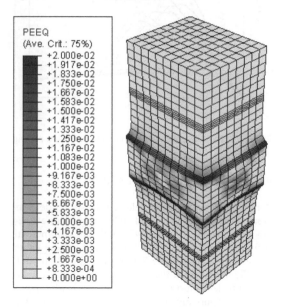

Figure 3. Deformed shape and equivalent plastic strain (deformation scale = 20).

PEEQ
(Ave. Crit.: 75%)
+2.000e-02
+1.917e-02
+1.833e-02
+1.750e-02
+1.667e-02
+1.583e-02
+1.500e-02
+1.417e-02
+1.333e-02
+1.250e-02
+1.167e-02
+1.083e-02
+1.000e-02
+9.167e-03
+8.333e-03
+7.500e-03
+6.667e-03
+5.833e-03
+5.000e-03
+4.167e-03
+3.333e-03
+2.500e-03
+1.667e-03
+8.333e-04
+0.000e+00

Table 2. Material properties (damage plasticity).

	Brick	Mortar
Modulus of elasticity (psi)	3.0×0^6	5.0×10^5
Poisson ratio	0.1	0.2
Uniaxial comp. strength (psi)	4,840	732
Uniaxial tensile strength (psi)	372	141
Tensile fracture energy G_I(lb/in)	0.46	0.34
Dilatancy angle Ψ	20°	20°
Friction angle	19°	19°
Deviatoric out-of-roundness K	0.7	0.7
Biaxial strength ratio f_{bo}/f_{co}	1.15	1.15
Vertex rounding ecc	0.1	0.1

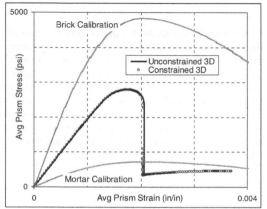

Figure 4. 3D model prism response.

specified with the fracture energy shown in Table 2. There was no damage considered in the brick and mortar elements in order to simplify the constitutive input of the damage-plasticity model in Abaqus.

The 3D model was analyzed with and without lateral restraints at the top and bottom faces of the prism. Figure 4 illustrates the average stress-strain response of both the constrained and unconstrained 3D prism models. Very little difference is evident in the two results. This may be due to the localized effect of the top and bottom boundary conditions. With the prisms under consideration having an aspect ratio of 3.4, these end effects become insignificant. Figure 4 also includes the input data of the compression calibration curves for the mortar and brick. As expected, the prism behavior falls between the mortar and brick response in axial compression.

Figure 5 shows the shear stress contours in the x-y plane at a center mortar joint immediately before plastic behavior began in the prism. The mortar is clearly being restrained from lateral expansion by the surrounding brick material due to interface shearing, which causes the brick to be in a state of bilateral tension-compression.

A comparison of the 3D numerical model and the analytical approximation in Equation (4) is displayed in Figure 6. The brick stresses were taken at the center of the middle brick, and the mortar stresses at the center of the adjacent mortar joint. Clearly the stresses in the mortar agree very well up to the peak prism stress. The stresses in the brick agree well at low stress levels, but start to diverge from the analytical solution before the peak prism stress is reached. This can be explained by the nature of the analytical approximation. Equation

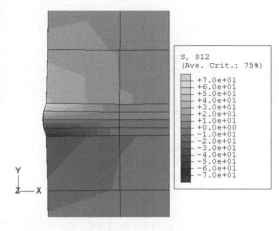

Figure 5. Shear stresses at mortar joint (Deformation Magnification = 500).

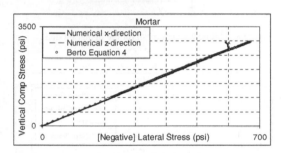

Figure 6a. A comparison of the numerical and analytical results (center of prism).

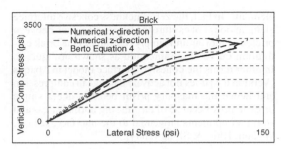

Figure 6b. A comparison of the numerical and analytical results (center of prism).

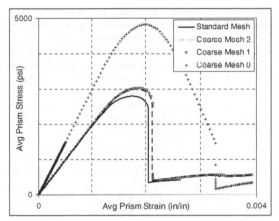

Figure 7. Mesh sensitivity study.

(4) represents the entire cross effect of axial compression in terms of two values, one for the mortar and one for the brick. This is a good approximation for the thin mortar layer, but it is a poor approximation for the much thicker brick with a large stress variation.

It is interesting to note that in the finite element model, the lateral stresses in the brick are no longer the same in the x- and z-directions as contrast to the analytical solution. An explanation for this emerges from the simplifying assumptions leading to Equation (4) in which the different geometry in the lateral directions of the prism is neglected. In reality the different lateral dimensions of the prism result in different response behavior in the x- and z-directions, as seen in the finite element results of the brick response which turns very pronounced close to failure.

5 MESH SENSITIVITY

The "standard" 3D mesh consists of brick units that are 6 elements high, 10 elements wide, and 8 elements deep, as can be seen in Figure 2. The mortar joints have 4 layers of solid elements through their thickness. In addition three coarser meshes were considered. The first, "coarse mesh 2", implements bricks of 3 elements high, 6 elements wide, and 4 elements deep, while 2 element layers comprise the thickness of the mortar joint. The second, "coarse mesh 1" uses the same mesh as "coarse mesh 2", except that the mortar joints are comprised of only 1 element layer through their thickness. The third, "coarse mesh 0" is the same as "coarse mesh 1", except cohesive interface elements replace the single layer of mortar elements. All 3D solid elements are trilinear 8 node hexahedral elements using full integration. The average stress-strain results for these three models are shown in Figure 7. The finite element models show close agreement with respect to mesh refinement in spite of the large drop of axial load resistance shortly after peak. Of particular interest are the nearly identical results for the single and double element layers used to idealize the mortar joint. This is due to the tension in the brick being the dominant source of prism failure. In contrast, the model using a single layer of cohesive interface elements for the mortar joint (designated as "coarse mesh 0") does result in a failure mechanism which is totally different from that of a single layer of finite thickness continuum elements. In fact, the zero-thickness cohesive interface element does not capture the lateral mismatch between brick and mortar and simply reproduces the brick calibration curve in compression as shown in Figure 4. In other terms the cohesive interface model in Abaqus [Version 6.5] does not activate damage in compression and hence the masonry prism behaves like a single brick mainly because of the missing normal stress and strain components tangential to the interface, see e.g. Willam, Rhee & Shing [2004].

6 THREE PLANAR MODELS

For meso-studies of full-scale masonry structures detailed 3D finite element models are still prohibitive in cost. Consequently 2D failure simulation models need to be assessed whether they are capable of reproducing the governing failure mechanism in the composite masonry prism, see e.g. Anthoine [1997]. Assuming plane stress, plane strain and generalized plane strain options all dimensional reductions are based on the standard mesh with 517 nodes and 460 elements, resulting in 1551 DOF, as illustrated in Figure 8. This leads to a large reduction of DOF by nearly tenfold and an even greater saving of computational resources because of the narrow bandwidth. The same damage-plasticity model is used for the material description as in the 3D analysis. The boundary conditions are laterally unconstrained at the top and bottom platen/specimen interfaces as before. The generalized plane strain option in Abaqus is based on the 3D formulation in which the out-of-plane motion is confined by two rigid bounding planes whose relative separation is controlled by a reference node implementing out-of-plane DOF 3. The in-plane DOF in the two bounding planes are subject to equality constraints eliminating the out-of-plane shear deformations but accounting for the out-of-plane normal strain. This is consistent with the observations by Li & Lim [2005] where a mixed variational formulation has been developed to capture generalized plane strain in terms of an additional DOF.

As can be seen from Figure 9, the plane strain and generalized plane strain models show a higher peak stress than the 3D model. The plane strain model reproduces the brittle in-plane failure mode present in the 3D model. The generalized plane strain model peaks closer to the 3D results than the plane strain model. However the very brittle post-peak behavior of the 3D mesh is reduced in favor of a more ductile response. The plane stress model shows dismal results, diverging from the linear elastic range far too early, and then failing to converge soon after. This premature failure is due to the change of the underlying failure mechanism from biaxial tension-compression in the brick units to biaxial compression in the mortar joints due to

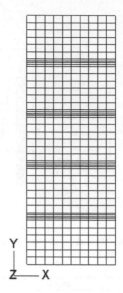

Figure 8.   Finite element mesh for the planar models.

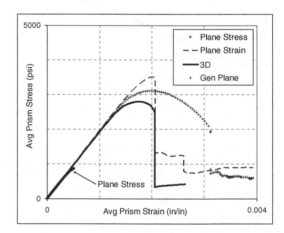

Figure 9.   A comparison of the four FEM models.

the lack of out-of-plane confinement. In short, the two dimensional reduction of the stress state introduces a dramatic change of the failure mode. Similarly to the 3D simulations the plane strain model provides lateral confinement in both components, in fact it prohibits out-of-plane failure in both brick and mortar, and henceforth leads to an upper bound. The generalized plane strain analysis provides an intermediate solution, the out-of-plane movement reduces the amount of plane confinement in the brick and in the mortar. For this reason, the generalized plane strain model is being used for the computationally intense parameter studies in the remainder of the paper.

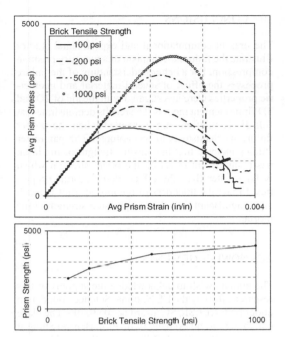

Figure 10. The effect of brick tensile strength on the prism compression strength.

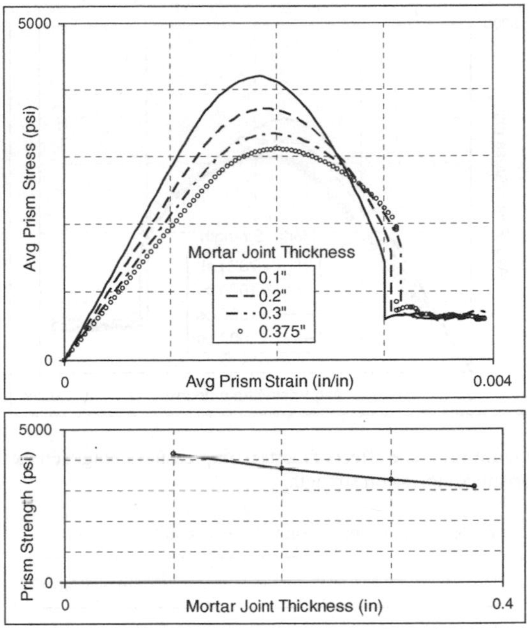

Figure 11. The effect of mortar joint thickness on the prism compression strength.

7 THE EFFECT OF BRICK TENSILE STRENGTH

With the biaxial tension-compression state of brick stress being considered to be the critical failure mechanism, one would expect that the brick tensile strength plays the dominant role in the overall strength of the masonry prism. For this reason the generalized plane strain model was used to study the effect of the tensile brick strength which was varied from $f_t^b = 100$ psi to $f_t^b = 1000$ psi, leaving all of the other parameters unchanged.

Figure 10 shows the results of this study. As expected, the increase in the brick tensile strength increases the compressive capacity of the prism by over 100%. The fact that the prism strength gradually approaches the axial compression capacity of the brick unit supports the observation that prism failure develops first in biaxial tension-compression in the brick for which the axial compression strength provides an upper bound.

8 THE EFFECT OF MORTAR JOINT PROPERTIES

In terms of composite analysis one would expect that the amount of mortar should play some role in the stiffness and strength properties of the prism. To this end the thickness of the mortar joint was varied from $0.1''$ to $0.375''$. The expectation is that a thicker mortar joint would reduce the stiffness of the prism. Because of the compliance mismatch in Equation (4) a higher tensile stress should develop in the brick and as a result the prism would fail at a lower compression stress. The results are displayed in Figure 11. As can be seen, decreasing the thickness of the mortar joint increases the stiffness of the prism and its compressive strength by over 30%.

In contrast Figure 12 illustrates the comparatively small effect of the mortar capacity on the prism strength which increases by less than 15% if one triples the mortar compression strength. As one would expect, a stronger mortar means a stronger prism. However, the mortar strength plays little role in the failure mechanism of the masonry prism which is governed by biaxial tension-compression in the brick unit rather than by triaxial compression in the mortar. However all failure simulations indicate that there is a residual strength of the prism after the large drop of load capacity has taken place which at first appears to be related to the compression strength of the mortar. However considering its invariance in Figure 12 it appears that both the peak as well as the residual value of compressive prism strength are closely related to the tensile capacity of the brick as shown in Figure 10.

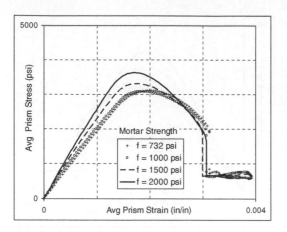

Figure 12. The effect of mortar compression strength on the prism compression strength.

Table 3. A comparison with test results.

	Prism strength
Experimental	2,490 psi
3D FE-Model	2,797 psi
Plane strain	3,514 psi
Generalized plane strain	3,113 psi
Plane stress	860 psi

9 SUMMARY OF PRISM STRENGTH

Table 3 compares the axial strength results of the 2D and 3D finite element models with the prism test data. The 3D model furnishes a prism strength which agrees well with the experimental result, it is 12% higher than the average of the experimental prism data with a COV of 9.8%. The plane strain peaks 41% higher, while the generalized plane strain peaks 25% higher. Due to the erroneous failure mode, the plane stress result yields a prism strength which is 65% lower than the test data.

The discrepancy among the test data and the numerical results may be in part due to the lack of test data to calibrate the large number of material parameters of the damage-plasticity model in Abaqus. In view of the experimental variations and the uncertainties in all input parameters the 3D model provides fairly close agreement, although the generalized plane strain model may be an acceptable alternative to capture the main features of masonry and to reduce computational effort.

10 CONCLUSIONS

The present computational and experimental studies illustrate the failure mechanism of a masonry prism in compression. For a better understanding of the failure processes the effect of several material and geometric properties were investigated with the aid of 2D and 3D finite element models. The results demonstrate that full 3D simulations are warranted when the proper failure mechanism is to be captured. The parameter studies indicate that the tensile strength of the brick has a significant effect on the prism strength which is far greater than the effect of the mortar properties. In fact the thickness of the mortar joint is much more significant than the compression strength of the mortar.

ACKNOWLEDGEMENTS

The authors wish to acknowledge the support of this research effort by the US National Science Foundation under NSF NEESR-SG award no. 0530709. The close collaboration and exchange with Prof. Benson Shing at UCSD and Prof. Sarah Billington at Stanford University are much appreciated on the joint project "Seismic Performance Assessment and Retrofit of Non-Ductile RC Frames with Infill Walls". Opinions expressed in this paper are those of the authors and do not necessarily reflect those of the sponsor.

REFERENCES

ABAQUS, inc., Abaqus version 6.5 finite element software, 2004, Providence RI, USA.

Anthoine, A.1997. Homogenization of Periodic Masonry: Plane Stress, Generalized Plane Strain or 3D Modelling?, Communications in Numerical Methods in Engineering 13: 319–326.

Berto, L., Saetta, A., Scotta, R., Vitaliani, R. 2005. Failure Mechanism of Masonry Prism Loaded in Axial Compression: Computational Aspects, Materials and Structures, 38: 249–256.

Lee, J., Fenves, G. 1998. A Plastic-Damage Concrete Model for Earthquake Analysis of Dams, Earthquake Engineering and Structural Dynamics, 27: 937–956.

Li, S., Lim, S. 2005. Variational Principles for Generalized Plane Strain Problems and their Applications, Composites: Part A, 36/3: 353–365.

Willam, K., Rhee, I., Shing, B. 2004. Interface Damage Model for Thermomechanical Degradation of Heterogeneous Materials, Computer Methods in Applied Mechanics and Engineering, 193: 3327–3350.

Fracture Mechanics of Concrete and Concrete Structures – High-Performance Concrete,
Brick-Masonry and Environmental Aspects – Carpinteri, et al. (eds)
© 2007 Taylor & Francis Group, London, ISBN 978-0-415-44617-4

A finite element study of masonry cracks

A.M. Fathy
Department of Properties of Materials, Faculty of Engineering, Ain Shams University, Cairo, Egypt

J. Planas
Departamento de Ciencia de Materiales, Universidad Politécnica de Madrid, Madrid, Spain

J.M. Sancho
*Departamento de Ingeniería de Edificación, Escuela Politécnica Superior, Universidad CEU-San Pablo,
Madrid, Spain*

ABSTRACT: Brick walls of ceramic without any mortar covering or paint are used extensively in building façades in Spain. Their good appearance and high resistance to environmental attacks are the reasons of the widespread use of such walls in Spain. One of the most used masonry wall system is based on nonbearing panels hanging partially, about two thirds of the brick width, on the edge beams of the structural skeleton. The edge beam is veneered with special thinner bricks to achieve the visual continuity of the façade. A considerable number of these walls show cracking. In this work, finite element simulations were performed in order to gain insight on the causes of cracking. A special finite element, based on the strong discontinuity analysis and the cohesive crack theory is used in the numerical simulations. The results agree with the overall cracking patterns observed.

1 INTRODUCTION

Walls are the most used single masonry element. From a structural point of view, walls are classified as either load bearing or non-load bearing. To this latter category belong closures and partitions, fences, parapets, and exterior wythes of multi-wythe walls that are subjected mainly to horizontal loads perpendicular to the plane of the wall (Casabbone 1994).

Most masonry units available can be classified in one of the following groups: concrete block, solid concrete brick, clay block, clay brick (solid or cored), clay tile, sand lime units or adobe units (Yamin & Garcia 1994). Masonry bearing walls with reinforced concrete slabs is of large use in residential buildings up to five stories in U.S. and Latin American countries (Casabbone 1994, Meli & Garcia 1994, Garcia & Yamin 1994, Gallegos 1994). There are different masonry systems of this type of walls as unreinforced masonry, confined masonry or reinforced masonry. In the Latin American countries, masonry units, following the Spanish construction tradition, were mainly solid clay bricks and the walls were reinforced. The reinforcement was concentrated in the perimeter of the wall, embedded in concrete elements, giving birth to the system called confined masonry (Fig. 1).

Figure 1. Confined masonry.

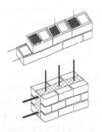

Figure 2. Reinforced masonry.

Reinforced masonry system is applied to masonry walls strengthened with distributed reinforcement along its length and height (Fig. 2).

In the last decades, the price of residential flats raised considerably in Spain. One of the consequences

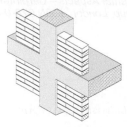

Figure 3. Masonry façade where columns and edge beams are seen.

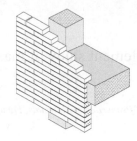

Figure 4. Masonry façade where columns and edge beams are covered.

was the reduction of the time employed in the building process. One of the most used construction systems for residential buildings is based on a skeleton of reinforced concrete or steel columns with reinforced concrete uni- or bidirectional slabs. Masonry walls are used as partitions. The façade is normally fabricated with double wythe wall with thermal insulation in between. The good quality and appearance of the ceramic masonry unit used lead to intensive use of this type of exterior walls without mortar cover or paint. The masonry wall can be built with totally or partially bearing over the end beams (Figs 3 & 4). Also, the exterior wythe may be fabricated out of the plane, which contain the end beams.

In this case, rigid steel angle is used to support the wall (Fig. 5)(Adell Argilés 2003). The edge beams in every floor are not in the same vertical plane. The horizontal eccentricities between those planes are in the same order as wall bearing depth over the edge beams of the partially bearing solution. So, steel angle is usually used in all the faced long to ensure the required bearing depth (Fig. 6)(Adell Argilés 2003).

Although the total bearing solution is the best, based on structural considerations, the last two solutions have the advantage of a better appearance and of leading to a larger effective area of the flats, which make these two solutions economically better in spite of their higher initial cost. The partial bearing of the wall is the most used solution in Spain. The masonry units are usually arranged with two thirds of the brick width supported on the edge beam. Dimensions of the customary brick units, Metric and DIN, are shown in figure 7 (Adell Argilés 2003, Hispalyt 1998).

There are no Spanish standards for this construction system. Usually the standards NBE_FL90 (Bearing walls) (Spanish standards 1990), NTE-FFL (External masonry walls of brickwork design) (Spanish Tecnical standards 1978) and NTE EFL (structural brickwork calculations) (Spanish Tecnical standards 1977) are used. The first gives instructions about the properties of the mortar and masonry units used in bearing walls and the last two standards provide rules for the structural design of bearing walls. A considerable number of these buildings show cracking in different zones of the façade walls. As a result of the lack of standards,

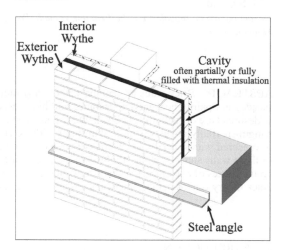

Figure 5. Exterior wythe built out of the plane of the edge beams.

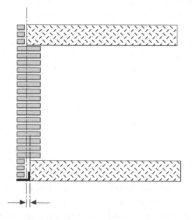

Figure 6. Horizontal gap between the vertical planes of the faces of the edge beams.

every construction company uses its own experience to prescribe the necessary recommendations to minimize the width and extension of the cracks. There are many possible load patterns that can cause such cracks.

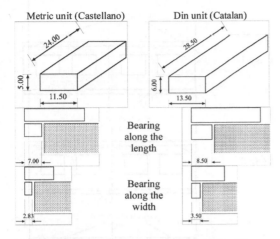

Figure 7.   Spanish masonry units (dimensions in cm).

Figure 8.   Cracked panels.

As a large number of flats were constructed in a short period, masonry units might be used before the necessary curing time. Most of the expected expansion in ceramic bricks develops in three or four years, but about 75 percent of it takes place in the first fifteen days after the fabrication (Adell Argilés 2003). The total expansion values and rates depend strongly on the type of the used clays.

Most of these façades were constructed without horizontal expansion joints. The walls are tied to columns with special metallic elements. The joint between the last row of masonry units and the edge beams is totally filled with mortar (Hispalyt 1998). This type of confinement and the volume changes of the masonry units due to water absorption or temperature variations can cause cracking.

Figure 8 shows two cracked panels. Partial bearing of the exterior wythe generates torsion moment on the edge beams. Rotation of edge beams can also cause cracking. In the USA, brick veneer walls are also intensively used, but the wall is totally constructed out

of the edge beam plane as shown in figure 5. The difference is that a horizontal expansion joint is left just below the steel angle that holds the wall (Memari et al. 2002, Drysdale et al. 1994).

The movement of the building as a result of wind loads is also a possible cause of cracking, especially in the last floor. The authors developed special finite element, based on the strong discontinuity approach and the cohesive crack theory, that is very effective in the numerical simulation of cracking in quasi brittle materials such as concrete and masonry The authors applied such element to the analysis of the cracks appearing in the foregoing masonry walls in order to have a better understanding of the cracking phenomena.

2   STRATEGY OF THE WORK

As a large number of walls were simulated, a preprocessor program with an easy user interface was developed to automatize the mesh generation process performed mainly using the ANSYS program. The special finite element was implemented in three different programs, i.e. as a user element in FEAP, as a special user material in ABAQUS and finally in an Object Oriented special purpose finite element code. Further details of this element can be found in (Sancho et al. 2004). In the study presently reported, the simulation was performed in two dimensions. The finite element is however capable of three-dimensional analysis.

In this study, a representative panel far from the boundaries of the façade is considered (i.e. it is assumed that there is large number of panels surrounding the studied panel in all directions). In a first series of analyses, a sensitivity analysis was carried out for each geometry to find the best set of computational parameters (such as mesh element density, load increment or calculation tolerance). The corresponding values were then kept constant in the remaining computations.

3   INPUT DATA

3.1   Geometric layout of the studied walls

Two different wall geometries were studied. Figure 9 shows the tested wall geometries. The layout was selected by a consulting office for a large number of construction companies in Spain. The two models have the same external dimensions but the window layout is different. A vertical expansion joint is assumed to exist every two panels. One panel is usually studied when there is a total symmetry. In the last load case two panels were studied, as shown next, because there is no symmetry.

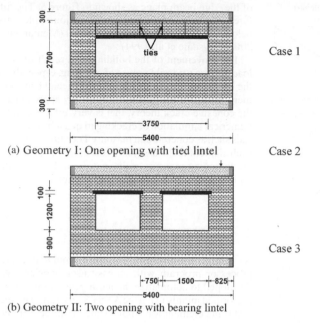

(a) Geometry I: One opening with tied lintel

(b) Geometry II: Two opening with bearing lintel

Figure 9.   Wall geometries used in the analyses. (All dimensions are in mm).

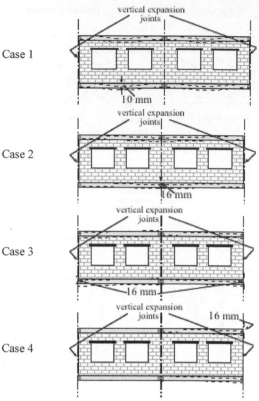

Case 1

Case 2

Case 3

Case 4

Figure 10.   Load cases.

3.2   Load cases

Four different load cases were used as depicted in figure 10. The first case assumes imposed deflection of 10 mm in the mid span of the beams with parabolic distribution. This is equivalent to a uniformly distributed load applied to the beam. The second case assumes a settlement of 16 mm in only one axis of the columns. This axis of columns is in the mid point between the expansion joints. In the third case, the settlement is applied to the two adjacent edges of the expansion joint. The last case is the same as the third but the settlement was applied to only one axis coinciding with an expansion joint.

3.3   Material properties assumed in the simulation

Every tested panel has two different materials, steel and bricks. Steel is used for beams, columns, lintel and their ties (in the first geometry). The modulus of elasticity of the steel is taken to be 200 GPa. There are many papers in the literature devoted to the behavior of the masonry units and mortar. Most of those papers deal with concrete blocks. In references (Beall 1993, Alcoser & Klingner 1994) it is shown that the overall strength of the wall depends primarily on the compressive strength of the unit, and very little on the mortar compressive strength. Other analytical papers (Zucchini & Lourenco 2004, Lee et al. 1996, Uva & Salerno 2005) uses the homogenization technique to give relations between the strength of the

masonry unit and mortar with the overall strength of the wall. The same concept was investigated experimentally (Khalaf et al.1994, Ramamurthy et al. 2000). As the masonry wall is a non-homogeneous material, several works investigate the strength of the masonry wall at varying directions with the bed joint (Khattab & Drysdle 1992). From all mentioned papers, it can be concluded that an equivalent homogeneous material can be used for masonry walls. The equivalent strength depends on the strength of the unit and mortar as well as on the direction of the loading with respect to the bed joint where the strength changes by about 20% (Khattab & Drysdle 1992). So, in this paper, homogenous isotropic material is assumed for the masonry wall with modulus of elasticity of 3 GPa, fracture energy of 100 N/m and tensile strength of 1.0 MPa.

3.4   Boundary conditions

For the first case of loading, the column lines coincide with vertical axes of symmetry for both the structure and the brick wall, as shown in figure 10. Therefore, for this case only one panel needs to be analyzed.

In the second and third cases, although there are two axis of symmetry along the two sides of the panel

1598

for the wall, there is only one axis of symmetry for the structure, which lies at the mid point between two adjacent expansion joints. The exact boundary conditions of the beams at the ends which coincide the joints is not known exactly because of the unknown stiffness of the remaining part of the structure at both sides of the panel. To make the calculation faster using the smallest possible number of nodes, one panel was used and two extreme conditions at the end of the beam were considered: In the first boundary conditions, it is assumed that the beams were hinged at one end, and fixed in the second. The exact condition lies in-between these two extreme conditions.

The last case is like the second and third cases with respect to the definition of the boundary conditions of the beams, but the axis of symmetry of the wall coincides with the line of columns on the right. In this case two panels were used. Fixed and hinged ends for the left end of the beams were considered as before.

Geometry I under load case one was used to make the sensitivity analyses to find suitable values for the calculation parameters.

3.5 Adaptation factor

To prevent the calculation block, the program permits that cracks with width lesser than $\alpha G_f/f_t$ can change the direction till found the correct one, where G_f is the fracture energy and f_t is the tensile strength. The factor α called adaptation factor Paper (Sancho et al. 2005a) shows that a value of 0.2 for the adaptation factor is sufficient for no blocking the calculation. This value is used in all the calculations.

4 RESULTS

4.1 Geometry I, load case 1

Triangular elements were used to mesh the wall and special elements with embedded crack were used for calculation. Quadratic four-node elements were used to mesh the beams and the lintel. Enhanced-strain quadrilateral elements developed by J. Simo and R.L. Taylor were used for the computations, since this type of element allows computing the stress and strain in bending accurately with few elements across the beam depth.

Figure 11 shows the resulting crack pattern for a deflection of 10 mm for a computation carried out at loading steps of 0.1 mm, with a tolerance of 1e-6 and a mesh size corresponding to 2100 elements. Black and grey crack lines correspond to crack openings respectively larger and lesser than 0.02 mm (corresponding to adaptation factor $\alpha = 0.2$ (Sancho et al. 2005a)). The wall appears to be extensively cracked for a deflection at mid-span of 10 mm; therefore in subsequent

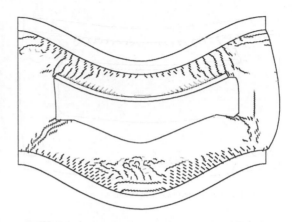

Figure 11. Crack pattern for geometry I, load case 1. (maximum deflection = 10 mm, deflection increment = 0.1 mm, Tolerance = 1e-6) Black and grey crack lines correspond to crack openings respectively larger and lesser than 0.02 mm.

computations the mid-span deflection was reduced to 2.5 mm.

To increase the accuracy, the deflection step was reduced to 0.01 mm. The resulting crack pattern is shown in the upper plot in figure 12. Next, to decrease the time of computation, the beams surrounding the concrete were eliminated and the parabolic vertical displacements were applied directly to the upper and lower edges of the masonry panel. The resulting crack distribution (lower plot in Fig. 12) was clearly different from the previous one, which demonstrates that the two ways of applying the boundary conditions are not equivalent and that the results are sensitive to the computational details. Moreover, the time required for the last type of computation was longer than for the one involving the beams, despite that the number of degrees of freedom were less, which is counter-intuitive.

Four possible sources for the observed differences were identified:

1. The meshes used for the masonry are very similar, but not identical
2. The horizontal displacements at the upper and lower edge of the masonry panel are not the same (in the second case the horizontal displacement is zero)
3. The results are sensitive to the tolerance
4. The results are sensitive to the step size.

Figure 13 shows that the positions of the nodes for the two cases in figure 12 are actually very close. However, to fully avoid this problem, in all subsequent computations, the beam and wall are always meshed and then the beams are removed.

To appropriately account for the horizontal displacements along the upper and lower edges of the masonry, the horizontal displacement of the lower and upper edges of the beams were computed using

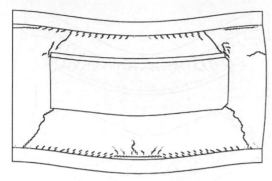

(a) Explicit modeling of edge beams

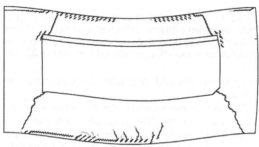

(b) Implicit computed boundary conditions

Figure 12. Crack pattern with and without edge beams. (max. deflection = 2.5 mm, deflection increment = 0.01 mm, Tol. = 1e-6).

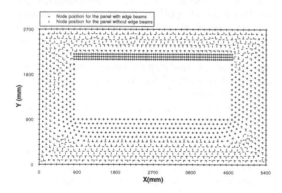

Figure 13. Node layout for computations with explicit modeling of edge beams and without beams. X, Y is horizontal and vertical distances to the bottom-left corner of the masonry panel.

classical beam theory and Navier's hypothesis for the cross-section of the beams. Figure 14 shows the horizontal and vertical components of the displacement at the upper and lower edges of the panel. Figure 15 shows that with this condition, the results are identical for computations with and without explicit inclusion of the beams.

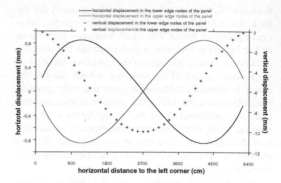

Figure 14. Computed horizontal and vertical displacements at the top and bottom edges of the masonry panel (according to the assumed deflection of the edge beams).

The influence of the step size is shown in figure 15. The results are essentially identical for step sizes equal or less than 0.01 mm, and thus the step size was set to that value in all subsequent calculations.

When reducing the tolerance down to 1e-12, the time of computation of the panels without the beams was less than those with the beams, as expected. The explanation for this behavior is that, in the panel with beams, the stiffness of the concrete is much larger than that of the masonry and thus the unbalance of nodal forces at the beginning of the step is larger, which means that identical relative tolerances (as used) correspond to larger absolute tolerances for the panel with beams.

Following this analysis, in the remaining of the research masonry panels with imposed parabolic displacements including computed horizontal components, with a step size of 0.01 mm and a tolerance of 1e-12 were used.

4.2 Geometry II, load case 1

Applying the load case 1 to the geometry II in figure 9 with the calculation parameters defined before, the results shown in figure 16 are obtained.

Figure 15 shows that for geometry I the dominant crack is the horizontal crack at the bottom of the masonry, while figure 16 for the geometry II shows that this type of crack doesn't appear because the of the compressive strut connecting the bottom and top strips. A vertical crack appears in the middle-upper part that is not present in geometry I. However, the most open cracks are the bottom-left and bottom-right cracks at the window corners.

4.3 Load case 2, 3 and 4

Computations were carried out to ascertain the cracking behavior of geometries I and II under loadings

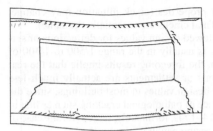

(a) Explicit modeling of edge beams, step inc. = 0.01.

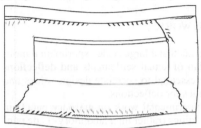

(b) Explicit modeling of edge beams, step inc. = 0.002.

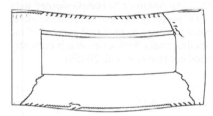

(c) Implicit computed boundary conditions, step inc. = 0.01.

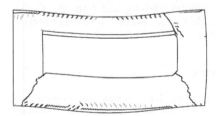

(d) Implicit computed boundary conditions, step inc. = 0.002.

Figure 15. Crack patterns for Geometry I under load case 1 The maximum deflection is 2.0 mm and the tolerance is 1e-12 in all cases.

2–4. Since the crack patterns were very similar for both geometries, only results for geometry I are given here (Figs 17, 18 & 19).

Figures 17 & 18 show the results for load case 2 and load case 3 respectively (As previously explained, two extreme boundary condition were assumed at the end opposite to the one experiencing the settlement: fixed end (left) or hinged end (right). As can be seen,

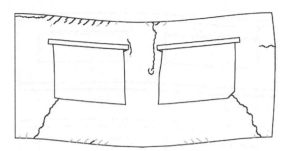

Figure 16. Crack pattern for Geometry II under load case 1. (Maximum deflection 2.0 mm, step increment 0.002 mm, tolerance 1e-12.).

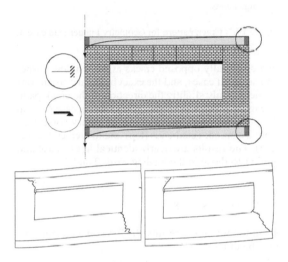

Figure 17. Crack pattern for Geometry I under load case 2.

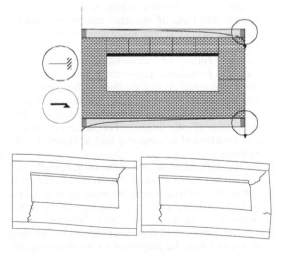

Figure 18. Crack pattern for Geometry I under load case 3.

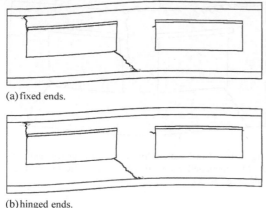

(a) fixed ends.

(b) hinged ends.

Figure 19. Crack pattern for Geometry I under load case 4.

two diagonally opposed cracks at the window corner appear in all cases, and the exact boundary conditions influence only slightly the direction of the crack path.

Figure 19 shows the results for loading case 4. As in the previous cases, two extreme boundary conditions were assumed: fixed ends (top) and hinged ends (bottom). The results are nearly identical in this case and similar to those in the loading cases 2 and 3.

5 CONCLUSIONS

From the foregoing results, the foregoing conclusions may be drawn

1. Finite elements with embedded cohesive cracks can describe the relative complex cracking patterns that arise in masonry façades, with multiple cracks growing simultaneously. As described elsewhere, this kind of elements can be essentially implemented in a general-purpose finite element code.
2. Careful fitting of computation parameters is required to achieve consistent results. A simple methodology to do so is outlined in this paper.
3. The influence of the geometry on the crack pattern is substantial only for load case 1 (excessive deflection of the beams). The remaining cases 2–4 (settlement of supports) lead to similar crack patterns.
4. For the deflections or settlements in the range 1/500 to 1/300 of the span (in the first intent of geometry one loaded with case one), the masonry appears to be fully cracked with crack openings of the same order of magnitude as the beam deflections.
5. Although the results shown correspond to a particular set of material properties for the masonry, the final crack pattern is similar for any set of realistic

material properties (crack initiation may change, but the final cracks are essentially the same).
6. Since allowed design values for deflections or settlements is usually in the range 1/500 to 1/300 of the span, the foregoing results implie that the real deflections or settlements are actually much less than the design values in most buildings, since the occurrence of pathological cracking such as that in figures 14–17 is, fortunately, scarce (for the case of totally bearing panels).

6 FUTURE WORK

1. This suggests that a large scale experimental monitoring plan of actual settlements and deflections would be essential to correlate design values and actual values for deflections.
2. Further experimental work on masonry panels in actual buildings (rather than laboratory-made panels) would be required to ascertain average and statistical properties of masonry façades.
3. The 2D computations must be complemented with 3D analysis to properly take into account the effect of eccentrical loading due to partial support of the masonry on the panels. Such extension is currently being carried out (Sancho et al. 2005b).

ACKNOWLEDGEMENTS

The authors gratefully acknowledge support for this research from the Spanish Ministry of Education under Grant CIT 380000-2005-40, and from the Comunidad de Madrid under Project S-0505/MAT-00155 (DUMEINPA).

REFERENCES

Adell Argilés, J. M. 2003. *Tratado De Construcción, Fachadas y Cubiertas*. Madrid, Spain: Editorial Munilla-Lería.

Alcoser, S.M. & Klingner, R.E. 1994. Masonry Research in Americas. *ACI SP-147-05, September*: 127–169.

Beall, C. 1993. *Masonry design and detailing for Architects Engineers, and Constructors*. Third edition, McGraw-Hill.

Casabbone, C. 1994. General Description of Systems and Construction Practices. *ACI SP-147-02, September*: 21–55.

Drysdale, R.G., Hamid, A.A. & Baker, 1994. *Masonry Structure, Behavior and Design*. New Jersey, USA: Prentice-Hall.

Gallegos, H. 1994. Masonry in Peru. *ACI SP-147-11, September*: 307–331.

Garcia, L.E. & Yamin, L.E. 1994. A Review of Masonry Construction in Colombia. *ACI SP-147-10, September*: 283–305.

Hispalyt, 1998. *Ejecución de fachadas con ladrillo cara vista*. Manual, Madrid, Spain: Editorial la sección de Ladrillo de Cara Vista de Hispalyt.

Khalaf, F.M., Hendy, A.W. & Fairbairn, D.R. 1994. Study of the Compressive Strength of Blockwork Masonry. *ACI Structural Journal, Vol. 91, No. 4, July–August*: 367–375.

Khattab, M.M. & Drysdle, R.G. 1992. Tests of Concrete Block Masonry Under Biaxial Tension-Compression. *Canadian Masonry Symposium, 15–17 June (1992)*: 645–656.

Lee, J.S., Pande, G.N., MIddleton, T.J. & KraIj, B. 1996. Numerical Modeling Of Brick Masonry Panels Subject To Lateral Loadings. *Computers and Structures, Vol. 61*: 735–745. Great Britain.

Meli, R. & Garcia, L.E. 1994. Structural Design of Masonry Buildings: The Mexican Practice. *ACI SP-147-08, September*: 239–262.

Memari, A.M., Burnet, E.F.P. & Brian, M.K. 2002. Seismic response of a new type of masonry tie used in brick veneer walls. *Construction and Building Materials, 6*: 397–407.

Ramamurthy, K., Sathish, V & Ambalavanan, R. 2000. Compressive Strength Prediction of Hollow Concrete Block Masonry Prisms. *ACI Structural Journal, Vol. 97, No. 1, January–February*: 61–67.

Sancho, J.M., Planas, J., Cendón, D.A., Reyes, E. & Gálvez, J.C. 2007. An embedded cohesive crack model for finite element analysis of concrete fracture. *Engineering Fracture Mechanics, Vol. 74*: 75–86.

Sancho, J.M., Planas, J., Cendón, D.A., Gálvez, J.C. & Reyes, E. 2005. Ventajas De La Fisura Cohesiva Adaptable En La Simulación Numérica De La Fractura De Materiales Cuasifragiles. *Anales de Mecánica de la Fractura, Vol. 22*: 553–558.

Sancho, J.M., Planas, J., Fathy, A.M., Gálvez, J.C. & Cendón, D.A. 2006. Three-dimensional simulation of concrete fracture using embedded crack elements without enforcing crack path continuity. *Int. J. Numer. Anal. Meth. Geomech.* in press, published online, September 4, 2006.

Spanish standards for bearing walls, 1990. *Muros resistentes de fábrica de ladrillo*. NBE FL-90.

Spanish Tecnical standards for buildings, 1977. *Structural brickwork calculations*. NTE-EFL.

Spanish Tecnical standards for buildings, 1978. *External masonry walls of brickwork design*. NTE-FFL.

Uva, G. & Salerno, P.B.G. 2005. Towards a multiscale analysis of periodic masonry brickwork: A FEM algorithm with damage and friction. *Solids and Structures, in press, received 2 June (2005)*.

Yamin, L.F. & Garcia, L.E. 1994. Masonry Materials. *ACI SP*-147-01, September: 1–20.

Zucchini, A. & Lourenco, P.B. 2004. A coupled homogenisation–damage model for masonry cracking. *Computers and Structures, 82*: 917–929. Great Britain.

*Fracture Mechanics of Concrete and Concrete Structures – High-Performance Concrete,
Brick-Masonry and Environmental Aspects – Carpinteri, et al. (eds)
© 2007 Taylor & Francis Group, London, ISBN 978-0-415-44617-4*

Fracture criterion for brick and mortar subjected to tri-axial state of stress

A. Litewka
Universidade da Beira Interior, Covilhã, Portugal

L. Szojda
Silesian University of Technology, Gliwice, Poland

ABSTRACT: The aim of the paper is to present experimental and theoretical study of fracture of brittle rock-like materials. To this end the tests of the specimens of brick and mortar subjected to various combinations of the tri-axial state of stress components were performed. The experiments made it possible to construct for both material tested the limit surfaces at material failure. The data obtained for brick and mortar subjected to uni-axial compression were used to calibrate the own theoretical model capable to describe the mechanical behavior of brittle materials. The limit surfaces obtained experimentally were compared with the theoretical predictions based on the fracture criterion proposed.

1   INTRODUCTION

Requirements of modern technology and progress in mechanics of solids and structures give rise to mutually interrelated extensive theoretical and experimental studies of mechanical properties of structural materials. Strong motivation for suitable oriented experiments exists in the case of brittle rock-like materials because of complexity of the phenomena that affect their mechanical response. Some results of experimental and theoretical studies of mechanical behavior of brittle rock-like materials have been previously reported mainly for uni-axial and biaxial loading of concrete (Kupfer 1973, Ehm & Schneider 1985, Thienel et al. 1991, Ligęza 1999). Relatively small amount of respective experimental data for such materials subjected to tri-axial state of stress is available. Some data can be found in mono-graphs on rock mechanics (Cristescu & Hunsche 1998, Derski et al. 1989, Goodman 1989) and on mechanics of concrete (Chen 1982, Neville 1995). Simultaneously new approach based on the methods of continuum damage mechanics has been used to formulate phenomenological models capable to describe the mechanical behavior of brittle rock-like materials in presence of oriented damage growth (Chaboche et al. 1995, Litewka et al. 1996, Murakami & Kamiya 1997, Halm and Dragon 1998). However, all those theoretical descriptions are based on limited experimental data, particularly for tri-axial state of stress and were verified for some specific cases of loading only. To obtain more realistic theoretical description of overall material response

that could account for oriented damage growth and development of damage induced anisotropy further extensive experimental studies are needed.

The aim of this paper is to supply experimental data on fracture of brick and mortar subjected to tri-axial state of stress as well as to show potentialities of own theoretical model (Litewka et al. 1996, Litewka & Dębiński 2003). A study of such a state of loading is a necessary first step towards analysis of complex conditions that are experienced by brick and mortar structures in practice during earthquakes or due to mining subsidence. That is why the tests were performed for relatively high values of the compressive mean normal stresses as well as for pure hydrostatic pressure.

2   EXPERIMENTS

The experiments presented here have been done as a continuation of those discussed by the authors in earlier paper (Litewka & Szojda 2006). The new results were obtained for specimens of the same types of mortar and brick that were tested earlier, and that is why the recent and older data could be compared. These two different series of the specimens are referred to as the specimens of Brick 1 and Mortar 1 for those reported in Litewka & Szojda (2006) and Brick 2 and Mortar 2 for these more recent presented here. The height and diameter of the cylindrical specimens used were equal to 12 cm and 6 cm, respectively. The specimens of brick were cut out from standard plain

Table 1. Experimental data and constants used in theoretical analysis of tri-axial state of stress.

Constant	Unit	Mortar	Brick
E_0	MPa	8730	2550
ν_0	–	0.173	0.103
f_c	MPa	−7.90	−10.85
A	MPa^{-2}	2095×10^{-5}	1249×10^{-5}
B	MPa^{-2}	62.55×10^{-5}	200.0×10^{-5}
C	MPa^{-1}	-0.9469×10^{-5}	-1.100×10^{-5}
D	MPa^{-1}	1.678×10^{-5}	3.754×10^{-5}
F	–	1.070	0.6900

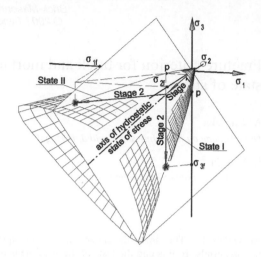

Figure 1. Limit surface at material fracture and loading paths for State I and State II: point corresponding to material fracture.

brick whereas those of mortar were prepared in special moulds. The details of the specimens preparation and the experimental procedure can be found elsewhere (Szojda 2001, Litewka & Szojda 2005).

The tests were performed for two cases of tri-axial compression referred to as State I and State II and also to pure hydrostatic pressure and uni-axial compression. The objective of the test performed under uni-axial compression was to calibrate the materials. That is why the initial Young modulus E_0 and Poisson ratio ν_0 as well as uni-axial compressive strength f_c were measured experimentally for both materials tested. The data shown in Table 1 were calculated as mean values measured for seven specimens of brick and seven for mortar. These values of standard constants E_0, ν_0, f_c and those for five other parameters A, B, C, D and F seen in Table 1 are necessary to employ the theoretical model proposed.

The objective of the tests under tri-axial state of stress was to measure the stresses at material fracture for prescribed loading programs. The tri-axial State I is a combination of uni-axial compression and hydrostatic pressure whereas the State II is a simultaneous action of hydrostatic pressure and uniform bi-axial compression. The respective stress tensor components that correspond to State I and State II are expressed by Equations (5), (6). Various combinations of the stress tensor components and at least two different loading paths are necessary to supply information on the shape of the limit surface at failure of the material subjected to tri-axial states of stress. The possible form of such a limit surface together with the loading paths for State I and State II of tri-axial compression is shown in Figure 1. It is seen from this figure that respective loading paths consisted of two stages. The Stage 1 was the same in both cases of tri-axial loading and consisted in a mo-notonic increase of hydrostatic pressure up to prescribed value p. In the Stage 2 of the first tri-axial state of stress (State I) the compressive vertical normal stress σ_V was increased up to material failure that occurs for $\sigma_{3f} = p + \sigma_V$. In the Stage 2 of the State II of tri-axial loading two compressive hori-zontal components σ_H of uniform bi-axial state of stress were

Table 2. Experimental and theoretical failure stress for mortar subjected to State I of tri-axial compression.

Material	Specimen	Hydrostatic pressure, p MPa	Failure stress σ_{3f}	
			Experiment MPa	Theory MPa
Mortar 1	ZC1*	−1.04	−10.90	−13.32
	ZD1*	−0.934	−12.19	−12.83
	ZC2*	−1.84	−17.55	−16.71
	ZD2*	−1.92	−16.03	−17.01
	ZC3*	−2.69	−18.74	−19.77
	ZD3*	−2.92	−20.01	−20.56
Mortar 2	M-H4-1	−3.41	−25.59	−22.14
	M-H4-2	−4.33	−23.97	−24.89
	M-H8-2	−8.45	−39.85	−35.51
	M-H8-3	−8.20	−36.05	−34.92
	M-H12-1	−12.28	−48.34	−43.98
	M-H12-2	−12.37	−49.53	−44.17
	M-H12-3	−11.95	−49.75	−43.29

* The data for these specimens of Mortar 1 were discussed in Litewka & Szojda (2006).

increased simultaneously up to material failure that corresponds to $\sigma_{1f} = \sigma_{2f} = p + \sigma_H$. To obtain several combinations of the stress tensor components the various levels of the hydrostatic pressure p were used. The respective numerical data presented earlier (Litewka & Szojda 2006) as well as new ones are shown in Tables 2–5. The new experiments performed for Mortar 2 and Brick 2 according to the program seen in Tables 2–5 made it possible to determine the stresses at material fracture for higher levels of hydrostatic pressure p than those in earlier tests done for Mortar 1 and Brick 1.

Table 3. Experimental and theoretical failure stress for mortar subjected to State II of tri-axial compression.

Material	Specimen	Hydrostatic pressure, p MPa	Failure stress $\sigma_{1f} = \sigma_{2f}$ Experiment MPa	Theory MPa
Mortar 1	ZA1*	−0.01	−7.19	−8.60
	ZB1*	−0.21	−9.40	−10.98
	ZA2*	−2.30	−15.86	−21.74
	ZB2*	−2.37	−16.83	−22.01
	ZA3*	−3.81	−23.55	−26.79
	ZB3*	−3.92	−24.87	−27.12
Mortar 2	M-V0-1	−0.14	−11.47	−10.21
	M-V0-2	−0.09	−7.92	−9.64
	M-V2-1	−1.87	−21.69	−20.07
	M V2 2	2.29	22.03	21.71
	M-V4-1	−3.85	−24.50	−26.89
	M-V4-2	−4.46	−29.06	−28.70
	M-V8-1	−7.83	−29.43	−37.50
	M-V8-2	−8.00	−30.67	−37.90
	M-V12-1	−12.17	−44.22	−47.23
	M-V12-2	−11.63	−45.89	−46.07
	M-V12-3	−12.20	−44.75	−47.30

* The data for these specimens of Mortar 1 were discussed in Litewka & Szojda (2006).

Table 4. Experimental and theoretical failure stress for brick subjected to State II of tri-axial compression.

Material	Specimen	Hydrostatic pressure, p MPa	Failure stress $\sigma_{1f} = \sigma_{2f}$ Experiment MPa	Theory MPa
Brick 1	CA1*	0	−10.49	−10.96
	CB1*	−0.06	−14.36	−11.02
	CA2*	−2.76	−21.94	−22.87
	CB2*	−2.51	−17.19	−22.13
	CA3*	−3.34	−24.70	−24.46
	CB3*	−3.63	−22.83	−25.22
Brick 2	B-V9-1	−9.09	−36.21	−37.50
	B-V9-2	−9.26	−34.83	−37.85
	B-V9-3	−9.53	−31.56	−38.40

* The data for these specimens of Brick 1 were discussed in Litewka & Szojda (2006).

3 FRACTURE CRITERION

The theoretical model of fracture and deformability of brittle rock-like materials employed in this paper, based on the assumption of tensorial nature of the material damage was presented in earlier papers

Table 5. Experimental and theoretical failure stress for brick subjected to State I of tri-axial compression.

Material	Specimen	Hydrostatic pressure, p MPa	Failure stress σ_{3f} Experiment MPa	Theory MPa
Brick 1	CC1*	−1.20	−15.27	−15.29
	CD1*	−0.93	−16.58	−14.29
	CC2*	−1.95	−20.86	−17.87
	CD2*	−2.13	−22.05	−18.45
	CC3*	−3.20	−22.23	−21.56
	CD3*	−3.03	−22.80	−21.09
Brick 2	B-H3-1	−3.14	−19.53	−21.40
	B-H3-2	−3.55	−21.56	−22.51
	B-H6-1	−6.64	−34.17	−29.87
	B-H6-2	−6.55	−33.66	−29.67
	B H6 3	6.30	31.83	29.12
	B-H9-1	−9.92	−47.26	−36.67
	B-H9-2	−8.66	−37.87	−34.13
	B-H9-3	−9.34	−40.68	−35.53

* The data for these specimens of Brick 1 were discussed in Litewka & Szojda (2006).

(Litewka et al. 1996, Litewka & Dębiński 2003, Litewka & Szojda 2006) and that is why the final form of the respective relations will be shown here. According to the rules of the continuum damage mechanics presented by Hayhurst (1983), Lemaitre (1984), Murakami (1987) and Krajcinovic (1995) the current state of the deteriorated material structure is described by the symmetric second rank damage tensor Ω_{ij} defined by Murakami & Ohno (1981) and Betten (1983). The explicit form of the relevant constitutive equations was found (Litewka et al. 1996, Litewka & Dębiński 2003) by employing the methods of the theory of tensor function represen-tations as applied to solid mechanics by Boehler (1987) and Betten (1988, 1998). The first equation of the theoretical model is the stress-strain relation for anisotropic elastic solid

$$\varepsilon_{ij} = -\frac{\nu_0}{E_0}\delta_{ij}\sigma_{kk} + \frac{1+\nu}{E_0}\sigma_{ij} + \\ + C\left(\delta_{ij}D_{kl}\sigma_{kl} + D_{ij}\sigma_{kk}\right) + 2D\left(\sigma_{ik}D_{kj} + D_{ik}\sigma_{kj}\right), \quad (1)$$

where ε_{ij} is the strain tensor and σ_{ij} is the stress tensor. Equation (1) contains the Kronecker delta δ_{ij}, the Young modulus E_0 and Poisson ratio ν_0 for an originally undamaged material, two constants C and D to be determined experimentally and the second order modified damage tensor D_{ij} responsible for the current state of internal structure of the material defined by Litewka (1989).

Deterioration of the material structure due to applied load was described by the damage evolution equation expressed in the form of the tensor function

$$\Omega_{ij} = As_{kl}s_{kl}\left(1 + H\det\right)^F \delta_{ij} + \\ + B\sqrt{\sigma_{kl}\sigma_{kl}}\left(1 + H\det\right)^F \sigma_{ij}, \tag{2}$$

where Ω_{ij} is a classical second order damage tensor formulated by Murakami & Ohno (1981) and Betten (1983), s_{kl} is the stress deviator, det σ is the determinant of the matrix σ of the stress tensor σ_{ij} and A, B, F are material parameters to be determined experimentally. The multiplier H explained by Litewka & Dębiński (2003) and Litewka & Szojda (2006) is a function of the stress tensor components that was expressed by the following function of the stress tensor invariants

$$H = \frac{227}{200\left|\det\ \right| + \left|\left(\sigma_{pp}\right)^3\right|} . \tag{3}$$

Equation (3) is a result of detailed analysis of possible form of such a function of the stress tensor invariants necessary to fulfill the physical conditions discussed by Litewka & Szojda (2006).

The damage tensor Ω_{ij} that accounts for the continuity of the material is not sufficient to describe directly the overall macroscopic properties of damaged material. That is why it was necessary to define a second order modified damage tensor D_{ij} capable to account for the strength and stiffness reduction of the damaged material. The relation

$$D_i = \frac{\Omega_i}{1 - \Omega_i}, i = 1, 2, 3 \tag{4}$$

between the principal values Ω_1, Ω_2 and Ω_3 of the damage tensor Ω_{ij} and the principal components D_1, D_2 and D_3 of the modified damage tensor D_{ij} contained in Equation (1) was formulated by Litewka (1989). The principal components of the modified damage tensor (4) increase to infinity for fully damaged materials and that is why could reduce to zero the stiffness of the material expressed by Equation (1).

Theoretical model used in this paper can also be used to determine the maximum stresses that can be sustained by the material subjected to multi-axial state of stress. To this end the appropriate fracture criterion for brittle material was formulated according to the rules of the damage mechanics. The physical background of this criterion was looked for in the results of experiments done and in the failure modes of broken specimens. It is seen from Figure 2a that the failure of brick specimen subjected to uni-axial compression occurs due to accumulation and growth of vertical cracks. In the case of tri-axial compression

shown in Figures 2b, c the material before its failure is totally crushed into separate tiny particles. This is well seen particularly in Figure 2c in the case of the specimen subjected to relatively high hydrostatic pressure p. In this specific case of loading the lower part of the specimen of brick is not seen in Figure 2c as it was completely crushed into powder whereas the other parts did not show so advanced degradation of the material structure. No photographs could be taken for three specimens of brick subjected to hydrostatic pressure $p = -9.92$, -8.66 and -9.34 MPa. These specimens were completely crushed into small particles and into powder. The similar failure modes were detected for the specimens of mortar where the total degradation of the internal structure occurred even for lower values of hydrostatic pressure. It means that tri-axial compression of brittle rock-like materials results in crack growth to such a state that at fracture the net cross section area on certain planes is reduced to zero. This full deterioration of internal structure of the material occurs when at least one of the principal components Ω_1, Ω_2 or Ω_3 of the damage tensor Ω_{ij} determined from Equation (2) reaches the limit value equal to unity.

To compare the experimental results with theoretical prediction the Equation (2) was expressed in terms of the stress tensors components

$$\sigma_{ij} = \begin{bmatrix} \sigma_{11} = p & 0 & 0 \\ 0 & \sigma_{22} = p & 0 \\ 0 & 0 & \sigma_{33} = \sigma_V + p \end{bmatrix} \tag{5}$$

for State I of tri-axial compression and

$$\sigma_{ij} = \begin{bmatrix} \sigma_{11} = \sigma_H + p & 0 & 0 \\ 0 & \sigma_{22} = \sigma_H + p & 0 \\ 0 & 0 & \sigma_{33} = p \end{bmatrix} \tag{6}$$

for State II of tri-axial compression. Taking into account the notation adopted in Equation (5) the relation

$$\Omega_1 = \Omega_2 = \left(\frac{2}{3}A\sigma_V^2 + Bp\sqrt{\sigma_V^2 + 2\sigma_V p + 3p^2}\right) \cdot \\ \cdot \left[1 + \frac{227(\sigma_V + p)p^2}{200\left|(\sigma_V + p)p^2\right| + \left|(\sigma_V + 3p)^3\right|}\right]^F = 1 . \tag{7}$$

was obtained for State I. The third principal component of the damage tensor Ω_3 does not decide in this case on the material fracture as it grows slower than Ω_1 and Ω_2. The State II of tri-axial compression expressed by Equation (6) is characterized by faster

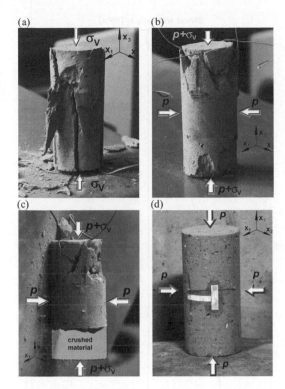

(a) (b) (c) (d)

Figure 2. Specimens of bricks after tests: (a) Specimen B-H0-2 tested under uni-axial compression, (b) Specimen B-H3-2 tested under tri-axial compression, p = −3.55 MPa, (c) Specimen B-H6-1 tested under triaxial compression, p = −6.64 MPa, (d) Specimen BC1 tested under pure hydrostatic compression.

growth of the principal component Ω_3 of the damage tensor and that is why the material fracture occurs when

$$\Omega_3 = \left(\frac{2}{3}A\sigma_H^2 + Bp\sqrt{2\sigma_H^2 + 4\sigma_H p + 3p^2}\right).$$
$$\cdot\left[1 + \frac{227(\sigma_H + p)^2 p}{200\left|(\sigma_H + p)^2 p\right| + \left|(2\sigma_H + 3p)^3\right|}\right]^F = 1. \quad (8)$$

In this case the growth of two others principal components Ω_1 and Ω_2 of the damage tensor is slower and that is why they do not decide about the onset of fracture.

Application of the fracture criterion proposed requires calibration of the material. The numerical va-lues of the constants A, B, C, D and F shown in Table 1 were obtained by using the stress-strain curves determined experimentally for uni-axial compression of brick and mortar seen in Figures 3, 4.

The details of the method used here to identify the material parameters have been described by Litewka &

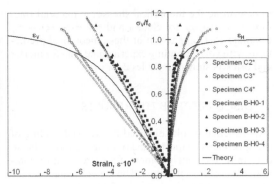

Figure 3. Experimental and theoretical stress-strain curves for brick subjected to uni-axial compression.
* The data for these specimens of Brick 1 were used in Litewka & Szojda (2006).

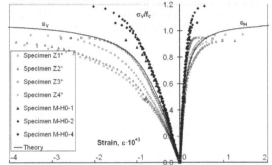

Figure 4. Experimental and theoretical stress-strain curves for mortar subjected to uni-axial compression.
* The data for these specimens of Mortar 1 were used in Litewka & Szojda (2006).

Dębiński (2003). The constant F that appears in Equations (2), (7), (8) was also determined experi-mentally and to do this, one point taken from one stress-strain curve obtained experimentally for tri-axial compression is sufficient. Theoretical stress-strain curves shown in Figures 3, 4 were obtained from the relations

$$\varepsilon_1 = \varepsilon_2 = \varepsilon_H = -\frac{v_0}{E_0}\sigma_V +$$
$$+ C\left[\frac{2A\sigma_V^3 - 3B\sigma_V^3}{3 - 2A\sigma_V^2 + 3B\sigma_V^2} + \frac{2A\sigma_V^3}{3 - 2A\sigma_V^2}\right] \quad (9)$$

$$\varepsilon_3 = \varepsilon_V = \frac{\sigma_V}{E_0} + (2C + 4D)\frac{2A\sigma_V^3 - 3B\sigma_V^3}{3 - 2A\sigma_V^2 + 3B\sigma_V^2}. \quad (10)$$

Equations (9), (10) were obtained by specifying the stress-strain relation (1) for uni-axial compression.

Equations (7) and (8) were used to calculate the values of σ_V and σ_H corresponding to material failure in State I and State II. These data made it possible to determine the theoretical stresses at material fracture $\sigma_{3f} = p + \sigma_V$ for State I and $\sigma_{1f} = \sigma_{2f} = p + \sigma_H$ for State II. Comparison of these theoretical predictions with corresponding experimental data for mortar and brick is shown in Tables 2–5.

4 DISCUSSION OF THE RESULTS

Experimental results obtained for tri-axial loading of the specimens of mortar and brick made it possible to determine the configuration of the stress tensor components at material failure. The experimental data for State I and State II shown in Tables 2–5 were used to determine the limit surfaces at material fracture. The form of such limit surface for brittle rock-like materials, constructed at the stress space (Willam and Warnke, 1975; Szojda, 2001) is shown in Figures 1, 5, 6. The axes of the coordinate system shown in Figure 1 correspond to the principal stresses σ_1, σ_2, σ_3 whereas those in Figures 5, 6 are defined by the mean stress

$$\sigma_m = \frac{\sigma_{11} + \sigma_{22} + \sigma_{33}}{3} \qquad (11)$$

and stress intensity

$$\sigma_i = \left\{ \frac{1}{2} \left[(\sigma_{11} - \sigma_m)^2 + (\sigma_{22} - \sigma_m)^2 + (\sigma_{33} - \sigma_m)^2 \right] + \tau_{12}^2 + \tau_{23}^2 + \tau_{31}^2 \right\}^{0.5}, \qquad (12)$$

where σ_{11}, σ_{22}, σ_{33}, τ_{12}, τ_{23} and τ_{31} are the stress tensor components. The meridians of the limit surface seen in Figures 5, 6 can be divided into three parts. Tri-axial tests presented in this paper made it possible to study experimentally the main almost rectilinear part of the meridian that corresponds to negative mean stresses. To this end the experimental results presented in Tables 2–5 for both materials subjected to State I and State II of tri-axial loading and also those obtained for the specimens of each material subjected to uni-axial compression were used. The experimental study of region of positive mean stresses where failure occurs in brittle manner by fracture of the material requires multi-axial tensile tests which are very difficult in the case of brittle rock-like materials. To obtain any experimental data for third region of large negative mean stresses where failure occurs by particle crushing, the tri-axial tests should be performed at very high values of hydrostatic pressure p combined with uni-axial compression (State I) or uniform bi-axial compression (State II).

Theoretical model (Litewka & Dębiński 2003, Litewka & Szojda 2006) applied in this paper can also be used to determine the form of the limit surface at material failure. The respective theoretical results

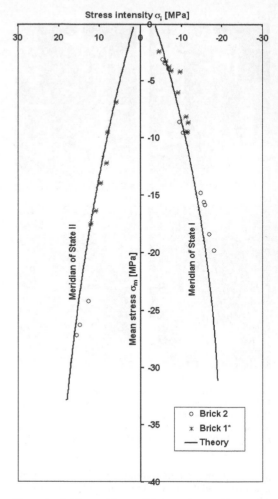

Figure 5. Main meridians of limit surface at fracture for brick.
*These data for Brick 1 were discussed in Litewka & Szojda (2006).

obtained from Equations (7), (8), (11), (12) were used to construct the meridians of the theoretical limit surfaces for brick and mortar. Fairly good agreement of the theoretical predictions with experimental data is seen in Figures 5, 6. The meridians of the limit surfaces for both materials become curvilinear for negative mean stresses larger than $3f_c$. It could suggest that the specimens subjected to very high values of pure hydrostatic pressure could failure and the limit surface might be closed for certain sufficiently large mean stresses.

To obtain any experimental evidence that the limit surface would really be closed for negative mean stresses it should be necessary to do the tests at extremely high hydrostatic pressure. The experimental study of this problem was also attempted here. To this

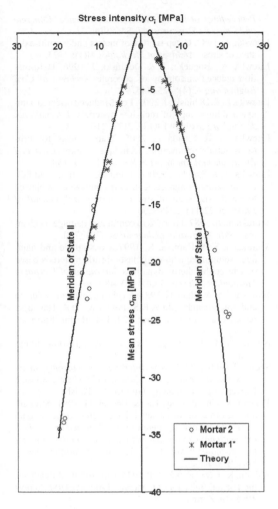

Figure 6. Main meridians of limit surface at fracture for mortar.
*These data for Mortar 1 were discussed in Litewka & Szojda (2006).

Table 6. Specimens of mortar and brick tested under pure hydrostatic pressure.

Mortar 2		Brick 2	
	Hydrostatic pressure, p		Hydrostatic pressure, p
Specimen	MPa	Specimen	MPa
MC1	−64.8	BC1	−65.9
MC2	−66.3	BC2	−66.9
MC3	−64.3	BC3	−65.1

suggest that this region of limit surface is out of the range of actually existing testing machines.

5 CONCLUSIONS

Experiments on behavior of brick and mortar under tri-axial loading presented in the paper were used to study the fracture of brittle materials subjected to higher values of negative mean stresses than those applied in earlier tests. The shape of theoretical limit surface at material failure was determined and compared with the experimental data obtained. Fairly good agreement of the experimental data and theoretical predictions was detected for both materials tested. Increasing compressive strength of brittle rock-like materials known from earlier experiments for specimens of rocks and soils subjected to confined axial compression was also observed in second tri-axial test used here. The experiments performed under pure hydrostatic pressure up to $p = 8f_c$ did not show any evidence that the limit surface for such a level of negative mean stresses might be closed. All these phenomena can also be explained theoretically within the mathematical model proposed. Thus, the experimental technique adopted and phenomenological model used in this paper proved to be accurate enough to study the shape of the limit surface at material failure.

ACKNOWLEDGEMENTS

This work was done within the F.C.T. Program C.E.C. U.B.I. and was financially supported by KBN Grant 5 T07E 028 25 and EC contract MTKD-CT-2004-509775.

REFERENCES

Betten, J. 1983. Damage tensors in continuum mechanics. *Journal of Méchanique Théorique et Appliquée* 2(1): 13–32.
Betten, J. 1988. Applications of tensor functions to the formulation of yield criteria for anisotropic materials. *International Journal of Plasticity* 4(1): 29–46.

end three specimens of the Brick 2 and three specimens of Mortar 2 were tested under pure hydrostatic pressure up to the limit capacity of the testing machine. For the machine used in these experiments the maximum hydrostatic pressure was equal to $p = −70$ MPa. The list of the specimens tested and respective maximum values of p applied is shown in Table 6. All the specimens during the process of loading did not show any symptoms of failure. Moreover, inspection of the specimens after unloading did not reveal any degradation of the internal structure. This can be seen in Figure 2d where the photograph of the specimen BC1 after loading up to $p = −65.9$ MPa and unloading is presented. The results of these experiments and also of those similar reported for concrete by Neville (1995)

Betten, J. 1998. Anwendungen von Tensorfunktionen in der Kontinuumsmechanik anisotroper Materialen. *Zeitschrift für Angewandte Mathematatik und Mechanik* 78(8): 507–521.

Boehler, J.P. 1987. *Applications of tensor functions in solid mechanics*. Wien: Springer-Verlag.

Chaboche, J.L., Lesne, P.M. & Maire, J.F. 1995. Continuum damage mechanics, anisotropy and damage deactivation for brittle materials like concrete and ceramic composites. *International Journal of Damage Mechanics* 4(1): 5–22.

Chen, W.F. 1982. *Plasticity of reinforced concrete*. New York: McGraw-Hill.

Cristescu, N.D. & Hunsche, U. 1998. *Time effects in rock mechanics*. Chichester: John Wiley & Sons,.

Derski, W., Izbicki, R., Kisiel, I. & Mroz, Z. 1989. *Rock and soil mechanics*. Amsterdam-Warsaw: Elsevier-P.W.N.

Ehm, C. & Schneider, U. 1985. Biaxial testing of reactor concrete. In *Transactions of 8th International Conference on Structural Mechanics in Reactor Technology*: Vol. H 349–354. Brusselles: North Holland.

Goodman, R.E. 1989. *Introduction to rock mechanics*. New York: John Wiley & Sons.

Halm, D. & Dragon A. 1998. An anisotropic model of damage and frictional sliding for brittle materials. *European Journal of Mechanics A/Solids* 17(3): 439–460.

Hayhurst, D.R. 1983. On the role of creep continuum damage in structural mechanics. In B. Wilshire & D.R.J. Owen (eds.), *Engineering Approaches to High Temperature Design*: 85–175. Swansea: Pineridge Press.

Krajcinovic, D. 1995. Continuum damage mechanics: when and how?. *International Journal of Damage Mechanics* 4(3): 217–229.

Kupfer, H. 1973. Das Verhalten des Betons unter mehrachsiger Kurzzeitbelastung unter besonderer Berucksichtigung der zweiachsiger Beanspruchung. In *Deutcher Ausschluss fur Stahlbeton*, 229: 1–105. Berlin: Wilhelm Ernst & Sohn.

Lemaitre, J. 1984. How to use damage mechanic. *Nuclear Engineering and Design* 80: 233–245.

Ligęza, W. 1999. Experimental stress-strain relationship for cement concrete under biaxial compression. In *Proceedings of the International Conference, Concrete and Concrete Structures*: 47–54, Zilina.

Litewka, A. 1989. Creep rupture of metals under multi-axial state of stress. *Archives of Mechanics* 41(1): 3–23.

Litewka, A., Bogucka, J. & Dębiński, J. 1996. Deformation induced anisotropy of concrete. *Archives of Civil Engineering* 42(4): 425–445.

Litewka, A. & Dębiński, J. 2003. Load-induced oriented damage and anisotropy of rock-like materials. *International Journal of Plasticity* 19(12): 2171–2191.

Litewka, A. & Szojda L. 2005. Triaxial tests for brittle materials: motivation, technique, results. *Civil and Environmental Engineering Reports* 1: 169–188.

Litewka, A. & Szojda, L. 2006. Damage, plasticity and failure of ceramics and cementitious composites subjected to multi-axial state of stress. *International Journal of Plasticity* 22(11): 2048–2065.

Murakami, S. 1987. Progress in continuum damage mechanics. *JSME International Journal* 30: 701–710.

Murakami, S. & Kamiya, K. 1997. Constitutive and damage evolution equations of elastic-brittle materials based on irreversible thermodynamics. *International Journal of Mechanical Sciences* 39(4): 473–486.

Murakami, S. & Ohno, N. 1981. A continuum theory of creep and creep damage. In A.R.S. Ponter &, D.R. Hayhurst, (eds.), *Creep in Structures*: 422–444. Berlin: Springer-Verlag.

Neville, A.M. 1995. *Properties of concrete*. Harlow: Longman.

Szojda L. 2001. *Analysis of interaction of masonry structures and deformable foundation*, PhD Thesis, Silesian University of Technology, Gliwice, (in Polish).

Thienel, K.-Ch., Rostasy, F.S. & Becker, G. 1991. Strength and deformation of sealed HTR-concrete under biaxial stress at elevated temperature. In *Transactions of 11th International Conference Structural Mechanics in Reactor Technology*: Vol. H 73–78. Tokyo: Atomic Energy Society of Japan.

Willam, K. J. & Warnke, E. P. 1975. Constitutive Models for the Triaxial Behavior of Concrete, *Proc. of IABSE Report 19*: 1–30, Zürich.

Fracture Mechanics of Concrete and Concrete Structures – High-Performance Concrete, Brick-Masonry and Environmental Aspects – Carpinteri, et al. (eds)
© 2007 Taylor & Francis Group, London, ISBN 978-0-415-44617-4

Repair of damaged multiple leaf masonry: Assessment by the Acoustic Emission technique

A. Anzani & L. Binda
Department of Structural Engineering, Politecnico di Milano, Italy

A. Carpinteri, G. Lacidogna & A. Manuello
Department of Structural Engineering and Geotechnics, Politecnico di Torino, Italy

ABSTRACT: Most historic centers in Italy are characterized by the presence of monumental buildings like churches and palaces, and also by a diffused built environment which itself constitutes an important part of the cultural heritage. The problem of safety of these historic masonry constructions started to attract increasing attention after the sudden collapse of ancient monuments happened in the last decades. In this work an experimental research has been carried out on three-leaf walls, using two stones of different characteristics trying to reproduce masonry typologies frequently used in Italian historic centres. Aim of the research was to understand the stress-strain behavior of this masonry under compression and shear actions and to find out suitable repair techniques. The application of the Acoustic Emission technique (AE), gave the possibility to interpret the damage mechanisms and to evaluate the effectiveness of the repair interventions, contributing to the determination of the most suitable methodology for their optimisation.

1 INTRODUCTION

Most historical centres in Italy are characterised not only by the presence of monumental buildings, but also by a man-made environment, which in itself constitutes an important part of the cultural heritage. Both monumental and minor architectural structures are often made of stonework masonry and were built using different techniques. In most cases they are made of the so-called multiple leaf stone masonry, the specific characteristics of which should be understood before undertaking a safety assessment on many kinds of structures. This is the case, for instance, when considering the seismic vulnerability of stonework buildings, but it also applies to the evaluation of the stability of ancient cathedrals, whose pillars have non-homogeneous cross-sections.

An experimental research was carried out on three-leaf walls built using two stones having different characteristics in an attempt to reproduce two masonry types frequently used in Italian historical city centres. The aim of the research was to understand the stress-strain behaviour of this masonry under shear actions and to investigate possible repair techniques. After shear testing, some specimens could be recovered and repaired by grout injections and steel rods confinement. The initial results obtained from the shear tests subsequently carried out on the repaired specimens to verify the effectiveness of the techniques adopted are discussed below.

For the localisation of damage, a sophisticated equipment for the analysis of AE signals was used. This equipment consists of six USAM® units for signal acquisition and processing. Each of these units analyses in real time, and transmits to a PC, all the characteristic parameters of an ultrasonic event. In this manner, each AE event is identified by a progressive number and characterized by a series of data giving the amplitude and the time duration of the signal and the number of oscillations. Each unit is equipped with a pre-amplified wide-band piezoelectric sensor (PZT), which is sensitive in a frequency range of 50 kHz to 800 kHz. The signal acquisition threshold can be set in a range of $100 \mu V$ to 6.4 mV. As known, the cracking process inside the material subjected to compression and shear takes place through the formation of a quantity of microcracks. With increasing loads, these coalesce into macrocracks, whose formation brings about a reduction in the bearing capacity of the structures (Carpinteri & Lacidogna 2006, Carpinteri et al. 2007). The statistical techniques derived from seismology, such as the Gutenberg-Richter (GR) law, and the analysis performed in terms of Self Organized Criticality (SOC), mated to AE monitoring, clarify the relationships between microstructural events and the macroscopic behaviour of structures during the damaging process (Carpinteri 1994, Carpinteri et al. 2006, Richter 1958, Chen et al. 1991).

2 EXPERIMENTAL DETAILS

2.1 Specimen preparation

Three-leaf stone walls, measuring $310 \times 510 \times 790\,\mathrm{mm}^3$, were built at DIS – Politecnico di Milano, by Spadaro srl Contractor – Rosolini, Sicily, so as to reproduce the morphology of the multiple-leaf walls and piers frequently encountered in historical centres (Fig. 1). The two external leaves, made of stone blocks with horizontal and vertical mortar joints, are connected to an internal rubble masonry leaf made of pebbles of the same kind of stone and connected with the same mortar as the outer leaves (Binda et al. 2003).

Two types of stones were chosen: the same limestone used for the Noto Cathedral (Noto stone) and a sandstone, frequently used in central and southern Italy (Serena stone). The characteristics of the two stones were obtained on cylindrical specimens, 80 mm in diameter and 160 mm high, cored from regularly cut stones. Compressive strength (σ_c) was determined in both directions, normal and parallel to the bedding planes, and tensile strength (σ_t) was determined in the same direction as the bedding planes. The values of the elastic modulus (E_a) and Poisson's ratio (ν_a), determined as an average over 3 specimens, are given in Table 1.

2.2 Pseudo-shear tests before and after repairs

The walls were tested to failure in pseudo-shear conditions according to the loading scheme shown in Figure 1. Further details on the complete experimental campaign can be found in (Anzani et al. 2004, Binda et al. 2003).

Once removed from the testing machine and stored in the laboratory, the walls were repaired and subsequently tested again. The repair was carried out by means of grout injections designed to fill the main cracks formed at the interface between the inner and the outer leaves during the test and by fitting steel bars. In particular, completely inorganic mixes based on microfine pozzolanic binders were used (Modena et al. 2002) and rods, having a diameter of 16 mm and characterized by an elastic modulus of $19500\,\mathrm{N/mm}^2$, were applied by tightening the nuts to a torque of 50 NM. This corresponds to a longitudinal action of 15.63 kN on each rod, corresponding in its turn to a total lateral confinement on each wall of $0.37\,\mathrm{N/mm}^2$.

2.3 Analysis of the results

The characteristic values reached through the shear tests before and after the repair and the techniques adopted to repair the specimens are given in Table 2.

PS4   PS6   PN4

Figure 1. Loading condition and aspect of the specimens once removed from the testing machine.

Table 1. Characteristics of the stones.

Stone σ_t	Load vs Bedding Planes Direction	σ_c	E_a	ν	
Noto	Normal	20,6	9476	0,10	2,06
Noto	Parallel	17,6	8526	0,09	–
Serena	Normal	104,2	18218	0,19	6,07
Serena	Parallel	89,0	23293	0,21	–

Compressive strength, elastic modulus and tensile strength are expressed in $\mathrm{N/mm}^2$.

Table 2. Characteristics of the walls tested before and after repair.

Wall	Repair [kN]	Virgin $P_{max,b}$ [μm]	δ_{max} [kN]	Repaired $P_{max,b}$ [μm]	δ_{max}
PS4	Injection	391,10	4056,84	485,33	2971,91
PS6	Steel Rods	415,89	4369,80	700,88	8462,52
PN4	Injection + Steel Rods	283,85	3631,31	520,17	6636,5

The results of the tests on the repaired specimens can be seen in Figure 2, 3 and 4, where the load vs. displacement diagrams are shown, compared to those obtained after the shear tests and before the repair.

Considering the different types of stone constituting the walls, the repair always caused an increase in load-bearing capacity, of 1.24 in the case of specimen PS4, made of Serena stone and repaired by grout injections only, more appreciable in the other cases.

In the two cases where the repair was made through injections (PS4 and PN4), the subsequent collapse of the two collar joints, that could be clearly observed in the diagrams plotted before the repair, is no longer visible (Fig. 2). This is scarcely evident in the case of the wall repaired only with steel rods (PS6), in which case the diagram shows a small decrease, corresponding to a displacement of 6000 μm.

In the case of specimen PS6, the most ductile one because of the steel rods and the quality of the stone,

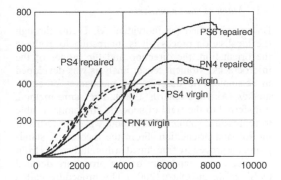

Figure 2. Results on virgin and repaired specimens.

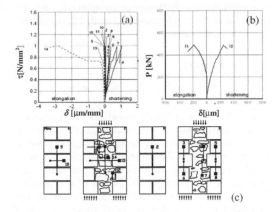

Figure 3. Results on specimen PS4.

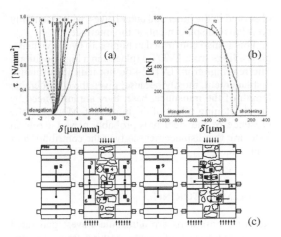

Figure 4. Results on specimen PS6.

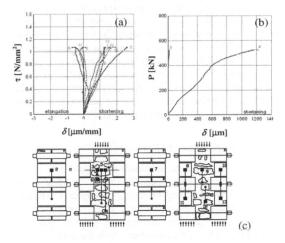

Figure 5. Results on specimen PN4.

the outer leaves did not crack. Failure took place through cracks propagation in the inner leaf, mostly involving the mortar at the interface with the stone pebbles, and at the interfaces between the inner and the outer leaves; cracks also propagated diagonally, indicating the diffusion of the load, and presented a concentration resulting in the detachment of material from the second stone layer below the top.

In the case of specimen PN4 a greater involvement of the outer leaf was observed at failure, when new cracks opened and, in some areas, delamination occurred, together with the propagation in the inner leaf of cracks that sometimes also cut the stone pebbles. In this case horizontal cracks were also visible, both in the outer leaf and in the inner leaf at the base level, due to the compression effect produced horizontally by the presence of the steel rods during the loading phase, given the low strength of the stone.

Figures 3a–5a show the shear stress vs. strain diagrams obtained from the readings of single LVDTs, whereas Figures 3b–5b show the load vs. relative displacement diagrams from LVDTs placed at the interface between the inner and outer leaves.

Conventionally, shortening was plotted as positive and elongation as negative; apart from some anomalous readings indicated below, vertical strains were always positive and horizontal ones were always negative.

In the case of wall PS6, repaired only with steel rods (Fig. 4b), the values recorded by both the LVDTs measuring the relative displacement were higher than in the other walls, consistently with the presence of previous cracks at the interface that had not been filled and in agreement with the compliant behaviour mentioned above. In the case of wall PN4 only LVDT 5, having a longer base, gave higher values than the other vertical LVDTs, which all showed very similar strains. This indicates that, in the case of Noto stone, which is characterised by a high porosity, a good adhesion developed, as borne out by the first series of test (Binda et al. 2003), during which similar strains were observed in the inner and outer leaves, with essentially no relative displacements at the interface.

This is also confirmed in Figure 3b by the virtually nil readings from LVDT 3 (placed in a high position). The readings from LVDT 4 do not indicate a relative displacement, mostly because, given the considerable length of the transducer base, they are influenced by the shortening effect of the inner leaf. Moreover, the presence of the steel rods probably played a role in reducing the displacements, given the contribution of frictions.

3 AE MONITORING

Acoustic emissions are ultrasonic waves; generated by a rapid release of energy, coming from disconti-nuities or cracks propagating in materials subject to states of stress or strain. In isotropic and homogeneous materials, the waves propagate through the damaged solid, according to straight paths, moving at the same rate v in every direction, until they reach the outer surface, where they are captured by ad hoc sensors (Carpinteri & Lacidogna 2006, Carpinteri et al. 2007). The process is similar to the one that takes place in seismology, where the elastic waves generated by the earthquakes reach the monitoring stations positioned on the surface of the earth (Richter 1958).

3.1 Localisation of AE sources

The first stage in the localisation method consists of recognising the data needed to identify the AE sources, followed by the triangulation procedure (Carpinteri et al. 2006). During the first stage, the groups of signals, recorded by the various sensors, that fall into time intervals compatible with the formation of microcracks in the volume analysed, are identified. These time intervals, of the order of microseconds, are defined on the basis of the presumed speed of transmission of the waves (P) and the mutual distance of the sensors applied to the surface of the mater-ial. It is common practice to consider the amplitude threshold of $100\,\mu V$ of the non-amplified signal to distinguish between P-wave and S-wave arrival times. In fact, P-waves are usually characterized by higher value signals. In the second stage, when the forma-tion of microcracks in a three-dimensional space is analysed, the triangulation technique can be applied if the signals recorded by at least five sensors fall into compatible time intervals. Thus, with this procedure it is possible to define both the position of the micro-cracks in the volume and the speed of transmission of P-waves. The localisation procedure can also be performed through numerical techniques using opti-misation methods such as the Least Squares Method (LSM) (Carpinteri et al. 2006).

3.2 Moment tensor analysis

Having identified the individual AE events through the localisation process, it is interesting to evaluate the orientation, the direction and the modes of the microcracks. Moment tensor analysis was developed in seismology to describe the mechanics of earth-quakes (Aki & Richards 1980). The same method, transferred to the field of acoustic emissions, is able to represent a source of ultrasonic waves in terms of motion and orientation. From the theoretical stand-point, the procedure fine-tuned in this study relies from the theoretical standpoint on the procedure defined by Shigeishi and Ohtsu (2001). This procedure, called SiGMA, characterised the AE signal by taking into account only the first arrival time of P-waves. In this procedure the moment tensor components m_{pq} are pro-portional to the amplitudes $A(\mathbf{x})$ of the first P-waves to reach the transducers:

$$A(\mathbf{x}) = \frac{C_s\, REF(\mathbf{t,r})}{R}(r_1\ r_2\ r_3)\begin{pmatrix} m_{11} & m_{12} & m_{13} \\ m_{21} & m_{22} & m_{23} \\ m_{31} & m_{32} & m_{33} \end{pmatrix}\begin{pmatrix} r_1 \\ r_2 \\ r_3 \end{pmatrix}. \quad (1)$$

C_s in Eq. (4) is a calibration coefficient of the acous-tic emission sensors, R is the distance between the AE source at point $\mathbf{y}$ and the sensor located at point $\mathbf{x}$. Vector $\mathbf{r}$ represents the components of the distances, R, obtained through the localisation procedure, and $REF(\mathbf{t,r})$ is the reflection coefficient of the sensitiv-ity of the sensor between vector $\mathbf{r}$ and direction $\mathbf{t}$. The moment tensor provides a general representation of the seismic source. The seismic moment tensor is a variety of stress tensor for an elastic medium, similar to the stress tensor used in elastomechanics. Since the moment tensor is symmetrical, to be able to represent it, it is necessary to determine the six independent unknowns m_{pq}. To this end, in order to determine the components of the moment tensor, the amplitude of the signal $A(\mathbf{x})$ must be received from at least six AE channels. From an eigenvalue analysis of the moment tensor, it is possible to determine the type of crack localised:

$$\frac{\lambda_1}{\lambda_1} = X + Y + Z, \frac{\lambda_2}{\lambda_1} = 0 - \frac{Y}{2} + Z, \frac{\lambda_3}{\lambda_1} = -X - \frac{Y}{2} + Z, \quad (2)$$

where, $\lambda_1, \lambda_2, \lambda_3$ are the maximum, medium and min-imum eigenvalues, respectively, X is the component due to shear, Y is the deviatoric tensile component, Z is the isotropic tensile component. Ohtsu classified an AE source with $X > 60\%$ as a shear crack, one with $X < 40\%$ and $Y + Z > 60\%$ as a tensile crack, and one with $40\% < X < 60\%$ as a mixed mode crack. More-over, from an eigenvector analysis, it is possible to determine the versors, l and n, which determine the direction of the displacement and the orientation of the crack surfaces (Shigeishi & Ohtsu 2001).

4 AE DATA ANALYSIS

4.1 Specimens PS6 and PN4

Masonry test pieces PS6 and PN4 were monitored with the AE technique throughout the loading test. Figure 6 shows, for the PS6 specimen, the diagrams of the AE cumulative and differential counts vs. time obtained during the test. Since the imposed displacements and times are proportional, it is possible to compare the diagram in Figure 2 with that in Figure 6. It can be seen that cumulative counts increase, very slowly at first and then proportionally to the load, reaching a peak in the proximity of the highest load reached in the test. The function representing the differential counts (AE count number per minute), instead, reaches its maximum value in the ascending branch of the load-displacement curves and reduces quickly to nil in the proximity of the ultimate load. From the diagrams of the differential counts of the AE signals, it is therefore possible to identify two phases in the loading process: a first phase during which the biggest cracks are formed and the material reaches a critical damage condition (phase 1), and a second phase during which the material tends to exhaust its bearing capacity (phase 2).

From the diagrams in Figure 6, it can also be seen that the material begins to release energy when the stress level, already reached in the uncracked specimen, is exceeded during the tests on the repaired specimen. In perfect agreement with the so-called *Kaiser effect* (Kaiser, 1950).

Figure 7 shows, on the A and D faces of specimens PS6, the projections of the AE sources identified during the loading test. In particular, during loading test 34, AE signal source points were identified (Fig. 7). The transmission speed of the ultrasonic signals was found to be ca 1350 m/s, the frequency ca 120 kHz. The sources determined during the first phase of the loading test (phase 1) are represented by triangles; the ones represented by squares were determined during phase 2, when the material tends to exhaust its bearing capacity (Table 3). During the load test on specimen

PN4, instead, 32 source points were identified from AE signals. The transmission speed of the signals was found to be ca 980 m/s, the frequency was ca 90 kHz.

Damage process zones are in good agreement with the considerations expressed in Section 2.3. The higher speed of the signals in specimen PS6 compared to specimen PN4 should probably be ascribed to the greater compactness of Serena stone vs. Noto stone. AE signals, in fact, undergo a considerable attenuation in porous media.

4.2 Estimating the volumes of the microcracks

Shigeishi and Ohtsu (2001) pointed out that the trace component of the moment tensor can be expressed with the following relationship:

$$m_{kk} = (3\lambda + 2\mu)l_k n_k \Delta V, \qquad (3)$$

where λ and μ are Lamè's constants, l_k and n_k are the components of the direction vector and the vector normal to the surface of the crack, m_{kk} is the trace component of the moment tensor, and ΔV is the volume expansion of each microcrack. As a rule, in elasticity theory, factor $(3\lambda + 2\mu) = K$ is considered as a constant specific to each material. Volume expansion may therefore be expressed as follows:

$$\Delta V = \frac{1}{K} \frac{m_{kk}}{l_k n_k}. \qquad (7)$$

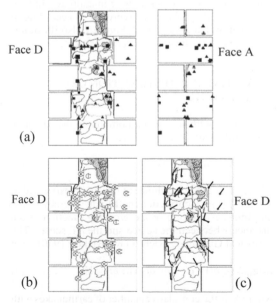

(a)

(b)   (c)

Figure 7. AE sources in specimen PS6. (a) Localization and magnitudes of relative crack volumes, (b) crack typology, (c) crack direction vectors.

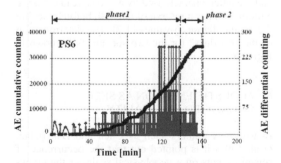

Figure 6. Diagram of cumulative and differential AE counts for the PS6 specimen.

Table 3. Markers of AE sources representing crack volume expansion in phase 1 and phase 2 of the test and labels identifying crack typology.

Phase 1	Phase 2	
▲ $\Delta V \leq 0.1 \, \text{mm}^3$	■ $\Delta V \leq 0.1 \, \text{mm}^3$	
▲ $0.1 \leq \Delta V \leq 0.2 \, \text{mm}^3$	■ $0.1 \leq \Delta V \leq 0.2 \, \text{mm}^3$	
▲ $\Delta V \geq 0.2 \, \text{mm}^3$	■ $\Delta V \geq 0.2 \, \text{mm}^3$	
Crack Typology (Mode)		
shear ⊗	tensile ①	mixed ⊛

For specimen PS6 in Serena stone, considering an average elastic modulus $E \cong 2000 \, \text{N/mm}^2$ and a Poisson coefficient of 0.2 (Table 1), we get $\lambda = Ev/(1+v)(1-2v) \cong 5700 \, \text{N/mm}^2$ and $\mu = Ev/2(1-2v) \cong 1700 \, \text{N/mm}^2$, which means it is possible to assume an approximate value of $K_S \cong 20000 \, \text{N/mm}^2$. Similarly, for specimen PN4 in Noto stone, the value of $K_N \cong 3800 \, \text{N/mm}^2$ can be determined.

Assuming term K to be approximately constant for both specimens throughout the loading tests, it can be seen that the highest magnitude volume expansion events are concentrated at the macrocracks identified. If the AE sources are identified by means of a graphic representation where individual crack magnitudes are symbolised, an interesting superposition is obtained between the AE sources with large crack volumes and the failure planes visible to the naked eye (Fig. 7). For specimen PS6, crack volume magnitudes, broken down by loading phase, are represented according to the notation listed in Table 3. Figure 7 also illustrates crack types and the crack direction vectors.

5 STATISTICAL DISTRIBUTION OF AE EVENTS

By analogy with seismic phenomena, in the AE field, magnitude is defined as follows:

$$m = Log_{10} A_{max} + f(r), \qquad (4)$$

where, A_{max} is the amplitude of the signal expressed in μV, and $f(r)$ is a correction coefficient whereby signal amplitude is taken to be a decreasing function of the distance r between the source and the AE sensor. The empirical Gutenberg-Richter law (Richter 1950) give:

$$Log_{10} N(\geq m) = a - bm \quad \text{or} \quad N(\geq m) = 10^{a-bm}, \qquad (5)$$

where N is the cumulative number of earthquakes with magnitude $\geq m$ in a given area and a specific time-range, whilst b and a are positive constants varying from one region to another. Eq. (5) has been used

successfully in the AE field to study the scaling laws of AE wave amplitude distribution. According to Eq. (5), the "b-value" stands for the slope of the regression line in the "log-linear" diagram of AE signal amplitude distribution. This parameter changes systematically at different times in the course of the damage process and therefore can be used to estimate damage evolution modalities.

Moreover, scale effects on the size of the cracks identified with the AE technique entail, by analogy with earthquakes (Carpinteri et al. 2006), the validity of the following relationship:

$$N(\geq L) = cL^{-2b}, \qquad (6)$$

where N is the cumulative number of AE generated by cracks having a characteristic size $\geq L$, c is the total number of AE events and $D = 2b$ is the noninteger exponent of the distribution. The cumulative distribution (6) is substantially identical to the one proposed by Carpinteri (1994), according to which the number of cracks with size $\geq L$ contained in a body is given by:

$$N^*(\geq L) = N_{tot} L^{-\gamma}. \qquad (7)$$

In Eq. (7), γ is an exponent reflecting the disorder, i.e., crack size scatter, and N_{tot} is the total number of cracks contained in the material. By equating distributions (6) and (7), we find that: $2b = \gamma$. At collapse, when the size of the largest crack is proportional to the characteristic dimensions of a structure, function (7) is characterised by an exponent $\gamma = 2$, corresponding to $b = 1$. In (Carpinteri 1994) it is also demonstrated that $\gamma = 2$ is a lower limit corresponding to the minimum value $b = 1$, observed experimentally when the bearing capacity of a structural member is exhausted.

By applying these concepts to the analysis of the "b-value" of specimens PS6 and PN4, it can be seen that the former exhausted its bearing capacity during the load test, with the formation of cracks of a size comparable to that of the specimen, along the interface between the filler and the outer wall (b-value $\cong 1$), and the latter, characterised by a widespread cracking pattern, still has a reserve of strength before reaching the final collapse (b-value $\cong 1.7$). The determination of the "b-values" for the two specimens is shown in Figure 8.

6 ACOUSTIC EMISSION AS SELF ORGANIZED CRITICALITY

In works by authors such as Chen and Bak (Chen et al. 1991) it is pointed out that the occurrence of seismic events on a geophysical scale, or the cracking of rock specimens subjected to stress conditions in laboratory tests, may be interpreted as instances

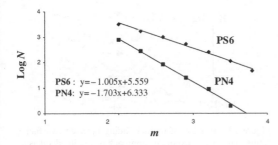

Figure 8. Determination of the "b-value" at the end of the loading test on the two specimens.

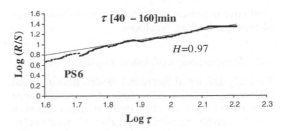

Figure 9. R/S analysis data sets for specimen PS6. The straight line shows the power law $R/S \propto \tau^H$, Eq. (9).

of Self Organized Criticality (SOC) phenomena. The concept of SOC implies that, in a system evolving in permanently critical conditions, the disturbances localised in a portion of the system may have cascade effects on the entire system. In this context, SOC state anomalies play the role of precursors of catastrophic behaviour. SOC behaviour is a permanent competition between the action of the process driving forces and the response of the system's dynamics. The determination of the time series power law exponent, referred to as Hurst exponent, can facilitate the identification of the self-organised nature of these phenomena (Hurst et al. 1965).

6.1 Hurst exponent of AE time series

The AE signal time series obtained from specimens PS6 and PN4 were interpreted by analysing the Hurst exponent. The starting consideration is that evolutive phenomena accompanied by a long-term memory of earlier states – such as a loading process culminating in the failure of a structural element – departs from a Gaussian random sequence type behaviour. In physics, this sequence is identified, for example, by the Brownian motion of a particle submerged in a fluid (Einstein 1916). Moving in an erratic fashion, this particle covers a distance, which, on average, is a function of the square root of time multiplied by a constant:

$$R = k\, t^{1/2}. \tag{8}$$

In Eq. (8), R is the distance covered, t is the time and k is a constant. In a generalised form, Eq. (8) can be rewritten as follows:

$$R/S \propto \tau^H, \tag{9}$$

where τ is the time associated with a certain time window and H is the Hurst exponent. $R(N)$ and $S(N)$ are defined as:

$$R(N) = \max_{1 \le j \le N}\left\{ \sum_{i=1}^{j}(A_i - \overline{A}) \right\} - \min_{1 \le j \le N}\left\{ \sum_{i=1}^{j}(A_i - \overline{A}) \right\}, \tag{10}$$

$$S(N) = \sqrt{\sum_{i=1}^{N}(A_i - \overline{A})^2}. \tag{11}$$

In Eqs. (10) and (11), $R(N)$ and $S(N)$ stand for the range and the standard deviation of the historical time series in question, while N and $\overline{A}$ are, respectively, the length and the mean value of the sequence A_i in a time window τ. The range, $R(N)$, is obtained as the difference between the maximum and the minimum value of the cumulated summation of the deviations from the mean of all sample observations. Hence, by determining the ratio between $R(N)$ and $S(N)$ it becomes possible to standardise the measurements and compare the results of different analyses (Rescaled Range Analysis).

In general, R/S increases with increasing τ, according to a power law, which is a function of H, and this exponent reflects the greater or lesser persistence of the phenomenon considered (Mandelbrot 1997). Moreover, H lies between the values 0 and 1; $H > 1/2$ corresponds to a persistent variation of the function: the greater is τ, the more probable is its further increase; $H \le 1/2$ characterizes an "anti-persistent" behaviour: the greater is τ, the more probable is a decrease in the function; $H = 1/2$ represents the Gaussian random sequence.

By analysing the AE data time series obtained from specimen PS6, for a time window of between 40 and 160 min (Fig. 9), i.e., extending from a few minutes of the start of the test till the end of the test, the Hurst exponent was found to be 0.97. This value remains virtually constant even for time windows of decreasing size, starting from the initial instant of the analysis. Similar results were also obtained for specimen PN4.

It may be inferred that imposed deformation tests carried out through the application of a constantly increasing load display a persistent behaviour on the different time scales. This conclusion is highly relevant to the interpretation of structural damage phenomena and the monitoring of structures, possibly on site, with the AE technique, where the persistence of the energy released during the damage process may be used for predictive purposes when dealing with self-organised phenomena leading to collapse.

6.2 Hurst exponent of rupture profiles

Recently developed theoretical models have acquired increasing relevance in the study of seismicity. Some authors (see Hallgass et al. 1996) assumed that fault profiles can be regarded as fractals, and, in particular, as statistically self-affine profiles $F_H(t)$, whose roughness depends on the time scale as $|F_H(t+\tau) - F_H(t)| \sim \tau^H$. The exponent $0 \leq H < 1$, in this case, controls the roughness of the fault where the standard Brownian profile corresponds to $H = 1/2$, and a differentiable limit curve corresponds to $H = 1$. More generally, the fractal dimension of the profile is well known to be $D_F = d - H$, in which the fractal ensemble is embedded in a d-dimensional Euclidean space. For a self-affine rough surface, the Euclidean reference dimension is $d = 3$. If one considers instead a representative profile of this surface we get $d = 2$. In either case, exponent H does not depend on the reference dimension (Zavarise et al. 2006).

The laboratory tests performed on multi-leaf specimens PS6 and PN4 entailed, due to the testing modalities, a phenomenon of mutual sliding between outer and inner leaves (Fig. 1). Hence, the failure mechanism involved is similar to that of facing faults as shown in Figure 10. In either mechanism, failure is accompanied by a decrease in the roughness of the profiles in contact. Based on the foregoing, the variation in the Hurst exponent can be seen as a significant indicator of the failure process.

The b-value of the GR law and the Hurst exponent can be correlated during the damage process. The article by Hallgass et al. 1996 gives the following relationship:

$$\frac{b}{c} = 1 - \frac{H}{d-1}. \quad (12)$$

In Eq. (12), c is the slope in the linear-log graph of the Richter law that correlates the magnitude of earthquakes with seismic energy (Richter 1958):

$$Log_{10}E = q + cm. \quad (13)$$

Equation (13) applied to the AE technique correlates the energy released by AE with magnitude m calculated as in Eq. (4). Accordingly, b-value and parameter c were computed by subdividing the test into homogenous time intervals. By introducing these values into Eq. (12) together with the Euclidean dimension, $d = 3$, we get the H vs. b-value diagram shown in Figure 11.

From Figure 11 it can be seen that for both specimens, when the b-value approaches 1, i.e., in the

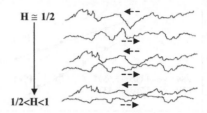

Figure 10. Process of mutual sliding between two facing faults. The Hurst exponent approaches 1 with decreasing profile roughness.

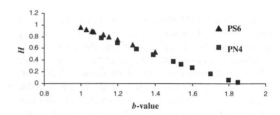

Figure 11. Exponent H vs. b-value from the tests performed on specimens PS6 and PN4.

proximity of critical conditions, the roughness exponent H approaches 1. The physical meaning of this relationship lies in the fact that the surfaces in contact of the specimen leaves become smoother and smoother, causing a critical sliding rate up to collapse. The values of H computed at the start of the tests, when the b-values are high, provide an indication of the initial roughness of the faults as well as of the volumetric energy dissipation. At the end of the tests, the maximum value of H reached in PS6 turned out to be greater than that reached in PN4. This result regarding the roughness of the surfaces in contact corroborates the observations made in Section 5.

7 CONCLUSIONS

An experimental research was carried out on repaired three-leaf walls reproducing the masonry types commonly used in Italian historical centres. The damage process zones, identified in the specimens through the AE technique, were in good agreement with the results of the tests performed by means of displacement transducers. The statistical techniques derived from seismology, such as the Gutenberg-Richter (GR) law, and the analysis performed in term of Self Organized Criticality (SOC), mated to AE monitoring, made it possible to interpret the damage mechanisms and to evaluate the efficiency of repair interventions, contributing to the identification of the most effective methodology for their optimisation.

ACKNOWLEDGEMENTS

The authors gratefully acknowledge the assistance provided by M. Brazzale, I. Mecca and S. Rampoldi in the
experimental activities, M. Antico and M. Iscandri for
the technical support received.

REFERENCES

Aki, K. & Richards, P.G. 1980. *Quantitative Seismology,
Theory and Method*. New York: W. H. Freeman.
Anzani, A. Binda, L. Fontana, A. Pina-Henriques, J. 2004.
An experimental investigation on multiple-leaf stone
masonry. *13th Int. Brick/Block Masonry Conference RAI
Amsterdam, 4–7 July 2004.*
Binda, L. Anzani, A. Fontana, A. 2003. Mechanical behaviour
of multiple-leaf stone masonry: experimental research.
*3-Day Int. Conf. Structural Faults & Repair, London, 1–3
July 2003*, MC Forde (Ed.), Engineering Technics Press,
Edinburgh, ISBN 0-947644-53-9, CD-ROM, Keynote
Lecture.
Carpinteri, A. 1994. Scaling laws and renormalization groups
for strength and toughness of disordered materials. *International Journal of Solids and Structures* 31: 291–302.
Carpinteri, A. & Lacidogna, G. 2006. Damage monitoring of
an historical masonry building by the acoustic emission
technique. *Materials and Structures* 20: 143–149.
Carpinteri, A. Lacidogna, G. Niccolini, G. 2006. Critical
behaviour in concrete structures and damage localization by acoustic emission. *Key Enineering Materials* 312:
305–310.
Carpinteri, A. Lacidogna, G. Pugno, N. 2007. Structural
damage diagnosis and life-time assessment by acoustic
emission monitoring. *Engineering Fracture Mechanics*
74: 273–289.
Chen, K. Bak, P. Obukhov, S. 1991. Self-organized criticality
of the fracture process in rock. *Phys Rev A* 43 625.
Einstein, A. 1916. *Investigations on the Theory of Brownian
Movement*. New York: Dover, essay I and II.
Hallgass, R. Loreto, V. Mazzella, O. Paladin, G. Pietronero,
L. 1996. Earthquakes statistics and fractal faults. *Physical
Review E* 56: 1346–1256.
Hurst, H.E. Black, R.P. Simaika, Y.M. 1965. *Long Term
Storage. An Experimental study*. London: Constable.
Kaiser, J. 1950. *An investigation into the occurrence of noises
in tensile tests, or a study of acoustic phenomena in tensile
tests*. Ph.D, Technische Hochschule München: Munich
FRG.
Mandelbrot, B. 1997. *Fractals and Scaling in Finance*, New
Yor: Springer-Verlag.
Modena, C. Valluzzi, M.R. Tongini Folli, R. Binda, L. 2002
Design Choices and Intervention Techniques for Repair-
ing and Strengthening of the Monza Cathedral Bell-Tower,
Construction Building Materials, ISSN 0950-0618, Spe-
cial Issue, 16(7): 385–395.
Richter, C.F. *1958 Elementary Seismology*. S. Francisco and
London: W.H. Freeman and Company.
Shigeishi, M. & Ohtsu, M. 2001. Acoustic emission moment
tensor analysis: development for crack identification in
concrete materials. *Construction and Buildings Materials*
15: 311–319.
Zavarise, G. Borri-Brunetto, M. Paggi, M. 2007. On the resolution dependence of micromechanical contact models.
Wear 262: 42–54.

Fracture Mechanics of Concrete and Concrete Structures – High-Performance Concrete,
Brick-Masonry and Environmental Aspects – Carpinteri, et al. (eds)
© 2007 Taylor & Francis Group, London, ISBN 978-0-415-44617-4

Numerical simulation of brick-masonry subjected to the double flat-jack test

A. Carpinteri, S. Invernizzi & G. Lacidogna

Dipartimento di Ingegneria Strutturale e Geotecnica, Politecnico di Torino, Torino, Italy

ABSTRACT: In the present paper, we describe the results obtained from double flat-jack tests performed varying the size of the masonry prism involved in the test. In addition, not only the deformations have been acquired, but also the acoustic emissions (AE), in order to get information about local cracking during the test. We present a meso-scale numerical model of the test, where every brick of the masonry is modeled in the details. Discrete cracks can arise both in the mortar joints and in the brick units. A good correlation is found between the amount of cracking simulated numerically and the experimental acoustic emissions for different prism sizes. The model is also able to catch the decrease in the compressive strength with increasing size. It is not possible to obtain an easy direct relation between the acoustic emission and the amount of cracking; nevertheless, it is possible to state that the two quantities are proportional to each other when increasing sizes are considered.

1 INTRODUCTION

Nondestructive and instrumental investigation methods are currently employed to measure and check the evolution of adverse structural phenomena, such as damage and cracking, and to predict their subsequent developments. The choice of a technique for controlling and monitoring reinforced concrete and masonry structures is strictly correlated with the kind of structure to be analyzed and the data to be extracted (Carpinteri & Bocca 1991; Anzani et al. 2000). For historical buildings, nondestructive evaluation (NDE) techniques are used for several purposes: (1) detecting hidden structural elements, such as floor structures, arches, piers, etc.; (2) determining masonry characteristics, mapping the nonhomogeneity of the materials used in the walls (e.g., use of different bricks during the life of a building); (3) evaluating the extent of the mechanical damage in cracked structures; (4) detecting voids and flaws; (5) determining moisture content and rising by capillary action; (6) detecting surface decay phenomena; and (7) evaluating the mechanical and physical properties of mortar and brick, or stone.

This study addresses some of the aforementioned problems deemed of special significance. The structural geometry was defined through the customary survey methods. Damage, cracking, and the evolution of these phenomena over time were assessed through a number of nondestructive techniques: tests with flat-jacks were conducted in order to evaluate the range of stresses affecting the structures; and at the same time,

the cracking processes taking place in some portions of the masonry structures were monitored using the acoustic emission (AE) technique.

The AE technique has proved particularly effective (Carpinteri & Lacidogna 2002, 2003, 2006), in that it makes it possible to estimate the amount of energy released during the fracture process and to obtain information on the criticality of the process underway. Strictly connected to the energy detected by AE is the energy dissipated by the structure being monitored. The energy dissipated during crack formation in structures made of quasibrittle materials plays a fundamental role in the behavior throughout their life. Strong size effects are clearly observed in the energy density dissipated during fragmentation. Recently, a multiscale energy dissipation process has been shown to take place in fragmentation, from a theoretical and fractal viewpoint (Carpinteri & Pugno 2002a,b, 2003). Based on Griffith's assumption of local energy dissipation being proportional to the newly created crack surface area, fractal theory shows that the energy will be globally dissipated in a fractal domain comprised between a surface and a volume in the Euclidean space. According to fractal concepts, an ad hoc theory is employed to monitor masonry structures by means of the AE technique. The fractal theory takes into account the multiscale character of energy dissipation and the strong size effects associated with it. With this energetic approach it becomes possible to introduce a useful damage parameter for structural assessment based on a correlation between AE activity in a structure and the corresponding activity recorded on masonry elements of different sizes, tested to failure by means of double flat-jacks.

2 NONDESTRUCTIVE EVALUATION TESTS

2.1 Flat-jack tests

The single flat-jack test concerns the measurements of in-situ compressive stress in existing masonry structures by use of a thin flat-jack device that is installed in a saw cut mortar joint of the masonry wall (ASTM 1991a). The method is relatively non-destructive. After the slot is formed in the masonry, compressive stress at that point causes the masonry above and below the slot to get closer. Inserting the flat-jack into the slot and increasing its internal pressure until the original distance between points above and below the slot is restored, can thus measure the compressive stress in the masonry. The slots in the masonry are prepared by removing the mortar from masonry bed joints, avoiding disturbing the masonry. Care must be taken in order to remove all mortar in the bed joint, so that pressure exerted by the flat-jack can be directly applied against the cleaned surface of the masonry units. The state of compressive stress in the masonry is approximately equal to the flat-jack pressure multiplied by factors which account for the ratio K_a of the bearing area of the jack in contact with the masonry to the bearing area of the slot, and for the physical characteristic of the jack K_m. In fact, the flat-jack has an inherent stiffness which opposes expansion when the jack is pressurized. Therefore, the fluid pressure in the flat-jack is greater than the stress that the flat-jack applies to masonry, and a conversion factor K_m is necessary to relate the internal fluid pressure to the stress really applied. The average compressive stress in the masonry, f_m, can be calculated as:

$$f_m = K_m K_a p,$$ (1)

where, p is the flat-jack pressure required to restore the gage points to the distance initially measured between them. We performed the tests using rectangular flat-jack 240 mm × 120 mm wide and 7 mm thick (by BOVIAR s.r.l., Italy). Their calibration factor was $K_m = 0.90$–0.92. The loading procedure was synchronized and the pressure was applied with a manual equipment (pressure range between zero and 60 bar). The usual coefficient of variation of this test method can be estimated equal to 20%; therefore, at least three tests have been carried out on each area of interest.

The double flat-jack test provides a relatively non-destructive method for determining the deformation properties of existing unreinforced solid-unit masonry (ASTM 1991b). The test is carried out inserting two flat-jacks into parallel slots, one above the

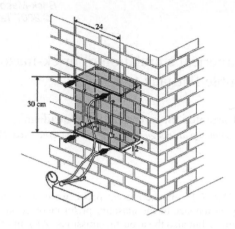

Figure 1. Typical set-up for in situ flat-jack test. The dimensions given are those of the specimen referred to as Vol. 1. (Reprinted from Gregorczyk and Lourenço 2000).

other, in a solid-unit masonry wall (Fig. 1). By gradually increasing the flat-jack pressure, a compressive stress is induced on the masonry comprised in between. The stress-strain relation can thus be obtained measuring the deformation of the masonry. In addition, the compressive strength can be obtained, if the test is continued to local failure. However, this may also cause damage to the masonry in the area adjacent to the flat-jacks. The tangent stiffness modulus at any stress interval can be obtained as follows:

$$E_t = \frac{\delta \sigma_m}{\delta \varepsilon_m},$$ (2)

where, $\delta \sigma_m$ is the increment of stress, and $\delta \varepsilon_m$ is the increment of strain. On the other hand, the secant modulus is given by:

$$E_s = \frac{\sigma_m}{\varepsilon_m},$$ (3)

where, σ_m and ε_m are the actual stress and strain in the masonry.

3 ACOUSTIC EMISSION MONITORING

Monitoring a structure by means of the AE technique makes it possible to detect the onset and evolution of stress-induced cracks. Crack opening, in fact, is accompanied by the emission of elastic waves that propagate within the bulk of the material. These waves can be captured and recorded by transducers applied to the surface of the structural elements (Fig. 2). The signal identified by the transducer (Fig. 3) is preamplified and transformed into electric voltage; it is then filtered to eliminate unwanted frequencies, such as

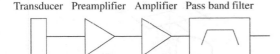

Transducer   Preamplifier   Amplifier   Pass band filter

Figure 2.   Acoustic emission measurement system.

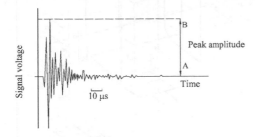

Figure 3.   AE signal identified by the transducer.

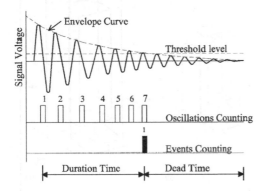

Figure 4.   Counting methods in AE technique.

the vibrations caused by the mechanical instrumentation, which are generally lower than 100 kHz. The signal is then analyzed by a threshold measuring unit which counts the oscillations exceeding a certain voltage value. This method of analysis is called ring-down counting (Pollock 1973; Brindley et al. 1973).

As a first approximation, the counting number, N, can be correlated to the quantity of energy released during the loading process. This technique also considers other procedures. For instance, by keeping track of the characteristics of the transducer and, in particular, of its damping, it is possible to consider all the oscillations produced by a single AE signal as unique events and to replace ring-down counting with the counting of events (Fig. 4).

3.1   AE data acquisition system

The AE monitoring equipment adopted by the writers consists of piezoelectric transducers fitted with a preamplifier and calibrated on inclusive frequencies between 100 and 400 kHz. The threshold level

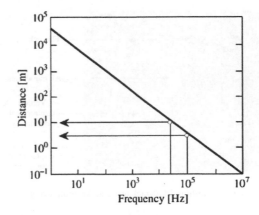

Figure 5.   Acoustic emission relationship between signal detection distance and signal frequency.

of the signal recorded by the system, fixed at 100 μV, is amplified up to 100 mV. The oscillation counting capacity is limited to 255 every 120 s of signal recording. In this way a single event is the result of two recorded minutes.

As specified in the literature (Ohtsu 1996), the maximum amplitude of direct non amplified signals is about 100 μV, hence, neglecting the attenuation by reducing to a few cm the distance of the transducer from the signal generation point, it can be assumed that the measuring system is able to detect the most meaningful AE events reflecting cracking phenomena in the masonry. Attenuation properties, in fact, depend on the frequency range: higher frequency components propagate in masonry with greater attenuation (Fig. 5). Based on experimental results, for a measuring area at a distance of 10 m, only AE waves with frequency components lower than 100 kHz are detectable (Carpinteri et al. 2005). With this system, the intensity of a single event is, by definition, proportional to the number N recorded in the time interval (event counting). Clearly, this hypothesis is fully justified only in the case of slow-crack growth (Holroyd 2000).

4   FLAT-JACK AND AE TESTS

Flat-jack testing is a versatile and powerful technique that provides significant information on the mechanical properties of historical constructions. The first applications of this technique on some historical monuments (Rossi 1982) clearly showed its great potential. The test is only slightly destructive, and this is why it is now widely accepted and used by monument monitoring and rehabilitation experts (Binda & Tiraboschi, 1999; Gregorczyk & Lourenço, 2000). When double jacks are used, this test works according to the same principle as a standard compressive test. The difference

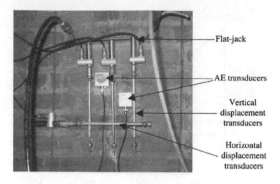

Flat-jack

AE transducers

Vertical
displacement
transducers

Horizontal
displacement
transducers

Figure 6. Combined flat-jack test and AE monitoring.

is that it is performed in situ and the load is applied by means of two flat-jacks instead of the loading platens. The test method is based on the following assumptions: the masonry surrounding the slot notches is homogenous; the stress applied to the masonry by the flat-jacks is uniform and the state of stress in the test prism is uniaxial.

In order to assess the extent of damage in the zone monitored using the AE technique, a compressive test was conducted on the masonry through the combined use of double jacks and AE sensors (Fig. 6). The tests were carried out with flat-jacks measuring $24 \times 12\,cm^2$. The cuts made into the masonry wall to obtain a smaller-sized specimen were made into two horizontal mortar joints spaced about 30 cm apart.

The minimum slenderness ratio of the specimens was $h/t = 2.5$, where h is the height of the prism comprised between the two flat-jacks and $t = 120\,mm$ the deepness of each flat-jack. This made it possible to reduce the friction effects on masonry behavior arising from the action of the flat-jacks.

During the tests, the stress-strain relationship of the masonry was determined by gradually increasing the pressure applied by the flat-jacks in the course of three loading-unloading cycles. Peak compressive strength was obtained from the load–displacement diagram, when the latter became highly nonlinear, denoting imminent failure. Compressive tests were performed on three different masonry portions. The prismatic masonry volumes tested in compression were delimited crosswise by vertical cuts (Fig. 7). Consequently, the in-situ test is equivalent to a compression test performed on specimens with different sizes, as shown in Figure 8. The tests were performed in keeping with the procedures specified in ASTM (1991b), other than for the vertical cuts produced in order to eliminate, in the cracked element, the influence of the adjacent masonry portions.

Figure 9 shows the results obtained from these tests for the intermediate element (Volume 2). Similar results were obtained for the other two elements. The figure also shows the three loading cycles performed

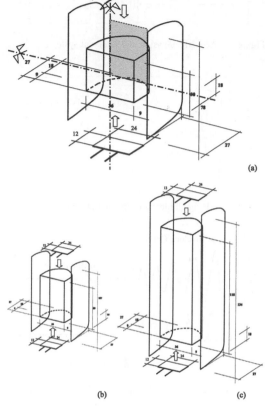

(a)

(b)   (c)

Figure 7. Schemes of the double flat-jack tests performed on different wall sizes.

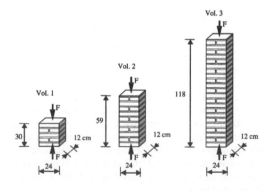

Figure 8. Equivalent masonry prisms tested in compression by means of double flat-jacks.

as a function of time and the diagram of the cumulative number of AE counts. From the AE diagram it can be clearly seen that the material releases energy when the stress level reached previously is exceeded

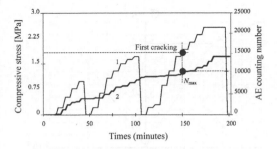

Figure 9. Double flat-jack test on Volume 2: cumulative number of AE events (2) versus cyclic loading (1).

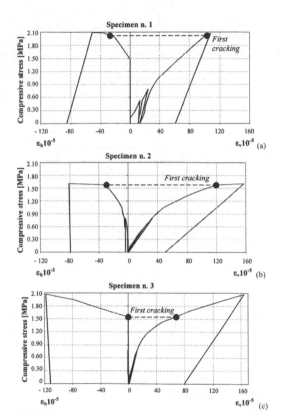

Figure 10. Experimental results obtained from the double flat jack tests.

Table 1. Experimental values obtained from flat-jack tests and AE measurements.

Specimen	Volume (cm³)	Peak stress (MPa)	N_{max} at σ_u
Vol. 1	8640	2.07	~6500
Vol. 2	16 992	1.61	~12 000
Vol. 3	33 984	1.59	~18 000

Table 2. Mechanical properties adopted in the analysis.

		Unit	Joint
Young's modulus	E	$6.0 \, 10^9$ Pa	$1.0 \, 10^9$ Pa
Poisson ratio	ν	0.15	0.15
Tensile strength	f_t	$3.0 \, 10^6$ N	$3.0 \, 10^5$ N
Fracture energy	G_f	50 N/m	10 N/m
Shear ret. Factor	β	0.01	0.01
Compress. strength	f_c	$3 \, 10^7$ Pa	$1 \, 10^7$ Pa

strength of the masonry, is deduced not only from a visual inspection during the test, but also monitoring when the horizontal strain suddenly increases.

5 NUMERICAL SIMULATION

The numerical model of the double flat-jack test was built exploiting the symmetry of the problem. Quadratic elements were used to represent both the brick units and the mortar joints. The failure of both components was assumed as ideal plasticity in compression and linear softening in tension. A fixed smeared crack model based on total deformation was used. All the analyses were performed with the Finite Element Software DIANA 9.1 (de Witte & Schreppers, 2005). The mechanical properties of the materials are summarized in Table 2.

Figure 11a shows the mesh used to model the smallest specimen. Taking advantage of the problem symmetry, only one quarter of the geometry has been discretized. Figure 11b shows details about the loading and boundary conditions. The following procedure has been applied. First, a displacement is imposed to the top of the specimen.

The amount of such displacement can be calculated from another model of the masonry wall ("uncut"), without the cut where the flat-jack is placed afterward. This corresponds to the in situ configuration before the test. The imposed displacement is determined such then the vertical stress equals the in-situ value.

Afterwards, the pressure load in both sides of the cut is applied incrementally. When the pressure reaches the in-situ value of the vertical stress, the deformation of the model approaches the configuration obtained from the "uncut" model, exactly like in the experimental procedure.

(Kaiser effect, Kaiser 1950). Moreover, from the diagram, we find that the cumulative number of AE counts at failure stress (i.e. immediately before the critical condition is reached) is $N_{max} \cong 12000$. The experimental results obtained on the three masonry elements are summarized in Table 1.

The stress-strain diagrams obtained from experiments are shown in Figure 10. The first cracking load, which reasonably corresponds to the compressive

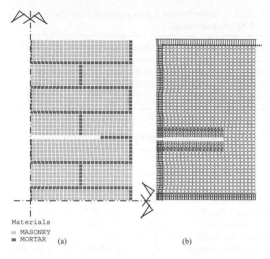

Materials
- MASONRY
- MORTAR   (a)   (b)

Figure 11. Finite element mesh adopted for Volume 1 exploiting symmetry (crf. Shaded area in Figure 7a). Mesh and materials (a); loads and boundary conditions (b).

If the load is increased further, the material comprised in between the two flat-jacks starts to damage. This behavior is caught correctly by the numerical model. Figure 12 shows the stress-strain diagrams obtained for the three different sizes. The arrows indicates the moment at which the horizontal strain suddenly increases, that corresponds to the first vertical cracking.

The compressive strength decreases with increasing the specimen size in a rather good agreement with the experimental tests. On the other hand, the stress-strain path in compression looks a bit stiffer than the experimental one, especially after the cracking occurs.

The crack pattern for the three sizes is shown in Figure 13. It slightly changes varying the size, probably due to the different aspect ratio.

In a previous work (Carpinteri & Lacidogna 2006), a statistical and fractal analysis of data from laboratory experiments was performed, considering the multiscale aspect of cracking phenomena. The fractal criterion takes into account the multiscale character of energy dissipation and the strong size effects associated with it. This makes it possible to introduce a useful energy-related parameter for the determination of structural damage (as used by Carpinteri et al. 2003, 2004, for reinforced concrete structures) by comparing the AE monitoring results with the values obtained on masonry elements of different sizes tested up to failure by means of double jacks.

Fragmentation theories have shown that, during microcrack propagation, energy dissipation occurs in a fractal domain comprised between a surface and the specimen volume V (Carpinteri & Pugno 2002a, b, 2003).

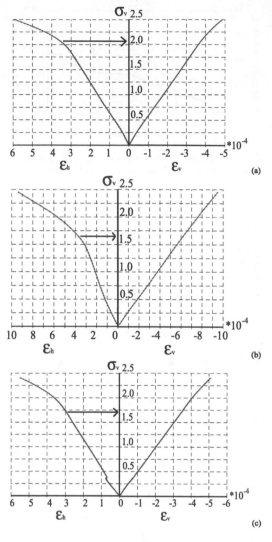

Figure 12. Stress-strain diagrams: Volume 1 (a); Volume 2 (b) and Volume 3 (c). The arrow indicates when first cracking spreads into the specimen.

This implies that a fractal energy density (having anomalous physical dimensions):

$$\Gamma = \frac{W_{\max}}{V^{D/3}},$$ (4)

can be considered as the size-independent parameter. In the fractal criterion of Eq. (4), $W_{\max}$ = total dissipated energy; Γ = fractal energy density; and D = fractal exponent, comprised between 2 and 3. On the other hand, during microcrack propagation, acoustic emission events can be clearly detected. Since the energy dissipated, W, is proportional to the number of AE events, N, the critical density of acoustic emission

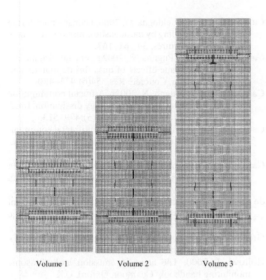

Figure 13. Crack patterns due to flat-jack pressure in the three specimens.

events, Γ_{AE}, can be considered as a size-independent parameter:

$$\Gamma_{AE} = \frac{N_{max}}{V^{D/3}}. \tag{5}$$

where Γ_{AE} = fractal acoustic emission energy density; and N_{max} is evaluated at the peak stress, σu. Eq. (5) predicts a volume effect on the maximum number of AE events for a specimen tested to failure.

The extent of structural damage can be worked out from the AE data recorded on a reference specimen (subscript r) obtained from the structure and tested to failure. Naturally, the fundamental assumption is that the damage level observed in the reference specimen is proportional to the level reached in the entire structure before monitoring is started.

From Eq. (5) we get:

$$N_{max} = N_{max,r}\left(\frac{V}{V_r}\right)^{D/3}, \tag{6}$$

from which we can obtain the structure critical number of AE events N_{max}. An energy parameter describing the damage level of the structure can be defined as the following ratio:

$$\eta = \frac{W}{W_{max}} = \frac{N}{N_{max}}, \tag{7}$$

N being the number of AE events currently recorded by the monitoring apparatus.

Now, we can assume that the number of AE is also proportional to the number of Gauss points subjected

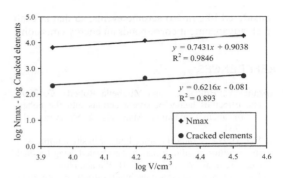

Figure 14. Volume effect on N_{max} and on the number of cracked finite elements.

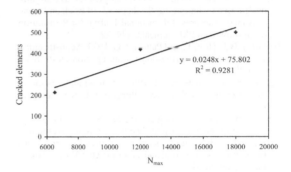

Figure 15. Linear dependency between N_{max} and the number of cracked finite elements.

to cracking in the finite element model. Therefore, the number of AE and the number of cracks in the finite element model should show the same exponent with respect to the considered volume. In fact, this is what we can substantially observe from Figure 14.

The linear relation between the number of cracked elements (or Gauss points) in the finite element model, and the AE is put into evidence also in Figure 15, where the two quantities are plotted in a direct comparison.

Finally, let us observe that the slope of this linear relation depends on the discretization of the finite element model. Nevertheless, refining the mesh (e.g. dividing by two the linear size of each element) does not change sensibly the exponent in Figure 14.

6 CONCLUSIONS

A numerical simulation of an innovative double flat-jack test combined with AE has been proposed. The numerical results agree rather well with the experimental evidences, both in terms of the estimated compressive strength and of the crack pattern. In addition, the number of Acoustic Emissions is put into relation with the number of Gauss points in the finite element model where cracking takes place. AE can be considered like micro seismic events, so that at each crack advancement corresponds an energy emission.

REFERENCES

Anzani, A., Binda, L., and Mirabella Roberti, G. 2000. The effect of heavy persistent actions into the behavior of ancient masonry. Materials & Structures, 33: 251–261.

ASTM. 1991a. Standard test method for in situ compressive stress within solid unit masonry estimated using flat-jack measurements. ASTM C1196-91, Philadelphia.

ASTM. 1991b. Standard test method for in situ measurement of masonry deformability properties the using flat-jack method. ASTM C1197-91, Philadelphia.

Binda L., and Tiraboschi, C. 1999. Flat-jack test as a slightly destructive technique for the diagnosis of brick and stone masonry structures. International Journal for Restoration of Buildings and Monuments, 449–72.

Brindley, B.J., Holt, J., and Palmer, I. G. 1973. Acoustic emission. III: The use of ring-down counting. Non-Destr. Test., 6(5):299–306.

Carpinteri, A., and Bocca, P. 1991. Damage and Diagnosis of Materials and Structures, Pitagora Editrice, Bologna, Italy.

Carpinteri, A., and Lacidogna, G. 2002. Structural monitoring and diagnostics by the acoustic emission technique: Scaling of dissipated energy in compression. Proc., 9th Int. Congress on Sound and Vibration (ICSV9), Orlando, Fla., Paper No. 166.

Carpinteri, A., and Lacidogna, G. 2003. Damage diagnosis in concrete and masonry structures by acoustic emission technique. J. Facta Univ., 3(13):755–764.

Carpinteri, A., and Lacidogna, G. 2006. Damage monitoring of a masonry building by the acoustic emission technique. Materials & Structures, 39:161–167.

Carpinteri, A., and Pugno, N. 2002a. Fractal fragmentation theory for shape effects of quasi-brittle materials in compression. Mag. Concrete Res., 54(6):473–480.

Carpinteri, A., and Pugno, N. 2002b. A fractal comminution approach to evaluate the drilling energy dissipation. Int. J. Numer. Analyt. Meth. Geomech., 26(5):499–513.

Carpinteri, A., and Pugno, N. 2003. A multifractal comminution approach for drilling scaling laws. Powder Technol., 131(1):93–98.

Carpinteri, A., Invernizzi, S., and Lacidogna, G. 2005. In situ damage assessment and nonlinear modeling of an historical masonry tower. Eng. Struct., 27:387–395.

de Witte F.C. & Schreppers G.J. DIANA Finite Element Analysis User's Manual, TNO DIANA BV, Delft, The Netherlands.

Gregorczyk P., and Lourenço, P.B. 2000. A review on flat-jack testing. Engenharia Civil UM, 9:39–50.

Holroyd, T. 2000. The acoustic emission and ultrasonic monitoring handbook, Coxmoor, Oxford, U.K.

Kaiser J. An investigation into the occurrence of noises in tensile tests, or a study of acoustic phenomena in tensile tests. Ph. D. dissertation. Munich (FRG): Technische Hochschule München; 1950.

Ohtsu, M. 1996. The history and development of acoustic emission in concrete engineering. Mag. Concrete Res., 48(177):321–330.

Pollock, A.A. 1973. Acoustic emission. II: Acoustic emission amplitudes. Non-Destr. Test., 6(5):264–269.

Rossi P.P. 1982. Analysis of mechanical characteristics of brick masonry tested by means of in situ tests. In: 6th Int. Brick/Block Masonry Conf., Rome, Italy.

Fracture Mechanics of Concrete and Concrete Structures – High-Performance Concrete, Brick-Masonry and Environmental Aspects – Carpinteri, et al. (eds)
© 2007 Taylor & Francis Group, London, ISBN 978-0-415-44617-4

Strengthening of unreinforced masonry walls using ECC for impact/blast resistance

M. Maalej
University of Sharjah, Sharjah, UAE

V.W.J. Lin, M.P. Nguyen & S.T. Quek
National University of Singapore, Singapore, Singapore

ABSTRACT: This paper presents the results of quasi-static and low-velocity projectile impact tests on unreinforced masonry (URM) walls retrofitted with a cementitious-based material known as Engineered Cementitious Composite (ECC). A total of 18 masonry wall panels were tested to assess the extent to which ECC can enhance the impact/blast resistance of the strengthened masonry walls by subjecting these series of panels to three types of load patterns namely patch load, uniformly-distributed load and impact load. Test results reported in this paper demonstrate the efficiency of the ECC-strengthening system in improving the ductility of URM walls, increasing their ultimate load-carrying capacity, enhancing their resistance against multiple low-velocity impacts and preventing sudden and therefore catastrophic failure.

1 INTRODUCTION

In recent years increasing research efforts have been devoted towards the development of blast-resistant and blast-retrofit designs for building structures. For these structures, the first real defense against the effect of blast loading is the building exterior, which is commonly made of masonry. However, unreinforced masonry elements are primarily designed to withstand in-plane compression loads and wind loads with little consideration of the forces generated in accidental events such as earthquakes and blasts. In the occurrence of such events, the unreinforced masonry elements experience in-plane and out-of-plane horizontal loads which they are not designed for, and thus, they will not be able to withstand these additional forces.

Many methods were proposed to enhance this out-of-plane resistance, one of which was to use composites like FRP to strengthen the masonry wall. Studies have shown that FRP reinforcement could be easily attached onto the masonry wall, thereby increasing its out-of-plane resistance significantly (Ehsani 1995, Triantafillou 1998, Gilstrap and Dolan 1998). However, masonry walls strengthened with FRP usually failed with little or no ductility.

In general, the material performance requirements for impact/blast applications include (a) fracture energy to resist spalling, scabbing and punching, (b) strength to prevent penetration and perforation,

(c) ductility for bending and residual strength and (d) damage tolerance to withstand multiple impacts (Maalej et al. 2005). In addition, material used for retrofit applications should provide ease of installation and should also have minimum shrinkage deformation after installation.

In a study conducted by Zhang et al. (2004), a hybrid-fiber ECC using proper volume ratio of high and low modulus fibers was developed to better meet the functional requirement for impact- and blast-resistant structures. This hybrid-fiber ECC mix containing 0.5% steel and 1.5% Polyethylene (PE) fibers (by volume), has displayed an optimal balance between ultimate strength and strain capacity. Typical uniaxial compressive and tensile stress-strain curves of the hybrid-fiber ECC material are presented in Figures 1 and 2, respectively. Maalej et al. (2005) studied the high velocity impact resistance of this hybrid-fiber ECC experimentally, demonstrating its high energy absorbing capacity and its ability to resist multiple impacts with little spalling, fragment ejection and small crater size on the impact face. It was concluded from the above study that ECC is a good material for impact and blast-resistant design of building structures.

Apart from the mechanical properties of ECC mentioned thus-far, recent investigations conducted by Li and Li (2006) have also verified ECC as an outstanding repair material. It was proven in the study that the

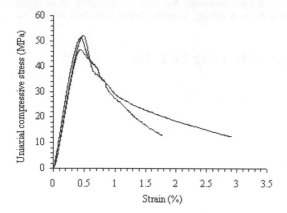

Figure 1. Typical hybrid-fiber ECC uniaxial compressive stress-strain curves.

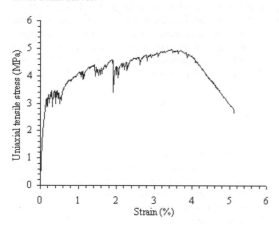

Figure 2. Typical hybrid-fiber ECC uniaxial tensile stress-strain curve.

ductility of the repair material is essential for achieving durability in the repaired structure. In particular, ECC has a high tensile ductility that relaxes potential stress build-up in the repair layer under drying shrinkage conditions. This behavior was accomplished by the multiple micro-crack damage that minimizes the delamination at the interface. Therefore, ECC is expected to be a good material for strengthening unreinforced masonry structures against impact/blast loading.

The main objective of this research work is to investigate experimentally the behavior and effectiveness of ECC-strengthened masonry wall against impact/blast loading. As conducting a field test on strengthened walls under blast loading is difficult and expensive, three series of laboratory tests were carried out in this study to assess the extent to which ECC can enhance the impact/blast resistance of the strengthened masonry walls. These series of panels were subjected to three types of load patterns namely patch load,

uniformly-distributed load and low-velocity projectile impact load.

2 EXPERIMENTAL PROGRAM

A total of 18 masonry wall panels were constructed and tested in the laboratory. Each wall panel measured 1000×1000 mm in plan and 100mm in thickness excluding the ECC layer. All walls were fabricated using solid clay bricks each having dimensions of $215 \times 100 \times 70$ mm. The brick units were laid in running bond with mortar layers each approximately 10mm in thickness. ASTM Type 1 cement and plastering sand were used to prepare the mortar mix, for which the cement: sand proportions were 1:4 by volume. Five $100 \times 100 \times 70$ mm bricks and fifteen 100mm mortar cubes were fabricated to determine its compressive strength.

The test specimens were grouped into three series, with Series I and II tests focusing on quasi-static loading, and Series III tests focusing on impact loading. Each series of test consisted of two unreinforced masonry walls (except Series III having one unreinforced masonry wall) to serve as control specimen and four strengthened masonry walls. Four strengthening configurations were studied, namely, (a) single-face of 34mm-thick ECC-strengthening layer (SE34), (b) double-face of 34mm-thick ECC-strengthening layer (DE34) each, (c) single-face of 34mm-thick ECC-strengthening layer with 8mm-diameter steel mesh (SD8) and (d) double-face of 34mm-thick ECC-strengthening layer with 8mm-diameter steel mesh (DD8) each. In Series I test, an additional reinforcing configuration, single-face of 20mm-thick ECC-strengthening layer (SE20) was included to investigate the effects of strengthening layer thickness variation. Each strengthened wall panel is identified using a combination of four to five characters. The first character, P (Patch), U (Uniformly distributed) or I (Impact) refers to the type of loading and the three to four characters that follow refer to the reinforcement configurations as describe above.

The ECC mix used was a hybrid-fiber mix containing 0.5% steel fibers and 1.5% high performance Polyethylene (PE) fibers. This was the mix proportion used in an earlier study on the performance of ECC under high velocity impact (Maalej et al. 2005); the only difference is that the diameter of the steel fiber used is 200 μm instead of 160 μm. The properties for the fibers used are shown in Table 1.

2.1 Quasi-static load test set-up

The test set-ups for Series I and II are shown in Figures 3 and 4, respectively. The test wall was laid horizontal with the leveled face or the reinforced face downwards, and simply-supported along four sides on

Table 1. Properties of fibers used.

Fiber type	Length (mm)	Diameter (μm)	Elastic mod. (GPa)	Tensile str. (MPa)
Steel	13	200	200	2500
Polyethylene	12	39	66	2610

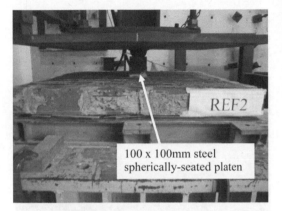

Figure 3. Quasi-static patch load test set-up.

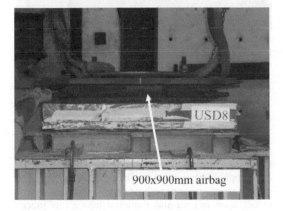

Figure 4. Quasi-static uniformly-distributed load test set-up.

round steel bar supports without edge restraint. This means that the corners of the test wall were free to uplift. The effective span of the panel in both directions was 900mm.

The test set-ups for Series I and II were similar except for their loading areas. For Series I test, a patch load of 100×100 mm was applied at the center of the test specimen through a spherically-seated platen by means of a MTS hydraulic jack head and the load was applied using displacement control at a constant rate of 0.1 mm/min. For Series II test, a uniformly-distributed load of 780×780 mm was applied at the center of the

Figure 5. Low-velocity projectile impact load test set-up.

test specimen using a Kevlar reinforced airbag of size 900×900 mm and the loading rate was controlled by the inflation of the airbag at an approximate rate of 1.5 kN/min.

2.2 Low-velocity projectile impact load test set-up

A drop weight impact facility similar to the one used by Ong et al. (1999) was adopted in this study (see Figure 5). This test facility consisted of a square steel frame welded on rigid columns that were bolted to the strong floor. Similarly, the specimen was laid horizontal with the leveled face or reinforced face downwards, and simply-supported along four sides on round steel bars welded to a square frame. The effective span of the slab in both directions was 900mm. To prevent the specimen from bouncing under the impact, the test wall was clamped with a steel strip at the top face using G-clamps along the support line.

The impact was achieved by dropping a projectile of mass 51 kg from a height of 4 m. This test facility made use of a manually-operated winch system to raise the hammer to the desired height. The impact point was locked at the center of the test specimen using a vertical aluminum square guide. The hammer was then allowed to slide freely through the use of smooth rollers, greased to minimize friction effects. The average velocity of the projectile measured from the experiment was 8 m/s.

3 EXPERIMENTAL RESULTS

3.1 Quasi-static loading test results

The tests carried out for Series I walls subjected to 100×100 mm patch loading and Series II walls subjected to approximately 780×780 mm uniformly-distributed loading revealed five possible failure modes. They were: (a) tensile failure, (b) compression

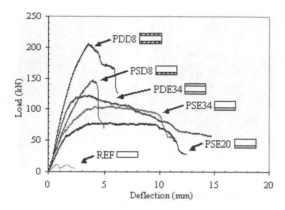

Figure 6. Load-deflection responses of wall components in Series I tests.

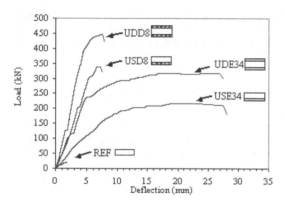

Figure 7. Load-deflection responses of wall components in Series II tests.

failure, (c) compression-induced buckling-bond failure (d) punching shear through the bricks and (e) shear de-bonding of the ECC-strengthening layer. These failure modes will be described and discussed in the analysis of results and discussion section.

The load-deflection responses of Series I and Series II test walls are shown in Figures 6 and 7, respectively. The applied load was recorded by the hydraulic actuator load cell, while the deflection of the specimen was measured at its center using a 100mm-range LVDT.

In general, the ECC-strengthening layer with or without steel mesh displayed very good stress distribution, having well-distributed cracks beneath the loading area stretching out towards the support in a radial pattern as shown in Figure 8. The only difference was that the cracks were less dense in the ECC-strengthening layer with the 8mm-diameter steel mesh indicating that the stresses were mostly distributed through the steel mesh. Lastly, it was interesting to note that as the steel mesh was pressing against the ECC layer, the shape of the steel mesh was visible from the crack patterns.

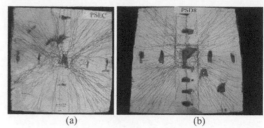

Figure 8. Typical cracking pattern of the ECC-strengthening layers (a) without steel mesh (b) with steel mesh.

Figure 9. Failure of unreinforced masonry wall under impact loading.

3.2 Low-velocity projectile impact loading test results

The importance of retrofitting an unreinforced masonry wall was demonstrated in this Series of tests when such wall exhibited sudden and therefore catastrophic failure upon its first impact loading. The unreinforced masonry wall was perforated by the projectile upon impact and shattered into a few pieces as shown in Figure 9. A large amount of fragment ejections were observed during the impact.

All ECC-strengthened masonry walls were able to withstand multiple impacts before perforation – five for ISE34, nine for ISD8, nine for IDE34 and eighteen for IDD8. Furthermore, during the impact, no fragments ejection was observed at the surface with the ECC-strengthening layer. This observation demonstrated the ECC's ability to prevent fragmentations due to the impact, which may help reduce human injuries in the event of blast/explosion.

The impact load versus time graphs obtained from the first 3 impact loadings applied to Series III test walls are shown in Figures 10–13. In general, two peak loads can be identified from each of the impacts.

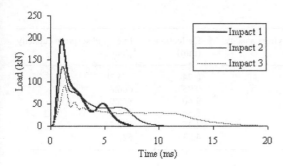

Figure 10. Load versus time graph for the first 3 impacts of ISE34.

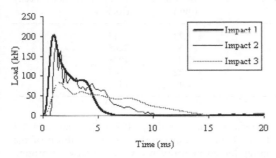

Figure 11. Load versus time graph for the first 3 impacts of ISD8.

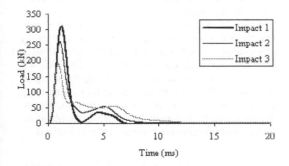

Figure 12. Load versus time graph for the first 3 impacts of IDE34.

The primary peak load occurs within 2 ms upon impact, corresponding to the instant when the projectile strikes the specimen accelerating it downwards, and the secondary peak load takes place when the accelerating specimen loses its kinetic energy and rebounds upward, increasing its contact pressure with the on-coming projectile.

On the whole, the duration of the impact (taken to be the time upon the impact until the load was smaller than 5 kN) ranges from 4.315 to 16.370 ms, depending on the interaction between the projectile and the specimen before the projectile rebounded.

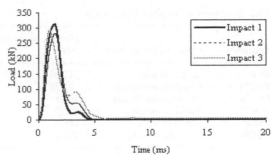

Figure 13. Load versus time graph for the first 3 impacts of IDD8.

4 ANALYSIS OF RESULTS AND DISCUSSION

4.1 Quasi-static loading tests

The results from the quasi-static tests have shown significant improvements in the ultimate load-carrying capacity and the ductility of the unreinforced masonry walls with the ECC-strengthening systems. The ultimate load-carrying capacity and deflection were used as comparison parameters to evaluate the performance of each reinforcing configuration. To be consistent in the comparison study, 90% post-ultimate load was selected as the cut-off point in the load-deflection curve, as most structures will still be able to function with 90% of the design strength. The results of Series I and II tests are summarised in Tables 2 and 3, respectively.

As mentioned earlier, five failure modes were observed from the experimental tests. Tensile failure was only observed in the unreinforced masonry walls in which the wall had little resistance against out-of-plane loading due to the low tensile strength of the masonry.

Compression failure was generally observed in the masonry walls that were singly-reinforced with ECC strengthening layer without steel mesh (PSE20, PSE34 and USE34). In this reinforcing configuration, the tensile ECC layer had a significantly-high strain capacity, while the strain capacity at the compressive face remained unchanged. Therefore, when the wall experienced out-of-plane loading, the applied strain at the compressive face exceeded the strain capacity of the masonry and hence crushing of the masonry took place.

Compression-induced buckling-bond failure of the ECC layer was observed in the masonry walls that were doubly reinforced with ECC-strengthening system without steel mesh (PDE34 and UDE34). In the case of uniformly-distributed loading, however, the large loading area delayed this mode of failure, contributing to increased load-carrying capacity of the test wall (Fig. 7). In addition, comparing the load-deflection curves for specimens PSE34 and PDE34

1635

Table 2. Summary of Series I tests results.

Specimen	Observed failure mode	Ultimate load-carrying capacity (kN)	Deflection (mm)	Energy absorption capacity (Joules)
REF	Tensile flexure	11.94	0.74	5.28
PSE20	Compression flexure	78.26	9.95	666.8
PSE34	Compression flexure	105.5	9.60	836.9
PDE34	Buckling-bonding	122.7	5.54	522.0
PSD8	Punching shear	146.9	4.35	398.7
PDD8	Shear de-bonding	206.6	4.53	621.9

Table 3. Summary of Series II tests results.

Specimen	Observed failure mode	Ultimate load-carrying capacity (kN)	Deflection (mm)	Energy absorption capacity (Joules)
REF	Tensile flexure	20.37	1.73	17.8
USE34	Compression flexure	217.9	27.5	3265
UDE34	Buckling-bonding	318.3	27.0	4978
USD8	Shear de-bonding	337.3	7.22	945
UDD8	Shear de-bonding	447.3	7.43	1456

on the one hand, and USE34 and UDE34 on the other hand, it seems that the large loading area in the latter case allowed Specimen UDE34 to preserve a high deflection capacity in comparison to Specimen PDE34.

Punching shear failure was observed in the masonry wall that was retrofitted using a single ECC-strengthening layer with steel mesh (PSD8). In this configuration, the additional 8mm-diameter steel mesh made the ECC-strengthening layer stiffer and harder to deform. In the case of patch loading, the inability of the reinforced wall to deform resulted in stress concentration near the loading area. Due to the brittle nature of the masonry material, at the moment when the load exceeded the punching shear capacity of the masonry layer, punching failure took place without warning. The stresses were then transferred to the ECC-masonry interface causing it to de-bond.

Shear de-bonding failure was generally observed in the masonry walls that were doubly-reinforced with ECC-strengthening layer with steel mesh (PDD8, USD8 and UDD8). Here again, the additional 8mm-diameter steel mesh made the ECC-strengthening layer stiffer and harder to deform. In these cases, high shear stresses were built up at the ECC-masonry interface due to the unequal deformation of the ECC and masonry layer, leading to bond failure.

The test results shown in Table 2 for Series I indicate that the ECC-strengthening system increased the failure loads and deflection capacities of the walls significantly by 6.5 to 17.3 times and 5.8 to 13.4 times, respectively, relative to those of the unreinforced masonry walls. Likewise, from Series II test results, the failure loads and deflection capacities were

significantly increased by 10.7 to 22 times and 4.2 to 15.9 times, respectively, relative to those of the unreinforced masonry walls. The energy absorption for Series I and II masonry walls were significantly increased as well from 75 to 158 times and from 53 to 279 times, respectively, relative to those of the unreinforced masonry walls.

Lastly, it was observed from Series I and II tests that the ECC-strengthening systems incorporating the 8mm-diameter steel mesh possessed a higher ultimate load-carrying capacity with lesser ductility. This was because the additional steel mesh made the composite panel stiffer, thereby enhancing its load-carrying capacity but reducing its ability to deform. Hence, depending on the nature of the retrofitting application, different reinforcing configuration can be utilized.

4.2 Thickness variation study of ECC-strengthening layer

Series I test results for the masonry walls incorporating single ECC-strengthening layers with 20mm- (PSE20) and 34mm- (PSE34) diameter steel mesh were used to study the effect of the thickness of the ECC layer on the load and deflection capacity. The load-deflection responses of the reinforced walls were similar as shown in Figure 6; the only difference was the ultimate load-carrying capacity. Despite being thicker by 70%, the ultimate load-carrying capacity of PSE34 was only 35% higher than that of PSE20. It was found from section analysis that the strain in the ECC-strengthening layers directly beneath the loading area of PSE20 and PSE34 was 3.54% and 2.43%, respectively (both below the tensile strain capacity of

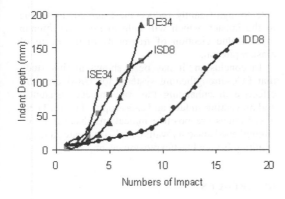

Figure 14. Plot of indent depth development.

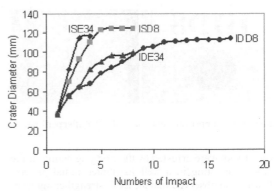

Figure 15. Plot of crater diameter development.

the ECC material). These observations suggest that in specimen PSE20 the ECC layer had a higher contribution to ultimate load-carrying capacity than in specimen PSE34.

4.3 Low-velocity projectile impact loading tests

The behavior and physical damage suffered by the test walls under low-velocity projectile impact loading differed with different ECC-strengthening systems applied. As mentioned earlier, all ECC-strengthened masonry walls were able to withstand multiple impacts, hence, the degree of damage inflicted in the specimens under repeated impacts was of greater interest compared to the energy absorption capacity of the specimen under single impact. The damage level is evaluated and characterized based on the average crater diameter, indentation depth, crack propagation as well as fragmentation. The first two parameters were measured directly from the specimens after each test, while the rest were observed qualitatively from high speed camera and digital camera recordings.

To evaluate the penetration resistance of each ECC-strengthening system, the indentation depths and crater diameters created on the impact face were plotted against the number of impacts as presented in Figures 14 and 15, respectively. From the two figures, it can be observed that the indentation depths as well as the crater sizes of the reinforced walls decreases with the addition of the 8mm-diameter steel mesh, for both singly and doubly-reinforced walls, at the same number of impacts. Double-faced reinforced masonry walls displayed higher penetration resistance with lesser indentation depth and smaller crater size compared with single-faced reinforced masonry walls. This is because the ductile ECC-strengthening layer at the impact face of the wall absorbed a significant part of the impact energy, thereby protecting the brittle masonry sandwich layer and resulting in a smaller degree of damage.

Figure 16. Impact and distal face of ISE34 after perforation.

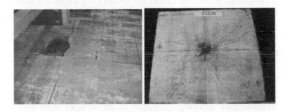

Figure 17. Impact and distal face of ISD8 after perforation.

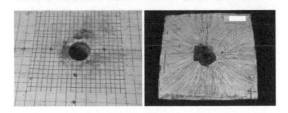

Figure 18. Impact and distal face of IDE34 after perforation.

Figures 16–19 show the impact and distal faces of the specimens at perforation. All the ECC-strengthened masonry wall panels remained structurally intact and showed only localized damage after being perforated. In general, all reinforced walls had cracks on the distal side that propagated from the impact point towards the support edges. Under point-load impact, the panel tried to bend close to a conical shape with high hoop stress. The cementitious matrix of the ECC material being brittle then fractured but

Figure 19. Impact and distal face of IDD8 after perforation.

the cracks were arrested by the bridging action of the fibers, resulting in a well-distributed radial pattern. It was also observed that the ECC-strengthening layer without steel mesh had very densely distributed cracks, while the ECC-strengthening layer with steel mesh had cracks that were sparsely distributed, indicating that the stresses were largely distributed by the steel mesh rather than the ECC-strengthening layer.

5 CONCLUSION

From the quasi-static loading tests, it was shown that The ECC-strengthening systems improved the out-of-plane resistance of the masonry walls significantly. In particular, the ultimate load-carrying capacity and maximum deflection of the masonry wall increased from 6.5 to 22 times and from 4.2 to 15.9 times for Series I and II tests, respectively. The steel mesh within the ECC-strengthening layer increased the ultimate load-carrying capacity of the reinforced wall by 40% to 68%, but reduced the deflection capacity of the specimen by 17% to 74%. The ability of ECC to strain-harden and develop multiple micro-cracks was also observed in this masonry retrofit application.

When subjected to low-velocity projectile impact, the ECC-strengthened masonry walls demonstrated an ability to resist multiple impact loadings, unlike URM which failed catastrophically when subjected to the first impact. The ECC-strengthening layers were also able to significantly reduce the fragmentations due to the impact, which will help to minimize human injuries due ejection of fragments in the event of blast/explosion.

In conclusion, it has been shown in this study that ECC-strengthening systems for URM are very effective in enhancing the wall impact resistance, and preventing sudden and catastrophic failure. These observations are positive indications that the ECC-strengthened masonry walls can also mitigate damages resulting from a blast/explosion event.

REFERENCES

Ehsani, M. R. 1995. Strengthening of Earthquake-Damage Masonry Structures with Composite Materials, in L. Taerwe ed. Non-metallic (FRP) Reinforcement for Concrete Structures. Spon, London. 680–687.

Gilstrap, J. M. and Dolan, C. W. 1998. Out-of-Plane Bending of FRP-Reinforced Masonry Walls, Composite Science and Technology, v58. 1277–1284.

Li, M. and Li, V. C. 2006. Behaviour of ECC/Concrete Layered Repair System under Drying Shrinkage Conditions. Journal of Restoration of Buildings and Monuments. v12 (2). 143–160.

Maalej, M., Quek, S. T. and Zhang, J. 2005. Behavior of Hybrid-Fiber Engineered Cementitious Composites Subjected to Dynamic Tensile Loading and Projectile Impact. Journal of Materials in Civil Engineering, ASCE. v17 (2). 143–152.

Ong, K. C. G., Basheerkhan, M. and Paramasivam, P. 1999. Resistance of Fiber Concrete Slabs to Low Velocity Projectile Impact, Cement and Concrete Composite. v21. 391–401.

Triantafillou, T. C. 1998. Strengthening of Masonry Structures Using Epoxy-Bonded FRP Laminates. Journal of Composites for Construction, ASCE, v2 (2). 96–104.

Zhang, J., Maalej, M. and Quek, S. T. 2004. Hybrid Fiber Engineered Cementitious Composites (ECC) for Impact and Blast-Resistant Structures, in Proceedings of First International Conference on Innovative Materials and Technologies for Construction and Restoration–IMTCR04. 6–9 June 2004, Lecce, Italy. 136–149.

Fracture Mechanics of Concrete and Concrete Structures – High-Performance Concrete,
Brick-Masonry and Environmental Aspects – Carpinteri, et al. (eds)
© 2007 Taylor & Francis Group, London, ISBN 978-0-415-44617-4

Structural analysis for the diagnosis of crackings in the Gothic masonry structure of the vaults of Trinidad Bridge in Valencia, Spain

A. Martinez Boquera & A. Alonso Durá
Architect, PhD Professor of the Polytechnic University of Valencia, Spain

V. Llopis Pulido
Architect

ABSTRACT: Trinidad bridge, built in 1402, the oldest bridge in Valencia for its Gothic ashlar masonry structure, is located in front of the old entrance with the same name in the walls of Valencia. In the preliminary research phase aimed at the Restoration of the Bridge, it is essential to know the historical-constructive evolution that provides the keys for the recognition of its current situation of deterioration, in order to accurately approach its restoration. The Bridge presents, as most noticeable structural pathology, cracks that run parallel to the arches in their joint with the vaults, mainly in the end spans of the bridge. In this phase of analysis described here, we aim at recognizing the structural behavior of the bridge and the current resistance conditions of masonry, by means of the theoretical-practical simulation, with the purpose of determining the coherence of the results and the conditions of the failure of material identified. The results of the numerical analysis performed, coincide, to a large extent, with the real cracking conditions of the bridge.

1   INTRODUCTION

In the area of consolidation of historical monuments, *structural stability* is a field with a high theoretical and practical interest that requires a specific evaluation and knowledge, closely related to architecture as discipline itself. It is very important then, to add to the architectonic restoration project that knowledge based on the relation between the historical evolution of the structural stability analysis and the current techniques related with the behavior of the masonry constructions, whether with stone or brick, and organize the development of the theory and critically analyze its validity within the framework of the modern *limit analysis* of masonry structures, that have been developed mainly by professor Jacques Heyman from the late 1960's. (Heyman J.1998).

The calculation of the thrust of arches and vaults has its first references in father Tosca's Treatise, (Tosca T.1794), one of the most cultured men of his time in Spain, and specially in volume 5 of his *Compendio* devoted to civil architecture in which he refers to these issues with enough detail to understand the context of statics at that time. His comments on the stability of masonry structures demonstrate a deep knowledge of their structural behaviour.

In the case of the historic Trinidad bridge of the city of Valencia, within the framework of the multidisciplinary team that has drafted the project of its complete restoration, starting from the graphic survey carried out in previous studies and based on a study from the constructive and geometric point of view, tests of materials, prospecting carried out in the cover, pilaster, backing, and land, as well as from the current state of the deteriorations observed in the masonry, we started to carry out the structural analysis of the bridge with the purpose of determining its stability, current masonry resistance conditions and with these data, be able to identify the causes of the damages detected and contribute to the most suitable solutions for its restoration.

Preliminary studies have revealed several damages in the stone, and as most noticeable structural

Figure 1.   Damage in the stone in the end bridge bays.

Figure 2.   Foundations of the bridge.

pathology, crackings that run parallel to the arches in their joint with the vaults, mainly in the end bridge bays. With the purpose of carrying out a diagnosis of those structural damages, two lines of work have been followed.

On one hand, surveys and geotechnical studies of the foundations of the bridge.

These results have revealed that those damages are not due to foundation problems, and at the same time a structural study of the bridges by means of a numerical model to study their behavior faced with different load hypotheses which can explain the more relevant cracking conditions of the bridge and plan the proposal of structural restoration. The description of the numerical procedure carried out is presented below.

2   CALCULATION MODEL OF THE STRUCTURE OF THE BRIDGES

The development of the calculation model has been carried out through the program EFCID, a finite element analysis program developed in the MMCYTE Department of the UPV that uses CAD environments for the definition of this calculation model from the made accurate laser scanner survey.

2.1   Analysis conducted

The following types of analysis have been conducted: (López J. 1998) (Luccioni B. 1996)

a Static linear
b Dynamic with a modal-spectral calculation for seismic actions
c Nonlinear Static with a damage isotropic model to characterize the breaking. (Lourenço P.B. 1996)

The objective is to determine the degree of safety against the actions planned on the bridge according to the current IAP standard, and to be able to determine the type of action that has produced the cracks in the lower part of the bridge vaults and parallel to the line of parapets.

The mortising on the bridge cover of installation conduits has been carried out.

2.2   Definition of the mesh of the calculation model

The generation of the finite elements mesh is carried out by the graphical procedures of the design environment that the calculation program used allows, on the graphical medium of CAD drawing. This causes the resulting meshing to adjust to the contour of the form of the bridges with high accuracy, since their survey has been made by scanner-laser.

Superficial and volumetric elements have been used in order to make up the mesh. The solid elements model the mass parts of the bridges: piers, vaults, lateral arches and backing. The superficial elements have been arranged to model the asphalt layer of the cover and they are also used to support the traffic load. The layer elements connect with the solids through the mesh of nodes in the upper plane of the covers.

The characteristics of the flat elements used are:

Triangular isoparametric elements of three nodes and quadrilateral ones of four:

The superficial elements have two work levels whose effects operate in a disconnected form. Membrane effect with deformations and requests in the plane of its xy surface, and plate effect with flexions in the perpendicular direction based on the local z-axis.

2.3   Membrane elements

C. Felippa optimized triangle consists of three degrees of freedom per node, two transfers and one rotation. It is characterized because the efforts and deformations operate in their plane and are stresses ($\sigma_X, \sigma_Y, \tau_{XY}$) and transfers d $_x$ and d$_y$ and round z. referred to its local axes.

2.4   Plate elements

Triangular superficial flexion element, with three knots, with three degrees of freedom per node (two rounds with respect to x–y and one transfer with respect to z), is the so-called DKT, discrete Kirchhoff triangle, based on Reissner-Mindlin plate model. Its characteristic efforts are bending moments M_x, M_y, M_{xy} and sharp T_x and T_y according to the local axes.

2.5   Lamina elements

Triangular superficial element of three nodes with six degrees of freedom per node. It is formed by the union of membrane elements. The quadrilateral elements are formed by the double partition by their diagonals in triangular elements.

Four parts corresponding to different constructive and mechanical characteristics have been distinguished in the mesh.

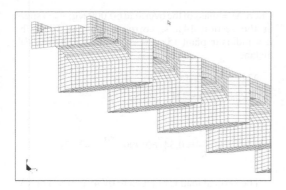

Figure 3. Model of the Trinidad Bridge.

1. Ashlar masonry with unions in which the loss and degradation of mortar leads them to be virtually joined to bone. They are rot in their outside facade, which corresponds to the vaults of the bridge bays, the lateral arches, parapets and side spandrel walls, as well as the outside surface of the piers of the walls of the bridge. It has been modeled with hexahedral elements.
2. Backing of the vaults and piers, made up of coffer-work, with a very resistant concrete and in very good state. Formalized with solid elements.
3. A layer of hexahedral solid elements that form the pavement that tops the backing.
4. The fine asphalt layer placed on the stone pavement has been modeled with quadrilateral superficial layer elements, on which the traffic loads are entered.

The mortising produced has been materialized on the cover of the bridge and next to the parapets due to the conduit for installations. The triangular and tetrahedral elements are used only in those zones of transition between meshes of different discretization. The model of the Trinidad bridge consists of:

Hexahedral solid elements of 8 nodes.	26,452
Plate elements of four nodes.	3,042
Number of Nodes	37,746
Constrained nodes	1,440
Degrees of freedom.	118,000

2.6 Characteristics of the materials

In previous studies test cylinders extracted from the bridge have been analyzed, corresponding to ashlar masonry structures. From the compression tests it has been obtained the load factor and the deformation module that is taken as parameter in the numerical model for the structural analysis. The values have been measured in two points of the ascending straight part of the chart of the simple compression tests. The results

for the test cylinders of the Trinidad Bridge are the following:

Test cylinder	Den-sity T/m^3	C. Breakage N/mm^2	$\Delta\sigma$ N/mm^2	$\Delta\varepsilon$	Module E N/mm^2
1	1.949	7.05	2.6738	0.00175	1527.885
2	1.914	7.13	3.0680	0.00200	1534.000
3	1.891	9.57	5.4858	0.00452	1213.672
Average	1.918	7.916			1425.186

The analysis of the results and the comparison with the values of test cylinders tested of the source quarry show that the ashlar masonry has undergone serious deterioration, increasing its porosity and consequently diminishing its density, its resistance and the deformation module. The values of the parameters of calculation for the backing material of the bridges have been obtained by comparison with data from the literature, since no test has been performed. (León J. 2001) (P.I.E.T 1971) (Eurocódigo 6 1997).

Properties of the materials considered in the numerical calculation model:

- Coffer work backing material.
  Density   2.3 T/m^3.
  Module of deformation.   9,000 N/mm^2
  Poisson's ratio   0.2
  Compressive force   15 N/mm^2
  Tensile strength   0.5 N/mm^2
- Material of the stone pavement layer.
  Density   2.4 T/m^3.
  Module of deformation.   11,000 N/mm^2
  Poisson's ratio   0.2
  Compressive force   20 N/mm^2
  Tensile strength   0.6 N/mm^2
- Material of the asphalt layer.
  Density   2.3 T/m^3.
  Module of deformation.   100 N/mm^2
  Poisson's ratio   0.2
- Ashlar masonry. Trinidad Bridge.
  Density   1.918 T/m^3.
  Module of deformation.   1,425 N/mm^2
  Poisson's ratio   0.2
  Compressive force   7.90 N/mm^2
  Tensile strength   0.15 N/mm^2

2.7 Evaluation of loads. Hypotheses considered

Although this is a urban historic bridge, the estimation of the calculation actions on the bridge that have been entered in the numerical model to evaluate their tense-deformational conditions have been taken in compliance with the Regulations on the actions to consider in the project of road bridges, IAP of the

Department of Public Works. This regulation specifies the actions and safety factors for the construction of new bridge. The Trinidad bridge was built centuries ago and throughout its history it has been under different types of loads and values, it is impossible to reproduce this history of loads. During the last century, it has been put progressively under traffic loads for which it was not intended.

The objective of the structural analysis performed is to determine with a degree reasonable of accuracy the conditions of the bridge under the actions undergone and to attempt to determine the causes of the structural pathologies that have appeared at present time. For that reason, the regulations in the IAP have been taken as reference.

The load hypotheses considered are the following:

Hypothesis 1. H1 Weight of the bridge. Determined according to the densities of the different materials that compose the bridges, which have been defined in the section on materials characteristics.

Hypothesis 2. H2 Overload of 10 KN/m^2 to be considered in roads with trucks according to the reference of NBE-88.

Hypothesis 3. H3 Overload of 4.0 KN/m^2 of vertical action all over the cover (IAP art.3.2.3.1.1.a1). It corresponds to the overload use as bridge for pedestrian use exclusively, or simultaneously used to the load of vehicles.

Hypothesis 4. H4 Loads of vehicles of 600 KN with parallel longitudinal axis to the road, divided into six loads of 100 KN according to IAP regulation art. 3.2.3.1.1.a2. In this hypothesis the higher number of vehicles will be placed in a row, one after another throughout all the bridge.

Hypothesis 5. H5 the same load than H4, but only vehicles are placed in the center of the space of each bridge bay.

Hypothesis 6. H6 is applied to the load of the H4 hypothesis, placing the vehicles on each one of the supports.

Hypothesis 7. H7 Applied to the alternation of loads by applying the loads of vehicles in the centers of the even spaces of the complete the bridge.

Hypothesis 8. H8 horizontal load of finishing and springing (IAP art. 3.2.3.1.1.b1.) It is applied in the direction of the axis of the platform of the board and operates at the level of the surface of the pavement as a uniformly distributed action.

Hypothesis 9. H9 inclined load of centrifugal action of the vehicles in the entrance and exit of each bridge. It is applied in the center of the initial and final span of the bridge. The load is evaluated according to IAP art.3.2.3.1.1.b2.

The centrifugal force Fc applied is:

$$F_c = K \cdot M \frac{V_e^2}{R}$$

where M = mass of the overload 60,000 Kg; V_e = speed in the section 14 m/s. corresponding to 50 Km/h; R = radius in plant 15 m; K = factor of adimensional distance.

$$\frac{231}{V_e^2 + 231} = \frac{231}{196 + 231} = 0.54$$

$$F_c = K \cdot M \frac{V_e^2}{R} = 0.54 \cdot 60,000 \frac{196}{15} = 423,360 N$$

The vertical load is reduced to $0.54 \cdot 600 = 324\ KN$ Hypothesis 10. H10 Seismic loading.

The seismic loading have been considered in accordance with Art.1.2.3 of the NCSE-02 standard and instruction (IAP art. 3.2.4.2.). A basic acceleration of ab = 0.06 g. is taken, specified for the city of Valencia. The construction is considered of Special Importance (Art. 1.2.2 of NCSE-02). and (IAP art. 3.2.4.2.1.). The calculation is performed by means of the analysis of response spectra. A three-dimensional model of the structure with 6 degrees of freedom per node is carried out, without restrictions nor simplifications.(Code UIC 1995) (ACI 1999) (Hendry A. 1998).

Given the special case of historical bridge and the high value of its "useful life" a coefficient of higher risk is taken by considering it as 1.5.

The combinations carried out are:

C1: 1H01 + 1H02
C2: 1H01 + 1H03 + 1H04
C3: 1H01 + 1H03 + 1HO5
C4: 1H01 + 1H03 + 1HO8
C5: 1H01 + 1H03 + 1HO9
C6: 1H01 + 1H03 (Combination for bridge without road traffic)
C7: 1H01 + 0.5H3 + 1H10

With combination 5, considered as the most unfavorable for the purpose of studying the cracking produced, the nonlinear analysis of damages has been performed. (Hanganu A.D.1997)

2.8 Analysis of results

The analytical study of the bridge, through the program of calculation by finite elements EFCID, from the calculation models, load hypotheses and combinations, and materials described above, has shown the following issues as the most outstanding features of the structural behavior:

The results of the behavior of the bridge, in its analysis with static gravitational loads, in any of their combinations, as well as earthquake dynamics, the latter taking into consideration a period of recurrence of 500 years, are acceptable values that do not imply possible structural deteriorations.

Figure 4.   Axial stresses X direction.

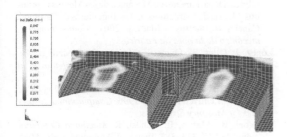

Figure 5.   Model of damages.

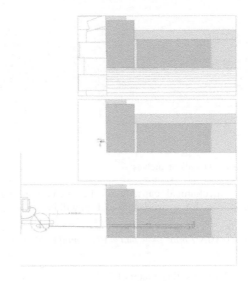

Figure 6.   Details of anchors.

In the linear analysis, the values of the compression membrane stresses, in any of the directions, are not relevant with any of the combinations of static load analyzed, with quite moderate values of the compression stresses in all of them, mainly below 0.3 N/mm^2, very distant from those allowed for the ashlar masonry tested.

The tensile stresses, under these same combinations of static load, are nonexistent or negligible, which does not justify any of the existing damages in the bridge.

The hypothesis of seismic loading, combined with gravitational statics, does not produce significant increases of the tensional state. From the nonlinear analysis with models of damages, for values of main stresses in ashlars that exceed 6 N/mm^2 in compression or 0.1 N/mm^2 in tensile strength, and values in coffer work backing of 15 N/mm^2 in compression and 0.5 N/mm^2 in tensile strength, shows the absence of significant damages in any of the combinations of static gravitational load analyzed as well as in the combination of gravitational loads plus seismic loading, therefore it can be concluded that the behavior under these actions is safe.

Regarding the traffic actions considered in hypotheses 8 and 9, of braking and centrifuge action in circular layouts respectively, combined with gravitational statics, described in Combinations 4 and 5, the tensions of traction in the inside face of the first vaults of the bridge, increase considerably, reaching values of up to 0.5 N/mm^2, which justifies cracking conditions that correspond with those appeared in the vaults of the bridge, mainly in the end spans.

These stress conditions values are coherent with the nonlinear analysis with damage models, ([12]) for such values of main stresses that in the previous combinations, that is, 6 N/mm^2 in compression in ashlar or 0.1 N/mm^2 in tensile strength, and values in coffer work backing of 15 N/mm^2 in compression and 0.5 N/mm^2 in tensile strength, which reveals the appearance of significant damages of around 0.8 in ahslar masonry of the bridge, within a scale that ranges between 0, absence of damages and 1 collapse of the material.

These stress conditions values, may be due to the way in which traffic affects the bridge, and that those horizontal stresses are transmitted to the joints of the ashlar masonry structure causing the most significant cracking of the vaults in its inside face.

From the above it can be concluded that, road traffic is the cause of the cracking in the vaults. The cracks that appear parallel to the arches in their union with the vaults, in the end spans of the bridges, are due to the horizontal actions caused by the road traffic in the curved layouts at the entrance and exit of the bridge.

With respect to the calculation of the anchors that can be arranged for the bonding of the existing cracks produced by the tensile stress in the masonry, its approximate value is the following:

The maximum tension produced in the zones of the crackings, for a situation of light traffic, is 0.10 N/mm^2 of traction in the inside surface of the vault, which is already a compression of 0.05 N/mm^2 at a distance of 0.5 m. over that surface, therefore the anchors are appropriate as repair system approximately arranged 20 cm. from the inside face of the voussoirs of the arches.

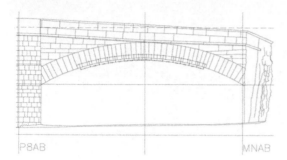

Figure 7.   Details of anchors.

The mechanical capacity of the anchors in the tensile area, for an arc-length of 1 m. will be:

$$F = 0.10N / mm^2 \cdot \frac{340}{2} mm = 17T / mm(1.7T / m)$$

Applying a safety factor of 3

$Ftotal = 5.1T / m$

This force has to be distributed between the number of anchors to be arranged in that surface based on the mechanical characteristics of the material selected.

REFERENCES

ACI 530–99.American Concrete Institute. 1999. *Building Code Requirements for Masonry Structure.*

Code UIC 778–3. Union International des Chemins de fer. 1995. *Recomendations pour l'evaluation de la capacité portante des ponts-voûtes existants en maçonnerie et beton.*

Eurocódigo 6.UNE-ENV 1996-1-1.Marzo 1997. *Proyecto de Estructuras de Fábrica.*

Hanganu, A.D.; Barbat, A.H.; Oñate, E. Metodología de evaluación del deterioro en estructuras de hormigón armado. Monografía CIMNE n° 39, Julio 1997. Pag. 183–203. *Aplicación del Indice Global de Daño al estudio de Construcciones Históricas.*

Hendry, A. Macmillan Press Ldt. 1998. *Structural Masonry.*

Heyman, Jacques. Cambridge University Press, 1998.*Structural Análisis. A Historical Aproach.*

León, J.; Martín-Caro, J.A.; Martínez, J.L. "Comportamiento mecánico de la obra de Fábrica". Monografía sobre el análisis estructural de construcciones históricas de fábrica. Departamento de Mecánica de los Medios Continuos y Teoría de Estructuras. E.T.S. Ingenieros de Caminos, Canales y Puertos . Madrid. 2001. *Comportamiento mecánico de la obra de Fábrica.*

López, J.; Oller, S.; Oñate, E.;Monografía CIMNE n° 46. Diciembre de 1998.*Cálculo del Comportamiento de la Mampostería mediante Elementos Finitos.*

Lourenço, P.B. Tesis Doctoral. Universidad Tecnológica de Delf. Delf University Press. 1996. *Computational Strategies for Masonry Structures.*

Luccioni, B.; Martín, P.; Danesi, R. *Mecánica Computacional,* Vol.17, 373–382. Tucumán. Argentina. 1996. *Modelo elastoplástico general para materiales ortótropos.*

P.I.E.T. 70. Consejo Superior de Investigaciones Científicas. Madrid. 1971. *Obras de fábrica. Prescripciones del Instituto Eduardo Torroja.*

Tosca, Thomas Vicente *"Tratados de arquitectura civil, montea y canteria, y reloxes".* Valencia: Oficina de los Hermanos Orga, 1794. . *"Compendio matemático en que se contienen todas las materias principales de las ciencias que tratan de la cantidad..."* Valencia: Antonio Bordazar, 1707–1715. 9 vols.

Fracture Mechanics of Concrete and Concrete Structures – High-Performance Concrete,
Brick-Masonry and Environmental Aspects – Carpinteri, et al. (eds)
© 2007 Taylor & Francis Group, London, ISBN 978-0-415-44617-4

Use of historical buildings to validate corrosion induced cracking models

A. Millard
CEA/DM2S/SEMT/LM2S, Saclay, France

V. L'Hostis
CEA/DPC/SCCME/LECBA, Saclay, France

R. Faquin
CUST, Clermont-Ferrand, France

ABSTRACT: In order to predict the long term evolution of reinforced concrete structures exposed to environmental conditions, the French Atomic Energy Agency (CEA) has launched a project, called CIMETAL, which aims at characterizing and predicting the effects of corrosion of the reinforcements induced by carbon dioxide penetration.

One of the main objectives of this study is to evaluate the capability of a cracking model to predict the time required for the formation of visible cracks on a old building. For this purpose, a 50 years old water tower, classified as historical building, and showing degradations, has been selected. Various measurements have been performed on it to characterize, as much as possible, some data required for the modelling, such as the accurate position of the rebars in the structure, the nature of the iron oxides, etc.

A particular cracked zone of the tower has been selected for the mechanical analysis. Bidimensionnal transient analysis have been performed, in which the concrete, the reinforcements as well as the corrosion products layers are modelled.

1 INTRODUCTION

Cracking of concrete due to corrosion is a very common pathology that can be observed on various structures. Corrosion is initiated by the penetration of chlorides or of carbon dioxide in most cases. Usually, the associated deterioration process follows three sequential stages (Tuuti, 1982) (figure 1):

- Initiation: during this long period, the corrosion rate is very low despite of the ingress of aggressive species from the environment to the steel.
- Depassivation: this step happens when the conditions required for the onset of corrosion are fulfilled thanks to the transport of aggressive species through concrete cover.
- Propagation: the reinforcement corrosion causes significant loss of section of the reinforcements. Internal micro cracking and spalling of the concrete cover appear. They are due to the high tensile stresses generated by the expansive volume of the corrosion products.

The mechanical consequences of corrosions are (i) the reduction of the resistive section of reinforcements

(ii) the creation of expansive products (commonly denoted rust) (iii) the fragilization of steel and finally (iv) the cracking of concrete (Andrade et al, 1996, Petre-Lazar, 2000).

As reinforced concrete is a common material used in the nuclear industry for the construction of power plants and nuclear waste storage facilities, the degradation of infrastructures has to be mechanically understood and modeled. In this context, the *CIMETAL* research program has been launched by the French *Commissariat à l'Energie Atomique* (CEA), for the prediction of the evolution of cement/metallic material systems in an open unsaturated environment. It deals with interactive studies dedicated to short term experimentations (corrosion (Huet et al, 2005) and mechanical behavior of structures), modeling to predict the corrosion (Huet et al, 2006) and the mechanical behavior of objects for several hundred years, and finally validate some hypotheses with analyses of old corrosion systems (ancient ferrous artifacts or archaeological analogues (Chitty et al, 2005)), and mechanical behavior of old reinforced concrete structures. The aim of this paper is precisely to show how a 50 years old water tower has been used to try to validate the modeling tools developed in the framework of the research program.

2 MODELING APPROACH FOR OLD REINFORCED CONCRETE STRUCTURES

2.1 General features

The overall process of corrosion development involves electro-chemical processes, heat and mass transport by convection and diffusion, and mechanical aspects. A totally predictive modeling requires a multi-physics approach, which is the final goal of the CIMETAL program, but which is not yet available. Therefore, at the present stage, the modeling approach is based on one hand on measurements and on the other hand on a sequential treatment of couplings.

First an accurate investigation of the steel-cementitious material interface is performed on the old reinforced concrete by means of various techniques, such as optical microscopy, spectrometry coupled to electronic microscopy, mercury porosimetry, Raman micro-spectroscopy, and X-ray diffraction. Of particular importance is the composition of the corrosion products, because the expansion coefficient may significantly differ between iron oxides.

Then, the average corrosion rate is determined from the measure of the iron oxides thicknesses coupled to a correction of the local density.

Finally, these data are used in a damage mechanical model, called CORDOBA, which can predict the consequences of the active corrosion phase, in terms of displacements, stresses and cracks pattern.

2.2 Composition of the long term corrosion products

The general pattern encountered on old corroded metallic reinforcements in binders is made up of a multi-layer structure (figure 1). The corrosion layout can be described as: the metallic substrate (M), the dense product layer (DPL), the Transformed Medium (TM) and the Binder (B). This corrosion pattern has been precisely described previously (Chitty et al, 2005). X-Ray Diffraction analyses reveal that Dense Product Layer is mainly made of iron oxy-hydroxides (goethite α-FeOOH, lepidocrocite γ-FeOOH, and akaganeite β-FeOOH), and iron oxides (maghemite γ-Fe$_2$O$_3$ and magnetite Fe$_3$O$_4$). The local structure of the Dense Product Layer has been studied by μXRD on some samples. Two different phases were noticed in the diffracted volume: goethite and magnetite (or/and maghemite). Indeed the fact that mixes of different phases are observed can be linked to the relatively large diffracted volume ($20 \times 20 \times 50\,\mu m^3$) compared to the size of the marbling observed by Scanning Electron Microscope and Optical Microscope. In order to

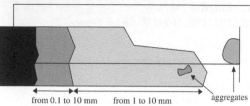

Figure 1. Macro photograph of a cross section (GR) and schematic section f the corrosion system (M: metal; DPL: Dense Product Layer; TM: Transformed Medium; B: Binder).

Table 1. Relative increase in volume for different iron oxides compared to the initial iron volume.

Iron oxide	Relative volume (oxide/Iron)
FeO	1,8
Fe$_3$O$_4$	2,0
Fe$_2$O$_3$	2,0
Fe(OH)$_2$	3,7
Fe(OH)$_3$	4,1
Fe(OH)$_3$, 3H$_2$O	6,2

study more in detail the structure of the marbling, the microscopic laser beam of the μRaman microscope was used. These analyses show that veins are only made of iron oxides i.e. maghemite and/or magnetite and that matrixes are made of iron oxy-hydroxides i.e. goethite, lepidocrocite and akaganeite.

The following table 1 gives the values of the expansion coefficient of the iron oxides, according to their composition.

2.3 Average corrosion rate

Systematic studies of corrosion of old reinforced concrete structures and archaeological analogues reveal characteristics that can be helpful for modelling purposes:

– Corrosion layers seem to be mainly composed of goethite that presents a volumetric expansion of 4 (according to values in table 1).
– The average corrosion rates estimated on archaeological analogues are about 4 μm/year.

In order to dispose of a more accurate evaluation of the average corrosion rate, the total iron quantity involved in the corrosion products can be measured by compositional analyses. Then, a density correction allows converting this quantity into metal loss.

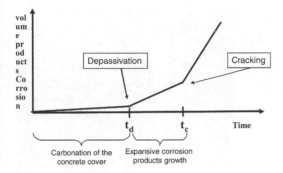

Figure 2. Tuutti's diagram for a reinforced concrete structure degradation.

For this purpose, average composition profiles lead to access on one hand to the evolution of the iron contents from the metal/oxide interface to the binder, and on the other hand to the thicknesses of the different layers (Dense Product Layer and Transformed Medium). A more precise description of this estimation can be found in Chitty et al, 2005.

2.4 Modelling of corrosion products growth

Average corrosion rates and nature of corrosion products identified during the characterization, can then be used as input data for the CORDOBA mechanical model.

Thanks to the Tuutti's model (Tutti, 1982) on stages of concrete damage due to the reinforcements corrosion (figure 2), two periods can be distinguished for the structure life. A first phase called "passive corrosion phase" when the unsaturated carbonation front penetrates into the concrete cover. We assume that no mechanical damage is induced by the passive layer growth. After the depassivation of the steel, expansive corrosion products growth leads to the mechanical damage of the concrete. The CORDOBA model is used for the estimation of this second period, until the appearance of the first through crack. This means that the predictions require the time of initiation of the depassivation as input data, which is unfortunately difficult to appraise.

The mechanical model mimics, at a macroscopic scale, the above depicted multi-layers structure, by assigning different material behaviour laws to the concrete (Binder and Transformed Medium), the reinforcements (Metallic Substrate) and the interfaces (Dense Product Layer).

For the concrete, a model capable of describing cracking must be used. Two kinds of such models have been tested: the Mazars' damage model (Mazars, 1984) formulated in an integral non-local framework, and the Ottosen's elasto-plastic model (Mersseman et al, 1994) using the Hillerborg's fracture energy based regularization technique.

For the reinforcements, since no plasticity is expected, a simple linear elastic material model is used.

The development of the corrosion products is simulated by means of special interface finite elements which are placed between the rebars and the concrete, and which can swell with time.

One difficulty in this model lies in the determination of the parameters of the interface, in terms of stiffness (Young's modulus for example) and swelling to reproduce the development of corrosion. Since the interface elements are composed of two superposed surfaces (or lines in a bidimensionnal case), their thickness must be given through their equivalent stiffnesses, k_n the normal one, and k_t the shear one, which are proportional to the Young's modulus of the rust.

The two stiffnesses are then calculated as:

$$k_n = \frac{E}{t_h} \quad \text{and} \quad k_t = \frac{E}{2(1+v) \times t_h} \qquad (1)$$

where v is the Poisson's ratio of the rust, and t_h is the thickness of the rust layer, calculated as the product of the expansion rate by the time. The expansion rate depends itself on the average corrosion rate as well as on the volumetric expansion of the corrosion products.

The Young's modulus of the iron oxides has been determined by small scale experiments, on cylindrical concrete samples containing an embedded reinforcement (Millard et al, 2004, Ouglova, 2004). According to the type of oxide formed (mainly goethite), the Young's modulus is found equal to 0.1 GPa.

It as to be mentioned that this first step of the development of the CORDOBA model does not consider any initial cracks within the concrete that could induce corrosion locally. Concrete is considered as mechanically sound when the active corrosion is initiated.

The model has been validated by means of experiments on centimetric concrete plates containing a reinforcement (Nguyen et al, 2006). The corrosion was due to the penetration of chlorides, accelerated by prescribed electric current. The developments of the cracks could be followed by image analysis and strain gauges, and compared to the numerical predictions.

The computations were performed using the CEA finite elements code CASTEM 2000 (Verpeaux et al, 1989). For plate geometries such as the one shown on figure 3, a good agreement is obtained up to the formation of a through-crack. The comparison of the strain measured across the through-crack, and the calculated one, is displayed on figure 4. For this purpose, equal reference lengths have been used in the measurement and in the calculation. Concerning the cracks formation, the image analysis technique revealed that the cracks first form close to the rebars, then others may

Figure 5. The two zones of the structure selected for simulation.

Figure 3. Comparison of cracks pattern, after 40 hours, between experiment and prediction.

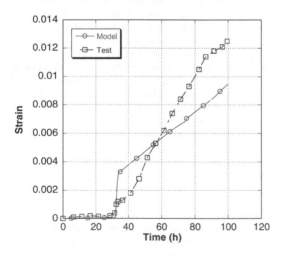

Figure 4. Comparison between experimental strain across the crack, and calculated one (Nguyen et al, 2006).

initiate on the outer boundary and finally coalesce with some initial one.

The next step consists in trying to validate the model on a real size structure, subjected to atmospheric corrosion.

3 APPLICATION TO A HISTORICAL BUILDING

3.1 Presentation of the structure

The chosen building is a water tower, built in 1950 at Saclay, France, and classified as historical. Before its repair in 2004, some degradations such as spalling have been recorded. They occurred mainly on some of the vertical columns supporting the tank, located on the side of the tower exposed to winds. On these columns, two representative damaged zones have been selected for the simulation (figure 5): One zone at the outer corner of the column, and the other close to the junction of the column with the wall . The section dimensions are 51 cm × 90 cm.

Unfortunately, construction reports and drawings were no longer available. Measurements on some specific zones of the tower performed prior to its repair in 2001 have shown significant variations on the concrete cover thickness as well as on the rebars diameters. Therefore, additional measurements have been conducted, using a Ferroscan tool, in the two zones selected. In particular, the real number of rebars and their positions have been precisely determined. It has also been found that the cover varies between 25 and 47 mm.

3.2 Finite element model

In view of the slenderness of the columns, a bidimensional analysis of an horizontal cross-section of the column is considered. In such a calculation, only the longitudinal reinforcements are considered. This simplification may lead to a non conservative prediction of the time to rupture, since the transverse frames also contribute to inception of corrosion induced cracking. However, it prevents the use of more expensive three-dimensional calculations.

A typical mesh is shown on figure 6. The mesh has been densified around the rebars to capture the cracks

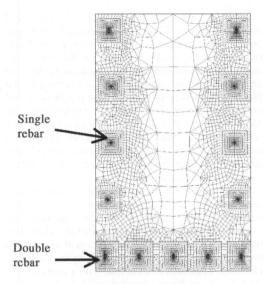

Single rebar

Double rebar

Figure 6.   Mesh of a transverse cross-section of a column.

Sane steel

Dense Product Layer

Figure 7.   View of interface cored from the tower column, and placed in resin.

formation. For symmetry reasons, only half of it is used in the calculations.

3.3   Determination of the corrosion data

As outlined above, input data such as the composition of the corrosion products, as well as the corrosion rates are required. During the repair of the tower, concrete samples containing reinforcement pieces have been cored. The analysis of the interface has revealed that corrosion products consist mainly of goethite. After 50 years, the corrosion products thickness reaches 100 μm (see figure 7). The figure shows that the rust thickness is not uniform around the rebar, the maximum being obtained where the cover is minimum. Nevertheless, in the computation it has been assumed uniform for simplicity.

In addition, an initial rust thickness of 30 μm has been measured from a reinforcement still in a passive corrosion phase. The remaining difficulty lies in the

Table 2.   Various cases studied in the parametric analysis.

Expansion law	Square root						Slow	Linear
	Fast							
ft (MPa)	2,5		3			2,5	2,5	3
rebars (mm)	10	10	12	10	12	10	10	10
Rear cover (mm)	40	30	40	40	40	40	40	40
Reference	M1	M2	M3	M4	M5	M6	M7	M8

determination of the starting time of the active corrosion phase. Unfortunately, for the water tower, concrete samples cored from the column were not long enough to localize the carbonation front. Since no reliable data was available, we decided to let the corrosion start from the beginning, thus producing the 70 μm in 50 years. From this assumption, and assuming a $t^{0.5}$ time evolution law, it is possible to derive the rust expansion rate. For the mechanical properties of the rust, Young's modulus has been taken as 0.1 GPa, and Poisson's ratio 0.2.

3.4   Concrete and steel properties

Some concrete properties have been measured in 2001:

– Young's modulus = 30 GPa
– Compression strength = 34 MPa

The other properties required by the non linear behaviour models have been either determined from usual recommendations or estimated from parametric investigations.

The rebars are smooth, their diameter is either 10 or 12 mm. They are considered as elastic. Their properties have been taken as:

– Young's modulus = 210 GPa
– Poisson's ratio = 0.3

3.5   Parametric investigations

Because of the uncertainties on the evolution of the corrosion products, various expansion laws have been compared: linear expansion with time using the average value of 4 μm/year recorded on archeological analogues, and expansions as square root of time, either fast (200 μm in 50 years) or slow (30 μm in 50 years).

In addition, the rebars diameter has been changed from 10 to 12 mm, the concrete traction strength from 2.5 to 3 Mpa, and the cover from 30 to 40 mm. The various cases studied are summarized in table 2.

The calculations are run over a 50 years period. Because of the uncertainties of the data, only the cracks pattern then obtained can be reasonably compared to the real observations. Nevertheless, the influence of the parameters can be appraised.

Table 3. Summary of predicted results.

Reference	Crack location	Time at cracking (years)	Rust thickness (μm)
M1	F	0.9	9.4
	R	–	–
M2	F	0.9	9.4
	R	1.8	13.3
M3	F	0.8	8.9
	R	2	14
M4	F	1.2	10.3
	R	–	–
M5	F	0.9	9.4
	R	–	–
M6	F	2.2	6.3
	R	–	–
M7	F	1.6	6.4
	R	–	–
M8	F	1.9	7.6
	R	–	–

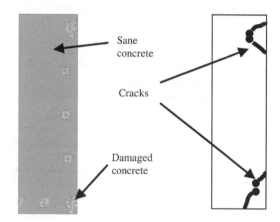

Figure 8. Comparison of predicted and observed cracks pattern.

3.6 Main results

The results presented here were obtained with the Mazars' concrete model. Depending on the parameters, the observed cracks at the front (F) or at the rear (R) of the column are predicted or not, after 50 years. It is also possible to compare the time of first cracking, and the corresponding thickness of the corrosion products. All these results are summarized in table 3.

For the case M2, corresponding to the smaller rear cover, the final cracks pattern is compared to the real one on figure 8.

The above results show that for all parameters combinations, cracking is always predicted at the front face of the column, which is not the case for the rear corner. As can be observed on figure 5, the real damage is more important on the front face.

It is well known that the durability of a reinforced concrete structure is mainly influenced by the concrete traction strength as well as the cover. This fact is recovered by the model. Beyond that, the model confirms that the increase of the rebars diameter leads to an earlier cracking, which can be interpreted by the fact that the volume of the corrosion products formed around a rebar with a large diameter, is more important. Moreover, through-craking is only observed where reinforcement is made of two adjacent rebars. This means that accounting for the transverse frames in a three-dimensional analysis would certainly lead to a more important damage.

Concerning the influence of the rust expansion rate, table 3 shows that cracking occurs for a rust thickness which highly depends on the rust expansion. One explanation could be that the rapid increase of the rust thickness is accompanied by a decrease of its stiffness, thus reducing the pressure exerted on the concrete. However, additional calculations with a constant rust thickness have shown a similar trend. A clarification of this issue will require additional studies.

4 CONCLUSION

The use of historical buildings to validate corrosion induced cracking models is not a straightforward exercise. Indeed, much information related to the initial state of the structure is not known. In the present case, it has not been possible to find maps and drawings and even with the recourse to techniques such as Ferroscan analysis, the real positions and diameters of the rebars are difficult to estimate. Moreover, the concrete properties evolve with time, and only some of them were measured in 2001.

Another important issue is the determination of the parameters associated with the corrosion products: beginning of the active corrosion phase, rust expansion rate, nature of the iron oxides, and initial rust thickness. It has been shown that some of them can be deduced from analysis on cored samples. However, the corrosion modeling approach proposed is based on simplifying assumptions, such as a square root of time expansion law, which might not account for the real complexity, for example induced by variable environmental conditions.

Finally, concerning the mechanical modeling, a more realistic calculation would require a three-dimensional description, including the transverse frames. Nevertheless, despite of all the above mentioned uncertainties, it has been possible to simulate the crack pattern observed on the 50 years old water tower, for a combination of realistic values of the parameters.

Of evidence, this does not constitute a validation of the model. For this purpose, it is most probably necessary to design long term large scale experiments, in the same spirit as the experiments performed on reinforced concrete beams by François et al, 1994.

REFERENCES

Almusallam A.A., Al-Gahtani A.S., Aziz A.R., and Rasheeduzzafart, 1995, *Const. and Build. Mat.*, **10**, 123–129

Andrade C., Alonso C., Molina F. J., 1993, *Mat. and Struc.*, **26**, 453–464.

Castel A., François R., Arliguie G., 2000, *Mat. & Struct.*, vol. **33**, 539–544.

Chitty W.J., Dillmann P., L'Hostis V., Lombard C., 2005, Long term corrosion resistance of metallic reinforcements in concretes – A study of corrosion mechanisms based on archaeological artefacts, Corrosion Science, 47 (6), pp. 1555–1581.

François R., Arliguie G., Maso J.C., 1994, Durability of reinforced concrete in chloride environment, Annales de l'ITBTP, 529, pp. 1–48.

Huet B., L'Hostis V., Miserque H., Idrissi H., 2005, Electrochemical behaviour of mild steel in concrete: influence of pH and carbonate content of concrete pore solution, Electrochimica Acta, 51 (1), pp. 172–180.

Huet B., L'Hostis V., Santirini G., Feron D., IDRISSI H., 2006, Steel corrosion in concrete: determinist modeling of cathodic reaction as a function of water saturation degree, Corrosion Science, in press.

Mazars J., 1984, Application de la mécanique de l'endommagement au comportement non linéaire et à la rupture du béton de structure, PhD Thesis, Paris 6, France (in french).

Mersseman (de) B., Millard A., 1994, Modèle Ottosen de comportement du béton en fissuration, CEA technical Report, CEA/DM2S, (in french).

Millard A, L'Hostis V., Beddiar K., Bethaud Y., Care S., 2004, Modelling the cracking of a reinforced concrete structure submitted to corrosion of steels – first validation of a damage model based on experimental tests, Proceedings of the OECD/NEA/CSNI-RILEM Workshop on use and performance of concrete in NPP fuel cycle facilities, Madrid, Spain.

Nguyen Q.T., Millard A., Care S., L'Hostis V., Berthaud Y., 2006, Fracture of concrete caused by the reinforcement corrsion products, proceedinds NUCPERF conference, Cadarache, France.

Ouglova A., 2004, Etude du comportement mécanique des structures en béton armé atteintes par la corrosion, PhD Thesis, Paris 6, France (in french).

Petre – Lazar I., 2000, Evaluation du comportement en service des ouvrages en béton armé soumis à la corrosion des aciers, PhD thesis, Université de Laval, Québec.

Tuuti K., 1982, 'Corrosion of steel in concrete', *Swedish Cement and Concrete Research Institute, Stockholm.*

Verpeaux P., Millard A., Charras T., Combescure A., 1989, A modern approach of computer codes for structural analysis. Proceedings of the SMIRT conference, Los Angeles, USA.

*Fracture Mechanics of Concrete and Concrete Structures – High-Performance Concrete,
Brick-Masonry and Environmental Aspects – Carpinteri, et al. (eds)*
© *2007 Taylor & Francis Group, London, ISBN 978-0-415-44617-4*

An application of sequentially linear analysis to settlement damage prediction for masonry façades

J.G. Rots & M. Boonpichetvong
Delft University of Technology, Delft, The Netherlands

B. Belletti*
University of Parma, Parma, Italy

S. Invernizzi†
Politecnico di Torino, Torino, Italy

ABSTRACT: In Amsterdam, the North-South metro-link is being constructed underneath the urban historical city centre. This implies that historical masonry facades and monuments face the risk of being damaged due to the settlements imposed by the subsurface building activities. The situation is critical because we have the combination of very soft soil, timber pile foundations and brittle un-reinforced masonry. A representative Amsterdam masonry façades from 18-19th century is analysed. A crucial aspect of large-scale masonry structural behaviour is that cracks may initiate, gradually propagate and then suddenly snap to a free surface, thus the final fracture is localized and often highly brittle. Existing Newton-Raphson based non-linear finite element techniques face difficulties in modelling this, whilst softening results in negative stiffness, sharp snap-backs and bifurcations. In this paper, the sequentially linear analysis is proposed as an alternative approach. The paper illustrates the potential of numerical models to gain insight in the risks and in the need for mitigating measures to preserve the architectural heritage.

1   INTRODUCTION

Ground movements may considerably affect the serviceability and safety of structures. In particular the behaviour of large-scale historical masonry buildings subjected to settlements induced by tunnelling activity has been analysed in this paper. The numerical difficulties encountered in modelling these full-scale buildings have been highlighted in previous works (Rots 2000, Boonpichetvong and Rots 2003) and are summarized in this paper where the case study of a representative Amsterdam masonry façades dated from 18–19th century is analysed. Analytical and empirical approaches are available for evaluating the cracking damage of surrounding structures subjected to settlement. Potts and Addenbrooke (1996) employed elastic beam elements to represent an overlying structure on a soil continuum. However, due to many provided assumptions, the direct application of these simplified methods in very sensitive cases such as fragile historical masonry buildings raise the need

for more detailed analysis. For quasi-brittle materials like concrete and masonry especially, the introduction of tension-softening models, fracture mechanics and nonlinear finite element (NLFE) methods have improved the possibilities for predicting cracks in structures. Several NLFE analyses were performed in the past with the DIANA finite element code in order to investigate how mesh refinement, element typology, constitutive law, convergence method and loading can affect the façade response (Boonpichetvong and Rots 2003). The main remark arising from the results of these studies is that the façade response strongly depends on the adopted constitutive models and convergence methods. These previous investigations, revealed a very brittle snap-back response associated with full fracture of the façade, which cannot be caught by smeared crack models. When the crack pattern develops the structural response obtained with smeared crack model shows a snap-through, due to the significant unbalance between high elastic energy stored in the façade and low fracture energy of masonry. For this reason the façade was re-analysed with discrete crack models and arc-length control technique. In this case snap-back response can be predicted, but practical use of this numerical procedure requires a lot of user's experience. In this paper it will be shown that the strong and sharp snap-back response, due to settlement, can be automatically detected by a sequentially linear analysis method. Sequentially linear analysis is an alternative to non-linear finite element analysis of structures when bifurcation, snap-back or divergence problems arise. The incremental-iterative procedure is replaced by a sequence of linear finite element analyses performed by decreasing Young's modulus and tensile strength at the integration point of elements, when their damage increases.

2 OVERALL EVENT-BY-EVENT PROCEDURE

The locally brittle snap-type response of many structures inspired the idea to capture these brittle events directly rather than trying to iterate around them in a Newton-Raphson scheme. A critical event is traced and subsequently a secant restart is made from the origin for tracing the next critical event. Hence, the procedure is sequential rather than incremental. The sequence of critical events governs the load-displacement response. To this aim, the softening diagram is replaced by a saw-tooth curve and linear analyses are carried out sequentially (Rots 2001). The global procedure is as follows. The structure is discretized using standard elastic continuum elements. Young's modulus, Poisson's ratio and initial strength are assigned to the elements. Subsequently, the following steps are sequentially carried out:

- Add the external load as a unit load.
- Perform a linear elastic analysis.
- Extract the 'critical element' from the results. The 'critical element' is the element for which the stress level divided by its current strength is the highest in the whole structure.
- Calculate the ratio between the strength and the stress level in the critical element: this ratio provides the 'global load factor'. The present solution step is obtained rescaling the 'unit load elastic solution' times the 'global load factor'.
- Increase the damage in the critical element by reducing its stiffness and strength, i.e. Young's modulus E_i and tensile strength f_{ti}, according to a saw-tooth constitutive law as described in the next section.
- Repeat the previous steps for the new configuration, i.e. re-run a linear analysis for the structure in which E_i and f_{ti} of the previous critical element have been reduced. Trace the next critical saw-tooth and element, repeat this process till the damage has spread into the structure to the desired level.

2.1 Saw-tooth diagram for masonry in tension

The way in which the stiffness and strength of the critical elements are progressively reduced constitutes

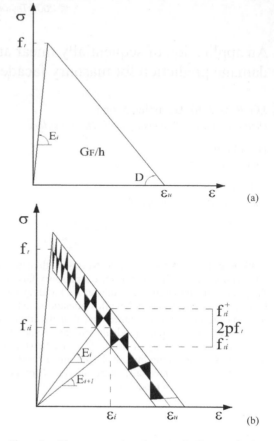

Figure 1. The stress-total strain curve for linear softening (a) and the consistent saw-tooth diagram (b).

the essence of the model. In other words, it is necessary to provide a saw-tooth approximation of the constitutive stress-strain relation. A very rough method would be to reduce E to zero immediately after the first, initial strength is reached. This elastic perfectly brittle approach, however, is likely to be mesh dependent as it will not yield the correct energy consumption upon mesh refinement (Bažant and Cedolin 1979). In the sequentially linear procedure the consecutive strength and stiffness reduction is based upon the concept of tensile strain softening, which is fairly accepted in the field of fracture mechanics.

Different approaches have been investigated in previous works (Rots and Invernizzi 2004), in the present paper a new generalized tooth size approach, already adopted for concrete (Rots et al. 2006), is presented and applied to masonry. The tensile softening stress-strain curve is defined by Young's modulus E, the tensile strength f_t, the shape of the diagram, and the area under the diagram, Fig. 1a. The area under the diagram represents the fracture energy G_f divided by the crack band width h, which is a discretisation parameter associated with the size, orientation and integration scheme of the finite element. Please note that the softening diagram is adopted only as a 'mother' or envelope curve that determines the consecutive strength reduction in sequentially linear analysis. For a linear softening diagram, the ultimate strain ε_u of the diagram reads:

$$\varepsilon_u = \frac{2G_f}{f_t h},$$ (1)

and the tangent to the tensile stress-strain softening curve is, see Fig. 1a:

$$D = \frac{f_t}{\varepsilon_u - \frac{f_t}{E}}$$ (2)

In a sequentially linear strategy, the softening diagram can be imitated by consecutively reducing Young's modulus as well as the strength. As shown in Fig. 1b, a strength range is set, as a percentage of the maximum tensile strength, p. As a result, a band or 'strip' is introduced into the softening diagram, delimited by two curves parallel and equidistant to the original diagram. The number of required teeth need to reach the complete damage of the element (N) and the values of Young's modulus (E_i) and tensile strength (f_{ti}) at the current stage i in the saw-tooth diagram are automatically obtained as values depending on this strength range, chosen by the user. We can interpret this as a pre-set 'ripple curve' on top of the base curve, so if i denotes the current stage in the saw-tooth diagram, the following relation can be imposed:

$$f_{ti}^- = f_{ti}^+ - 2pf_t$$ (3)

The reduced strength f_{ti}^+ corresponding to the reduced Young's modulus E_i is taken in accordance with the envelope softening stress-strain curve:

$$f_{ti}^+ = \varepsilon_u^+ E_i \frac{D}{E_i + D}$$ (4)

where:

$$\varepsilon_u^+ = \varepsilon_u + p \frac{f_t}{D}$$ (5)

$i + 1$ denotes the next stage in the saw-tooth diagram and a_{i+1} is a variable which increases the damage of the element by reducing Young's modulus:

$$E_{i+1} = \frac{f_{ti}^-}{\varepsilon_i} = \frac{E_i}{a_{i+1}}$$ (6)

The value of a_{i+1} can be obtained from Eqs. 3 and 6, see Fig. 1b:

$$a_{i+1} = \frac{E_i}{f_{ti}^-} \varepsilon_i = \frac{f_{ti}^+}{f_{ti}^-} = \frac{f_{ti}^+}{f_{ti}^+ - 2pf_t}$$ (7)

Note that this is the softening curve in terms of stress versus *total* strain, i.e. the sum of elastic strain and crack strain of an imagined cracked continuum. The model always provides a solution: the secant saw-tooth stiffness is always positive, so that ill-conditioning or divergence does not appear in sequentially linear analysis. An advantage of this new formulation is that no special techniques are required to handle mesh-size objectivity.

3 SETTLEMENT DAMAGE PREDICTION OF HISTORICAL MASONRY FAÇADE

3.1 Historical masonry façade

Presently, in the Netherlands, bored tunneling is planned to be driven in soft soil adjacent to historical masonry buildings founded on timber pile foundations. An example for an Amsterdam historical masonry building is as shown in Fig. 2a. The structure of these historical masonry buildings normally consists of un-reinforced façades and bearing walls. The layout of the selected masonry façade as in Fig. 2b is made of a block of three house units. A uniform thickness of one brick (220 mm) is adopted here for the whole façade. The opening pattern shows two large openings at ground floor and a regular pattern of three window openings at the three floors above. Above the window openings at ground floor, lintels in the form of steel beams are present to distribute the vertical load to either side of the opening. The load-bearing and house-separating walls perpendicular to the façade have not been included in the model. Also their connections to the façade and other 3D effects have been omitted. Although it is well understood that 3D effects can modify the damage pattern, the context of this study will be limited to 2D response.

3.2 Overview of modeling approach

The currently available design practice for evaluating settlement effects on the surrounding structures due to soft-ground tunneling can be classified into two types, namely uncoupled analysis and coupled analysis. In the first class, the greenfield settlement determined from empirical equations is directly imposed under building models, while the second class allows for the full interaction between above-ground structure and underlying soil. To account for the situation of historical masonry buildings founded on fragile piles, Rots

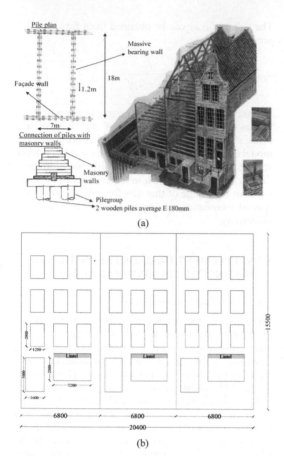

(a)

(b)

Figure 2. An example of historical masonry buildings in Amsterdam (after Netzel and Kaalberg, 2001) (a); the selected historical masonry façade (all dimensions in mm) (b).

(2000) introduced the semi-coupled scheme in which so-called bedding interface elements were employed to represent a simplified "smeared" model for fragile pile foundation. This strategy is adopted in the present study and the hogging situation i.e. the building sits at the point of inflection of ground settlement profile, is investigated here as being a common case in practice. The estimated settlement trough is taken in the form of Guassian distribution curve (Mair et al. 1996) for any percentage of expected ground loss (V):

$$S_v = S_{max} \exp\left[\frac{-x^2}{2i^2}\right] \qquad (8)$$

where $S_{max} = 0.31VD^2/i$, i is the horizontal distance from the tunnel centerline to the point of inflection on the settlement trough (i = KZ_o). K is the trough width parameter for tunnels and is taken as 0.5 here. Z_o is the depth to the tunnel axis, which is 20 m herein. Tunnel diameter (D) is 6.5 m.

4 FINITE ELEMENT MODELLING AND RESULTS

4.1 Smeared crack analysis and results

The façade is modeled by eight-node quadratic plane stress elements. A three by three integration scheme is adopted. The non-linear behavior of masonry is simulated by using a multiple fixed crack version of crack band model. Young's modulus (E) and Poisson's ratio for the façade material are 6000 N/mm² and 0.2 respectively. A linear tension softening has been adopted with tensile strength (f_t) equal to 0.3 N/mm² and fracture energy (G_f) equal to 0.05 N/mm. Constant shear retention value of 0.01 is adopted. A mesh with element size of 500 by 400 mm² is used. The crack bandwidth is chosen as 225 mm as a half of an average of the mesh dimension because the strain field in a quadratic element is likely to be lumped on one side. The loading scheme starts with the activation of self-weight and live load of 5 kN/m at each floor. Mass density of masonry is 2400 kg/m³. The normal stiffness of the no-tension bedding interface elements is taken as 0.15 N/mm³ by smearing out the stiffness of foundation system in an average manner. Finally, the settlement trough is applied incrementally. Fig. 3a illustrates the relation between the angular distortion (Boscardin and Cording 1989) and the maximum crack width in the façade. The maximum crack width is the maximum value that occurs somewhere in the façade. The position of the maximum crack width may differ with increasing the angular distortion. Smeared crack response reveals that initially the crack width remains very low, indicating that the façade behaves approximately elastically with only minor crack. The maximum crack width gradually increases when the cracks at the corner of windows start to propagate. In correspondence of a critical value of angular distortion equal to 1/800 the top band of the façade is suddenly fully cracked and the maximum crack width suddenly increases very rapidly, while the angular distortion remains almost constant. The moment of full through-depth cracking occurs when the top band of the façade is not longer able to carry the tensile stress in that region.

Then, a separation occurs into two unloaded parts that rotate, Fig. 3b. The sudden breakage of the façade is accompanied by a sharp snap-back in the structural response.

With displacement control, a snap-through response is captured in which, for a slight increase of the angular distortion, the dominant vertical crack shows an abrupt jump in the magnitude along the second branch of the façade response. Along this branch, the magnitude of maximum crack-width is rapidly increased and the quality of observed convergence performance in the nonlinear analysis becomes dramatically poor. The difficulty to achieve a good convergence is partly

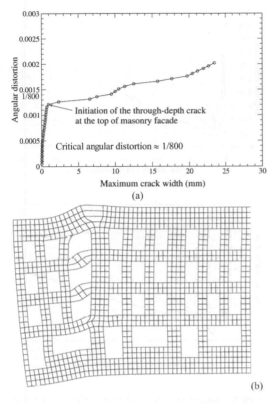

(a)

(b)

Figure 3.   The façade response in smeared crack analysis (a) localization shown at the top of façade (b).

due to the significant residual stress in the fully open crack zone and mainly due to the significant unbalance between high elastic energy stored in the façade and low fracture energy of masonry.

4.2   Discrete crack analysis and results

The same façade is now re-analyzed by placing six-noded quadratic interface elements at the vertical line of the final crack predicted in the former smeared crack analysis. The rest of masonry continuum is now assumed to be linearly elastic. Interface elements, having the same fracture properties of masonry in section 4.1, are placed at the vertical line of the final crack predicted by smeared crack analyses. Lobatto integration scheme is adopted to avoid the stress oscillation problem. Instead of steering the analysis by displacement control, the arc-length control technique is pursued here hopefully to reveal a stable crack propagation path of this façade. Switching to this arc-length control requires the modification of the previously prescribed displacement degrees of freedom at the bottom of the façade into the free displacement degrees of freedom. Note that as minimum constraints are still required

to avoid the rigid body movement of the facade, few nodes at the right bottom of the façade are inevitably made as fixed supports. The effect of the live load and self-weight of masonry will be excluded here as it is found that this initial stresses result in severe false-cracking when a tying condition of the free displacement degrees of freedom along the bottom nodes of the façade is imposed. In addition, the effect of self-weight and live load is supposed to be negligible with respect ground settlements when the potential vertical crack line develops. A stable response can be predicted with the discrete crack corrector analysis and selecting into the constraint equation of the arc-length control procedure only the incremental displacements at the top of the façade. This approach, controlling the parameter which dominates the crack opening displacement, gives good convergences during all the course of loading. In fact, it is found that the façade snaps back during the stable cracking process along the corner of window openings at each floor, see Fig. 4a. With the classical incremental-iterative displacement control, the revisit of smeared crack analysis by the other types of smeared crack models, e.g., by total-strain rotating crack model (Feenstra et al. 1998) as enclosed in Fig. 4b cannot reveal this phenomena. The strong and sharp snap-back response is indeed the brittle-failure characteristics for large-scale masonry structures. However, even with the discrete crack analysis, much effort is required and the approach is still somewhat not attractive. Application of the sequentially linear continuum concept is appealing and will be investigated in the following section.

4.3   Sequentialyt linear analysis and results

The masonry façade is described with four-node membrane elements having size of 200×250 mm. From mesh refinement test presented in Boonpichetvong and Rots (2003), it came out that the structural response remains the same if 400×500 eight node elements or 200×250 mm four-node elements are used. So the results obtained with smeared or discrete crack models can be in any case compared with the ones obtained with sequentially linear analysis. The final crack predicted by smeared crack analyses, indicated the position of interface elements for discrete crack analysis carried out with the arc-length control technique. For the sequentially linear analysis, this same final crack pattern is also achieved by pre-defining the bands of elements to crack. A linear tension softening characterised by the same mechanical properties (Young's modulus, tensile strength and energy fracture) chosen for smeared crack model have been adopted for these predefined elements, while the remaining part of the façade is made linear elastic. According to section 2.1 a strength percentage p equal to 10% has been fixed for the saw-tooth diagram.

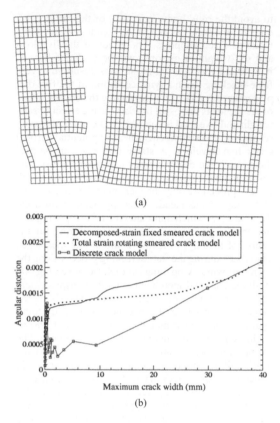

(a)

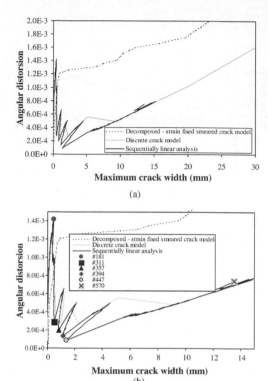

(a)

Figure 4. Incremental deformation during snap back response in discrete crack analysis (a); Computed angular distortion versus maximum crack graph (b).

Figure 5. Angular distortion versus maximum crack width curves: complete curves (a); zoom-in (b).

Consequently, the total number of teeth needed in order to reach the complete damage is automatically evaluated (N = 19). An isotropic reduction of stiffness was assumed. It means that Young's modulus was (saw-tooth wise) reduced in all directions. No-tension interface elements, previously inserted with the aim to ignore tensile resistance of soil boundary underneath buildings, are still included but now assumed to behave elastically. From the smeared crack analyses, these interface elements are shown to be in compressive mode, so this simplification can be made here without loosing the accuracy. Settlement is imposed as a displacement of nodes at the bottom of the façade.

As in the case of NLFE analysis carried out with discrete crack models and arc-length control technique, self-weight and live load are not considered.

Figure 5 shows the curves in terms of angular distortion versus maximum crack width obtained with smeared crack model, discrete crack model and sequentially linear analysis. In the case of sequentially linear analysis the crack width is determined as the

maximum principal strain times the crack band width h, assumed equal to 200 mm.

Smeared crack models proved inadequate tools for predicting such structural response and the results detected a snap-through where the convergence was very poor. This problem was circumvented by using discrete crack models, which display a very sharp snap-back beyond peak and subsequently a rising branch again. On the other hands non-proportional loadings cannot be applied and several numerical difficulties related to the arc-length procedure must be overcome in order to obtain this curve. The response given by sequentially linear analysis fits very well to the one obtained with the discrete crack model, but the easiness of the input data implementation required by sequentially linear analysis represents one of the main advantages of this numerical procedure. In Fig. 5b, where a zoom-in of the curve is shown, are also indicated the reference numbers of sequentially linear analyses corresponding to the sudden jumps which characterize the snap-back of the structural response. In Fig. 6 the level of element damage in correspondence of these jumps is reported. The damage level is evaluated in terms of number of teeth reached in the saw-tooth diagram providing an indication of crack

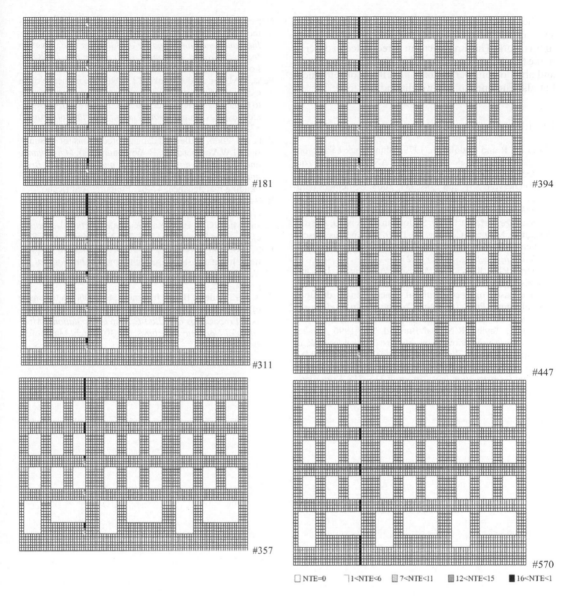

#181

#394

#311

#447

#357

#570

☐ NTE=0 ⌐1<NTE<6 ▓ 7<NTE<11 ▓ 12<NTE<15 ■ 16<NTE<1

Figure 6. Level of damage in correspondence of jumps in the structural response.

pattern and, if required, it can be transformed to crack width values.

5 CONCLUSIONS

The smeared crack models allow taking into account non-proportional loading (first the dead weight and live load, subsequently settlement are applied), but the very poor convergence performance makes the results untrustworthy. The advantage of smeared crack models is that cracks can occur anywhere in the mesh in any direction, while, with the discrete crack concepts, interface elements have to be predefined in the mesh. In the latter case snap-back response can be predicted, but practical use of this arc-length control technique requires a lot of user's experience. The response given by sequentially linear analysis fits very well to the one obtained with discrete crack model and no special user's skill or experience are needed to predict with good accuracy the structural response. In this way, contact can be made between the fracture mechanics "science community" and the practical engineering design world.

The implementation of non-proportional loading in sequentially linear procedure should be the next step of this work. This improvement is requested in order to make this proposed numerical procedure more general and able to consider not only the effects of settlement but also the ones of self-weight and live load.

REFERENCES

Bažant, Z.P. and Cedolin, L. 1979. Blunt crack band propagation in finite element analysis. ASCE *J. Engineering Mechanics Division*, 105(2), p. 297–315.

Boonpichetvong, M. and Rots, J.G. 2003. Settlement damage modelling of historical buildings. *Proceedings of EURO-C 2003 Computational Modelling of Concrete Structures*, March 17–20, St. Johann im Pongau, Austria.

Boscardin, M.D. and Cording, E.J. 1989. Building response to excavation induced settlement. *ASCE Journal of Geotechnical Engineering*, 115(1): 1–21.

Feenstra, P.H., Rots, J.G., Arnesen, A., Teigen, J.G. and Høiseth, K.V. 1998. A 3D constitutive model for concrete based on a co-rotational concept, In R. de Borst et al. (eds), *Proc. Int. Conf. Computational Modeling of Concrete Structures*: 13–22. Rotterdam: Balkema.

Mair, R.J., Taylor, R.N. and Burland, J.B. 1996. Prediction of ground movements and assessment of risk of building damage due to bored tunnelling. In R.J. Mair and R.N. Taylor (eds), *Geotechnical aspects of underground construction in soft ground*: 713–718. Rotterdam: Balkema.

Netzel, H. and Kaalberg, F.J. 2001. Settlement risk assessment of the North-South Metroline in Amsterdam, *Proc. Int Conference on Response of Buildings to Excavation Induced Ground Movement (Full paper in CD-ROM)*, London, UK.

Potts, D.M. and Addenbrooke, T.I. 1996. The influence of an existing surface structure on the ground movements due to tunnelling, In R.J. Mair and R.N. Taylor (eds), *Geotechnical aspects of underground construction in soft ground*: 573–578. Rotterdam: Balkema.

Rots, J.G. 2000. Settlement damage predictions for masonry, In L.G.W. Verhoef and F.H. Wittmann (eds), *Maintenance and restrengthening of materials and structures – Brick and brickwork*, Proc. Int Workshop on Urban heritage and building maintenance: 47–62. Freiburg: Aedificatio.

Rots J.G. 2001. Sequentially linear continuum model for concrete fracture. In de Borst R, Mazars J, Pijaudier-Cabot G, van Mier JGM, Balkema AA (eds), *Fracture Mechanics of Concrete Structures*, Lisse: The Netherlands, 831–839.

Rots, J.G. and Invernizzi, S. 2004. Regularized sequentially linear saw-tooth softening model. *Int. Journal for Numerical and Analytical Methods in Geomechanics*, 28, p. 821–856.

Rots, J.G., Belletti, B. and Invernizzi, S. 2006. On the shape of saw-tooth softening curves for sequentially linear analysis. *Proceedings of EURO-C 2006 Computational Modelling of Concrete Structures*, 27–30 March, Mayrhofen, Tyrol, Austria.

Fracture Mechanics of Concrete and Concrete Structures – High-Performance Concrete, Brick-Masonry and Environmental Aspects – Carpinteri, et al. (eds)
© 2007 Taylor & Francis Group, London, ISBN 978-0-415-44617-4

Selection of repair methods for rendered façades

J. Lahdensivu
Tampere University of Technology, Tampere, Finland

ABSTRACT: The repair methods of rendered facades are selected primarily on the basis of the technical condition of the structure. The information on the deterioration level of structures from a condition investigation allows the selection of appropriate repair methods and an assessment of related risks and the service life of the repair. The content of a condition investigation is to be such that set goals are achieved. Usually the aim is to determine the repair need and safety of structures. To achieve that, the deterioration and performance defects of structures need to be established. Rendered facades can be subject to several types of deterioration. This requires establishing the existence, scope, location, degree, impacts and future progress of deterioration in each case. The large variation in the level of deterioration between different buildings, and the fact that the most significant damage is not visible until it has progressed quite far, make a thorough condition investigation necessary in most facade repair projects.

1 INTRODUCTION

Prestigious and durability are the attributes commonly associated with rendered facades. The appearance requirements specified for the architectural design of buildings and facades of this type are usually high. The original facade makes demands also for the final result of facade repair. For the selection of repair method appears primarily damages and technical condition of the structure among the appearance requirements of facades.

Rendering was popular in the Finnish buildings facades until 1960s, when concrete structures and pre-cast concrete facades replaced rendering in new buildings facade. A small stock of older buildings with rendered facades consists mainly of churches, castles and other public buildings. The best known rendered buildings in Finland are probably those around the Senate Square in Helsinki including the Senate Building, Helsinki Cathedral and buildings of the University of Helsinki.

Rendering has been popular again during last ten years. Especially in the repair of those pre-cast concrete facades, which were made during 1960's and −70's. Also new buildings has now been build with rendered facade, but the construction technique and materials are different than old days. Insulating rendering is the main rendering system in these days in Finland.

Despite the young age of most of the building stock, some facades have had to be repaired after only 10 years' service life. Structures are damaged by various degradation phenomena whose progress is influenced by many structural, exposure and material factors. Consequently, the service lives of different structures vary a lot in practice. In the Finnish climate, frost weathering is the major deteriorating factor for porous materials.

2 STRUCTURES AND MATERIALS

The substrates of old rendered facades are mainly of burnt brick and mortar. In new buildings there is quite lot variation on substrates, materials and structures. The foundations of the buildings have varied by construction site. A common feature is a plinth of natural stone laid under the bearing brickwork. An advantage of a high natural stone plinth is that it prevents capillary transfer of moisture from the ground to the brickwork.

2.1 *External wall structure*

In old rendered buildings the mortar was applied onto solid masonry walls. Such external wall structures were common in Finnish rendered buildings until the 1960's. The thickness of the walls have varied over time (Neuvonen et al. 2002).

In the beginning of the 19th century bricks were usually manufactured on or nearby the building site. The quality of bricks varied a lot because they were fired in temporary kilns (Lahti 1960). Many brick factories were founded in Finland in the middle of the 19th century. In 1900 there were still 173 small brick factories in Finland, and 85% of the bricks were made manually (Kuokkanen & Leiponen 1981).

At the turn of the century each factory produced bricks of non-standard sizes with the moulds it had. An agreement on standard brick sizes was reached already in 1897, but standardised brick production was not started until the 1920's. Only those external walls that were faced with natural stone or rendering were built of domestic brick. Imported bricks were used for fair-faced external walls due to their higher quality. The dimensions and rates of firing of domestic bricks varied significantly which directly affected their durability (Lahti 1960).

The properties of bricks made a hundred years ago depended mainly on two things: the quality of the clay used and the firing. The temperature was not uniform throughout the kiln but fluctuated above and below 1,000°C. The range in firing temperature, therefore, resulted in over-burnt to under-burnt bricks. Over-burnt bricks are dark, almost black, durable and impermeable. Under-burnt bricks are typically light in colour, weak, highly porous and partly stratified.

From 1980's in small houses rendered external wall structure has consist on blocks of autoclave aerated concrete (AAC) or breeze-blocks which has an insulation layer between two breeze-block. These kind of wall structure is still quite popular in rendered small houses in Finland.

In the beginning of this century insulation rendering has taken a big part of rendering market in Finland. Traditionally (1980's) insulation rendering has been made on mineral wool with lime- or lime-cement mortars and the tree coat rendering has been fasten with mechanical anchor to buildings framework.

In new insulation rendering systems insulation is mineral wool or expanded polystyrene (EPS), mortars are cement-based and usually polymer modificated mortars. And rendering is fixed to insulation only with mortar, there is usually no mechanical anchor.

2.2 Foundations

Brick-buildings are brittle compared, for instance, to concrete ones and are easily damaged due to uneven settling of foundations and frost heave. The heavy weight of masonry buildings and their brittleness place demands on items such as, soil bearing capacity and type of foundation.

The type of foundation has naturally depended on the construction site and its soil composition. There have been three main foundation systems: rock-based, soil-based and pile-based.

Until the 1890's the typical foundation walls were made of dry set boulders (no mortar). At the beginning of the 20th century mortar or sand was normally placed between the boulders (Tawast 1993). The internal surface of the foundation, which formed the cellar wall, was generally lined with brick and later cast concrete. In the 1910's concrete foundation walls began to be used. The foundations were wider at the bottom in order to distribute the loads over a larger area. Bitumen or coal tar was used as waterproofing between the stacked boulders and the interior brickwork lining. Asphalt was used instead of mortar for tanking as well as in laying the brickwork of the interior surface (Neuvonen et al. 2002).

2.3 Mortars and rendering methods

Slaked lime has been used as an ingredient of mortar worldwide for thousands of years, and for hundreds of years also in Finland. The skill of mortar making was brought to Finland around 1100 by Swedish, Central European and Eastern Baltic masters representing the Germanic school. As late as the early 1900's Finnish builders used primarily pure lime mortars for masonry work and rendering – cement was added only occasionally (Perander et al. 1985).

In the 19th century most Finnish cement was imported, although the first cement factory already started operating in Savio in 1869 (Hurme et al. 1991).

The appearance, structure and colouring of rendered facades has varied quite extensively over time according to architectural trends. In the 16th and 17th centuries rendering was used to imitate natural stone surfaces being a cheaper alternative. Sandstone was imitated by pale yellow rendering, marble slabs by white rendering, and granite by grey rendering. A rendered surface was easy to colour and had no joints which allowed drawing lines on it to imitate stone slabs. Mortar was also used for mouldings. (Anon. 1999). The rendering jobs of old involved the application of three coats across the entire substrate.

Modern substrates like AAC or breeze-blocks demands different mortars for rendering than those old bricks. That because the strength and the suction of substrates are very much different than in burned bricks. In modern renderings there are used usually two coat renderings. It means that we use two different mortar. Base mortar can be sprayed in two layers with one mortar and surface also sprayed in two layers (Anon. 2005). Mortars are cement-based polymer modified mortars.

In insulation renderings mortars are developed especially for this use. Mortars should have very good adhesion to insulation material and it must be still workable. Those mortars are usually cement-based but they includes a lot of polymers.

3 LOADS AND DETERIORATION MECHANISMS RELATED TO RENDERED FACADES

The damage to rendered facades as structures age is mainly due to weathering. Deterioration may be fast enough to be harmful if used materials or the quality of work have been poor or structural solutions have been defective or perform unsatisfactorily. Weathering sets off several parallel deterioration phenomena which means that usually more than one cause contributes to the degradation of the facade. Initially deterioration phenomena proceed slowly, but as the damage increases the rate of deterioration also normally accelerates.

Deterioration of rendered facades can be divided into three basic categories:

- structure-induced deterioration of rendering
- deterioration of materials
- damage resulting from moisture.

3.1 Structure-induced deterioration

The damage caused by structures to rendered facades typically consists of cracks. Harmful cracks in masonry are generally the result of uneven settling of foundations. A crack in the masonry substrate also always leads to cracking of the rendering layer. In brick walls the cracks normally run along the interfaces between mortar and masonry units. A condition investigation must be carried out to determine the causes of cracks and their movement. The cause can often be determined from the direction and location of cracks. Repair of active cracks without first determining the cause will usually lead to renewed cracking.

Drying of structure and mortar leads to shrinkage. Shrinkage is characteristic value for each material. In mortars a plastic shrinkage can be avoid by careful after-care, but characteristic shrinkage happens when rendering dries during time. Both of these shrinkage mechanisms can cause cracking on rendering. In most cases these cracks are quite narrow and they are usually only aesthetical problems. I cases, when rendering will be coated with organic paints or coatings which form an impermeable skin over the rendering, it might be very harmful.

Rainwater can penetrate into the rendering and the wall structure through cracks in facades resulting in moisture and frost-weathering damage.

3.2 Deterioration of materials

Frost weathering due to a high moisture load is the most common reason for the deterioration of the rendering layer, especially with weak, old lime mortars. Rendering mortar is a porous material whose pore system may, depending on the conditions, hold varying amounts of water. As the water in the pore system freezes, it expands about 9% by volume which creates hydraulic pressure in the system (Pigeon and Pleau 1995). If the level of water saturation of the system is high, the overpressure cannot escape into air-filled pores and thus damages the internal structure of the concrete resulting in its degradation.

Probably the most widely known frost damage theory is the hydraulic pressure theory by Powers published in 1948. Accordingly, damage occurs as freezing water expands creating hydraulic pressure within the pore structure of a porous material. The pressure is created when part of the water in a capillary pore freezes and expands forcing thereby the unfrozen water out of the pore. The migration of water causes localised internal tensions in the material whereby its strength may fail resulting in cracking (Powers and Brownyard 1949).

The cause of the rapid weathering is generally a damaged or non-performing structural detail or a structure connected to the rendering surface. Frost weathering is manifested as reduction in strength of the rendering, loss of adhesion, or crazing or chipping off of the surface. In Finnish climatic conditions frost weathering is the most common cause of damage.

The degree of frost weathering may vary in different parts of the wall surface – depending on, for instance, the load and variation in material properties – as well as thorough the thickness of the rendering. Weathering due to high local moisture load may affect only a very limited area. On the other hand, an improper surface treatment may result in deterioration across most of the wall surface.

3.3 Defects in moisture performance

Rendered surfaces have traditionally been painted with inorganic lime paints. These coatings are porous, allowing water and water vapour to pass through in both directions.

Organic paints form a uniform, almost impermeable skin over the rendering. In principle, impermeable surface treatment prevents rainwater from penetrating into the rendering layer. However, in practice water always migrates into the rendering layer through permeable points in the coat of paint, structural cracks, etc. The wall structure cannot dry out fast enough as a result of the impermeable coating which can lead to frost weathering. The deteriorating effect of organic facade has been widely recognised.

One function of flashings, eaves, etc. is to prevent rainwater from entering structures through seams and joints and to allow them to dry to avoid any detrimental effects of moisture within. The performance of seams and joints play a major role in the durability of the overall structure. Moreover, the performance of flashings has a major impact on how dirty the rendering gets.

4 REPAIR ALTERNATIVES

Selection of suitable repair alternative for each case has several demands:

- the repair must be cost effective
- aesthetic requirements

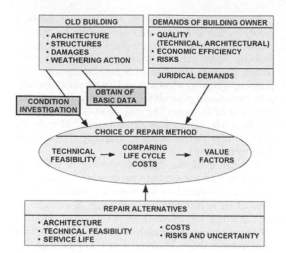

Figure 1. Different factors which has influence on decision of repair method.

- degree of different deterioration mechanisms
- service life requirements of repair
- requirements concerning the reliability of the repair.

The following repair methods are used based on the degree and extent of the deterioration:

- light coating repair
- patching and coating repair
- rendering removal and re-rendering
- repair of entire facade, for instance, with an insulating rendering.

All the requirements of selection of repair method has not the same value. The repair methods are selected primarily on the basis of the technical condition of the structure. The one that best meets the architectural and financial criteria of the project is then selected from among those that meet the technical requirements. In figure 1 it has been shown the principles of selection of repair method.

The technical condition of rendering and the service life which facade has left can be found by systematic condition investigation. The information on the deterioration level of structures from a condition investigation allows the selection of appropriate repair methods and an assessment of related risks and the service life of the repair. Thus, the investigator must be highly knowledgeable about the repair methods for rendered facades and the demands on them.

5 SYSTEMATIC CONDITION INVESTIGATION

The object of the condition investigations of facades is to produce information for building owners and designers which can be used to evaluate the need for repair. They will also help select the most suitable repair methods.

The content of a condition investigation should be such as to allow reaching the set aims. Usually the goal is to determine the repair needs and safety of structures. This requires establishing the damage to structures as well as their performance defects. For that, the existence, extent, location, degree, cause, impacts and future progress of damage must be determined (Anon. 2002).

The content of a condition investigation is determined by the type of structure and materials used, the climatic conditions and visible deterioration as well as the set goals. Thus, the investigator should be highly knowledgeable about the performance of structures and the deterioration phenomena which affect both structures and materials.

6 SYSTEMATIC CONDITION INVESTIGATION

Investigation of the rendering itself is only a part of the condition investigation of facade rendering. The focuses are:

- the substrate and deterioration of the substrate
- type, strength and deterioration of rendering
- deformation, movements and cracking
- coatings
- moisture behaviour
- thermal performance
- condition and attachment of ornaments
- special issues, such as earlier repairs.

6.1 Substrate and its deterioration

The structural performance of the rendering substrate is determined during the condition investigation. Construction drawings are very helpful in determining structural performance. The materials of the substrate, their strength and attachments are also determined.

The substrate may have been damaged by, for instance, weathering and steel corrosion in addition to cracking. The condition investigation should determine the reason for, extent and degree of the deterioration. Moreover, renovation planning must also consider any possible impacts on bearing capacity, slenderness of the building structure and bending of beams from deterioration or corrosion.

6.2 Type, strength and deterioration of rendering

Determination of the strength and adhesion of rendering is an essential part of the condition investigation. The investigation of facades involves assessment of the adhesion of rendering and deterioration of different mortar layers, the degree of deterioration, and the extent and location of various types of deterioration on the facades. Large, uniform areas of loose rendering influence the selection of the type of repair. Small, solid loosened areas can in some cases be left unrepaired.

The strength of the rendering layer is more significant when coated with organic substances. Removal of the coatings requires higher strength of the rendered surface than cleaning and re-coating of facades with permeable coatings.

The condition investigation should also determine the type, materials and thicknesses of rendering layers.

6.3 Deformation, movements and cracking

The cracks visible on facades have most often been caused by uneven settling of the building foundations, loading of the building frame, or restraint actions or rendering deformations. The condition investigation must determine the reason for cracking before the type of repair is selected.

The condition investigation must also determine the extent of cracking and its impact on the selection of the type of repair and the useful life of the repairs.

6.4 Coatings

The coatings of rendered facades can be divided roughly into two groups:

– organic paints and coatings which form an impermeable skin over the rendering
– inorganic paints and coatings which do not substantially alter the moisture performance of the rendering surface.

The condition inspection determines the type and condition of the coating. As a rule, impermeable organic coatings should be removed during repairs. Certain organic coatings have contained, for instance, asbestos which means that the asbestos-content of the coating must be established.

6.5 Moisture behaviour

Old rendered facades subjected to local moisture loads will deteriorate prematurely. The moisture load on facades and their various sections are assessed in the condition investigation.

The moisture behaviour of a facade is affected by, for instance, eaves, effectiveness of rainwater drainage, joints with various structural elements, and the condition of flashings. The condition investigation evaluates the visible impacts of defects in moisture performance.

Figure 2. This kind of cracks in facade are usually a sign for uneven settling of the building foundations.

Defects in moisture performance are the most important factor causing local deterioration of rendering.

6.6 Thermal performance

The thermal performance of an external wall structure generally needs to be assessed in buildings made of thin LECA blocks or aerated concrete masonry units. It is possible to improve the energy performance of the building by, for instance, applying an insulating rendering should the existing rendering be in bad enough condition to require re-rendering to the substrate.

Old buildings with solid brick walls do not meet today's thermal insulation standards. Yet, they are not generally required to have additional insulation installed because of architectural considerations.

A condition investigation of insulating renderings should determine its the type, moisture content, the attachment method and the quality of the bond to the substrate.

6.7 Condition and attachment of possible ornaments

Sometimes a rendered facade is adorned with numerous plaster ornaments. The bonding reliability of plaster ornaments is determined during the condition investigation. Small and flat decorations usually adhere to the plaster with the help of plaster mortar. Large ornaments which protrude from the facade are mechanically fastened by, for instance, wrought iron nails and laths.

The condition investigation of plaster ornamentation should generally be conducted by an expert. Actual breakage of an ornament can easily be noticed during a general condition investigation.

6.8 Special issues including earlier repairs

Areas of earlier patch repairs of rendering are often distinguishable from the rest of the facade. Different mortars have often been used for the repair. The condition investigation should determine deterioration and adhesion to the substrate of the patching.

The timing of rendering repairs must also take into account the repair of adjoining building elements. Repairs that affect rendering repairs include window replacement and repair of balconies and eaves.

The condition investigation of a rendered facade should also evaluate the need for repair of adjoining structures – at least visually.

7 SAMPLE SIZE AFFECTS RELIABILITY

The condition investigation of an old structure always involves a degree of uncertainty, because it is not usually possible to investigate all facades and all structures. The information describing the condition of structures is collected as samples and the condition and properties of structures vary in different sections of the facade. Systematic condition investigation attempts to collect parallel information from as many sources as possible. This makes the evaluation of results easier and increases the reliability of conclusions.

Information on the potential problems occurring in structures and the state and progress of deterioration can be collected, for example, from the building's design documents, through visual observations on site, by various field research methods and by sampling and laboratory tests.

The deterioration mechanisms of rendered facades are generally such that a sufficiently reliable view of the condition of the facade can generally be formed by careful visual inspection and simple field research methods. Measurements requiring special equipment and laboratory testing of material samples are mainly needed in special cases.

8 CONCLUSIONS

The repair methods are selected primarily on the basis of the technical condition of the structure. The information on the deterioration level of structures from a condition investigation allows the selection of appropriate repair methods and an assessment of related risks and the service life of the repair. Thus, the investigator must be highly knowledgeable about the repair methods for rendered facades and the demands on them.

The large variation in the level of deterioration between different buildings, and the fact that the most significant damage is not visible until it has progressed quite far, make a thorough condition investigation necessary in most facade repair projects.

REFERENCES

Anon. 1999: *Rendering book. Helsinki.* BY 46. 90 p. (in Finnish)

Anon. 2002: *Condition Investigation Manual for Concrete Facade Panels.* Helsinki. Concrete Association of Finland BY 42 178 p. (in Finnish)

Anon. 2005: *Rendering book 2005.* Helsinki. BY 46. 158 p. (in Finnish)

Hurme Riitta, Häyrynen Maunu, Penttala Vesa, Putkonen Lauri, Soini Eero 1991: *Concrete in Finland 1860–1960.* Jyväskylä, Suomen Betonitieto Oy. 195 p. (in Finnish)

Kuokkanen Rauno, Leiponen Kauko 1981: *History of Finnish Brick Industry.* Helsinki, Suomen Tiiliteollisuusliitto r.y. and Tiilikeskus Oy. 553 p. (in Finnish)

Lahti Matti J. 1960: *How Helsinki has been Built.* Rakentajain Kustannus Oy. 336 p. (in Finnish)

Neuvonen Petri, Mäkiö Erkki, Malinen Maarit 2002: *Multistore Apartment Buildings 1880–1940.* Helsinki, Rakennustietosäätiö RTS. 192 p. (in Finnish)

Tawast Ismo 1993: *Underpinning.* Tampere, Tampere University of Technology, Institute of Geotechnics Publication 26, Institute of Structural Engineering Publication 59. 199 p. (in Finnish)

Perander Thorborg et al. 1985: *Mortars in Historic Buildings.* Espoo, VTT Research Reports 341. 148 p. (in Finnish)

Pigeon M. and Pleau R. 1995: *Durability of Concrete in Cold Climates.* Suffolk. E & FN Spon. 244 p.

Powers T.C. and Brownyard T.L. 1948: *Studies of the Physical Properties of Hardened Portland Cement Paste.* Bulletin 22. Portland Cement Assosiation Bulletin. Chicago. IL.

Fracture Mechanics of Concrete and Concrete Structures – High-Performance Concrete,
Brick-Masonry and Environmental Aspects – Carpinteri, et al. (eds)
© 2007 Taylor & Francis Group, London, ISBN 978-0-415-44617-4

Acoustic emission in asphalt mixtures at low temperature

X. Li, M.O. Marasteanu & J.F. Labuz
Department of Civil Engineering, University of Minnesota, Minneapolis, USA

ABSTRACT: The objective of the research was to investigate the use of acoustic emission to analyze the microstructural phenomena of damage and the corresponding macroscopic behavior in asphalt mixtures tested at low temperature. An acoustic emission system with eight channels of recording was used to monitor the specimens tested in creep and in strength testing. The source location and the AE event count were used to illustrate the relationship between the micro damage and macroscopic behavior for the two different test conditions. The analysis indicates that a damage zone develops in both the creep and the strength tests. The damage zone changes with the test temperature and the loading level applied during the creep test.

1 INTRODUCTION

It is well recognized that asphalt mixtures have complex temperature-sensitive behaviors. The response to a given loading is strongly dependent on temperature and loading path. A change of a few degrees in temperature induces dramatic changes in behavior of the asphalt mixture. The behavior can vary from relatively ductile at higher temperature to brittle at lower temperature.

Low temperature cracking is the most significant distresses in asphalt pavements built in areas with cold climates. Low temperature cracking is attributed to tensile stresses induced in the asphalt pavement as the temperature drops to extremely low values. The accumulation of tensile stress to a certain level is associated with the formation of microcracks, which release energy in the form of elastic waves called acoustic emission (AE). The current AASHTO specification for asphalt mixture low temperature characterization consists of two tests: the indirect tensile creep test (ITC) and the indirect tensile strength test (ITS). For both tests, cylindrical specimens of 150 mm diameter × 50 mm thickness are loaded in compression across the diametral plane.

This paper investigates the use of AE to analyze the microstructural phenomena and the corresponding macroscopic behavior in asphalt mixtures tested at three low temperatures. For each test temperature, two different loading levels were used in the creep test and one constant stroke rate of 1 mm/min was employed in the strength tests. This was done to identify the damage development with loading levels and to compare the damage zone in creep and failure tests.

2 BACKGROUND

AE methods represent a well-documented and widely used tool to characterize microscopic fracture processes and therefore to evaluate damage growth in brittle materials. An acoustic emission is defined as a transient elastic wave generated by the sudden release of energy from localized damage processes within the stressed material. This energy release causes the propagation of stress waves that can be detected at the surface of the material. Acoustic emissions result from microcracking, dislocation movement, phase transformation, and other irreversible changes in the material.

Due to its capability of detecting internal damage, AE has been used for many years to study the behavior of materials such as rock and concrete. By studying the occurrence of AE events, the investigations are generally focused on the cumulative number of AE events, the rate of occurrence, amplitude distribution, energy and frequency distribution.

Compared to the vast number of studies performed on Portland cement concrete and rocks, a literature search on the use of AE in asphalt materials results in a very limited number of references. This is due to the fact that for most service temperatures asphalt mixtures display viscous and ductile behavior and the development of defects is gradual and does not produce emissions that can be detected.

In one of the first efforts to document the use of AE to characterize the fracture of asphalt mixtures under thermal loading conditions at low temperatures, Valkering & Jongeneel (1991) used a single piezoelectric transducer mounted on the surface of the specimen. The cumulative event counting was analyzed to determine how damage accumulates during loading. Their work was followed by a small number of studies performed by different authors who also limited their analyses to counting of AE events.

AE techniques were also investigated during the Strategic Highway Research Program (SHRP) in an effort to better characterize the cohesive, adhesive and thermal properties of asphalt binders (Chang, 1994; Wang, 1995; Qin, 1995). However, this work was performed at higher temperatures that considerably limited the analysis of the AE results. Sinha (1998) reported on the AE activity in unrestrained asphalt mixture samples exposed to thermal cycling at low temperatures to show that micro cracking occurs in asphalt mixtures due to the difference between the thermal contraction coefficients of the aggregate and of the asphalt binder. Hesp et al. (2001) investigated AE activity in restrained asphalt mixture samples undergoing various temperature cycles and Cordel et al. (2003) used two transducers to detect the AE events in asphalt mixture direct tension tests performed at low temperatures.

In a recent study by Li & Marasteanu (2006), the accumulated AE events obtained from an AE system with eight channels of recording were analyzed to illustrate the relationship between the micro damage and macroscopic behavior of the asphalt mixtures at different loading levels. In addition, the initiation and propagation of the cracking were observed using the localization of the event source and the fracture process zone in the asphalt mixtures was measured by observing the distribution of microcracks with different energy levels.

3 EXPERIMENTAL PROCEDURE

The asphalt mixture used in this study was prepared using the Superpave design procedure outlined in SP-2. One asphalt binder with performance grade (PG) 58–34 and modified by styrene-butadiene-styrene (SBS) was used. Granite aggregate was selected to prepare the mixture. Cylindrical specimens 150 mm by 170 mm were compacted using the Brovold gyratory compactor. A 4% target air voids was achieved after compaction. The compacted samples were then cut into 3 slices with 50 mm in thickness.

The indirect tensile test set-up is shown in Figure 1. A sample with dimensions of 150 mm diameter by 50 mm height was loaded in static compression across its diametral plane. Different load levels were applied in the creep test to investigate the effect of load levels on the development of the micro damage during the creep test. A constant loading rate of 10 kN/s was used at the beginning of the creep test. Following the creep test, an indirect tensile strength test was performed at the same temperature. A constant stroke rate of 1 mm/min was applied during the strength test until failure.

An AE system with eight channels of recording was used to monitor the asphalt mixture specimens tested in creep and strength tests. The AE event signals were recorded using four DAQ cards (Model PCI-5112, National Instruments). Each card had two independent channels which acquired AE signals detected by eight piezoelectric sensors (Model S9225, Physical Acoustics Corporation). Four sensors were mounted on each side of the specimen using M-Bond 200, a modified alkyl cyanoacrylate. The preamplification of the AE signals was provided by eight preamplifiers (Model 1220C, PAC) with a gain set to 40 dB. One of the sensors was used as a trigger, which was often the one closest to the tip of the initial notch. Trigger level was set at 10 mV in this research. Once the recording was triggered, signals were band-pass filtered (0.1–1.2 MHz) and sampled at 20 MHz over 200 microseconds. Considering the ringing of the resonant sensor, a sleep time of 9 milliseconds between two consecutive events was prescribed during which the system could not be triggered. The velocity of propagation of the longitudinal waves was determined by generating an elastic wave by pencil lead (0.5 mm diameter) breakage on the opposite side of the specimens.

Tests were performed in a MTS servo-hydraulic testing system. The TestStar IIs control system was used to set up and perform the tests and to collect the data. The software package MultiPurpose TestWare was used to custom-design the tests and collect the data. All tests were performed inside an environmental chamber. Liquid nitrogen tanks were used to obtain the required low temperature. The temperature was controlled by MTS temperature controller and verified using an independent platinum RTD thermometer.

Three test temperatures, −12°C, −24°C and −36°C were selected based on the PG lower limit of the

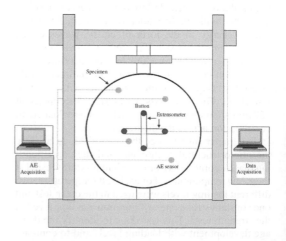

Figure 1. Schematic of experimental setup.

asphalt binder. Two specimens, one replicate for two different loading levels, were fabricated for the creep testing at each temperature. The specimen used for the creep test at the lower loading level was also used for the strength test, 30 minutes after the creep test.

the highest temperature ($-12°C$), the asphalt mixture is more ductile and it has lower peak load and larger displacements. At the lowest temperature ($-36°C$), the material is brittle. At $-24°C$ temperature, the mixture exhibits an intermediate behavior.

4 DISCUSSION OF RESULTS

The mechanical response for all three temperatures during the strength tests are shown in Figure 2. At

4.1 AE event count

The event count curves under different temperatures follow a similar pattern. Very few AE events were detected at the beginning of the test but soon after the AE rate was constant at 7 to 8 events per second. The black circle on the event count curve shows the loading level in the creep test performed before the strength test. For all strength plots under different temperatures, more than 100 events were recorded before the loading level applied in the creep test was reached. This appears to indicate that the Kaiser effect, which describes the phenomenon that a material under load emits acoustic waves only after a primary load level is exceeded, is not applicable to the material investigated in this study.

A typical plot of loading and AE event count versus time for the creep test is shown in Figure 3. The two creep tests with different load level under the same

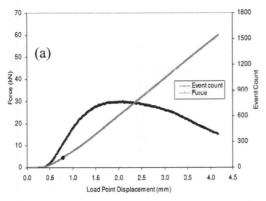

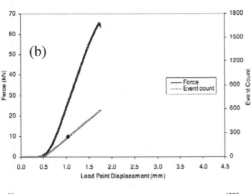

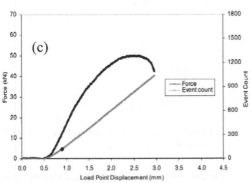

Figure 2. Load and AE count versus displacement from strength test: (a) $-12°C$; (b) $-24°C$; (c) $-36°C$.

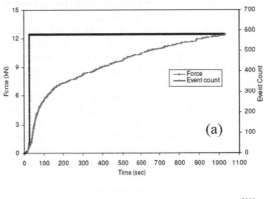

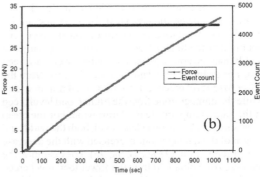

Figure 3. Typical plot for load and AE count versus time from creep test: (a) 12.5 kN at $-24°C$; (b) 30 kN at $-24°C$.

temperature show a similar trend with time. AE events were recorded with a fast event rate immediately after the creep load level was applied in a short time and the event rate decreased with time. It was found that more events were recorded in the creep test with higher loading level.

4.2 AE source location

The source location of the AE events can be inferred by investigating the differences of the first time of arrival among the transducers placed at different locations on the specimen. Therefore, it is necessary to determine the arrival time of the elastic waves for each sensor from the recorded AE event. The correct picking of the first arrival strongly affects the accuracy of the source location (Labuz et al., 1997). The mean amplitude and standard deviation of the noise during the pre-trigger period was calculated and a threshold was set at three times the standard deviation from the mean. By measuring the time at which the signal passes this threshold value for each sensor, the relative arrival time can be obtained.

Once the arrival times of the event were determined from the sensor records, the location of the AE event source can be estimated. It should be noted that not all events can be located. It was found that only 143 events were located with 10 mm error for the creep test under the lower loading level at −24°C. Therefore, event locations with 143 events are selected and plotted for the comparison. A rectangle was drawn to contain 90% of these 143 events. Figure 4 shows the first 143 AE locations obtained from the strength test under different temperatures. It is found that the size of the damage zone, which is defined here as the area with microcracking inside the rectangle, changes with temperature for the strength test. The width of this zone decreases with the decrease in test temperature. While no clear difference is found for the length of the damage zone for the two higher temperatures, the damage zone at the lowest temperature is obviously longer than the damage zones at the two higher temperatures.

Figures 5, 6 and 7 show the locations for the last 143 events during the creep test for all temperatures. It is observed that for all temperatures the damage zone obtained at the higher load was significantly larger in width than the zone obtained from the lower load creep test. For example, the damage zone obtained from the lower load level creep test at −12°C is 24 mm in width, while the damage zone from the higher load level creep at the same temperature is 29 mm wide. For the creep tests with relatively lower load level, both the width and length of the damage zone decreased with the decrease of test temperature (Figs 5a, 6a, 7a). For example, the damage zone obtained from the lower load level creep at −24°C is 21 mm by 60 mm, while the size of damage zone from the higher load level creep at −36°C is 18 mm by 55 mm.

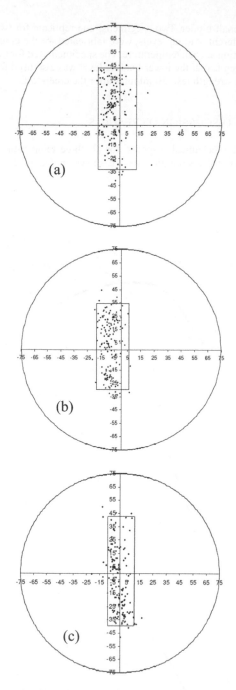

Figure 4. Event locations and damage zone for strength test: (a) −12°C; (b) −24°C; (c) −36°C.

The comparison for the damage zone obtained from both creep test and strength test shows that for all temperatures the size of damage zone obtained from strength test is very close to the size of damage zone

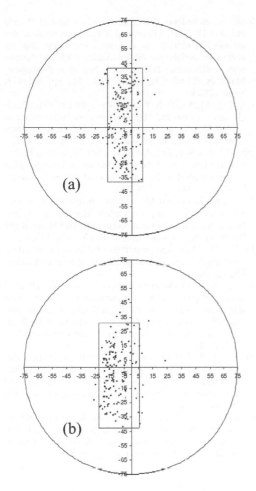

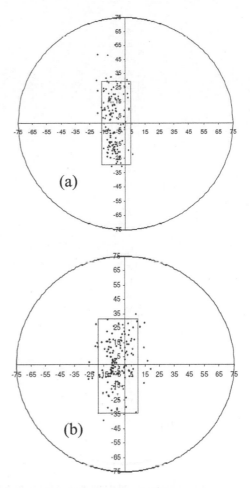

Figure 5. Event locations and damage zone for creep test at −12°C: (a) 10 kN (33% of strength); (b) 15 kN (50% of strength).

Figure 6. Event locations and damage zone for creep test at −24°C: (a) 12.5 kN (25% of strength); (b) 30 kN (60% of strength).

from the creep test with higher loading level, which was approximately equal to 50% to 60% of the strength value (Figs 5b, 6b, 6c).

5 CONCLUSION

Indirect tensile creep tests with two different loading levels, and strength tests were performed under three different test temperatures for one asphalt mixture. An acoustic emission system with eight channels of recording was used to monitor the development of microcracking in the specimens for both tests.

The experimental data show that the asphalt mixture presented more brittle behavior with decrease in test temperature. Few events were recorded at the

beginning of the strength test but the AE rate was constant at 7 to 8 events per second after certain loading level. A similar event rate was found immediately after the creep load was applied, but the event rate decreased with time. Creep tests at higher load levels were found to produce more events than the tests with lower load levels for all test temperatures. The Kaiser effect was not applicable for the tested specimens.

It was observed that the width of the damage zone developed during strength tests decreased with the decrease of temperature and the damage zone obtained from the higher loading level creep test was significantly larger in width than the zone obtained from the lower loading level creep test. The size of the damage zone was found to decrease with temperature decrease for the creep tests. The damage zone obtained from strength test was approximately equal to that from

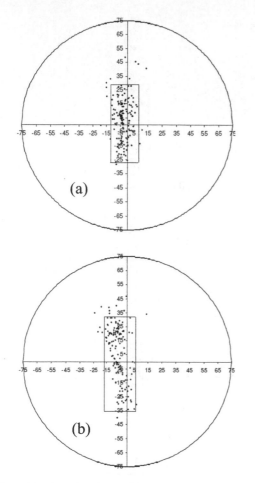

Figure 7. Event locations and damage zone for creep test at −36°C: (a) 25 kN (38% of strength); (b) 40 kN (61% of strength).

creep test with a load level about 50% to 60% of the strength value.

REFERENCES

Chang W. V. 1994. Application of Acoustic Emission to Study the Cohesive and Adhesive Strength of Asphalt. Strategic Highway Research Program, SHRP-A-682, pp. 81–148.

Cordel S., Benedetto H.Di., Malot M., Chaverot P., and Perraton D. 2003. Fissuration à basse température des enrobés bitumineux -essai de retrait thermique empêché et émission acoustique. 6th International RILEM Symposium, Performance Testing & Evaluation of Bituminous Materials (PTEB 03), Editor Partl, M., pp. 465–472, Zurich, 2003.

Hesp S. A. M., Smith B. J. 2001. The Effect of Filler Particle Size on Low and High Temperature Performance of Asphalt Mastics and Mixtures. Journal of the Association of Asphalt Paving Technologists, Vol. 70, pp. 492–542.

Labuz J. F., Dai S-T., and Shah K. R. 1997. Identifying Failure Through Location of Acoustic Emission. Transportation Research Record 1526, TRB, National Research Council, Washington, D.C., pp.104–111.

Li X., and Marasteanu M. O. 2006. Investigation of Low Temperature Cracking in Asphalt Mixtures by Acoustic Emission. International Journal of Road Material and Pavement Design, Vol. 7-No. 4/2006, pp.491–512.

Qin X. 1995. Adhesion properties of polymeric materials. Ph.D. thesis, University of Southern California, Chemical Engineering.

Sinha N. K. 1998. Acoustic emission is asphalt subjected to thermal cycling at low temperature. In Acoustic emission/microseismic activity in geologic structures and materials, Proceedings of the Six Conference, by Reginald Hardy, Jr., H., Pennsylvania State University, Vol. 21, 109–120.

Valkering C. P., and Jongeneel D. J. (1991). Acoustic emission for evaluating the relative performance of asphalt mixes under thermal loading conditions. Journal of the Association of Asphalt Paving Technologists, vol. 60, 160–187.

Wang H.-C. 1995. Ultrasonic and acoustic emission in nondestructive evaluation of viscoelastic solids – elastomer, human cornea and asphalt. Ph.D. thesis, University of Southern California, Chemical Engineering.

Fracture Mechanics of Concrete and Concrete Structures – High-Performance Concrete,
Brick-Masonry and Environmental Aspects – Carpinteri, et al. (eds)
© 2007 Taylor & Francis Group, London, ISBN 978-0-415-44617-4

Some mechanisms for the genesis of fractures in sedimentary rocks

C. Putot & D. Quesada
Institut français du pétrole, Rueil Malmaison, France

D. Leguillon
LMM, CNRS UMR 7607, Université Pierre et Marie Curie, Paris, France

ABSTRACT: Understanding the spatial and temporal distribution of fluid flow in the subsurface is of fundamental importance to the successful management of groundwater and hydrocarbon resources. The analysis is restricted to open mode fracture sets, usually recognized for important contributors to permeability in the case of low porosity reservoirs. We investigated, on a mechanical basis, several field representations likely to occur when jointing and fracture clustering take place. Clustering cannot be explained by effective horizontal tensional conditions resulting in fluid driven fractures, in view of the screening effect between very closely spaced fracture planes, even if we assume sub-critical propagating conditions. Systematic joints form at depth, as a result of tectonic compression in combination with high pore pressure. Unconfined effective conditions arise to produce low rock strength and brittle deformation and load parallel extension fractures are the rule.

1 INTRODUCTION

The importance of fractures and faults in hydrocarbon entrapment, migration and flow has just been recently recognized. The previously common attitude that most basins and reservoirs have no fractures or faults is now impossible: in particular, tight reservoirs discourage any attempt to get a reasonable view of gas or oil recovery by conventional models.

Three common structural types may be considered:

- dilatant fractures (joints, veins, dykes and sills also known as hydraulic natural fractures)
- contraction/compaction structures (compaction bands)
- shear fractures (faults)

1.1 Joints and faults

Joints are a distinct mode of geologic fracture, distinguished from faults in that the displacement that occurs across the fracture interface is a dilation. Because joints often occur in parallel trending sets of closely-spaced fractures, they can control the mechanical and hydraulic properties of the enclosing rock mass. Consequently, joints affect the productivity of oil and natural gas reservoirs.

1.2 Objectives of this presentation

The objective of this presentation is to get a synthetic view of what is possible in the arrangement of joints (opening mode fractures) on a sound mechanical basis, spatially and temporally speaking.

More precisely the matter is about statements assessing the evolution of fractures during genesis of the reservoir. It means a complex loading time history coming with diagenetic processes during lithification, tectonics. A great number of unknown variables may explain the variety of situations; rock constitutive relationships, structural heterogeneities, abnormal pore pressure, triaxial stress tensor at the time of fracturing.

1.3 Pore pressure and tectonics

Historically, it was commonly believed that open fractures could not exist at depth. However, as the role of pore pressure in the reduction of effective stress began to be recognized, open fractures at depth began to be considered not only possible, but entirely likely under certain conditions.

Many attempts have been made in the past to justify the existence of effective tractions orthogonal to joint plane (horizontal tractions for most of the cases). Excessive pore pressure has been generously invoked, joints propagating as natural hydraulic fractures. In the most common version of the assumed mechanism, abnormal formation pore pressure is supposed to exceed the least compressive horizontal principal stress, causing failure within the rock. But many arguments such as fracture morphology are telling against this interpretation.

Figure 1. Systematic vertical jointing.

Figure 2. Fracture swarm.

– for clustering (Fig. 2) there is no satisfactory answer of this type.

According to Bessinger et al. (2003) compression driven tensile fracturing mechanisms during jointing has not yet been recognized as important. Evidence for joint–parallel compressive stresses in the vicinity of small or large scale heterogeneities has been demonstrated by field observations and measurements.

Oil industry (Fonta et al. 2005) is indirectly aware of such behaviour: non-tectonic fractures – mainly early diagenetic features – are differentiated from tectonic fractures, with only the latter having a real potential effect on production.

1.4 Joint size and shape

Engineering studies have examined joint spacing distributions, but because they typically do not separate genetically distinct joints, their measurements contribute little to the scientific understanding of joint development (Narr & Suppe 1991). The standard engineering technique involves measuring the spacing between joints along a borehole of arbitrary orientation.

Assuming for simplification parallel joint sets, we distinguish between:

– systematic jointing (Fig. 1) scaled by layer thickness
– multilayer jointing encouraging to define a "mechanical unit" larger than the stratigraphic unit (Cooke & Underwood 2001)

Aspect ratio, length over spacing, is a useful parameter for capturing typology of fracture.

– systematic jointing, scaled with stratigraphic or mechanical units has been classically explained using remote tension perpendicular to fracture;

2 EARLY IDEAS

2.1 Importance of pore pressure

Joint propagation occurs when appropriate failure criteria are met; they are often specified in terms of states of stress considered as function of depth of burial, variation of rock properties during consolidation and diagenesis, stress history and pore-fluid pressures. Of particular interest is for Engelder (1985) the "evidence" that joints propagate as natural hydraulic fractures under the influence of abnormal pore pressure. According to him, two types of joints may be distinguished: those propagating while burial is in progress and those propagating during erosion and uplift. Abnormal pore pressures are required in the former case whereas thermal-elastic contraction is primarily responsible for the latter case. Hydraulic joints are those caused by abnormal pore pressure during burial under restricted pore-water circulation (these joints form at depths in excess of 5 km). Tectonic joints are distinguished from hydraulic joints in that they form at depth (less than 3 km) under the influence of high pore pressure which developed only during tectonic compaction (no overpressure during burial). Active compression of the host rocks is needed in this case to account for abnormal pore pressure. Unloading and release joints form in response to the removal of overburden during erosion (Nur, 1982).

2.2 Modelling fracture sets

Although joint spacing is found to be roughly proportional to layer thickness in many studies, data are not always consistent with each other (Wu & Pollard 1995) and they have lead to contradictory conclusions about the jointing process. One possible explanation is that the data collection methods introduces significant bias. Another possibility is that the 2D linear mechanical relationship between spacing and thickness is too simplistic.

Indeed, the development of a joint set is a complex 3D process with possible changes during the history for physical conditions and loading. Joint fractography and petrophysical properties may help to find the timing of fracturing. These properties allow sometimes to identify depth of burial and variations of pore pressure.

2.3 Experimental modelling

Experimental models based on brittle coating techniques share some of the kinematics features such as lateral propagation parallel to bedding. Trends about sensitivity of spacing to parameters are studied by changing the thickness of the brittle coating. Observing joint sets on bedding planes are a necessity when propagation is dominantly parallel to bedding; spatial distribution of these joints is not visible in layer cross-sections.

As far as geometry of fractures sets is concerned, two kinds of joints sets are distinguished on bedding surfaces (Wu & Pollard 1995): a *poorly-developed* set represents the early stages of development when typical joint lengths are less than typical spacing; a *well-developed* set represents later stages when lengths are much greater than spacing.

Rives et al. (1992) have studied how the frequency distribution of spacing depends on the stage of development of fracture sets.

2.4 Modelling geometry of fracture set in a semi-infinite medium: extension cracking

The first attempts of mechanical analysis are found in Lachenbruch (1961) who connects brittle cracking in the direction normal to the component of maximum tensile stress with many conditions of geologic interest. He points out the interest of Irwin's modification of Griffith's theory and considers useful to describe initiation in terms of "tensile strength" which is the macroscopic average tension under which small flaws start to grow and coalesce whereas macroscopic size cracks are more relevant to an energy criterion. Examples are thermal contraction cracks in cooling basalt, desiccation cracks in mud, tension cracks on the convex sides of flexures. The crack is generally initiated at a surface of great stress (often at or near the ground surface) and is propagated toward the interior of the medium where the tension decreases and ultimately passes in compression.

Depth and spacing give information about the mechanical conditions under which the cracks formed.

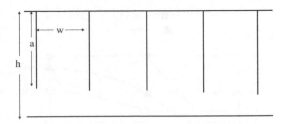

Figure 3. Fracture spacing w and length (height) a in a bed of thickness h.

Scale is partly provided by Irwin's internal length l_0 and partly by an implicit geometric parameter.

$$l_0 = \frac{EG_c}{\sigma_c^2} \tag{1}$$

2.5 Single layer configuration: jointing by bending

Kemeny & Cook (1985) present a similar model to that of Lachenbruch, except that geometry is limited according to a finite layer thickness h assumption. Boundary conditions consistent with the process of *tectonic uplift* at some depth below the surface are assumed, with *bending* along *gravity* as loading parameters. Initiation length of cracks is characterised as the actual depth reached at the stable equilibrium point, whereas average spacing is determined by a principle of least amount of energy.

2.6 Aspect ratios for single layer jointing

As mentioned before, average spacing w between parallel growing cracks and length a (Fig. 3) are parameters of great concern.

Reduced toughness κ allows to condense in an appropriate way what is rock relevant and what is connected to geometry (h is here the layer thickness as the relevant scaling parameter). We write (Putot et al. 2001):

$$\kappa = \left(\frac{l_0}{h}\right)^{1/2} \tag{2}$$

κ parameter collects synthetically all information needed to express results in terms of:

w average spacing between fractures
a fracture length (or height)
h layer thickness ; might be considered as the independent leading parameter.

Following equivalencies between geometric mean values have been obtained (Putot et al. 2004):

$$\frac{w}{h} \approx \kappa^3 \quad \frac{w}{a} \approx \kappa^2 \quad \frac{a}{h} \approx \kappa \tag{3}$$

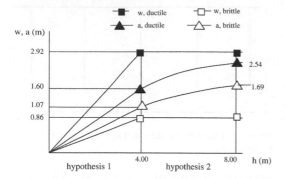

Figure 4. Results of the numerical model, spacing w and fracture length a as a function of layer thickness h for conditions of low hydrostatic pressure (predominant bending).
$K_{IC} = 1 MPa\sqrt{m}$ $\sigma_c = 3.3 MPa$ ductile rock
$K_{IC} = 1 MPa\sqrt{m}$ $\sigma_c = 5 MPa$ brittle rock.

2.7 Some trends

Describing variations of w and a with layer thickness h requires statement of hypotheses on our mixed failure criteria:

1st hypothesis: we speculate (usual formulation, Ladeira & Price, 1981) that toughness EG_c is constant with layer thickness h but σ_c is decreasing with h according to $\sigma_c \approx h^{-1/2}$

For these conditions l_0 is proportional to h and κ is constant when h is varied. Geometrical characteristics are proportional to thickness of the layer h according to:

$$w \approx a \approx h \qquad (4)$$

2nd hypothesis: toughness is still constant with h but $\sigma_c \approx h^{-1/6}$

For these conditions, l_0 is proportional to $h^{1/3}$ such that κ varies according to $h^{-1/3}$; w is constant

$$(w \approx h^0) \text{ and } a \approx h^{2/3} \qquad (5)$$

This last hypothesis is in better accordance with observations when dealing with large layer thicknesses (also Ladeira and Price, 1981). Spacing is no longer depending on bed thickness scaling.

The example presented in Figure 4 shows results of the numerical model described in (Putot et al. 2001) for superficial geologic settings and rock types (predominantly bending).

Carpinteri et al. (2005) provide a statistical model to the size effect on grained materials tensile strength and fracture energy very close to our second assumption.

3 MULTI LAYER JOINTING AND FRACTURE SWARMS

3.1 Some evidence for multi layer jointing

Models have attempted to assess the processes and parameters important in determining joint spacing. For most of them, as was pointed out in the last section, the computed joint spacing depends on the thickness of the jointed layer, on a contrast in physical properties between the jointed layer and adjacent beds and on layer-parallel extensional strain. Actual joint spacing distributions seem to differ somewhat from these models: consistent bed-thickness to fracture-spacing relationship can be demonstrated for evenly bedded lithologies, but it deteriorates rapidly as bedding thickness increases.

Narr & Suppe (1991) consider that spacing of joints should be referred to "mechanical layer" thicknesses rather than to individual "bed" thickness, to emphasize the fact that joints are confined to mechanically determined layers, which may nevertheless contain significant bedding planes and sedimentary laminations that are cross-cut by the joints.

Strata are either "brittle", meaning they sustain a well-developed joint system, or else are relatively soft and have poorly-developed joint systems. The "brittle" rocks are harder and more cohesive, the softer are principally mudstone and shale.

All these considerations seem to point out that some kind of interbedding (rock properties, shale thickness) is more prone to arrest fractures, the reason why it is suggested to investigate the bed/interbed coupling.

3.2 Vertical initiation and propagation; horizontal propagation

The most fundamental features of a joint surface include an initiation point and the associated hackle, which have been termed a *plumose structure*. Typically, initiation points are almost always located at bedding interfaces. *Hackle*, the slightly curved topographic feature formed parallel to the local propagation direction and perpendicular to joint front, is remarkably well developed on fine grained rocks. Growth kinematics of successive joint fronts can be analyzed in order to elucidate history of jointing. Joints initiated at the bottom of the layer propagate vertically upward. Reaching the upper interface with shale, further vertical propagation seem inhibited. Propagation then proceed laterally in both directions until conditions for joint propagation are no longer met (Fig. 5).

Composite joint features in several siltstone contiguous layers have been analyzed: each individual layer has its own plumose structure, indicating apparent independent sequential initiation and propagation of each layer.

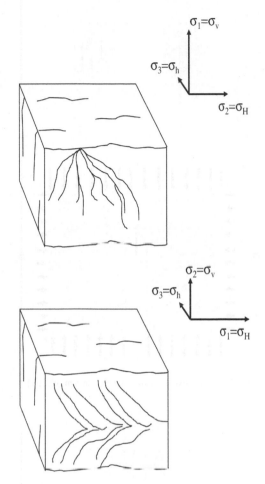

$\sigma_1=\sigma_v$

$\sigma_3=\sigma_h$

$\sigma_2=\sigma_H$

$\sigma_2=\sigma_v$

$\sigma_3=\sigma_h$

$\sigma_1=\sigma_H$

Figure 5. Schematic representation of plumose structure for vertical and horizontal propagation: non tectonic and tectonic case.

3.3 Fracture swarms (clusters)

Full development of joints, for which lengths are much greater than spacing, have been studied for fractures confined to *one single bed*. Brittle coating techniques seem then appropriate for an analogue experimental model. Nevertheless, we notice that spacing is regular and with order of magnitude close to thickness of the coating.

The geometry of fracture swarms or clusters appear very different: length of fractures is well developed in *two directions*, the order of magnitude being several hundred meters length, not only parallel to bedding but also perpendicular, exhibiting a large degree of "persistency" across layers.

No adequate explanation has been found yet for fracture swarms which are nevertheless of great importance and interest for petroleum geologists, particularly for tight gas reservoirs.

3.4 Olson's picture of clustering

A step towards understanding the process has been numerically achieved by Olson (2004). He considers the dynamics of pattern development for large populations of *layer-confined fractures*. His position is very close to that of brittle coating techniques. He assumes that the initial starter flaws, no matter how short (10 cm in practice), extend across the full thickness of the layer (8m for the presented example), neglecting the initiation and initial vertical propagation. The simulation results demonstrate the apparent important role of the sub-critical crack growth characteristic (expressed by a power-law relationship) in the control of the joint spacing to bed thickness ratio. The idea is that a propagating joint causes the stresses ahead of the tip of a blunted crack to be more tensile, promoting the growth of nearby fractures in a manner similar to the process zone often observed around igneous dykes, where the density of dykes-parallel joints is found to be very high close to the dyke.

A similar analysis is developed by the authors (Picard 2005) when considering the cluster as a large blunted crack (see section 4.4) and discriminating between propagation of the fracture swarm as a whole or involving initiation of a crack as a prerequisite first step (see section 4.4 and Figure 8).

4 FURTHER INVESTIGATION TRACKS

4.1 The inheritance of quasi-brittle materials formulations

Quasi-brittle behaviour seems to be the most relevant frame for studying propagation of large size fractures at the reservoir's scale. Failure of brittle materials under triaxial compression has been an important issue for many years.

We proposed an approach which allows to predict crack initiation at V-notches, interfaces with contrasting mechanical properties and various geometrical non singular concentrators such as circular cavities (Picard et al. 2006, Leguillon et al. in press).

The analysis is based on a two-scale asymptotic approach in plane strain elasticity. Far and near representations of the stress and displacement fields are matched in accordance to remote loading and geometry of the microstructure. A mixed criterion involving an energy balance and a maximal stress allows to determine the crack jump at initiation. This length depends on material properties and on the dimension of the local geometry around the stress concentration point (Leguillon 1993).

Interpreting the geologic history from vertical outcrop patterns of fractures requires consideration of tectonics and stratigraphy, which can both produce variations in the fracture pattern.

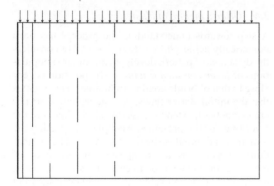

Figure 6. Final state of the fracture swarm.

Modelling fracture mechanisms and possible arrests at interfaces follows a similar approach.

4.2 A fracture swarm model, considered as representative of a low confinement effective situation

A brief presentation of the model developed in Putot et al. (2001) is made in the following. It allows some features of the complex phenomenon of failure in compression in a low confinement situation (Kendall 1978) to be addressed. It is similar to situations discussed by Vardoulakis (1986) considering spalling mechanisms in the vicinity of a stress-free surface. The use of such a model is justified by the very low effective horizontal stress demonstrated to be realis tic for generation of regional fractures (see section 5).

A figurative view of the idealised fracture swarm is presented Figure 6 ; the axis of symmetry is on the left. The important facts are:

When considering fractures close to the assumed free surface (confining parameter equals zero) spacing w reaches a well defined limit depending on fracture properties and aspect ratio a/w remains unspecified: any ratio is possible and long fractures are statistically the most likely.

The dual proposition concerns the fracture swarm's margin, where a/w ratio is reaching a specified limit whereas spacing w remains unspecified (no bedding in this model as a scale parameter).

4.3 Joint propagation across interbedding

Three types of fracture behaviour have been investigated: fracture transection through bed contacts, termination at interface and step-over. Conclusions have been inferred, considering strength parameters at the interface coupled with geometrical and loading parameters (Picard et al. 2004a, b). Size effects have also been investigated (Leguillon et al. in press)

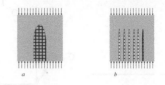

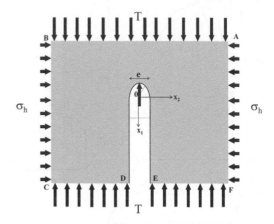

Figure 7. Figurative view of a fracture swarm (cluster).

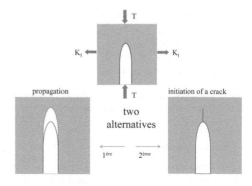

Figure 8. Stationary fracture swarm propagation or initiation of a crack at the notch tip.

4.4 Modelling a cluster as a process zone

The formation of closely spaced fractures involves the process of fracture set initiation, propagation and arrest. The only loading likely to lead to a reproduction in exactly the same shape – as a steady state process – seems to be a compressive stress parallel to the fracture planes (Figs 7, 8). Actually, whatever the density of clustering, such a loading is able to make cracks ignore each other. More exactly, some interaction may exist, especially when spacing is reduced, but propagation does not prevent the compression to operate and maintain propagation. Moreover, it is necessary to consider realistic fractures, with a finite thickness, (Picard et al. 2006), in order to take into account the propagation of clusters in the plane of fracturing (Picard 2005).

5 DISCUSSION

Systematic fracture sets have been studied for over a century but divergent opinions and sometimes inconsistent ideas cloud the issue. This section is an attempt to synthesize the most credible mechanisms.

5.1 Regional fractures, fold related systems and others

The first idea is the necessity to separate regional fracture systems – the only ones with great economic significance – from fold related systems. Data from both categories have been mixed and some confusion results in interpretations (Lorenz et al. 1991).

It seems realistic that regional fractures sets formed during relatively early deformation, prior to the accumulation of stress that produced flexure. Frequently, horizontal compression apparently fractured the flat-lying strata prior to their incorporation into the fold and thrust belts.

Similarly, cross fracturing systems occur later by stress release during erosion and uplift and are characteristic of younger effects.

5.2 Load parallel extension fracture

The load-parallel extension fracture concept is supported by fracture data and morphology; it has been overshadowed by the concept of natural hydraulic fracturing which is entirely inappropriate in this context. It requires a sustained pressure differential between the fluid within a propagating fracture and the fluid in the pores of the rock, not compatible with usual permeability properties.

The relevant mechanism includes rather the formation, at depth, of regional fractures, as the result of tectonic compression in combination with high pore pressure. Thrusting creates anisotropic horizontal stresses and ultimately fractures in relatively undisturbed strata adjacent to the thrust belt.

5.3 Triaxial stress state, abnormal pore pressure and tectonics in reservoir

Fractures propagate preferentially in the horizontal direction if the maximum compressive total stress is horizontal, and vertically if the maximum stress is vertical (Fig. 5). The maximum compressive stress is commonly the vertical overburden stress in undeformed sedimentary adjacent basins. It characterizes the initiation of the tectonic process. Horizontal stress anisotropy requires the horizontal tectonic movement of rocks, the maximum compressive stress being horizontal within thrust belts.

The importance of pore pressure in fracturing is well known, but the mechanical effects of pore pressure are misunderstood and overextended. It is generally implicitly assumed that the total minimum horizontal stress remains constant during increase in pore pressure ; however, it is not the case. As pore pressure is increased, at the same time, an increasing compressive component is added to the total stress. As a consequence, tensile fracturing resulting from vanishing minimum horizontal effective stress is totally excluded, as will be demonstrated in the next section.

5.4 Proof of unlikely limit case

Assuming classical effective stress law:

$$\sigma_{min} = f(v, E)\sigma_{max} \qquad (6)$$

where v and E are Poisson ratio and Young modulus, respectively, f is a function of rock properties which describes how the overburden is transmitted to the horizontal, and Biot relationship between effective stresses and pore pressure, where α is near unity:

$$\sigma' = \sigma - \alpha p \qquad (7)$$

the hypothetic limit case:

$$\sigma_{min} = \sigma_{max} = 0 \Leftrightarrow p = \frac{\sigma_v}{\alpha} = \frac{\sigma_h}{\alpha} \qquad (8)$$

means that the effective horizontal stresses will not reach zero until the pore pressure reaches the overburden value. Note that at failure $\sigma_{max}/\sigma_{min}$ becomes large, although all effective stresses decrease ; the effective stress ratio becoming large is a prerequisite to fracture growth.

Should the limit be attained, this would mean that the rock would be unconfined in effective stress (all compressive stresses equal, with systematic orientation jointing unlikely).

Convergence of pore pressure and overburden stress is nevertheless confirmed in field data, showing that pressure commonly approaches the local overburden stress, but never exceeds it. This proves the unlikely occurrence of zero horizontal effective horizontal stress and tensile fracturing.

5.5 Brittle behaviour

Because the strata are almost unconfined (in effective stresses) the rocks yield in load parallel extension fractures at stresses well below those necessary to create folds, faults or shear fractures in the laboratory ; failure could occur with differential stresses of only few MPa under geologic conditions ; the brittleness of rocks at reservoir depths is enhanced by high pore pressure.

5.6 Low strain rates, presumably sub-critical

The fractures propagate at quasi-static rates, compatible with geologic phenomena (Atkinson 1984), and probably at rates in accordance with thrust belt tectonic. Some fractographic features such as closely spaced arrest lines are clearly indicative of slow and stable fracture propagation, as opposed to forked terminations indicative of instability.

5.7 Subsequent flexure

Existing regional fractures may be reactivated during subsequent bending of strata, in which case the fracture system may become more permeable. It is not always clear if inherited fractures on structures are or not related to regional stresses that contributed to bending but bending is certainly not the main factor for fracturing.

6 CONCLUSIONS

The genesis of fractures in reservoirs is a complex problem raising questions about loading and scaling conditions in space and time.

Pore pressure exerts a major influence on the effective compressive stresses taken by rocks when fracturing. Nevertheless, balance between pore pressure and compressive stresses is never reached.

Quasi brittle conditions dominate with low effective confinement and relatively low effective failure stresses.

Modeling techniques well suited to structural heterogeneities representation, such as matched asymptotics between near and far fields are useful.

Understanding fracture swarms need further investigation. Several tracks have been proposed. We think that low rate processes at the crack tip combined with extended fracture mechanics are good candidates for a satisfactory analysis, but full validation requires in-situ evidences.

REFERENCES

Atkinson, B.K. 1984. Subcritical crack growth in geological materials, J. Geophys. Res., 89: 4077–4114.

Bessinger, B., Cook, N.G.W., Myer L., Nakagawa S., Nihei K., Benito P. & Suarez-Rivera R. 2003 The role of compressive stresses in jointing on Vancouver Island, British Columbia Journal of Structural Geology., 25: 983–1000.

Carpinteri, A., Cornetti, P. & Puzzi S. 2005. Size effects in grained materials: extreme value theory approach, ICF11, 11th International Conference on Fracture, Turin, march 20–25, 2005.

Cooke, M.L. & Underwood, C.A. 2001. Fracture termination and step-over at bedding interfaces due to frictional slip and interface opening, Journal of Structural Geology, 23: 223–238.

Engelder, T. 1985. Loading paths to joint propagation during a tectonic cycle: an example from the Appalachian plateau, USA. Journal of Structural Geology, 7: 459–476.

Fonta, O., Al-Ajmi H., Verma N.K., Matar S., Divry V. & Al-Qallaf H. 2005. The fracture characterization and fracture modeling of a tight-carbonate reservoir. SPE 93557 Middle East Oil and Gas Show and Conference, Bahrain.

Kemeny, J. & Cook, N.G.W. 1985. Formation and stability of steeply dipping joint sets, 26th US Symposium on Rock Mechanics/Rapid City.

Kendall, K. 1978. Complexities of compression failure, Proc. R. Soc. Lond. A., 361: 245–263.

Lachenbruch, A.H. 1961. Depth and spacing of tension cracks, Journal of Geophysical Research, 66(12): 4273–4292.

Ladeira, F.L. & Price, N.J. 1981. Relationship between fracture spacing and bed thickness, Journal of Structural Geology, 3(2): 179–183.

Leguillon, D. 1993. Asymptotic and numerical analysis of a crack branching in non-isotropic materials, Eur. J. Mech. A/Solids, 12(1): 33–51.

Leguillon, D., Quesada D., Putot C. & Martin E. in press. Size effects for crack initiation at blunt notches or cavities, to appear in Engineering Fracture Mechanics.

Lorenz, J.C., Teufel L.W. & Warpinski N.R. 1991. Regional fractures: A mechanism for the formation of regional fractures at depth in flat-lying reservoirs, The American association of Petroleum Geologists Bulletin, 75(11): 1714–1737.

Narr, W. & Suppe, J. 1991. Joint spacing in sedimentary rocks, Journal of Structural Geology, 13(9): 1037–1048.

Nur, A. 1982. The origin of tensile fracture lineaments, Journal of Structural Geology, 4(1): 31–40.

Olson, J.E. 2004. Predicting fracture swarms – the influence of subcritical crack growth and the crack-tip process zone on joint spacing in rock. From Cosgrove, J.W. & Engelder, T., (eds) The Initiation, Propagation, and Arrest of Joints and Other Fractures. Geological Society, London, Special Publications, 231.

Picard, D., Putot, C. & Leguillon, D. 2004a Z-99 A model for joint propagation across layer interfaces, EAGE 66th Conference & exhibition, Paris, 7–10 juin 2004

Picard, D., Putot, C. & Leguillon, D. 2004b A020 Joint propagation in relatively undeformed bedded sedimentary rock through compression failure mechanisms, ECMOR IX, Cannes 30 août- 2 septembre 2004.

Picard, D., 2005. Modèle de représentation mécanique de la formation des fractures naturelles d'un réservoir pétrolier, Thèse de doctorat, Université Pierre et Marie Curie (Paris VI), 22 septembre 2005.

Picard, D., Leguillon, D. & Putot, C. 2006. A method to estimate the influence of the notch-root radius on the fracture toughness measurement of ceramics, *Journal of the European Ceramic Society*, 26: 1421–1427.

Putot, C., Chastanet, J., Cacas M.C. & Daniel, J.M. 2001. Fracturation naturelle d'un massif rocheux: diaclase et couloir de fracturation *Oil and gas Science and Technology Rev IFP*, 56(5): 431–439.

Putot, C., Picard, D., Leguillon D. & Daniel, J.M. 2004. Joint propagation in bedded sedimentary rock through compression failure mechanisms, *Euroconference on Rock Physics and Rock Mechanics (Scaling laws in space and time)*. Potsdam, Germany.

Rives, T. 1992. Mécanismes de formation des diaclases dans les roches sédimentaires: Approche expérimentale et comparaison avec quelques exemples naturels, *Thèse de IIIème cycle,* Univ. Montpellier II.

Vardoulakis, I. & Mulhaus, H. B. 1986. Technical note: local rock surface instabilities, *Int. J. Mech. Min. Sci. & Geomech. Abstr.* , 23(5): 379–383.

Wu, H. & Pollard, D. D. 1995. An experimental study of the relationship between joint spacing and layer thickness, *Journal of Structural Geology*, 17: 887–905.

Fracture Mechanics of Concrete and Concrete Structures – High-Performance Concrete,
Brick-Masonry and Environmental Aspects – Carpinteri, et al. (eds)
© 2007 Taylor & Francis Group, London, ISBN 978-0-415-44617-4

Sub-size fracture testing of FY sea ice

C.A. Totman, O.E. Uzorka & J.P. Dempsey
Clarkson University, Potsdam, USA

D.M. Cole
US Army ERDC-CRREL, Hanover, USA

ABSTRACT: In the Fall of 2004, a total of 14 sea ice beams were harvested off the coast of Antarctica in McMurdo Sound to be used for lab-scale ice fracture testing at Clarkson University. A sub-size test program was initiated, with a view to developing a lab-scale ice fracture test methodology for first-year sea ice and to compare the results with large-scale in-situ experiments. The geometry chosen for the lab-scale experiments was the single-edge-notched, self-weight compensated beam (SENB-SWC) loaded in the three-point bend configuration (as per the Level II method described in the ACI Committee 446 report on "Fracture Toughness Testing of Concrete"). Closed loop, constant CMOD rate control was used to achieve stable crack growth during the experiment. Although it was difficult to achieve stable crack growth in these experiments, recent success suggests that the sub-size testing program under way will be successful.

1 INTRODUCTION

1.1 *Background*

The need for an accurate and reliable testing methodology to measure and characterize the fracture properties of ice is of great importance, but difficult to realize considering the range of experiments performed and the variability found in the results of these experiments. In-situ testing, though extremely important and necessary, is laborious, difficult and expensive. It is therefore desirable to develop an acceptable lab-scale test method so that a more systematic approach can be taken to acquire the fracture properties of ice. This information can then be used to more accurately develop and validate ice fracture models.

A rigorous instructional laboratory testing component was added to a fracture class taught at Clarkson University, Totman & Dempsey (2003), based on the ACI Committee 446 report on "Fracture Toughness Testing of Concrete." The results from the tests conducted by the students were very encouraging and it seemed a logical step to apply the test method to the fracture testing of ice. The Level II procedure described in the ACI 446 report was selected for the fracture testing of ice to establish its applicability. This procedure uses a single-edge-notched beam that is self-weight compensated (SENB-SWC) in the three-point bend configuration. Closed loop, constant rate crack mouth opening displacement (CMOD) control is used to load the specimen. By using CMOD control,

the rate of crack opening is constant throughout the test and thus eliminates the variation in CMOD opening rate inherent with other loading methods. This is a very important aspect since the effect of loading rate is a critical parameter in the testing of ice.

Several initial experiments were performed using freshwater ice beams grown and harvested from an ice sheet in a basin situated in a cold room at Clarkson University. The results of the initial experiments showed that the test method was indeed suitable for ice. In the fall of 2004, a total of 14 first-year sea ice beams harvested off the coast of Antarctica in McMurdo Sound were shipped back to Clarkson for testing.

This paper assumes that the fracture of sea ice can be described by a cohesive or fictitious crack model (Hillerborg 1985, Mulmule & Dempsey 1997, 1999). The Level II method in the ACI 446 Report (2002) is followed, although the cohesive stress versus crack opening curve does not have the same characteristics as for concrete (Dempsey et al. 1999).

1.2 *Experimental setup*

The Level II ACI 446 method was used as a general guideline for the lab scale sea ice fracture experiments, however, some modifications were made based on the first few experiments. The general arrangement is shown in Figure 1. The beams were cut oversize from the ice sheet in the field using a chainsaw and further cut to the proper size using a band saw. The typical size of the beams was 150 mm × 150 mm × 950 mm. The

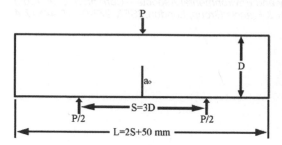

Figure 1.   Three-point-bend SENB-SWC.

loading span used for the experiments was 450 mm. This results in a span-to-depth ratio (S/D) of 3. The overall length of the beam is 50 mm longer than twice the span to ensure the weight of the beam does not contribute to loading. The notch in the beam is centrally located and cut with a radial arm saw such that the initial notch length (a_o) is ½ the beam depth. After cutting the notch, it is then sharpened by using the corner of a razor-type tool to initiate microcracks at the notch tip (DeFranco et al. 1991, Wei et al. 1991).

A total of 13 LVDTs and one load cell were used to measure displacements and load during the experiment. Figures 2 and 3 show the arrangement of the transducers attached to the sea ice beam. Both sides of the beam have an identical arrangement of instruments. Two ranges of LVDTs were attached to the bottom of the beam and both are used for CMOD control during the test. This was found to be necessary since the larger range LVDT did not provide suitable resolution for stable crack growth just before and after the peak load. A shorter range LVDT was used for control until the peak load occurs, at which time control is switched electronically to the longer range LVDT for the remainder of the test.

The four load point displacement (LPD) gages, two on each side of the beam, also have a coarse and fine range so that small events around the peak load region do not go unnoticed. The LPD gages are mounted to the specimen by means of a reference frame that is supported by the beam directly above the lower reaction supports. This ensures that any deformation at the supports is not measured by the LPD gages. One additional LPD gage is mounted to the frame of the test machine to compare with the LPD gages mounted on the reference frame.

Six more LVDTs are used to monitor the crack opening displacement (COD) and displacements ahead of the pre-cut notch, so-called fictitious crack (F) gages. These gages provide a means to measure the crack propagation and cohesive zone during the test.

The bend fixture used in this experiment was designed and fabricated at Clarkson University and is very robust such that the compliance of the fixture is insignificant compared to the sea ice. The lower

Figure 2.   Detail of gage mounting locations.

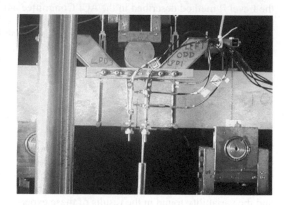

Figure 3.   SENB-SWC transducer configuration.

reaction support rollers are mounted to the bend fixture using roller bearings to minimize friction effects at the supports. Rigid bearing blocks are placed between the sea ice beam and the rollers so that crushing or melting at the supports do not inhibit the rotation of the rollers. One lower support and the upper fixture are free to rotate parallel to the long axis of the beam to reduce torsional effects due to any minor geometric misfits. Four springs were attached from the center point-loading fixture to the beam in order to maintain good contact throughout the experiment. The test machine used for the experiments was an Instron 8500 series, 250 kN capacity servohydraulic load frame that provides good response and rigidity.

1.3   Data acquisition and control

Sampling of each analog transducer channel during the tests was performed using 16-bit National Instruments data acquisition hardware. The scan rate on

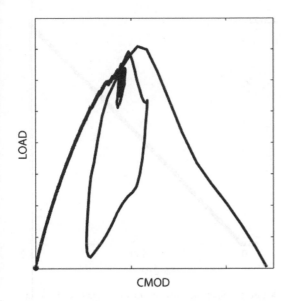

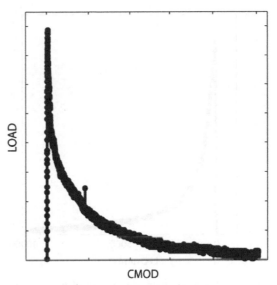

Figure 4. Result of tuning on a notched beam with intact ligament.

Figure 5. Result of tuning on a pre-tested beam.

each channel was 1000 samples per second, which was sufficient to fully characterize the peak load region as discussed in the results section.

Closed-loop feedback control of the CMOD was performed using an Instron 8800 series digital controller. The controller features a minimum control loop update rate of 250 microseconds, user selectable filters on all input channels and programmable control mode transfer. The high performance of this controller made it possible to obtain very tight control throughout the tests and maintain the constant CMOD rate. The bumpless, automatic control mode transfer capability of this controller made it possible to achieve a constant CMOD rate without interruptions for the duration of the experiment. As mentioned above, the key to success was using a high resolution, short range LVDT to control the CMOD rate until just after the peak load and then switching control to a longer range LVDT for the rest of the test.

Selecting appropriate tuning parameters for each feedback loop proved to be a difficult aspect of these experiments. Tuning was initially performed using the beam specimens with the pre-cut notch and the intact ligament. However, it was found, through trial and error that this resulted in a control system that overreacted to small fracture events. Since the compliance of the specimen decreases constantly throughout the test and changes drastically as the crack begins to propagate, it was observed that the control system would actually close the CMOD when crack propagation occurred and sometimes go into a resonance condition. Figure 4 shows a sample test result due to an under-damped control system.

To avoid the oscillations associated with an under-damped system, control loop tuning was performed on a beam that had been tested and the ligament fractured. To all intents and purposes, the self-weight compensation, springs and a small cohesive stress were holding the tuning specimen together. This resulted in tuning parameters that produced the proper level of damping for the specimen compliance. The outcome of this tuning method (shown in Fig. 5) was a stable, constant rate CMOD throughout the bend test. Unfortunately, overcoming this tuning issue consumed about half of the sea ice beams shipped back from Antarctica. Moreover, since stable fracture had been achieved, the displacements on subsequent tests exceeded initial estimates such that the displacement transducers were going out of range before the conclusion of the test. This led to a few more experiments with incomplete data and even fewer beams to work with. Longer range LVDTs were fitted to the setup in an effort to capture the complete unloading behavior.

2 RESULTS & CALCULATIONS

2.1 Discussion

The full analysis of this set of experiments is still in progress but some of the initial calculations and results will be presented here. All the experiments were run at a constant CMOD rate of 10 microns per second and isothermal temperature of $-2.5°C$. The temperature of each beam was monitored using thermocouples near the surface and at the core of the beam. Following each experiment, salinity and density measurements were

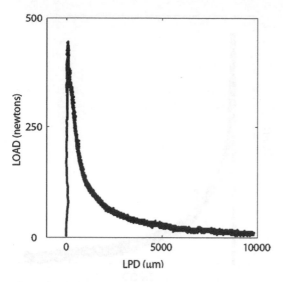

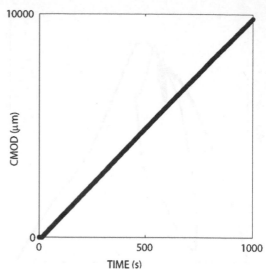

Figure 6. Specimen 4B measured work of fracture with residual load subtracted.

Figure 8. Specimen 4B record of constant CMOD rate control.

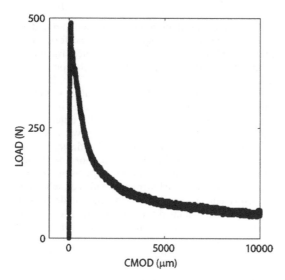

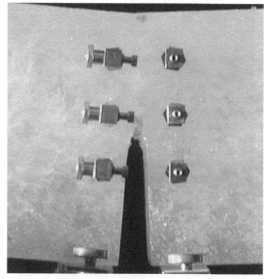

Figure 7. Specimen 4B record of load vs. CMOD showing residual load.

Figure 9. Fracture after 10 mm CMOD illustrating the extent of bridging.

taken. Thin sections of each specimen were obtained and photographed to determine grain size distribution and the extent and direction of crystallographic alignment. Additionally, post-test photographs were taken of thin sections of the fracture profile to conduct a fractal analysis of the fracture surfaces.

2.2 Work of fracture and fracture energy

The measured work of fracture (W_{Fm}) is calculated as the area under the load versus LPD record. Typical

results of the successful tests are shown in Figures 6–8. The area was computed using trapezoidal approximation directly from the data. As per the ACI Committee 446 report, the residual load at the end of the test is subtracted from the complete load record when determining the work of fracture. For these tests this value ranged from 60 to 90 Newtons at 10 mm CMOD.

Some specimens did not completely fracture, as illustrated in Figure 9 where a post-test view of the

Table 1. Summary of test results.

Specimen ID	W_F (J)	G_F (N/m)
4B	0.350	37
5B	0.379	39

beam is shown after 10 mm CMOD. Clearly, more experiments need to be conducted to capture the complete unloading.

Following directly from the ACI Committee 446 Report, once the W_{Fm} is determined from the load versus LPD record the total work of fracture, W_F, is estimated based on the behavior of the far tail region of the load-CMOD record. The corrected load, P_1, is found by subtracting the residual load at the end of the test from all points in the measure load record. The quantity X is computed from the CMOD record, according to Equation 1, corresponding to all points in the post peak P_1 record that are less than or equal to 5% of the corrected peak load. The far tail constant, A, is then found by least squares fitting of the quadratic equation in Equation 2. The far tail constant is then used in Equation 3 to compute the total work of fracture. Finally, the fracture energy, G_F, is found from Equation 4. Work of fracture and fracture energy results for two of the most successful and complete experiments are summarized in Table 1.

It should be noted that the sea ice used in these experiments was very highly drained and may not be representative of the in-situ conditions. The cold storage facility at Clarkson was not in an operational state at the time the beams arrived from Antarctica and, as such, the ice was stored in the same cold rooms that the preparation and testing were conducted.

The cold rooms where kept at approximately -7^oC except when testing was conducted and the temperature adjusted to -2.5^oC. The storage conditions and time spent in storage resulted in specimens that were highly drained with low salinity and density values. Large continuous and discontinuous drainage channels were present throughout the sea ice beams.

Another factor affecting the results was the design of the reaction support rollers. As mention previously, they are mounted on roller bearings to minimize friction. However, this particular fixture was designed for use with concrete at room temperature and the bearings did not perform as well in the sub-freezing test conditions. A new fixture is currently being designed and fabricated at Clarkson specifically for conducting the ice fracture tests that will eliminate this issue.

2.3 Peak load region fracture behavior

From the test records shown in Figures 6 & 7, the fracture behavior of the first year sea ice at a relatively

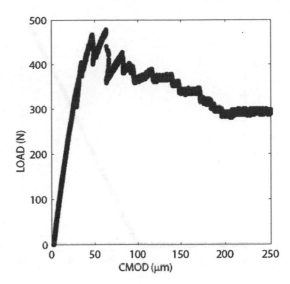

Figure 10. Short range CMOD gage record to illustrate the fracture behavior in the fracture behaviour in the peak load region.

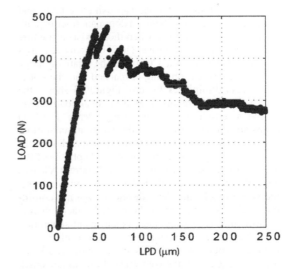

Figure 11. Record of short range LPD to verify the fracture behavior.

warm temperature appears to exhibit a brittle behavior followed by an extensive region of diminishing cohesive stress. However, if the peak load region is shown in greater detail using the load and the short-range CMOD gage, Figure 10, the fracture events can be characterized by a series of stress releases. Figure 11 is the same experiment showing the load and the short range LPD and Figure 12 shows the short range CMOD gage versus time to verify that the fracture events were not due to deviations in CMOD control.

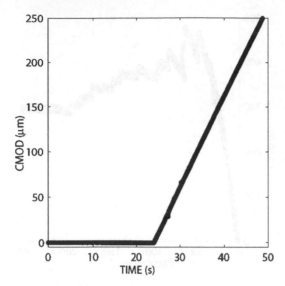

Figure 12. CMOD vs. time record to show constant rate control in the peak load region.

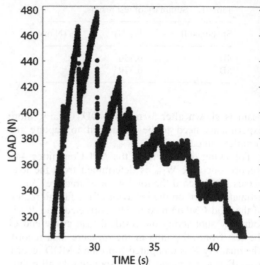

Figure 13. Load vs. time record in the peak load region.

It is well known that ice can exhibit a large and complex damage zone before a dominant fracture path develops. However, even when there exists a fracture path, it may not be exclusive. There is extensive bridging that occurs and the COD must be relatively large before the critical opening is reached and the cohesive stress diminishes to zero. Figure 13 shows the load versus time record around the peak load region to give a sense of the rate of energy release. The load drops are sudden with very little change in separation, which suggests only small, hairline cracks developed through the ligament in the more brittle phases of the ice microstructure. Moreover, since the stress releases are followed by a rise in the load before another release, the small cracks that have developed are not continuous. The coalescence of these small cracks occurs a short time later when there is sufficient separation such that a more dominant fracture path develops. Following the development of the initial, major crack, the process zone continues to grow and the dominant crack continues to advance in a stable manor.

$$X = \frac{1}{\left(w_M - w_{MA}\right)^2} - \frac{1}{\left(w_{MR} - w_{MA}\right)^2} \quad (1)$$

Where w_M = recorded CMOD, w_{MA} = CMOD at zero P_1 for the rising part of the curve, w_{MR} = CMOD at the end of test.

$$P_1 = X(A + KX) \quad (2)$$

Where P_1 = corrected load, A = far tail constant, K = unused coefficient.

$$W_F = W_{Fm} + \frac{2A}{\delta_R - \delta_A}\left(\frac{S}{4D}\right)^2 \quad (3)$$

Where W_F = total work of fracture, W_{Fm} = measured work of fracture, δ_R = LPD at end of test, δ_A = LPD at zero P_1 for the rising part of the curve, S = reaction support span, D = beam depth.

$$G_F = \frac{W_F}{Bb} \quad (4)$$

Where G_F = fracture energy, W_F = total work of fracture, B = beam thickness, b = uncracked ligament length.

Further information concerning crack propagation can be gathered from the load versus displacement records of the F gages located ahead of the precut notch. Using simple beam theory and taking the neutral axis to be half the length of the uncracked ligament, the first gage ahead of the notch, F1, is positioned below the neutral axis in the tensile region. The next gage, on the same side of the beam, F3, is located above the initial neutral axis in the compressive region. Figures 14 and 15 show the record of the F1 and F3 gages, respectively.

Observing the record of the F1 gage, located in the tensile region of the ligament, it exhibits similar behavior as the CMOD, as would be expected. However, the F3 gage, in the initial compressive region, shows a quite different response. From zero to the peak load, the separation closed a few microns confirming this

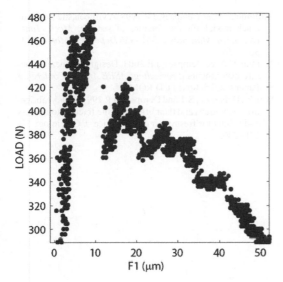

Figure 14. F1 gage located below the initial neutral axis in the tensile region.

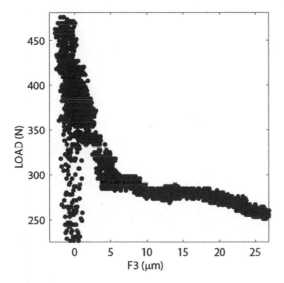

Figure 15. F3 gage located above the pre-tested beam's neutral axis in the initial compressive region of the ligament.

region of the ligament was in compression until the peak load is reached and the crack begins to advance. Moreover the "spiky" behavior is absent and there is simply one dominant separation that occurs before the fracture activity transitions to a cohesive dominated failure.

From these results it can be concluded that numerous small cracks developed at different locations in the fracture process zone before coalescing into a dominant failure path. At this time, the main crack was then able to advance ahead of the F3 gage location transitioning the stress in this region from compressive to tensile.

3 CONCLUSIONS

The outcome of the initial round of fracture testing of sea ice beams using the Level II method described in the ACI Committee 446 report has yielded successful results warranting further testing and detailed analysis. The SENB-SWC geometry along with CMOD control provides an effective means to characterize the fracture behavior and extract relevant fracture properties of the sea ice.

A more thorough and controlled round of testing is scheduled to begin in the spring of 2007 when 40 first-year sea ice beams will arrive. These beams were harvested off the coast of Antarctica in McMurdo Sound during a round of in-situ testing conducted during the fall of 2006.

Since the first series of sea ice beam tests, the cold storage room has been repaired so that the beams can be stored at a constant $-20^{o}C$ until testing. Fabrication of a custom bend fixture designed specifically for ice testing is underway so that unwanted external influences on the results will be eliminated. The first series of tests reported here provided numerous lessons and insights that will be applied to the incoming test program so that the testing will be more efficient and productive.

Ongoing application of the cohesive crack model to the in-situ Antarctic sea ice tests is able to track the stable growth of the crack and back-calculate associated changes to the stress separation curve, the fracture energy and the process zone size. The fracture energy determined from the in-situ tests is an increasing function with extent of stable crack growth. The values determined to date are approximately a third of the values reported in Table 1. Unfortunately, the in-situ ice and the ice finally tested at Clarkson University differed considerably in composition, salinity and porosity. With better control of the temperature history, and the custom bend fixture, it will be interesting to see if the values become more compatible.

ACKNOWLEDGEMENT

The financial support of the US National Science Foundation through the Antarctic Oceans and Climate Systems Program, Award # 0338226 and the Division of Undergraduate Education, Award # 0311075 is gratefully acknowledged.

REFERENCES

ACI Committee 446 Report 2002. *Fracture Toughness Testing of Concrete*.

DeFranco, S.J., Wei, Y. & Dempsey, J.P. 1991 Notch-acuity effects on the fracture toughness of saline ice. *Annals of Glaciology* Vol. 15: 230–235.

Dempsey, J.P., Adamson, R.M. & Mulmule, S.V. 1999. Scale effects on the in-situ tensile strength and fracture of ice. Part II: First-year sea ice at Resolute, N.W.T. *Int. J. Fract.* 95: 347–366.

Hillerborg, A. 1985. The theoretical basis of a method to determine the fracture energy G_F of concrete. Materials and Structures 18: 291–296.

Mulmule, S.V. & Dempsey, J.P. 1997. Stress-separation curves for saline ice using the fictitious crack model. *ASCE J. Eng. Mech.* 123: 870–877.

Mulmule, S.V. & Dempsey, J.P. 1998. A viscoelastic fictitious crack model for the fracture of sea ice. *Mech. Time-Dependent Materials* 1: 331–356 *Dependent Materials* 1: 331–356.

Totman, C.A. & Dempsey, J.P. 2003. Design of concrete fracture experiments. *Proceedings ASEE Annual Conference* Paper # 2215, 8p (in CD ROM).

Wei, Y., DeFranco, S.J. and Dempsey, J.P. 1991. Crack fabrication techniques and their effects on the fracture toughness and CTOD for freshwater columnar ice. J. Glaciology 37: 270–280.

Fracture Mechanics of Concrete and Concrete Structures – High-Performance Concrete,
Brick-Masonry and Environmental Aspects – Carpinteri, et al. (eds)
© 2007 Taylor & Francis Group, London, ISBN 978-0-415-44617-4

Experimental and theoretical investigation of fracture in glass matrix composites reinforced by alumina platelets

M. Kotoul, J. Pokluda & P. Šandera
Faculty of Mechanical Engineering, Brno University of Technology, Brno, Czech Republic

I. Dlouhý & Z. Chlup
Institute of Physics of Materials, Academy of the Czech Republic, Brno, Czech Republic

A. Boccaccini
Department of Materials, Imperial College London, London, United Kingdom

ABSTRACT: Experimental and theoretical investigation of fracture in the borosilicate glass/Al_2O_3 platelets composite was performed. This composite is a perspective structural material for many applications due to its low production expenses and satisfactory properties even at elevated temperatures. The fractographical analysis was employed to reveal vitality of toughening mechanism with increasing content of reinforcement. Possible synergy between crack deflection and other toughening mechanisms was examined.

1 INTRODUCTION

Glass is known as a relatively cheap and easy to fabricate material with satisfactory properties. However, it is a material with very brittle behaviour (a typical fracture toughness value is around of $0.6\,MPa.m^{1/2}$). The low fracture toughness is a limiting factor for employing such material in design of loaded components. Therefore, an extensive research dedicated to the improvement of mechanical properties of inherently brittle materials including glass has taken place. There are many possible ways how to increase the fracture resistance. The possible synergy of more than one toughening mechanism is apparently advantageous. Particle/matrix interface decohesion and particle pull out, accompanied by deflection of crack trajectory provide the typical synergistic toughening effect.

A successful example of ceramic platelet reinforcement of glass is the borosilicate glass/Al_2O_3 platelets composite developed by Boccaccini et al. (2003). They demonstrated a better mechanical behaviour of the composite over that of the unreinforced glass matrix in terms of hardness, Young's modulus, fracture strength and fracture toughness. By means of a detailed experimental investigation, the mechanical properties enhancement was ascribed to three concurrent phenomena: the Young's modulus increment resulting from the platelets addition, the presence of a compressive residual stress in the glassy matrix, and the crack deflection mechanism.

The paper aims to analyze a relationship between reinforcement volume fraction as well as surface roughness and mechanical properties especially fracture toughness both experimentally and theoretically.

2 EXPERIMENTAL

The experimental glass ceramic composite was fabricated via powder technology and hot-pressing, as described in a previous study (Boccaccini & Trusty, 2003). Alumina platelets (TS100, Lonza-Werke, Waldshut-Tiengen, Germany) of hexagonal shape and with major axes between 5 and 25 μm and axial ratio of 0.2 were used. A commercially available borosilicate glass (DURAN, Shott Glass, Mainz, Germany) was selected for the composite matrix. Samples containing 0, 5, 10, 15 and 30 vol.% of platelets were considered in this study.

As was presented elsewhere (Boccaccini & Trusty, 2003, Tood et al., 1999) the composite microstructure exhibits a dense glass matrix where the platelets are distributed homogenously. The existence of a strong bond between the matrix and platelets was confirmed by transmission electron microscopy (Winn et al., 1997). The thermal expansion mismatch between matrix and reinforcement cause presence of internal residual stresses. The thermal expansion coefficient of the borosilicate glass matrix is much lower than that of the alumina platelets, which results in net tangential compressive and radial tensile stresses in the matrix upon cooling from the processing temperature. The measurement of these residual stresses was conducted by fluorescence spectroscopy technique, as reported in (Tood et al., 1999).

Fracture toughness values were obtained using the chevron notch technique. Test pieces of standard cross-section (3×4 mm) were cut from the round shaped plates of diameter 40 mm and thickness of 4 mm by precise diamond saw. The chevron notch with top angle of $90°$ was machined by ultra thin diamond blade into each test piece. A Zwick/Roell electromechanical machine was used for loading in three point bend test with a span of 20 mm. Crosshead speed of 0.1 mm/min was used for loading. The samples were tested at room temperature and at $500°$C. The elevated temperature has been selected just below the temperature of viscous flow of the glass matrix. The Maytec high temperature furnace was used to conduct tests at elevated temperatures. Load-deflection traces were recorded and the fracture toughness was calculated from the maximum load (F_{max}) and the corresponding minimum value of geometrical compliance function (Y_{min}^*) using the equation

$$K_{IC} = \frac{F_{max}}{B\sqrt{W}} \, Y_{min}^*, \qquad (1)$$

where B and W stand for the width and height of the specimen, respectively. The calculation of the geometrical compliance function was based on Bluhm's slice model (Bluhm, 1975). Reliability of this technique for composite materials was reported elsewhere (Boccaccini et al., 2003, Dlouhy & Boccaccini, 2001).

Scanning electron microscopy (SEM) was used for fractographic analyses of fracture surfaces of tested chevron notched specimens.

Fracture surface roughness was measured by profilometer MicroProf FRT using a chromatic aberration method for z-axis measurement. The FRT Mark III software was applied for analysis of measured fractured surfaces and 3D surface reconstructions.

A plot of fracture toughness values on the volume content of alumina platelets in borosilicate glass matrix is shown in Figure 1.

The combination of toughening mechanisms puts into effect during crack propagation which has an influence on fracture surface formation. Therefore the fracture surface characteristics indicate the employment of toughening mechanisms during different stages of fracture process. Figure 2 shows a dependence of relative surface roughness on alumina platelets content accompanied by fracture toughness data at room temperature. The surface roughness is linearly increasing with rising amount of alumina platelets in the borosilicate glass matrix up to approximately 15 vol.%. At higher reinforcement content

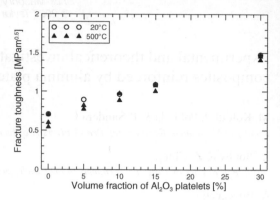

Figure 1. Dependence of fracture toughness on alumina platelets volume fraction in glass matrix at room and elevated temperature.

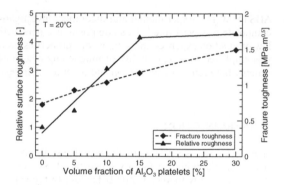

Figure 2. Dependence of relative surface roughness and fracture toughness on alumina platelets volume content in glass matrix at room temperature.

the roughness increase is slowing down. Figure 2 compares the change of surface roughness and the evolution of fracture toughness with platelets volume fraction.

The two main toughening mechanisms responsible for roughness (pull out and crack deflection) are weakened at the highest reinforcement volume fraction and therefore the increase of fracture toughness increase is lower in comparison to lower content of platelets. On the contrary, the higher content of alumina platelets, which are tougher than the glass matrix, is acting against the weakening of key toughening mechanisms. Typical examples of reconstructed fracture surface obtained from the profilometric measurement conducted on the fracture surfaces of chevron notch test pieces for both 0% and 30% of alumina platelets volume content in borosilicate glass matrix are shown in Figure 3. The corresponding scanning electron images are displayed in Figure 4. It is evident that the fracture surface roughness has been significantly increased

a)

b)

Figure 3. Reconstructed fracture surface for (a) 0% and (b) 30% volume fraction of alumina platelets in borosilicate matrix.

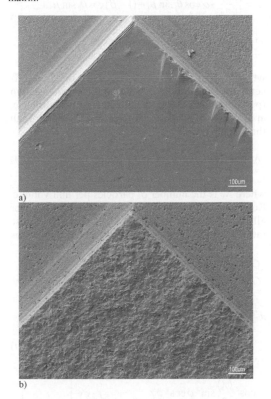

a)

b)

Figure 4. Fracture surface for a) 0% and b) 30% volume content of alumina platelets in borosilicate glass matrix (SEM).

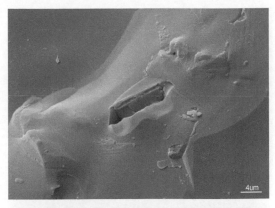

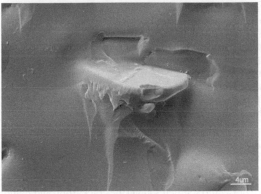

Figure 5. Example of toughening mechanisms evidence in the SEM image of fracture surface.

when reinforcement is incorporated into the borosilicate glass. The connection between surface roughness and reinforcement volume content was proved however the roughness will certainly depend on the shape and dimension of the platelets as well as on the bonding between platelets and matrix.

The fractographical analysis proved presence of several toughening mechanisms. The micrographs supplying the evidence of examples of crack deflection and particle pull out are shown in Figure 5.

At the highest volume fraction of alumina platelets in borosilicate glass matrix, not all particles interacting with the crack front (present at the fracture surface) contribute to toughening effect as shown in Figure 6. Platelets cluster are observed even though a desirable degree of homogenous particles distribution is reached.

3 THEORETICAL

The most comprehensive model for describing toughening by crack deflection has been developed by Faber & Evans (1983). This model uses a strain energy

Figure 6. Clustering of alumina platelets when high reinforcement content is present.

release rate approach where the ratio between the average strain energy release rate at the deflected crack front $\langle G \rangle$ and the strain energy release rate for the undeflected crack front G_{Im} gives the relative toughening. Crack advance is assumed to be governed by the strain energy release rate

$$G = \frac{1}{E}\left[k_1^2\left(1-\nu^2\right)+k_2^2\left(1-\nu^2\right)+k_3^2\left(1+\nu\right)\right], \qquad (2)$$

where k_1, k_2, and k_3 are the local stress intensities for the deflected segments along the crack front, and E and ν are the Young modulus and Poisson's ratio, respectively. The toughening increment G_c is predicted as

$$G_c = \frac{G_{Im}}{\langle G \rangle}G_{Imc}, \qquad (3)$$

where G_{Imc} stands for the critical energy release rate of the matrix. We have corrected some errors in the expression for the strain energy release rate derived by Faber & Evans, (1983) and obtained

$$\frac{G}{G_{Im}} = \cos^2\frac{\lambda}{2}\left(2\nu\sin^2\phi+\cos^2\frac{\lambda}{2}\right)^2\cos^4\phi+\cos^2\phi\times$$

$$\times\sin^2\frac{\lambda}{2}\cos^4\frac{\lambda}{2}+\frac{\cos^2\frac{\lambda}{2}\sin^2\phi\cos^2\phi}{1-\nu}\left(2\nu-\cos^2\frac{\lambda}{2}\right)^2, \qquad (4)$$

where λ is a tilt angle and ϕ is a twist angle. Observe that the expression in Equation 4 possesses the required limiting properties, i.e.

$$\lim_{\phi\to\pi/2}\frac{G}{G_{Im}}\to 0, \quad \lim_{\phi\to 0}\frac{G}{G_{Im}}\to\cos^4\frac{\theta}{2}. \qquad (5)$$

Faber and Evans considered cracks deflected by spheres, discs, and randomly oriented short rods (whiskers). In the case of disc shaped particles, there is necessary to describe the disc orientation with respect to the crack front and to adjacent discs. The effective tilt angle $\lambda \equiv \theta$ was introduced, and, by virtue of the particle geometry, the average tilt angle $\langle \lambda \rangle$ can be expressed as

$$\langle \lambda \rangle = \frac{\alpha/2\theta_1\sin\theta_1+(1-\beta)/2\theta_2\sin\theta_2}{\alpha\sin\theta_1+(1-\beta)\sin\theta_2}, \qquad (6)$$

where the angles θ_1 and θ_2 describe the tilt of neighbouring discs with respect to the plane xz occupied by a planar crack, α, $\beta \in \langle 0;1\rangle$ are relative locations at which he crack plane intercepts the discs. The twist angle ϕ of the crack between two adjacent discs is

$$\phi = \arctan\left[\frac{\alpha\sin\theta_1+(1-\beta)\sin\theta_2}{\Delta'}\right], \qquad (7)$$

where

$$\Delta' = \left\{\left[\frac{\Delta}{2r}-\alpha\cos\theta_1\sin\mu_1+(1-\beta)\cos\theta_2\sin\mu_2\right]^2+\right.$$

$$\left.+\left[\alpha\cos\theta_1\cos\mu_1+(1-\beta)\cos\theta_2\cos\mu_2\right]^2\right\}^{1/2}, \qquad (8)$$

where μ_1 and μ_2 are angles by which discs are offset with respect to the direction of crack propagation (parallel to the x-axis), Δ is the interparticle distance, approximated by the point-to-point spacing through the volume

$$\langle \Delta \rangle \sim N_v^{-1/3}, \qquad (9)$$

where N_v is the number of particles per unit volume. Replacing λ by $\langle \lambda \rangle$ in Equation 4 and integrating over all possible configurations, the strain energy release rate due to twist of the crack front can then be written as

$$\frac{\langle G \rangle^T}{G_{Im}} = \frac{4}{\pi^4}\int_{-\frac{\pi}{2}}^{\frac{\pi}{2}}\int_{-\frac{\pi}{2}}^{\frac{\pi}{2}}\int_0^1\int_0^1\int_{-\frac{\pi}{2}}^{0}\int_0^{\frac{\pi}{2}}d\theta_1\,d\theta_2\,d\alpha\,d\beta\,d\mu_1\,d\mu_2\times$$

$$\times\frac{\eta}{2}\left\{\cos^2\frac{\langle\lambda\rangle}{2}\left(2\nu\sin^2\phi+\cos^2\frac{\langle\lambda\rangle}{2}\right)^2\cos^4\phi+\right.$$

$$+\cos^2\phi\times\sin^2\frac{\langle\lambda\rangle}{2}\cos^4\frac{\langle\lambda\rangle}{2}+$$

$$\left.+\frac{\cos^2\frac{\langle\lambda\rangle}{2}\sin^2\phi\cos^2\phi}{1-\nu}\left(2\nu-\cos^2\frac{\langle\lambda\rangle}{2}\right)^2\right\}, \qquad (10)$$

1694

where already the modification by the amount of crack front subject to twist was introduced via the factor η

$$\eta = \frac{\dfrac{\Delta}{2r} - \alpha\cos\theta_1\sin\mu_1 + (1-\beta)\cos\theta_2\sin\mu_2}{\sqrt{\Delta'^2 + \left[\alpha\sin\theta_1 + (1-\beta)\sin\theta_2\right]^2}}. \quad (11)$$

η is the ratio of the undeflected to twisted crack front lengths.

For θ_1 and θ_2 of like-sign, the resultant tilted crack, occurring along one-half of the crack front has a driving force normalized with respect to the length of undeflected crack

$$\frac{\langle G\rangle^t}{G_{Im}} = \frac{4}{\pi^4}\int_{-\frac{\pi}{2}}^{\frac{\pi}{2}}\int_{-\frac{\pi}{2}}^{\frac{\pi}{2}}\int_0^1\int_0^1\int_{-\frac{\pi}{2}}^{\frac{\pi}{2}}\frac{\xi}{2}\cos^4\frac{\bar\lambda}{2}\,d\theta_1\,d\theta_2\,d\alpha\,d\beta\,d\mu_1\,d\mu_2,$$

$$(12)$$

where ξ is the ratio of the undeflected to tilted crack front lengths

$$\xi = \frac{\dfrac{\Delta}{2r} - \alpha\cos\theta_1\sin\mu_1 + (1-\beta)\cos\theta_2\sin\mu_2}{\sqrt{\Delta'^2 + \left[\alpha\sin\theta_1 - (1-\beta)\sin\theta_2\right]^2}} \quad (13)$$

and

$$\bar\lambda = \frac{1}{2}\left[\arctan\left(\frac{\tan\theta_1}{\cos\mu_1}\right) + \arctan\left(\frac{\tan\theta_2}{\cos\mu_2}\right)\right] \quad (14)$$

is the average tilt angle across the tilted plane.

The total strain energy release rate in the presence of discs is

$$\langle G\rangle = \langle G\rangle^t + \langle G\rangle^T. \quad (15)$$

The toughening increment derived from Equations 3 and 15 by numerical integration is plotted in Figure 7. Apparently, the crack deflection model fairly predicts the toughening increment up to the volume fraction of Al_2O_3 platelets about of 10 vol.%. However, for the volume fraction of 30 vol.% the deflection alone underestimates the experimental data by about of 50%.

Other toughening mechanism in this material could be considered: a contribution of residual stresses to toughening, an increase in fracture toughness due to the increase in Young's modulus resulting from the platelets additions (Table 2), and the crack trapping (Xu et al., (1998)). Note that no extensive crack bridging by Al_2O_3 platelets was observed.

One of several possible events may occur whenever the crack meets the particles. If the particle toughness exceeds that of the matrix, and the particles are strongly bonded to the surrounding material, then the crack is trapped by the particles. This process can significantly improve the strength of a brittle solid.

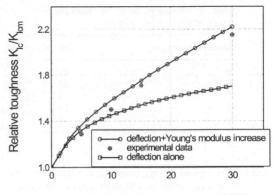

Figure 7. Relative toughness predictions base upon crack deflection model and upon the combination of the crack deflection+Young's modulus increase model.

Table 1. Thermomechanical properties of the composite constituents.

	E[GPa]	G[GPa]	ν	$\alpha\,[10^{-6}/^\circ C]$
Glass matrix	63	26	0.22	3.3
Al_2O_3 platelets	402	248	0.23	8.9

However, for trapping to be effective, the particle toughness must be at least three times that of the matrix. If the particles have a low toughness, the crack breaks through them, and the toughness of the composite is little better than that of the matrix. Observe that for the composite investigated, the ratio of the particle fracture toughness K_{Ic}^{part} and the matrix fracture toughness K_{Icm} is about of 3.5. More generally, Bower & Ortiz, 1993 suggested that for bridging particles to form the particle critical strain energy release rate G_{Ic}^{part} should exceed

$$\frac{G_{Ic}^{part}}{G_{Imc}} \geq \left(2,1 + 4,8\frac{R}{b}\right)^2, \quad (16)$$

where R denotes the particle radius and b is the particle spacing. If this is not the case, the crack cuts through the particles: the maximum possible toughness of the composite is then (Rose, 1975)

$$\frac{G_{Ic}^{eff}}{G_{Imc}} = 1 + 2\frac{R}{b}\left(\frac{G_{Ic}^{part}}{G_{Imc}} - 1\right). \quad (17)$$

If G_{Ic}^{part} is comparable to G_{Imc}, very little improvement in toughness is observed. This is the case of the composite investigated due to a high difference of Young's modulus, see Table 1.

In the case that $\alpha_f > \alpha_m$ (where α_f and α_m are the coefficients of thermal expansion of the second phase

Table 2. Young'modulus of borosilicate glass containing Al_2O_3 platelets.

Platelet content [%]	E[GPa]
0	63
5	65
10	70
15	79
30	102

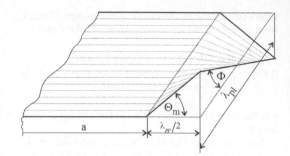

Figure 8. Scheme of the pyramidal element periodically approximating the tortuous crack front.

and matrix, respectively), compressive hoop stresses and tensile radial stresses will exist around the second phase, and the crack (growing perpendicular to trajectories of maximum tensile stress) will deflect around the particle.

It was predicted (Tood et al., 1999) that the existence of local compressive stresses between the particles would decrease the stress intensity factor and hence contribute to toughening. However, in the direction normal to the disc-shaped inclusion, there is no constraint and the residual thermal tractions are very low in both the matrix and the inclusion. Thus, the crack prefers to deflect along the platelet interfaces while the twisted crack front between platelets is shortened due to a decrease of twist angle leading to higher energy release rate. As a result, both terms in Equation 3, G_{Imc} and $\langle G \rangle$ respectively, effectively increase and the net toughening increment does not change.

The effect of Young's modulus is easily to introduce using Equation 3. Namely, it holds

$$\frac{K_{Ic}}{K_{Imc}} = \sqrt{\frac{E}{E_m}\frac{G_c}{G_{Imc}}} = \sqrt{\frac{E}{E_m}\frac{G_{Im}}{\langle G \rangle}}, \qquad (18)$$

where E is Young's modulus of the composite, see Table 2. The toughening prediction based upon Equation 18 is plotted in Figure 7, too. It is seen that the agreement with experimental data is very good.

4 FRACTURE TOUGHNESS ASSESSMENT BASED UPON SURFACE ROUGHNESS ANALYSIS

For a sufficiently precise assessment of the roughness-induced shielding effect (RIS) the following steps must be undertaken:

(i) Construction of a real-like model of the crack front based on a 3D determination of the surface roughness;
(ii) Calculation of local stress intensity factors k_1, k_2 and k_3 along the crack front;
(iii) Calculation of the effective stress intensity factor K_{eff}.

The first step can be achieved by means of a 3D reconstruction of fracture morphology. This can be carried out by means of the MicroProf FRT based on the optical chromatography. The second problem can be solved using a numerical program system FRANC3D based on the boundary element method. The third step is solvable by a standard mathematics. The nearly exact numerical solution by means of the FRANC3D is, however, connected with extremely high time consumption, more or less inadequate to the efficiency of the results obtained. Therefore, a simple pyramidal model of the crack front was proposed (Pokluda et al., 2004) for approximate analytical estimations. This model is based on a pyramid-like periodical approximation of the tortuous crack front, which is characterized by respective tilt and twist angles Φ and Θ_m towards the macroscopic crack plane, see Figure 8. The profile roughness R_L (measured along the crack front) and the periodicity λ_{pl} (λ_{pp}) measured parallel (perpendicular) to the crack front are associated with the angles Φ and Θ_m (the highest twist angle of the pyramidal band) by following simple equations:

$$\lambda_{pp} \tan \Theta_m = \lambda_{pl} \tan \Phi, \ R_L = \cos^{-1} \Phi. \qquad (19)$$

The characteristic periodicities λ_{pp} and λ_{pl} are usually determined by Fourier analysis of the roughness profiles measured at appropriate locations on the fracture surface. The effective stress intensity factor k_{eff} for the pyramidal front (normalized by the remote K_I factor) can be calculated using the following approximate analytical expressions for local stress intensity factors:

$$k_1 = \cos\left(\frac{\Theta}{2}\right)\left[2\nu \sin^2 \Phi + \cos^2\left(\frac{\Theta}{2}\right)\cos^2 \Phi\right],$$

$$k_2 = \sin\left(\frac{\Theta}{2}\right)\cos^2\left(\frac{\Theta}{2}\right), \qquad (20)$$

$$k_3 = \cos\left(\frac{\Theta}{2}\right)\sin \Phi \cos \Phi \left[2\nu - \cos^2\left(\frac{\Theta}{2}\right)\right].$$

1696

Table 3. Computed characteristics of the pyramidal model related to measured specimens.

Sample	Al$_2$O$_3$ [%]	R_L	λ_{pp} [μm]	λ_{pl} [μm]	θ_m	k_{eff}
E6	0	1.011	373	114	0.0455	0.983
C2	5	1.053	412	171	0.3178	0.924
S1	10	1.199	102	32	0.2040	0.763
D3	15	1.115	341	170	0.2410	0.719
B6	30	1.229	102	128	0.7311	0.714

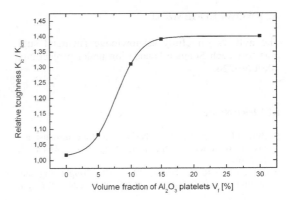

Figure 9. Influence of the roughness-induced shielding to fracture toughness as function of the percentage of Al$_2$O$_3$ particles.

The results calculated according to Equation 20 are sufficiently accurate provided that $\lambda_{pp} << 2a$, where a is the precrack length. The global normalized effective factor k_{eff} for the pyramidal model of the crack front can then be computed as

$$k_{\mathrm{eff}}^2 = \frac{\pi - 2}{2\Theta_m(2R_L + \pi - 4)} \int_{-\Theta_m}^{\Theta_m} \left(k_1^2 + k_2^2 + \frac{k_3^2}{1-\nu} \right) d\Theta \,.(21)$$

Comparison of results obtained by means of the pyramidal model and the FRANC3D code revealed that, in the whole range of both the surface roughness and the roughness periodicity typical for real surfaces, the difference lies within the error band of 10%, (Pokluda et al., 2004). Thus, the pyramidal model is used hereafter for the assessment of the contribution of RIS to fracture toughness in investigated materials.

4.1 Results and analysis

The measurements of 3D fracture surface morphology were performed by means of the MicroProf FRT. The obtained profiles were subjected to the Fourier analysis in order to determine the characteristic periodicities λ_{pp} and λ_{pl} and, simultaneously, the corresponding profile roughness R_L were established. The measured values for all specimens are displayed in Table 3. Note that the values of λ_{pp} are an order lower than the double-length of the precrack ($2a = 4$ mm) which ensures a reasonable validity of the pyramidal model. Determined roughness characteristics were then used for computation of the normalized effective factor k_{eff} according to Equations 19, 20, and 21; the results are also shown in Table 3.

The dependence of the reciprocal value of k_{eff}, i.e. of the ratio K_{Ic}/K_{Imc} (predicted relative fracture toughness ratio) on the volume fraction of Al$_2$O$_3$ platelets is shown in Figure 9. It is clear that the RIS effect raises the fracture toughness of about 40%.

However, the saturation of the RIS effect starting at about 15 vol.% of Al$_2$O$_3$ is, most probably, associated with the resolution limit of the MicroProf FRT device. Namely, the angles of surface micro-facets higher than

nearly 40% cannot be measured and, therefore, the roughness values are underestimated. The number of such facets rapidly increases for specimens with the content of Al$_2$O$_3$ platelets beyond 10 vol.%. Consequently, the real maximal contribution of the RIS effect to the enhancement of the fracture toughness can be, in fact, substantially higher than 50%. Observe that the theoretical prediction of the fracture toughness increment based upon Equation 3 is nearly 70%.

5 CONCLUSIONS

The fracture toughness values were determined at both room and elevated temperatures using the chevron notch technique. The increase in fracture toughness values by incorporating 30 vol.% of alumina platelets into borosilicate glass matrix was of about 1.5 MPa.m$^{0.5}$. This value is more than two times higher than the fracture toughness of plain borosilicate glass. Surface roughness of all fractured chevron notch test pieces was analysed with the aim to establish a relationship between the fracture resistance and the surface roughness. The surface roughness increases linearly up to 15 vol.% alumina platelets content in borosilicate glass matrix. This onset of roughness is followed by a steady state where changes in roughness are negligible. However, the analysis based upon the pyramid-like periodical approximation of the tortuous crack front (Pokluda et al., 2004) revealed that the saturation of the RIS effect starting at about 15 vol.% of Al$_2$O$_3$ is, most probably, associated with the resolution limit of the MicroProf FRT device. Theoretical calculations of the fracture toughness enhancement based upon corrected crack deflection model developed by Faber & Evans, (1983) combined with the influence of the increase in Young's modulus resulting from the platelets additions were found to be in good accordance with experimental data.

ACKNOWLEDGEMENT

The authors gratefully acknowledge financial support to Czech Science Foundation under project No. 106/06/0724.

REFERENCES

Bluhm, J.I. 1975. Slice synthesis of a three dimensional "work of fracture" specimen. *Eng. Fract. Mech.* 7: 593–602.

Boccaccini, A.R. & Trusty, P.A. 1996. Toughening and strengthening of glass by Al_2O_3 platetels, *J. Mat. Sci. Lett.* 15: 60–62.

Boccaccini, A.R. & Winkler, V. 2002. Fracture surface roughness and toughness of Al_2O_3 platelet reinforced glass matrix composites. *Composites. Part A* 33:125–131.

Boccaccini, A.R. et al. 2003. Reliability of the chevron-notch technique for fracture toughness determination in glass. *Materials Science and Engineering* A347: 102–108.

Bower, A.F. & Ortiz, M. 1993. The influence of grain size on the toughness of monolithic ceramic. *J. Engn. Mat. Technology* 115: 228–236.

Breder, K. 2000. *Comprehensive Composite Materials* 4: 77–93.

Cannillo, V. et al. 2002. Numerical models for thermal residual stresses in Al_2O_3 platelets/borosilicate glass matrix composites. *Materials Science and Engineering* A323:246–250.

Chou, Y.S. & Green, D.J. 1993. Silicon-carbide platelet alumina composites. 3. Toughening Mechanisms. *Am. Ceram. Soc.* 76: 1985–1992.

Dlouhy, I. & Boccaccini, A.R. 2001. Reliability of the chevron notch technique for fracture toughness determination in glass composites reinforced by continuous fibres, *Scripta Mater.* 44: 531–537.

Evans, A.G. & McMeeking, R.M. 1986. On the toughening of ceramics by strong reinforcements. *Acta met.* 34: 2435–2441.

Faber, K.T. & Evans, A.G. 1983. Crack deflection processes—I. Theory. *Acta Met.* 31:565–576.

Janssen, R. & Heussner, K.H. 1991. Platelet-reinforced ceramic composite materials, *Powder Metall. Int.* 23: 242–245.

Huang, X. & Nicholson, P.S. 1993. Mechanical and fracture-toughness of α-Al2O3-platelet-reinforced Y-PSZ composites at room and high-temperatures. *J. Am. Ceram. Soc.* 76: 1294–1301.

Pokluda, J. et al. 2004. Statistical approach to roughness-induced shielding effects. *Fatigue Fract. Engng. Mater. Struct.* 27: 1–17.

Rawlings, R.D. 1994. Glass-ceramic matrix composites, *Composites* 25: 372–379.

Rose, L.R.F. 1975. Toughening due to crack front interaction with a second phase dispersion. *Mechanics of Materials* 6: 11–15.

Tood, R.I. et al. 1999. Thermal residual stresses and their toughening effect in Al_2O_3 platelet reinforced glass. *Acta Mat* 47: 3233–3240.

Winn, A.J. et al. 1997. Examination of microhardness indentation-induced subsurface damage in alumina platelet reinforced borosilicate glass using confocal scanning laser microscopy. *J. Microscopy*, 186: 35–41.

Xu, G. et al. 1998. The influence of crack trapping on the toughness of fiber reinforced composites. *J. Mech. Phys. Solids* 46: 1815–1833.

Fracture Mechanics of Concrete and Concrete Structures – High-Performance Concrete,
Brick-Masonry and Environmental Aspects – Carpinteri, et al. (eds)
© 2007 Taylor & Francis Group, London, ISBN 978-0-415-44617-4

Reinforced glass concept: An experimental research

P.C. Louter, G.J. Hobbelman & F.A. Veer
Faculty of Architecture, Delft University of Technology, The Netherlands

ABSTRACT: The Glass & Transparency research group has developed a safety concept for structural glass beams which shows some analogy with reinforced concrete. Annealed float glass beams are reinforced by adhesively bonding a stainless steel section to the layout of the beam. Current research focuses on the effect of different reinforcement layouts and different adhesive types on the post-failure behaviour of reinforced glass beams. For this research 30 specimens with a length of 1.5 m have been subjected to a four-point bend test. Three reinforcement layouts and two adhesives (DELO GB368 and Araldite 2013) have been tested. The test results show differences in fracture patterns, failure behaviour and failure mechanisms for the tested reinforcement layouts and applied adhesives. Regarding structural qualities and consistency layout III and Araldite 2013 performed best. However, DELO GB368 provides some major advantages at the manufacturing process. Combining the advantages of both adhesives into one adhesive seems preferable.

1 INTRODUCTION

In contemporary architecture there is an increasing demand for transparent buildings and structures. Glass is desired as a load bearing material for structural components such as columns and beams. However, due to its brittleness and sudden failure behaviour, glass is considered a structurally unsafe material. At the faculty of Architecture, Delft University of Technology, the Zappi Glass & Transparency research group focuses on the development of transparent components and structures with safe failure behaviour. The research group has developed a safety concept for structural glass beams which shows some analogy with reinforced concrete. Annealed float glass beams are reinforced with a stainless steel section which is integrated in the layout of the beam and rigidly bonded to the glass.

Upon overloading the glass will crack but crack propagation will be limited due to the dissipation of fracture energy by deformation of the reinforcement. Furthermore the stainless steel section will act as a crack bridge carrying the tensile forces after the glass has cracked. Together with the compression force in the (un-cracked) compression zone an internal couple will be generated and the beam will still be able to carry load, see figure 1. In practice this will provide bystanders time to flee or to take measures. This concept has been developed in preceding research by Veer (2005) and successfully tested in previous beam designs up to a length of 7.2 m by Louter (2005).

The post-failure behaviour of a reinforced glass beam is highly dependent on the bond between glass

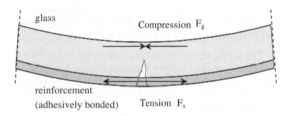

Figure 1. Reinforced glass beam concept: schematic overview of distribution of forces after glass failure.

and reinforcement. Current research focuses on the effect of different reinforcement layouts and different adhesive types on the post-initial failure behaviour of reinforced glass beams. For this research 30 specimens with a length of 1.5 m have been subjected to a four-point bend test. Three different reinforcement layouts (I, II and III) and two different adhesives (DELO GB368 and Araldite 2013) have been tested.

2 REINFORCEMENT LAYOUTS

The tested reinforcement layouts are displayed in figure 2. For each layout an annealed float glass beam of $1500 * 115 * 10$ mm is applied.

– Layout I consists of two stainless steel sections (each $2 * 9$ mm) which are bonded to the side panes of the glass beam.
– Layout II consists of a stainless steel box section ($10 * 10 * 1$ mm) which is bonded to the edge of the glass beam.

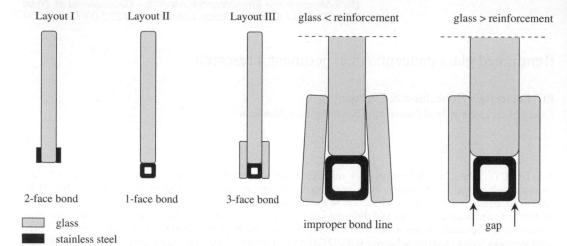

Figure 2.   Tested reinforcement layouts.

Figure 3.   Effects of dimensional inaccuracies for layout III.

– Layout III consists of a stainless steel box section (10 * 10 * 1 mm) which is bonded to the edge of the glass beam and encapsulated by two additional outer layers (each 40 * 6 mm). These outer layers are bonded to the side panes of the glass beam (using the transparent DELO GB368 adhesive).

The amount of steel in the section is equal for each layout; the area of the box section (36 mm^2) is equal to the area of both full sections ($2 * 2 * 9$ mm $= 36$ mm^2). The developed reinforcement layouts differ in two important aspects:

(a) bond area (*1-, 2- or 3*-face bond)

The interaction between glass and reinforcement is fully dependent on the adhesive bond. Forces are transferred via shear in the adhesive layer. Upon glass failure large tensile forces (in the reinforcement) have to be transferred from and to the glass. The bond area of the tested reinforcement layouts differ in 1-, 2- or 3-face bond, see figure 2. Enlarging the bond area will reduce shear stresses in the adhesive bond and will prevent from premature detachment of reinforcement. In this respect layout III, which has the largest bond area, should perform best.

(b) capacity to adapt to dimensional inaccuracies

The strength of an adhesive bond is, amongst other aspects, dependent on the thickness of the bond layer. Generally a thin adhesive layer will result in a strong bond. For reinforcement layout I and II the thickness of the adhesive bond can be controlled during the bonding process. By clamping the reinforcement to the glass any adhesive surplus will be pressed out. However, for reinforcement layout III this can not be done since the reinforcement is bonded *between* two glass sheets. In this case the thickness of both vertical bond layers depends on to what extent the dimensions of the reinforcement correspond with the thickness of the inner

glass layer. Any dimensional inaccuracy in glass and/or reinforcement will result in a deviating bond thickness. In case the glass is *thinner* than the reinforcement an improper bond line occurs, see figure 3. In case the glass is *thicker* than the reinforcement the bond layer will have to be thicker and the applied adhesive will have to be capable of filling this gap while maintaining its strength. According to the European standard EN 572-2 the allowed tolerances for float glass with the applied glass thickness of 10 mm are ±0.3 mm.

3   ADHESIVE TYPES

Each reinforcement layout has been tested for two different adhesives, DELO GB368 (abr. GB368) and Araldite 2013 (abr. AR2013). The most important properties of these adhesives are listed in table 1. A short description of both adhesive is provided in the following paragraphs.

3.1   *DELO Photobond GB368*

The acrylic based photo-initiated curing adhesive DELO GB368 has been developed for glass-glass and glass-metal bonding. In previous reinforced glass beam designs, which consist of multiple glass layers and stainless steel reinforcement, this adhesive has already been applied for both glass-glass and glass-reinforcement bonding.

For the bonding process of (multi-layer) reinforced glass beams DELO GB368 provides three main *advantages*;

(a) Polymerization is only initiated by UV-radiation, which enables an accurate positioning of the substrates without being rushed by curing times,

Table 1. Key properties of DELO GB 368 and Araldite 2013, according to the manufacturer's datasheets.

		DELO GB 368	Araldite 2013
		Acrylic based	Two-component epoxy
Shear strength	[MPa]	23 (glass-glass) 23 (glass-alu)	18
Viscosity	[mPas]	5700	Thixotropic (mixed) Gap filling capacity up to 5 mm
Colour		Transparent	Grey
Curing time		Minimum curing time 15 seconds	4 hours at 23°C to reach 1 MPa. 10 hours at 23°C to reach 10 MPa.

(b) The adhesive cures within 30 seconds which limits production time,

(c) Glass-glass and glass-reinforcement bonding can be executed simultaneous, since the same adhesive is applicable for both purposes.

The main *disadvantages* of this adhesive for the reinforced glass concept are:

(a) The adhesive has to be cured with a maximum thickness of 0.1 mm and is not able to fill gaps caused by dimensional inaccuracies,

(b) Preceding experimental research showed the adhesive is only limited resistant to shock loads which occur upon glass failure, causing premature collapse of the beam.

3.2 *Araldite 2013*

Although the two-component epoxy Araldite 2013 has been developed as a metal bonding adhesive it is also suitable for bonding other materials such as ceramics, rubbers, rigid plastics and glass. Although the Araldite has not yet been applied in the reinforceced glass conept this adhesive might provide some *advantages*:

(a) According to the manufacturer's datasheets Araldite 2013 has a filling capacity up to 5 mm, which should make it suitable for taking up any dimensional inaccuracies,

(b) The Araldite is a rather tough adhesive, which should make it more resistant to the shock loads, which occur upon glass failure.

The main *disadvantages* for the reinforced glass concept of the Araldite 2013 seem to be:

(a) Since this adhesive is not suitable for glass-glass bonding (due to its grey colour) the reinforced

glass bonding process has to be divided in two stages; the glass-glass bonding has to be executed with a different (transparent) adhesive than the glass-reinforcement bonding, which will increase production time.

(b) The adhesive has a limited handling time of about 1 hour after the two components have been mixed; accurate positioning of the substrates has to be completed within this time span,

(c) At room temperature the adhesive has a curing time of 4 hours to reach a light handling strength of 1 N/mm^2, and a curing time of 10 hours to reach 50% of the final shear strength. Due to this long curing time the substrates have to be clamped for several hours, which will increase production time.

4 TEST RESULTS

Of each layout (LI, LII and LIII) and for both adhesives (GB368 and AR2013) 5 specimens have been made, which results in a total of 30 specimens. All specimens have been subjected to a 4-point bend test to validate their structural behaviour. For this test a Zwick 100 kN test rig was provided with a steel hinge-and-roller-support rig. Supports were 1400 mm apart, loads were 400 mm apart and lateral (anti-buckling) supports were 550 mm apart. The specimens were loaded at a rate of 1 mm/minute and loading was continued until total destruction. Load and vertical displacement were monitored. The specimens were provided with a grid, which enables visual monitoring of crack heights. All tests were captured on video.

4.1 *General failure behaviour*

First of all the general failure process of all specimens will be described in this paragraph. The specific failure behaviour for each layout and adhesive type will be discussed in proceeding paragraphs.

The stress-displacement diagrams for each layout and adhesive type are given in figures 5, 6, 9, 10, 13 and 14. A general and schematic stress-displacement diagram is given in figure 4. Three general stages/phases can be distinguished:

(a) Linear elastic behaviour

All specimens showed a linear elastic behaviour until at a global tensile bending stress at the lower edge of the glass of 30–65 N/mm^2 a first crack occurred. This crack originated at the lower edge of the glass beam and ran about 2/3 of the total beam height before being stopped in the upper compression zone. Due to the sudden increase in vertical displacement of the specimen the load dropped.

(b) post-failure behaviour/residual strength

As loading was continued the load started to rise again until at a second peak-load another crack

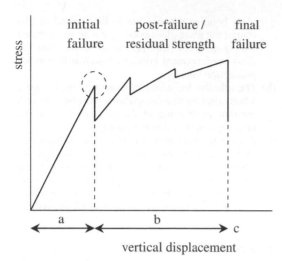

Figure 4. Schematic stress-displacement diagram for all reinforcement layouts.

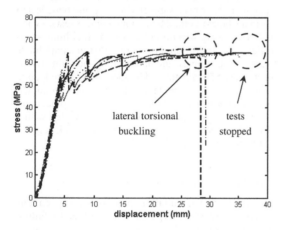

Figure 5. Stress-displacement diagram of layout I - GB368 specimens.

occurred causing a drop in load. This process might be repeated one or several times. The stress-displacement diagram shows a decrease in beam stiffness after each peak-load. For most specimens the residual strength exceeded the initial failure load.

(c) total failure
Final failure (point c) occurred due to either,

– Progressive detachment of reinforcement; adhesive failure, or
– Lateral torsional buckling; lateral instability of the compression zone due to increasing crack growth and increasing compression forces.

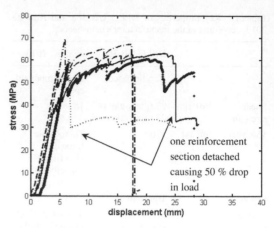

Figure 6. Stress-displacement diagrams of layout I - AR2013 specimens.

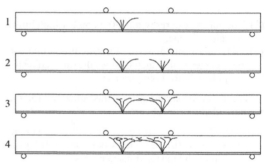

Figure 7. Schematic overview of crack propagation, at different time steps (1–4), for L I – GB368 specimens.

4.2 Layout I

The stress-displacement diagrams of layout I specimens for both adhesives are given in figures 5 and 6. Regardless of the applied adhesive, all specimens of layout I showed comparable bending stiffness in the linear elastic phase. After initial failure the specimens showed a small increase in load of−15 to +64% and a rather large increase in vertical displacement of 214–675%.

4.2.1 L I – GB368 – specimens
Figure 7 gives a schematic overview of the crack propagation at different time steps (1–4) for layout I-GB368 specimens. Upon overloading a V-shaped crack with a height of 75–90 mm occurred (1). As loading was continued a second or even third V-shaped crack occurred (2). Subsequently the cracks started to propagate horizontally and started to grow towards each other (3). Although smaller cracks occurred at the outline of the compression zone a compression zone of 10–15 mm

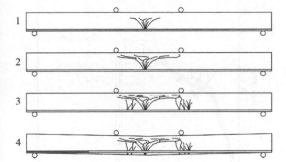

Figure 8. Schematic overview of crack propagation, at different time steps (1-4), for L I – AR2013 specimens.

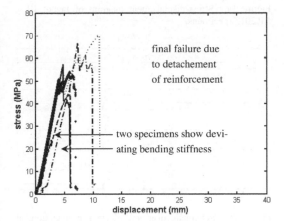

Figure 9. Stress-displacement diagram of layout II – GB368 specimens.

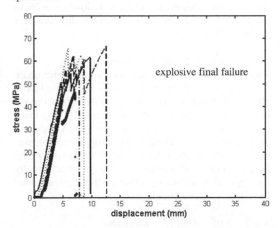

Figure 10. Stress-displacement diagram of layout II - AR2013 specimens.

remained un-cracked (4) until the beam failed due to lateral torsional buckling. For three specimens the test was stopped due to extensive vertical displacements, see figure 5.

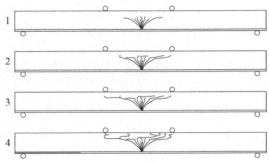

Figure 11. Schematic overview of crack propagation, at different time steps (1-4), for L II – GB368 specimens.

4.2.2 L I – AR2013 – specimens

Figure 8 gives a schematic overview of the crack propagation at different time steps (1–4) for layout I-AR2013 specimens. At the initial failure load a V-shaped crack with a height of 80 mm occurred (1) which started to propagate horizontally (2) as loading was continued. Subsequently multiple V-shaped cracks occurred (3). Gradually the reinforcement started to detach at mid-span and the glass started to slide past the reinforcement (4). For two specimens one of the reinforcement sections detached at one beam end, causing a 50% drop in load, see figure 6.

Finally the specimens failed due to detachment of both reinforcement sections at one beam end. For two specimens the residual strength did not exceed the initial failure load.

4.3 Layout II

The stress-displacement diagrams of layout II specimens for both adhesives are given in figures 9 and 10. Except for two LII-GB368 specimens all specimens show comparable bending stiffness. After initial failure the specimens show an increase in load of −25 to +94% and a relatively small increase in vertical displacement of 27–225%.

4.3.1 L II – GB368 – specimens

Figure 11 gives a schematic overview of the crack propagation at different time steps (1–4) for layout II-GB368 specimens. Upon overloading a V-shaped crack with a rather dense fracture pattern and a height of 70–80 mm occurred (1). As loading was continued a second crack occurred for some specimens. Subsequently the V-shaped crack(s) started to propagate in a horizontal manner (2/3). Finally all specimens failed due to detachment of reinforcement at one beam end. The specimens showed a 'bull-bar' shaped crack pattern (4).

For one specimen the residual strength did not exceed the initial failure load.

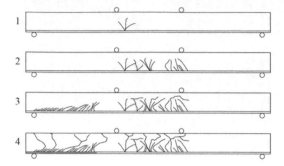

Figure 12. Schematic overview of crack propagation, at different time steps (1-4), for L II – AR2013 specimens.

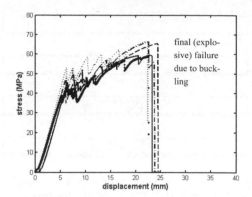

Figure 14. Stress-displacement diagram of layout III – AR2013 specimens.

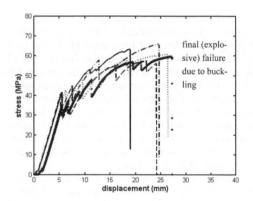

Figure 13. Stress-displacement diagram of layout III – GB368 specimens.

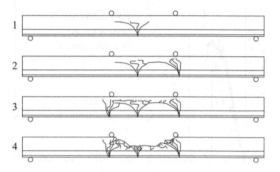

Figure 15. Schematic overview of crack propagation, at different time steps (1–4), for L III – GB368 specimens.

4.3.2 L II – AR2013 – specimens

Figure 12 gives a schematic overview of the crack propagation at different time steps (1–4) for layout II-AR2013 specimens. Upon overloading a V-shaped crack with a height of 70–80 mm occurred (1). This crack has fewer branches as was observed for LII-GB368 specimens. As loading was continued multiple V-shaped cracks occurred (2). The cracks seemed to be 'non-related' and they did not grow towards each other. Subsequently small diagonal/sloped cracks occurred towards the beam end (3). Finally the specimens failed rather explosive. At the beam end large cracks were observed and the reinforcement had largely been torn from the glass (4). However, small glass particles remained attached to the reinforcement.

For one specimen the residual strength did not exceed the initial failure load.

4.4 Layout III

The stress-displacement diagrams of layout III specimens for both adhesive are given in figures 13 and 14. All specimens show comparable bending stiffness and large vertical displacements. After initial failure the specimens show an increase in load of +26 to +84% and a large increase in vertical displacement of 240–401%.

4.4.1 L III – GB368 – specimens

Figure 15 gives a schematic overview of the crack propagation at different time steps (1–4) for layout III – GB368 specimens. Upon overloading a V-shaped crack with a height of 80–90 mm occurred (1). As loading was continued a second or even a third V-shaped crack occurred (2). These cracks propagated horizontally and tended to overlap (3). At the upper edge of the beam a zone of 30 mm remained un-cracked. Finally the upper zone at mid-span failed rather explosive due to lateral torsional buckling (4).

4.4.2 L III – AR2013 – specimens

Figure 17 gives a schematic overview of the crack propagation at different time steps (1–4) for layout III – AR2013 specimens. Upon overloading one or multiple V-shaped cracks with a height of 90–95 mm occurred (1). As loading was continued the existing cracks remained rather stable and successively multiple cracks occurred (2/3). At the upper edge of the

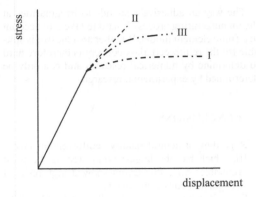

Figure 16. Schematic stress-displacement diagram of layout I, II and III.

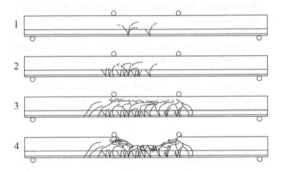

Figure 17. Schematic overview of crack propagation, at different time steps (1–4), for L III – AR2013 specimens.

beam a zone of 30 mm remained un-cracked. Finally the specimens failed explosive due to lateral torsional buckling.

5 DISCUSSION

The test results show differences in fracture patterns, failure behaviour and failure mechanisms for the tested reinforcement layouts and the applied adhesives. The results will be discussed by layout and adhesive type.

5.1 Layouts

Figure 16 shows a schematic stress-displacement diagram/tendency of all three tested reinforcement layouts.

The layouts show different post-failure trajectories:

Layout I shows an elastic ideal plastic behaviour. After initial failure bending stiffness is strongly reduced. The specimens show a large deformation capacity, but only a small capability of carrying increasing loads. This might be caused by local detachment of reinforcement, which allows for large

deformations since the glass can 'slide' past the reinforcement. Local detachment of reinforcement or even full detachment of one reinforcement section did not lead to a full collapse of the specimens, see figure 6. Due to the large bond area and the application of two reinforcement sections a preferable redundancy has been built in, which contributes to safe failure behaviour.

Layout II shows the most brittle behaviour. After initial failure a limited decrease in bending stiffness is observed and the beams are able to carry increasing loads. For this layout all specimens failed due to detachment of reinforcement (adhesive failure). This can be explained by the limited bond area between glass and reinforcement, which leads to high shear stresses. For layout I only a 1-face bond is applied, which excludes any redundancy. Upon failure of this single bond face fails there is no second bond face to limit this failure or to carry the forces. In this respect this layout option is the least preferable.

Layout III shows an elastic/strain hardening behaviour. After a first crack occurs bending stiffness gradually decreases, but the beams are still able to carry extensive and increasing loads. The large bond area in this beam layout (3-face bond) prevents from premature detachment of reinforcement. If one of the bond faces might fail (partially), there are still 2 bond faces left to carry the tensile forces in the reinforcement, which provides redundancy and safe failure behaviour. Due to the proper bond detachment of reinforcement did not occur and the specimens were able to carry increasing loads. As the compression force in the glass increased lateral instability of the compression zone became critical. Due to increasing crack growth the upper (un-cracked) compression zone detaches from the lower part and becomes susceptible to buckling. All specimens of layout III finally failed due to buckling. This failure mechanism has been observed in preceding research by Louter (2006) and is a determining factor in dimensioning (reinforced) glass beams according to Belis (2005).

The layout III-specimens showed consistent results. For all specimens the final failure load exceeded the initial failure load and the final failure mechanism was equal for each specimen. Cause of its consistency, built-in redundancy and safe failure behaviour this layout seems the most preferable regarding structural qualities.

5.2 Adhesives

The test results show a difference in fracture pattern for the AR2013 and GB368 specimens. Regardless of layout the GB368-specimens show *few*, but *large* cracks whereas the AR2013-specimens show *many*, but *small* cracks. This difference becomes most distinct for failure stage 3 of both LIII-AR2013 and

LIII-GB368 specimens, see figures 15 and 17. The AR2013-specimens show a more dense fracture pattern than the GB368-specimens. This difference in fracture pattern can be explained by a difference in toughness of both adhesives. For GB368-specimens local de-bonding of reinforcement was observed at the crack tips/origin. The shock load which occurs upon glass fracture causes the adhesive to fail for several centimetres on either side of the crack tip. This local de-bonding of reinforcement allows for large crack opening displacements and extensive crack propagation. Due to the higher toughness of the AR2013-adhesive local de-bonding occurs to a lesser extend for the AR2013-specimens. Crack opening displacement and crack propagation are limited and stresses are more evenly/equally (re)distributed. The crack itself remains stable and new cracks will occur next to the existing crack.

Both adhesives have only been tested for single layer glass beams. In a multi-layer glass beam the occurrence of *few* but *large* cracks (as for GB368-specimens) might be more advantageous than the occurrence of *many* but *small* cracks (as for AR2013-specimens). In a multi-layer glass beam lateral instability due to overlapping cracks is more likely to occur for a dense (many small cracks) fracture pattern. In this respect the GB368-adhesive seems advantageous, but it is noted that further research is recommended.

The AR2013-specimens show more consistent results. For instance for layout III the scatter in ultimate vertical displacement of AR2013-specimens is less than of GB368-specimens. Due to its gap filling quality, the AR2013-adhesive is able to take up any dimensional inaccuracy, which leads to a more consistent structural quality. The structural performance of the AR2013-adhesive seems less dependent on irregularities at the manufacturing process than the GB368-adhesive. In this respect the AR2013-adhesive seems more preferable and generates more predictable and consistent failure behaviour than the GB368-adhesive. However, AR2013-adhesive has more limitations at the bonding process as has been noted in paragraph 3.2. A combination of both adhesives seems ideal: a transparent adhesive which is rapidly cured by UV-light, tough, resistant to shock loads, not sensitive to irregularities at the manufacturing process and applicable for both glass-glass and glass-reinforcement bonding.

The way an adhesive responds to irregularities at the manufacturing process is hard to take into account in a finite element model. Whether an adhesive is suitable for the reinforced glass concept is therefore hard to determine by numerical research and can only be determined by experimental research.

6 CONCLUSIONS

– Regarding structural quality reinforcement layout III, which has the largest bond area, is the most preferable, since this layout showed high residual strength and consistent results.
– Regarding structural quality adhesive Araldite 2013 is the most preferable adhesive for the reinforced glass concept, since this adhesive can adapt to dimensional inaccuracies and provided the most consistent results.
– Regarding the production process DELO GB368 is preferable because of its short curing time and its applicability for both glass-glass and glass-reinforcement bonding.
– Combining the structural and production advantages of both tested adhesives is preferable.

REFERENCES

Belis, J. 2005. Kipsterkte van monolithische en gelamineerde glazen liggers. *Dissertation, Laboratory for Research on Structural Models, Ghent University*.
Veer, F.A. 2005. 10 years of Zappi research. *Proceedings of 9th Glass Processing Days, Tampere, Finland*
Louter, P.C. & Belis, J. & Bos, F.P. & Veer, F.A. & Hobbelman G.J. 2005. Reinforced Glass Cantilever Beams. *Proceedings of 9th Glass Processing Days, Tampere, Finland*.
Louter, P.C. & Schetters L. & Veer, F.A. & Van Herwijnen, F. & Romein, T. 2006. Experimental research on scale 1:8 models of an 18 m reinforced glass beam. *Proceedings of the 2nd International Symposium on the Architectural Application of Glass. Munich, Germany*.
European Standard 2004. EN 572-2 Glass in Building – Basic Soda lime silicate glass products – part 2: Float glass.

Part XII
Environmental Issues

Fracture Mechanics of Concrete and Concrete Structures – High-Performance Concrete,
Brick-Masonry and Environmental Aspects – Carpinteri, et al. (eds)
© 2007 Taylor & Francis Group, London, ISBN 978-0-415-44617-4

Strength and stiffness of concrete under heating and cooling treatments

J.S. Lee, Y. Xi, & K. Willam
Department of Civil, Environmental, and Architectural Engineering at University of Colorado Boulder, USA

ABSTRACT: Understanding the properties of concrete under high temperature is essential to enhance the fire resistance of reinforced concrete structures (RCS) and to provide accurate information for fire design of RCS. During and after a fire, different parts of a concrete structure experience different heating and cooling scenarios. Hence the remaining stiffness and strength properties in the concrete structure need to be assessed before a decision can be made as to whether the structure can be repaired or must be rebuilt. The purpose of this study is to investigate the strength and stiffness performance of concrete subjected to various temperature scenarios with the aid of residual compressive strength testing. In addition, the temperature distribution in the concrete cylinders, color changes, and cracks are investigated.

1 INTRODUCTION

Under a rapid heating condition, Portland cement concrete experiences a large volume change resulting from thermal dilatation of coarse and fine aggregates as well as from shrinkage of the cement paste. When the heating rate is high, spalling damage may occur in concrete due to high thermal stresses and pore pressure build-up. Therefore, it is very important to assess the residual mechanical properties after the concrete has been subjected to high temperature at different heating and cooling rates. The results will provide essential information to the concrete industry for improving the fire resistance of concrete.

Extensive experimental studies on this important topic were performed in the past. The important experimental parameters included maximum temperature, heating rate, types of aggregates used, various binding materials, and mechanical loads under high temperature conditions. The experimental studies have mainly concentrated on the strength of concrete. Poon et al. conducted experimental studies on the strength of normal and high strength concretes. However their studies did not consider the effects of heating and cooling rates which are very important factors on the degradation of concrete due to fire. From the strength point of view, the reduction of concrete strength is affected strongly by the heating rate as well as the cooling rate.

The purpose of this study is to investigate the residual compressive strength and stiffness of concrete subjected to various cooling regimes. The test variables are the maximum temperatures and cooling regimes under the constant heating rate of 2°C/min and holding time of 4 hrs at the maximum temperatures.

Table 1. Moderate strength mix design of normal weight concretes (by weight)

Moderate strength mix design – kN/m^3				
Cement	Water	Fine agg.	Coarse agg.	Strength (Mpa)
3.49	1.75	8.32	10.12	30

Additionally, the temperature distribution in the concrete cylinder, color changes in concrete, and crack patterns are also investigated.

2 SPECIMEN PREPARATION, HEATING EQUIPMENT, AND TEST VARIABLES

2.1 Mix design and specimen preparation

In this experimental study granite was used as the coarse and fine aggregate in all concrete specimens. In the initial mix design the water/cement (w/c) ratio was 0.50 (see Table 1).

However the adjusted w/c ratio was 0.71, which is very high (see Table 1). This was due to the dry aggregates and contents of very dry, fine aggregates in crushed granite sand. Different water/cement ratios were examined for workability and a w/c ratio of 0.71 was the lowest value with which the proper workability with slump value 38.10 mm could be obtained. The moisture content of the aggregates was measured using the test method B in ASTM D 2216. The absorption capacity of aggregates was 0.78%. By subtracting the

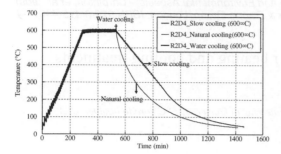

Figure 1. Temperature histories subjected to different cooling regimes.

water absorption capacity of the aggregates from the initial water content a final w/c ratio of 0.67 by weight is obtained. The maximum size of coarse aggregate was 19.05 mm.

Concrete cylindrical specimens with dimensions of 101.6 × 203.2 mm were made for the experimental study. The size of the specimen was determined by the internal size of the furnace used in the study, which is 198.12 mm × 289.56 mm × 167.64 mm. The specimens were cured under the standard condition (Temperature 23°C and RH 93%) for 8 weeks in a fog room.

2.2 *Heating equipment and test variables*

An electrically heated furnace (Model RHF 15/8 of CARBOLITE) designed for maximum temperature up to 1600°C was used. The temperature history inside the furnace was measured and recorded using type K-thermocouples. Model OM-CP-OCTTEMP produced by OMEGA was the data logger for the thermocouples.

The effect of three cooling conditions was examined in the study, and they were called slow cooling, natural cooling, and water cooling. Four maximum temperatures (target temperatures) were tested: 200, 400, 600, and 800°C. The heating rate was 2°C/min, and the holding time at the maximum temperature was 4 hrs.

In the case of natural cooling, the specimen was left in the furnace and the temperature change was recorded over time. For slow cooling, the cooling rate was controlled as shown in Fig. 1, slower than natural cooling. For water cooling, the specimen was taken out of the furnace and put in a tank of water with initial temperature of 20°C. The furnace was heated using two groups of three heating elements on either side of the specimen. In order to provide a more uniform thermal condition inside the concrete specimen, the concrete specimen was placed inside a hollow tube made of mullite with a maximum operational temperature of 1700°C.

Figure 1 shows the temperature histories using a heating rate of 2°C/min, a hold time of 4 hrs at the maximum temperature of 600°C, and three different cooling conditions. The ambient temperature

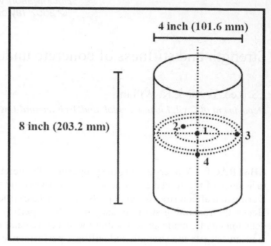

Figure 2. Locations of thermocouples and specimen geometry.

history was recorded between the mullite tube and the specimen with the help of a K-thermocouple.

3 TEMPERATURE DISTRUBUTION IN THE CONCRETE

Figure 2 shows the concrete cylinder used to investigate the temperature distribution at elevated temperature.

One thermocouple was used to measure ambient temperature between the specimen and mullite hollow tube, and four additional K thermocouples were installed inside and on the surface of the specimen to measure the temperature distribution in the concrete cylinder. Figure 2 shows the specimen geometry and the locations of the thermocouples. The heating rate was 1°C/min, the maximum temperature was 900°C with a hold time of 2 hrs, and the cooling method was natural cooling.

Figure 3 shows temperature histories over time measured at each thermocouple location. One can see that in the heating phase the surface temperature is higher than the internal temperature and in the cooling phase the surface temperature is lower than the internal temperatures, which are expected experimental results. The two vertical drops in the cooling phase in Figure 3, near 540°C at the center and 330°C at mid-radius, are due to the damage of the two thermocouples in position 1 and position 2.

Figure 4 shows the temperature difference between the surface and center of the specimen at different temperature ranges. One can see that the temperature difference varies during the entire testing period, which means that a steady state condition in the concrete was never reached. This is mainly due to the phase

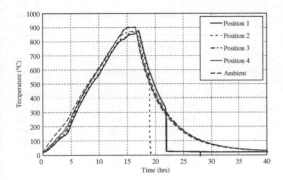

Figure 3. Transient temperature at each position in the concrete cylinder.

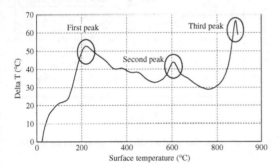

Figure 4. ΔT vs. Surface temperature (during heating).

transformations taking place at different temperatures in the concrete. The three peaks shown in the figure are related to the micro-structural changes due to complex physicochemical transformations in the concrete under different high temperatures. The peaks shown in Fig. 4 may be explained by the test results of DTA (Differential Thermal Analysis) conducted by Lankard. The first peak occurs at about 200°C due to evaporation of free water and dehydration of calcium silicate hydrate (C-S-H). The second peak appears between 550°C and 650°C and is related to the decomposition of calcium hydroxide (CH) and calcium silicate hydrate (C-S-H). The third peak occurring at 850°C is likely due to the decomposition of calcium carbonate ($CaCO_3$) which was observed in the DTA experiments by Lankard.

Figure 5 shows the temperature profiles measured at positions 1, 2, and 3, when the temperature at the center of specimen reaches 600°C. The surface temperature corresponding to each temperature profile is indicated in the figure. One can clearly see that while undergoing natural cooling, the temperature difference between the center and the surface of the concrete is larger than that during the heating period. In general, a larger temperature difference represents a higher temperature gradient, which leads to a higher damage. Therefore,

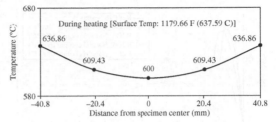

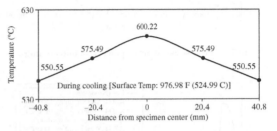

Figure 5. Temperature profiles during heating and cooling.

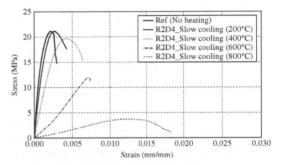

Figure 6. Axial stress-strain response (Slow cooling).

this test data means that the damage of the concrete is strongly affected by cooling method.

4 RESIDUAL COMPRESSION TEST

Axial strains were measured using two extensometers (MTS model 632.94E-20). The average of the two strain readings was taken as the axial strain. Displacement control with a rate of 0.0001 inch/sec was used for the compression test.

Figure 6, 7, and 8 plot stress versus strain for the specimens that were subjected to slow, natural, and water cooling, respectively. As can be seen from the figures both strength and stiffness properties of concrete decrease significantly with increasing maximum temperature.

The elastic response behavior is fairly nonlinear for those specimens subjected to slow or natural cooling and maximum temperatures of 600°C and beyond. The initial slope of the axial stress-strain response is lower

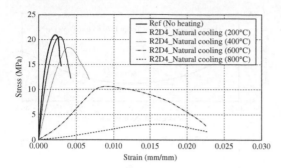

Figure 7. Axial stress-strain response (Natural cooling).

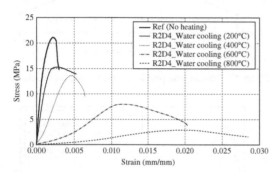

Figure 8. Axial stress-strain response (Water cooling).

than the slope in the middle part, indicating that the concrete is hardened or rather reconsolidating during mechanical loading. For the case of water cooling, the same trend is shown in the stress-strain curves for the range of maximum temperatures of 400°C and beyond. This hardening behavior may be caused by the closure of thermally induced cracks formed during high temperature heating and cooling treatments. In the initial stage of loading the cracks are open as the load gradually increases, resulting in a lower slope at the beginning. After the cracks are closed, the slopes of the curves gradually increase and follow the general trend of a compression strength test.

Table 2 summarizes the ultimate strength values from the residual compression tests. Figure 9 depicts the residual strength normalized by the reference strength (25°C) for the three cooling methods as a function of the target temperature. The residual strength rapidly drops beyond 400°C, whereby the specimens subjected to water cooling decrease more rapidly than the specimens subjected to other cooling methods. The strength of specimens exposed to a target temperature of 600°C is less than 57% of the reference strength at room temperature. For 800°C, the strength is less than 18% of the reference strength.

Table 3 summarizes the test results for the initial tangent modulus from the residual compression test.

Table 2. Ultimate strength results of residual compression test.

Temp. (°C)	Ultimate strength (Mpa)		
	Slow	Natural	Water
25	21.16	21.16	21.16
200	21.15	20.77	15.32
400	19.69	18.59	13.64
600	11.96	10.72	8.02
800	3.71	3.19	2.93

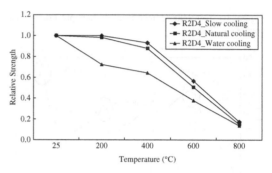

Figure 9. Relative residual strength vs. maximum temperature.

Table 3. Initial tangent modulus from residual compression test.

Temp. (°C)	Initial tangent modulus (Mpa)		
	Slow	Natural	Water
25	2.48E+04	2.48E+04	2.48E+04
200	1.61E+04	1.44E+04	1.12E+04
400	7.86E+03	7.17E+03	2.43E+03
600	1.32E+03	9.51E+02	3.97E+02
800	3.03E+02	1.68E+02	8.96E+01

Figure 10 shows the values of the initial tangent modulus with different temperatures, normalized by reference specimen tested at room temperature.

The basic trend of the variation of initial tangent modulus shown in Figure 10 is similar to that of the residual strength shown in Figure 9. However, the initial tangent modulus of concrete is more sensitive to elevated temperature than the compression strength. As shown in Figure 10, the drop in stiffness at lower temperatures is larger than the drop in strength shown in Figure 9. In contrast, the drop in strength in the higher temperature range is larger than the drop in stiffness shown in Figure 10. Specifically, the initial tangent moduli of the specimens exposed to a maximum temperature of 600°C are less than 5.3% of

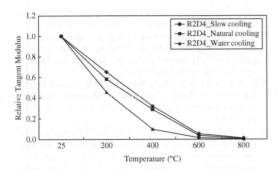

Figure 10. Relative initial tangent modulus vs. maximum temperature

the initial reference tangent modulus. For 800°C, the moduli are less than 1.3% of the reference values. In addition to the deterioration of stiffness and strength due to maximum temperature, we also observe that both heating as well as cooling rates have significant effects on the mechanical properties. In terms of cooling rate, faster cooling contributes to a more severe decrease of both the strength and stiffness properties of concrete.

5 COLOR CHANGES AND CRACKS

Color changes and crack patterns in concrete are important features for assessing the fire damage of the concrete. They can be used to evaluate the degree of damage in concrete and to estimate the maximum temperature experienced by the concrete specimen.

In the concrete exposed to high temperature, the exposed maximum temperature can be predicted from some changes in color. The chemical composition of granite used as the aggregates in the study is composed mostly of silica (SiO_2) and alumina (Al_2O_3). Pink or red spots were observed on every specimen exposed to a maximum temperature of 400°C and above. The colors of the specimens subjected to slow and natural cooling were generally pink or red, while the colors of the specimens subjected to water cooling were dark pink or dark red due to water absorption during cooling in water.

The distributed cracks were visually observed on every surface of the specimens exposed to a maximum temperature of 600°C and above. Particularly, it is likely that the distributed cracks between the aggregate and cement paste matrix are deeply related to the differences in the thermal deformations between the two. The crack widths of the specimens exposed to a maximum temperature of 800°C were visibly wider than those of the specimens exposed to a maximum temperature of 600°C. Also, in the specimens exposed to maximum temperature 800°C, the crack widths of the specimens heated quickly (15°C/min)

were significantly larger than those of the specimens heated slowly (2°C/min).

6 CONCLUSIONS

1. A systematic experimental study was conducted for concrete specimens under different maximum temperatures and cooling rates. Residual strength and stiffness, color changes, and crack patterns were observed and reported. Also temperature profiles inside the specimens were measured and recorded.
2. Temperature differences between the center and surface of concrete cylinders at steady state did exhibit distinguishing features. There are three distinct peaks in the temperature difference within the range of room temperature and 850°C. The three peaks are related to the micro-structural changes due to complex physicochemical transformations in the concrete under high temperature. The first peak at about 200°C is associated with the dehydration of calcium silicate hydrate (C-S-H) and the evaporation of free water. The second peak at 550°C is related to the decomposition of calcium hydroxide (CH) and calcium silicate hydrate (C-S-H). The third peak at 850°C may be due to the decomposition of calcium carbonate.
3. Complete stress-strain curves were obtained for evaluating residual mechanical properties of the concrete. Strength and stiffness properties decrease significantly as the maximum temperature and the cooling rate increase. The present test data demonstrate that the cooling rate is an important parameter for the degradation of mechanical properties of concrete subjected to high temperatures.
4. The crack width increases with increasing temperature and heating rate which means both contribute and enlarge the damage of concrete.

REFERENCES

Abrams, M.S., "Compressive strength of concrete at temperatures to 1600°F", American Concrete Institute (ACI) SP 25, Temperature and Concrete, Detroit, Michigan, 1971.

Castillo, C., and Durani, A.J., "Effect of transient high temperature on high-strength concrete", ACI Material Journal, V. 87, No. 1, Jan/Feb. 1990, pp. 38–67.

Cheng, F.P., Kodur, V.K.R., and Wang, T.C., "Stress-strain curves for high strength concrete at elevated temperature", Journal of Materials in Civil Engrg., ASCE, V. 16, No. 1, Jan/Feb. 2004, pp. 84–90.

Felicetti, R., and Gambarova, P.G., "Heat in concrete: special issues in materials testing", Studies and Researches, Graduate School in Concrete Structures-Fratelli Pesenti Politecnico di Milano, Italy, Vol. 24, 2003.

Furumura, F., Abe, T., and Shinohara, Y., "Mechanical properties of high strength concrete at high temperature", Proceedings of the Fourth Weimar Workshop on High Performance Concrete: Material Properties and Design,

Hochschule fuer Architektur und Bauwesen (HAB), Weimar, Germany, Oct. 1995, pp. 237–254.

Guo, J.S., and Waldron, P., "Development of the stiffness damage test (SDT) for characterization of thermally loaded concrete", Materials and Structures, V. 33, Oct. 2000, pp. 483–491.

Joshi, R.C., Chatterji, S., Achari, G., and Mackie, P., "Technical note : re-examination of ASTM C1202", September, 1999.

Khoury, G.A., Sullivan, P.J.E., and Grainger, B.N., "Strain of concrete during first heating to 600°C under load", Magazine of Concrete Research, V. 37, No. 133, 1985, pp. 195–215.

Khoury, G.A, Sullivan, P.J.E., and Grainger, B.N., "Transient thermal strain of concrete: literature review, conditions within specimen and individual constituent behavior", Magazine of Concrete Research, V. 37, No. 132, 1985, pp. 131–144.

Khoury, G.A., Grainger, B.N., and Sullivan, P.J.E., "Strain of concrete during first cooling to 600°C under load", Magazine of Concrete Research, V. 38, No. 134, 1986, pp. 3–12.

Kodur, V.K.R., and Sultan, M.A., "Effect of temperature on thermal properties of high-strength concrete", Journal of Materials in Civil Engrg, ASCE, V. 15, No. 2, March/April. 2003, pp. 101–107.

Lankard, D.R., " The dimensional instability of heated Portland cement concrete", Ph.D Dessertation, Ohio State University, Cleveland, OH, 1970.

Leithner, D., "Experimental investigation of concrete subjected to fire loading", Diploma Thesis, Technical University Vienna, Austria, 2004.

Ludirdja, D. Berger, R.L., and Young, F., "Simple method for measuring water permeability of concrete". ACI material journal, V. 86, No. 55, Sep–Oct, 1989, pp. 433–439.

Luo, X., Sun, W., Chen, S.Y.N., "Effect of heating and cooling regimes on residual strength and microstructure of normal strength and high-performance concrete", Cement and Concrete Research, V. 30, 2000, pp. 379–383.

Phan, L.T., "Fire performance of high-strength concrete: A report of the state-of-the art, Building and Fire Research Laboratory", National Institute of Standards and Technology, NISTIR 5934, Dec. 1996.

Phan, L.T., and Carino, N.J., "Review of Mechanical Properties of HSC at Elevated Temperature", Journal of Materials in Civil Engineering, ASCE, V. 10, No. 1, Feb. 1998, pp. 58–64.

Phan, L.T., Lawson, J.R., and Davis, F.L., "Effect of elevated temperature exposure on heating characteristics, spalling, and residual properties of high performance concrete", Materials and Structures (RILEM), V. 34, March. 2001, pp. 83–91.

Phan, L.T., "High-strength concrete at high temperature-an overview", Utilization of High Strength/High Performance Concrete, 6TH International Symposium, Leipzig, Germany, V. 1, June. 2002, pp. 501–518.

Poon, C.S., Azhar, S., Anson, M., and Wong, Y.L., "Comparison of the strength and durability performance of normal- and high-strength pozzolanic concretes at elevated temperatures", Cement and Concrete Research, V. 31, 2001, pp 1219–1300.

Schneider, U. "Concrete at high temperatures-A general review", Fire Safety Journal, V. 13, 1988, pp. 55–68.

Ravindrarajah, R.S., Lopez, R. and Reslan, H., "Effect of elevated temperature on the properties of high-strength concrete containing cement supplementary materials', 9TH International Conference on Durability of Building Materials and Components, Australia, March. 2002.

Reis, M.L.B.C., Neves, I.C., Tadeu, A.J.B., and Rodrigues, J.P.C., "High-temperature compressive strength of steel fiber high-strength concrete", Journal of Material in Civil Engrg., ASCE, V. 13, No. 3, May/June. 2001, pp. 230–234.

Fracture Mechanics of Concrete and Concrete Structures – High-Performance Concrete,
Brick-Masonry and Environmental Aspects – Carpinteri, et al. (eds)
© 2007 Taylor & Francis Group, London, ISBN 978-0-415-44617-4

On the tensile behaviour of thermally-damaged concrete

P.F. Bamonte & R. Felicetti
Department of Structural Engineering, Politecnico di Milano, Milan, Italy

ABSTRACT: The mechanical properties of thermally-damaged concrete have been the subject of many investigations in the last fifty years. Nonetheless, a noteworthy impulse in this field has been promoted in recent times by the advent of high-performance concrete, because the wide range of materials at issue makes the well-established references difficult to be generalized. This is particularly true for the tensile response, due to the challenging experimental conditions and to the still not standardized test methods. In this paper the results collected by the authors on a number of different concrete mixes are drawn together with the twofold objective to clarify the relations among the most common direct- and indirect-testing techniques and to sketch any possible general trend in the tensile properties of ordinary and special concretes exposed to high temperature.

1 INTRODUCTION

Concrete is known to exhibit a good behaviour at high temperature, owing to its incombustible nature and low thermal diffusivity, that guarantee a slow propagation of thermal transients within the structural members. Nevertheless, a series of chemo-physical transformations occur in the material at increasing temperature (Khoury, 2000): the physically-combined water is released above 100°C; the silicate hydrates decompose above 300°C and the portlandite is dehydrated above 500°C; some aggregates begin to convert or to decompose at different temperatures (release of bound water, $\alpha \rightarrow \beta$ SiO$_2$-conversion of the quartz, decomposition of the limestone). At the same time, the conflicting strains of the aggregate (which expands) and of the cement paste (which shrinks as a consequence of drying) activate a diffuse set of micro-cracks in the transition zone between these two phases. In some cases the thermal strain and the vapour pressure build-up lead to spalling, i.e. the sudden expulsion of concrete chips, that has the effect of exposing deeper layers of concrete to fire, thereby increasing the rate of heat transmission.

The mechanical response of the material is considerably weakened by the above-mentioned phenomena and the compressive strength decreases slowly below 450–500°C and rapidly above 500°C (RILEM, 1985a). Given the irreversible nature of the cited transformations, the decay keeps unchanged after cooling down to room temperature and only minor differences usually occur between the "hot" and the "residual" mechanical properties.

Concerning the other constitutive parameters, a more marked decrease of the Young's modulus is generally observed, whereas the tensile strength often exhibits the most temperature-sensitive behaviour (RILEM, 1985a). However, the decay may be partly offset by the increased ability of thermally-damaged concrete to undertake inelastic strains, by the smoother softening of the cohesive stress along the cracks and by the almost constant fracture energy resulting from increased crack roughness and branching (Felicetti & Gambarova, 1998). These latter favourable effects are of particular interest when the resistant mechanisms governed by the tensile behaviour of the material are at issue, as in the case of the shear capacity of beams and slabs (Beltrami et al., 1999; Felicetti & Gambarova, 2000), of the bond strength of rebars and fasteners (Bamonte et al., 2006) and of the sensitivity to spalling (Gawin et al., 2006).

Nonetheless, the beneficial influence of the reduced brittleness should also be considered in the characterization of the material tensile behaviour via the indirect testing techniques (bending, splitting, etc), because the actual strength decay may be masked to some extent by the better "structural" response of the small-size concrete specimens (Figure 1). For this reason, well-controlled direct tests are often preferred for material characterization. However, since testing in direct tension at high temperature is hardly possible, these tests are performed in residual conditions (Felicetti et al., 2000).

These general trends are liable to significant variations depending on the mix-design, to the point that the strength of some concretes may be negligible at

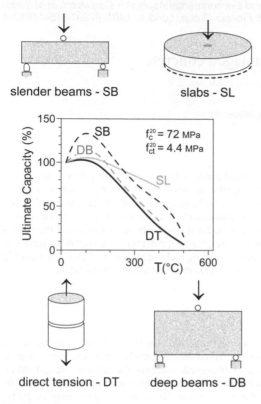

Figure 1. Decay of the residual capacity exhibited by different plain-concrete specimens.

500°C (Felicetti & Gambarova, 1998), whereas other concretes still keep a remarkable share of their initial performance (Khoury, 2000). Furthermore, the advent of High-Performance Concretes is leading to a variety of different behaviours: high-strength concrete is usually more temperature-sensitive, because of its high stiffness, low porosity and smaller defects; light-weight concrete has the advantage of the soft and thermally-stable aggregate, but it is prone to explosive spalling; self-compacting concrete is more susceptible to drying and creep strains in the cement paste.

In this paper the main experimental issues concerning the characterization of thermally-damaged concrete in tension will be discussed, in order to recognize the possible connections among the different testing techniques. Then, the results collected by the authors in the last decade will be summarized in a rather harmonized form, with the aim to outline any general relationship between the fracture parameters and the thermal damage experienced by the material.

2 MATERIALS AND TESTING TECHNIQUES

In the last decade a number of research projects have been carried out at Politecnico di Milano, on different types of thermally-damaged concrete. Since the objective was in all cases to investigate the material constitutive behaviour, an important issue was to rule out all possible "structural" effects, such as thermal gradients and spalling. This is the reason why the thermal cycles were always slow enough to allow an almost uniform distribution of the thermal damage inside the specimen (heating rate $v_h = 0.5$–$2°C/min$).

Given the minor differences from the "hot" response, most of the tests were performed in the residual state, in order to allow a complete control of the testing conditions and an accurate monitoring of the results. The different types of concrete herein considered are as follows:

A. *High-Performance Concretes for massive structures*: two concretes were examined in order to check their viability in the construction of the secondary containment shells of nuclear power plants (in the framework of the EC project "CONT", Felicetti & Gambarova, 1998). Owing to the relatively-small content of slow-reacting cement, these concretes are characterized by a low hydration heat and are suitable for thick structures. Another distinctive feature is the highly-siliceous aggregate (mostly flint consisting of quartz, opal and calcedonium) which is commonly used in certain regions of Central Europe for its good mechanical properties. However, the relatively-high content of zeolitically-bound water makes flint-based concrete very sensitive to high temperature, since the water is slowly expelled between 100 and 600°C, with relevant volume changes and subsequent splitting.

B. *Normal- and High-Performance Light-Weight Concretes*: besides a reference ordinary concrete (B1), the main interest in this case was to ascertain the role played by light-weight aggregates in fire conditions, both in a normal-strength (B2, expanded clay) and in a high-performance concrete (B3, dense expanded clay) see Felicetti et al., 2002.

C. *High- and Ultra-High Strength Concretes*: these concretes were studied in the framework of the Project Brite-Euram HITECO III, including a high-strength concrete (hyposiliceous gabbro aggregate) and two steel fibre-reinforced Ultra-High Strength Concretes (a Compact Reinforced Concrete and a Reactive Powder Concrete, not reported in this paper). Owing to the fruitful cooperation with the research team at the Imperial College of London, it was possible to run the first-ever reported direct tensile tests in hot conditions (Felicetti et al., 2000).

D. *Ordinary concretes for fastener testing*: the properties of these concretes are not very peculiar, but particular attention was devoted to the characterization of their tensile behaviour, since the research

Table 1. Main characteristics of the concretes considered in this paper.

	TA	d_{max} [mm]	a/f	w/b	f_c^{20} [MPa]	E_c^{20} [GPa]	G_f [J/m^2]	f_t	$f_{t,fl}$	$f_{t,sp}$	symbol
A1	flint	25	4.35	0.43	72	47	245	×	×		◇
A2	flint	25	4.17	0.30	95	53	229	×	×		◆
B1	siliceous	12	5.88	0.67	30	25	64	×		×	+
B2	sil. +exp. clay	12	4.76	0.63	39	16	64	×		×	×
B3	sil. +exp. clay	12	3.13	0.33	56	17	65	×		×	✳
C	gabbro	16	3.13	0.29	92	44	161	×			✿
D1	siliceous	28	8.33	0.80	20	19	97		×		□
D2	siliceous	25	4.35	0.41	63	33	146		×	×	■
E1	normal	16	3.57	0.50	52	36	83		×	×	○
E2	normal	16	2.78	0.35	83	39	132		×	×	●
E3	normal	16	2.56	0.33	90	41	142		×	×	△
E4	siliceous	4.5	2.33	0.31	125	42	42		×		▲

project was focused on the residual capacity of undercut fasteners (Bamonte et al., 2006).

E. *Self-Compacting Concretes of different grades*: not many research projects have been carried out so far on the fire behaviour of this type of concrete (Persson, 2004), especially if the tensile properties are at issue.

The previous list does not include several results on steel fiber-reinforced concrete, because in this case the main resistant mechanism is governed by the pull-out of the fibers rather than by the fracture properties of the concrete matrix, that are the object of this paper.

Table 1 summarizes the most significant factors (RILEM, 1985a) influencing the performance at high temperature of the 12 concretes examined in this paper:

- type of aggregate (TA);
- maximum size of the aggregate (d_{max});
- aggregate/fines ratio by mass (a/f);
- water/binder ratio (w/b);
- compressive strength at ambient temperature (f_c^{20}).

With reference to the type of aggregate (TA), its influence on the performance of concrete in tension at high temperature is two-fold. The nature of the aggregate, i.e. its mineralogical composition, usually influences its mechanical decay at high temperature. On the other hand, the structure of the aggregate (crushed or natural round) can influence its fracture properties, especially the fracture energy.

Other important parameters are the maximum size of the aggregate (d_{max}) and the aggregate/fines ratio (a/f): the former because of its role in the bridging effect across a crack, the latter because of its influence on the roughness of the crack surface. Moreover, the aggregate/fines ratio seems to have a significant effect on the strength decay of concrete exposed to high temperature, since the relative decay is smaller for lean mixes than for rich mixes (RILEM, 1985a).

Finally, the last four columns of Table 1 summarize the different mechanical characteristics investigated in the various concretes and the symbols used in the plots throughout the paper: reference is made to the direct- and indirect- (bending and splitting) tensile strengths, taht are the most common in concrete testing (RILEM, 1985a).

Figure 2a summarizes the decay of the compressive strength as a function of the temperature for the 12 concretes under examination. Most of the plots lie below the Eurocode 2 decay curves as a possible consequence of the relatively high values of initial strength. Moreover, the results appear to be very dispersed, because the temperature is not the best parameter to quantify the thermal damage, since the same temperature might be more or less detrimental, depending on the characteristics of the concrete. If concretes with different characteristics are to be compared, as it is the case in this paper, it is more appropriate to refer to a damage variable D, defined on the basis of the decay of the elastic modulus:

$$D(T) = 1 - E(T)/E^{20} \qquad (1)$$

Figure 2b confirms that the compressive strength tends to be less sensitive to high temperature than the elastic modulus, since all the diagrams are in the upper part of the plot for almost the whole range of thermal damage.

3 EXPERIMENTAL ISSUES IN TENSILE TESTING OF THERMALLY DAMAGED CONCRETE

As is generally recognized, the most straightforward experimental technique to ascertain the tensile response of concrete is the direct-tension test. Though very simple in principle, this test is greatly influenced by the end restraints of the specimen, since

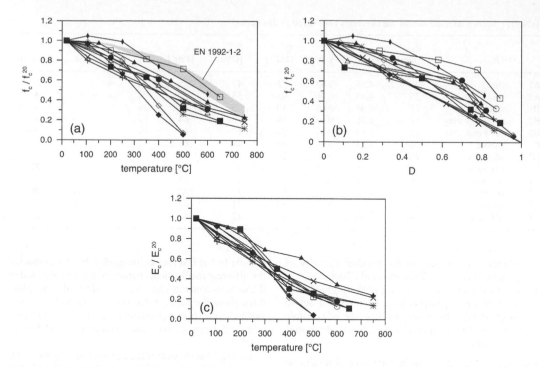

Figure 2. (a) Compressive strength decay of the different concrete mixes herein investigated and (b) its relation with the thermal damage, namely the decay of the elastic modulus, (c).

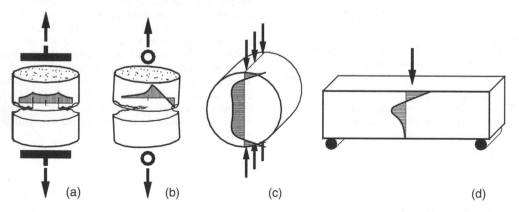

Figure 3. Outline of the stress patterns in the usual tensile tests: direct-tension with (a) restrained and (b) hinged ends; (c) splitting test; and (d) bending test.

preventing – or not preventing – the end rotations affects the inherent tendency of the specimen to bend, due to the buckling instability triggered by the softening behaviour of the material (Bazant & Cedolin, 2003).

As a result, the crack is likely to start opening along one side of the fracturing cross-section, while the opposite side turns to compression (Figure 3b). This effect is particularly evident in slightly-damaged high-strength concretes (Figure 4a), that combine

the heterogeneity ascribable to the thermally-induced flaws with a still remarkable brittleness (namely a steep softening behaviour).

The phenomenon can be prevented by forcing an almost uniform crack opening via specially-designed test setups and rather short specimens (Felicetti & Gambarova, 1998; Felicetti et al., 2000), but they are hardly practicable inside a hot furnace (RILEM, 2000). Nonetheless, the effect of the testing environment is generally less pronounced than the influence of

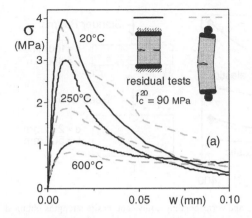
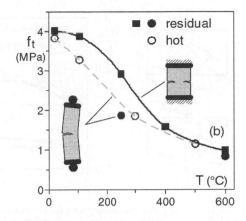

Figure 4. Effect of specimen shape and boundary conditions in direct-tension tests: (a) residual stress-crack opening curves of a thermally-damaged high-strength concrete (type C); and (b) tensile strength decay in hot and residual tests.

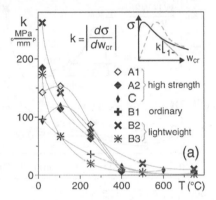
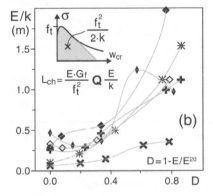

Figure 5. (a) Reduction of the slope k of the softening branch in thermally-damaged concretes; and (b) relationship of k with the material deformability (elastic modulus E); k is the mean slope of the descending branch past the peak and up to the counterflexure point.

specimen shape and boundary conditions (Figure 4b). Hence, the restrained-end residual test seems at present the most objective way to determine the fracture properties of concrete in the whole range, from pristine to completely damaged material.

The restrained-end direct-tension test not only makes it possible to accurately evaluate concrete tensile strength (that coincides with the stress at the peak of the stress-crack opening curve, Figure 4a), but allows also to determine the slope of the softening branch, which is a measure of the cohesive-stress reduction associated with the increase of crack opening. As already stated, the thermal damage makes fracture more branched and tortuous, and then remarkably smoothens the post-peak response (Figure 5a).

One consequence of the milder response is the better correspondence between the restrained- and the hinged-ends direct-tension tests in the case of severely-damaged concretes (Figure 4b). Moreover,

the improved cohesion of the cracks generally prevails over the increasing tendency to open (as a result of the increased deformability of the damaged material). A clear indication on the mutual roles of these conflicting trends is provided by the increasing values of the ratio of the elastic modulus E to the softening slope k at increasing damage levels (Figure 5). The ratio is equivalent to the Hillerborg's characteristic length, i.e. the span of a totally-developed fracture process zone, where the crack is assumed to be able to open but still capable to transfer the cohesive stresses.

This parameter is of crucial importance in the interpretation of the results of the indirect test methods, given the variable width of the fractures activated in such tests (Figure 3). In the case of bending, the equivalent strength of the material is usually worked out from the moment at peak M_{max} under the limit assumption of an elastic-brittle tensile response, which seems reasonable for pristine concrete (Figure 6). As a matter of fact, some stable propagation occurs beyond

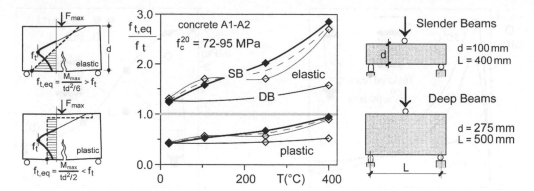

Figure 6. Possible extreme assumptions for the material response and corresponding equivalent tensile strengths obtained from three-point bending tests on specimens of different size.

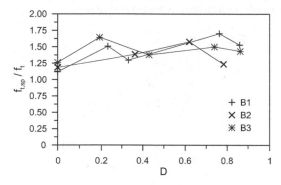

Figure 7. Effect of the thermal damage on the ratio between the splitting and direct tensile strengths.

the incipient formation of the crack and the equivalent strength is higher than the true tensile strength. The divergence rises up to a ratio of about 3 for severely-damaged small concrete specimens, denoting a deeply-propagated process zone subjected to an almost uniform cohesive stress. Under these conditions, the opposite assumption of perfect plasticity in tension seems closer to the real stress pattern within the fractured cross section.

Furthermore, the bending response is markedly affected by beam depth and the benefit of the increased material ductility is definitely more pronounced in laboratory specimens than at the scale of real structural members (e.g. the deep beams of Figure 6). For this reason, the bending test appears to be not an objective method for the assessment of concrete strength decay, unless more complex models are adopted in the interpretation of the results.

Things are expected to improve if the splitting tests are considered, owing to the better uniformity of the crack opening at the load peak, especially in the case of short cylinders, which are less affected by fracture propagation through specimen thickness. Also

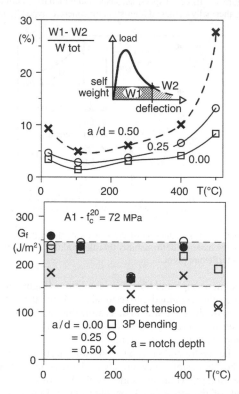

Figure 8. Work of fracture in three-point bending tests: (a) possible error in lack of counterweight system; and (b) influence of the notch depth.

in this case the equivalent material strength is generally worked out under the assumption of an elastic material, though some inelastic strains occur in the stressed regions of the specimen, be they in tension or in compression. The comparison with the actual material strength (Figure 7) shows a slightly variable ratio at increasing temperature, with a minor influence of the

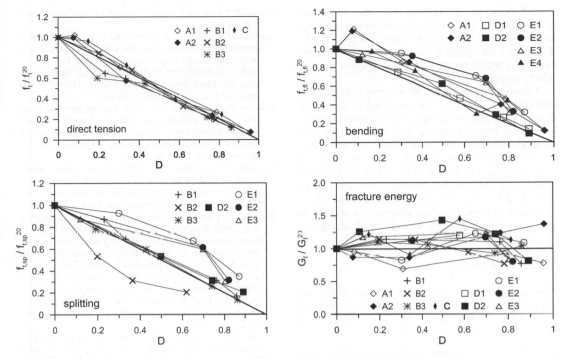

Figure 9. Plots of the materials properties as a function of the thermal damage.

improved ductility on the assessment of the strength decay.

Besides the above considerations, the bending test still remains a viable method for determining the total work of fracture of the material, provided that special attention is paid to the design of the test setup. One critical aspect to be considered is the need to cut a notch at mid-span, whose optimum depth (generally in the range 0.2–0.3d) comes from the requirement to localize the failure in a single fracture, while preserving the extent of the ligament and preventing the formation of sizcable cracks during the thermal exposure. A second aspect to be considered is the accurate measurement of the net beam deflection, because of the increasing relevance of the inelastic local settlement over the loading rollers. Finally, the work performed by the specimen self-weight should be properly taken into account, since under increasing temperatures the specimen weakens, the displacements prior to failure become sizable, the work done by the self-weight is no longer negligible and the self-weight may even trigger the anticipated collapse of the specimen. Concerning this latter point, a series of weight-compensated bending tests (Felicetti & Gambarova, 1998) showed that the work W1 performed by the self-weight before specimen breaking into two pieces does not coincide with the area W2 below the tail of the load-deflection curve (Figure 8a), as postulated in the 80's (RILEM, 1985b). Then, a suitable counterweight system should be adopted in order to perform a fully-controlled test up to the complete breaking-up of the specimen.

If all these requirements are properly fulfilled, the works associated with the fracture assessed via the bending test and the direct-tension test are practically the same (Figure 8b) and they can be both adopted for determining the dissipation capacity of thermally-damaged concretes.

4 EXPERIMENTAL RESULTS

Figure 9 shows a summary of the test results concerning the tensile properties of the concretes.

The direct tensile strength exhibits the same relative decay as the elastic modulus: this comes to no great surprise, since any possible non-linearity or redistribution due to structural effects is ruled out in this case, where all tests were performed in direct tension with fixed platens. The only exception is representted by the concretes B1 and B3, which seem to suffer a larger decay in terms of the tensile strength for low values of the thermal damage.

On the contrary, the afore-mentioned structural effects become evident should one consider the tensile strength in bending: most of the concretes seem to take advantage from this test modality, which allows some stress redistribution on the cross-section and thus leads to higher apparent values of the tensile strength. The same happens for the splitting strength, even though in this case the test set-up certainly allows less stress redistribution if compared to a bending test. Even in the case of the splitting tests, concrete B2 exhibits a behaviour which seems to be very different from the other concretes.

A further explanation of the increase in the bending and splitting strengths is to be found in the dependence of the fracture energy on the thermal damage: unlike the other properties, and even for the great variety of concretes herein examined, fracture energy does not decrease monotonically with temperature, but even increases in a certain range of values of the thermal damage D (generally between 150 and 350°C). This trend is responsible for the enhanced ductility of structural specimens at high temperature and for the subsequent increase of the indirect tensile strength.

5 CONCLUSIONS

This paper aims to give some general indications regarding the experimental techniques commonly used to measure the effects of the thermal damage on concrete. The principal conclusions are:

- the best way to measure the tensile strength is surely by testing in direct tension with restrained platens: in case of free rotation of the specimen ends, the tensile strength may be underestimated, especially for slightly-damaged concrete;
- the ratio between the splitting and the direct tensile strengths is almost constant throughout a broad range of temperatures, thus confirming the brasilian test to be a sound alternative to assess the relative strength decay;
- the tensile strength in bending, on the contrary, is far too much affected by the increasing ductility of the small-size bent specimens;
- the fracture energy can be equally-well measured in a direct as well as in a bending test, provided that some particular experimental arrangements are taken.

REFERENCES

Bamonte, P., Gambarova, P.G., Gorla, A. & Niglia, A. 2006. Residual Capacity of Undercut Fasteners Installed in Thermally-Damaged Concrete. roc. of the 2nd International fib Congress, June 2006, Naples (Italy), Vol.1; 234–235 (12-page paper on CD).

Bazant Z.P., Cedolin L., 2003. Stability of structures. Dover Publ.Inc., p1011.

Beltrami C., Felicetti R. & Gambarova P.G., 1999. Ultimate behavior of thermally damaged HSC deep beams: test results and design implications, Proc. 5th Int. Symposium on Utilization of High Strength/High Performance Concrete, Sandefjord (Norway), Ed. by Holand I. and Sellevold E.J., June 20–24, 1:137–146.

Felicetti R., Gambarova P.G., 1998. On the Residual Properties of High Performance Siliceous Concrete Exposed to High-Temperature, Special Volume in honor of Z.P. Bazant's 60th Anniversary, Prague, March 27–28, Ed. Hermes (Paris), 167–186.

Felicetti R., Gambarova P.G., 2000, On the residual behavior of HPC slabs subjected to high temperature, Proc. PCI/FHWA/FIB Int. Symp. On HPC and 46th Annual PCI Convention, Orlando (Florida), Sept. 25–27, 598–607.

Felicetti R., Gambarova P.G., Natali Sora M.P. & Khoury G.A., 2000, Mechanical behaviour of HPC and UHPC in direct tension at high temperature and after cooling. Proc. 5th Symposium on Fibre-Reinforced Concrete BEFIB 2000, Lyon (France), September 13–15, p. 749–758.

Felicetti R., Gambarova P.G., Silva M., Vimercati M., 2002, Thermal Diffusivity and Residual Strength of HPLWC Exposed to High Temperature, Proc. of the 6th Int. Symposium on Utilization of HSC/HPC, Leipzig (Germany), Vol.2, pp.935–948.

Gawin D., Pesavento F., Schrefler B.A., 2006. Towards prediction of the thermal spalling risk through a multi-phase porous media model of concrete. Comput. Methods Appl. Mech. Engrg. 195: 5707–5729.

Khoury G.A., 2000. Effect of Fire on Concrete and Concrete Structures. *Progress in Structural Engineering Materials*, 2: 429–447

RILEM-Committee 44-PHT, 1985. Behaviour of Concrete at High Temperatures. Technical Report Ed. by U. Schneider, Dept. of Civil Engineering, Gesamthochschule, Kassel Universität, Kassel (Germany), 122 pp.

RILEM TC 129-MHT *Test methods for mechanical properties of concrete at high temperatures*, 2000. Tensile strength for service and accident conditions, Materials and Structures, V.33, May, p. 219–223.

RILEM-Committee 50-FMC, 1985. "Determination of fracture energy of mortar and concrete by means of three-point bend tests on notched beams", Draft Recommendation, Materials and Structures, Vol.18, N.106, pp. 285–290.

Fracture Mechanics of Concrete and Concrete Structures – High-Performance Concrete,
Brick-Masonry and Environmental Aspects – Carpinteri, et al. (eds)
© 2007 Taylor & Francis Group, London, ISBN 978-0-415-44617-4

Experimental investigation on spalling mechanisms in heated concrete

M. Zeiml[1] & R. Lackner[2]

[1]*Institute for Mechanics of Materials and Structures, Vienna University of Technology, Vienna, Austria*
[2]*Computational Mechanics, Technical University of Munich, Munich, Germany*

ABSTRACT: Some recent results obtained during an experimental campaign on concrete spalling in fire conditions are presented in this paper. Since spalling was visually recorded by means of a high-speed camera, the slow-motion sequences provided an insight into the size, shape, and velocity of the spalled-off pieces which made it possible to evaluate the released energy during the acceleration of the concrete chips and to clarify the causes of spalling. The two commonly-accepted theories about the causes of concrete spalling are also recalled, taking advantage of the test results. The roles of the various thermal, mechanical and hydral processes controlling concrete spalling are discussed.

1 INTRODUCTION

In case of fire loading, concrete structures are affected by various physical, chemical, and mechanical processes, leading to degradation of the material parameters of concrete and spalling of near-surface concrete layers. Especially in case of tunnel fires, with the thermal load characterized by a steep temperature increase during the first minutes of the fire and a maximum temperature exceeding 1200°C, spalling significantly reduces the load-carrying capacity of the structure. In the literature, different types of concrete spalling due to fire loading are defined (Kalifa et al. 2000; Schneider and Horvath 2002; Hertz 2003; Khoury and Majorana 2003). Depending on its origin, spalling can be divided into

1. aggregate spalling (splitting of aggregates),
2. corner spalling (i.e., corners of columns or beams fall off), and
3. surface spalling (surface layers of concrete fall off or burst out of the structural element).

Moreover, depending on the underlying physical mechanisms, spalling can be divided into

1. progressive spalling (or sloughing-off, where concrete pieces fall out of the structural element) and
2. explosive spalling (violent burst-out of concrete pieces characterized by sudden release of energy).

Two phenomena are considered to be the main causes for spalling. On the one hand, restrained thermal dilation results in biaxial compressive stresses parallel to the heated surface which lead to tensile stresses in the perpendicular direction (Bžzant 1997; Ulm et al. 1999). This type of spalling in consequence of thermo-mechanical processes can be referred to as thermal-stress spalling (Khoury and Majorana 2003; Khoury 2006). On the other hand, the build-up of pore pressure in consequence of vaporization of water (thermo-hydral processes) results in tensile loading of the microstructure of concrete (Meyer-Ottens 1972; Anderberg 1997; Consolazio et al. 1997; Kalifa et al. 2000; Schneider and Horvath 2002; Hertz 2003), which can be referred to as pore-pressure spalling (Khoury and Majorana 2003; Khoury 2006).

In this paper, results of recently-conducted fire experiments are presented where spalling was recorded visually by means of a high-speed camera. The recorded images and sequences are used to investigate the velocity of the spalled-off pieces, yielding estimates for the released energy, and to highlight the processes responsible for spalling and their respective influence.

2 SPALLING EXPERIMENTS

Within the underlying fire experiments, reinforced concrete slabs with the dimensions $0.60 \times 0.50 \times 0.12$ m made of concrete C30/37 and C60/75 (water/cement-ratios of $w/c = 0.35$ and 0.55, respectively; limestone aggregates) were subjected to fire loading. In selected batches, air-entraining agents and/or polypropylene (PP) fibers were added to the mix design. The PP-fibers ($18\,\mu$m in diameter and

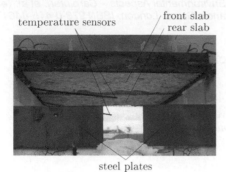

temperature sensors   front slab / rear slab

steel plates

Figure 1. Experimental setup used within spalling experiments.

3 or 6 mm long, respectively) were introduced in order to investigate the resulting improvement of the spalling behavior via a decrease of the pore pressure due to vaporization of water. Two slabs at a time were subjected to pre-specified temperature histories, i.e., the ISO fire curve (prEN1991 1-2 2002) or the HCI (hydrocarbon fire with $T_{max} = 1300°C$). Figure 1 shows the experimental setup. In order to cover the bright and reflective oven walls, steel plates were placed in areas visible on the movie sequences. During the fire experiments, the temperature history was recorded in the oven and at selected depths from the heated surface. Spalling was recorded visually by (i) a video camera, producing real-time movies, and (ii) a high-speed camera recording selected spalling events at a rate of 250 frames per second.

Since spalling (especially explosive spalling) is a very quick process, only the slow-motion sequences could be used to investigate the size/shape and to compute the velocity of the concrete pieces that got detached from the heated surface. The intensity of certain spalling events could be assessed from the acoustics of the event and the from time sequences before and after certain spalling events, accessible via the real-time movies.

Within the experiments, all previously mentioned types of spalling were observed. Whereas the most violent type of spalling (explosive spalling) was observed mostly as surface spalling, corner spalling was rather slow (i.e., progressive spalling with low velocity). Moreover, the size of some pieces originating from corner spalling was considerably bigger than the size of all pieces from surface spalling. The size (and, hence, the mass) was found to be inversely-proportional to the velocity of distinct pieces, with smaller velocities for bigger pieces. Aggregate spalling was observed to be of explosive as well as progressive type. In some events, no distinct spalled-off pieces could be identified but a cloud of chips was visible.

$t = 0$

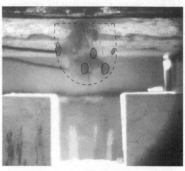

$t = 8$ ms

$t = 16$ ms

Figure 2. High-speed camera images from spalling experiments (C60/75): position of spalling front for three time instants.

Figure 2 shows three screen shots of a selected spalling event (surface spalling). The dashed lines mark the location of the spalling front at selected time instants, whereas the solid lines mark distinct spalled-off pieces. It can be seen that the spalling front (i.e., the fastest small-sized chips) moves faster than the distinct spalling pieces and also that the pieces move with different velocities. For the depicted spalling event, the velocity of the spalling front was calculated as $v_{front} = 12$ m/s, whereas the velocity of the five distinct pieces lied within the range of $5.6 \leq v_{piece} \leq 12$ m/s.

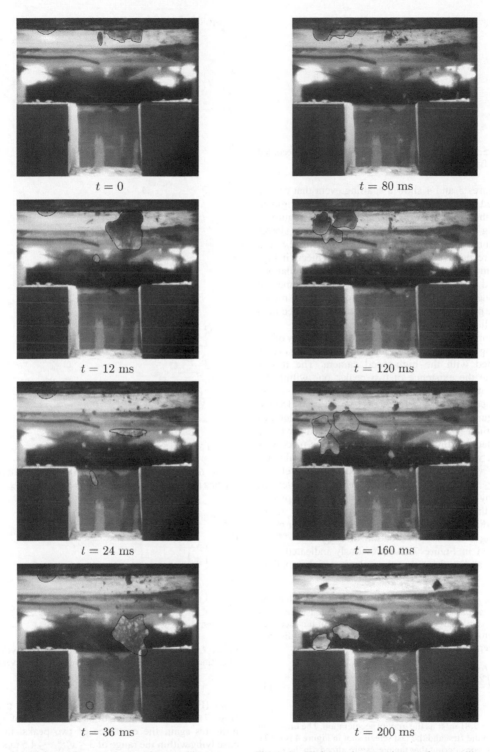

$t = 0$

$t = 80$ ms

$t = 12$ ms

$t = 120$ ms

$l = 24$ ms

$t = 160$ ms

$t = 36$ ms

$t = 200$ ms

Figure 3.   High-speed camera images from spalling experiments (C30/37): concrete pieces with different size and velocity.

Figure 4.   High-speed camera images from spalling experiments (C30/37): concrete pieces in free fall, no further acceleration.

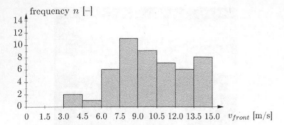

Figure 5. Distribution of velocity of spalling front recorded within fire experiments.

Figures 3 and 4 show a spalling event that can be defined as corner spalling since the pieces originate from the front edge of the specimen. In Figure 3, the big plate-like piece is apparently much slower than the small piece, with velocities being 2.8 and 6 m/s, respectively. The marked concrete pieces in Figure 4 move with an even lower velocity. Calculations showed that the velocity of the marked pieces may be explained exclusively by gravity acceleration[1], meaning the pieces were simply detached from the specimen and fell downwards after that.

Figure 5 shows the distribution of the velocity of the spalling front observed during selected spalling events recorded with the high-speed camera. The results indicate the existence of two peaks: one within the range of $7.5 \leq v_{front} \leq 9$ m/s and a second not so distinct peak within the range of $13.5 \leq v_{front} \leq 15$ m/s. In Figure 6, the minimum and maximum velocities of distinct spalled-off pieces are displayed for every recorded spalling event. In case only one piece was visible within the spalling event, minimum and maximum velocities are equal. Pieces with a velocity $v_{piece} \leq 1.5$ m/s were considered as being accelerated only by gravity forces, denoted as free-fall pieces. From the identified minimum and maximum velocity of distinct spalled-off pieces determined for every spalling event, the distribution of v_{piece}^{min} and v_{piece}^{max} is depicted in Figure 7. As previously indicated, Figures 5 and 7 show that the velocity of the spalling

---

[1] Figure 4 shows three pieces in free fall. The length of the path these pieces are visible in the slow-motion sequence is $L = 13$ cm. Starting at the bottom surface of the specimen with zero velocity, the time span for a piece to move a distance $L = 13$ cm is given by

$$t = \sqrt{\frac{2L}{g}} = \sqrt{\frac{2 \cdot 0.13}{9.81}} = 0.16 \text{ s}, \tag{1}$$

with $g = 9.81$ m/s^2 as the gravity acceleration. The time span between the first and the last screen shot in Figure 4 is 0.12 s which – considering that the pieces are already in the downward motion in the first screen shot in Figure 4 – corresponds well to the situation of free fall.

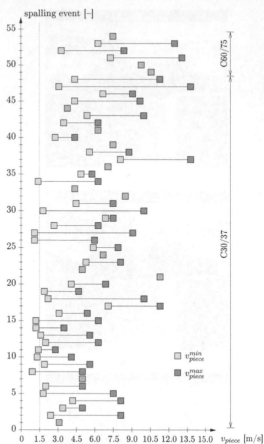

Figure 6. Minimum and maximum velocity of distinct spalling pieces determined for every recorded spalling event.

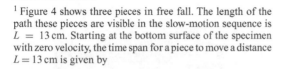

Figure 7. Distribution of minimum and maximum velocity of distinct spalling pieces within recorded spalling events.

front (Fig. 5) is greater than that of single pieces (Fig. 7). The distribution of the minimum of v_{piece} indicates again the existence of two peaks, in this case lying within the range of $3 \leq v_{piece}^{min} \leq 4.5$ m/s and $6 \leq v_{piece}^{min} \leq 7.5$ m/s, respectively. As in case of v_{front}, the second peak is smaller. The maximum of v_{piece}, on the other hand, exhibits one peak, lying within the range of $6 \leq v_{piece}^{max} \leq 7.5$ m/s.

3 DISCUSSION

The origin of spalling is still open to debate (see, e.g., (Meyer-Ottens 1972, Anderberg 1997, Bažant 1997; (Consolazio et al. 1997; Ulm et al. 1999; Kalifa et al. 2000; Schneider and Horvath 2002; Hertz 2003; Khoury and Majorana 2003; Khoury 2006)). As already indicated in Section 1, two phenomena are considered to cause spalling, namely thermo-hydral processes and thermo-mechanical processes.

When investigating the governing processes involved in concrete spalling, the released energy during a spalling event is considered (Gawin et al. 2006). Hereby, the kinetic energy of the spalled-off concrete piece is calculated. In case of thermo-mechanical processes, this kinetic energy is equal to the difference between the elastic strain energy stored in the piece prior to spalling and the fracture energy consumed during its dislocation. In case of thermo-hydral processes, this kinetic energy can be set equal to the performed work associated with the expansion of water vapor when the concrete piece is dislodged.

Numerical studies (Gawin et al. 2006) have shown that both the released elastic energy and the performed work during vapor expansion can result in a velocity in the range of $4 \leq v_{piece} \leq 5$ m/s. When a combination of the two described processes is considered, the resulting velocity becomes 7 m/s. In (Gawin et. al. 2006), both thermo-mechanical and thermo-hydral processes are considered to influence the stress state within the concrete member, whereas the former (thermo-mechanical processes) are considered to initiate cracking and, hence, spalling. The latter (thermo-hydral processes) are not considered to cause spalling without the former, i.e., the assistance of stresses in consequence of restrained thermal dilation, which are present in every concrete structure subjected to heating. Thermo-hydral processes, however, are considered to substantially contribute to the acceleration of the spalled-off piece, depending on the magnitude of the gas pressure within the concrete structure. In (Gawin et al. 2006), the ratio between kinetic energies resulting from thermo-mechanical and thermo-hydral processes lies within the range of 1:1 to 1:6.

According to (Bažant 2005), thermo-hydral processes are not the major source for explosive spalling. They, nevertheless, contribute to the triggering of fracture and crack opening. Furthermore, after cracking and during the crack opening, the pore pressure in the crack drops to zero almost instantly since the available volume increases by several orders of magnitude. Therefore, thermo-mechanical processes are considered as the major source for explosive spalling.

It is agreed upon the fact that in case of heating of concrete during fire loading, a combination of thermo-mechanical and thermo-hydral processes causes spalling (see, e.g., (Bažant 2005, Gawin et al. 2006; Khoury 2006)). Whether the former or the latter is the main driving process has not been clarified yet. In any case, the relative influence of these two processes depends on numerous factors, such as concrete strength, moisture content, heating rate, etc.

For the underlying spalling experiments, the contribution of the two described processes to the total amount of kinetic energy available for accelerating the spalled-off piece was investigated following the scheme outlined in (Gawin et al. 2006). As mentioned in (Gawin et al. 2006), the proposed quantitative analysis has only a simplified character and represents a rough estimate for the kinetic energy due to the fact that several assumptions are made (e.g., adiabatic vapor expansion, geometry of fracture and spalled piece) and some physical phenomena are neglected. Nevertheless, a trend could be extracted from investigating the selected spalling events depicted in Figures 2 to 4 in Section 2: the results indicate that the importance of thermo-mechanical and thermo-hydral processes depends on the water/cement-ratio and, hence, the concrete strength. In case of concrete with $w/c = 0.35$ (C60/75, see selected spalling event depicted in Figure 2), the ratio between kinetic energies associated with thermo-mechanical and thermo-hydral processes turned out to be approximately 1:1. In case of concrete with $w/c = 0.55$ (C30/37, see selected spalling event in Figure 3), on the other hand, thermo-hydral processes had a higher influence with the ratio being approximately 1:4. The velocities determined from the so-obtained kinetic energy and the corresponding mass of the spalled-off piece were found in the range of the measured velocities reported in Section 2.

4 CONCLUSIONS

Spalling of near-surface concrete layers considerably influences the stability of concrete structures subjected to fire loading. Two types of processes are considered to be mainly responsible for spalling: (i) thermo-mechanical and (ii) thermo-hydral processes. Within recently conducted fire experiments, spalling was recorded visually, giving insight into the size/shape and velocity of spalled-off pieces, allowing assessment of the released kinematic energy during acceleration of the concrete pieces.

Within the fire experiments, different types of spalling were observed ranging from (i) explosive spalling with velocities of up to 14 m/s and (ii) progressive spalling with smaller velocities to (iii) fall-off of concrete pieces with the gravity as the only source of acceleration.

Both thermo-mechanical and thermo-hydral processes are considered to contribute to spalling, the relative importance of the two processes has not been clarified yet and is subject of discussion. Within the underlying spalling experiments, the resulting kinetic energy from these two processes was estimated using the simplified analysis scheme proposed in (Gawin et al. 2006). The results indicated that the ratio between the kinetic energies originating from thermomechanical and thermo-hydral processes, respectively, is dependent on the w/c-ratio and, hence, the concrete strength. In case of concrete with a lower w/c-ratio (resulting in a higher concrete strength), thermo-mechanical and thermo-hydral processes were equally important, whereas thermo-hydral processes contributed to a larger amount to the total kinetic energy in case of concrete with a higher w/c-ratio.

ACKNOWLEDGMENTS

The authors wish to thank Ulrich Schneider, Heinrich Bruckner, Johannes Kirnbauer, Günter Sinkovits, and Michael Baierl from Vienna University of Technology, Vienna, Austria, for the fruitful cooperation and assistance within the described fire experiments. Financial support by the Austrian Science Fund (FWF) via project P16517-N07 "Transport processes in concrete at high temperatures" is gratefully acknowledged.

REFERENCES

Anderberg, Y. (1997). Spalling phenomena in HPC and OC. In L. T. Phan, N. J. Carino, D. Duthinh, and E. Garboczi (Eds.), *Proceedings of the International Workshop on Fire Performance of High-Strength Concrete*, Gaithersburg, Maryland, pp. 69–73. NIST.

Bažant, Z. P. (1997). Analysis of pore pressure, thermal stress and fracture in rapidly heated concrete. In L. T. Phan, N. J. Carino, D. Duthinh, and E. Garboczi (Eds.), *Proceedings of the International Workshop on Fire Performance of High-Strength Concrete*, Gaithersburg, Maryland, pp. 155–164. NIST.

Bažant, Z. P. (2005). Concrete creep at high temperature and its interaction with fracture: recent progress. In G. Pijaudier-Cabot, B. Gérard, and P. Acker (Eds.), *Proceedings of the 7th International Conference on Creep, Shrinkage and Durability of Concrete and Concrete Structures*, London, pp. 449–460. Hermes Science.

Consolazio, G. R., M. C. McVay, and J.W. Rish III (1997). Measurement and prediction of pore pressure in cement mortar subjected to elevated temperature. In L. T. Phan, N. J. Carino, D. Duthinh, and E. Garboczi (Eds.), *Proceedings of the International Workshop on Fire Performance of High-Strength Concrete*, Gaithersburg, Maryland, pp. 125–148. NIST.

Gawin, D., F. Pesavento, and B. A. Schrefler (2006). Towards prediction of the thermal spalling risk through a multi-phase porous media model of concrete. *Computer Methods in Applied Mechanics and Engineering 195*, 5707–5729.

Hertz, K. D. (2003). Limits of spalling of fireexposed concrete. *Fire Safety Journal 38*, 103–116.

Kalifa, P., F.-D. Menneteau, and D. Quenard (2000). Spalling and pore pressure in HPC at high temperatures. *Cement and Concrete Research 30*, 1915–1927.

Khoury, G. (2006). Tunnel concretes under fire: Part 1 – explosive spalling. *Concrete (London) 40*(10), 62–64.

Khoury, G. and C. E. Majorana (2003). Spalling. In G. Khoury and C. E. Majorana (Eds.), *Effect of Heat on Concrete*, Udine, pp. 1–11. International Centre for Mechanical Science.

Meyer-Ottens, C. (1972). *Zur Frage der Abplatzungen an Betonbauteilen aus Normalbeton bei Brandbeanspruchung [Spalling of normal-strength concrete structures under fire loading]*. Ph. D. thesis, Braunschweig University of Technology, Braunschweig, Germany. In German.

prEN1991 1-2 (2002). Eurocode 1 – Actions on structures – Part 1–2: General actions – Actions on structures exposed to fire. European Committee for Standardization (CEN).

Schneider, U. and J. Horvath (2002). Abplatzverhalten an Tunnelinnenschalenbeton [Spalling of concrete for tunnel linings]. *Beton- und Stahlbetonbau 97*(4), 185–190. In German.

Ulm, F.-J., O. Coussy, and Z. Bažant (1999). The "Chunnel" fire I: chemoplastic softening in rapidly heated concrete. *Journal of Engineering Mechanics (ASCE) 125*(3), 272–282.

Fracture Mechanics of Concrete and Concrete Structures – High-Performance Concrete,
Brick-Masonry and Environmental Aspects – Carpinteri, et al. (eds)
© 2007 Taylor & Francis Group, London, ISBN 978-0-415-44617-4

Ultimate capacity of undercut fasteners installed in thermally-damaged high-performance concrete

P.F. Bamonte & P.G. Gambarova
Department of Structural Engineering, Politecnico di Milano, Milan, Italy

M. Bruni & L. Rossini
MS Engineers, Milan, Italy

ABSTRACT: Medium-capacity metallic undercut fasteners installed in thermally-damaged concrete has been recently the subject of a research project in Milan ($\varnothing = 10$ mm = net shank diameter). The results of the third phase of this project, concerning a high-performance concrete (HPC, $f_c - 60$–65 MPa) are presented in this paper, but the results obtained in the two previous phases are systematically recalled (low-strength concrete-LSC, $f_c = 20$–25 MPa and normal-strength concrete-NSC, 50–55 MPa). Beside room temperature, five "reference" temperatures (between 200 and 450°C) are considered, to represent as many values of the fire duration prior to the installment of the fastener. The investigation covers four values of the installment depth ($h/h_N = 0.45$, 0.60, 0.80 and 1.00, where h_N = nominal installment depth $= 10\varnothing$), and includes the mechanical and thermal characterization of the concrete. In all cases the failure was due to the damaged concrete, with the formation of a conical crack. The systematic formation of this crack was instrumental in formulating a relatively simple model based on limit analysis and on the assumption that the conical crack forms in Mode 1. The model allows to quantify the further loss of capacity that occurs in actual fire situations, where the heating rate is one order of magnitude higher that in the electric furnaces generally used in a lab.

1 INTRODUCTION

Post-installed mechanical fasteners are increasingly used in the prefabrication industry, in most industrial plants, and in a variety of structures and infrastructures often exposed to extreme load situations. As a result of the extensive use of this technology, several valuable technical documents and papers have been published in the last ten years (CEB 1994 & 1997; ACI 2001; Cook et al. 1992; Eligehausen & Ozbolt 1998; Reick 2001; Cattaneo & Guerrini 2004), to allow the designers and the technicians to make the best use of the many types of fasteners available on the market. However, limited attention has been devoted so far to certain severe environmental conditions, like – for instance – high temperature and fire (Eligehausen et al. 2004; Bamonte & Gambarova 2005).

This is the case of tunnels, whose concrete lining may be damaged during a fire, making it necessary to assess the residual capacity of the fasteners after the fire or even to replace the fasteners, to guarantee the safety of suspended pipes (for fresh-air delivery and gas expulsion) and mechanical devices (for electrical and ventilation systems). In both cases, information about the residual capacity provided by the thermally-damaged concrete is a must, since not only the thermal damage in the concrete is irreversible, but may even increase during the cooling phase. On the contrary, the steel shank of a pre-installed fastener recovers most of its original strength after cooling and is mechanically similar to the shank of a newly-installed fastener.

In this research project, medium capacity post-installed non-predrilled undercut fasteners (nominal shank diameter $\varnothing = 10$ mm) are tested, in order to investigate their behavior, after being installed in thermally-damaged concrete (Bamonte & Gambarova 2005). The attention is focused on undercut fasteners, since they are more efficient than expansion fasteners in resisting high temperatures (the bearing action of the concrete is developed farther from the heated surface). Other fasteners, like grouted and adhesive fasteners, should be mentioned as well, but they are non-mechanical devices, and their technology is totally different.

Three concrete grades ($f_c = 20$–25, 50–55 and 60–65 MPa), 6 reference temperatures (200, 250, 300, 350, 400 and 450°C at the distance $h^* = 8\varnothing_N$ from the heated surface) and 4 installment depths ($h = 45$, 60, 80 and 100 mm) are investigated, beside room

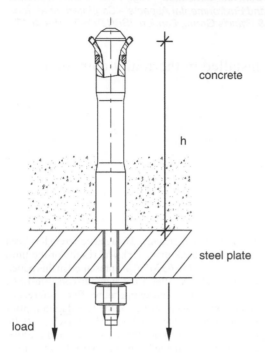

concrete

h

steel plate

load

Figure 1. Typical post-installed mechanical fastener, with non-predrilled undercut.

temperature (20°C). Because of the combinations of fastener depth and diameter, fastener failure is mostly controlled by concrete failure and not by shank yielding, except in the virgin concrete.

Other objectives of the project are: (a) the formulation of a relatively-simple limit-analysis model for the description of fastener failure due to concrete fracture (prefixed conical surface, linear crack-opening distribution at the crack interface and cohesive stresses dependent on crack opening); and (b) the definition of a reduction factor to be applied to the ultimate capacity of a fastener installed in slowly-heated concrete, in order to introduce the further damage ensuing from the high thermal gradients caused by an actual fire (fast heating).

During the experimental campaign, a third objective emerged: it became very clear that the well-known C-C Method (Concrete-Capacity Method) (CEB 1994) had to be reformulated in order to explain the extra-capacity of the fasteners installed in the low-grade concrete used in this study (LSC, $f_c = 20$ MPa), with respect to the fasteners installed in the high-grade concretes used in the first and third phase of this research project (NSC, $f_c = 52$ MPa and HPC, 65 MPa, see Bamonte & Gambarova 2005).

Sixty fasteners (Figure 1) were post-installed in as many thermally-damaged concrete blocks, simulating 5 values of the fire duration. The ultimate capacity turned out to be a highly-decreasing function of the

reference temperature. Preliminarily, the mechanical and thermal properties of the concrete mixes were measured, and the thermal properties were implemented into a FE code, in order to predict the temperature distribution inside each specimen.

The thermal field was then compared with the temperature values measured inside the first specimen of each concrete grade, by means of a set of thermocouples, and the agreement was very good.

Since the thermal damage in the concrete mostly depends on the maximum temperature reached during the heating process (RILEM 1985; Felicetti & Gambarova 1998; Phan & Carino 2002; Cheng et al. 2004), the residual capacity of a fastener is a good indication of its high-temperature capacity.

As for the formulation of a model to describe the further damage induced by fast heating (as in actual fires), the roughly conical fractures typical of the tests on headed studs indicate a major role for crack opening and cohesion (Mode I: only past the peak load does some slip occur). For such a reason, a simple model based on the transmission of normal stresses is under develop-ment, and the necessary checks are still in progress.

2 EXPERIMENTAL PROGRAM

2.1 Test philosophy and temperature-related problems

As already mentioned, this project is focused on non-predrilled undercut fasteners of medium capacity, since this type of fasteners is interesting in terms of fire-sensitivity and pull-out capacity. As a matter of fact, the large depth of large fasteners makes these devices less sensitive to concrete damage, for any reasonable fire duration, while small fasteners are of less importance for their limited capacity. A single diameter was considered (nominal shank diameter $\emptyset_N = 10$ mm; net diameter $\emptyset = 8.6$ mm; outside diameter of the whole body $\emptyset_0 = 18$ mm; suggested nominal depth of the drilled hole $h_N = 10\emptyset_N = 100$ mm).

The effective depth of the fasteners was limited in most cases to $8\emptyset_N$ to represent a situation where the heavily-damaged concrete layer exposed to the fire has been removed after the fire, to be replaced with a new concrete or mortar layer (thickness h_0), that has no structural relevance. In such a case, even if the depth of the fastener is given the nominal value (h_N), the effective depth is smaller ($h = h_N - h_0 = 0.8h_N$ in most of the cases investigated in this study).

The value $h^* = 80\% \ h_N (= 80$ mm) has thus been adopted as a reference for the maximum temperature reached inside each specimen, and as a criterion upon which comparisons are made between "slow heating" (in the electric furnace) and "fast heating" (typical of both standard fires and real fires). Consequently,

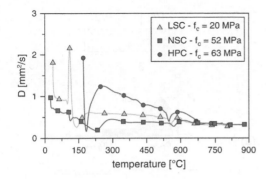

Figure 2. Plots of the thermal diffusivity of the concrete measured in this study and adopted in the analysis.

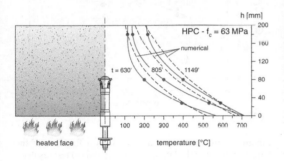

Figure 3. Temperature profiles: measured values (full curves) and thermal analysis (dashed curves) for three different values of the heating duration.

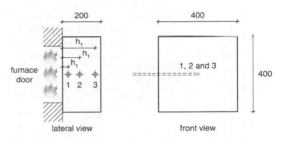

Figure 4. Typical preliminary specimen used to determine the temperature distribution inside the concrete blocks.

Figure 5. Loading rig during a typical pull-out test.

in each test reference was made to the temperature reached at the depth h^* (T = 200, 250, 300, 350, 400 and 450°C).

In one specimen of each concrete grade, the temperature profile was monitored by means of 3 thermocouples installed at approximately 25, 80 and 175 mm from the heated surface (specimen thickness — 200 mm, Figure 4). In Figure 2, the diffusivity of the concretes adopted in this study is plotted as a function of the temperature; the measurement of the thermal diffusivity allows to perform numerical simulations to evaluate the temperature profiles at any given heating duration. An example is shown in Figure 3, where the measured temperature profiles in the HPC specimen are compared with the corresponding numerical simulations: note that, although the overall trend of the temperatures is correctly described, sizeable differences are observed at the exposed and unexposed surfaces.

In the following, the thermal profiles used in the interpretation of the test results will be taken from the experimental measures, using a third-degree interpolating function along the depth of the specimen. The numerical simulation will be used only to work out the thermal profiles ensuing from the ISO-834 Fire Curve.

Sixty concrete blocks – or specimens – were cast (Figure 4: dimensions $400 \times 400 \times 200$ mm in 55 blocks and $400 \times 300 \times 200$ mm in the remaining 5 blocks). All blocks were fastened to the front face of the chamber of the furnace (Figure 4), with the door open, in order to have one of the face of the specimen exposed to high temperature. After the cooling process, a hole was drilled in the centroid of the heated face of each specimen, up to the required depth. Then the fastener was installed and tested in the next few days (maximum one week past the cooling). The loading set-up was a simple steel rig (Figure 5) consisting of a loading ring, three inclined legs, a reaction plate and a hydraulic actuator. One LVDT was used to measure the displacement of the fastener with respect to the undisturbed concrete mass, and to control the test. All tests were displacement-controlled, and the displacement rate was $ds/dt = 0.5$ mm/min (1.0 mm/min in the final part of the test, at the end of the softening branch).

In each case (corresponding to a given couple of values of the reference temperature and of the installment depth), at least two nominally-identical specimens were tested for repeatability.

The diagrams of concrete mechanical and physical properties are plotted in Figures 6 and 7 (cylindrical compressive strength, direct tensile strength, stabilized

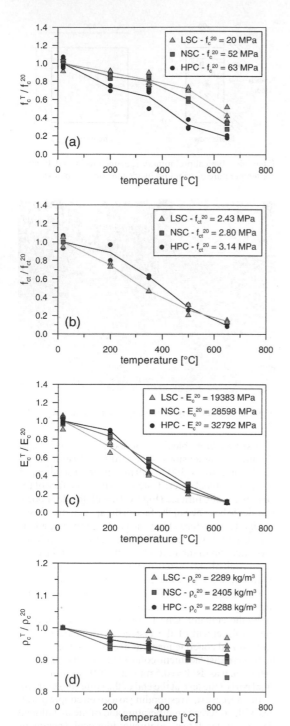

Figure 6. Residual mechanical and physical properties as a function of the temperature: (a) cylindrical compressive strength; (b) direct tensile strength; (c) stabilized elastic modulus; and (d) mass per unit volume.

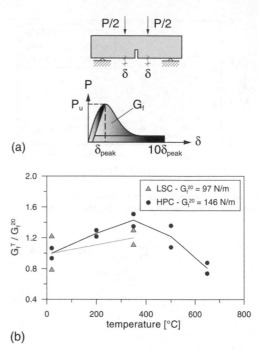

Figure 7. Residual fracture energy as a function of the temperature: (a) calculation procedure; and (b) test results.

elastic modulus, mass per unit volume and fracture energy) as a function of the temperature.

With reference to the fracture energy (Figure 7), only two reference temperatures were considered in the case of low-strength concrete (T = 20 and 350°C, for want of specimens, Fig. 7b), compared to 5 values in the case of high-performance concrete (T = 20, 200, 350, 500 and 600°C, Figure 7b). The direct tensile strength (f_{ct}) was evaluated by multiplying the indirect tensile strength measured in three- (LSC) or four-point bending (HPC) ($f_{ct,fl}$: size of the prisms $600 \times 150 \times 150$ mm) by the size-dependent factor $\{[1.5 \cdot (h/h_o)^{0.7}]/[1 + 1.5 \ (h/h_o)^{0.7}]\}$, where h is the actual depth of the section of the prisms and $h_o = 100$ mm is a reference depth (MC90). The fracture energy was evaluated as the area enveloped by the load-displacement curve minus the energy dissipated during nonlinear cracking phenomena taking place during the loading phase prior to the attainment of the maximum load (Figure 7a). Contrary to the other mechanical properties, fracture energy increases up to 300–400°C and then starts decreasing (at 500–600°C the values are very close to those in the virgin conditions, see also Zhang & Bicanic 2002).

2.2 Test results

Considering the various parameters coming into play and the difficulties encountered during the experimental campaign, it was possible to perform a total of 60 successful tests, Specimen's failure was always controlled by concrete fracture in the thermally-damaged specimens and in LSC virgin specimens, while NSC and HPC virgin specimens failed because of shank yielding. (In NSC and HPC virgin specimens, the ultimate load ensuing from concrete fracture was evaluated by means of the CC* method, as explained in the following). As a rule, two nominally-identical specimens were tested in each case for repeatibility, except in one case, where the scattering of the test results required a third test.

In most cases a clear-cut conical fracture occurred, with and without thin radial cracks (Bamonte et al. 2006). In some cases, the conical fracture was accompanied by the subdivision of the specimen into 2–4 blocks, but in most of these cases the overall cracking was the result of the simultaneous formation of the conical crack and of cross-shaped radial cracks, as if the two mechanisms were activated at the same time. Moreover, the formation of these cracks and of the ensuing failure mechanism was observed after the attainment of the peak load: thus, it is reasonable to assume that the bearing capacity of the different specimens was not affected by the evolution of cracking past the peak load and specifically by the radial cracks, that appeared only in the final stage of the loading process.

2.3 Interpretation of the test results at 20°C

A number of preliminary tests at room temperature were performed, for different values of concrete grade and embedment depth, in order to study the influence of the different parameters coming into play. As it is well known, the original C-C Method is based on the following formulation for the ultimate capacity:

$$P_u = k_1 f_c^{0.5} h^{1.5} \tag{1}$$

Eq.(1) is valid for a given family of concretes, since all the parameters (maximum aggregate size d_a, aggregate type, cement type and content, hydraulic adjuncts, . . .) are lumped into a single coefficient (k_1).

However, if different mixes are to be compared as in this project, with very different fracture properties, the C-C Method proves to be inadequate, and Eq.(1) should be reformulated. In the following, in Eq.(1) the tensile strength of the concrete ($f_{ct} \approx f_c^{0.5}$) is replaced by the fracture energy G_f, that is given the formulation proposed in MC90 (1991):

$$G_f = 0.2 \cdot (10 + 1.25 d_a) f_c^{0.7} \tag{2}$$

By introducing the fracture energy it is possible to take care of the maximum aggregate size (d_a) that controls crack roughness and cohesion:

$$P_u = k_2 G_f h^{1.5} \tag{3}$$

Table 1. Main parameters of the tests at 20°C considered in Figure 8 for the evaluation of the parameter k_2 in eq.(3).

	h [mm]	d_a [mm]	f_c [MPa]	author
▣	60	22	26	Eligehausen et al.
+	65	19	27	prelimary test
⊠	65	20	52	NSC, 1st phase
☎	60	19	27	preliminary test
▣	60	28	20	LSC, 2nd phase
×	45	25	63	HPC, 3rd phase

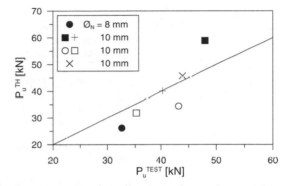

Figure 8. Fitting of the test results at 20°C by means of the proposed reformulation of the CC-Method (CC*-Method, Eq.(3)).

where k_2 is a parameter whose value mostly depends on fastener type ($k_2 = 1.01$ is the mean value in the tests carried out by the authors and by Eligehausen & Ozbolt 1998). Note that Eqs.(1) and (3) take care of the so called size effect ($\approx h^{-0.5}$, CEB 1997; Eligehausen & Ozbolt 1998). Since the pull-out of the fasteners (except in the case of shank yielding) is controlled by concrete fracture, introducing concrete fracture energy in the formulation of the pull-out load looks very reasonable, as shown by the somewhat better agreement between the test results and the theoretical predictions. (However, a rather large scattering cannot be ruled out, since parameters other than d_a and f_c come into play).

Moreover, in Eq.(3) the possible advantage of explicitly introducing the aggregate is weakened by the enhanced dependence on the compressive strength compared to Eq.(1). A further improvement could be the explicit introduction of the role of the aggregate size by means of the bilinear expression contained in Eq. (2), together with the dependence on the square root of the compressive strength, as in Eq. (1).

In Figure 8 the results obtained in this project at 20°C (1st, 2nd and 3rd phases), some test results by other authors and the values of the ultimate capacity predicted by Eq.(3) are compared for $d_a = 19$–28 mm, $f_c = 20$–63 MPa.

All the test results obtained in this study are gathered together in Figure 9, that clearly shows that the

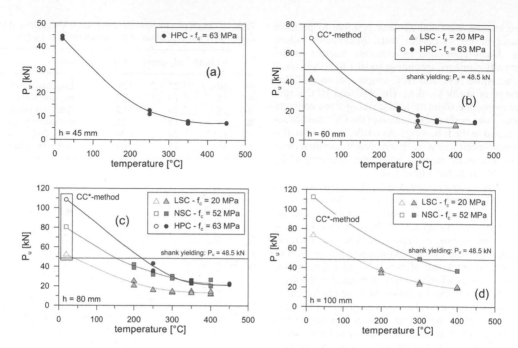

Figure 9. Residual pull-out loads as a function of temperature: (a) h = 45 mm; (b) h = 60 mm; (c) h = 80 mm; and (d) h = 100 mm.

failure mode is markedly affected by the temperature. Even if in the virgin specimens (T = 20°C), at medium and large installment depths, and for the highest-grade concretes, the pull-out failure is caused by shank yielding, the ultimate load-temperature curves show a regular descending trend, a marked "flattening" above 300°C and a tendency towards a smaller role for concrete strength with increasing temperature (there are marginal differences between NSC and HPC above 250°C, Fig. 9c, and limited differences between LSC and HPC above 300°C, Fig. 9b).

For instance, Figure 9c shows that after reaching 400°C at h/h_N = 0.8, the pull-out capacity is roughly 1/5 of the capacity of the virgin specimens – if reference is made to concrete fracture –, and from 1/5 (LSC) to 1/3 (NSC and HPC) – if reference is made to shank yielding.

As previously observed in Bamonte & Gambarova (2005) with reference to the C-C Method, both the C-C Method and the Modified C-C Method (C-C*) are totally unable to describe the mechanical decay of fasteners installed in thermally-damaged concrete. Of course, the average temperature of the concrete between the head of the fastener and the heated surface should be introduced, but even in this case the afore-mentioned models tend to overestimate the capacity of a fastener by up to 3 times.

There are at least two reasons: firstly the tensile strength is more affected by the temperature than

the compressive strength, and secondly the thermally-induced microcracks tend to damage the crack interface to the detriment of the roughness induced by the aggregates. However, further tests and deeper studies are needed, since concrete fracture energy is an increasing function of the temperature up to 300°C (Figure 7).

2.4 High and low heating rates

The heating rate in an electric furnace generally does not exceed 10–15°C/minute, which is an order of magnitude less than in real fires. Consequently, an answer should be given to the question to what extent the capacity measured under low heating rates (as in the tests performed in this project) is overestimated compared to that under high heating rates.

In order to compare the ultimate capacities of a fastener installed in a slowly-heated or rapidly-heated concrete block, a theoretical model is under development, with the following assumptions:

(a) the ultimate capacity is reached when the opening of the conical crack has the critical value under the head and zero value along the heated surface; as a matter of fact, the fitting of the test results shows that the ultimate resistance is reached before the attainment of the critical value w_{cr}^T under the head. The coefficient β takes care of such fact (β = 0.25 for LSC; = 0.15 for NSC and = 0.05 for

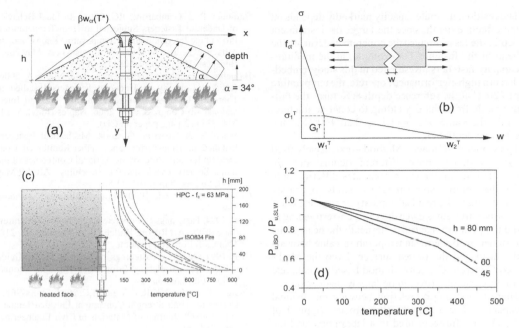

Figure 10. (a) Limit analysis model; (b) bilinear softening law in tension; (c) temperature profiles for fast heating (ISO 834, dashed curves) and slow heating (these tests, continuous curves), for the same temperature at the reference depth (h* = 80 mm, T = 200, 300 and 400°C); and (d) ultimate capacity under fast heating normalized to that under slow heating, as a function of the temperature (HPC).

HPC). Since for $\beta = 0$ there is no softening, β may be considered as a "toughness parameter" (as it is well known, the lower the grade, the greater the toughness for the same aggregate size);

(b) the profile of the opening is linear (see "w" in Figure 9a);

(c) the stress-crack opening law of the material is a bilinear decreasing function, where the tensile strength and the fracture energy are a function of the temperature (Figure 10b; $\sigma_1^T = f_{ct}^T/4$, $w_1^T = 0.75 G_f^T/f_{ct}^T$, $w_2^T = 5 G_f^T/f_{ct}^T$);

(d) the fracture energy is variable with the temperature;

(e) the pull-out force is the integral of the cohesive stresses transmitted at the crack interface, where the local stress is related to the local crack width, through the temperature-dependent stress-crack width law.

The temperature profiles (Figure 10c) and the ensuing temperature values along the conical crack in the case of a standard ISO-834 Fire were evaluated by means of numerical simulations, taking advantage of the diffusivity measured in the preliminary tests (Figure 2). The criterion used to compare the exposures to fires of different heating rates was the temperature reached at the reference depth of h* = 80 mm. With reference to HPC ($f_c = 63$ MPa), Figure 10c shows the temperature profiles in the case of a slow-heating process (continuous curves) compared to the temperature profiles ensuing from standard ISO-834 Fire (dashed curves).

Modeling the various situations (h = 45, 60, 80 mm, low heating rate and standard ISO-834 fire), leads to the curves shown in Figure 10c, where the ratio between the capacity after a standard fire and that after a slow heating process is plotted as a function of the reference temperature reached at the depth h*. For T = 400°C, the capacity after the standard fire is 75%, 65% and 60% of the capacity after a slow heating, depending on the installment depth (h = 80, 60, 45 mm).

3 CONCLUDING REMARKS

In spite of the limits set by the technology used in the tests and by the choice of the specific fasteners adopted in this project, some general remarks can be formulated, to help producers, designers and technicians in their day-to-day activity aimed at structural safety:

• Mechanical undercut fasteners installed in thermally-damaged concrete exhibit a strong temperature sensitivity, that seems to be larger in low-grade concretes than in high-grade concretes.

1735

- The residual ultimate capacity markedly depends on the effective depth, since the larger the installment depth, the less damaged the concrete underneath the head of the fastener. For example, in the medium-capacity fasteners investigated in this study, embedded in a high-performance concrete, the temperature of $220°C$ at the reference depth $8\emptyset_N$ turns the failure mode from shank yielding to concrete fracture, while the same occurs at less than $100°C$ at the effective depth $6\emptyset_N$.
- The Concrete-Capacity Method – extensively used when fasteners have different depths and are installed in similar virgin concretes differing only for the compressive strength – tends to grossly overevaluate the residual capacity, if concrete compressive strength is given the value corresponding to the temperature reached underneath the head of the fastener or to the mean temperature value between the head and the heated surface. Even the Modified Concrete-Capacity Method based on concrete fracture energy proves to be largely inadequate.
- There is a strong correlation between the residual capacity after a slow heating process (typical of the electric furnaces used in a laboratory) and the fast heating process associated with the standard fire. The latter situation leads to definitely-lower capacities. However, slow heating processes are instrumental in avoiding concrete spalling in a lab. For instance, after reaching $350°C$ at the reference depth $h = 0.8\,h_N$, depending on the actual fastener depth the capacity during the standard fire is from 44% to 66% of that under slow heating for the high-performance concrete investigated in this study.

ACKNOWLEDGEMENTS

The financial support by Italcementi Group – C.T.G. (Bergamo, Italy) and the supervision of CIS-E (Int. Center for the Constructions in Europe) are gratefully acknowledged (Research Project "Experimental and Theoretical Investigation on the Mechanical Behavior of Post-Installed Undercut Fasteners Installed in Thermally-Damaged Plain and Fiber-Reinforced High-Performance Concretes", March 2004).

The technicians Antonio Cocco and Paolo Broglia of the Laboratory for Building Materials should also be praised for their commitment to the success of this project.

REFERENCES

ACI. 2001. Evaluating the Performance of Post-Installed Mechanical Fasteners in Concrete and Commentary. ACI 355.2/ACI 355.2R, Concrete International, February 2001: 106–136.

Bamonte, P. & Gambarova, P.G. 2005. Residual Behavior of Undercut Fasteners Subjected to High Temperatures. Proc. of the International fib Symposium "Keep Concrete Attractive", edited by Balázs G, Borosnyói A, May 2005, Budapest (Hungary); Vol.2: 1156–1163.

Bamonte, P., Gambarova, P.G., Gorla, A. & Niglia, A. 2006. Residual Capacity of Undercut Fasteners Installed in Thermally-Damaged Concrete. Proc. of the 2nd International fib Congress, June 2006, Naples (Italy), Vol.1; 234–235 (12-page paper on CD).

Cattaneo, S. & Guerrini, G. 2004. Mechanical Fasteners Installed in High-Performance, Fiber-Reinforced Concrete (in Italian). Proc. of the National Conference of the Italian Society for R/C and P/C Structures – AICAP, May 2004, Verona (Italy): 101–112.

CEB-FIB Model Code 1990. Thomas Telford ed., London, 1991.

CEB. 1994. Fastenings to Concrete and Masonry Structures. State-of-the-Art Report, Bulletin d'Information No. 216, Thomas Telford ed., London.

CEB. 1997. Design of Fastenings in Concrete. Design Guide – Parts 1,2,3, Bulletin d'Information No. 233, Thomas Telford Ed., London.

Cheng, F.P., Kodur, V.K.R. & Wang, T.C. 2004. Stress-Strain Curves for High Strength Concrete at Elevated Temperatures. ASCE Journal of Materials in Civil Engineering; 16(1): 84–90.

Cook, R.A., Collins, D.M., Klingner, R.E. & Polyzois, D. 1992. Load-Deflection Behavior of Cast-in-Place and Retrofit Concrete Anchors. ACI Structural Journal; 89(6): 639–649.

Eligehausen, R. & Ožbolt, J. 1998. Size Effect in Design of Fastenings. Special Volume on "Mechanics of Quasi-Brittle Materials and Structures", edited by Pijaudier-Cabot G, Bittnar Z, Gérard B, HERMES, Paris: 95–118.

Eligehausen, R., Kožar, J., Ožbolt, J. & Periskic, G. 2004. Transient Thermal 3-D FE Analysis of Headed Stud Anchors Exposed to Fire. Proc. of the International Workshop "Fire Design of Concrete Structures: What now? What next?", edited by Gambarova PG, Felicetti R, Meda A, Riva P, December 2004, Milan (Italy): 185–198.

Felicetti, R. & Gambarova, P.G. 1998. Effects of High Temperature on the Residual Compressive Strength of High-Strength Siliceous Concretes, ACI Materials Journal; 95(4): 395–406.

Phan, L.T. & Carino, N.J. 2002. Effects of Test Conditions and Mixture Proportions on Behavior of High-Strength Concrete Exposed to High Temperatures. ACI Materials Journal; 99(1): 54–66.

Reick, M. 2001. Fire Behavior of Fasteners Embedded in a Concrete Mass and Subjected to a Pull-out Force. PhD Dissertation, News of the Dept. of Building Materials – IWB; V.2001/4, University of Stuttgart, 166 pp.

RILEM-Commitee 44-PHT. 1985. Behavior of Concrete at High Temperatures. Edited by Schneider U, Department of Civil Engineering, Gesamthochschule Kassel, Kassel (Germany), 122 pp.

Zhang B. and Bicanic N. 2002. Residual Fracture Toughness of Normal- and High-Strength Gravel Concrete after Heating to 600°C, ACI-Materials Journal, V.99, No.3, pp. 217–226.

Fracture Mechanics of Concrete and Concrete Structures – High-Performance Concrete,
Brick-Masonry and Environmental Aspects – Carpinteri, et al. (eds)
© 2007 Taylor & Francis Group, London, ISBN 978-0-415-44617-4

Thermal stress analysis and fatigue strength measurement of the interface between concrete and polymer cement mortar

T. Matsumoto & T. Mahaboonpachai
Department of Civil Engineering, University of Tokyo, Tokyo, Japan

Y. Inaba
Kajima Technical Research Institute, Kajima Corporation, Tokyo, Japan

R. Rumbayan
Polytechnic of Manado, Manado, Indonesia

ABSTRACT: This paper mentions two major studies on the delamination of external wall tile structures. External wall tile structures are usually made of tile, adhesive mortar, and concrete, and the interfaces between layers are considered to be a weak location against delamination. First, from the viewpoint of reaction, the heating experiment of a small scale tile structure was conducted. The temperature distribution in the through-thickness direction was measured with a thermography, and, based on the measured temperature, the stresses at the interface are estimated with simple analytical models. Next, from the viewpoint of resistance, the measurement of interfacial fracture energy and fatigue strength between concrete and polymer cement mortar were carried out. The interfacial bond strengths and their S-N diagrams are obtained. Based on the studies of reaction and resistance, the mechanisms of interfacial failure observed in external wall tile structures are discussed in order to achieve the durability of the interface and the tile structures.

1 INTRODUCTION

This paper mentions preliminary studies under a research program that has been conducted for the purposes of understanding the mechanisms of interfacial failure observed in external wall tile structures and achieving the durability of the interface and the tile structures.

The external wall tile structure of a building is composed of three different materials: ceramic tiles, adhesive mortar, and concrete. Therefore, bi-material interfaces are inevitable in the tile structure. Under service conditions, these interfaces are subjected to environmental actions such as temperature cycles, dry and wet cycles, UV radiation, chemical pollutants, and so on, and it is observed that the interfaces are the location where the failure of the tile structure initiates (Kumagai 1991).

Among these environmental actions, temperature cycles, which lead to the cycles of interfacial stresses in the normal and shear direction, are considered to be important. If these interfacial stresses are significant compared to their static strength, it is possible that interfacial delamination takes place and its fatigue propagation leads to the failure and fall-off of the tile

structure. Therefore, in order to understand the mechanisms of interfacial failure, the current study conducts the thermal stress analysis of the interface via temperature measurement with a thermography, and also conducts the experimental measurement of interfacial fracture energy under static loading and of fatigue strength under fatigue loading. Namely, these studies are conducted in order to clarify the "reaction" and "resistance" of the interface in the tile structure.

First, the heating experiment of a tile structure will be explained. The experiment was carried out in order to obtain the temperature distribution of a tile structure in the thickness direction. Using the measured temperature distribution, thermally induced interfacial stresses are calculated based on simple analytical models.

Next, the fracture energy and the fatigue strength measurement of the bi-material interface between concrete and polymer cement mortar will be explained. Based on interfacial fracture mechanics, bi-material interface specimens were tested under the mixture of normal and shear stress, and the interfacial fracture energy was measured by taking into account the mismatch parameters of two materials. The interfacial fracture energy is also analyzed with finite element analysis so that the tensile and shear bond strengths can be calculated. Furthermore, the fatigue test was carried out for the same type of bi-material interface specimens in order to grasp the fatigue strength characteristics in terms of S-N diagrams.

Finally, based on the thermal stress estimate as reaction and the bond strengths and the S-N diagrams as resistance, the possible failure mechanisms of the bi-material interface in the tile structures will be discussed.

2 THERMOGRAPHIC MEASUREMENT OF THE TILE STRUCTURE AND THERMAL STRESS ANALYSIS AT THE INTERFACE

First, this chapter explains the thermographic measurement of temperature distribution of the tile structure under monotonic thermal loading, where the thermal loading was given simulating the real building environmental conditions.

Second, in this chapter, using the measured temperature distribution, thermal stresses at the interface are estimated based on simple analytical models for shear and normal stress.

2.1 Material and specimen

The tile structure is made of three layers: concrete substrate, adhesive mortar, and tiles. A small scale specimen was fabricated for the current measurement purpose (Figure 1). The specimen has the thickness of 110 mm, and its tile surface side has the area of 200 mm by 100 mm.

The concrete layer of the specimen is made of normal strength concrete. The adhesive mortar of the current experiment is polymer cement mortar, the mix proportion of which is given in Table 1. The thickness of the adhesive mortar is 5 mm, which is commonly used in practice. The tiles are glazed porcelain with gray color, and they are 45 mm in length and width and 5 mm in thickness.

First, the adhesive mortar was mixed with water by hoe until achieving buttery consistency. Secondly, the fresh adhesive mortar was applied on a concrete plane. After five minutes of opening time, the glazed porcelain tiles were laid in. The open time may lead to reducing the wettability and also make the fresh adhesive mortar more effective in spreading over. Finally, the specimen was stored for 28 days at room temperature before thermographic temperature measurement.

Since the specimen has to represent a real tile structure in a building where heat flows only in the wall through-thickness direction, the four side surfaces of the specimen were covered by a heat insulation material. The heat insulation material is made of glass

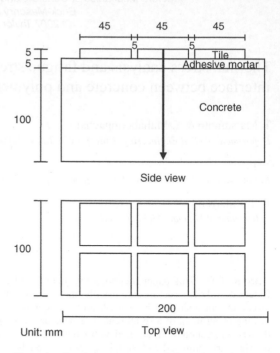

Figure 1. Specimen geometry and dimensions. The arrow shows the line along which the surface temperature is measured.

Table 1. Mix composition of adhesive mortar. W: water, C: cement, S: sand, P: Ethylene-vinyl acetate copolymer, MC: Methylcellulose (kg/m^3).

S/C	W/C	W	C	S	P	MC
50	30	335	1117	559	45	1.69

wool, and it prevents the heat from neither entering nor leaving the specimen. This was confirmed by comparing the heating experiment with and without the heat insulation. The results showed that the current heat insulation was satisfactory enough to achieve its purpose.

2.2 Heating experiment

Monotonic thermal loading was applied to the tile surface of the specimen by using two lamps. The tile surface of the specimen was heated from room temperature, about 29°C, up to around 50°C, and the measurement of temperature distribution was done with a thermography.

For the thermographic measurement, one side of the heat insulation material was opened for five seconds (the side shown in Figure 1), and the measurements were carried out at the interval of ten minutes, so that the heat loss could be minimized.

Figure 2. Thermal image of the specimen side surface at 40 minutes after heating.

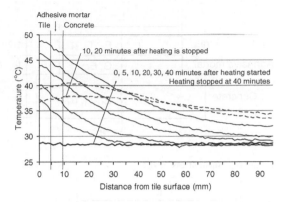

Figure 3. Temperature profile in the tile structure in the thickness direction.

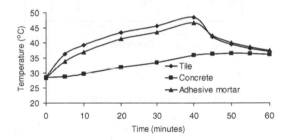

Figure 4. Average temperature of each layer.

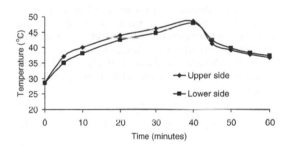

Figure 5. Tile temperature at both sides.

For the justification of these measurement procedures, a measurement was carried out without opening the side until the tile surface reached 50°C, and its result was compared with that of the current method. The difference of the measured temperature was less than 1°C. This implies that the error of the current method is less than around 2%, and also supports that the measurement on the side surface substitutes that of inside specimen.

Figure 2 shows the thermal image of the specimen side surface at 40 minutes after heating. It shows uniform enough temperature distribution horizontally. The thermographic measurements were done before heating was applied, and at 5, 10, 20, 30, and 40 minutes. At 40 minutes, heating was stopped, and the measurements were continued at 10 and 20 minutes after heating was stopped.

Figure 3 shows the temperature profile on the side surface of the specimen in the thickness direction (along the arrow in Figure 1). Before heating, the temperature was around 28.5°C through the thickness. It can be observed that the tile surface temperature quickly increased up to 37°C at 5 minutes after heating was applied. From 5 to 40 minutes, the temperature of concrete increased due to the heat conduction. At 40 minutes, the heating was stopped, and soon the tile temperature dropped from 48.6 to 41.2°C. By 20 minutes after the stop of heating, the heat conduction took place in the reverse way, resulting in the temperature decrease of concrete on the tile surface side.

For the later thermal stress analysis, the average temperature of each layer, i.e. tile, adhesive mortar, and concrete, is calculated and plotted with time in Figure 4. Also, the temperature at both sides of the tile is plotted with time in Figure 5. In Figure 4, it is seen that the temperature difference is more significant between adhesive mortar and concrete, and, in Figure 5, the maximum difference between both sides appears at 5 minutes, although it is rather small. These figures will be used for shear and normal stress estimate in the next section, respectively.

2.3 Analytical models for thermal stress analysis

Stresses arise at the interfaces due to the mismatch of elastic and thermal properties of three layers. In this section, shear and normal stress are estimated by using simple analytical models proposed by Kumagai (1994).

The shear stress model consists of three linear elastic materials: tile, adhesive mortar, and concrete (Figure 6). The main assumptions of the model are no bending moment and no shear deformation in tile and concrete.

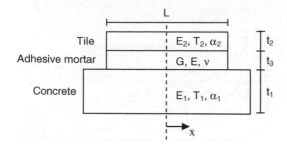

Figure 6. Shear stress model.

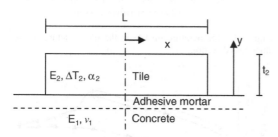

Figure 7. Normal stress model.

The shear stress, τ, can be expressed as

$$\tau = \frac{C(\alpha_1 T_1 - \alpha_2 T_2)}{\left(\dfrac{1}{E_1 t_1} + \dfrac{1}{E_2 t_2}\right) \cosh \dfrac{\beta_s}{2}} \sinh \frac{\beta_s x}{L} \qquad (1)$$

where $\alpha_1 =$ thermal expansion coefficient of concrete; $\alpha_2 =$ thermal expansion coefficient of tile; $T_1 =$ average temperature of concrete; $T_2 =$ average temperature of tile; $E_1 =$ Young's modulus of concrete; $E_2 =$ Young's modulus of tile; $t_1 =$ thickness of concrete; and $t_2 =$ thickness of tile. C and β_s are given as follows:

$$\beta_s = C \times L = \sqrt{\frac{G}{t_3}\left(\frac{1}{E_1 t_1} + \frac{1}{E_2 t_2}\right)} \times L \qquad (2)$$

where $G =$ shear modulus of adhesive mortar which can be obtained with $E =$ Young's modulus of adhesive mortar and $v =$ Poisson's ratio of adhesive mortar and where $t_3 =$ thickness of tile and $L =$ the longer side length of tile. Details of the derivation can be found elsewhere (Kumagai 1994).

The normal stress model consists of two layers: tile and lamped layer of adhesive mortar and concrete (Figure 7). It is assumed that the adhesive is very thin compared to concrete thickness and that no bending exists in concrete, meaning that tile bends over elastic foundation.

The normal stress, σ, can be expressed as

$$\sigma = yK \qquad (3)$$

Table 2. Material properties.

	Concrete	Adhesive mortar	Tile
Young's modulus (GPa)	30.7	21.6	80.0
Poisson's ratio	0.207	0.212	–
Thermal expansion coefficient (/°C)	0.000006	–	0.000008

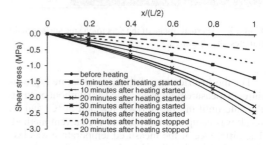

Figure 8. Shear stress distribution.

where

$$y = \frac{\cosh \beta_n x \cos \beta_n x + \phi \sinh \beta_n x \sin \beta_n x}{2\beta_n{}^2\left(\sinh \dfrac{\beta_n L}{2} \sin \dfrac{\beta_n L}{2} - \phi \cosh \dfrac{\beta_n L}{2} \cos \dfrac{\beta_n L}{2}\right)}$$
$$\frac{\alpha_2 \Delta T_2}{t_2} \qquad (4)$$

and

$$\phi = \frac{\sinh \dfrac{\beta_n L}{2} \cos \dfrac{\beta_n L}{2} + \cosh \dfrac{\beta_n L}{2} \sin \dfrac{\beta_n L}{2}}{\sinh \dfrac{\beta_n L}{2} \cos \dfrac{\beta_n L}{2} - \cosh \dfrac{\beta_n L}{2} \sin \dfrac{\beta_n L}{2}} \qquad (5)$$

ΔT_2 is temperature difference between upper and lower side of tile. Furthermore, K is normal stiffness of the lamped layer per unit length and can be obtained as

$$K = \frac{E_1}{mLd_2\left(1 - v_1{}^2\right)} \qquad (6)$$

where d_2 is the shorter side length of tile and m is tile shape coefficient, which is equal to 0.95 for square shape. Again, details can be found elsewhere (Kumagai 1994).

2.4 Thermal stress analysis

Using the temperature distribution by a thermography and material properties in Table 2, shear and stress distribution is plotted with time in Figure 8, and normal stress distribution in Figure 9.

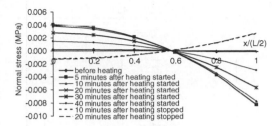

Figure 9. Normal stress distribution.

It is observed that the maximum shear stress occurs always at the edge of the tile. The shear stress rapidly increases up to 20 minutes, and slowly increases after 20 minutes. This coincides with the temperature distribution in Figure 4. Namely, the temperature difference between tile and concrete produces the shear stress in the observed way. Heating was stopped at 40 minutes, and the observation continued up to 20 minutes after heating stopped. The decrease of the shear stress after 40 minutes is fast, again in relation to the decrease of the temperature difference between tile and concrete. During the observation, the maximum shear stress is estimated to be 2.625 MPa for the tile surface temperature 48.64 degrees Celsius.

For the normal stress, the maximum value in compression is observed at the edge of the tile, while the maximum value in tension is at the center of the tile. Both maximum values happen at 5 minutes.

This again coincides with the temperature difference between both sides of the tile, which can be seen in Figure 5. Heating increases the temperature difference up to 5 minutes, but it soon decreases the difference after 5 minutes. Following this behavior, the normal stress becomes close to zero along the interface at 40 minutes. Furthermore, the stress direction changes after 40 minutes, meaning that the tile bends upwards. This is due to the fact that the lower side of the tile becomes hotter than the upper side in cooling process, as is seen in Figure 5. During the observation, the maximum normal stress is estimated to be 0.008 MPa in compression at the edge and 0.004 MPa in tension at the center. Both happen at 5 minutes for the tile surface temperature 37.12°C.

Although it has to be admitted that the current simple models have a lack of accuracy due to their assumptions and that a more rigorous analysis such as finite element analysis is necessary to analyze this kind of crack problem, the current models together with thermographic measurement yield the estimate of interfacial stresses under thermal loading. It estimate the shear stress on the order of 1 MPa and the normal stress on 10^{-3} MPa. The interface is subjected more severely to shear stress, if we assume the strengths in shear and normal are also on the order of 1 MPa. With this estimate of stresses and the fatigue strength characteristics.

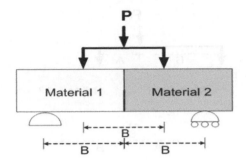

Figure 10. Symmetric four point loading.

3 MEASUREMENT OF FRACTURE ENERGY AND FATIGUE STRENGTH OF BI-MATERIAL INTERFACE

3.1 Determination of interfacial fracture energy under mixed mode fracture

The evaluation of bond at the interface is usually made either in tension or flexure.

The tensile bond test is common, and is widely used for the evaluation of the interfacial bond strength evaluation between adhesive mortar and concrete (Austin et al. 1995). The interfacial bond is evaluated as tensile bond strength of the bonded area.

The evaluation method in flexure is also reported (Kunieda 2000). In this method, beam specimens made of two kinds of materials are fabricated, and a notch is introduced at the bi-material interface (Figure 10). The beam specimens are loaded under four point flexure, and the observed interfacial fracture is treated as a mode I fracture problem. Therefore, the interfacial bond is evaluated in terms of fracture energy and bridging stress-crack opening displacement relation.

The current study aims at the evaluation of interfacial fracture under the combination of tension and shear stress, since the bi-material interface of a tile structure is subjected to the various combination of tension and shear stress depending on the environmental conditions. Hence, in addition to symmetric four point loading in Figure 10, asymmetric four point loading in Figure 11 was applied.

Under the mixture of tension and shear stress, the interfacial fracture has to be treated as a fracture under the combination of mode I and II, and the interfacial fracture energy is expressed as a function of phase angle, which represents the relative proportion of shear to normal stress at the interface.

Following O'Dowd (1992), interfacial fracture energy, Γ, and phase angle, Ψ, can be calculated as follows:

$$\Gamma(\psi) = \frac{1-\beta^2}{E_*}\left|K_c^2\right| \quad \text{and} \tag{7}$$

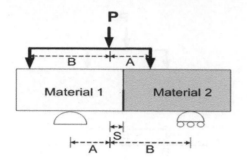

Figure 11. Asymmetric four point loading.

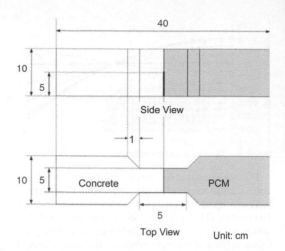

Figure 12. Specimen dimensions.

$$\psi = \psi_0 + \varepsilon \, ln\left(\frac{l}{a}\right), \tag{8}$$

where K_c = critical stress intensity factor; E_* = effective modulus of the bi-material system; a = crack length; and $\varepsilon = (1/2\pi)ln[(1-\beta)/(1+\beta)]$. β is one of Dundurs parameters, which represent the mismatch of elastic properties (Dundurs, 1969). l is arbitrary reference length (Rice 1988), and normally 200 μm is used for the interface between cementitious materials (Lim 1996).

The critical stress intensity factor and the effective modulus of the bi-material system can be determined as

$$K_c = YT_c\sqrt{a} \quad \text{and} \tag{9}$$

$$\frac{1}{E_*} = \frac{1}{2}\left[\frac{1}{\overline{E_1}} + \frac{1}{\overline{E_2}}\right] \tag{10}$$

where Y = geometric correction factor; T_c = critical nominal stress occurred at the interface; and

$$\overline{E_i} = E_i/(1-\nu_i) \tag{11}$$

for plane strain elastic modulus of material i.

For the symmetric loading in Figure 10,

$$T_c = \frac{P_c}{t}\frac{3B}{2W^2}, \tag{12}$$

$$Y = \sqrt{f_1^2 + (2\varepsilon g_2)^2}, \text{ and} \tag{13}$$

$$\psi_0 = arctan\left(\frac{2\varepsilon g_2}{f_1}\right). \tag{14}$$

For the asymmetric loading in Figure 11,

$$T_c = \frac{P_c}{tW}\left[\frac{B-A}{B+A}\right], \tag{15}$$

$$Y = \sqrt{Y_1^2 + Y_2^2}, \text{ and} \tag{16}$$

$$\psi_0 = arctan\left(\frac{Y_2}{Y_1}\right). \tag{17}$$

In Equation 12–17, P_c is the maximum load, which takes place at the onset of fracture and is obtained from the experiment; t is thickness of the beam at the interface; W is height of the beam; A and B are explained in Figure 10 and also in Figure 11; $Y_1 = (6sf_1/W) - 2\varepsilon g_1$; $Y_2 = f_2 + 12(s\varepsilon g_2/W)$; s is the loading offset that is the distance between the interface and the loading line as shown in Figure 11; and f_1, f_2, g_1, and g_2 are the calibration factors given in O'Dowd (1992). By having symmetric as well as asymmetric loading cases, the phase angle can be covered from 0 (mode I, pure tension) to 80 degrees (mixed mode, but mostly shear).

3.2 Fracture energy measurement

Fracture energy was measured based on the procedures in the previous section.

Figure 12 shows specimen dimensions. The specimens have a notch of 5 mm height at the interface, and they have a reduced cross section in the middle so as to avoid the unwanted cracks in other locations. The materials are concrete and polymer cement mortar (PCM) to simulate the bi-material interface of a tile structure.

The specimens were fabricated in the following manner. First, the concrete part was cast, and the specimens were cured under water for 28 days and in air for 7 days. The notch was made with a plastic tape, and the interface was roughened with water jet, causing the undulation of the interface to be 0.5 mm at maximum. Next, the PCM part was cast to complete the specimens, and they were cured under water for 28 days. The material properties are shown in Table 3.

Table 3. Material properties.

Material	Young's modulus (GPa)	Poisson's ratio
Concrete	30.7	0.207
PCM	21.6	0.212

The Dundurs parameters are calculated with these values.

The phase angles tested in the current study were 0, 30, and 60 degrees. For 0 degree, symmetric loading was applied, and, for 30 and 60 degrees, asymmetric loading was applied. All the loadings were executed under the displacement control of 0.005 mm/sec.

By using the equations in the previous section, interfacial fracture energy is calculated. All the experimental results are plotted in Figure 13. At 0 degree, fracture propagated along the interface. On the other hand, at 30 and 60 degrees, fracture initially propagated along the interface, but in the middle of the interface, it kinked out to PCM. Since the initial fracture was along the interface, these cases can also be considered as interfacial fracture.

Figure 13 shows the increasing trend of interfacial fracture energy with phase angle. As a reference, the interfacial fracture energy between concrete and fiber reinforced concrete (FRC) is taken from Lim (1996) and plotted together.

For the phase angle from 0 to 60 degrees, the interfacial fracture energy between concrete and PCM was found to vary in the range of 0.5 to 11.2 J/m², while Lim (1996) shows that the interfacial fracture energy between concrete and FRC varies in the range of 2–24 J/m² for the phase angle from 0 to 70 degrees. Although these bi-material systems are different from each other, it can be observed that the interfacial fracture energy of the current system is lower than the one between concrete and FRC, especially at high phase angle.

Based on the fracture energy measurement, finite element analysis is conducted where the interface is modeled with interface elements (Mahaboonpachai & Matsumoto 2005, Mahaboonpachai et al. 2006). Interface element is a four node element with zero thickness, and its constitutive law is defined in normal and shear direction. According to the finite element analysis, the tensile bonding strength is calibrated as 2.5 MPa, and the shear bond strength 5.6 MPa. These values with the thermal stress analysis in the previous section confirm that the interface is subjected more severely to shear stress.

3.3 Fatigue strength measurement

Fatigue strength was also measured by testing the specimens of the same dimensions and fabrication processes.

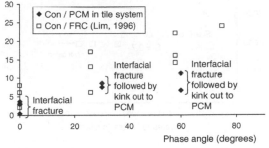

Figure 13. Interfacial fracture energy between concrete and PCM plotted with that between concrete and FRC measured by Lim (1996).

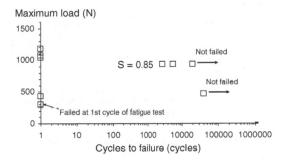

Figure 14. S-N diagram for phase angle = 0 degree.

The phase angles are also 0, 30, and 60 degrees. Under either symmetric or asymmetric loading, whichever is appropriate for each angle, fatigue loading was applied under load control at the frequency of 0.1 Hz. The ratio of maximum load to ultimate load was varied from 0.65 to 0.85, where ultimate load had been already obtained in the interfacial fracture energy measurement as P_c. In all the fatigue tests, the ratio of minimum load to maximum load was kept 0.2.

As a result of fatigue tests, S-N diagrams can be obtained, and they are shown in Figures 14, 15, and 16 for the phase angle of 0, 30, and 60 degrees, respectively. Ultimate loads are plotted at 1 cycle, and fatigue life of specimens is plotted in the manner of S-N diagram. In the case of 30 and 60 degrees, one specimen was observed to fail prematurely at 1st cycle of fatigue test. Also, some specimens did not fail before the prescribed number of cycles, and they are labeled with "Not failed" on the plot at the prescribed number of cycles.

Although there are not enough data points to perform a regression analysis, the following two observations can be made generally.

First, the higher the phase angle, the shorter the fatigue life. At the phase angle of 60 degrees, it is seen

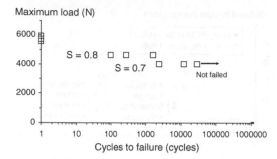

Figure 15. S-N diagram for phase angle = 30 degrees.

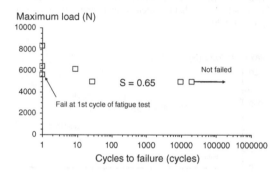

Figure 16. S-N diagram for phase angle = 60 degrees.

that the average fatigue life is on the order of 100 cycles for the ratio of 0.8, and 10,000 cycles for 0.7. On the other hand, at 0 degree, fatigue life ranges from 1000 to more than 10,000 cycles, exhibiting a longer fatigue life at higher load level. And, at 60 degrees, fatigue life is possibly shorter than that at 30 degrees. This implies that fatigue strength is weaker under shear dominant loading conditions.

Second, although it is considered that bi-material interface is generally a weak location, the S-N diagram of the current system is not significantly different from those of cementitious materials. It is generally observed that fatigue life data points of most cementitious materials lie along a line which connects ultimate load level at 1st cycle and half the ultimate load level at one million cycles. The data points of fatigue life at 30 degrees also lie along this line. Therefore, it can be said that the current bi-material interface has similar fatigue strength characteristics in terms of S-N diagram.

4 CONCLUDING REMARKS

This paper presented preliminary studies for the purposes of understanding the mechanisms of interfacial failure observed in external wall tile structures and achieving the durability of the interface and the tile

structures. The studies are conducted from the viewpoints of reaction and resistance of the interface in the tile structure.

For the former part, the temperature measurement in the through-thickness direction has been conducted with a thermography, and the interfacial stresses under thermal loading are estimated with simple analytical models. It estimates the shear stress to be 2.625 MPa and the normal stress in compression 0.008 MPa.

For the latter part, the bi-material interface between concrete and polymer cement mortar was tested to measure interfacial fracture energy and also to measure fatigue strength.

The measurement of interfacial fracture energy shows that, for the phase angle from 0 to 60 degrees, the interfacial fracture energy between concrete and PCM varies from 0.5 to 11.2 J/m^2. The values are smaller than the literature values of the interface between concrete and fiber reinforced concrete, especially at high phase angle.

The interfacial fracture energy was analyzed with finite element analysis in order to obtain the shear and tensile bond strength. The obtained strength values are 5.6 and 2.5 MPa, respectively.

Fatigue strength characteristics of the bi-material interface under different phase angles are summarized in terms of S-N diagrams. The results show that higher phase angle seems to exhibit shorter fatigue life and that the S-N diagram is similar to those of other cementitious materials.

Overall conclusions are as follows. The bi-material interface of a tile structure under thermal loading is subjected more severely to shear stress direction, and the shear stress is on the same order of shear strength. On the other hand, the S-N diagram shows that the interface is not so weak and is comparable to other cementitious materials, while the diagram shows that the interface is slightly shorter life at higher phase angle.

In order to understand the interfacial failure of a tile structure, a finite element analysis is necessary to estimate the interfacial stresses more accurately. The finite element analysis should include the constitutive law of the interface elements in fatigue, and it should conduct the delamination propagation analysis under thermal cyclic loading.

REFERENCES

Dundurs, J. 1969. Edge-bonded dissimilar media. *Journal of Applied Mechanics* 36: 650–652.
Kumagai, T. 1991. Separation failure analysis of ceramic tile applications on external walls. *Journal of Structural and Construction Engineering, Architectural Institute of Japan,* 422: 15–25. (In Japanese).
Kunieda, M., Kurihara, N., Uchida, Y. & Rokugo, K. 2000. Application of tension softening diagrams to evaluation

of bond properties at concrete interfaces. *Engineering Fracture Mechanics* 65(2–3): 299–315.

Lim, Y.M. 1996. Interface fracture behavior of rehabilitated concrete infrastructures using engineered cementitious composites. *Ph.d. Thesis*. Ann Arbor: University of Michigan.

Mahaboonpachai, T. & Matsumoto, T. 2005. Investigation of interfacial resistance between concrete and polymer-cement mortar and development of constitutive material model for the interface. *Journal of Applied Mechanics, Japan Society of Civil Engineers* 8: 977–985.

Mahaboonpachai, T., Inaba, Y. & Matsumoto, T. 2006. Investigation of interfacial resistance of the interface between concrete and polymer–cement mortar in tile applications.

Proceedings of the second international conference on structural health monitoring of intelligent infrastructure, Shenzhen, 16–18 November, 2005, 2: 1481–1487.

O'Dowd, N.P., Shih, C.F. & Stout, M.G. 1992. Testing geometry for measuring interfacial toughness. *International Journal of Solids and Structures* 29(5): 571–589.

Rice, J.R. 1988. Elastic fracture concepts for interfacial cracks. *Journal of Applied Mechanics* 55: 98–103.

Rumbayan, R., Mahaboonpachai, T. & Matsumoto, T. 2006. Thermographic measurement and thermal stress analysis at the interface of external wall tile structure. *Journal of Applied Mechanics, Japan Society of Civil Engineers*, 9: 1069–1076.

Fracture Mechanics of Concrete and Concrete Structures – High-Performance Concrete,
Brick-Masonry and Environmental Aspects – Carpinteri, et al. (eds)
© 2007 Taylor & Francis Group, London, ISBN 978-0-415-44617-4

Sensitivity analysis of a drying model concerning concrete delayed strains in containment vessels by means of a non-intrusive method

M. Berveiller, Y. Le Pape & B. Sudret
Electricité de France, R&D Division, Site des Renardières, Moret-sur-Loing, France

ABSTRACT: Drying impacts the delayed behavior of the containment vessels. Many tests of loss of mass are available in the data base of *Electricité de France* or in the literature and make it possible to evaluate profiles of water content in a test tube. The macroscopic retiming of the drying parameters and the uncertainty level on the latter force to carry out a sensitivity analysis over the lifespan of the vessel. The ranking of the input parameters (sensitivity analysis) was carried out using an expansion of the model response onto the polynomial chaos basis (non intrusive regression method) and computing the Sobol' indices analyticaly from the polynomial chaos coefficients. One is more particularly interested in the delayed vertical and tangential strains. Initially, the parameters of Mensi's law and the water contents actually play the most important role. Then, the parameters of the desorption curve desorption and more particulary the initial water content are the most important parameters.

1 INTRODUCTION

Concrete drying drives the delayed behavior of containment vessels. Many tests of loss of mass are available in the data base of *Electricité de France* or in the literature and make it possible to evaluate profiles of water content in a test-tube or a work. The macroscopic retiming of the drying parameters and the level of uncertainty on the latter force to carry out a sensitivity analysis. It means that we want to quantify the relative importance of each input parameter. In our case, the response is expanded onto the polynomial chaos basis. The coefficients are evaluated from a series of deterministic finite element analysis (regression method). Sobol' indices are computed analytically from the obtained coefficients of the response surface.

2 PRESENTATION OF THE MODEL FOR THE DELAYED STRAIN OF CONCRETE IN CONTAINMENT VESSEL

The containment of French pressurized water reactors is ensured by two concrete vessels. The inner containment is made of reinforced prestressed concrete. Drying impacts the delayed behavior of containment vessels. The constitutive law, implemented in EDF's Finite Element Code, *Code_Aster*[1], is the result of

the previous works by L. Granger (Granger 1996) and F. Benboudjema (Benboudjema 2002). The total strain is a function of the temperature T, the hydration degree β, the relative humidity h and the macroscopic stress σ. The conventional strain rate decomposition reads:

$$
\begin{aligned}
\dot{\varepsilon}(T, \beta, h, \sigma) = \ & \dot{\varepsilon}_e(\dot{\sigma}) + \dot{\varepsilon}_{th}(T) \\
& + \dot{\varepsilon}_{as}(\dot{\beta}) + \dot{\varepsilon}_{ds}(h, T) \\
& + \dot{\varepsilon}_{bc}(\sigma, h) + \dot{\varepsilon}_{dc}(\sigma, \dot{h})
\end{aligned}
\tag{1}
$$

with ε_e: elastic strain, dependent on σ ; ε_{th}: thermal dilation/contraction ; ε_{as}: autogenous shrinkage, being a function of the hydration degree $\beta \in [0; 1]$; ε_{ds}: drying shrinkage, dependent on the drying process controlled by the evolution of h; ε_{bc}: basic creep, naturally being a function of σ, but also, of the hydrous state of the material – basic creep of predried specimen exhibits some dependency with the equilibrated relative humidity–; ε_{dc}: drying creep. The model proposed by Bažant and Chern is assumed (Bažant and Chern 1985). In the sealed specimens (autogenous shrinkage and basic creep), the stress state remains homogeneous. Therefore, the computation is done analytically. Conversely, the drying specimens (loss of weight, drying shrinkage and creep) exhibit a humidity gradient responsible for an heterogeneous stress state. Their analysis requires a numerical simulation performed with *Code_Aster*. The calibration of the parameters is performed on the basis

---

[1] This code can be downloaded for free at http://www.code-aster.org

of the previous experimental results and follows the procedure:

1. the drying process is modeled through a non linear thermal analogy, where the water diffusion coefficient D is a function of the water content C, *i.e.* $D(C) = a \cdot \exp(b.C)$. With the back-analysis of the time evolution of the specimen loss of weight, the parameters a and b are retrieved.
2. The autogenous shrinkage is fitted on a simple hyperbolic law: $\varepsilon_{as} = K_{as}\beta$, with $\beta = t/(t + t_{1/2})$.
3. The basic creep constitutive law assumes the full uncoupling of the spherical and the deviatoric strains. It requires the knowledge of the strain tensor derived from the experimental test.
4. The drying shrinkage is assumed to be proportional to the loss of water content: $\dot{\varepsilon}_{ds} = -K_{ds}\dot{C}$. The influence of the auto-induced creep due to the stress gradient was found insignificant while calibrating K_{ds}.
5. The drying creep is ultimately modeled once all the other parameters are known. The intrinsic drying creep constitutive equation reads $\dot{\varepsilon}_{dc} = |\dot{h}|\sigma/\eta_{dc}$.

3 NUMERICAL SIMULATION

On the basis of the previously calibrated parameters, some numerical computations (Le Pape, Toppani, and Michel-Ponnelle 2005) are performed on a so-called Representative Structural Volume. This RSV is located approximately at an equal distance from the dome and the raft of hte containment vessel and far enough from the equipment hatch and the tendon buttresses, so that homogeneous strain states may be applied in the directions of the prestressing. ZC1450 is a simplified model; the thickness of the wall is discretized to account for the gradient of humidity. A single element is used in the vertical and the tangential direction due to the total strain homogeneity. The prestressing is introduced by external forces. Iterations on the effective applied prestress are computed over the non linear calculation to account for the loss of prestress induced by the concrete creep. The temperature, humidity and prestress evolutions follow the scenario: 15°C-60%RH inside and outside during building period, 35°C–45%RH inside the containment building when the reactor is in-service. The prestressing is applied graduously following the building stages. Figure 1 presents the mesh used.

4 SENSITIVITY ANALYSIS METHOD

As the numerical model is quite large, a response surface model has to be used to minimize the computational cost of the sensitivity analysis. The expansion onto the polynomial chaos (Berveiller 2005; Ghanem

GIBI FECIT

Figure 1. Mesh of the Representative Structural Volume.

and Spanos 1991) is relevant for sensitivity analysis (Sudret 2006a). The non intrusive regression method presented in the sequel only requires deterministic finite element anlyses and an analytical postprocessing of the results, which makes it appealing for the non linear coupled problem under consideration.

4.1 Non intrusive regression method

The non intrusive method presented in this communication is based on a least square minimization between the exact solution and the solution approximated using the polynomial chaos (Isukapalli 1999; Berveiller 2005). First the input random variables (gathered in a random vector $\underline{U}$ whose joint PDF is prescribed) are transformed into a standard uniform vector $\underline{\xi}$ (*i.e.* a vector whose components are uniformly distributed over $[-1;1]$). In our case, we suppose that all variables $\underline{U}$ are uniform. If these M variables are independent, the one-to-one mapping reads:

$$\xi_i = \Phi^{-1}(F_i(U_i)) = -1 + 2\frac{x - L_{low}}{L_{up} - L_{low}} \quad (2)$$

where Φ is the standard uniform CDF and $\{F_i(U_i), i = 1, \cdots, M\}$ are the marginal CDF of the U_i's wgich expand as in Eq.(2). Suppose now that we want to approximate a response quantity $\underline{S}$ by the truncated series expansion:

$$\underline{S} \approx \underline{\tilde{S}} = \sum_{j=0}^{P-1} \underline{S}_j \Psi_j(\underline{\xi}) \quad (3)$$

where $\{\Psi_j, j = 0, \cdots, P - 1\}$ are P multidimensional Legendre polynomials of $\underline{\xi}$ whose degree is less or equal than p. Note that the following relationship holds:

$$P = \frac{(M + p)!}{M! \, p!} \quad (4)$$

Let us denote by $\{\underline{\xi}^{(k)}, k = 1, \cdots, n\}$ n outcomes of the standard uniform random vector $\underline{\xi}$. For each outcome

$\underline{\xi}^{(k)}$, the isoprobabilistic transform yields a vector of input random variables $\underline{U}^{(k)}$ (Eq.(2)). Using a classical finite element code, the response vector $\underline{S}^{(k)}$ can be computed. Let us denote by $\{s^{(k),i}, i = 1, \cdots, N_{ddl}\}$ its components. Using Eq.(3) for the i-th component, one gets:

$$\tilde{s}^i(\underline{\xi}) = \sum_{j=0}^{P-1} s_j^i \Psi_j(\underline{\xi}) \tag{5}$$

where (s_j^i) are coefficients to be computed. The regression method consists in finding for each degree of freedom $i = 1, \cdots, N$ the set of coefficients that minimizes the difference:

$$\Delta s^i = \sum_{k=1}^{n} \left[s^{(k),i} - \tilde{s}^i(\underline{\xi}^{(k)}) \right]^2 \tag{6}$$

These coefficients are solution of the following linear system:

$$
\begin{pmatrix}
\sum_{k=1}^{n} \Psi_0(\underline{\xi}^{(k)})\Psi_0(\underline{\xi}^{(k)}) & \cdots & \sum_{k=1}^{n} \Psi_0(\underline{\xi}^{(k)})\Psi_{P-1}(\underline{\xi}^{(k)}) \\
\vdots & \ddots & \vdots \\
\sum_{k=1}^{n} \Psi_{P-1}(\underline{\xi}^{(k)})\Psi_0(\underline{\xi}^{(k)}) & \cdots & \sum_{k=1}^{n} \Psi_{P-1}(\underline{\xi}^{(k)})\Psi_{P-1}(\underline{\xi}^{(k)})
\end{pmatrix}
$$
$$
\begin{pmatrix} s_0^i \\ \vdots \\ s_{P-1}^i \end{pmatrix} =
\begin{pmatrix} \sum_{k=1}^{n} s^{(k),i}\Psi_0(\underline{\xi}^{(k)}) \\ \vdots \\ \sum_{k=1}^{n} s^{(k),i}\Psi_{P-1}(\underline{\xi}^{(k)}) \end{pmatrix} \tag{7}
$$

Note that the $P \times P$ matrix on the left hand side may be evaluated once and for all. Moreover it is independent on the mechanical problem under consideration. The crucial point in this approach is to properly select the regression points, *i.e.* the outcomes $\{\underline{\xi}^{(k)}, k = 1, \cdots, n\}$. Note that $n \geq P$ is required so that a solution of (7) exist. (Isukapalli 1999; Berveiller 2005) choose for each input variable the $(p+1)$ roots of the $(p+1)$-th order Legendre polynomial, and then built $(p+1)^M$ vectors of length M using all possible combinations. Then they select n outcomes $\{\underline{\xi}^{(k)}, k = 1, \cdots, n\}$ out of these $(p+1)^M$ possible combinations. (Sudret 2006a) presents a method to obtain the minimum number of points that allows the left matrix to be invertible.

4.2 Computation of the Sobol' indices

Global sensitivity analysis aims at quantifying the uncertainty in the model output due to the uncertainty in the input parameters. More precisely, the so-called *ANOVA techniques* aim at decomposing the variance of the output as a sum of contributions of each input variable, or combinations thereof. Many papers have been devoted to this topic in the last twenty years. A good state-of-the-art of the techniques is available in (Saltelli, Chan, and Scott 2000).

Consider a scalar response quantity s^k (*i.e.* a component of $\underline{S}$ in Eq.(3)). The Sobol' decomposition of its variance D reads (Sobol' 1993):

$$D \equiv \mathrm{Var}\left[s^k\right] = \sum_{i=1}^{n} D_i + \sum_{1 \leq i < j \leq n} D_{ij} + \cdots + D_{12...n} \tag{8}$$

where each term $D_{i_1 i_2 ... i_s}$ represents the part of the variance associated with the combination of variables $\{i_1, i_2, \ldots, i_s\}$. The Sobol' indices are nothing but the normalized version of these partial variances:

$$\delta_{i_1 i_2 ... i_s} = D_{i_1 i_2 ... i_s}/D \tag{9}$$

They sum up to 1 and thus represent the fraction of the response variance that may be attributed to the combination of input variables $\{i_1, i_2, \ldots, i_s\}$. Monte Carlo estimates of these so-called partial variances are usually used, leading to an unaffordable computational cost when the reponse is the result of a finite element analysis.

Alternatively, (Sundret 2006a; Sudret 2006b) shows that the decomposition of the total variance is straightforward once the model response has been expanded onto the polynomial chaos basis.

First remember that each multivariate polynomial in Eq.(3) is completely defined by a list of M nonnegative integers $\{\alpha_1, \cdots, \alpha_M\}$ as follows:

$$\Psi_j(\underline{\xi}) \equiv \Psi_\alpha(\underline{\xi}) = \prod_{i=1}^{M} P_{\alpha_i}(\xi_i) \quad , \quad \alpha_i \geq 0 \tag{10}$$

where $P_q(.)$ is the q-th Legendre polynomial. Let us denote by $\mathcal{I}_{i_1, \ldots, i_s}$ the set of $\underline{\alpha}$ multi-indices such that only the indices $(i_1, \ldots, i_s)$ are non zero:

$$\mathcal{I}_{i_1, \ldots, i_s} = \left\{ \underline{\alpha} : \begin{array}{llll} \alpha_k & > & 0 & \forall k = 1, \ldots, n, & k \in (i_1, \ldots, i_s) \\ \alpha_j & = & 0 & \forall k = 1, \ldots, n, & k \notin (i_1, \ldots, i_s) \end{array} \right\} \tag{11}$$

Note that $\mathcal{I}_i$ corresponds to the polynomials depending only on parameter x_i. Using this notation, the $P-1$ terms in Eq.(3) corresponding to the polynomials

Table 1. Presentation of input random variables.

Parameter	Notation	Type of distribution	Lower limit	Upper limit
Initial humidity	C_0	Uniform	116.4 l/m^3	135.8 l/m^3
Outside humidity	C_{EXT}	Uniform	65.61 l/m^3	80.19 l/m^3
Inside humidity	C_{INT}	Uniform	49.12 l/m^3	54.28 l/m^3
Thermal activation energy	Q	Uniform	4465 K	4935 K
Parameter a of the drying process	a	Uniform	$1.05 \cdot 10^{-13}$ m^2/s	$1.95 \cdot 10^{-13}$ m^2/s
Parameter b of the drying process	b	Uniform	0.06118 m^3/l	0.06762 m^3/l
Desorption abscissa	ABS_H	Uniform	0	0.6

$\{\Psi_j, j = 1, \ldots, P-1\}$ may now be gathered according to the parameters they depend on:

$$s^k = s_0^k + \sum_{i=1}^{n} \sum_{\underline{\alpha} \in \mathcal{I}_i} s_{\underline{\alpha}}^k \Psi_{\underline{\alpha}}(x_i)$$

$$+ \sum_{1 \le i_1 < i_2 \le n} \sum_{\underline{\alpha} \in \mathcal{I}_{i_1,i_2}} s_{\underline{\alpha}}^k \Psi_{\underline{\alpha}}(x_{i_1}, x_{i_2}) + \ldots$$

$$+ \sum_{1 \le i_1 < \cdots < i_s \le n} \sum_{\underline{\alpha} \in \mathcal{I}_{i_1,\ldots,i_s}} s_{\underline{\alpha}}^k \Psi_{\underline{\alpha}}(x_{i_1}, \ldots, x_{i_s})$$

$$+ \cdots + \sum_{\underline{\alpha} \in \mathcal{I}_{1,2\ldots,n}} s_{\underline{\alpha}}^k \Psi_{\underline{\alpha}}(x_1, \ldots, x_n)$$

(12)

Thus the Sobol-PC sensivity indices of the k-th component of the response vector:

$$\delta_{i_1 \ldots i_s}^k = \sum_{\underline{\alpha} \in \mathcal{I}_{i_1,\ldots,i_s}} (s_{\underline{\alpha}}^k)^2 \, \mathrm{E}\left[\Psi_{\underline{\alpha}}^2\right] / \mathrm{Var}\left[s^k\right]$$

(13)

where the total variance of s^k is easily obtained from the PC coefficients:

$$\mathrm{Var}\left[s^k\right] = \sum_{j=1}^{P-1} (s_j^k)^2 \, \mathrm{E}\left[\Psi_j^2\right]$$

(14)

In the sequel, the first order Sobol-PC sensitivity indices are computed.

5 SENSITIVITY ANALYSIS OF THE DRYING MODEL ON THE DELAYED STRAIN OF CONCRETE IN CONTAINMENT VESSEL

The macroscopic retiming of the drying parameters and the level of uncertainty on the latter force to carry out a sensitivity analysis. It means that we want to quantify the relative importance of each input parameter.

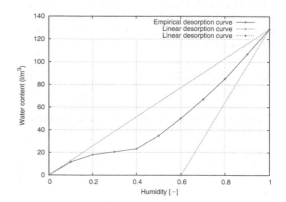

Figure 2. Desorption curve.

5.1 Presentation of the random variables

In this sensitivity analysis, we take into account 7 random variables (gathered in Table 1). The humidity in the intersapce between the two concrete vessels is called outside humidity. The desorption curve (cf. Figure 2) relates the water content to the humidity. It is usually difficult to get data about this curve. Thus it is important to take this parameter into account in the sensitivity analysis. The empirical curve on Figure 2 is not that of the concrete of the vessel under consideration. It is difficult to model this parameter by a random variable. We make the choice to enclose this curve by two lines, which are defined with two random variables: the initial humidity C_0 and the absorption abscissa ABS_H.

The first step in the sensitivity analysis is to expand the evolution of delayed strain onto the polynomial chaos basis. As all input random variables are uniform, the polynomial chaos is built from Legendre polynomials. (Sudret 2006a) shows that an order 2 gives the best ratio accuracy/efficiency for computing Sobol' indices. In our case, which involves 7 random variables, we have 36 coefficients to compute. $n = 56$ finite element analyses are enough to obtain the whole expansion of delayed strain onto the polynomial chaos by using the non intrusive regression method presented in section 4.1.

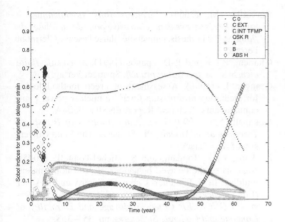

Figure 3. Evolution of Sobol' indices for tangential strains.

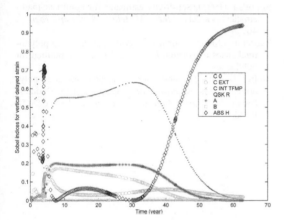

Figure 4. Evolution of Sobol' indices for vertical strains.

5.2 Results of the sensitivity analysis

Figures 3 and 4 show the evolution of the Sobol' indices for each input parameter versus time on the tangential and vertical strains. One can first of all notice that for $t \in [0; 5]$ years, the Sobol' indices vary a lot. This corresponds to the date when the installation of prestressing is modelled. The parameters Q, C_{INT} and C_{EXT} have a relatively low index both for the tangential and the vertical strains. On the other hand, the other parameters play a considerable role. As it can be seen on each figure, the reduction of the importance of C_0, a and b (all three directly influencing drying) coincide with the increase in importance of ABS_H. This could mean that when drying reaches a certain level, it is the desorption curve which plays the most important role. This means that the role of drying process decreases in time and that its variability has no influnece on the amplitude of the delayed strains after a certain time. This implies that the very fine modeling of the kinetics

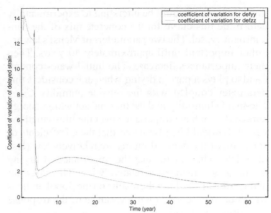

Figure 5. Evolution of coefficient of variation of tangentials and vertival strains.

of drying is not necessary when one is only interested in the delayed strains in the long run.

The figure 5 presents the temporal evolution of the coefficients of variation of the tangential and vertical deformations. One notices that these coefficients of variation are relatively low (lower than 4%). That means that the importance of drying on the differed deformations is relatively low.

6 CONCLUSION

This paper presents a sensitivity analysis of concrete drying in the delayed behavior of containment vessels. Only the parameters influencing drying are described by random variables. The analysis of sensitivity was carried out using the non intrusive method of regression, which makes it possible to have a stochastic response surface of the delayed strains. The Sobol' indices, which allow to rank the input variables according their weight in the response variance, are calculated in an analytical way from the coefficents of the response surface rank. The input random parameters are:

- Initial humidity;
- Outside humidity;
- Inside humidity;
- Thermal activation energy;
- Two parameters of Mensi's law;
- The curve of desorption.

The response variables are the tangential and vertical delayed strain computed over the time range [0–60 years]. As far as the tangential and vertical delayed strains are concern, it appears that the parameter which has the most importance is the curve of desorption, which utilizes two random variables, in particular the initial humidity. However we have little information on this curve, so it would be intersting to experimentally determine this curve for the concrete mix of the containment vessel. The two parameters of Mensi's law are rather important until approximately 40 years. Then their importances decrease. The initial water content C_0 also plays a part in drying when one considers this parameter coupled with the outside humidity C_{EXT}. The inside humidity and the thermal activation energy almost do not have importance over the lifespan of the vessel. It should be also noted that the coefficients of variation of the delayed strains are relatively low (less than 4%). That means that the importance of drying on the delayed strains is relatively low. This does not deteriorate in anything the results on the Sobol' indices which do not depend on the variance of the response quantity considered.

Finally, the analysis shows that it is necessary to accurately determine the desorption curve and the parameters of Mensi's law to get accurate prediction of the delayed strains.

REFERENCES

Bažant, Z. and J. Chern (1985). Concrete creep at variable humidity: constitutive law and mechanism. *Materials and Structures 18*(103), 1–20.

Benboudjema, F. (2002). *Modélisation des déformations différées du béton sous sollicitations biaxiales. Application aux enceintes de confinement de bâtiments réacteurs des centrales nucléaires*. Ph. D. thesis, Université de Marne la Vallée.

Berveiller, M. (2005). *Eléments finis stochastiques : approches intrusive et non intrusive pour des analyses de fiabilité*. Ph. D. thesis, Université Blaise Pascal – Clermont Ferrand.

Ghanem, R.-G. and P.-D. Spanos (1991). *Stochastic finite elements – A spectral approach*. Springer Verlag.

Granger, L. (1996). Assessment of creep methodologies for predicting prestressing forces in nuclear power plant containments. Technical Report ENSIGC9604A.

Isukapalli, S.S. (1999). *Uncertainty Analysis of Transport-Transformation Models*. Ph. D. thesis, The State University of New Jersey.

Le Pape, Y., E. Toppani, and S. Michel-Ponnelle (2005). Analysis of the delayed behaviour of NPP containment building. In G. Pijaudier- Cabot, B. Gérard, and P. Acker (Eds.), *Proceedings of the seventh International Conference CONCREEP, Creep, Shrinkage and Durability of Concrete and Concrete Structures*, pp. 353–358.

Saltelli, A., K. Chan, and E. Scott (Eds.) (2000). *Sensitivity analysis*. J. Wiley & Sons.

Sobol', I. (1993). Sensitivity estimates for nonlinear mathematical models. *Math. Modeling & Comp. Exp. 1*, 407–414.

Sudret, B. (2006a). Global sensitivity analysis using polynomial chaos expansion. *Rel. Eng. Sys. Safe.* submitted for publication.

Sudret, B. (2006b). Global sensitivity analysis using polynomial chaos expansion. In P. Spanos and G. Deodatis (Eds.), *Proc. 5th Int. Conf. on Comp. Stoch. Mech (CSM5)*. Rhodos.

Fracture Mechanics of Concrete and Concrete Structures – High-Performance Concrete,
Brick-Masonry and Environmental Aspects – Carpinteri, et al. (eds)
© 2007 Taylor & Francis Group, London, ISBN 978-0-415-44617-4

Evaluation of adhesion characteristics of joint in concrete by tension softening properties

K. Yamada, A. Satoh & S. Ishiyama
Akita Prefectural University, Yurihonjo, Japan

ABSTRACT: This study intends to reveal some clues for improving mechanical properties of joint in concrete. The authors conducted fracture mechanics test of nine types of specimens as the models of vertical construction joint. Along with the investigation on fracture mechanics parameters, SEM analysis was made from the samples on detached and fractured surfaces of specimens. The resulted fracture mechanics parameters and tension softening diagram showed clearer difference of performance in joint than flexural strength does. There are many pores and fragmental layers of $Ca(OH)_2$ observed in smooth part of ligament after the test, which could be the main cause of detaching without fracturing of the surface.

1 INTRODUCTION

1.1 Construction joint

Every concrete structure has inevitably construction joint that is a discontinuous plane of concrete produced during construction. The joint in concrete induces many types of deterioration to performances, such as decreased tensile strength that also makes shear strength lower (Hamazaki 2003), a higher possibility of water penetration (Tanaka & Shin 2000) and a higher tendency of carbonation through the joint than monolithic concrete with no joint (Yamamoto 2001). Also horizontal joint suffers dry out that changes the pore structure, eventually making the durability performance lower (Yuasa 1998).

The construction joint is also a good example of researching adhesion performance of repair in concrete structure, because the interfacial adhesion is the most critical issue for both of construction joint and repaired surface. There are surging needs for improving interfacial adhesion of concrete (Sakami 2006) as repair and retrofitting of existing concrete structure is a major market of construction industry in developed countries (Sakai 2006).

1.2 Previous studies

Many previous studies revealed that additional placing of concrete should be well advanced before hardening of previously placed one, if discontinuity of concrete should be avoided (Yamamoto 2001; Sugata 2003). After hardening, tensile strength between two bodies (a previously placed body and an additionally placed one) decreases depending on many conditions, such as time after the previous one was placed, roughness of the surface where additional one is placed, direction and thickness of concrete layer downward which is related to an amount of bleeding water (Yamamoto 2001; Hamazaki 2003), and so on.

It was common that the adhesion strength of joint was evaluated with bending strength through bending test (Yamamoto 2001; Hamazaki 2003), and there was few study which evaluated tension-softening properties of joint.

Kurihara et al. (1996) investigated 5 types of concrete prism specimens with the results that fracture energy calculated by tension softening diagram (TSD) denotes clear difference in adhesion performance of joint in the specimen. Other than this study, there is no research that employs TSD for an evaluation of adhesion performance.

1.3 Purpose of the research

Previous studies do not tell the cause of the decrease of adhesive performance through joint. There have been no studies observing the joint surface with a scanning electron microscope (SEM). Also there is no research discussing the difference of TSD characteristics with varied types of joints; such as cast surface with a delay of 24 hours, or 48 hours, with a mortar layer and with a permanent form made of fiber reinforced cement composites (FRCC).

There is a difference of conditions between horizontal and vertical joint. It is well known that the horizontal joint surface of concrete suffers dry-out and bleeding, which is more complicated than vertical joint surface. Vertical joint surface scarcely suffers such complex combination of condition for concrete, which eventually makes the researching focus on adhesion performance sharp.

Then the authors employed specimens with a vertical joint at the center of them for evaluation of adhesion performance with various types of joint in concrete, and discuss the cause of the decrease from the results of TSD and SEM observation.

2 EXPERIMENT

2.1 Specimens

Table 1 and Figure 1 show the attribute and illustration of specimens, and Table 2 the mix proportion in which following materials were used. Cement is ordinary Portland cement. Gravel is crushed stone with the size under 20 mm. Sand is natural pit sand. FRCC board is made of PVA (polyvinyl alcohol) fiber and cement, produced through Hatcheck machine before pre-curing at 50 degree, whose thickness is 6 mm and

Table 1. Attribute of specimens.

Specimen	Attribute of specimens
R-1	Reference with no joint
D-1	Separated with a piece of dried FRCC board
Dr-1	Separated with a piece of wet FRCC board
F-1	Cast after 24 hours on the surface (slightly roughened with wire brush)
F-2	Cast after 24 hours on the surface (strongly roughened with wire brush)
Fs-1	Cast after 48 hours on the surface whose form was painted steel
Fs-2	Cast after 48 hours on the surface of F-2 specimen
I-1	Cast after 24 hours on the surface whose form was joint sheet
J-1	Cast immediately after on the surface of mortar attached after 24 hours on previously cast concrete

Cast simultaneously as a reference	FRCC board
R-1	D-1 and Dr-1

Cast previously	Roughening Cast afterwards	Cast previously	Joint sheet Cast afterwards
F-1, F-2, Fs-1 and Fs-2		I-1	

Cast previously	Mortar Cast afterwards
J-1	

Note: A half depth notch is provided and then the remained part of joint is a ligament.

Figure 1. Detail of specimens.

tensile strength is 12.8 MPa. Joint sheet is made of plastic with many cones on the surface and ordinarily used for vertical joint to give enhanced shear strength (Civil Eng. Res. C. 2002). (See Figure 2).

Table 3 shows the mechanical properties of concrete and mortar, in which the strengths were measured with cylinder type specimens (diameter is 100 mm for concrete and 50 mm for mortar). Other than those specimens, each specimen for fracture toughness test was a prism with a section of 100 mm by 100 and a length of 400 mm. At center of the specimen, a half-depth (50 mm) notch was provided prior to fracture toughness test.

2.2 Tension softening diagram

Standard curing in water was applied for 28 days from the cast of the latter concrete. After that, fracture toughness test was executed with observing JCI's standards (Izumi 2004). The only test method that differs from the standard is that the authors provided counter weights at both ends of the specimen to cancel the weight of the specimen. The load was applied at the center of the span and at a speed of 0.06 mm/min with measuring load, deflection and crack mouth opening

Table 2. Mix proportion of concrete and mortar.

Mixture	Weight of materials				Super Plasticizer kg/m³
	Water kg/m³	Cement kg/m³	Sand kg/m³	Gravel kg/m³	
Concrete*	177	344	739	1010	1.72
Mortar	209	523	1569	–	–

* W/C = 51.4%, s/a = 43%, Air = 3%, Slump = 16.4 cm.

Above: Section of joint sheet. The height of cone is 8mm.

Left: Plan of joint sheet. Spacing between cones is 30mm.

Figure 2. Detail of joint sheet.

Table 3. Mechanical properties of concrete and mortar.

Mixture	Density g/cm³	Compressive strength MPa	Tensile strength MPa
Concrete*	2.31	42.70	3.83
Mortar*	2.20	41.45	3.74

* Cure = 28 days of standard curing in water.

Table 4. Resulted fracture mechanics parameters.

Name	F_b MPa	F_t MPa	G_F N/m	K_{1C} MN/m$^{3/2}$	G_c N/m
R-1	6.58	6.82	91.1	0.693	17.73
D-1	1.87	2.50	5.6	0.196	1.31
Dr-1	2.06	2.30	6.2	0.216	1.73
F-1	1.56	1.66	3.1	0.160	0.87
F-2	4.29	4.70	23.6	0.454	7.61
Fs-1	2.87	3.33	17.5	0.299	3.31
Fs-2	3.92	3.75	39.8	0.413	5.82
I-1	4.22	5.40	35.1	0.383	5.01
J-1	4.22	5.10	36.6	0.439	7.04

F_b: Flexural strength, F_t: Tension softening initial stress.
G_F: Fracture energy, K_{1C}: Stress intensity factor.
G_c: Energy release rate.

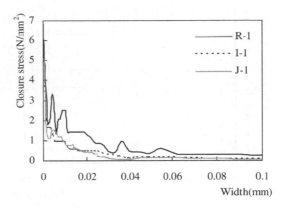

Figure 3A. TSDs for R-1, I-1 and J-1.

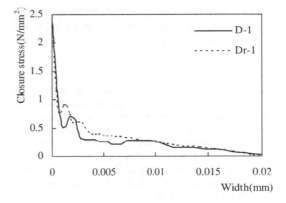

Figure 3B. TSDs for D-1 and Dr-1.

displacement (CMOD). The loading speed was controlled with a feed-back system to be exactly the same at anytime.

Inverse analysis was adapted to the resulted load-deflection curve to calculate TSD with observing JCI's standards (Izumi 2004). The finite element (FEM) model used in the analysis has 389 elements in half of the specimen and 41 nodes in ligament.

3 RESULTS

3.1 Fracture mechanics parameters

Table 4 shows the resulted fracture mechanics parameters, in which F_b is flexural strength, F_t is tension softening initial stress, G_F is fracture energy, K_{1c} is stress intensity factor calculated with equation (1) and G_c is energy release rate calculated with equation (2). The values for R-1 are similar to the previous studies (Ohgishi 1988, Kitsutaka 1997).

$$K_{1C} = \frac{PS}{BW^{\frac{3}{2}}} \left\{ 2.898\left(\frac{a}{W}\right)^{\frac{1}{2}} - 4.613\left(\frac{a}{W}\right)^{\frac{3}{2}} + 21.8\left(\frac{a}{W}\right)^{\frac{5}{2}} \right.$$
$$\left. -37.7\left(\frac{a}{W}\right)^{\frac{7}{2}} + 38.74\left(\frac{a}{W}\right)^{\frac{9}{2}} \right\} \quad (1)$$

$$Gc = \frac{K_{1C}^2}{E} \quad (2)$$

There are 2 groups recognized in Table 4 except for R-1. One group is F-2, Fs-2, I-1 and J-1, which has large values of F_b and F_t. Other group is D-1, Dr-1, F-1 and Fs-1, which has smaller values than group 1. Though F_b and F_t are almost equal within group1 or group2, fracture mechanics parameters are different ranging from double to triple or moreover. This result is the same as the one from a research by Kurihara (1996).

3.2 Tension softening diagram

Figure 3A, 3B and 3C show TSDs. Figure 3A tells that the major difference between R-1 and other two is the closure stress between 0.01 mm and 0.05 mm. It is suggested that this range of closure stress is the major cause of difference for fracture energy. Figure 3B tells that wet surface of the layer of FRCC is essential for improving adhesion performance. Figure 3C tells that the joint with the different roughening produces the different TSD.

The ratio of G_F divided by that of weakest joint reaches moreover 10 (F-1 vs. Fs-2) in Table 4, indicating the adequate roughening is very essential for the enhanced performance of the joint. In the roughened joints, the order of G_F is F-1 < Fs1 < F-2 < Fs2, meaning cast after 48 hours has good results.

FRCC permanent form did not have good results in this test, but it can be pointed out that pre-wetting of the form should be the requisite for improved adhesion because D-1 < Dr-1 in G_F.

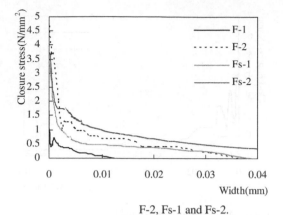

F-2, Fs-1 and Fs-2.

Figure 3C. TSDs for F-1, F-2, Fs-1 and Fs-2.

The interesting finding is that the better the G_F becomes, the larger the critical width when closure stress becomes zero (Figure 3A–3C).

3.3 Observation of fracture surface

There are two types of fracture surfaces at the ligament of the specimen after fracture toughness test. One is a fractured part, the other is a detached part. The authors made careful observation on the surface, and the smooth surface was determined as a detached part and rough surface as a fractured part.

Figure 4 depicts the map of them. The dashed area indicates fractured part while blank area detached smooth part in ligament.

4 DISCUSSIONS

4.1 F_b and fractured area

If the detached part in ligament of the specimen does not contribute to the flexural strength, F_b should be proportional to the moment of inertia calculated only with the fractured part. At the beginning of the first crack when stress of the tensile edge becomes F_b, the neutral axis can be assumed to be the center of ligament. Then the moment of inertia calculated within the tensile fractured part would be proportional to the F_b.

These assumptions are described in following equations (3) and (4), and depicted in Figure 5. In Figure 5, x-axis represents J_r/J_A and y-axis F_b.

$$M_{cr} = \int \sigma y dA = \frac{\sigma}{y} J_r = \sigma_0 J_r \qquad (3)$$

$$F_b = \frac{M_{cr}}{J_A} \frac{h}{2} = \sigma_0 \frac{h}{2} \frac{J_r}{J_A} = k \frac{J_r}{J_A} \qquad (4)$$

Where M_{cr} = bending moment at cracking, σ_0 = stress at unit height from the neutral axis, k = constant,

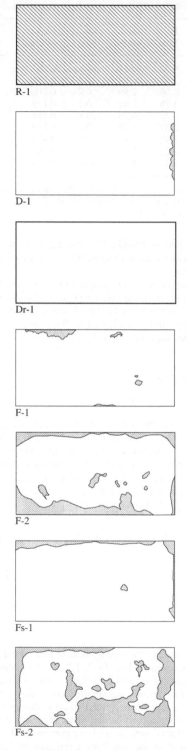

Figure 4. Map of fractured part and detached part.

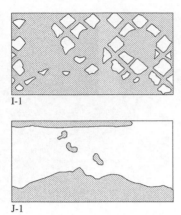

I-1

J-1

Figure 4. (Continued)
Note: Dashed area indicates fractured rough part while blank area indicates detached smooth part.

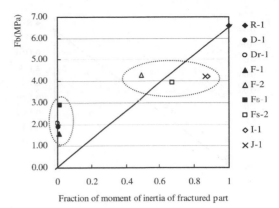

Figure 5. Relationship between flexural strength and fractured area.

J_r = moment of inertia for only fractured tensile part and J_A = moment of inertia for all tensile part.

The location of the symbols above the solid line indicates that the smooth part that was not considered as effective should be considered as effective for adhesion strength. On the contrary, the location of symbols below the solid line tells vice versa, and also the possibility of the weaker strength for the fractured part than that of reference.

There are two groups in this graph. One group gathers near y-axis, which tells detached part has some contribution to F_b. The other group has relatively high F_b (4 MPa) and they locates near the solid line that connect origin and R-1, telling that these assumptions can be correct for them.

4.2 G_F and fractured area

If total area in ligament is available for G_F, it can be calculated with equation (5). If fractured part distributes uniformly within the section, G_F only by the fractured part of specimen (G_{F-part}) should be proportional to the fraction of fractured part to total area in ligament (ϕ), resulting in equation (6).

$$G_F = \frac{E(w) + E'(w)}{A_{lig}} \quad (5)$$

$$G_{F-part} = G_F \phi \quad (6)$$

Where $E(w)$ = consumed energy in load-deflection curve until crack width is w, $E'(w)$ = differentiated $E(w)$ with respect to w, A_{lig} = total area of ligament, ϕ = fraction of fractured area to total area, k = some constant and $G_{F-part} = G_F$ of specimen that has a smaller width than a full width.

Equation (5) derives from equation (7) that is well known equation usually employed for calculation

of closure stress with modified J-integral method (Uchida 1991).

$$\sigma(w) = \frac{wE''(w) + 2E'(w)}{A_{lig}} \quad (7)$$

Where $\sigma(w)$ = closure stress and $E''(w)$ = differentiated $E'(w)$ with respect to w.

In Figure 6, x-axis represents ϕ and y-axis represents G_F. Almost all symbols (except for Fs-1 and I-1) locate near the solid line that connects origin and R-1, telling that these assumptions can be correct for them.

The distance of the same symbol from the solid line is different between Figure 5 and 6, which suggests that the governing cause of F_b and G_F is different. It means that even detached part contributes to F_b (even smooth surface could bear stress by chemical bond), whereas visibly rough surface is necessary for consuming energy like G_F.

The reason for the poor performance in both F_b and G_F produced from fractured part in the case of I-1 may be the weakness of the adhesion strength, which may derive from the produced $Ca(OH)_2$ by plastic joint sheet.

4.3 SEM observation in detached part

The authors cut a sample of 1 cm square from the surface of each specimen. After platinum spattering on it, SEM observation was done. Figure 7A–7F show typical observations of a surface of concrete that was cast afterwards on the surface of previously cast one.

Many pores are seen on the surface that contacted to FRCC board in Figure 7A. The air is considered to come from FRCC board, because the surface of the board is rough enough to entrap large pores on the surface. Smooth surface in Figure 7B is $Ca(OH)_2$ because element analysis told Ca = 73.0% and Si = 22.5%. This was produced during the hydration on the surface of form and remained after

Figure 6. Relationship between fracture energy and fracture area.

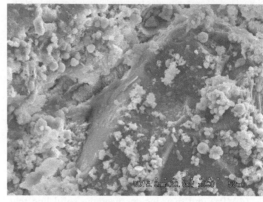

Figure 7C. SEM photo from fractured part in R-1 (×1500).

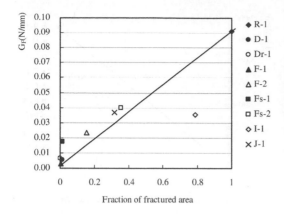

Figure 7A. SEM photo from detached part in D-1 (×30).

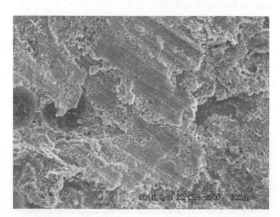

Figure 7D. SEM photo from fractured part in Fs-1 (×900).

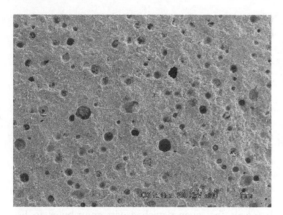

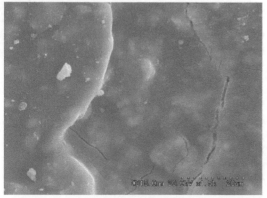

Figure 7B. SEM photo from detached part in Fs-2 (×1500).

Figure 7E. SEM photo from detached part in F-2 (×200).

wire brushing of the surface. Rough surface is the usual observation in fractured part like Figure 7C in R-1 specimen. On the other hand, smooth surface in Figure 7D is aggregate because element analysis

told that Na = 13.5%, Al = 18.2%, Si = 61.3% and Ca = 4.2%.

From Figure 7D, it is suggested that transient layer (Uchikawa 1993, Kobayashi 1998) on the aggregate

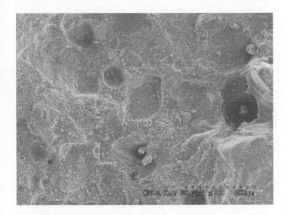

Figure 7F. SEM photo from fractured part in J-1 (×70).

near the surface of joint may have fractured and the aggregate appeared.

The surface that have scratches in Figure 7E is made of $Ca(OH)_2$ because element analysis told $Ca = 78.1\%$ and $Si = 21.9\%$. Thus $Ca(OH)_2$ remains on the surface of previously cast concrete after roughening by wire brush.

Figure 7F shows fractured surface of mortar layer on J-1 specimen. There are some holes where grains of sand torn off and some grains of sand remained on the surface.

In summary, there are many pores and a layer of $Ca(OH)_2$ observed in detached part, which could be the main cause of detaching. On the other hand, rough surface of CSH gel and transient layer around aggregate are observed in fractured part. There are grains of sand and holes where grains of sand torn off along with gravels (that are avoided for SEM observation).

5 CONCLUSIONS

Fracture mechanics parameters and tension softening diagram of nine types of specimens for vertical construction joint were examined along with SEM analysis from the samples on detached and fractured surfaces.

The findings are as follows.

[1] Even though flexural strengths at ligament of nine specimens are not substantially different from each other, fracture energies of them are substantially different from each other. The authors point out that the key to improve the structural performance of joint should be the enhancement of fracture energy.

[2] Smooth surface of detached part that appears on specimen after fracture toughness test can contribute to flexural strength, whereas it cannot contribute to fracture energy. This suggests that the mechanism of determining the strength and the fracture energy is different with each other.

[3] There are many pores and layer of $Ca(OH)_2$ observed in detached part, which could be the main cause of detaching. Then, some of the keys to enhance the performance of joint are thoroughly removing a layer of $Ca(OH)_2$, or to find mix proportions or conditions with which $Ca(OH)_2$ does not produce in joint.

REFERENCES

Civil Eng. Res. C. (eds) 2002. *Evidence report on inspection of construction technology No. 0123* Tokyo: Civil Eng. Res. C.

Hamazaki, J. 2003. Study on evaluation and prevention method of discontinuous joint of concrete. *Annual report of Building Research Institute of Japan* 2001: 71–72.

Izumi, I. (eds) 2004. *Test method for fracture energy of plain concrete (Draft) in JCI standards* Tokyo: JCI.

Kitsutaka, Y. & Oh-oka, T. 1997. Influence of short cut fiber on fracture parameters of high strength mortar matrix. *J. Struct. Constr. Eng. by AIJ* 497: 1–8.

Kobayashi, K. et al. 1996. Characters of interfacial zone of cement paste with additives around aggregate. *J. Mat. Sci. Japan* 45(9): 1001–1007.

Kurihara, T. et al. 1996. Evaluation of adhesive performance of construction joint in concrete by tension softening diagram. *Proc. annual meeting of JCI* 18(2): 461–466.

Ogawa, A. et al. 2005. PVA-fiber reinforced high performance cement board, *Proc. Int'l workshop on high performance fiber reinforced cementitious composites in structural applications (Hawaii)* Task Gr. C: 1–8.

Ohgishi, S. & Ono, H. 1988. Influence of testing factors on fracture toughness, G1c and fracture energy, G_F of plain and fiber reinforced concrete. *Concrete technology* 26(2): 103–118.

Sakai, E. 2006. Methods of cross sectional restoration. *Cement and concrete* 713: 39–44.

Sakami, S. (eds) 2006. *Check list for construction joint in RC structures* Tokyo: Gihodo.

Sugata, N. et al. 2003. Improvement of union of concrete by placing process. *Cement science and Concrete technology* 57: 186–192.

Tanaka, K. & Shin, Y. 2000. Permeability and pore structure of placing joint of cement mortar in casting. *J. Struct. Constr. Eng. by AIJ* 529: 7–12.

Uchida, Y. et al. 1991. Determination of tension softening diagrams of concrete by means of bending tests. *Journal of JSCE in V-1* 426: 203–212.

Uchikawa, H. et al. 1993. Estimation of the thickness of transition zone in hardened mortar and concrete, and investigation of the relationship between their thickness and strength development. *Concrete research and technology* 4(2): 1–8.

Yamamoto, Y. (eds) 2001. *Counter plans and problems for cold joint in concrete structure: Concrete library 103 by JSCE* Tokyo: Maruzen.

Yuasa, N. et al. 1998. Inhomogeneous distribution of the moisture content and porosity from the surface layer to internal parts. *J. Struct. Constr. Eng. by AIJ* 509: 9–16.

Fracture Mechanics of Concrete and Concrete Structures – High-Performance Concrete,
Brick-Masonry and Environmental Aspects – Carpinteri, et al. (eds)
© 2007 Taylor & Francis Group, London, ISBN 978-0-415-44617-4

Quantification of freezing-induced damage in reinforced concrete

K. Zandi Hanjari

Dept. of Civil and Environmental Engineering, Chalmers University of Technology, Göteborg, Sweden

P. Kettil

Dept. of Applied Mechanics, Chalmers University of Technology, Göteborg, Sweden

K. Lundgren

Dept. of Civil and Environmental Engineering, Chalmers University of Technology, Göteborg, Sweden

ABSTRACT: The paper presents a methodology to analyze the mechanical behavior of reinforced concrete structures with an observed amount of freezing-damage at a given time. It is proposed that the effect of internal freezing damage can be modeled as change of material properties, and that surface scaling can be modeled as change in geometry. The change in material properties was examined, and it was found that relations between compressive and tensile strength commonly used for undamaged concrete could not be directly applied to freezing damaged concrete. A modified relation was suggested. The proposed methodology was tested on concrete beams affected by internal freezing damage, using non-linear finite element analyses based on fracture mechanics in the program Diana, and the results were compared with available experimental results. The results indicated that an uncertainty in the analyses was the Young's modulus for damaged concrete, and that this influenced the results to a rather large extent.

1 INTRODUCTION

There is a growing need for reliable methods of assessing deteriorated structures, since an optimized maintenance and repair method involves the capability to predict the load-carrying capacity and remaining service life of deteriorated structures. In an ongoing research project, load-carrying capacity of damaged bridges is studied on the structural and component level. Mainly damage due to environmental impacts will be considered, such as corrosion, damage due to freezing, splitting of covers, and damaged bond between the concrete and reinforcement. The part of the project presented here is mainly focused on damage due to freezing, which is one of the major causes of deterioration in reinforced concrete structures and is still a pending research topic in assessment of old bridges.

While previous research has been chiefly concerned with the causes and mechanisms of freezing deterioration, relatively little attention has been given to the problem of assessing the residual load-bearing capacity of deteriorated reinforced concrete structures. This has exposed the need for improved understanding of freezing damage effects upon structural integrity. The aim of this research is to develop a method to quantify the damage caused by freezing of reinforced concrete.

Freezing damage in concrete is caused by the volume expansion of freezing water in the concrete pore system. If the expansion cannot be accommodated in the pore system, but is restrained by the surrounding concrete, it induces tensile stresses in the concrete. The tensile stresses cause cracks, which affect the strength, stiffness, and fracture energy of the concrete as well as the bond strength between the reinforcing bar and surrounding concrete in damaged regions; see Powerst (1945) and Shih (1988).

Two types of freezing damage can be distinguished, Fagerlund (2004):

1. Internal freezing damage caused by freezing of moisture inside the concrete. This may cause cracking and substantial reduction of strength and stiffness, Fagerlund (2004).
2. Surface scaling, which is usually caused by freezing of salt water in contact with the concrete surface. This damage usually results in spalling of the concrete surface, while the remaining concrete is mainly unaffected, Fagerlund (2004) and Gudmundsson (1999).

This paper presents a methodology to analyze the mechanical behavior (e.g. stiffness and strength) of reinforced concrete structures affected by freezing damage. At this stage, the methodology is restricted to the prediction of the mechanical behavior for a structure with an observed amount of damage at a given time. The freezing-damage development over time is not included.

The paper is organized as follows: Chapter 2 presents the proposed methodology. In Chapter 3, the methodology is tested on concrete beams affected by internal freezing damage. Finally, Chapter 4 concludes the work.

2 MODELING OF FREEZING DAMAGE

In the following, a methodology to analyze the mechanical behavior of reinforced concrete structures affected by freezing damage is proposed. The methodology is based on the assumption that the usual method of structural analysis applies. Furthermore, it is assumed that the effect of internal freezing damage can be modeled as change of material properties; i.e. reduction of strength and stiffness, and that surface scaling can be modeled as change in geometry; i.e. reduction in dimensions.

2.1 *Change of material properties*

Internal freezing damage is here modeled as change of material properties; i.e. reduction of strength and stiffness.

2.1.1 *Compressive strength*
Measurements of the compressive strength for internal freezing damage have been reported in e.g. Fagerlund (1994). Internal frost-damage is caused by freezing of water inside the concrete. The damage caused by freezing is always extended to such parts of the concrete where the degree of water saturation, S, exceeds the critical value. The critical degree of saturation, S_c, is a material property and independent of the number of frost cycles and freezing rate. As reported in Fagerlund (1994), in order to reach different degree of saturation, certain fractions of air bubbles, captured inside the pore system of concrete during casting, had to be water-filled. This could only be achieved by dissolution of the entrapped air and replacement by water to different extent. Vacuum treatment to different residual pressure (of 2, 20, and 50 mm Hg), have been used to empty the initially air-filled pores in the concrete and make it possible for water to rapidly fill the pore system of concrete including the air pores.

So far, based on a study of the reported results, it seems not possible to directly relate the reduction of compressive strength to the extent of freezing damage observed. Hence, it must be concluded that at least the compressive strength must be measured in each individual case by e.g. compression tests on a few drilled cores, supplemented with non-destructive testing such

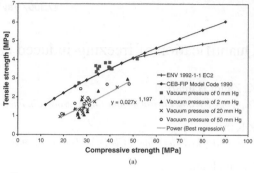

(a)

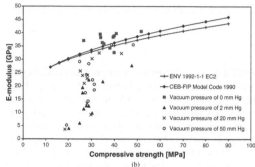

(b)

Figure 1. Relations between (a) compressive strength and tensile strength, and (b) compressive strength and Young's modulus for undamaged and freezing damaged concrete; test results from Fagerlund (1994).

as with Schmidt-hammer, to determine the extent of the damaged region.

2.1.2 *Other strength and stiffness properties*
For undamaged concrete, relations between compressive strength and other properties such as tensile strength, stiffness, and fracture energy are well established and widely used. For practical reasons, to reduce the required amount of testing down to compressive tests only, it would be useful if similar relations could be found for freezing damaged concrete. Therefore, in Figure 1 test results for tensile strength and Young's modulus for freezing damaged concrete are compiled and plotted versus measured damaged compressive strength. This is compared with relations for and test results of undamaged concrete.

All test results are from Fagerlund (1994). It should be noted that the tests of the compressive strength have been done on three specimens for each case. The measured compressive strengths were recalculated from 100 mm cube to 150 mm cube by multiplying with a factor of 0.96, according to Neville (2003), and then to standard 150*300 mm cylinder that commonly are used in different codes according to Ljungkrantz *et al.* (1994). The tensile strengths were obtained from measured splitting tensile strength by multiplying with a factor of 0.9 according to CEB (1993). Finally, Young's modulus was obtained from dynamically measured Young's modulus, reported by Fagerlund (1994), by multiplying with a factor of 0.83, according to Neville (2003).

As can be seen in Figure 1, the relations for undamaged concrete cannot be used directly for freezing damaged concrete. The tensile strength of the damaged concrete is markedly lower than would be estimated from the relations for undamaged concrete. By curve fitting, although the scatter is large, the following relation for damaged concrete is suggested:

$$f_{ct} = 0.027 f_{cc}^{(1.197)} \qquad (1)$$

where f_{ct} is the tensile strength of the damaged concrete and f_{cc} is the measured compressive strength of concrete using standard 150*300 mm cylinder. The suggested relation is also shown in Figure 1.

For Young's modulus, the results are unclear; while many results indicate lower values than would be expected from the relations for undamaged concrete, some even indicate higher values. Due to this large scatter, it was not considered possible to give any suggested relation between compressive strength and Young's modulus.

2.1.3 Stress-strain curve and failure envelope
In absence of experimental investigations, the only, but yet reasonable, possibility is to assume that the shapes of the stress-strain relation and failure envelope of the freezing damaged concrete are the same as for undamaged concrete.

2.2 Change of geometry

Surface scaling freezing damage is suggested to be modeled as change of geometry; i.e. reduction in dimensions. The remaining concrete is assumed to be unaffected, according to Gudmundsson (1999). The extent of the damaged region and the depth of surface scaling must be measured on site. The geometry of the structural analysis model must be updated accordingly.

3 FE ANALYSIS – COMPARISION WITH TESTS

In the following, the methodology proposed in Chapter 2 was tested on concrete beams affected by internal freezing damage. No tests on the structural effects of scaling damage were found in the literature; therefore no such comparisons could be done.

3.1 Experimental test setup

Four-point bending tests of frost damaged reinforced concrete were reported by Hassanzadeh (2006).

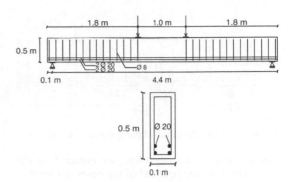

Figure 2. Example of geometry of beams tested, redrawn from Hassanzadeh (2006).

A short summary of the test setup relevant for the finite element (FE) analysis is given in the following.

In total 14 beam tests have been carried out, comprising two different geometries, varying reinforcement content (bending reinforcement content ratio $\omega/\omega_b = 64 - 114\%$, and with and without stirrups), and different climate exposure (L = laboratory climate, S and V varying forms of freezing exposure). In addition, the compressive strength, splitting tensile strength and fracture energy have been measured on cubes, cylinders and RILEM test beams. The compressive and tensile strength of the concrete exposed to freezing have been determined from cores drilled out from concrete blocks stored in the same conditions as the beams. The specimens exposed to freezing showed typical internal freezing damages. Surface scaling did not occur.

3.2 Finite element model

One beam type tested by Hassanzadeh (2006), beam type 1s, was modeled and analyzed by finite element method (FEM) using the program Diana. This beam type was chosen because freezing damage changed the failure mode from yielding of the reinforcement to bending compression failure. Both the undamaged beam and the beam exposed to freezing (internal freezing damage) were modeled and analyzed. The effect of the internal freezing damage was modeled as a change of material properties in accordance with the methodology proposed in Section 2.1.

The beam was modeled in 2D, see Figure 3. Due to symmetry only half of the beam was modeled. In the tests, steel plates and roller bearings have been used at the supports. In the FE-model, the steel plate was modeled as infinitely stiff by constraint equations, see left end of the beam in Figure 3. The FE nodes along the plate were tied to the centre node, thus forcing the nodes to remain in a straight line, but allowing for rotation. The centre node was supported for displacement in the y-direction. Also for modeling of the loading plates on the top of the beam, the nodes were

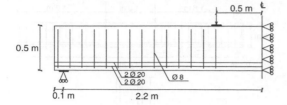

Figure 3. 2D model of half of the symmetric beam.

Table 1. Concrete properties.

	f_{cc} [MPa]	f_{ct} [MPa]	G_F [N/m]	E_c [GPa]
Undamaged	36.2	2.41	139	31.26
FEA-damaged-1	16.8	0.9	163	31.26
FEA-damaged-2	16.8	1.59	163	24.23
FEA-damaged-3	16.8	0.68	163	4.8
FEA-damaged-4	16.8	0.68	163	7.0
FEA-damaged-5	16.8	0.68	163	15.0

Table 2. Reinforcement properties.

	f_y [MPa]	E_s [GPa]
Reinforcement	670	196

tied to remain in a straight line, see Figure 3. At the symmetry line, see right end of the beam in Figure 3, all nodes were fixed in the x-direction.

For the concrete, 4-node plane stress solid elements were used. The rebar layers were modeled by 2-node truss elements. To model slip between rebar and concrete, interface elements were used. The stirrups were modeled by an option called "embedded reinforcement", corresponding to full interaction between the concrete and the steel.

The concrete was modeled with a constitutive model based on non-linear fracture mechanics using a smeared rotating crack model based on total strain; see DIANA (2006). The crack band width was assumed to be equal to the element size, 50 mm. For the tension softening, the curve by Hordijk *et al.* was chosen, as described in DIANA (2006). In compression, an ideal plastic behavior was used. The bond-slip relation was based on the CEB/FIP Model Code for confined concrete with good bond conditions, see CEB (1993).

Since the reinforcement type used in the experiments had not been reported in Hassanzadeh (2006), the reinforcement steel was in the finite element analyses modeled as elastic-perfect plastic and the yield stress was calibrated so that the maximum load agreed with the experimental result for the undamaged beam. The same yield stress was used for the rest of the analyses.

The material properties used in the analyses are shown in Tables 1–2. For the undamaged case, the material properties used in the analysis were based on the measured compressive strength and fracture energy in Hassanzadeh (2006) and calculated tensile strength and E-modulus using ENV 1992-1-1 EC2 and CEB-FIP Model Code 1990 respectively. For the frost-damaged case, several analyses were run with varying material properties due to the large scatter of the measured material properties. The fracture energy in all analyses were assumed to be the same as the measured fracture energy, 163 N/m, for the damaged concrete by Hassanzadeh (2006). In FEA-damaged-1 analysis, all measured material properties were used. It should be noted that the compressive strength was recalculated from 100*200 mm cylinder to standard 150*300 mm cylinder by multiplying with a factor of 0.96 according to Ljungkrantz *et al.* (1994), and the tensile strength was obtained from measured splitting tensile strength

by multiplying with a factor of 0.9 according to CEB (1993). As the E-modulus had not been measured, the calculated undamaged E-modulus was used in this case. In FEA-damaged-2 analysis, all damaged properties were calculated from measured compressive strength using relationships for undamaged concrete. In this case the tensile strength and E-modulus were calculated using ENV 1992-1-1 EC2 and CEB-FIP Model Code 1990 respectively. For FEA-damaged-3 to 5 analyses, the tensile strengths were calculated from measured compressive strength using proposed damaged tensile strength in Equation 1. Since the scatter of E-modulus reported by Fagerlund (1994) is quite high, Figure 1(b), and the reported damaged compressive strength by Hassanzadeh (2006) is rather low, the minimum reported E-modulus for FEA-damaged-3 analysis and two higher E-modulus for FEA-damaged-4 and 5 analyses were used.

An incremental static analysis was performed using a Newton-Raphsson iterative scheme to solve the non-linear equilibrium equations. First, the self-weight gravity load was applied. Then, the external load was gradually applied as prescribed displacement at the loading point.

3.3 Results

The total load versus midpoint deflection graphs from the experiments and FE-analyses for both the undamaged and freezing damaged beams are shown in Figure 4. Figures 5-7 show the deformed shape and the crack distribution (in terms of maximum contour plots of the tensile strain) for both the undamaged and frost-damaged beams from the analyses. Figures 8–10 show the compressive stress and strain in concrete and the stress in the reinforcement versus beam deflection.

The load-deflection graph shows good agreement between the analysis and test for the undamaged case,

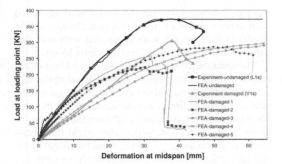

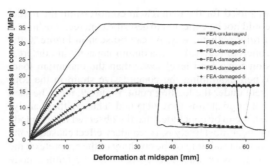

Figure 4. Load-displacement curves from analyses and experiments.

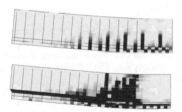

Figure 5. Deformed shape and crack development (maximum tensile strains) for undamaged beam before and after failure (5 and 59 mm deformation at midspan) from FEA-undamaged analysis.

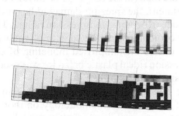

Figure 6. Deformed shape and crack development (maximum tensile strains) for frost damaged beam before and after failure (5 and 59 mm deformation at midspan) from FEA-damaged-1 analysis.

Figure 7. Deformed shape and crack development (maximum tensile strains) for frost-damaged beam before and after failure (5 and 59 mm deformation at midspan) from FEA-damaged-3 analysis.

Figure 8. Compressive stress in concrete under the loading point versus beam deflection at midspan.

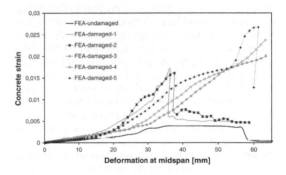

Figure 9. Strain in concrete in x-direction under the loading point versus beam deflection at midspan.

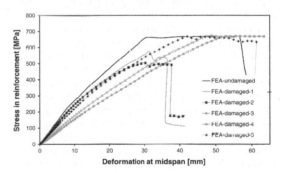

Figure 10. Stress in bottom reinforcement at midspan versus beam deflection at midspan.

see Figure 4. For the damaged case the agreement between the test and the analyses are less good, which may depend on the large scatter of the measured material properties. The analyses of the damaged case are discussed in the following.

The analyses cases 1 and 2 are initially too stiff compared to the test, but fails at a lower load level (−15% and −30%, respectively), see Figure 4. The deformed shape and crack distribution, Figure 6, indicate that a shear failure occurred in these analyses. However, it should be noted that as an ideal plastic behavior was used for the concrete in compression, the model could not be expected to describe concrete crushing in a fully realistic way. As can be seen in Figure 8, the concrete reached its maximum capacity in compression for a load level lower than the maximum load. When examining the compressive strain in the concrete under the loading point, Figure 9, it can be seen that large strains were obtained. Thus, in these analyses, it can be concluded that the observed failure mode in shear most likely is a secondary effect caused by the limited modeling of the crushing of the concrete on the compressive side. This agrees with the failure mode reported from the test, which was concrete crushing in bending.

In the analyses cases 3 to 5 where the tensile strengths were calculated using proposed relationship in equation(1) and Young's modulus were varied, the stiffness prediction improves compared to the experiment. Case 4 gives the best agreement for the stiffness. Concerning the failure mode, the reinforcement yielded in these analyses, see Figure 10. However, it should be noted that the concrete reached its maximum capacity in compression before the reinforcement yielded, and that large compressive strains in the concrete under the loading point were obtained, Figure 10 and 9. Thus, again, most likely the analyses should not be trusted all the way to maximum load, due to the simplified modeling of concrete in compression. Hence, by varying the Young's modulus, the stiffness could be better predicted compared to the test, but the failure load and failure mode could not be described properly.

To enable a better description of the concrete compression failure mode, it would have been necessary to include the softening of concrete in compression, and also to give a descending branch of the stress-strain curve. When this is done, localization of the deformations in a compressive failure needs to be taken into account. Van Mier (1984) showed that the compression softening behavior is related to the boundary conditions and the size of the specimen. One problem when modeling this is that the number of elements in which the compressive region will localize is not known when the analysis is started. While in tension, it seems reasonable to assume that a crack will localize in one element, an assumption that is not so obvious for compression. This complication is the reason why the simplified modeling in compression was chosen in the analyses presented here.

4 CONCLUSIONS

The paper has presented a methodology to analyze the mechanical behavior of reinforced concrete structures affected by freezing damage. The proposed methodology was tested on concrete beams affected by internal freezing damage, using non-linear FE-analyses based on fracture mechanics in the program Diana 9.1. The results of the analyses were compared with available experimental results from Hassanzadeh (2006).

The undamaged beam failed due to yielding of the bending reinforcement. As expected, good agreement between the analysis and the experiment was obtained, both regarding stiffness and strength. For the freezing damaged case, the agreement between the test and the analyses are less good, which also could have been expected due to the large scatter of the measured material properties. Therefore, several analyses were run with varying material properties. With a proper choice of the elastic modulus the stiffness of the damaged beam could be reproduced. However, the analyses gave a low failure load (in the range −5%–30%). Further, in the experiment, the freezing-damaged beam changed failure mode to bending compression failure. This failure mode could not be predicted in a correct way in the analyses, due to the simplified modeling of concrete in compression (ideal plastic behavior was used for the concrete in compression).

Future research should focus on refinement and testing of the proposed methodology on more experimental set-ups, including a better modeling of the compressive behavior. In particular, the correlation between compressive strength and other parameters need to be examined by more tests.

REFERENCES

CEB (1993). *CEB-FIP Model Code 1990*. Lausanne, Switzerland: Bulletin d'Information 213/214.

DIANA (2006). *DIANA Finite Element Analysis, User's Manual, release 9.1*. TNO Building and Construction Research.

Fagerlund, G. (2004). *A service life model for internal frost damage in concrete*. Lund: Univ.

Fagerlund, G. & Janz, M. & Johannesson, B. (1994). *Effect of frost damage on the bond between reinforcement and concrete*. Div. of Building Materials, Lund Institute of Technology.

Gudmundsson, G. & Wallevik, O. (1999). Concrete in an aggressive environment. *Frost damage in concrete (International RILEM Workshop), Pro 25*, 1999.

Hassanzadeh, M. & Fagerlund, G. (2006). Residual strength of the frost-damaged reinforced concrete beams. *III European Conference on Computational Mechanics Solids, Structures and Coupled Problems in Engineering*, Lisbon, Portugal, 2006.

Neville, A.M. (2003). *Properties of concrete*. London: Pearson prentice Hall.

Powers, T.C. (1945). A working hypothesis for further studies of frost resistance of concrete. *Journal of American Concrete institute* 16 4: 245–271.

Shih, T.S. & Lee, G.C. & Chang, K.C. (1988). Effect of freezing cycles on bond strength of concrete. *Journal of Structural Engineering* 114 3: pp. 717–726.

Fracture Mechanics of Concrete and Concrete Structures – High-Performance Concrete,
Brick-Masonry and Environmental Aspects – Carpinteri, et al. (eds)
© 2007 Taylor & Francis Group, London, ISBN 978-0-415-44617-4

Determination of the degradation level in fire-damaged RC constructions

A. Dufka & F. Khestl

Brno University of Technology, Faculty of Civil Engineering, Institute of Technology of Building Materials and
Components, Brno, Czech Republic

ABSTRACT: Degradation of material as a consequence of synergistic action occurs at the fire affected reinforced concrete constructions of both physical (temperature shocks, expansion pressures of water steam etc.), and physically chemical mechanisms (decomposition of cement swage, modification changes in aggregate etc.). With regard to these facts it is obvious that correct judgement of state respectively measure of disruption of constructions interfered this way makes demands on measure of tests and analysis realized in the frame of constructional technical research. The article presented treats problems of diagnostic procedures indispensable for valuable judgement of state of fire-affected constructions.

1 INTRODUCTION

Fire is one of the factors that might be the cause of considerable failure on constructional objects and in extreme cases they might cause the collapse of the whole construction. It is possible to state generally that extreme temperatures connected with the outbreak of fire degraded building materials by contribution of both physical (expansion pressures of water steam, linear respectively volume changes etc.), and physically chemical principles (degradation of material swage, modification changes etc.). This synergy of negative mechanisms will be applied understandably in case of fire affected reinforced concrete constructions.

The fact that reinforced concrete elements are degraded in case of fire by contribution of different principles must be naturally taken into account while choosing diagnostic methods respectively while putting together the conception of constructional technical research whose aim is to judge succinctly the state of building respectively measure of damage caused by fire.

This article is focused on problems with choosing diagnostic procedures that are required for correct judgement of state of reinforced concrete constructions. It is possible to state that these problems (i.e. problems of diagnostic of reinforced concrete constructions affected by fire) are a scientific field that is somehow neglected nowadays by expert public.

One of the key aspects which are indispensable to accept while choosing diagnostic procedures whose aim is to judge the state of constructions affected by fire is taking into account the mechanisms that cause degradation of reinforced concrete. In this

chapter, attention is drawn on principles of degradation of reinforced concrete.

2 THE PRINCIPLES OF DEGRADATION OF REINFORCED CONCRETE CONSTRUCTIONS IN ACTION OF EXTREME TEMPERATURES

We can summarise the negative effects in consequence of which failures may occur after the outbreak of fire, possibly entire loss of load-bearing capacity by the following:

Rapid growth in temperature occurs at the outbreak of fire, that is, the surface of construction is exposed to considerable changes of temperature. The speed of growing temperature of construction depends primarily on intensity of fire and character of construction. Destruction of concrete is a consequence of such shocking temperature stress when breaking of the surface concrete layers in thickness up to several centimeters occurs. This kind of failure is called as so-called "blasting of concrete" (spalling). The consequence of this type of failure is among others the fact that the top layer of concrete above the reinforcement was weakened; eventually the reinforcement is exposed to direct action of fire. A considerable heating of reinforcement occurs, accompanied by loss of its bastion, which can threaten in principle the structural analysis of the whole construction.

From the point of view of decrease of physically mechanical parameters of concrete on a basis of Portland cement, which occur as a result of action of higher temperature, the processes that occur in microstructure of cement swage have crucial importance. Mainly these facts are important:

The dehydrating of calciumhydrosilicates respectively calciumhydroaluminates occurs at temperatures upto c. 400°C. These reactions are to a certain extent reversible. The phases stated above do not create as a rule major binder component of cement stone.

The dissociation of portlandite in cement stone happens in the interval of temperatures 460 until 560°C. Portlandite is one of the materials that significantly affect binder capabilities of cement stone.

Modification change of silica that as a rule creates the majority part of concrete aggregate (gravel) occurs in temperature of concrete 573°C. Change of β silica to modification γ that happens at the temperature stated is accompanied by significant volume changes and is therefore a cause of genesis of considerable expansion pressures in the structure of concrete.

The decomposition of compact-grained calcium carbonate (that is, vaterite and aragonite) occurs in approximate range of temperatures 700–820°C. The dissociation of large-grained calcium carbonate (mainly calcite) occurs in the interval of temperatures c. 820–940°C. The modification of calcium carbonate belongs to phases that considerably influence adhesive force of cement swage.

It is obvious from the above stated that decrease of mechanical features of concrete exposed to extreme temperatures created by fire is caused by synergic action both physical principles (i.e. destruction as a consequence of "temperature shock") and physically chemical processes that proceed in microstructure of concrete (i.e. decomposition of concrete swage, modification changes in gravel etc.). From the point of view of load-bearing capacity the changes in features of reinforcement respectively decrease of mechanical bond are very important.

Another aspect that can further increase the development of degradation of constructions evaluated is chilling during fire fighting.

3 DIAGNOSTICS OF FIRE AFFECTED CONSTRUCTIONS

In the previous text the fact was stated that mechanisms, which cause the development of degradation on reinforced constructions affected by fire, are not trivial. This reality understandably determines the fact that if the state respectively measure of degradation on reinforced construction affected by fire has to be judged correctly, a complex approach to this problem is indispensable.

A complex of activities whose aim is to judge succinctly the state of fire affected construction respectively judge the measure of degradation on reinforced constructions by extreme temperatures, can be designated as constructional technical research.

Methodological approaches applied in the frame of diagnostics of constructions affected by fire, must be understandably adapted to character and state of construction. The following text is looking at the problems.

Visual inspection is a source of initial information about the extent of damage caused by fire and measure of disturbance of individual construction elements. In the frame of visual inspection, the whole state of construction is evaluated primarily, that is, an inquiry is made if some of the elements collapsed, deviation from column is then watched in vertical elements, deflexion was watched in roof girders etc. (gained knowledge this way can be specified geodetically). Furthermore, in the frame of visual inspection, attention is directed on judgement on state of surface of individual elements (destruction of surface layers, presence of breaks etc.), disturbance of decking above reinforcement is watched, measure of bare reinforced steel etc. Gained knowledge in visual inspection enables us to gain source information about damage caused by fire that, understandably, must be completed and expanded in a considerable way for correct judgement of state of construction.

To judge succinctly the state of reinforced constructional elements, in the frame of constructional technical research, determinations are made whose aim is to monitor physically mechanical (that is, especially bastion) and physically chemical (i.e. especially the state of cement swage) parameters of concrete.

3.1 *Physically mechanical parameters of concrete:*

- Determination of compression strength – destructively on core holes (ČSN EN 12390-3) eventually non-destructively (for instance, ČSN 73 1373),
- Determination of bastion of concrete in simple tension (ČSN 73 1318),
- Determination of tensile strength of surface concrete layers (ČSN 73 1318),
- Determination of elongation modulus (ČSN 731371).

3.2 *Physically chemical analysis*

The aim of physically chemical determinations is to judge the state respectively measure of disturbance of cement swage of concrete, eventually to analyse the state of aggregate in concrete. It is possible to use these analyses in advantage for discovering of these facts:

- X-ray diffraction analysis (RTG analysis) – qualitative analysis whose aim is to determine mineralogical composition of swage, eventually aggregate. The procedure of this analysis is defined by methodological approach of VUT FAST in Brno, no. 30-33/1,

- Differential thermal analysis (DTA analysis) – quantitative analysis. Its aim is to quantify especially the volume of phases that create concrete swage. The procedure of this analysis is defined by methodological approach of VUT FAST in Brno, no. 30-33/1.

Another phenomenon that is also monitored in the frame of research of fire-affected constructions is the features of reinforced steel (tensile strength, characteristic strength).

The places for realisation of making tests respectively places of offtake of core holes are chosen as a rule the way that both the localities in which visual inspection proved obvious intensive affection of evaluated elements by fire would be involved and the places in which this disruption is not visible by visual inspection. The aspect applied when judging the state of concrete in separate elements is then comparison of characteristics monitored determined on samples taken in places markedly affected by fire and from constructions in which the traces of action of fire were not obvious.

Another criterion that can be taken into account when judging the state of constructions researched is comparison of features of concrete evaluated with facts declared by project documentation (mainly with declared concrete grade of bastion etc.).

4 THE PROBLEMS THAT COMPLICATE THE INTERPRETATION OF GAINED KNOWLEDGE

Problems of judging of the state of reinforced constructions affected by fire require a complex approach. While judging the state of real constructions, the facts can occur that can significantly complicate the interpretation of gained knowledge in the frame of constructional technical research. The best way will be to document this statement by experience and examples that were gained while evaluating of state of reinforced concrete constructions affected by fire.

We made a constructional technical research of one storey prefabricated reinforced concrete hall, created by prefabricated columns, beams and bearers; membrane roofing was then created by reinforced concrete TT panels.

Gained knowledge at visual inspection stated the fact that most of constructional elements of this hall (that is, columns, bearers and membrane roofing) were quite markedly struck by action of temperatures emerging during fire. A collapse of joining balks and membrane roofing occurred in some parts of the hall.

Some of the joining balks did not collapse entirely, but their deflexion was considerable. There were a lot of cracks on joining balks. In some localities, decking of concrete on joining balks was entirely destroyed

Figure 1.   Fire affected hall.

Figure 2.   A collapse of joining balks and membrane roofing.

and reinforcement was disclosed. Membrane roofing was also struck by significant failures.

In localities where joining balks collapsed, a collapse of membrane roofing occurred understandably. There were places on columns in which traces of fire were striking, that is, it concerns mainly destruction of decking above the reinforcement. Presence of cracks was also identified on columns. Some of the columns deviated from perpendicular, which could be seen already in visual inspection.

Figure 3. Local places with totally burned through membrane roofing.

Gained knowledge in visual inspection were naturally completed by a complex of determination both physically mechanical and physically chemical characterisations of concrete. It is possible to summarize this knowledge by the following:

It was found that bastion characterisation of concrete evaluated range in very wide interval.

In localities where intense action of fire was obvious, tensile strength of surface layers was significantly in decrease and in some cases it actually equaled zero.

A rather different situation was found in case of compressive strength. In some elements in which evident traces after massive action of fire were detected by visual inspection, from the point of view of compression strength concrete corresponded with facts noted by project documentation. On the contrary, in some cases the compression strength of concrete was lower than concrete grade of bastion declared by project documentation. From the point of view of judgement of state of separate constructional elements the fact that in some cases decomposition of elements occurred during offtake of core holes was important too. Destruction of core holes was caused by presence of cracks in concrete. In the majority of cases these were cracks that on the surface of construction were only hair cracking, whereas in concrete, which creates the "inner mass" of the element, their opening was wider. The decomposition of core holes during offtake occurred in some cases as well as when core holes were taken from components in which traces of action of fire were not visible in visual inspection.

The samples of concrete were put through physically chemical determinations that enable them to analyse their composition for judgement of measure of degradation. One of the aspects that were taken into account in evaluation of results of physically chemical analyses was comparison of gained knowledge for separate samples with results of analyses made on sample that was taken off collapsed joining balk, i.e. element that was very strongly affected by fire. It was stated that cement swage is very intensely degraded by action of high temperatures in samples taken off collapsed joining balks. Cement swage of these samples was practically wholly decomposed including phases of calcium carbonate. The modifications of calcium carbonate (calcite, vaterite and aragonite) are minerals that decompose in relatively high temperatures (c. 800 to 950°C). In this case the results of physically chemical analyses proved very intense destruction of cement swage and this fact entirely corresponded with very low bastion of concrete that was found. Intense development of degradation on cement swage by high temperatures was found as well in some other constructional elements. Generally, it can be stated that elements, in which the decomposition of cement swage was found, this fact was accompanied by decrease (and as a rule very considerable) of bastion characterisation of concrete.

Indeed, the results of physically chemical analyses in some of the elements evaluated indicate a relatively low measure of degradation on cement swage by high temperatures, which did not respond to the results of physically mechanical determinations, when a considerable decrease of bastion parameters of concrete was found. This disproportion between gained knowledge by physically chemical analyses and results of tests of bastion characterisation of concrete can be justified be these principles respectively by their synergy:

- In consequence of "temperature shocks" emerging at the outbreak of fire respectively even while putting it out conditions for development and propagation of cracks are created in the structure of concrete. The cracks can also develop as a consequence of cumulation of water steam in the structure of concrete (spalling). The development of cracks can occur even in relatively low temperatures. These cracks do not result in significant changes in mineralogical composition of cement swage.
- The changes happening in aggregate, which creates filler in concrete, also take part in lowering of mechanical parameters of concrete. Aggregate with relatively high volume of silica was used in this case. Modification changes occur in increase of temperatures in this material (change of β silica to γ modification) accompanied by volume changes. However, these processes are reversible, that is, after lowering of temperature silica in modification β dominates absolutely again in the structure of concrete. It is obvious from the stated above that detection whether modification changes in concrete filler occurred as a consequence of fire is (with regard to reversibility of these changes) relatively difficult.
- In action of high temperatures on cement swage decomposition reactions of phases that create this swage occur. It is mainly the decomposition of, for instance, calciumhydrosilicates, ettringite,

portlandite, phases of calcium carbonate etc. It was proved that some of these decomposition reactions are reversible in real time horizon. As an example of this fact, we can state repeated creation of portlandite in cement paste exposed in temperatures that go beyond the temperature of dissociation of portlandite.

5 CONCLUSION

This article treats diagnostic processes whose aim is to judge succinctly the state of reinforced concrete constructions struck by fire and especially securing the data for statical judgement of constructions evaluated.

Generally, it can be stated that while monitoring the state of constructions struck by fire, de facto analogical characteristics are watched, as in case of constructional technical research of objects exploited in "common" conditions. Thus both physically mechanical parameters of concrete are watched (especially its bastion characterisation) and physically chemical analyses are made (with an aim to judge the state of cement swage) and the state of steel reinforcement is monitored as well. It is indispensable to respect mechanisms that caused destruction of reinforced concrete constructions evaluated (i.e. the fact that synergy of physical and physically chemical principles is the cause of creation of failures) for relevant interpretation of gained knowledge by a complex of these tests.

On the basis of knowledge found at evaluating of state of reinforced concrete constructions struck by fire, it is possible to state that one of the phenomenons which significantly take a share in many cases on intense decrease of bastion characterisation of concrete is creation and development of cracks. The creation of cracks is determined mainly by physical principles (as for instance, changes in temperature by shock, expansion pressure by water steam).

Another piece of knowledge which was gained in the frame of diagnostic procedures realised on fire affected constructions, is the fact that the quantity which very precisely describes measure of interference on reinforced concrete element by extreme temperatures, is tensile strength of surface layers of concrete.

Generally, it is possible to state that action of extreme temperatures significantly decreases tensile strength of concrete than compression strength is. This fact is caused by creation of microscopic cracks in cement swage, by creation of defects in the interface of swage – grains of aggregate etc. It is obvious that tensile strength of surface layers is a quantity which describes mainly the state respectively measure of disturbance in full the surface of element evaluated. We recommend unambiguously expanding the results of statement of tensile strength of surface layers of concrete by carrying out tests of concrete tensile strength (tests on core holes) for a more complex judgement of state of constructional elements.

Physically chemical analyses whose results in an effective way complete the gained knowledge by physically mechanical tests are an integral part of complex of analyses realised in research of reinforced concrete constructions struck by fire. It is possible to specify more closely the temperature at which separate elements were exposed and thus formulate an assumption about change of bastion parameters of concrete on a basis of determination of mineralogical composition of cement swage in concrete etc.

ACKNOWLEDGEMENTS

Knowledge was stated in the article that was gained in the frame of solution in "GP103/04/P006 *Increasing immunity of cement composites against extreme conditions*" and with support of "MSM 0021630511 *Progressive Building Materials with Utilization of Secondary Raw Materials and their Impact on Structures Durability*".

REFERENCES

Gluekler, E.L. 1979. Local thermal and structural behavior of concrete at elevated temperatures. *Transaction of the 5th International Conference on Structural Mechanics in Reactor Technology vol. A.* International Congress Center Berlin. p. 13–17 H8/4.

Malhotra, H.L. 1956. The effect of temperature on the compressive strength of concrete. *Mag. Conc. Res. 18 23* p. 85–94.

Matoušek, M., Drochytka, R. 1998. Atmosférická koroze betonů (The atmospheric corrosion of concrete). IKAS Praha.

Fracture Mechanics of Concrete and Concrete Structures – High-Performance Concrete,
Brick-Masonry and Environmental Aspects – Carpinteri, et al. (eds)
© 2007 Taylor & Francis Group, London, ISBN 978-0-415-44617-4

Homogenization-based modelling of reinforced concrete in the context of durability-oriented analyses

E. Rumanus & G. Meschke

Institute for Structural Mechanics, Ruhr University Bochum, Bochum, Germany

ABSTRACT: In the context of a poromechanics framework developed for durability-oriented analysis, the paper is mainly concerned with a constitutive model for reinforced concrete. A continuous concrete matrix and two different sets of steel reinforcement characterize the three-phase composite material. For each phase, the nonlinear pre- and post-peak behavior is described separately, while considering interactions between the reinforcement bars and the concrete. The material behaviour of the matrix material is formulated by a combined multisurface elasto-plastic damage model. A classical J_2–plasticity flow rule describes the elasto-plastic response of the steel reinforcement. Based on continuum micromechanics, the Mori-Tanaka homogenization technique is considered as a suitable approach to derive the homogenized (macromechanical) constitutive relations of the composite material and related local (micromechanical) field information which is essential for durability analyses. Dowel action between the reinforcement bars and the concrete is implicitly captured by the chosen approach. The model performance is demonstrated by selected numerical and experimental studies.

1 INTRODUCTION

Corrosion of the reinforcement constitutes one of the major limiting factors for the durability of reinforced concrete structures. Expansive radial pressures induced by corrosion products (rust) along the interface between the steel bars and the concrete may cause cracking and spalling of cover concrete. Gradual loss of bond strength between the reinforcement and the surrounding concrete and the reduction of reinforcement cross-sectional area amplify the deterioration mechanisms caused by corrosion. On the structural level, a reduction of the structural stiffness and of the load carrying capacity affects the (residual) service life-time of reinforced structures and may lead to premature failure. Since the nature of steel corrosion is physico-chemical and its evolution is strongly moisture dependent, life-time oriented structural analyses require consideration of transport mechanisms of moisture and of corrosive substances such as chloride ions or calcium hydroxide leading to the depassivation of the reinforcement bars. Equally relevant is a suitable model for reinforced concrete allowing to represent the interacting mechanisms between the reinforcement bars and the surrounding concrete such as bond slip and dowel action as well as the corrosion-induced degradation of these interaction properties.

In this study, the main focus is laid on the latter aspect: Reinforced concrete is modelled as a three-phase composite material consisting of a continuous matrix and two different sets of rebars. This idea has been suggested recently by Pietruszczak and Winnicki 2003) and (Linero et al. 2006) using the classical *mixture theory* in order to obtain macroscopic properties of the composite material. In the present paper, the macroscopic behavior of this composite material is obtained by employing homogenization schemes to a representative volume element (RVE), in which the necessary conditions for the application of homogenization are fulfilled (Zaoui 2002). This model for reinforced concrete is being implemented within a multiphase model for partially saturated concrete accounting for heat and moisture transport and the relevant interactions observed on the nano- and micro-level between cracking, drying and creep (Meschke and Grasberger 2003; Grasberger and Meschke 2004).

While for mainly unidirectional loading a 1D-modelling of the reinforcement within an embedded approach would be sufficient (Linero et al. 2006), shear stresses transmitted by the rebar in case of cracked concrete (dowel action) suggest a homogenization approach using a fully 3D representation of the steel reinforcement within the considered RVE. Effects such as dowel-action are therefore captured automatically without any additional specifications of the residual shear stiffness in cracked zones.

The mechanical response of composite materials is highly influenced by the morphology of the microstructure, the properties of the constituents and by micromechanical interactions within the composite. The volume fraction, the aspect ratio, the orientation and the shape of constituents with correlated interactions have to be taken into account in order to describe the structural response accurately. To avoid modelling of the complex heterogeneous microstructure, adequate homogenization techniques are performed. Based on continuum micromechanics (Zaoui 2002) the MORI-TANAKA homogenization scheme is employed in this study in order to provide an estimate of the constitutive relations of reinforced concrete described as a three-phase composite material. The adopted micromechanical model ensures the continuity of the surrounding matrix and accounts for interactions between the embedded inhomogeneities (Mori and Tanaka 1973).

The main goal of this study is to derive a reliable macroscopic model for reinforced concrete as a composite material which provides information about the stress and strain fields of the individual constituents (the concrete matrix and the reinforcement bars). This information is essential when degradation mechanisms originating from mechanical, physical and physico-chemical processes have to be estimated accurately. Since details on the coupled hygro-mechanical model for concrete have been already presented elsewhere (Meschke and Grasberger 2003; Grasberger and Meschke 2004), the focus of this paper lies on the formulation of reinforced concrete based on homogenization micromechanical approach. Since this work is in progress, in a first version of the proposed model, bond-slip between the rebar and the surrounding concrete is not yet accounted for.

2 CONSTITUTIVE MODELS

2.1 Concrete matrix material

For describing damage and creep of cementitious materials subjected to external loading and changing hygral conditions an elasto-plastic damage model (Meschke and Grasberger 2003, Grasberger and Meschke 2004) formulated within the framework of the BIOT-COUSSY theory (Coussy 2004) is employed. Concrete is assumed to consist of a continuous matrix and pores, which, depending on the environmental conditions, are in general partially filled by liquid water and by an ideal mixture of water vapour and dry air. Based on the considered theory, the individual phases formed by the matrix phase and the pores are represented as a homogeneous material according to their volume fraction in each material point. The related constitutive relations capturing the main physical processes acting on the nano- and microlevel are obtained by means of defining an appropriate expression for the free energy of the thermodynamic system together with macroscopic coupling coefficients which are obtained from relating micro- and macroscopic quantities and exploiting symmetry conditions of the macroscopic energy function Ψ_m

$$\Psi_m = \mathcal{W}(\varepsilon_m - \varepsilon_m^p - \varepsilon_m^f, m_l - \rho_l \phi_l^p, \psi, \gamma_f, T) + \mathcal{U}(\alpha_R, \alpha_{DP}) \qquad (1)$$

for the matrix material (Grasberger and Meschke 2004). The index m refers to the matrix. The linearized strain tensor ε_m within the matrix is assumed to be small and can therefore be decomposed into elastic strains ε_m^e, plastic strains ε_m^p and long-term creep strains ε_m^f i.e.

$$\varepsilon_m = \varepsilon_m^e + \varepsilon_m^p + \varepsilon_m^f. \qquad (2)$$

Moisture distribution is described by the liquid mass content m_l and by liquid, with a density of ρ_l, occupying the non-recoverable portion of the porosity ϕ_l^p. The integrity ψ captures the isotropic damage state of the poromechanical material. Viscous slip γ_f, associated with relative motions within gel-pores, causes creep deformations observed on the macroscopic scale (Bažant et al. 1997; Grasberger and Meschke 2003). The thermal field is described by the absolute temperature T. The hardening and softening law, specifying the material behaviour beyond the elastic domain, is governed by the internal variables α_R for tension cracking and by α_{DP} for compression damage.

A multi-surface fracture energy-based damage-plasticity theory is employed to characterize the behaviour of concrete in tension and compression (Meschke et al. 1998). Degradation mechanisms and inelastic deformations are controlled by four threshold functions f_k defining a region of admissible stress states in the space of plastic effective stresses σ'_m

$$\mathbf{E} = \{(\sigma'_m, q_k) | f_k(\sigma'_m, q_k(\alpha_k)) \leq 0, \ k = 1, .., 4\}. \qquad (3)$$

Cracking of concrete is accounted for by means of a fracture energy based Rankine criterion, employing three failure surfaces perpendicular to the axes of principal stresses

$$f_{R,A}(\sigma'_m, q_R) = \sigma'_A - q_R(\alpha_R) \leq 0, \quad A = 1, 2, 3 \qquad (4)$$

with $q_R(\alpha_R) = -\partial \mathcal{U}/\partial \alpha_R$ denoting the softening parameter and the index A refers to the principal direction. The ductile behaviour of concrete subjected to compressive loading is described by a hardening/softening Drucker-Prager plasticity model

$$f_{DP}(\sigma'_m, q_{DP}) = \sqrt{J_2} - \kappa_{DP} I_1 - \frac{q_{DP}(\alpha_{DP})}{\gamma_{DP}} \leq 0 \qquad (5)$$

with $q_{DP}(\alpha_{DP}) = -\partial \mathcal{U}/\partial \alpha_{DP}$ as the hardening/softening parameter and the values κ_{DP} and γ_{DP} are obtained from the compressive strength of the material.

The mechanical behavior for the matrix in case that no moisture and heat transport is considered, is characterized by the stress field of the matrix σ_m which is equal to the plastic effective stress tensor σ'_m (Grasberger and Meschke 2004) and is obtained from equation (1) and (2) as

$$\sigma_m = \psi\, \boldsymbol{C}_m : \varepsilon^e_m = \psi\, \boldsymbol{C}_m : (\varepsilon_m - \varepsilon^p_m - \varepsilon^f_m) \qquad (6)$$

with $\boldsymbol{C}_m$ as the undamaged elasticity tensor of the matrix material.

2.2 Reinforcement

Similar to equation (2), the total strain tensor of the rebar ε_s can be decomposed into an elastic ε^e_s and a plastic part ε^p_s

$$\varepsilon_s = \varepsilon^e_s + \varepsilon^p_s. \qquad (7)$$

The subscript s refers to the deformations of the steel. Since the distribution of the stiffness within a reinforced concrete structure is discontinuous, the strains of the matrix given in equation (2) may differ from the strains of the reinforcement even when full bonding between reinforcement and matrix is assumed. The stress-strain relationship is obtained from the stored energy function Ψ_s as

$$\sigma_s = \partial\, \Psi_s / \partial\, \varepsilon^e_s = \boldsymbol{C}_s : \varepsilon^e_s = \boldsymbol{C}_s : (\varepsilon_s - \varepsilon^p_s), \qquad (8)$$

where $\boldsymbol{C}_s$ denotes the isotropic tensor of elasticity. The admissible stress field σ_s within the rebar is described by a classical J_2-plasticity model (Simo and Hughes 1998) and the non-linear regime beyond the yield stress σ_y is governed by an isotropic linear hardening law based on the von Mises yield condition

$$f(\mathbf{s}, \alpha) = \|\mathbf{s}\| - \sqrt{2/3}\,[\sigma_y + K\,\alpha] \leq 0. \qquad (9)$$

The evolution of the isotropic hardening is governed by the internal variable α and by the constant isotropic hardening plastic modulus K.

3 CONTINUUM MICROMECHANICS

3.1 Composite material (reinforced concrete)

In this study, the considered three-phase composite consists of a continuous matrix formed by concrete and by two sets of straight rebars representing the steel reinforcement forming the reinforcement-layer. The direction of the rebars and the geometry of the cross section may be arbitrary within the 2–3-plane of the reinforcement layer. Figure 1 contains an illustration of the composite material "reinforced concrete".

Such a configuration is typical for reinforced shell-like as well as beam structures. Besides steel

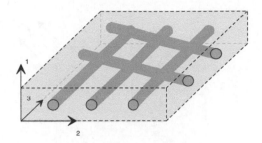

Figure 1. Illustration of the composite material.

reinforcement also textile materials are frequently used as fiber-reinforcement in concrete structures. Such composite material may also be described within a micromechanical framework (Richter 2005). Therefore, the proposed micromechanical model presented in the following sections is formulated in a rather general format in order to consider a broad class of reinforcing materials.

3.2 The representative volume element (RVE)

A widely used approach in continuum micromechanics is based on the consideration of a representative volume element (RVE) representing an arbitrary material point of a structure. Thereby, the complex morphology of the microstructure is captured in a simplified manner by the RVE in order to estimate the related effective (macroscopic) response by means of an averaging procedure. To confirm the representative character of the RVE, the considered size l has to be large enough in order to ensure a statistical distribution of the constituents with a characteristic size d and at the same time it has to be essentially smaller than a length of the structure L

$$d \ll l \ll L. \qquad (10)$$

In Figure 2 the assumed RVE is depicted schematically. The considered microstructure is governed by two straight steel rebars with the tensor of elasticity $\boldsymbol{C}_1$, $\boldsymbol{C}_2$ and the related angles α_1, α_2, which are embedded in a continuous matrix (concrete) with the material tensor $\boldsymbol{C}_m$. Depending on the position and volume fraction of each rebar, the expected effective mechanical response of the RVE is in general anisotropic or transversal isotropic. Hill's condition requires the equality of the energy on the micro and macro level independently of the constitutive law. This condition is a priori fulfilled by homogeneous strain boundary conditions applied by prescribing linear displacements at the boundary of the RVE (Zohdi and Wriggers 2005)

$$\boldsymbol{u}(\boldsymbol{x}) = \varepsilon^* \cdot \boldsymbol{x}, \quad \boldsymbol{x} \in \partial V, \qquad (11)$$

where ε^* defines the macroscopic (constant) strain tensor.

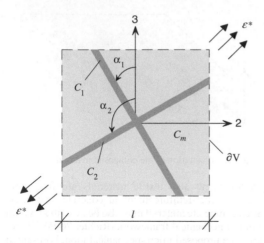

Figure 2. Representative volume element (RVE).

3.3 Micro-macro mapping

The local strain and stress fields within the RVE are averaged over the total volume V of the RVE in order to evaluate the homogenized values of the strains

$$< \varepsilon >_V = \frac{1}{V} \int_V \varepsilon(x) \, dV = \sum_{i=1}^{n} c_i < \varepsilon >_i \quad (12)$$

and of the stresses

$$< \sigma >_V = \frac{1}{V} \int_V \sigma(x) \, dV = \sum_{i=1}^{n} c_i < \sigma >_i . \quad (13)$$

Since the local averaged field values are assumed to be constant within each phase ($\sigma_i = < \sigma(x) >_i$ and $\varepsilon_i = < \varepsilon(x) >_i$) they can be summed up according to the volume fraction $c_i = V_i/V$, whereby V_i is the total volume of the phase i within the RVE. Hence, the volume of the RVE of the considered three-phase composite is assumed to be filled completely by all phases i.e. $c_1 + c_2 + c_m = 1$. According to the *average strain theorem*, for any perfectly bonded heterogeneous body the averaged strains $<\varepsilon>_V$ given by equation (12) can be identified as the macroscopic strain tensor ε^* applied on the RVE i.e. $<\varepsilon>_V = \varepsilon^*$, which is independent of the considered constitutive laws (Zohdi and Wriggers 2005). For a three-phase composite, a reformulation of equation (12) and (13) leads to

$$\varepsilon^* = < \varepsilon >_V = c_1 \varepsilon_1 + c_2 \varepsilon_2 + c_m \varepsilon_m \quad (14)$$

describing the homogeneous macroscopic strains applied onto the composite material and to

$$\sigma^* = < \sigma >_V = c_1 \sigma_1 + c_2 \sigma_2 + c_m \sigma_m \quad (15)$$

for the macroscopic stresses representing the composite stress field. The related local stress tensor $\sigma_i(\varepsilon_i)$

is calculated according to each constitutive law given by equations (6) and (8). The unknown local strain fields ε_i have to estimated by means of the forth-order localization (concentration) tensor $\mathcal{A}_i$ which relates the homogenized macroscopic strains ε^* to the local strains within each phase

$$\varepsilon_i = \mathcal{A}_i : \varepsilon^*, \quad i = 1, 2, m. \quad (16)$$

The tensor $\mathcal{A}_i$ of each phase accounts for the morphology of the microstructure by considering the elasticity, the volume fraction, the aspect ratio, the orientation and the shape of each constituent. It should emphasized that $\mathcal{A}_i$ relates micro and macro quantities and depends therefore on the theory chosen for the micromechanical model. If a three-phase composite is considered only two concentration tensors have to be known. The third one can be determined from the average value

$$< \mathcal{A} >_V = c_1 \mathcal{A}_1 + c_2 \mathcal{A}_2 + c_m \mathcal{A}_m = \mathbb{1}, \quad (17)$$

with $\mathbb{1}$ denoting the forth-order unit tensor. Due to different orientation and shape of the inhomogeneities, which is captured by $\mathcal{A}_i$, the mechanical response of the related homogenized stiffness tensor $\mathcal{C}^*$ is in general anisotropic even if all constituents are isotropic. As long as all constituents are in elastic regime, the mechanical constitutive relation for a composite material is defined by

$$\sigma^* = \mathcal{C}^* : \varepsilon^*, \quad (18)$$

where $\mathcal{A}^*$ can be derived from the localization tensors of each phase

$$\mathcal{C}^* = < \mathcal{C} : \mathcal{A} >_V = \sum_{i=1}^{n} c_i \mathcal{C}_i : \mathcal{A}_i. \quad (19)$$

In the post-cracking range of the matrix or in the yielding regime of the rebars, however, the macroscopic tangent stiffness tensor of the composite $\mathcal{C}^{*,tan}$ needs to be computed according to

$$\mathcal{C}^{*,tan} = d\sigma^* / d\varepsilon^*. \quad (20)$$

Depending on the considered micromechanical model the macroscopic tangent tensor $\mathcal{C}^{*,tan}$ is obtained from linearization of each constitutive law

$$\mathcal{C}_i^{tan} = d\sigma_i / d\varepsilon_i \quad i = 1, 2, m. \quad (21)$$

3.4 Three-phase Mori-Tanaka approach

An appropriate homogenization scheme to derive the effective mechanical response of a RVE is provided by the widely used Mori-Tanaka approach (Mori and Tanaka 1973). This micromechanical model ensures continuity of the matrix phase and accounts for mechanical interactions between the inclusions in an average manner. According to this homogenization scheme, the reference material playing the predominant morphological role of the composite is the continuous matrix. The inclusions and their states of strain and stress are directly affected by the matrix material. Within this approach, which is also denoted as *effective field theory*, the two limit cases (no inclusions exist ($c_i = 0$) and no matrix phase is considered ($c_m = 0$)) are covered by $\boldsymbol{C}^* = \boldsymbol{C}_m$ when $c_i = 0$ and $\boldsymbol{C}^* = \boldsymbol{C}_i$ when $c_m = 0$. In the following, the MORI-TANAKA equations for a non-linear three-phase composite are presented.

The relation between the strains of the phases ε_i and the applied macroscopic strains ε^* on the boundary of the RVE is formulated in the general format

$$\varepsilon_i = \boldsymbol{A}_i^{MT} : \varepsilon^*, \quad i = 1, 2, m. \tag{22}$$

In order to identify the fourth-order concentration tensor $\boldsymbol{A}_i^{MT}$ of each phase, the related assumptions of the MORI-TANAKA approach have to be taken into account. As mentioned before, the average strains of the inclusions ($\varepsilon_1, \varepsilon_2$) are defined by the average strains of the matrix ε_m

$$\varepsilon_1 = \boldsymbol{T}_1 : \varepsilon_m \quad \text{and} \quad \varepsilon_2 = \boldsymbol{T}_2 : \varepsilon_m. \tag{23}$$

Based on ESHELBY's equivalent inclusion approach, the fourth-order tensor $\boldsymbol{T}_i$ of each phase can be estimated by re-formulating the inclusion inhomogeneity problem as a homogeneous problem with eigenstrains (Eshelby 1957). The solution for a single elastic inhomogeneity with an ellipsoidal shape perfectly bonded to a surrounding homogeneous matrix is given by

$$\boldsymbol{T}_i = \left[\mathbb{1} + \boldsymbol{S}^i : (\boldsymbol{C}_m^{-1} : \boldsymbol{C}_i - \mathbb{1}) \right]^{-1} \quad i = 1, 2. \tag{24}$$

For an ellipsoidal geometry of the inclusions the fourth-order ESHELBY tensor $\boldsymbol{S}^i$ for each phase is solely dependent on the aspect ratio of the inclusion and on the Poisson's ratio ν_m of the surrounding isotropic matrix $\boldsymbol{C}_m$. Since a cylindrical shape can be regarded as an ellipsoidal geometry with a special aspect ratio, the solution of the ESHELBY tensor in a local coordinate system $\boldsymbol{S}^{loc}$ for the considered straight rebars can be computed (Eshelby 1957). In Figure 3, the cylindrical inhomogeneity representing a single rebar is illustrated. In this Figure, over-bars are used to characterize the local coordinate system. Note that the cross section of the rebar may have an elliptical or circular shape depending on the aspect ratio $s = a_2/a_1$. The ESHELBY tensor is transformed from the local to the global coordinate system within the 2–3-plane by means of the rotation tensor $\boldsymbol{Q}(\alpha)$

$$S_{ijkl} = Q_{im}(\alpha) Q_{jn}(\alpha) Q_{ko}(\alpha) Q_{lp}(\alpha) S_{mnop}^{loc}, \tag{25}$$

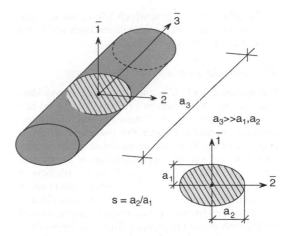

Figure 3. Representation of an inhomogeneity.

where α stand for the angle of rotation of the considered bar with respect to the positive x_1-axis (see Figure 2). For each set of rebars with the orientation α_i the related ESHELBY tensor $\boldsymbol{S}^i$ has to be obtained according to formula (25). The coordinate transformation of the stiffness tensors $\boldsymbol{C}_m$ of the matrix and $\boldsymbol{C}_i$ of the steel rebars is, because of the invariant property of isotropic tensors, not required.

As soon as the mechanical response of the matrix becomes inelastic, the related stiffness required for equation (24) is defined according to the actual damage state. In the post-cracking regime, the stiffness of the matrix phase degenerates to $\psi \boldsymbol{C}_m$ where ψ is the remaining integrity, while the elastic stiffness of the steel reinforcement remains unchanged in the post-yielding regime. It should be noted that $\boldsymbol{T}_i$ given by equation (24) also represents the localization tensor resulting from the dilute approach, where no interactions between inclusions are considered. The application of the dilute approach is limited to composites with very small volume fractions of the inclusions.

For the present three-phase composite the concentration tensors $\boldsymbol{A}_i^{MT}$ introduced in equation (22) related to each phase can be identified from combining equation (19) for the homogenized effective material tensor $\boldsymbol{C}^*$, equation (17) for the concentration tensors $\boldsymbol{A}_i^{MT}$ together with equation (14) for the homogenized macroscopic strains ε^* and with the microscopic strains ε_i given by equation (23) as

$$\boldsymbol{A}_1^{MT} = \left[c_1 \mathbb{1} + c_2 \boldsymbol{T}_2 : \boldsymbol{T}_1^{-1} + c_m \boldsymbol{T}_1^{-1} \right]^{-1} \tag{26}$$

$$\boldsymbol{A}_2^{MT} = \left[c_1 \boldsymbol{T}_1 : \boldsymbol{T}_2^{-1} + c_2 \mathbb{1} + c_m \boldsymbol{T}_2^{-1} \right]^{-1} \tag{27}$$

$$\boldsymbol{A}_m^{MT} = \left[c_1 \boldsymbol{T}_1 + c_2 \boldsymbol{T}_2 + c_m \mathbb{1} \right]^{-1}. \tag{28}$$

Since $\boldsymbol{A}_i^{MT}$ is a function of $\boldsymbol{T}_1$ and $\boldsymbol{T}_2$, it is obvious that the strains in each phase are affected by the other constituents, which allows for the consideration of micromechanical interactions within the MORI-TANAKA strategy.

3.5 Consideration of sets of reinforcement

The assumed RVE illustrated in Figure 2 considers two straight inhomogeneities (rebars) embedded in a continuous matrix. If the distance between neighboring rebars is relatively large, the assumed mechanical response of the RVE containing single rebars in each direction is a suitable approximation of the macroscopic effective shear stiffness. Since the stiffness of the composite material in longitudinal direction of the rebars is manifested by a parallel system (VOIGT-boundary conditions), the adopted micromechanical model reproduces the proper stiffness in longitudinal direction independent of the distance between adjacent rebars. If sets of reinforcement are considered, however, depending on the distance between neighboring reinforcement bars, the effective shear stiffness may differ from the effective shear stiffness of a RVE whose size is in accordance with the requirements described in section 3.2.

The stiffening effect provided by the rebar in shear mode (dowel action) is frequently accounted for by increasing the effective shear stiffness of the composite according to beam bending mechanisms (Pietruszczak and Winnicki 2003; Linero et al. 2006). In the present approach, however, a consistent modification of the ESHELBY tensor $\boldsymbol{S}$ is performed in order to capture the effective shear stiffness when reinforcement sets embedded in a matrix are considered. To this end, the aspect ratio parameter $s = a_2/a_1$ specifying the cross section of the rebar is modified in order to reproduce the correct effective shear stiffness. It should be emphasized, that this modification primarily affects the shear contribution while the macroscopic stiffness in longitudinal direction of the rebars remains unchanged. In Figure 4 the mapping from the standard RVE of a single set system to the modified RVE of sets of reinforcement bars is depicted schematically.

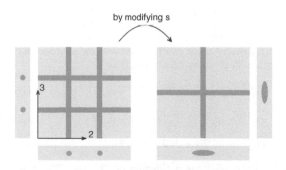

by modifying s

Figure 4.   Consideration of reinforcement sets.

3.6 Homogenized mechanical response

For the proposed three-phase composite material, the macroscopic (homogenized) free energy Ψ^* can be additively decomposed into the matrix part Ψ_m and the part Ψ_i associated with the rebars according to their volume fractions c_m and c_i, respectively

$$\Psi^* = c_m \Psi_m \mathbf{1} : \mathcal{A}_m^{MT-1} : \mathbf{1} + \sum_{i=1}^{2} c_i \Psi_i \mathbf{1} : \mathcal{A}_i^{MT-1} : \mathbf{1}. \quad (29)$$

Since the applied macroscopic strain tensor ε^* differs from the strains within each phase, the energetic consistency is ensured by taking the concentration tensor $\mathcal{A}_i^{MT}$ of the MORI-TANAKA scheme into account in equation (29). This approach can easily be confirmed by deriving the macroscopic stress tensor σ^* in equation (15) from the homogenized free energy Ψ^*

$$\sigma^* = \frac{\partial \Psi^*}{\partial \varepsilon^*} = c_m \frac{\partial \Psi_m}{\partial \varepsilon^*} : \mathcal{A}_m^{MT-1} + \sum_{i=1}^{2} c_i \frac{\partial \Psi_i}{\partial \varepsilon^*} : \mathcal{A}_i^{MT-1}$$

$$= c_m \frac{\partial \Psi_m}{\partial \varepsilon_m} : \frac{\partial \varepsilon_m}{\partial \varepsilon^*} : \mathcal{A}_m^{MT-1} + \sum_{i=1}^{2} c_i \frac{\partial \Psi_i}{\partial \varepsilon_i} : \frac{\partial \varepsilon_i}{\partial \varepsilon^*} : \mathcal{A}_i^{MT-1}$$

$$= c_m \sigma_m : \underbrace{\mathcal{A}_m^{MT} : \mathcal{A}_m^{MT-1}}_{\mathbb{II}} + \sum_{i=1}^{2} c_i \sigma_i : \underbrace{\mathcal{A}_i^{MT} : \mathcal{A}_i^{MT-1}}_{\mathbb{II}}$$

$$= c_m \sigma_m + \sum_{i=1}^{2} c_i \sigma_i \overset{!}{=} <\sigma>_V . \quad (30)$$

The homogenized non-linear tangent operator $\mathcal{C}^{*,tan}$ relating macro-strains to the macro-stresses is identified as

$$\mathcal{C}^{*,tan} = \frac{d\sigma^*}{d\varepsilon^*} = c_1 \frac{d\sigma_1}{d\varepsilon^*} + c_2 \frac{d\sigma_2}{d\varepsilon^*} + c_m \frac{d\sigma_m}{d\varepsilon^*}$$

$$= c_1 \frac{d\sigma_1}{d\varepsilon_1} : \frac{d\varepsilon_1}{d\varepsilon^*} + c_2 \frac{d\sigma_2}{d\varepsilon_2} : \frac{d\varepsilon_2}{d\varepsilon^*} + c_m \frac{d\sigma_m}{d\varepsilon_m} : \frac{d\varepsilon_m}{d\varepsilon^*}$$

$$= c_1 \mathcal{C}_1^{tan} : \mathcal{A}_1^{MT} + c_2 \mathcal{C}_2^{tan} : \mathcal{A}_2^{MT} + c_m \mathcal{C}_m^{tan} : \mathcal{A}_m^{MT}$$

$$= \mathcal{C}_m^{tan} + c_1 (\mathcal{C}_1^{tan} - \mathcal{C}_m^{tan}) : \mathcal{A}_1^{MT} +$$

$$c_2 (\mathcal{C}_2^{tan} - \mathcal{C}_m^{tan}) : \mathcal{A}_2^{MT} . \quad (31)$$

The tangent stiffness of each phase $\mathcal{C}_i^{tan}$ depends on the damage or yielding state and is calculated according to the adopted material model for each phase. Note, that in the non-linear regime, also the concentration tensor $\mathcal{A}_i^{MT}$ is affected by the damage or yielding state, which is manifested by $\mathcal{T}_i$ given in equation (24). As long as all constituents are within the elastic range however, equation (31) coincides with equation (19) and the classical micromechanical laws of elastic composites are valid.

3.7 Numerical study in the elastic regime

To illustrate the influence of the volume fraction and the orientation of the rebar on the effective stiffness $\mathcal{C}^*$

within the elastic regime of both constituents, a simple benchmark test is performed. To this end, one single rebar with Young's modulus $E_s = 76,000$ N/mm^2 and Poisson's ratio $\nu_s = 0.2$ is embedded within a matrix with $E_m = 30,000$ N/mm^2 and $\nu_m = 0.2$. The shape of the rebar is assumed to be cylindrical (aspect ratio $s = 1$). For different angles α of the rebars, the effective Young's modulus in 3-direction $E_3^* = 1/C_{33}^{*-1}$ is computed for three different volume fractions of the rebar ($c = 1/5/10\%$).

Two homogenization procedures – the *mixture theory* and the MORI-TANAKA approach – are used. Since in the *mixture theory* besides the volume fraction no micromechanical information is considered, the related equations are directly obtained by assuming $\mathcal{A} = \mathbf{11}$ for the concentration tensor leading to an isotropic homogenized stiffness tensor $\mathcal{C}^*$. The mechanical response obtained with the MORI-TANAKA technique, however, is manifested by an anisotropic stiffness even if all constituents are isotropic. Figure 5 illustrates the isotropic response based on the *mixture theory* (dashed lines) and the more realistic anisotropic, orientation-dependent response obtained from the MORI-TANAKA approach (solid lines). Since the *mixture theory* is based on a parallel system of the constituents, resulting in uniform (constant) strains within the RVE, the approximated macroscopic stiffness constitutes an upper bound (VOIGT approximation). In addition to the orientation also the volume fraction of the rebar c strongly affects the homogenized effective stiffness.

3.8 Homogenization within multi-physics

In this section, the extension of the adopted (mechanical) homogenization technique for reinforced concrete to a multi-phase model for concrete is briefly addressed. Since the free energy of the porous matrix Ψ_m given in equation (1) is governed by the moisture content m_l and by the absolute temperature T, the related state equations of the liquid pressure p_l and of the entropy S

$$p_l = \rho_l \, \partial \Psi^* / \partial m_l = \rho_l \, c_m \partial \Psi_m / \partial m_l \, \mathbf{1} : \mathcal{A}_m^{MT-1} : \mathbf{1} \quad (32)$$

$$S = -\partial \Psi^* / \partial T = -c_m \partial \Psi_m / \partial T \, \mathbf{1} : \mathcal{A}_m^{MT-1} : \mathbf{1} \quad (33)$$

can derived from the free energy of the composite material Ψ^* given in equation (29). It is implicitly assumed that moisture and heat transport are not affected by the embedded reinforcement. The expressions for $\partial \Psi_m / \partial m_l$ and $\partial \Psi_m / \partial T$ can be found in (Grasberger and Meschke 2004).

4 EXPERIMENTAL VERIFICATION

A re-analysis of a shear test of a bi-directionally reinforced panel tested experimentally (Collins et al. 1985) is performed in order to validate the proposed model for reinforced concrete. The considered panel PV27 with dimensions $890 \cdot 890 \cdot 70 \, mm^3$ is reinforced homogeneously with a volume fraction of 1.785% in each orthogonal direction. For this panel, the finite element discretization contains elements with identical material parameters for the composite. While the experiment has been performed load-controlled the finite element analysis is performed displacement-controlled. A relatively large amount of reinforcement and a low concrete strength have been chosen to provoke structural concrete shear failure (Collins et al. 1985). The material parameters for the concrete matrix material and for the reinforcing steel are collected in Table 1, with f_m^c and f_m^t representing the compression and tensile strength, respectively and G_f is the fracture energy. For the steel an ideal elasto-plastic behaviour is assumed. At the boundaries of the investigated panel, the concrete and the reinforcing rebars are perfectly bonded. The stiffening effect of the reinforcing steel sets is captured according to section 3.5 by modifying the aspect ratio of the cross section s.

The structural shear-stress versus the equivalent shear-strain behaviour is depicted in Figure 6. The

Table 1. Mechanical material parameters.

Concrete	Steel
$E_m = 20,000$ Mpa	$E_s = 20,000$ Mpa
$\nu_m = 0.2$	$\nu_s = 0.3$
$f_m^c = 20$ Mpa	$\sigma_y = 442$ MPa
$f_m^t = 1.0$ MPa	$K = 0$ MPa
$G_f = 0.1$ N/mm	$s = 100$

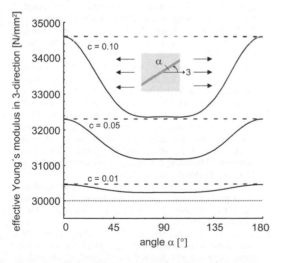

Figure 5. E_3^* for different volume fractions c (numerical).

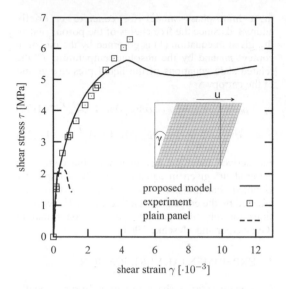

shear stress τ [MPa]

proposed model ——
experiment □
plain panel – – –

shear strain γ [$\cdot 10^{-3}$]

Figure 6. Analysis of an orthogonally reinforced shear panel (Collins et al. 1985).

structural shear-strains are computed from the prescribed displacements at the top of the panel, and the structural shear-stresses are obtained by averaging the shear-stresses calculated also at the top at the panel. The comparison between the experimental and numerical result shows a satisfactory agreement. The onset of cracking within the matrix is well predicted by the proposed model. The maximum shear-stress, however, is slightly underestimated in the numerical analysis by $\sim 10.8\%$. As reported in (Collins et al. 1985) and also confirmed by the numerical analysis, structural failure of the panel originates from concrete crushing. No yielding of reinforcement is observed. In order to illustrate the contribution of the rebars to the global shear stiffness, the same concrete panel without consideration of reinforcement is also analyzed numerically. The respective shear stress-shear strain curve is included in Figure 6. Since the stiffening effect provided by the reinforcement is missing, the maximum load capacity is controlled by matrix cracking. The maximum capacity of the plain concrete panel is approximately one third of the reinforced panel (Figure 6).

5 CONCLUDING REMARKS

In this paper, a constitutive model for reinforced concrete based on continuum micromechanics is presented. Reinforced concrete is represented as a three-phase composite material, characterized by a continuous concrete matrix and two different sets of steel reinforcement rebars. The MORI-TANAKA homogenization scheme is considered as a suitable approach to derive the homogenized constitutive relations of

the composite material and to obtain related local (micromechanical) field information. Dowel action between the reinforcement bars and the concrete is implicitly captured by the chosen approach. The proposed model is formulated within a poromechanics framework developed for durability-oriented numerical analysis of reinforced concrete structures. However, in the present paper the main focus has been laid on the purely mechanical response of reinforced concrete. Since the work is still in progress, debonding mechanisms between concrete and the embedded reinforcement (bond slip) as well as corrosion have not yet been taken into account.

ACKNOWLEDGMENT

Financial support was provided by the German National Science Foundation (DFG) in the framework of the project A9 of the collaborative research center SFB 398. This support is gratefully acknowledged.

REFERENCES

Bažant, Z. P., A. B. Hauggaard, S. Baweja, and F.-J. Ulm (1997). Microprestress-solidification theory for concrete creep. I: Aging and drying effects. *Journal of Engineering Mechanics (ASCE)* 123(11), 1188–1194.

Collins, M. P., F. J. Vecchio, and G. Mehlhorn (1985). An international competition to predict the response of reinforced concrete panels. *Canadian Journal of Civil Engineering* 12, 624–644.

Coussy, O. (2004). *Poromechanics*. Chichester, England: Wiley.

Eshelby, J. D. (1957). The determination of the elastic field of an ellipsoidal inclusion, and related problems. *Proc. Roy. Soc. London, Series A, 241*, 376–396.

Grasberger, S. and G. Meschke (2003). Drying shrinkage, creep and cracking of concrete: From coupled material modelling to multifield structural analyses. In R. De-Borst, H. Mang, N. Bićanić, and G. Meschke (Eds.), *Computational Modelling of Concrete Structures*, pp. 433–442. Balkema.

Grasberger, S. and G. Meschke (2004). Thermo-hygromechanical degradation of concrete: From coupled 3D material modelling to durability-oriented multifield structural analyses. *Materials and Structures 37*, 244–256.

Linero, D. L., J. Oliver, A. E. Huespe, and M. D. G. Pulido (2006). Cracking modeling in reinforced concrete via the strong discontinuity appoach. In G. Meschke, R. de Borst, H. A. Mang, and N. Bićanić (Eds.), *Computational Modelling of Concrete Structures*, London, pp. 173–182. Taylor & Francis Group.

Meschke, G. and S. Grasberger (2003). Numerical modeling of coupled hygromechanical degradation of cementitious materials. *Journal of Engineering Mechanics (ASCE) 129*(4), 383–392.

Meschke, G., R. Lackner, and H. A. Mang (1998). An anisotropic elastoplastic-damage model for plain concrete. *International Journal for Numerical Methods in Engineering 42*, 703–727.

Mori, T. and K. Tanaka (1973). Average stress in the matrix and average elastic energy of materials with misfitting inclusions. *Acta Metall. 21*(5), 571–574.

Pietruszczak, S. and A. Winnicki (2003). Constitutive Model for Concrete with Embedded Sets of Reinforcement. *Journal of Engineering Mechanics 129*(7), 725–738.

Richter, M. (2005). *Entwicklung mechanischer Modelle zur analytischen Beschreibung der Materialeigenschaften von textilbewährtem Feinbeton*. Ph. D. thesis, TU Dresden, Germany.

Simo, J. C. and T. J. R. Hughes (1998). *Computational inelasticity*. Berlin: Springer.

Zaoui, A. (2002). Continuum Micromechanics: Survey. *Journal of Engineering Mechanics 128*(8), 808–816.

Zohdi, T. I. and P. Wriggers (2005). *Introduction to Computational Micromechanics*. Springer.

Fracture Mechanics of Concrete and Concrete Structures – High-Performance Concrete,
Brick-Masonry and Environmental Aspects – Carpinteri, et al. (eds)
© 2007 Taylor & Francis Group, London, ISBN 978-0-415-44617-4

Damage mechanics based modelling of the relation between capillary pores and compressive strength of concrete

S. Akyuz, Y. Akkaya, H.O. Yazan & M.A. Tasdemir
Istanbul Technical University, Civil Engineering Faculty, Istanbul, Turkey

ABSTRACT: The amount and geometry of pores in hardened cement paste phase, that affect the mechanical behaviour of concrete, have been considered in the form of two different parameters: porosity and capillary sorptivity coefficient (or water/cement ratio) where the latter represents the geometry of pores in hardened concrete. A mathematical model based on the damage mechanics approach has been developed to relate the compressive strength of concrete to these two parameters. Equations developed in this model have been applied to the values obtained in an available experimental work, in which porosities, sorptivity coefficients and strengths were measured for 40 different concrete mixtures. For indirect determination of the water-cement ratio in both hardening and hardened concretes, a petrographic examination method using optical fluorescence is presented. Additionally, five concrete mixtures with different water-cement ratios were used to determine their fluorescence intensities, sorptivity coefficients and compressive strengths. It is shown that the flourescence intensity is directly affected by the water-cement ratio of concrete. Good agreement has been found between experimental and calculated results.

1 INTRODUCTION

Concrete is an extremely complex system of solid phases, pores, and water, with a high degree of heterogeneity. It has been shown that the pore size distribution in the cement paste in concrete and in mortar is different from that of plain cement paste, which does not contain aggregate (Winslow & Cohen 1994). Additional large pores occur at the interfacial zones surrounding each aggregate. The interface between cement paste and coarse aggregate particles is the weakest zone, and the use of particles, such as silica fume, is important for densification and for the improvement of the stability of fresh concrete, thus enhancing overall durability and strength. For material modeling purpose, concrete can be considered as a three-phase composite material consisting of hardened cement paste, aggregate, and the interfacial zone between aggregate and cement paste (Tasdemir et al. 1998). To gain benefits from the inclusions of fine particles, good dispersion within the concrete system is necessary and is provided by means of high range water reducing admixtures (Sawicz & Heng 1996, Detwiller & Mehta 1989, Nehdi et al. 1996). Recent studies have shown that measurements of permeability of concrete can be used as an indication of durability. Dinku & Reinhardt (1997) have shown that the gas permeability is sensitive to changes in curing duration, water/cement ratio, age of testing and moisture history

of concrete; according to their research, it is possible to predict the gas permeability from the capillary sorptivity measurements. According to Martys & Ferraris (1997), the sorptivity coefficient is essential to predict the service life of concrete as a structural material and to improve its performance. Since the pore structure has a dominant factor on the performance of concrete, the main objective of this work is to develop a combined approach for quality control of concrete using optical fluorescence microscopy, sorptivity measurements, and damage mechanics based modeling. It is thought that this approach will help to fill the need in bringing the theoretical concepts to implementation for the better understanding of the material behaviour.

2 THEORETICAL BACKGROUND

The close neighbourhood of a crack can be accepted as not being able to carry the load, i.e. in this region the stress is zero. v_0 is the volume which is able to carry load and v is the total volume of the solid body. $v - v_0$ and $v - v = H$ show the dead volume of the solid body under unloaded and loaded conditions, respectively. v is the total volume of the solid body under load. There is a well known relation of geometry between the area of crack surface A and the total dead volume H given as

$$A^{3/2} = kH = k(v - v) \qquad (1)$$

where k is a constant (Akyuz et al. 2000). When $v=v_0$, A is A_0, hence A/A_0 can be expressed as

$$\frac{A}{A_0} = \left(\frac{\mathcal{V}-v}{\mathcal{V}-v_0}\right)^{2/3} \tag{2}$$

Here A_0 is the area of crack surface in the solid when $U = V = 0$ (U is the work done by the external loads, V is the elastic component of the stored energy, Akyuz et al. 2000). Let us consider the stress-strain curve of concrete under uniaxial compressive loading and if the total volume of the concrete specimen is $\mathcal{v}$, the work done by the external forces can be written as

$$U = \left\{\int_0^\varepsilon \sigma(\varepsilon)\,d\varepsilon\right\}\mathcal{V} \tag{3}$$

where σ and ε are uniaxial stress and strain, respectively. The reversible (elastic) work is

$$V = \frac{1}{2}\cdot\frac{\sigma^2}{E_{ef}}\cdot\mathcal{V} \tag{4}$$

where E_{ef} is the effective modulus of elasticity which is slightly less than the dynamic modulus of elasticity, E_d. For the simplicity in calculations, it is assumed that $E_{ef} \approx E_d$. From Equations 3 and 4, the following equation can be written (Akyuz et al. 2000)

$$\int_0^\varepsilon \sigma(\varepsilon)\,d\varepsilon - \frac{1}{2}\cdot\frac{\sigma^2}{E_d} = \gamma_F\cdot\frac{A_0}{\mathcal{V}}\cdot\left(\frac{A}{A_0}-1\right) \tag{5}$$

where γ_F is the amount of energy required to create a unit area of fracture surface. If we take the differential of both sides, then we have

$$\sigma - \alpha\sigma\dot{\sigma} = \beta\frac{d}{d\varepsilon}\left(\frac{a}{a_0}-1\right) \tag{6}$$

where $a = A/\mathcal{v}$, $a_0 = A_0/\mathcal{v}$, $\beta = a_0\gamma_F$ and $\alpha = 1/E$. The damage function can be defined as

$$\psi = \frac{v}{v_0} = \begin{cases} 1, & \varepsilon = 0 \\ \dfrac{v}{v_0}, & 0 < \varepsilon < \varepsilon_m \\ 0, & \varepsilon = \varepsilon_m \end{cases} \tag{7}$$

ε_m is defined as the strain at compressive strength. As shown from Equation 7, ψ is a decreasing function and as result $\dot{\psi} < 0$. From Equation 2 a/a_0 can be written as

$$\frac{a}{a_0} = \left(\frac{\dfrac{\mathcal{V}}{v_0}-\dfrac{v}{v_0}}{\dfrac{\mathcal{V}}{v_0}-1}\right)^{2/3} = \left(\frac{q-\dfrac{v}{v_0}}{q-1}\right)^{2/3} \tag{8}$$

where $q = \mathcal{v}/v_0 > 1$ which is a constant. In this case, the following equation can be expressed as

$$\frac{d}{d\varepsilon}\left(\frac{a}{a_0}-1\right) = -\frac{2}{3}\cdot\frac{1}{(q-1)^{2/3}(q-\psi)^{1/3}}\cdot\frac{d\psi}{d\varepsilon} \tag{9}$$

Substitution of Equation 9 into Equation 6, the kinetic equation of damage can be obtained as

$$\dot{\psi} = \frac{d\psi}{d\varepsilon} = -K(q-\psi)^{1/3}(\sigma-\alpha\sigma\dot{\sigma}) \tag{10}$$

where $K = (3/2)(q-1)^{2/3}/\beta$. In a quasi-static case, the damage does not increase under a constant stress σ_0 for $\sigma_0 < f_c$, where f_c is the compressive strength of concrete. It is assumed that, the failure occurs near the peak point under a constant stress $\sigma_0 \approx f_c$. In this case, the differential equation becomes

$$\frac{d\psi}{(q-\psi)^{1/3}} = -Kf_c\cdot d\varepsilon \tag{11}$$

Taking the integration of Equation 11 with $\psi = 1$ at $\varepsilon = 0$ and $\psi = 0$ at $\varepsilon = \varepsilon_m$, then f_c can be written as

$$f_c = \gamma_F\frac{a_0}{\varepsilon_m}\left[\frac{1}{\left(1-\dfrac{1}{q}\right)^{2/3}}-1\right] \tag{12}$$

On the other hand, if we take a_m as the maximum area of crack surface per unit volume, then $a_0/a_m = (1-1/q)^{2/3}$. It is known that when ε_m increases, the area of crack surface also increases. From this point of view, we assume that $\varepsilon_m = a_m\eta$, thus we have

$$f_c = \frac{\gamma_F}{\eta}\left[1-\left(1-\frac{1}{q}\right)^{2/3}\right] \tag{13}$$

We assume that G is the volume of concrete excluding the crack. Hence, we can write the inequality of $v_0/\mathcal{v} \le G/\mathcal{v} \le 1$. On the other hand, as shown in Figure 1, when $G/\mathcal{v} = 0$, then $v_0/\mathcal{v} = 0$ and when $G/\mathcal{v} = 1$, then $v_0/\mathcal{v} = 1$. Thus, $v_0/\mathcal{v}$ can be written as

$$\frac{v_0}{\mathcal{v}} = \left(\frac{G}{\mathcal{v}}\right)^m, m \ge 1 \tag{14}$$

Substituting of $v_0/\mathcal{v}$ given in Equation 14 into Equation 13, the following equation can be obtained:

$$f_c = M\left\{1-\left[1-(1-p)^m\right]^{2/3}\right\} \tag{15}$$

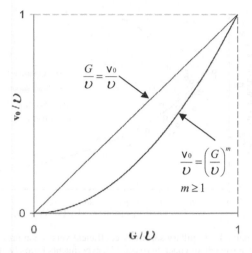

$$\frac{G}{\upsilon}=\frac{v_0}{\upsilon}$$

$$\frac{v_0}{\upsilon}=\left(\frac{G}{\upsilon}\right)^m$$

$$m \geq 1$$

Figure 1. v_0/υ versus G/υ.

where $M = \gamma_F/\eta$ and $m = 1 + \theta C$, in which C is the capillary coefficient or $m = 1 + \theta(w/c-0.23)$, in which w/c is the water/cement ratio. In Equation 15, M is the highest compressive strength of the material which corresponds to the zero porosity. As shown in Figure 1, m is a parameter which represents pore geometry. In case of spherical pore, there is no region without stress around the pore, thus $k = G/\upsilon = v_0/\upsilon$ and $m = 1$. As the sphere type of pore turns into an elliptical shape, m takes the values greater than one and the stress concentration increases, as a result compressive strength of concrete decreases. Thus, pore geometry can be determined by means of capillary pores as given above. According to the proposed model depending on the equation chosen from m, the compressive strength of concrete (f_c) can be written as

$$f_c = M_1\left\{1-\left[1-(1-p)^{1+\theta_1 C}\right]^{2/3}\right\} \qquad (16)$$

or

$$f_c = M_2\left\{1-\left[1-(1-p)^{1+\theta_2(\frac{w}{c}-0.23)}\right]^{2/3}\right\} \qquad (17)$$

where M_1 and M_2 are constants which represent the theoretical compressive strength of concrete without porosity (maximum strength), p is the porosity, C is the capillary coefficient as in Equation 16 and w/c is the water/cement ratio as in Equation 17 and θ_1 and θ_2 are the constants of these equations.

3 EXPERIMENTS

3.1 Materials

The test results used in this work were obtained at the Istanbul Technical University (ITU). The

combined study of experiments, the damage mechanics based approach explained above, prediction of the water-cement ratio using the optical fluorescence microscopy, capillary sorptivity and compressive strength of concrete are elaborated.

Locally available sea sand of the Istanbul area, a Portland cement (CEM I-42,5), crushed limestone and its fines were used in the mixtures. The grading and maximum aggregate size of concrete were kept constant. Mix proportions of sand, limestone fines and crushed limestone were 35%, 15%, and 50% respectively. Five concrete batches were made with the same Portland cement, sand and limestone, and its fines.

At least three specimens of each concrete mixture were tested at 126 days.

3.2 Sorptivity tests

Three test specimens for sorptivity test, cut from the cylinders of 100 mm diameter, were prepared for each mixture. Measurement of capillary sorption were carried out using specimens pre-conditioned in the oven at about 50°C until constant mass.

Test specimens were exposed to the water on the surface of 7854 mm^2 by placing in a pan. The water level in the pan was maintained at about 5 mm above the base of the specimens during this experiment. The lower areas on the sides of the specimens were coated with epoxy to achieve unidirectional flow. At certain times, the masses of the specimens were measured using a balance, then the amount of water absorbed was calculated and normalized with respect to the cross-section area of the specimens exposed to the water at various times such as 1, 2, 3, 5, 10, 20, 30, 60, 120, 180, 240, 300 and 360 minutes (Figure 2).

The sorptivity coefficient (k), was obtained by using the following expression:

$$\left(\frac{Q}{A}\right)^2 = k \cdot t \qquad (18)$$

where Q = the amount of water absorbed [cm^3]; A = the cross-sectional area of specimen that was in contact with water [cm^2]; k = the sorptivity coefficient of the specimen [cm^2/min]; t = time [min].

To determine the sorptivity coefficient, $(Q/A)^2$ was plotted against the time (t), then, k was calculated from the slope of the linear relation between $(Q/A)^2$ and t.

As seen in Figure 3, 4 and 5 the sorptivity coefficient of concrete decreases with increasing compressive strength of concrete. Similar trend was obtained by Tasdemir (2003), as test results indicate that the sorptivity coefficient of concrete decreases as the compressive strength of concrete increases. It was also shown that the sorptivity coefficient of concrete is very sensitive to the curing condition and

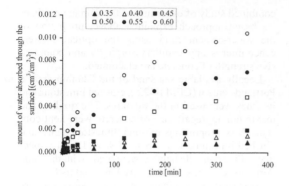

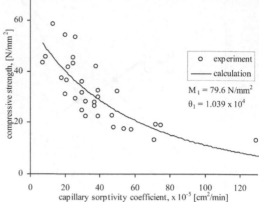

Figure 2. Time versus amount of water absorbed through the surface.

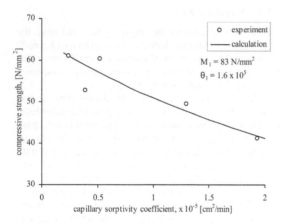

Figure 3. Sorptivity coefficient versus compressive strengths of concrete (p = 0.10).

Figure 4. Capillary sorptivity coefficient versus compressive strength of concrete (p = 0.13), experiments from Uyan (1974).

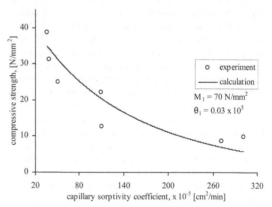

Figure 5. Capillary sorptivity coefficient versus compressive strength of concrete (p = 0.19), experiments from Uyan (1974).

the sorptivity coefficient of concrete is higher in low strength concrete.

Thus, concrete has larger capillary pores and lower compressive strength when a higher capillary sorption in concrete is obtained. In Figures 3, 4 and 5, the parameters M_i and θ_i can be calculated using the data obtained. Figures 6 and 7 show the relation between w/c ratio and compressive strength of concrete.

Equations developed by Akyuz et al. (2000) above were tested using the experimental data obtained from 40 different concrete mixtures. In each mixture the total porosity, the capillary coefficient and the compressive strength of concrete were measured (Uyan 1974).

In the first 31 different mixtures, porosity was in between 0.11 and 0.15, with an average value of 0.13. For the rest of the mixtures, the porosity values was between 0.17 and 0.21 and the average was 0.19.

The parameters M_i and θ_i were calculated using the data given by Uyan (1974). The curves obtained by substituting these parameters in Equation 16 and

17 were given in Figures 4, 5 and 7 together with the parameters. Good agreement obtained between the experimental and theoretical results shows that the amount of porosity as well as the geometry of pores can be related to the compressive strength of concrete.

3.3 Determination of w/c ratio using optical fluorescence microscopy

In hardened concrete, it is not possible to determine the water-cement ratio directly. However, there is an indirect method for determining the water-cement ratio in hardening concrete for both quality control at the early ages and the forensic investigation of deteriorated concrete (Jacobsen et al. 2000).

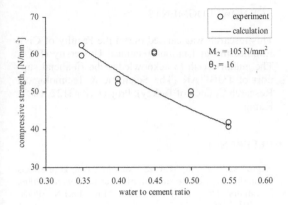

Figure 6. Variation of compressive strength with w/c ratio ($p = 0.10$).

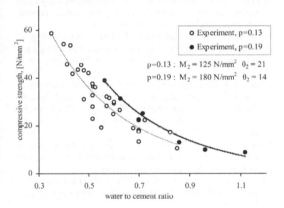

Figure 7. Variation of compressive strength with w/c ratio, experiments from Uyan (1974).

Modern concrete contains a variety of organic admixtures, supplementary cementitious materials, manufactured aggregates, special cements, and various types of fibers. These materials are added to the mixture to improve the properties of the concrete, enhance its performance, or control specific properties such as setting time, heat of hydration, shrinkage and workability (Powers 2006). In recent years, optical fluorescence microscopy of concrete has been used as a tool to investigate the internal structure of hardened concrete. Several attempts have been made to determine the water-cement ratio in concrete using optical fluorescent microscopy (Jacobsen et al. 2000; Henrichsen & Laugesen 1995, Laugesen 1993, Mayfield 1990).

The method used in concrete petrography has been borrowed from geology and it is based on vacuum impregnation of concrete using a yellow fluorescent epoxy. During impregnation the capillary porosity, cracks, voids, and defects in the concrete are filled

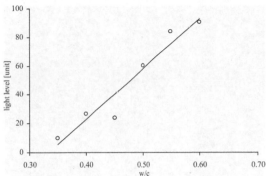

Figure 8. Relation between light level in the thin section and w/c ratio of concrete.

with epoxy. After impregnation, thin sections are prepared from slices of concrete that are attached to a glass slide, and then ground to a thickness of about 20 μm to 30 μm. These thin sections allow the concrete petrographer to identify the material constituents and predict their proportions, air content, water-cement ratio, paste homogeneity, paste volume, aggregate volume, effectiveness of curing, and examine the relationships between the various constituents.

The impregnation depth of the fluorescent epoxy into the sample is dependent on the capillary porosity of the cement paste (Laugesen 1993). Large cracks, cement paste-aggregate debonding, interconnected porosity, and air voids can be easily impregnated.

In Figure 8, as the water-cement ratio of concrete increases, the fluorescence intensity increases significantly. After obtaining such a calibration curve in laboratory conditions, water-cement ratio of concrete in-situ can be indirectly predicted providing sufficient measurement.

In Figure 9, as the water-cement ratio of concrete increases from 0.40 to 0.60, the lightness increases substantially. Hence, the increase in the capillary porosity at the higher water-cement ratios results in high flourescence intensity. However, at the lower water-cement ratios, the thin section becomes darker.

4 CONCLUSIONS

The dependence of compressive strength on the capillary coefficient or water-cement ratio is succesfully represented by the proposed model, based on damage mechanics. The model described here not only depends on the effect of pore volume, but also on the geometry of pores on the compressive strength of concrete.

In petrographic examination of concrete air-void system parameters, paste content and crack widths can be measured. The water-cement or water-binder ratio, air content, extent of cement hydration, paste and aggregate volumes can be predicted. Especially,

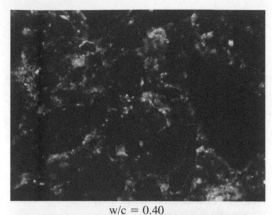

w/c = 0.40

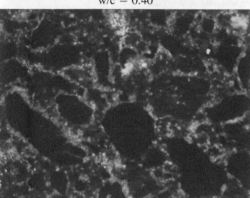

w/c = 0.50

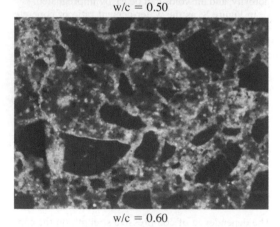

w/c = 0.60

Figure 9. Photographs of samples taken under the fluorescent light with water-cement ratios of 0.40, 0.50, and 0.60.

after obtaining the relationship between the light level in the thin section and the water-cement ratio in the laboratory condition, the quality control of concrete in-situ can be realized using an optical fluorescence microscopy.

ACKNOWLEDGMENTS

This research was carried out in the Faculty of Civil Engineering at Istanbul Technical University (ITU). The authors wish to acknowledge the financial support of TUBITAK (The Scientific & Technological Research Council of Turkey): Project:106G122, 1007-Kamu.

REFERENCES

Akyuz, S., Tasdemir, M.A. & Uyan, M. 2000. Pore effects on the compressive strength of concrete, *Concrete 2000*, Editors: Dhir, R.K. and Jones M.R., E&FN SPON: 1407–1416.

Detwiller, R.J. & Mehta,P.K. 1989. Chemical and physical effects of silica fume on mechanical behavior of concrete, *ACI Materials Journal*, 86: 609–614.

Dinku, A. & Reinhardt, H.W. 1997. Gas permeability coefficient of cover concrete as a performance control, *Materials and Structures*, 30: 387–393.

Henrichsen, A. & Laugesen, P. 1995. "Monitoring of Concrete Quality in High Performance Civil Engineering Constructions," *MRS Symposium Proceedings* 370: 49–56.

Jakobsen, U.H., Laugesen P. & Thaulow N. 2000. Determination of water to cement ratio in hardened concrete by optical fluorescence microscopy. *ACI SP-191, Water–cement ratio and other durability parameters—-techniques for determination*: 27–42.

Laugesen, P. 1993. Effective w/c ratio of cement paste in concrete. *Proc. 4th Euroseminar on Microscopy Applied to Building Materials, Wisby, Sweden.*

Martys, N.S. & Ferraris, C.F. 1997. Capillary transport in mortars and concrete, *Cement and Concrete Research*, 27: 747–760.

Mayfield, B. 1990. The quantitative evaluation of the water/cement ratio using flourescence microscopy. *Magazine of Concrete Research*, 150: 45–49.

Nehdi, M., Mindess,S. & Aïtcin, P.C. 1996. Optimization of high strength limestone filler cement mortars, *Cement and Concrete Research*, 26: 883–893.

Powers, L.J. 2006. The power of petrography, *Structure Magazine*, January: 25–28.

Sawicz, Z. & Heng, S.S. 1996. Durability of concrete with addition of limestone powder, *Magazine of Concrete Research*, 48: 131–137.

Tasdemir, M.A., Tasdemir, C., Akyuz, S., Jefferson, A.D., Lydon, F.D. & Barr, B.I.G. 1998. "Evaluation of Strains at Peak Stresses in Concrete: A Three Phase Composite Model Approach", *Cement and Concrete Composites*, 20: 301–318.

Tasdemir, C. 2003. Combined effects of mineral admixtures and curing conditions on the sorptivity coefficient of concrete, *Cement and Concrete Research*, 33: 1637–1642.

Uyan, M. 1974. Capillary in concrete. PhD thesis, Faculty of Civil Engineering, Istanbul Technical University: 180p (in Turkish with English summary).

Winslow, D.N. & Cohen, M.D. 1994. Percolation and pore structure in mortars and concrete, *Cement and Concrete Research*, 24: 25–37.

Fracture Mechanics of Concrete and Concrete Structures – High-Performance Concrete,
Brick-Masonry and Environmental Aspects – Carpinteri, et al. (eds)
© 2007 Taylor & Francis Group, London, ISBN 978-0-415-44617-4

Experimental study on the combined effect of temperature and mix parameters on the premature cracking of cement-based materials

A. Pertué, P. Mounanga, A. Khelidj & D. Fournol
GeM, UMR CNRS 6183, Research Institute in Civil Engineering and Mechanics, IUT St Nazaire, France

ABSTRACT: Cracking can occur in early-age concrete when thermal, hydrous or physico-chemical deformations are restrained. In addition to a deterioration of the aesthetic aspect, this premature cracking generates a loss of durability of the structure. In this article, the autogenous cracking of cement-based materials is studied using the ring test method. A specific experimental system based on this method was developed. It enables to keep the specimens in autogenous and temperature-controlled conditions during the whole test duration. The influence of both temperatures between 20°C and 40°C and the degree of restraint on the first autogenous cracking age of CEM I cement pastes with water-to-cement ratio W/C = 0.3 and 0.4 has been examined. Increasing the curing temperature and the degree of restraint and decreasing the W/C ratio leads to a shortening of the cracking age. Moreover, the level of maturity (hydration degree) corresponding to the age of the first autogenous cracking decreases when the curing temperature increases.

1 INTRODUCTION

At an early age, the volume changes of cement-based materials are the combined result of the physico-chemical evolution, the hydration heat release and the drying of the material. An inaccurate estimation of the amplitude and the rate of these volume variations can constitute an important cause of concrete durability problems.

In low water-to-cement ratio concretes (W/C < 0.5), the high cement proportioning results in an increase in the mechanical strength but also induces an augmentation of both autogenous and thermal deformations, in particular at a very early age. These deformations generate, when restrained, microscopic cracking (around the aggregates) or crossing cracks.

The knowledge of the deformation evolution in free conditions is not sufficient to precisely predict the risk of early-age cracking. Indeed, in restrained conditions, the relaxation phenomenon of the hydrating matrix attenuates the amplitude of autogenous shrinkage: the strain evolution is then different from that recorded in free conditions. It is therefore necessary to develop specific devices enabling the quantification of deformation in restrained conditions, taking into account the thermal history of the material. The parameters conditioning the development of early-age deformations are numerous: the cement type, the concrete's composition, the geometry and dimensions of the specimen, the curing temperature, the drying conditions, etc.

The effects of some of these parameters have been investigated in previous studies (Grzybowski et al. 1990, Weiss et al. 1999, Weiss & Ferguson 2001, Hossain & Weiss 2006), but early-age autogenous cracking and the temperature effect on this type of cracking did not receive much attention.

Several testing techniques exist for the determination of restrained shrinkage and cracking risk of cement-based materials (Bentur & Kovler 2003). Among them, an easy and simple method is the ring test. It consists in casting a specimen of cement-based material around a metal ring and in measuring the dimensional variations of this ring. The restrained shrinkage causes the development of tensile stresses in the material. As soon as they become higher than the material tensile strength, a transversal crack appears in the specimen.

The ring test method has been used in many research works (Table 1) for the quantification of the cracking risk of cement-based materials. A bibliography and an analysis of these studies have been recently presented by Radlinska et al. (2006).

The present work focuses on the early-age autogenous cracking in cement paste. The effects of three parameters are measured: the isothermal curing temperature, the degree of restraint and the initial amount of water (W/C ratio). The ring test device used in this study is equipped with a peripheral hermetic thermal regulation system, which enables to ensure quasi-isothermal and autogenous conditions during the test.

Table 1. Recent studies using the ring test method.

References	Mixture parameters			Test conditions	
	C	W/C	A/C	T (°C)	RH (%)
Grzybowski & quad Shah 1990	1	0.5	2	20	40
Weiss & Ferguson 2001	1	0.5	1.5	21	50
Weiss & Shah 2002	NSC	0.5	2.8	30	40
Hossain & Weiss 2006	1	0.3–0.4	1.5–1.4	23	50
Shah & Weiss 2006	1	0.5	1.8	23	50

C: cement type; NSC: normal strength concrete; W/C: water-to-cement ratio; A/C: aggregate-to-cement ratio; T: curing temperature; RH: relative humidity.

2 EXPERIMENTAL PROGRAM

2.1 Materials

French CEM I cement is used for all the experiments. According to Bogue's formula, it is composed of 60.4% of C_3S, 12.2% of C_2S, 8.8% of C_3A, 8.0% of C_4AF and 7.4% of gypsum; its specific surface is 3390 cm^2/g.

Two water-to-cement ratios (W/C = 0.3 and 0.4) and three curing temperatures (T = 20°C, 30°C and 40°C) are studied.

2.2 Quasi-isothermal ring-test device

The ring test device is schematized in Figures 1a and 1b. Figure 2 presents an overall photograph of the experimental system.

Each device is composed of two concentric rings: a PVC external ring and a metallic central ring. Three different metals (brass, steel and stainless steel) are used. The characteristics of these central rings are given in Table 2. Each cement paste specimen cast between the two rings has a cross section of 40 × 40 mm.

The strain of the central annulus due to the cement paste shrinkage is measured by four strain gauges (Fig. 1b). Four additional gauges are placed inside the central ring for thermal compensation.

In order to carry out the tests at a controlled temperature, each device is equipped with a peripheral thermal regulation system made of a copper tube network joined on two aluminium plates and fed by a thermostated water bath (±0.1°C) via a pump (Fig. 2). To avoid thermal losses, insulating foam is placed on the copper tubes and each device is set in a wooden box (Fig. 1a).

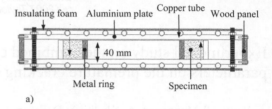

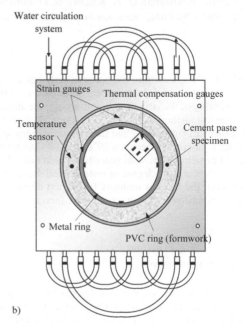

Figure 1. Diagrams of the quasi-isothermal ring-test device: a- side view and b- top view.

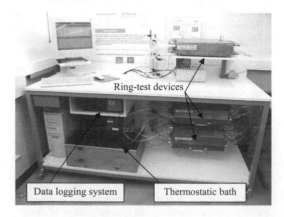

Figure 2. Photography of the experimental system.

2.3 Testing protocol

The water circulation in the thermal regulation system is activated two hours before the casting of cement

Table 2. Characteristics of the metal rings.

Designation	R_{IS} mm	R_{OS} mm	E_m MPa	υ -
Brass	95	105	100 000	0.33
Steel	95	105	211 000	0.30
Stainless Steel	95	105	200 000	0.30

R_{IS}: inner metal radius; R_{OS}: outer metal radius; E_m: elastic modulus; υ: Poisson coefficient.

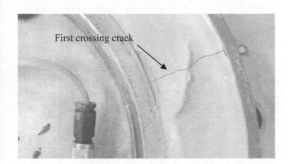

Figure 3. Transversal cracking in the cement paste annulus.

paste in order to stabilize the temperature of the rings. Form oil is applied on the rings and on the bottom of the device to limit friction between the specimen and the ring walls.

Each cement paste is prepared by mixing cement with water for 3 minutes. The specimen is then immediately cast around the central metal ring. A thermocouple, embedded in the sample, makes it possible to follow the temperature of the material during the test. The wooden box is then sealed and the automatic measurement acquisition begins. The time lag between the end of mixing and the first strain measurement does not exceed 15 minutes. The deformation of the central ring and the temperature of the specimen are recorded by a data logging system every 20 minutes. The test is stopped at the appearance of the first crossing crack (Fig. 3) detected by a brutal jump on the strain curve of the metal ring.

3 TEST RESULTS

3.1 Validation tests

3.1.1 Temperature control

The hydration heat can cause a significant elevation of the cement paste temperature during the first hours after mixing. It is essential to control these temperature variations in order to eliminate potential problems of thermal deformations.

Figure 4 shows the typical temperature curve of specimens during the tests. The beginning of each

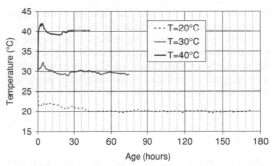

Figure 4. Temperature control of the cement paste specimens.

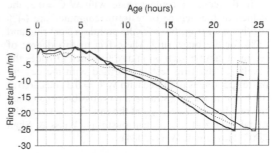

Figure 5. Repeatability of the ring tests (Brass ring, W/C = 0.3, T = 40°C).

curve is marked by a temperature variation of about ±2°C explained by the strong hydration heat release and the thermal inertia of the regulation system. This thermal instability lasts approximately 6 hours. Beyond this period, a maximal thermal deviation of 0.5°C is observed. The efficiency of the thermal regulation system can thus be considered as correct.

3.1.2 Repeatability tests

Figure 5 provides the result of three repeatability tests carried out with the brass ring and the W/C = 0.3-cement paste. The curing temperature is 40°C.

The repeatability level of the test method is acceptable. The strain curves follow the same evolution. The absolute difference in maximal strain amplitude and in cracking age is about 2 μm/m and 2.3 hours, respectively.

3.2 Parametric study

In the three next paragraphs, the effects of the ring nature, the water-to-cement ratio and the temperature on the appearance of the first crack will be analysed.

3.2.1 Effect of the degree of restraint

Recently, Hossain & Weiss (2006) studied the effect of the degree of restraint by measuring the influence of the ring thickness on the age of the first crack in drying conditions. They noted that the increase of the degree of restraint causes a shortening of the cracking age.

In the present work, the effect of the degree of restraint is investigated by using metal rings with different elastic moduli (brass, stainless steel and steel). The degree of restraint R is computed using the expression proposed by See et al. (2003):

$$R = \frac{A_m \cdot E_m}{A_m \cdot E_m + A_c \cdot E_c} \qquad (1)$$

where A_c and A_m are the section areas of the cement paste specimen ($A_c = 16\,cm^2$) and the metal ring ($A_m = 4\,cm^2$), respectively. E_c and E_m are the elastic moduli of the cement paste and the metal, respectively.

In the case of cement paste with W/C = 0.3, the values of the degree of restraint computed are 64% for the steel ring, 63% for the stainless steel ring and 46% for the brass ring. The effect of this degree of restraint can be seen on Figure 6, which provides the results of the ring tests performed with the cement paste specimens prepared with W/C = 0.3 and cured at 30°C. From the ring strain ε_m, it is possible to calculate the residual stress $\sigma_{residual}$ at the interface between the metal ring and the cement paste as (Weiss & Shah 2002, Hossain & Weiss 2004):

$$\sigma_{residual}(t) = -\varepsilon_m(t) \cdot E_m \cdot C_R \qquad (2)$$

where C_R is the ring geometry coefficient:

$$C_R = \frac{R_{OS}^2 + R_{OC}^2}{R_{OC}^2 - R_{OS}^2} \times \frac{R_{OS}^2 - R_{IS}^2}{2R_{OS}^2} \qquad (3)$$

with R_{OS} = outer metal radius; R_{OC} = outer cement paste ring radius; and R_{IS} = inner metal radius.

Figure 7 shows the evolution of the residual stress as a function of time for the three metal rings.

The three curves plotted in Figure 6 (and in Fig. 7) experience the same type of evolution: firstly a period of about 24 hours with low ring deformations (residual stress), then a faster augmentation of the strains (of the residual stress) and finally a brutal jump (drop) corresponding to the cracking of the specimens.

Figures 6 and 7 demonstrate that the nature of the central ring influences both the age of cracking and the maximal ring strain measured. The test performed with the brass ring, which has the lower rigidity, shows the later the age of cracking, the higher the ring strain and the residual stress: a reduction of 13% of the degree of restraint leads to double the amplitude of the ring strain and to increase the maximal residual stress of about 15%.

A difference of about 10 h is recorded between the age of cracking obtained with the steel ring and that measured with the stainless steel. This difference cannot be explained only by the slight difference existing

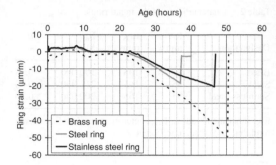

Figure 6. Ring strain for cement paste with W/C = 0.3 at T = 30°C.

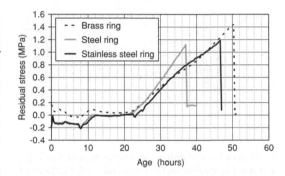

Figure 7. Residual stress for cement paste with W/C = 0.3 at T = 30°C.

Figure 8. Surface quality of the rings (enlargement = ×50).

between the elastic moduli of these two rings. The surface quality of the rings, which induces more or less friction at the interface with the cement paste specimen, could also play an important role in this cracking age difference. The metal surface in contact with the specimens was observed with a video-microscope for each ring (Fig. 8). Different scratch thicknesses have been observed for the three rings. It can thus be supposed that the intensity of friction forces will not be the same for the three types of rings. However at the current state of our knowledge, it is still difficult to quantify the effect of the surface quality on the difference in cracking age. Tests are in progress to determine precisely the depth of roughness of each ring.

3.2.2 Curing temperature effect
The second parameter studied is the curing temperature. Figure 9 gives the ring test results obtained for the specimens with W/C = 0.4 at 20, 30 and 40°C.

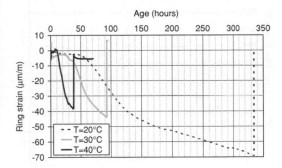

Figure 9. Brass ring strain for cement paste with W/C = 0.4 at T = 20, 30 and 40°C.

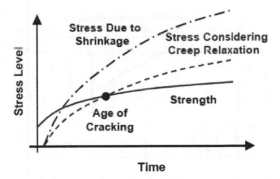

Figure 10. Competition between internal stress evolution and strength development in hydrating cementitious systems (Weiss 1999).

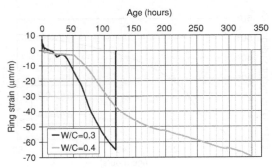

Figure 11. Brass ring strain for cement pastes with W/C = 0.3 and W/C = 0.4 at T = 20°C.

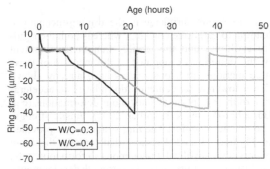

Figure 12. Brass ring strain for cement pastes with W/C = 0.3 and W/C = 0.4 at T = 40°C.

A preliminary point, which has to be underlined, is that, in the range considered, the temperature did not significantly modify the elastic modulus of metals: for instance, for stainless steel, it is 200 GPa at 20°C and 199 GPa at 40°C. In the following, the influence of this variation will be thus neglected.

Figure 9 shows that a higher curing temperature causes an acceleration of the early-age cracking: from 20 to 40°C, the cracking age is divided by 8.8. This phenomenon can be attributed to the thermo-activation of the cement hydration process at a very early age: when the temperature increases, the deformations of the cement paste develop faster (Turcry et al. 2002, Mounanga et al. 2006) and the value of cracking tensile stress is reached more quickly.

In fact, this acceleration is the result of the competition between several complex phenomena: on one hand, the increase of internal tensile stress due to autogenous shrinkage (partially counterbalanced by the creep relaxation of the hydrating matrix) and, on the other hand, the development of the material tensile strength. When the internal stress becomes higher than the tensile strength, the specimen cracks (Fig. 10). Temperature accelerates both the internal stress rate and the tensile strength evolution. According to our results, the fact that early-age cracking occurs sooner when the curing temperature is higher seems to demonstrate that the thermo-activation of the internal stress rate is more important than the thermo-activation of the tensile strength evolution.

Besides, a second effect of the increase of curing temperature is the diminution of the maximal ring strain measured just before the cracking of the specimen. Between 20°C and 30°C and for all the specimens investigated, a significant diminution of the maximal ring strain is recorded when temperature increases whereas the difference between the amplitudes of the maximal ring strain obtained at 30°C and at 40°C is much lower (Fig. 9).

3.2.3 W/C ratio effect

Figures 11 and 12 show the effect of the W/C ratio on the brass ring test results at 20°C and 40°C, respectively. The initial amount of water in the material influences both the rate of the ring strain and the age of cracking. A difference of cracking age of about 215 hours is observed between the two cement pastes (W/C = 0.3 and W/C = 0.4) at 20°C. This difference becomes lower at 40°C, because of the thermo-activation of early age cracking discussed in the previous section.

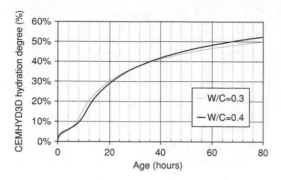

Figure 13. CEMHYD3D hydration degree of cement pastes with W/C = 0.3 and W/C = 0.4 at T = 20°C.

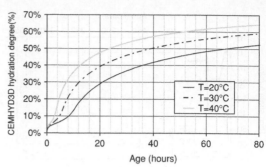

Figure 14. CEMHYD3D hydration degree of cement pastes with W/C = 0.4 at T = 20, 30 and 40°C.

Autogenous shrinkage is the driving-mechanism leading to early-age cracking, whereas both the rigidity evolution and the tensile strength development of the material are opposed to the increase of these deformations. Previous studies have already shown that a lower water/cement ratio induces an acceleration of the evolution of autogenous deformations (e.g. Baroghel-Bouny & Kheirbek 2001), tensile strength and elastic modulus. The ring strain curves provided in Figures 11 and 12 show that, among the main phenomena involved (autogenous deformation rate, evolution of rigidity and tensile strength development), the preponderant one remains the autogenous strain rate, since the gain of both rigidity and tensile strength due to the diminution of the initial water amount did not enable to slow down the occurrence of early-age cracking.

4 AUTOGENOUS EARLY-AGE CRACKING AND THE HYDRATION DEGREE CONCEPT

The test results discussed in the previous sections are expressed as a function of time. This mode of presentation is not optimal when comparing materials of different compositions since it does not enable to directly relate the cracking age to the physico-chemical evolution of the material.

In the following, the ring test results are plotted as a function of the maturity level of the cement pastes expressed in terms of the hydration degree.

The evolution of the cement hydration degree has been computed with a program developed by the NIST, CEMHYD3D (Bentz 2005).

The input data are the particle size distribution, the chemical composition and the apparent activation energy of cement. For this latter parameter, the value considered is 40 kJ/mol (Mounanga et al. 2006).

Figures 13 and 14 show the influence of the W/C ratio and the curing temperature on the hydration

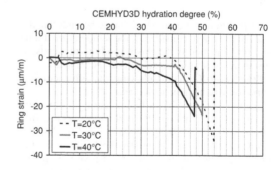

Figure 15. Brass ring test results as a function of hydration degree for cement pastes with W/C = 0.3 at T = 20, 30 and 40°C.

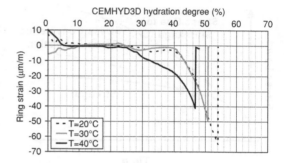

Figure 16. Stainless steel ring test results as a function of hydration degree for cement pastes with W/C = 0.3 at T = 20, 30 and 40°C.

degree of the cement pastes. In the range of the W/C ratio investigated, the effect of this parameter remains weak at a very early age. On the contrary, a significant acceleration of the hydration rate is observed when the temperature increases, in accordance with the thermo-activation principle.

Figures 15 to 18 present the evolution of the ring strain as a function of the cement hydration degree for different cement pastes.

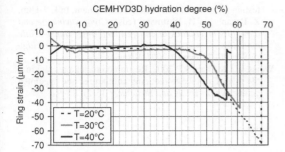

Figure 17. Brass ring test results as a function of hydration degree for cement pastes with W/C = 0.4 at T = 20, 30 and 40°C.

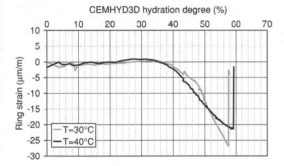

Figure 18. Stainless steel ring test results as a function of hydration degree for cement pastes with W/C = 0.4 at T = 30 and 40°C.

It can be noted that the augmentation of the curing temperature results in a decrease of the maturity level corresponding to the first autogenous cracking time. These results show that, when considering a relatively large temperature range (20 to 40°C), the thermal effect on the first early-age autogenous cracking is not only a kinetic one : temperature influences both the age of cracking and the maturity level corresponding to this cracking age.

5 CONCLUSIONS

From the experimental results presented in this paper, the following points can be underlined, concerning:

- *The curing temperature.* A higher curing temperature causes an acceleration of the early age autogenous cracking. This phenomenon can be attributed to the thermo-activation of hydration, which induces an increase of the autogenous shrinkage rate and therefore of the internal tensile stress. Temperature also influences the maximal amplitude of ring strain recorded just before the cracking.
- *The degree of restraint.* It was observed that the higher the degree of restraint the sooner the

autogenous cracking age. Besides, it seems that the surface quality of the ring has an impact on the early age cracking. Supplementary tests are in progress in order to better quantify the effect of this parameter.
- *The water/cement ratio.* A diminution of the initial water amount leads to faster cracking of the cement pastes. This effect has been explained by the fact that, in low water-to-ratio cement-based systems, the autogenous shrinkage rate, responsible for the development of internal tensile stress, is higher.

Finally, from computation results obtained with CEMHYD3D, it was shown that, for a given value of the W/C ratio and in the temperature range studied, the level of maturity (hydration degree) corresponding to the autogenous cracking age decreases when the curing temperature increases.

REFERENCES

Baroghel-Bouny, V. & Kheirbek, A. 2001. Effect of mix-parameters on autogenous deformations of cement pastes – Microstructural interpretations. *Concrete Science and Engineering* 3(9): 23–38.

Bentur, A. & Kovler, K. 2003. Evaluation of early age cracking characteristics in cementitious systems. *Materials and Structures* 36(3): 183–190.

Bentz, D.P. 2005. *CEMHYD3D: A three-dimensional cement hydration and microstructure development modelling pack-age Version 3.0.* NIST Report 7232. U.S. Department of Commerce.

Grzybowski, M. & Shah, S.P. 1990. Shrinkage cracking of fi-ber reinforced concrete. *ACI Materials Journal* 87(2): 138–148.

Hossain, A.B. & Weiss, J. 2004. Assessing residual stress development and stress relaxation in restrained concrete ring specimens. *Cement and Concrete Composites* 26(5): 531–540.

Hossain, A.B. & Weiss, J. 2006. The role of specimen geometry and boundary conditions on stress development and cracking in the restrained ring test. *Cement and Concrete Research* 36(1): 189–199.

Mounanga, P., Baroghel-Bouny, V., Khelidj, A. & Loukili, A. 2006. Autogenous deformations of cement pastes – Part I: temperature effects at early age and micro-macro correlations. *Cement and Concrete Research* 36(1): 110–122.

Radlinska, A., Moon, J.H., Rajabipour, F. & Weiss, J. 2006. The ring test: a review of recent developments. In O.M. Jensen, P. Lura, K. Kovler (eds), *Volume changes of hardening concrete; Rilem Proc. intern. symp., Lyngby (Denmark), 20–23 August 2006.*

See, H.T., Attiogbe, E.K. & Miltenberger, M.A. 2003. Shrink-age cracking characteristics of concrete using ring specimens. *ACI Materials Journal* 100(3): 239–245.

Shah, H.R. & Weiss, J. 2006. Quantifying shrinkage cracking in fiber reinforced concrete using the ring test. *Materials and Structures* 39(9): 887–899.

Turcry, P., Loukili, A., Barcelo, L. & Casabonne, J.M. 2002. Can the maturity concept be used to separate the auto-genous shrinkage and thermal deformation of a cement paste at early age? *Cement and Concrete Research* 32(9): 1443–1450.

Weiss, W.J. 1999. Prediction of early age shrinkage cracking in concrete. PhD Dissertation, Evanston, Illinois.

Weiss, W.J., Yang, W. & Shah, S.P. 1999. Factors influencing durability and early-age cracking in high-strength concrete structures. *ACI SP 189-22 High Performance Concrete: Re-search to practice, Farmington Hills MI.*

Weiss, W.J. & Ferguson, S. 2001. Restrained shrinkage testing: the impact of specimen geometry on quality control testing for material performance assessment. In F.-J. Ulm, Z. Bazant & F.H. Wittmann (eds), *Creep shrinkage and durability mechanics of concrete and other quasi-brittle materials (Concreep 6)*; *Proc. intern. symp., Cambridge (USA), 20–22 August 2001*. Elsevier.

Weiss, W.J. & Shah, S.P. 2002. Restrained shrinkage cracking: the role of shrinkage reducing admixtures and specimen geometry. *Materials and Structures* 35(2): 85–91.

Fracture Mechanics of Concrete and Concrete Structures – High-Performance Concrete,
Brick-Masonry and Environmental Aspects – Carpinteri, et al. (eds)
© 2007 Taylor & Francis Group, London, ISBN 978-0-415-44617-4

Cracking due to leaching in cementitious composites: Experimental investigation by means of X-ray microtomography and numerical modeling

T. Rougelot & N. Burlion
Laboratoire de Mécanique de Lille, UMR CNRS 8107, Villeneuve d'Ascq, France

D. Bernard
Institut de Chimie de la Matière Condensée de Bordeaux, UPR CNRS 9048, Pessac, France

F. Skoczylas
Laboratoire de Mécanique de Lille, UMR CNRS 8107, Villeneuve d'Ascq, France

ABSTRACT: Chemical degradation of cement based materials leads to significant degradation of their physical properties. A typical scenario is a calcium leaching due to water (water with very low pH compared with that of porous interstitial fluid). The main objective of this paper is to evaluate the evolution of microstructure induced by leaching of a cementitious composite using synchrotron X-ray microtomography, in the particular way of identification of cracking induced by leaching. After a brief description of the degradation mechanism and the X-ray synchrotron microtomographic analysis, we propose a numerical simulation performed in order to prove that cracking due to leaching is induced by an initial pre-stressing of concrete. This pre-stressing is due to endogenous shrinkage. After leaching, the tensile strength of cement paste is dramatically reduced leading to microcracking.

1 INTRODUCTION

Mortars and concretes are non-homogeneous materials whose macroscopic physical properties depend on their local characteristics. For instance, porosity and permeability of concrete are intimately related to the microstructure. Hence it is of importance to link local and macroscopic scale in order to study and model the mechanical behaviour and the transport properties of cementitious materials. This requires the knowledge of the 3D micro-geometry. In order to visualize this microstructure, various innovative techniques are currently developed such as acoustic emission analysis, infrared thermography or X-ray computed microtomography (XCMT).

Synchrotron XCMT finds applications in the study of sample microstructure without damaging it. The principle, similar to the medical scanner, consists of acquiring digital images of the material's X-ray absorption. This acquisition is undertaken at various angles: a three-dimensional image is then obtained by numerical reconstruction from the set of 2D-images.

Concrete leaching is often the result of a fluid attack (pure water or water with very low pH compared to that of the pore fluid), and leads to the hydrolysis of cement paste hydrates, important increase in porosity and permeability, and important decrease in mechanical properties. Existing models often use a damage behaviour in order to capture the main mechanical characteristic of the leached material. Using synchrotron XCMT as a non-destructive characterization method under accelerated leaching with ammonium nitrate solution leads to possible microtomographic analysis on the same specimen during the dissolution. It is then possible to determine the degradation kinetics, the leaching front position and the porosity increase without interfering with the material (Burlion et al. 2006). It is also possible to show cracking of material during the process. The cementitious materials containing aggregates (or glass spheres in the present case) are auto-stressed materials: indeed, the endogenous shrinkage of the cementing matrix causes a contraction around these aggregates. This contraction remains generally limited to values lower than the cement tensile strength. On the other hand, in the particular case of leaching, chemical attack will lead to a drastic decrease of tensile strength of the matrix. A microcracking will appear mainly around the aggregates, what results in a very notably weakening of the mechanical capacities of the material. As regards confirmation or not of these assumptions, a numerical study was led. It aims to highlight the preferential areas of cracks nucleation, by the use of a finite element code in nonlinear mechanics. The selected model has been chosen for its simplicity, and constitutes a first numerical approach of mechanical behaviour modelling of leached samples.

First we will briefly describe mortar degradation mechanism, and the principles of synchrotron X-ray computed microtomography. In a second part, we propose a numerical simulation of mortar leaching: the results obtained are described, demonstrating the capacity of the method to validate our hypothesis about the concrete cracking during leaching.

2 EXPERIMENTAL TOOLS AND MEASUREMENTS

2.1 X-ray microtomography used for leaching analysis

High resolution microtomographic acquisitions were performed on the BM05 beam line of the European Synchrotron Radiation Facility (ESRF, Grenoble, France). In our case a monochromatic beam with energy of *30 keV* was used. The mortar sample was mounted on a rotative table and the acquisition consisted in the recording of 900 two-dimensional (2D) radiographs at equally spaced angles between 0° and 180° (Figure 1). A scintillator is set behind the sample to convert, as efficiently as possible, X-rays to visible light. Contrast obtained on the 2D projections results from the difference in X-ray absorption by the phases/features encountered by X-rays in the specimen. A mirror and the optical set-up selected for the experiment directs the light to the detector. The FRELON CCD camera, developed at ESRF, comprises *2048 × 2048 pixels*. A pixel size of *5.1 × 5.1 μm²* is then obtained. The 2D radiographs were then exploited to reconstruct the volume of the samples using a conventional filtered backprojection algorithm.

2.2 Material choice and specimens design and Accelerated leaching process

The size of the sample was imposed by the microtomographic analysis. The maximal size was of 2048 × 5.1 μm (about 10 mm). Cylindrical cores of 8 mm diameter have been directly obtained from classical prism 40 × 40 × 160 mm³ (Figure 2). Their lengths varied between 20 to 30 mm. To perform this study, different cementitious composites have been done: the key idea has been to reproduce cementitious composites initially proposed by Bisschop and

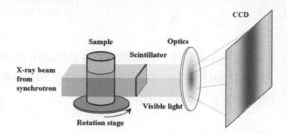

Figure 1. Sketch of the experimental setup for microtomography mapping.

Figure 2. Different cementitious composite specimens (*φ*8 mm) after various degradation processes (leaching, thermal exposure).

van Mier to study effects of drying on microcracking (Bisschop & van Mier 2002, Shiotani et al. 2003). We will focus here only on a composite constituted with 35% of glass spheres completed by a cement paste (cement CEM II/B 32,5R with water by cement ratio equal to 0.5). The glass sphere diameter is 2 mm. Furthermore, a sample of 8 mm diameter and 20 mm long can be considered as a representative volume of the material. Before microtomographic analysis and leaching process, samples were preserved from desiccation to avoid any risk of microcracking induced by drying.

The objective of experiments is to analyze different leaching states of the cementitious composite. This requires an accelerated test, able to reproduce material response that characterizes the long-term behavior of cementitious materials. The degradation process chosen for this experimental study has to fulfill the following imperative: because of time assigned for the microtomographic analysis, the material degradation must take at most 6 days. It was thus chosen, considering the geometry of the specimens, to leach the material by means of an accelerated test with ammonium nitrate solution ($NH_4NO_3 - 480 g/kg_{H2O}$). Such a leaching process has a very high kinetics (about 300 times the kinetics of leaching by deionised water) and that the ammonium nitrate-based calcium leaching leads to the same mineral end products in the cementitious material (Carde et al. 1996).

3 PRINCIPLES OF THE NUMERICAL SIMULATION

3.1 Experimental procedure and numerical simulation

The numerical simulation is linked to observations made on several cross-sections of a sample at different stages of leaching. The chosen cross-section has been selected because of its equal distance between the top and the bottom of the sample. The leaching front is then not influenced by end effect: the hypothesis of plane strains can be done in our calculations and would be well verified.

After a first image acquisition done on the sound sample, it has been immersed for 11 hours into the ammonium nitrate solution. This solution was regularly stirred during the total leaching period (96 h for the considered material). A new acquisition is then made, and the sample is dipped back again into the aggressive solution. New image acquisitions are performed periodically during leaching. The experimental advantage is that it is possible to compare data obtained with the same sample, making unnecessary statistical analysis. The analysis and the comparison of the various degradation stages are directly made without interference between the leaching process and the experimental device, therefore leading to accurate measurement of porosity evolution, of degradation kinetics or of the development of microcracks.

After numerical 3D reconstruction by back filtered projection, 3D-maps are obtained from the sample's X-ray absorption coefficient μ at different stages of leaching. The X-ray absorption coefficient is directly proportional to the density of material: dense material appears in white and porosity in black. Microcracking can be easily detected if their opening is higher than the resolution of the microtomographic analysis (here 5.1 micrometers). Any cross section through the sample can be visualized. Figure 3.a shows a cross-section of the sound sample: one can easily recognize on this picture the glass spheres and the cement matrix. On the left part of the figure, small black discs are visible: these are big porosity in which Portlandite crystals have developed. The studied cross-section has been chosen because of the geometrical distribution of the glass spheres (interactions ball-ball), and also because of the possible interaction between balls and the sample surface. In this section, the glass diameters observed are well representative of the entire sample. This cross-section is then numerically modelled: in our simulation, the glass balls and the interfaces sphere-matrix will be described, while the cement matrix is assumed homogeneous. For the sake of simplicity, all the glass spheres will not be taken into account. Figure 3.b shows the same cross-section (same distance from the sample top, Fig. 3.a) after an 18-hour period of leaching (2nd step of leaching). The cement matrix

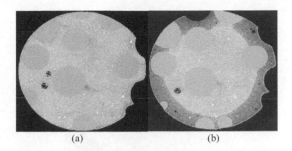

(a)   (b)

Figure 3.   Cross-section of the sample: (a) sound sate before leaching; (b) after 18 hours of leaching.

is no longer homogeneous: exterior part of the sample appears darker. It is due to the decrease in calcium content and to the increase in porosity after the passage of the dissolution front (Burlion et al. 2006). Due to leaching, microcracks occur, near the glass spheres. The leaching front induces a drastic drop of cement mechanical strength: as the zones around aggregates are stressed, the decrease in tensile strength leads to microcrack openings. A numerical simulation will be performed very simply: each step of leaching is mechanically reproduced with the assumption that each leached ring (named CP1 to CP4 in §4.1) has its mechanical properties reduced compared to sound material.

An example of a 3D-reconstruction is given Figure 3.c, in which a slab (about 0.25 mm) is presented after 18 hours of leaching. Aggregates are grey, sound cement light grey and leached cement dark grey. It is possible to distinguish the crack in 3D which connects the edge of the glass sphere at the surface sample. This crack occurs clearly during leaching process.

We have to notice that this microstructure analysis is not able to detect very small microcracks: visibility of microcracks depends on the resolution of the X-ray technique (here a pixel size of 5.1 μm), which means that cracks with width larger than 10 μm can be distinguished with certainty. Many cracks can be rather small, with a width smaller than 5 μm, and may be undetected in the present experiments.

3.2 How to model the evolution of the leaching front?

The finite element code used is CESAR LCPC v4, and mainly CLEO2D which solves two-dimensional problems. As mentioned previously, the hypothesis of plane strains is supposed to be verified for the studied samples.

The progression of the leaching front is numerically modelled by creating several layers in the sample. Thus, their mechanical properties will decrease gradually, taking in consideration effects of leaching process over cement paste. To remain rather qualitative than

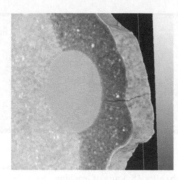

Figure 3.c. 3D-reconstruction of a slab of the leached sample during 18 hours – zoom on the right side of the figure 3.b.

quantitative, only one leaching state has been considered. Actually, strength decrease is about *70 %* when Portlandite (Carde et al. 1996), first dissolved hydrate, disappears. Moreover, a study on accelerated leaching on mortars shows the same tendency with a dramatic decrease in physical properties (Agostini et al. 2006).

In a first approach, the cementitious matrix will be supposed to be in one of two states: sound or leached. Figure 3.b, which represents a slice of the specimen after an 18-hour leaching process, clearly shows the border between these two areas thanks to a contrast difference. The leached area is less dense than the sound one and consequently more permeable to X-ray penetration. Glass spheres, made of silica, are not degraded: their mechanical properties will remain unchanged during leaching.

3.3 Mechanical behaviour of each constituent

On the one hand, the constitutive law used for glass spheres (considered isotropic) is linear elastic, since stresses reached in this material in the numerical model remain in the elastic behaviour.

On the other hand, the cementitious matrix is supposed to be isotropic-elastic perfectly plastic, modelled by Mohr-Coulomb criterion. It is widely used in numerical models for many materials, such as concrete (Camborde et al. 2000).

Besides, it appears not to be mandatory to implement a strain localization limiter with Mohr-Coulomb (Gerard et al. 1998).

3.4 Analyzed values

At each step of leaching, the stress field and the plastic strain norm is observed and studied. It will allow to represent either preferential areas of plasticization or evolution of stresses in the sample. In addition, the cracking pattern if a damage model, where damage variables are directly linked to plastic ones (Frantziskonis & Desai 1987), had been used.

Moreover, it should thus be possible to observe crack closing by means of stresses relaxation during the progression of the leaching front, particularly studying an absence of evolution of plastic strains, but a decrease in elastic strains or stresses. The norm of plastic strains $||\varepsilon_p||$ is defined by Equation 1.

$$\left\| \overline{\varepsilon_p} \right\| = \sqrt{\varepsilon_{Ip}^{\ 2} + \varepsilon_{IIp}^{\ 2}} \tag{1}$$

where ε_{Ip} and ε_{IIp} are main plastic strains.

Analyzed stresses are the main stresses which, when $||\varepsilon_p|| = 0$, allow to know volume variation around a node of the mesh, and so to determine whether cracks are closing or opening per Equation 2:

$$\frac{\Delta V}{V} = tr(\varepsilon_e) = \frac{(1 - 2v)(\sigma_I + \sigma_{II} + \sigma_{III})}{E}. \tag{2}$$

with ε_e the tensor of elastic strains, σ_I, σ_{II} and σ_{III} the main stresses, $\frac{\Delta V}{V}$ the volume variation, E the Young's modulus and v the Poisson's ratio.

3.5 Effect of maturation before leaching

The sample is cored in a 40 * 40 * 160 mm prismatic beam that has been cured in lime-saturated water for 28 days, and then protected of desiccation for 6 months by a self-adhesive aluminium film. It is thus only submitted to endogenous shrinkage. Indeed the prismatic beam is hydraulically isolated from its environment. Measured shrinkage depends on increase in capillary pressure due to relative humidity decrease because of water consumption by not hydrated cement. Shrinkage is monitored periodically with a displacement transducer. The linear length variation which stands for linear endogenous shrinkage, supposed to be isotropic, of cement paste, is measured.

To model this effect, cement paste will be initially submitted to a thermal strain $\varepsilon_{thermal}$ equivalent to the strain caused by endogenous shrinkage $\varepsilon_{endogenous}$, the finite element code used implying this analogy explained in Equation 3 to take into account this phenomenon.

$$\varepsilon_{endogenous} \equiv \varepsilon_{thermal} = \alpha.\Delta T \tag{3}$$

where α is the thermal dilatation ratio and ΔT the equivalent thermal variation.

4 MODEL DATA

4.1 Geometry

The numerical sample is composed of seven spheres (#1 to #7) located as may be seen in our reference slice before leaching process (Fig. 3.a). To focus our

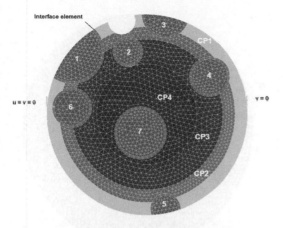

Figure 4. Mesh used in the model (u and v are respectively horizontal and vertical displacements, aggregates are numbered from 1 to 7 and cement layers from CP1 to CP4).

analysis on representative cases, only these 7 aggregates have been selected, since they are representative of many geometrical disposition possibilities (Fig. 4). For instance, an aggregate far from the surface, (#7), or on the contrary close to the surface (#4 and #6), an aggregate partially out of the sample (#1, 3 and 5), neighbour aggregates (#1, 6, 2 and 3), and finally under an asperity of the surface (#2). This geometry will be adequate to simulate crack openings and besides, to extend these results to other geometrical distribution of aggregates.

These glass spheres are inclusions in a cementitious matrix, which has been divided up in 4 layers (approximately 1 mm-thick for CP1 and CP2 layers, 2 mm for CP3, CP4 being the central part of the specimen).

4.2 Finite element mesh

The generated mesh is composed of about 3,850 6-nodes triangles (quadratic interpolation). At each interface between glass aggregate and cement paste, 6-nodes interface elements are added, for a total of 151 elements. They will allow to model the interface behaviour, and as for cement paste layers, they can be sound or leached, to take into account the calcium dissolution. Indeed, the chemical composition of this interface leads to a preferential leaching process around aggregates, due to its high concentration in calcium.

4.3 Mechanical characteristics

Table 1 recapitulates the values of parameters that are necessary for this model: glass spheres (Bridge et al. 1983), sound and leached cement paste (Heukamp et al. 2003, Carde & Francois 1999). Internal friction angle ϕ is deduced from friction coefficient δ with the

Table 1. Mechanical parameters used for each constituent.

	E (MPa)	v	C (MPa)	ϕ (°)
Sound cement paste	22800	0.24	17.1	54.9
Leached cement paste	3600	0.24	1.3	34.1
Glass aggregates	73000	0.17	–	–

approximation $\delta = sin\ \phi$, C is the cohesion of the material and E its Young's modulus. The Poisson's ratio v of the sound cement paste is 0.24, in accordance with a range of values generally reported in articles varying between 0.2 and 0.25 (Boumiz et al. 1996, Haecker et al. 2005). Due to a lack of experimental data, Poisson's ratio is assumed to be not affected by the leaching process. Intuitively, we can suppose that it will increase with degree of leaching, but it is not the hypothesis made here.

4.4 Interface elements

Interface elements allow to include in the model the paste-aggregate interface (also named Interfacial Transition Zone) which is, mechanically speaking, very weak compared to cement paste or glass. Indeed, the surface of the aggregates is smooth and rounded, avoiding creation of a strong link between these constituents, as noticed in several publications (Bisschop & van Mier 2002, Shiotani et al. 2003). Besides, glass is a non-porous material, reinforcing the interface weakness, as strong colloidal bridges are almost impossible.

That will be taken into consideration, assuming there is a Coulomb's friction at this transitional zone, with a low-value for the friction threshold. As explained previously, two cases are considered: a so-called sound interface around an aggregate when the front of leaching has not yet attacked more than half of its perimeter, and a degraded interface in the other case.

Values for these parameters, modelling behaviour of the interface zone, are extrapolated from cement paste data obtained in the literature. Young's modulus of interface is approximately 50% lower than the matrix's (Hashin & Monteiro 2002, Yang 1998, Lutz et al. 1997). Tensile strength (f_t) is supposed to be low: 1.5 MPa, then 1 MPa after leaching. To model the weakness of the interface, and its minor contribution to the mechanical behaviour of the studied specimen, the values of the two parameters of the Coulomb friction (C and ϕ) are also very low. Table 2 sums up all the values of parameters used for interface elements.

Friction, separation and non-interpenetrability conditions are verified, the simulation is done for a

Table 2. Properties of interface elements.

	E (MPa)	f$_t$ (MPa)	C (MPa)	$\phi(°)$
Sound interface	14400	1,5	2	35
Leached interface	1800	1	1	25

maximum of 1000 iterations with a tolerance of 0.1% for convergence of the solution.

4.5 Boundary conditions

The specimen is fixed in two diametrically opposed nodes at its surface, respectively by a zero-horizontal and vertical displacement for one node, and a zero-vertical displacement for the other node. These boundary conditions well represent the specimen when it is in ammonium nitrate solution.

Besides, we can notice that the plane deformation hypothesis induces stresses in Z-axis (in the axial direction of the specimen) since axial strains are supposed to be null. These over-imposed z-stresses can be interpreted as modelling the effect of supposed other aggregates in an upper or lower section of the specimen.

4.6 Shrinkage due to maturation

An endogenous linear shrinkage has been measured on 40 * 40 * 160 mm prismatic specimens from which studied samples submitted to leaching have been cored. This value of −300 micro-strains is supposed isotropic. Strains due to maturation are consequently of −300 micro-strains in each direction and each point of the cement paste.

By analogy (cf. §3.6), the equivalent thermal stresses, induced by endogenous shrinkage, initially applied are modelled by a temperature variation of −30°C of the cement paste with $\alpha = 10^{-5}$ as the dilatation coefficient.

5 ANALYSIS OF NUMERICAL RESULTS

5.1 Locations of sections

To clearly analyse results of the simulations, we have chosen to study several sections (Fig. 5) where we will be able to observe evolutions in cement paste and glass spheres:

– in the area where the most important plastic strains are supposed to occur (sections CC' and FF')
– around an aggregate just under an asperity of the surface and close to another aggregate (section DD')

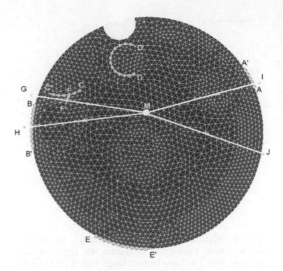

Figure 5. Localization of section lines (sections AA' to FF' and GM to JM). Arrows indicate the orientation of each section.

– on the surface of the sample, beginning in plain cement paste and ending close to an aggregate (section EE')
– on the surface for two aggregates close to the surface, one being an isolated aggregate (Section AA') and the other not (section BB')
– four section lines in the direction of the progression of the leaching front, one in plain cement paste far from inclusions (section JM), one in cement paste close to inclusions (aggregates #1 and #6) (section GM), and two through cement and glass aggregates (sections HM and IM)

These sections allow to study most of all existing configurations, and also to give a global view of the evolution of stresses and strains during leaching. The orientation of these sections is given by arrows as seen on the Figure 5. The origins of sections are the first point of the section (D is the origin for the section DD'), and the last point is the end of the section. For section lines, it is the same principle: M is the end of sections GM, HM, IM and JM. On curves that will be presented, the term "position on the section" stands for the X-coordinate in the local reference mark so defined and oriented.

5.2 Initial endogenous shrinkage

The sound sample mature is modeled by taking into account thermal strain, which is equivalent to endogenous strain during maturation. The numerical simulation showed that, even before leaching, the sample is submitted to some severe stresses, close to the aggregates. Figure 6 shows this, representing plane main stresses in each node of the mesh. The sample, due

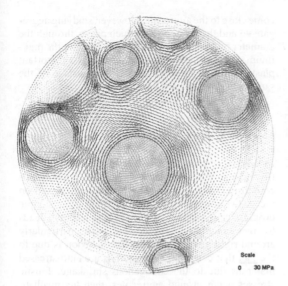

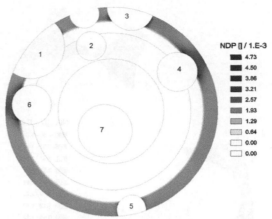

Figure 7. Isovalues of the norm of plastic strains (NDP) after leaching of the layer *CP1*. Scale is given at the right of the schema (from 0 to $4,73.10^{-3}$ strains).

Figure 6. Representation of the main stresses in each node of the mesh (in black, tensile stress and in grey compressive stress, length of segments is proportional to intensity of stress). Axial stresses are not represented.

to thermal restrain and so endogenous shrinkage, is prestressed.

An important part of the cementitious matrix is in a tensile state, since rounded aggregates avoid free shrinkage during maturation, while aggregates are in a compressive state, because of endogenous shrinkage which acts as a pressure all around them. Physically, the difference in Young's modulus between cement paste and glass explains this phenomenon. Indeed, shrinkage is higher in cement paste than in aggregates. However, there is no plastic zone in the matrix. This is a confirmation of visual observations made in Figure 3.a, where no crack can be detected at a mesoscale.

5.3 Leaching of the first layer

The first external layer (*CP1*, thickness = *1mm*) is now considered as totally leached. This case corresponds to an 11-hour leaching process. Its mechanical properties are supposed uniformly degraded. Results presented further show the location of zones where plastic strains are important (Fig. 7), a darker colour meaning a higher plastic strain norm.

They are located close to aggregates in the periphery of the specimen. Indeed, as underlined in the previous paragraph, the areas submitted to tension (which is the most unfavourable solicitation for cement paste) are around aggregates, where free strains are not possible. Moreover, strength of cement paste decreases with leaching, accentuating this phenomenon. So,

plasticization, or damage if a damage model had been used, are mainly likely to appear in these areas.

Aggregates in subsurface of the sample (#4 and #6) are surrounded by the highest plastic strains during leaching of the first external layer. This is logical, because the thickness of the cement paste is very low.

5.4 Leaching of the second layer

In the same way as for leaching of the *CP1* layer, *CP2* layer is now considered as being totally leached. That corresponds to 18-hour leaching process in ammonium nitrate solution, and almost to the state observed in Figure 3.b. Figure 8 presents the cartography of plasticized zones, and mainly those close to aggregates #1, #6 and #2 which are the most interesting ones. We can already notice that, concerning crack openings, there is a concordance between numerical simulation and experimental observations on Figure 3.b. The remark (previous paragraph) as regards repartition of plasticized zones around the aggregates is confirmed by the leaching of *CP2* layer, in which the same phenomenon appears.

A zoom onto the aggregate #6 (Fig. 9) allows to make a comparison between numerical simulation (Fig. 8) and reality obtained by means of X-ray microtomography. A good agreement exists between zones where plastic strains are important and the localization of cracks. In particular, the crack (b) of the Figure 9 perfectly coincides with numerical model (Fig. 8). In addition, if we study more precisely aggregates #1 and #6, the maximum of plastic strain deformation norm in the sample is located on the shortest distance between their surfaces (crack (a) in Figure 9). The two aggregates prevent the cement paste to freely shrink, and their proximity leads to an overlapping of the plastic

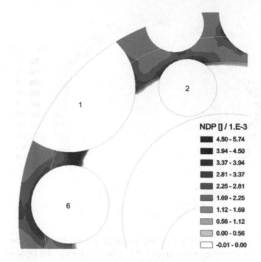

NDP [] / 1.E-3
- 4.50 - 5.74
- 3.94 - 4.50
- 3.37 - 3.94
- 2.81 - 3.37
- 2.25 - 2.81
- 1.69 - 2.25
- 1.12 - 1.69
- 0.56 - 1.12
- 0.00 - 0.56
- -0.01 - 0.00

Figure 8. Isovalues of the norm of plastic strains close to aggregates #1, #2 and #6 after leaching of *CP2* layer. Each level of grey corresponds to an interval of values.

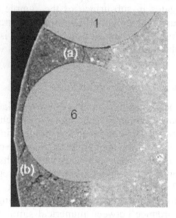

Figure 9. Zoom on aggregate #6 after an 18-hour leaching: (a) cracking between two neighbour aggregates, (b) cracking between subsurface aggregate and surface of the specimen.

areas that they each generate. This overlapping leads to a dramatic increase in values of the norm of plastic strains. If aggregates are farther from each other, plasticization becomes lower, and this could be observed looking attentively at aggregates #1 and #2, or #2 and #3 for instance.

Another remark can be drawn about occurring of plasticity close to aggregate #4 (the section IM passes through the highest area of plastic strain around aggregate #4). Indeed, propagation of plasticity begins from the surface of the aggregate, towards the periphery of the sample, and not the opposite phenomenon. This could be explained since stresses become higher as we come close to the aggregate. However, studying aggregate #6 and the section HM which passes through the diameter of the glass sphere and not through the maximum of plastic strain around it, the most important plastic strain is close to the external surface of the sample, and not around the aggregate.

6 CONCLUSIONS

In this paper, a new experimental approach to identify microcracking due to leaching of cementitious composites is presented. This technique is based on the X-ray microtomographic analysis of a sample progressively leached. We show that leaching leads to microcracking of the cement paste, particularly around rigid aggregates. This phenomenon is due to the fact that cementitious materials are auto-stressed materials due to the endogenous shrinkage. Tensile stresses occur around aggregates, then the mechanical properties of cement are reduced and lead to microcracks.

Some numerical simulations of the leaching process are performed in order to confirm this hypothesis. Experimental observations on the cement paste with glass spheres composite are confirmed by using a perfect elastoplastic model leached. As result, apparitions of high plastic strains areas are highly influenced by rounded glass aggregates, which will prevent free strains of the matrix. Moreover, the more at the surface of the specimen this aggregate is, the more important plastic strains are. High plastic strains, and so cracks, due to these rigid inclusions begin close to the aggregate surface, and then propagate inside cement paste. As a conclusion, damage modelling of leaching processes will be well adapted to numerical simulation of durability problems of concrete structures.

ACKNOWLEDGEMENT

We acknowledge the European Synchrotron Radiation Facility (ESRF) for provision of synchrotron radiation facilities.

REFERENCES

Agostini, F. Lafhaj, Z. Skoczylas, F. & Loodsveldt, H. 2007. Experimental study of accelerated leaching on hollow cylinders of mortar, *Cement and Concrete Research* (in press)

Bisschop, J. & van Mier, J.G.M. 2002. Effect of aggregates on drying shrinkage microcracking in cement-based composites, *Materials and Structures 35*: 453–461.

Boumiz, A. Vernet, C. & Cohen Tenoudji, F. 1996. Mechanical properties of cement pastes and mortars at early ages

: evolution with time and degree of hydration , *Advanced Cement Based Materials 3*: 94–106.

Burlion, N., Bernard D. & Chen D. 2006. X-ray microtomography: application to microstructure analysis of a cementitious material during leaching process,*Cement and Concrete Research 36*: 346–357

Bridge, B. Patel, N.D. & Waters, D.N. 1983. On the Elastic Constants and Structure of the Pure Inorganic Oxide Glasses, *Physical Status Solidi A 77*: 655.

Camborde, F. Mariotti, C. & Donzé, F.V. 2000. Numerical study of rock and concrete behaviour by discrete element modelling, *Computers and Geotechnics 27*: 225–247.

Carde, C. François, R. & Torrenti, J.M. 1996. Leaching of both calcium hydroxide and C-S-H from cement paste: modelling the mechanical behaviour, *Cement and Concrete Research 26*: 1257–1268.

Carde, C. & Francois, R. 1999. Modeling the loss of strength and porosity increase due to leaching of cement paste, *Cement and Concrete Research 21*: 181–188.

Frantziskonis, G. & Desai, C.S. 1987. Constitutive model with strain softening, *International Journal of Solids and Structures 23*: 733–750.

Gerard, B. Pijaudier-Cabot, G. & Laborderie, C. 1998. Coupled diffusion-damage modelling and the implications on

failure due to strain localisation, *International Journal of Solids and Structures 35*: 4107–4120.

Haecker, C.J. Garboczi, E.J. Bullard, J.W Bohn, R.B. Sun, Z. Shah, S.P.& Voigt, T. 2005. Modeling the linear elastic properties of Portland cement paste, Cement and Concrete Research 35: 1948–1960.

Hashin, Z. & Monteiro, P.J.M 2002. An inverse method to determine the elastic properties of the interphase between the aggregate and the cement paste, *Cement and Concrete Research 32*: 1291–1300.

Heukamp, F.H. Ulm,F.J. & Germaine, J.T. 2003. Poroplastic properties of calcium-leached cement-based materials, *Cement and Concrete Research 33*: 1155–1173.

Lutz, M.P. Monteiro, P.J.M. & Zimmerman, R.W. 1997. Inhomogeneous interfacial transition zone model for the bulk modulus of mortar, *Cement and Concrete Research 27:* 1113–1122.

Shiotani, T. Bisschop, J. & van Mier, J.G.M. 2003 Temporal and spatial development of drying shrinkage cracking in cement-based materials, *Engineering Fracture Mechanics 70:* 1509–1525.

Yang, C.C. 1998. Effect of the transition zone on the elastic moduli of mortar, *Cement and Concrete Research 28*: 727–736.

Fracture Mechanics of Concrete and Concrete Structures – High-Performance Concrete,
Brick-Masonry and Environmental Aspects – Carpinteri, et al. (eds)
© 2007 Taylor & Francis Group, London, ISBN 978-0-415-44617-4

An experimental study on the roles of water saturation degree and aggregate size in the mechanical response of cement-based composites

M. Szcześniak, N. Burlion & J. Fu Shao
Laboratoire de Mécanique de Lille, UMR CNRS 8107, Villeneuve d'Ascq, France

ABSTRACT: Desiccation of concrete induces the formation of microcracks leading to mechanical damage of concrete structures. Damage concentrates in the interfacial contact zone (ICZ), like also between aggregate particles. This damage, induced by drying, can be characterized in function of the 'Saturation level degree' variations. The main purpose of this study is concerned with specification of mechanical reactions and basic material constants for studied cement based composites during the desiccation process. In order to reduce number of analyzed factors, it was decided to use as aggregate particles glass and polystyrene spheres. This type of aggregate allows the characterization of almost exact ICZ. Each series of composite is a mixture of cement paste and 35% of spheres. In this first approach, only one diameter of aggregate particles is placed in one series of composite. Mechanical tests performed were compression tests: uniaxial, hydrostatic and quasi-triaxial, as well as non-direct tension tests.

1 INTRODUCTION

1.1 *Physical background, aim of the study*

Advances in understanding of crack and shrinkage mechanisms (Witmann 1982, Yurtdas et al. 2004) combined with numerical modeling procedures for cement based materials (Schlangen et al. 2007, Konderla et al. 2006, Koenders et al. 1997) and the falling rate of computation time have encouraged present study. All of presented experimental efforts aim for quantitative description of aggregate size and saturation level influences on damage caused by drying and autogenous shrinkage.

Verification of hydro-mechanical reactions for mono-diameter aggregate has a practical meaning. Firstly, during fabrication of concrete it is common, in some national norms obligatory, to determine granulometric curves. On this basis and knowledge of aggregate size effect, some conclusions on the future material durability and performance can be formed. Secondly, during the insertion of fresh concrete into forms and later wrong vibration time or technique segregation of aggregate particles can appear. It leads to bigger concentrations of larger aggregate particles in the lower element sections and appearance of cement 'milk' on the surface. It is inconvenient, because on the level of modeling, we were supposing homogenous material. It is embarrassing because in the absence of decor surface, cement paste 'milk' on the surface will shrink in absence of restraining aggregate. Finally, bigger aggregate particles are connected with

bleeding mechanism, which is responsible for macroscopic porosity and lower cement paste densities below the aggregate. If we consider beam element in its common steel reinforcement mostly concentrated in the lower tensile sections, it can further lead to easier access of humidity lead to corrosion.

In order to reduce number of analyzed factors, it was decided to use as aggregate particles glass and polystyrene spheres. Each series of composite is a mixture of cement paste and 35% of spheres, like composites proposed by Bisschop and van Mier (Bisschop et al. 2001). In first approach, only one diameter of aggregate particles is placed in one series of composite. This type of composite is not commonly used in practical engineering, however it will be further proposed how to make a transition to standard concrete with ellipsoidal natural aggregates. Not mentioned previously but not less important are measuring methods which allow analysis of moisture and cracks distributions at different stages of drying (Villain et al. 2006, Dela et al. 2000, Andrade et al. 1999).

In the first part of this study, we will present the objectives of the experimental campaign. In the second part, mechanical results are shown. Unixial compression, triaxial compression with 5 and 15 MPa of confining pressure have been performed. The effect of water saturation degree and microcracking during drying are put in light. The competition between the both phenomena will control the failure process of cementitious composites.

Figure 1. Four diameters of aggregate spherical particles were used 6, 4, 2, 1 mm.

Table 1. ICZ information.

Particle diam. mm	Surface of particle mm^2	Volume of particle mm^3	No. of part. –	Total ICZ surface T_{icz} mm^2	T_{icz}/V_s mm^2/mm^3
1	3,14	0,52	48989	153903	2,10
2	12,57	4,19	6124	76951	1,05
4	50,27	33,51	765	38476	0,53
6	113,10	113,10	227	25650	0,35
Mix	–	–	–	73745	1,01

Sample volume (V_s) 73287 mm^3.

1.2 Problems considered

On the basis of our observations, like also those presented in literature (Bisschop et al. 2001) for spherical glass particles, contact between aggregate and cement paste is considerably influenced by 'debonding' mechanism. For the further usage we adopt ICZ expression (Interfacial Contact Zone) which differs from ITZ (Interfacial Transition Zone) in a way that we consider cement paste as isotropic material. Debonding is more clearly observed for bigger aggregate dimensions; however it is only a matter of scale that after enlarging crack patterns for smaller diameters similar or even exact crack distributions appear.

For the reason of further modeling basic information is ratio between aggregate diameter and the smallest side of reference volume. In present case it is respectively 1/6, 1/9, 1/18, 1/36.

Table 1 contains information about total and particular surfaces and volumes of aggregate particles for one separate sample. In order to obtain 35% of aggregate volumetric ratio, sufficient number of spheres is presented in the fourth column of Table 1. As a derivative of the first four columns, we can compute average value of ICZ for each series of composites. This information, like also proportion of T_{icz} to total sample volume, brings us to the conclusion that the surface of ICZ cracks increases with decreasing radius of aggregate particles. The 'Mix' notion is used for proportional (25%) volumetric mixture of four previously mentioned aggregate diameters in order to obtain 35% of aggregate in total.

If we consider that hydric damage, which appears during drying in composite, is a sum of damages due to cracks around the aggregates and between particles.

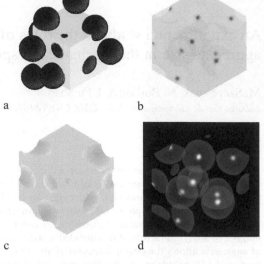

Figure 2. Four stages of numerical analysis. (a) RVE verification, (b) Aggregate cut off, (c) Matrix model, (d) Inclusion/Interface representation.

The following hypothesis can be then formulated. Total surface of cracks, space which can no longer maintain load for tensile stresses, reaches the highest level of 153903 mm^2 for 1 mm (in diameter) particles and is respectively: two, four and six times smaller for 2, 4, 6 mm diameter aggregates. T_{icz} for the mixture of four diameters obtains value of 73745 mm^2 what places it just near to 2 mm particles.

2 EXPERIMENTAL PROCEDURE

2.1 Types of tests performed

Series of samples previewed for compression were systematically mechanically tested on four levels of water saturation degree – roughly 100%, 66%, 33% and 0%. Because of the specific character of saturating and drying of cement based materials, we are adopting the testing procedure as first saturated samples and then dried ones. It prevents restarting hydration process for 0 to 100% saturation testing procedure. On the other hand, one should be aware of fragile difference between what's an effect of direct influence of different water saturation degree, what comes from cracks which are derivative of hydric gradients, and finally coupling between what's direct and what's indirect. It is not an easy task, and in order to partially restrain effect of cracks, triaxial tests with initial confining stress of 5 and 15 MPa in spite of uniaxial compression tests, were performed.

In order not to induce significant damage due to temperature gradient it was decided to use 30°C and 30 ± 5% RH for drying, like Bisschop and van Mier

Table 2. Constituents of the composite.

Constituent	Quantity
Glass spheres	638.75 kg/m³
Ultracem	52.5
Cem I 52.5 N CP2	800 kg/m³
Water	400 kg/m³
Water/cement ratio	0.5

Table 3. Properties of water at different stages of material preparations and maturation.

Preparation			Maturing		
		Units		mg/l	mg/l
Temp	19.51	°C	Al	1.791	K	452.515
El cond	0.998	mS/cm	B	0.225	Na	134.725
pH	7.26	–	Ba	0.013	Ni	0.006
O_2	5.7	mg/l	Ca	1.858	P	0.478
Redox	330	mV	Cr	0.020	Pb	0.037
O_2	61.8	%	Cu	0.022	Si	21.283
					Zn	0.031

(Bisschop et al. 2001). Meaningful time, to achieve 0% of saturation, influences maturation level. Testing additional 'witness' samples at the time of main tests solved this problem. 'Witness' samples were prepared and kept in exact atmosphere conditions, but protected by aluminium from drying.

For the reasons of statistical verification, but restrained by time on the other hand, it was decided to make three representations for each uniaxial and 'witness' compression test and two representations for all triaxial compression tests.

2.2 Composition of composites and samples conservation

Composition of the cementitious composites, presented in Table 2, used for the present material was based on the need to obtain considerably high level of shrinkage, like also on being in reference to (Bisschop et al. 2001).

For the mechanical study, cylindrical samples were prepared in forms of stainless steel. Each sample was 36 mm in diameter and 100 mm in height initially. After 24 hours, it was demouled and placed in water for 28 days. Chemical properties of water used for samples preparation and storing are presented in Table 3. After 28 days in water height of samples was reduced to 72 mm.

For the hydro-mechanical study prismatic samples were prepared in steel forms. Each sample was $40 \times 40\,mm^2$ in cross section and 160 mm in height initially. After 24 hours it was demouled and placed in water for 28 days.

2.3 Experimental devices

Uniaxial compression tests were performed using a hydraulic press INSTRON® with capacity 500 kN. Hydrostatic and triaxial test were carried out by additional usage of triaxial cell with capacity 60 MPa. Displacement rate used for vertical loading was 2 μm/s. Injection speed of hydraulic oil was about 2 ml/min. All of instrumentation was kept in $20 \pm 1°C$ and $45 \pm 5\% RH$.

3 RESULTS

3.1 Basic assumptions and hypothesis

Thus expression for water saturation degree level Sw is used, we are making first assumption. For each series of composites, six samples are dried in $90 \pm 1°C$ conditions. Final mass loss is taken as reference $m_{90}\ [g]$.

$$ml_{90} = \frac{m_0 - m_{90}}{m_0} \quad [-] \tag{1}$$

$$Sw = 100 - 100 \cdot \frac{m_0 - m_i}{ml_{90} \cdot m_0} \quad [\%] \tag{2}$$

where $m_0 =$ initial mass of a sample [g], $m_i =$ current mass of a sample [g]. It must be stated uniform saturation idea is quite reasonable for 100% saturation level degree only. Due to basic mechanisms of mass transportation and heat transfer, it is well known that drying provokes hydric gradients, which are one of the damage initiators, and sample core remains wet even in late stages of drying.

3.2 Results from mechanical tests

Before we go to the analysis of material parameters, i.e. strengths in uniaxial and triaxial compression, some comments should be made on Young's modulus, which at the present stage of study is assumed as the most important material reference.

Figure 3 shows variation of secant and tangent Young's modulus as a function of water saturation degree for strain-calculated by gages responses and, on the other hand, from crossheads displacements. The values represented in Figure 3 concern glass spheres with diameter 4 mm. The measures from crossheads displacements give big difference in comparison to those based on strain gages. It is natural because displacement of crossheads is a sum of sample displacement, load cell, grips, coupling between specimen and grips and the machine frame. Recent work of Chawla (2005), presenting how to make a transition between crosshead and sample displacement, encouraged us not to neglect these results. The modulus of elasticity decreases with the desiccation: thus,

1809

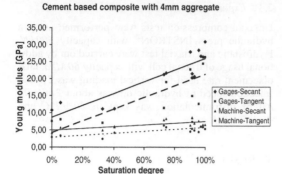

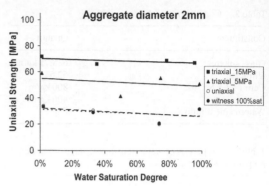

Figure 3. Variation of Young's Modulus in function of saturation degree, based on the strain gages (real value), based on the total displacement of machine cross head (real value + dependence of whole instrumentation stiffness).

Figure 5. Variation of strength versus water saturation degree for uniaxial and triaxial compression tests with hydrostatic pressure *5 MPa* and *15 MPa*. Aggregate diameter considered *2 mm*.

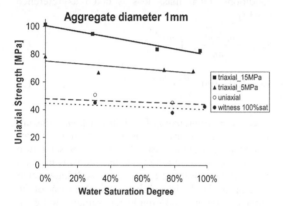

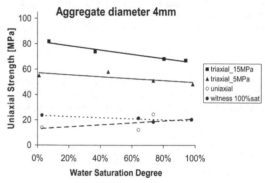

Figure 4. Variation of strength versus water saturation degree for uniaxial and triaxial compression tests with hydrostatic pressure *5 MPa* and *15 MPa*. Aggregate diameter considered *1 mm*.

Figure 6. Variation of strength versus water saturation degree for uniaxial and triaxial compression tests with hydrostatic pressure *5 MPa* and *15 MPa*. Aggregate diameter considered *4 mm*.

induced microcracking implies a reduction of the sample stiffness.

Figures 4–7 present in the short way all of the mechanical tests results for four aggregate diameters used. For triaxial tests results, some authors (instead of our 'y' axis descriptor – Uniaxial Strength, Fig. 4 to 7) prefer the deviatoric stress '$\sigma_1 - \sigma_3$'. However in our considerations it means the same. In order to read maximum average stress in the cross-section (in the loading direction), readers should add respectively *15, 5, 0* and *0 MPa* to 'triaxial_15 MPa', 'triaxial_5 MPa', 'uniaxial' and finally 'witness 100%sat' series. Whenever value from uniaxial test is presented, it is an average from 3 tests. For triaxial tests, because of rather extended research program and time necessary for one test, it was decided to make an average from 2 tests for each value. One have to notice that the 'witness' sample are always with a saturation degree equal to

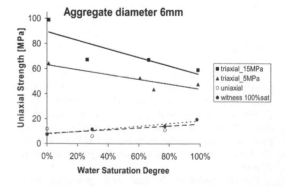

Figure 7. Variation of strength versus water saturation degree for uniaxial and triaxial compression tests with hydrostatic pressure *5 MPa* and *15 MPa*. Aggregate diameter considered *6 mm*.

100%. For a better comparison, we have represented the results on witness specimens at the same level of the uniaxial compression test performed on composites submitted to drying: it means that the compression test was done on the witness sample at the same date as the dried sample.

It should be recalled that 'witness' samples were in contact with air only during few seconds of demoulding, few minutes of aluminium covering and finally few minutes of the test. Results obtained for composites with *1 mm* aggregate particles allow us to make the following remarks (Fig. 4):

- Concerning uniaxial tests on 'witness' samples

  - First sample was tested after *53 days* after preparation, the last one about *43 days* later. During those *43 days*, the composite was exposed to two basic mechanisms, maturation and autogenous shrinkage. Total outcome of those reactions was about *10%* of growth in compression strength.

- Concerning uniaxial tests on dried samples

  - Dried samples are also exposed to maturation and autogenous shrinkage. In addition the shrinkage influences intensity and orientation of cracks in the material. Nevertheless *8.5%* of increase in the resistance is observed.

- Concerning triaxial tests on dried samples with *5 MPa* of confinement

  - Samples restrained to initial hydrostatic compression subjected to the same mechanisms of resistance variation. Samples close to zero saturation level have final resistance *12.5%* bigger then those fully saturated

- Concerning triaxial tests on dried samples with *15 MPa* of initial hydrostatic compression

  - Drying seems to strength material and final strength gain is around *20.7%*.

Uniaxial compression tests performed on samples in different stages of drying together with samples protected from water evaporation show similar kinematics of resistance variation. For the samples initially confined by hydrostatic compression, the influence of desaturation level becomes considerable with increase of confining pressure. For *1 mm* of aggregate diameter, resistance increase in composite, due to decrease of saturation level, becomes more important for bigger initial hydrostatic compression values. At the level of *15 MPa* of hydrostatic compression, the increase of strength due to desaturation exceeds that due to maturing process.

These various evolutions can be explained if one highlights 2 phenomena which control the failure process. The first phenomenon is the mechanical microcracking induced by the external loading. This cracking will propagate all the more easily as the material is initially damaged by hydrous microcrackings. The second phenomenon, concomitant with the first, is the effect of the capillary pressure which will lead, during the desiccation, to an increase in mortar prestressing. This prestressing will induce an increase in multiaxial compression strength of material. According to the preponderance of these 2 phenomena, the failure process is different. On Figure 4, one can see that the process which controls the evolution of strength in uniaxial compression according to the desiccation is mechanical cracking. The evolutions of strength's sample are about the same for specimens submitted to drying or not. On the other hand, in multiaxial compression, the capillary pressure plays an important role due to the fact that mechanical microcracking is limited by confinement. It is thus noted that the resistance decrease with drying (or remains quasi-stable) when mechanical microcracking can propagate, therefore in uniaxial compression. Resistance increases with drying in triaxial compression because of effect of capillary pressure. The more important the hydrous microcracking will be, the more the failure process by mechanical damage will be dominating. The more the material will be dried, the higher the capillary pressure will be and the larger its resistance will be (in the absence of microscopic crack).

Results obtained for composites with *2 mm* aggregate particles allow us to formulate following statements:

- Concerning uniaxial tests on 'witness' samples and drying samples

  - *15.6%* of growth in resistance is observed with no distinction for drying and witness samples

- Concerning triaxial tests on drying samples

  - Samples with *5 MPa* of initial hydrostatic compression have final resistance *10.5%* bigger then those fully saturated
  - Samples with *15 MPa* of initial hydrostatic compression have final resistance *4.7%* bigger then those fully saturated

Uniaxial compression tests performed on samples in different stages of drying together with samples protected from water evaporation show similar kinematics of resistance growth. What differs '*2 mm*' composites from '*1 mm*' is almost exact values of strength at all drying stages. It can imply that for this diameter of aggregate saturation level do not affect resistance.

Results obtained for composites with *4 mm* aggregate particles allow us to formulate following statements:

- Concerning uniaxial tests on 'witness' samples

  - *18%* of growth in resistance is observed in time due to maturation

- Concerning uniaxial tests on drying samples
  - For the first time drop of *6%* in resistance is observed with falling saturation level
- Concerning triaxial tests on drying samples
  - Samples with *5 MPa* of initial hydrostatic compression have final resistance *13%* bigger then those fully saturated
  - Samples with *15 MPa* of initial hydrostatic compression have final resistance *20%* bigger then those fully saturated

Uniaxial compression tests performed on samples in different stages of drying together with samples protected from water evaporation show that for the first time mechanism of drying shrinkage is more important than combined maturing and expected positive low saturation level effect. Samples become considerably cracked even in the early stages of drying. As a result values obtained in compression tests decrease with lower saturation, like also results became 'crack positions' dependent. Triaxial compression tests are less susceptible to cracks, because of its ability to close them in the initial loading stage.

Results obtained for composites with *6 mm* aggregate particles allow us to formulate following statements:

- Concerning uniaxial tests on 'witness' samples and drying samples
  - *6%* of decrease in resistance is observed with desiccation
- Concerning triaxial tests on drying samples
  - Samples with *5 MPa* of initial hydrostatic compression have final resistance *30.6%* bigger then those fully saturated
  - Samples with *15 MPa* of initial hydrostatic compression have final resistance *38.2%* bigger then those fully saturated

Uniaxial compression tests performed on samples in different stages of drying together with samples protected from water evaporation show that even samples subjected only to autogenous shrinkage have high damage level because of provoked cracks. As a result only for '*6 mm*' composites, the decrease in strength both for 'witness' and drying samples was observed. It should be underlined that negative effect of the biggest particle on bigger intensity of cracks formation seems to vanish for higher hydrostatic compression levels.

Table 4 summarizes the effect of drying on material strength, in function of aggregate diameter and mechanical loading condition. Presentation of the results by distinction between type of mechanical tests verifies the effect of aggregate size on the material strength. Figures 8 to 10 summarize this effect.

Table 4. Change rate in strength during drying from *100%* to about *0%* saturation level.

Type of test	Aggregate diameters			
	1 mm %	2 mm %	4 mm %	6 mm %
Uniaxial	+8.5	+15.6	−6.0	−6.0
	(4 MPa)	(4.9 MPa)	(−7.8 MPa)	(−7.6 MPa)
Uniaxial	+10.0	+15.6	+18.0	−6.0
'witness'	(4.5 MPa)	(4.9 MPa)	(4.3 MPa)	(−7.6 MPa)
Triaxial	+12.5	+10.5	+13.0	+30.6
5 MPa	(9.4 MPa)	(5.7 MPa)	(7.6 MPa)	(19.3 MPa)
Triaxial	+20.7	+4.7	+20	+38.2
15 MPa	(20.9 MPa)	(3.3 MPa)	(16.8 MPa)	(34.1 MPa)

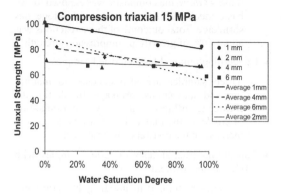

Figure 8. Variation of strength versus water saturation degree for triaxial compression tests with hydrostatic pressure *15 MPa* for different aggregate diameters.

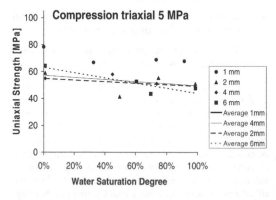

Figure 9. Variation of strength versus water saturation degree for triaxial compression tests with hydrostatic pressure *5 MPa* for different aggregate diameters.

There is a general tendency of resistance growth, for each of the mechanical tests performed, with decreasing diameter of aggregate. Moreover decreasing of water saturation degree seems to reinforce composites

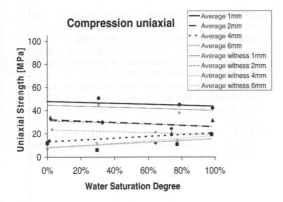

Compression uniaxial

— Average 1mm
— Average 2mm
- - Average 4mm
······ Average 6mm
— Average witness 1mm
- - - Average witness 2mm
····· Average witness 4mm
······ Average witness 6mm

Figure 10. Variation of uniaxial strength versus water saturation degree for uniaxial compression tests for different aggregate diameters.

with smaller aggregate. This tendency is contrary for bigger aggregate diameters. It is supposed, that damage mechanism, induced by weak interface contact, is more significant at that case.

It seems, from the Figure 8, *1 mm* composite has bigger strength at all saturation levels. Lasting diameters *2, 4* and *6 mm* start from approximately same strength level but, during drying, gain in strength varies. Kinematics of that growth increases with diameter.

For smaller initial hydrostatic pressure we observe similar difference between *1 mm* composites and the rest. At all level of saturation about *20 MPa* of difference is noticeable. As a difference to Figure 8 we can see smaller differences in kinematics of strength growth between bigger particles and only 6 mm composites have noticeable *30%* of strength growth.

What differs uniaxial results from triaxial one is general distinction between composites with monodiameter spherical particles. Bigger aggregate is used, smaller uniaxial strength is obtained at all levels of saturation. Within *1 mm* composites it is observed drying samples are more resistant to loading then protected from drying samples. For *2 mm* aggregate, this tendency vanishes and no significant difference between previously mentioned series is noticeable. As early as *4 mm* aggregate composites, decrease in strength during drying is observed. Only for *6 mm* composites decrease in strength for protected samples was observed.

4 CONCLUSIONS AND PERSPECTIVES

The present work contains first results from experimental part of a study directed for better understanding of cement based composites behavior at different stages of drying.

We did not find ICZ influencing compression strengths in uniaxial and triaxial compression

experiments. Table 1 allowed us to formulate hypothesis of bigger contact crack surfaces for smaller aggregate dimensions. However, Figures 8, 9 and 10 do not confirm that and composite with aggregate *1 mm* reaches the biggest strengths at each stage of drying. It seems that only tensile and bending strengths can be affected by ICZ cracks and it must be verified in the separate study.

Mechanical influences of hydrous microcracking have been shown. The initial sate of mortar, before loading, is crucial to understand its failure process. In particular, if hydrous damage is big – i.e. the diameter of aggregate is large, the failure of cementitious material during desiccation is piloted by mechanical damage only. If the effect of initial microcracking is reduced (for example by a confining pressure), the evolution of multiaxial compression strength is related to capillary pressure increase.

All of the data acquired, to be useful for further research, need to be generalized. Outlines of the numerical model, concerning this type of composite, were proposed recently (Konderla et al. 2006) and are forming the enclosure of the coherent study. It must be repeated, this work, concerning 'model' composite, was encouraged by recent achievements in finding analytical solutions for Eshelby problem, like also it has been recently proposed (Torquato et al. 2002) how to make a transition from ideal spherical particles to more complicated combinations of polydimensional spheres, which can state for ellipsoidal natural river aggregate.

Additional mechanical tests should be performed on composite with the same volume fraction of aggregate, but multi diameter sets of spheres. Within the text authors used notation *1, 2, 4,* and *6 mm* as for aggregate diameter. It seems more reasonable to use as reference notation D/L (inclusion diameter / smallest length of the whole sample). In the present work this notation implies values respectively 1/6, 1/9, 1/18, 1/36. If we consider now results from Points 3.3 and 3.4, where for the biggest aggregate dimension standard deviation of the results used for each series was generally always the biggest, it can be derived 1/36 proportion may yield not representative results for the type of composite considered.

REFERENCES

Andrade C., Sarria J., Alonso C., 'Relative humidity in the interior of concrete exposed to natural and artificial weathering', Cement and Concrete Research, 29, 1999, 1249–1259.

Bisschop J., Pel L., van Mier J.G.M., 'Effect of aggregate size and paste volume on drying shrinkage microcracking in cement-based composites', *Creep, Shrinkage & DurabilityMechanics of Concrete and other Quasi-Brittle Materials*, Proc. of CONCREEP-6@MIT, eds F. Ulm,

Z.P. Bažant and F.H. Wittmann, MIT, Boston, USA, 2001, p.75–80.

Dela B.F., Stang H., 'Two-dimensional analysis of crack formation around aggregates in high shrinkage cement paste', Eng. Fract. Mech., 65, 2000, 149–164 .

Koenders E.A.B., van Breugel, K. 'Numerical modeling of autogenous shrinkage of hardening cement paste', Cement and Concrete Research, 27, (10), 1997, Elsevier

Konderla P., Szcześniak M., 'Zagadnienie homogenizacji kompozytu o sprężysto-plastycznych składnikach' (in Polish) 'Homogenization of composite with elastic-plastic constituents', Polish Society of Theoretical and Applied Mechanics (PTMTS), Conference materials, Karpacz 2006.

Schlangen E., Koenders E.A.B., K. van Breugel, 'Influence of internal dilatation on the fracture behaviour on multi-phase materials', Eng. Fract. Mech., 74, 2007, 18–33.

Torquato S., Random 'Heterogeneous Materials: microstructure and macroscopic properties', Interdisciplinary applied mathematics; v.16, Springer-Verlag, 2002.

Villain G., Thiery M., 'Gammadensimetry: A method to determine drying and carbonation profiles in concrete' NDT&E International, 39, 2006, 328–337.

Witmann, F.H., 'Creep and shrinkage mechanisms', in 'Creep and shrinkage in concrete structures', edited by Z.P. Bazant and F.H. Witmann, (J.Wiley and Sons, 1982) 129–161.

Yurtdas I., Burlion N., Skoczylas F., 'Experimental characterisation of the drying effect on uniaxial mechanical behaviour of mortar', Mater. Struct., 37, (267) 170–176.

Influence of the saturation degree and mix proportions on the behavior of concrete under high level of stresses

X.H. Vu, Y. Malecot & L. Daudeville
3S-R, UJF INPG CNRS, Grenoble, France

ABSTRACT: The aim of the present study is to identify the concrete behavior under severe dynamical loading (explosions or ballistic impacts). This paper presents the effect of both the saturation degree and the water-cement ratio on the concrete behavior under static triaxial loading. The tests are done using a press with high capacities: confining pressure up to 650 MPa and axial stress up to 2.3 GPa. The test results show that the saturation degree has a major influence on the concrete deviatoric behavior. The strength of dried concrete strongly increases with the confining pressure whereas it is a constant for saturated samples beyond a confining pressure of 100 MPa. The results also show that the concrete behaves like a granular stacking under high confinement without any influence of the cement paste strength.

1 INTRODUCTION

Right after the cement setting, an ordinary concrete is a quasi-saturated material. Most of the time, it is then submitted to a lower environmental relative humidity, so that a drying process occurs in the concrete. As the pore network of the cement matrix is very thin, this moisture transport is a very slow process which can be described using a diffusion-like equation (Baroghel-Bouny et al. 1999). The time required to reach the moisture equilibrium varies with the square of the built structure thickness. As most of concrete sensitive infrastructures such as dams or nuclear reactors are very massive, they might remain quasi-saturated in core most of their life-time whereas their facing dries very fast.

The concrete drying effect on its shrinkage or cracking has been intensely studied. It is well known that the saturation degree of the hardened concrete has a significant effect on its static uniaxial behavior (Burlion et al. 2005). On the other hand, there are very few results about the effect of the water on the behavior of concrete when it is subjected to extreme dynamical loadings (explosions or ballistic impacts). This lack of knowledge is due to the difficulty of reproducing such loadings experimentally with a simultaneous control of the concrete moisture content. If we consider for example the impact of a missile on a concrete structure, we observe three phases of triaxial behavior, each one associated with different damages but sometimes occurring simultaneously (Zukas, 1992). The validation of concrete behavior models taking into account simultaneously the phenomena of brittle damage and irreversible strain such as compaction thus needs new test results reproducing the complex loading paths described previously. The majority of the available experimental results in literature only relate to triaxial loadings with moderate confinement pressure. They notably allowed us to understand the transition of brittle-ductile behavior which is a characteristic of floating cohesive materials (Li et al. 1970). Numerous studies show that dynamic tests performed on concrete, for example by means of split pressure Hopkinson bars (Zhao et al. 1996), are difficult to realize essentially because of the brittle feature of material that leads to a rupture in the transient stage of loading. The inhomogeneous character of the stress state in the sample, the very limited control of the load path and relatively poor instrumentation lead to a delicate test result exploitation.

We present in this paper an experimental study on the concrete mechanical behavior under high confinement pressure, using a static triaxial press, called "GIGA". This press allows us to attain homogeneous, static and well controlled stress levels of the order of one Giga Pascal. The static characterization of a behavior model with a view to predicting dynamic behavior is not a new practice in the study of geomaterials. The rheological behavior of concrete under compression seems to slightly depend on the deformation rate for dry specimens (Bischoff et al. 1991). The very strong dependence on the loading rate in traction can be mainly explained by the influence of defects (Hild et al. 2003). Similar experimental studies

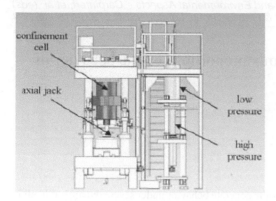

Figure 1. General scheme of the press.

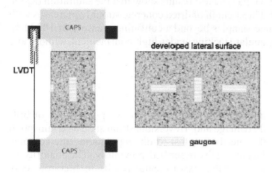

Figure 2. Scheme of strain measurement.

were carried out previously. They were limited to small mortar samples (Burlion et al. 2001). The aim of the present study is to extend this practice to the study of "true" concretes (centimetric aggregate size). This paper presents the effect of both the saturation degree and the water-cement ratio on the concrete behavior under extreme static triaxial loading.

2 EXPERIMENTAL DEVICE

GIGA press is a large capacity triaxial press which has been specifically designed and developed for this study (Thiot, 2004). With this press, cylindrical concrete specimens of 7 cm in diameter and 14 cm in length with a confining pressure of up to 0.85 Gpa and with a 2.3 GPa maximum axial stress can be tested. Figure 1 shows a general scheme of the press (see Vu et al. 2006, Gabet et al. 2006, for more details).

The concrete specimen, surrounded by a membrane impermeable to the confinement fluid, is positioned between caps made of tungsten carbide. The specimen is also instrumented with a axial displacement sensor of type Linear Variable Differential Transformers (LVDT), one axial gauge and two orthoradial gauges

Table 1. Concretes mix proportion and mechanical properties.

Mix proportion (kg/m^3)	R30A7	EC08	EC04
Aggregate D 0,5/8	1007	1007	1007
Sand D0/2	838	838	838
Cement CEMI52,5	263	226	352
Water	169	181	137
Water reducing admixture	0	0	4.57
Fresh concrete properties	R30A7	EC08	EC04
Slump test (cm)	7	15	7
Entrapped air volume (%)	3.4	5.0	4.1
Water-cement ratio W/C	0.64	0.8	0.4
Cement paste volume Vp (%)	0.286	0.286	0.286

(Figure 2). The axial stress applied to the specimen and the pressure inside the cell can be determined by means of a force sensor and a pressure one.

The porous feature of the concrete required the development of a protective multilayer membrane surrounding the specimen, preventing the confining fluid from infiltrating through the specimen (VU X. H. et al. 2006).

3 MATERIAL DESCRIPTION

3.1 Standard concrete R30A7 (dried, partially or "completely" saturated state)

The standard R30A7 concrete is formulated for a resistance in simple compression at age of 28 days (fc28) and a slump of 30 MPa ($\pm$ 2 MPa) and of 7 cm ($\pm$ 1 cm) respectively. The aggregate size of the concrete is such that it must pass through the 8 mm sieve when sieving. The composition and the mechanical properties of the concrete R30A7 are presented in table 1.

The concrete is preserved during 28 days in saturated surroundings inside waterproof bags immersed in water to insulate the concrete both physically and thermally. The specimens are then conserved in a drying oven at 50°C for the dry specimens, or in water for the saturated specimens. A concrete sample is considered "100% dried" or "100% saturated" when the difference in its mass between two consecutive weighing during 24 h is less than 1%. The partially saturated samples are conserved in water and then a few days in air. The saturation degree of these samples is determined by regular weighing and a drying kinetic of the saturated sample during preparation.

3.2 Two other concretes (EC04 and EC08)

In the composition of a concrete, the water-cement ratio (E/C) plays a very important role as it represents the cement paste strength. In the aim of studying of the water-cement ratio effect, two other concretes (EC04 and EC08) have been formulated at water-cement ratios of 0,4 and 0,8 respectively. Both concretes EC04 and EC08 have the same cement paste volume (total volume of water and cement) and the same aggregate composition as the standard concrete R30A7 (E/C = 0.64). The composition and the mechanical properties of the concrete EC04 and EC08 are also presented in table 1. The procedure of specimen fabrication for these concretes is similar that of the dried concrete R30A7.

4 INFLUENCE OF SATURATION DEGREE

In this paper, compressive stresses and strains are assumed to be positive, some following symbols are used: σ_x is the principal stress, p is the pressure inside the cell $\sigma_m = \sigma_x + 2p$ is the mean stress, $q = \sigma_x - p$ is the principal stress difference (deviator), ε_x is the axial strain measured by LVDT (or axial strain gauge), ε_θ is the orthoraxial strain measured by orthoradial strain gauge and $\varepsilon_v = \varepsilon_x + 2\varepsilon_\theta$ is the volumetric strain.

A series of triaxial tests on dried and saturated concrete R30A7 samples have been carried out to study the influence of saturation degree. There are also a few tests which concern intermediate saturation degree. Because the drying kinetic of a saturated sample during its preparation occurs quickly, a saturated sample loses in general 20% of the water mass in its volume after 24 h of exposure to air. Because of the complex sample instrumentation procedure before a test, the preparation of a concrete sample with strain gauges glued on its surfaces requires at least 24 hours, so that the majority of the "completely" saturated samples become partially saturated samples with a saturation degree of approximately 80% at the beginning of the test. Some triaxial tests have been carried out on "completely" saturated samples without strain gauges.

The results of triaxial tests carried out on dried and partially saturated concrete samples at confining pressure of 0 (simple compression), 50, 100, 200, 400, 650 MPa are presented as principal stress versus strains (Figures 3a and 3b) and as mean stress versus volumetric strains (Figure 5).

In the figures 3a and 3b, the hydrostatic part of each test shows a reproducible and isotropic behaviour because the two components of strain (ε_x, ε_θ) are very close. For the dried sample, the load-carrying capacity increases significantly with the increase of confining pressure. Whereas the saturated samples seem to have a perfectly plastic behaviour regardless of the confining pressure.

Figure 5 shows a similar compaction behaviour between saturated and dried samples during the hydrostatic phase. For saturated samples after a significant compaction phase, an abrupt dilatancy appears when

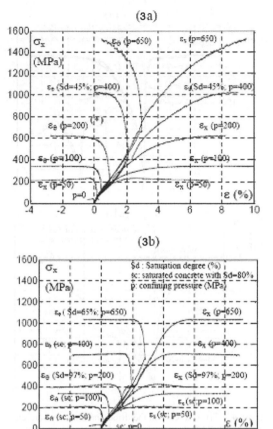

Figure 3. Triaxial tests on the dried concrete R30A7 (3a) and the partially saturated concrete R30A7 (3b) : stress/strain curves Sd: Saturation degree (Sd = 0% for "completely" dried concrete and Sd = 100% for "completely" saturated concrete); sc: saturated concrete with Sd = 80%; p: confining pressure in the triaxial test (MPa).

the maximal axial stress is reached. This is also the case for dried samples at low confining pressure, but at higher confining pressure, the stress level continues to increase after the appearance of the dilatancy.

Figures 4a and 4b show the deviatoric behaviour of both dried and saturated concrete. For confining pressure of 0 MPa, 50 MPa (Figure 4a), the dried concrete is stiffer than the saturated one. At 100 MPa confining pressure, the response of the saturated or dried samples are almost identical. This response may be associated with the response of granular stacking because the cement paste of both concretes may be almost destroyed at this confining pressure and the saturation degree of the saturated concrete is less than 100%. At a confining pressure of 200 MPa and above (Figure 4b), the strength of the dried samples strongly increases with confining pressure. This phenomenon

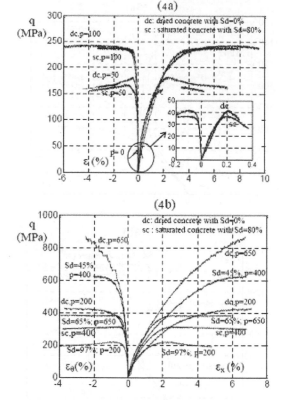

(4a)

(4b)

Figure 4. Triaxial tests on the dried and partially saturated concrete R30A7: principal stress difference versus strain curves with a confining pressure from 0 MPa to 100 MPa (5a), and from 200 MPa to 650 MPa (5b).

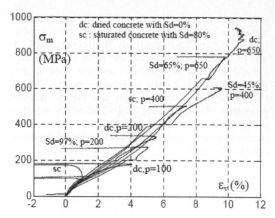

Figure 5. Triaxial tests on the dried and partially saturated concretes R30A7: volumetric behaviour.

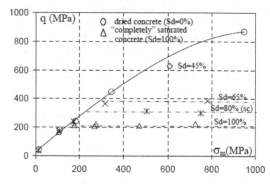

Figure 6. Limit states of the dried and saturated concrete in the stress invariant space (σ_m, q_{max}).

may be explained by the increase of the material density with confining pressure. For saturated samples, the maximum deviatoric stress increases only slightly under high confining pressure.

Figure 6 shows the maximum deviatoric stress versus the mean stress of all triaxial tests on dried and saturated concrete. The maximum deviatoric stress is associated with the dilatancy in most the cases except for the dried samples at higher pressure level. Some triaxial tests at a confining pressure of 200 MPa and above has been carried out on "completely" saturated concrete with only LVDT measurement, and the results do not show in increase in the maximum deviatoric stress. This phenomenon may be explained by a pore pressure effect in "completely" saturated concrete. The tests with intermediate saturation degrees show a limit stress state which is between the dried and the saturated samples, indicating that the limit state curve may be a function of the saturation ratio. At low pressure level, if there is still air in the concrete, the limit state of the partially saturated concrete is the same as that of dried concrete, whereas at higher pressure level if all

the dried porosity has been closed, it will behave as a saturated sample. This is an hypothesis which need to be confirm with more tests.

5 INFLUENCE OF WATER-CEMENT RATIO

To determine the influence of the water-cement ratio, a series of uniaxial compression tests (triaxial tests without confining pressure) and triaxial tests with confining pressures of 100 MPa and 650 MPa on concretes EC08 (E/C = 0,8), R30A7 (E/C = 0,64), EC04 (E/C = 0,4) have been performed. Results from the uniaxial compression tests are shown as stress/strain curves in Figure 7. Age of the concretes EC08, R30A7, EC04 are 203, 197, 197 days respectively. The uniaxial compression test results show that the higher the ratio, the more weak and porous the concrete.

Results from the triaxial tests with confining pressure of 650 MPa are also shown as stress/strain curves in figure 9a and as principal stress difference/strain curves in figure 9b. The hydrostatic part of figure 9a

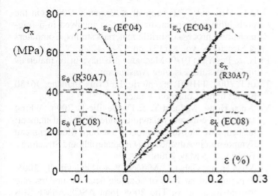

Figure 7. Uniaxial compression tests.

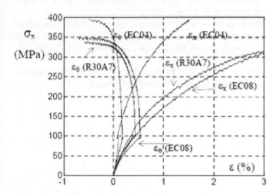

Figure 8. Triaxial tests with p = 100 MPa.

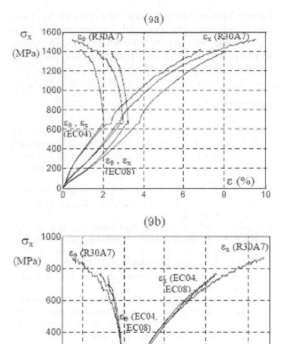

Figure 9. Triaxial tests on the concretes EC08, R30A7, EC04 with p = 650 MPa: stress/strain curves (a), principal stress difference/strain curves (b).

shows that the higher the ratio, the more the concrete compacts under hydrostatic stress. But what is more surprising is that the deviatoric responses of the three different concretes (Figure 9b) seem to be very close for each component (ε_x and ε_θ).

At the confining pressure of 100 MPa, Figure 8 also shows that the deviatoric responses of the concretes EC08 and R30A7 are also very close, as for the confining pressure of 650 MPa. This can be explained by the quasi-destruction of the cement paste matrix under high confinement (100 MPa and more), resulting a deviatoric response of both concretes similar to that of granular stacking. On the other hand, Figure 8 shows the concrete EC04 is harder than the others during both hydrostatic and deviatoric loading. This can be explained by the a good quality of cement paste which means that the cement paste matrix is not yet completely destroyed at the confining pressure of 100 MPa and therefore the concrete behaves as a matrix of cement paste and granular. These results confirm that the concrete behaviour at low confining pressure is strongly linked to the cement paste strength, whereas the behaviour is more like a granular stacking under high confinement.

6 CONCLUSION

The test results show that the saturation degree has a major influence on the concrete deviatoric behavior of concrete. The strength of dried concrete strongly increases with the confining pressure whereas it is limited for saturated samples. Some supplementary tests on partially saturated concrete will be carried out to confirm that a limit state curve may be a function of the saturation ratio.

In addition, the results concerning the water-cement ratio of the concrete mixture confirm that the concrete behavior at low confining pressure is strongly linked to the cement paste strength whereas it behaves more like a granular stacking under high confinement.

ACKNOWLEDGMENTS

The GIGA press was implemented in Laboratoire 3S in the framework of a cooperation agreement with the French Délégation Générale pour l'Armement (DGA,

French Ministry of Defence). This research has been developed with the financial support of the Centre d'Études Techniques de Gramat (CEG, DGA).We would like to thank Dr. Éric Buzaud (CEG) for giving technical and scientific advice.

REFERENCES

Baroghel-Bouny V., Mainguy M., Lassabatere T. & Coussy O. 1999. Characterization and identification of equilibrium and transfer moisture properties for ordinary and high-performance cementitious materials, Cem Concr Res 29, pp. 1225–1238.

Bischoff P.H. & Perry S.H. 1991. Compressive bahaviour of concrete at high strain rates, Materials and Structures, vol. 24, p. 425–450.

Burlion N. 1997. Compaction des bétons: éléments de modélisation et caractérisation expérimentale, PhD thesis, ENS Cachan, France.

Burlion N., Bourgeois F. & Shao J.-F. 2005. Effect of desiccation on mechanical behaviour of concrete, Cem. Concr. Comp. 27, pp. 367–379.

Gabet T., Malecot Y. & Daudeville L. 2006. Ultimate strength of plain concrete under extreme combined stresses : triaxial and proportional stress paths, Revue Europénne de Génie Civil vol.10 No 3, pp. 375–390.

Hild F., Denoual C., Forquin P. & Brajer X. 2003. On the probabilistic-deterministic transition involved in a fragmentation process of brittle materials strains, Computers & Structures, vol. 81(12), pp. 1241–1254.

Li H. & Pugh D. 1970. Mechanical behaviour of materials under pressure, Elsevier, Amsterdam.

Thiot P. 2004. THIOT Ingénierie, La Croix Blanche, 46130 Saint Michel Loubejou, France.

Vu Xuan Hong, Malecot Y. & Daudeville L. 2006. A First analysis of concrete behaviour under high confinement: influence of moisture content, First Euro Mediterranean Symposium in Advances on Geomaterials and Structures, Tunisia., 3–5 May, 2006.

Vu Xuan Hong, Gabet T., Malécot Y. & Daudeville L. 2005. Experimental analysis of concrete behavior undersevere triaxial loading ≫, The 2005 Joint ASCE/ASME/SES Conference on Mechanics and Materials, McMat 2005 Mechanics and Materials Conference Baton Rouge, Louisiana, 1–5 June, 2005.

Zhao H. & Gary G. 1996. On the use of SHPB techniques to determine the dynamic behavior of materials in the range of small strains, Int. J. Solids Structures, vol. 33(23), pp. 3363–3375.

Zukas J.A. 1992. Penetration and perforation of solids, Impact Dynamics, Krieger Publishing Company.

Fracture Mechanics of Concrete and Concrete Structures – High-Performance Concrete,
Brick-Masonry and Environmental Aspects – Carpinteri, et al. (eds)
© 2007 Taylor & Francis Group, London, ISBN 978-0-415-44617-4

Self-healing of cracked concrete: A bacterial approach

H.M. Jonkers & E. Schlangen

Delft University of Technology, Delft, The Netherlands

ABSTRACT: Crack occurrence in reinforced concrete should be minimized for both durability and economical reasons as crack repair is costly. Autogenous repair, or self-healing, of concrete would save a substantial amount of money, as manual inspection and crack repair could be minimized. Thus, a reliable self-healing mechanism for concrete would not only result in more durable structures, but would also be beneficial for the global economy. This study exploited the potential to apply calcite-precipitating bacteria as a crack-healing agent in concrete. The potential of different species to precipitate calcite, produce endospores, survive concrete-production, and heal cracks by sealing them with calcite was investigated. Furthermore, the mechanical properties of 'bacterial concrete' were tested. ESEM studies showed that alkali-resistant spore-forming bacteria embedded in the concrete matrix can precipitate substantial amounts of calcite. The bacterial approach thus seems a highly promising mechanism to mediate self-healing in concrete structures.

1 INTRODUCTION

Cracks can occur in concrete structures due to multiple reasons such as autogenous shrinkage, freeze-thaw reactions, mechanical compressive- and tensile forces. Although micro-cracks do not necessarily result in significant strength loss of concrete, the ingress of water and other reactive chemicals such as chloride and water may pose a thread to the steel reinforcement as these strongly enhance its corrosion rate. Thus for durability reasons and potential repair costs, crack occurrence should be minimized or, alternatively, occurring cracks should ideally be healed directly after formation by an autonomous repair mechanism. Different autonomous repair systems are feasible. One such a self-healing mechanism could involve secondary hydration reactions of still present but not fully reacted cement particles. Although a high percentage of non-reacted cement particles within its matrix may result in a concrete with a substantial self-healing capacity, the material characteristics of the initial concrete structure may not be satisfactorily as it may be more brittle and initially weaker as wanted. Another self-healing mechanism could be based on the addition of a self-healing agent that would make up a part of the concrete matrix without or insignificantly affecting its structural and mechanical characteristics. In this study the potential of bacteria to act as a self-healing agent in concrete is investigated. Although the idea to use bacteria and integrate them in the concrete matrix may seem odd at first, it is not from a microbiological viewpoint. Bacteria naturally occur virtually everywhere on earth, not only on its surface but also deep within, e.g. in sediment and rock at a depth of more than 1 km (Jorgensen & D'Hondt 2006). Various species of so-called extremophilic bacteria, i.e. bacteria that love the extreme, are found in highly desiccated environments such as deserts (Dorn & Oberlander 1981; DeLaTorre et al 2003), but also inside rocks (Fajardo-Cavazos & Nicholson 2006) and even in ultra-basic environments (Pedersen et al 2004; Sleep et al 2004) which can be considered homologous to the internal concrete environment. Typical for many desiccation- and/or alkali-resistant bacterial species is their ability to form endospores. These specialized cells are characterized by an extremely low metabolic activity, are known to be able to resist high mechanically- and chemically induced stresses (Sagripanti & Bonifacino 1996) and are viable for periods up to 200 years (Schlegel 1993). In some previously published studies the application of bacteria for cleaning of concrete surfaces (DeGraef et al 2005) and strength improvement of cement-sand mortar (Ghosh et al 2005) was reported. Furthermore, in some studies the crack-healing potential by mineral-precipitating bacteria on degraded limestone (Dick et al 2006) and ornamental stone surfaces (Rodriguez-Navarro et al 2003) as well as on concrete surfaces (Bang et al 2001; Ramachandran et al 2001) was investigated and reported. Although promising results were reported, the major drawback of the latter studies was that the bacteria and compounds needed for mineral precipitation could only be applied externally on the surface of the structures after crack-formation had occurred.

This methodological necessity was mainly due to the limited lifetime (hours to a few days) of the (urease-based) enzymatic activity and/or viability of the applied bacterial species. In the present study the application of alkali-resistant spore-forming bacteria to enhance the self-healing capacity of concrete is investigated. Tensile- and compressive strength characteristics of reference (no bacteria added) and bacterial concrete are quantified. Furthermore, the viability of bacteria immobilization in concrete is quantified and, finally, calcite precipitation potential of bacterial concrete is demonstrated by ESEM analysis.

2 METHODS

2.1 Cultivation of alkali-resistant spore-forming bacteria

Four strains of alkaliphilic spore-forming bacteria were purchased from DSMZ (German Collection of Microorganisms and Cell Cultures), Braunschweig, Germany: *Sporosarcina pasteurii* DSM 33; *Bacillus cohnii* DSM 6307; *Bacillus halodurans* DSM 497 and *Bacillus pseudofirmus* DSM 8715 and cultivated according to the suppliers recommendations (medium DSMZ-2 for *S.pasteurii* and DSMZ-31 for the others).

Endospore-forming potential was determined in mineral medium. This medium contained per liter of Milli-Q ultra pure water: $0.2\,g\,NH_4Cl$, $0.02\,g\,KH_2PO_4$, $0.225\,g\,CaCl_2$, $0.2\,g\,KCl$, $0.2\,g\,MgCl_2.6H_2O$, 1 ml per liter trace elements solution SL12B, 0.1 g yeast extract, 6.45 g citric acid trisodium salt and 8.4 g sodium bicarbonate. The pH of this medium was 9.2. Aerobic batch cultures were incubated in 2-l Erlenmeyer flasks on a shaker table at 150 rpm. Growth was monitored by microscopy and cell numbers and percentage of sporulating cells were quantified by microscopy using a Burger-Turk counting chamber.

2.2 Preparation and strength characteristics of bacterial concrete

Concrete bars with and without (control) added bacteria were prepared for tensile- and compressive strength determination. The aim of these tests was to check whether the strength of the concrete was not negatively affected by the bacteria. Firstly, for the preparation of bacterial concrete, a dense culture of *S.pasteurii* was obtained after growth in medium DSMZ-2. Total cell number was quantified by microscopy using a Burger-Turk counting chamber. Subsequently, cells were washed twice by centrifugation (20 min × 10000 g) and resuspension of the cell pellet in tap water. Washed cells were finally resuspended in a 20-ml aliquot of tap water. This cell suspension was applied as part of the needed water for concrete bar preparation.

Table 1. Cement, water and aggregate composition needed for the production of 9 concrete bars of dimensions 16 × 4 × 4 cm used for tensile- and compressive strength characterization of bacterial- and control (no bacteria added) concrete. For bacterial concrete, the 20-ml cell suspension was part of the total water volume needed.

Compound/Aggregate size (mm):	Weight (g):
Cement (ENCI CEM I 32.5)	1170
Water	585
Aggregate size fraction:	
4–8	1685
2–4	1133
1–2	848
0.5–1	848
0.25–0.5	730
0.125–0.25	396

Concrete bars for tensile- and compressive strength determination were prepared as follows. Two sets (bacterial concrete and control concrete without bacteria) of nine bars each (bar dimensions 16 × 4 × 4 cm) were made using ordinary portland cement (ENCI CEMI 32.5R), a water-cement ratio of 0.5 and aggregate composition (sand and gravel) as listed in Table 1. The bars were initially cured for 24 hours in plastic foil-sealed molds at room temperature, subsequently uncased and further cured in tap water-filled separate plastic containers at room temperature. Subsets of three bars each were tested for flexural tensile- and compressive strength after 3, 7 and 28 days curing following the procedure according to EN 196-1 Standard Norm.

2.3 Viability of concrete-immobilized spores

The viability (ability to germinate) of spores of the alkaliphilic bacterial species *B.cohnii*, *B.halodurans* and *B.pseudofirmus* immobilized in cement stone was determined. Cultures of the respective species were firstly grown in mineral medium (see above). These cultures were washed twice by centrifugation (20 min × 10000 g) and resuspension of the cell pellet in tap water after the number of spores formed was quantified by microscopy. Obtained spore suspensions were divided in two parts, one part was stored in a fridge at 4°C and served as non-concrete immobilized control for determination of spore viability during storage (see below), and one part was used for cement stone sample preparation. For the latter, the spore suspension was used as part of the make up water, and bacterial and control (no bacterial spores added) cement stone specimen were prepared. Number of spores added to cement stone as determined by microscopic counting was $10^9\,cm^{-3}$. Ordinary portland cement (ENCI CEMI 32.5R) and a water-cement ratio of 0.5 was used for the preparation of cement stone disks (4 cm diameter, 1 cm height), cast in plastic vials closed with a plastic lid. After 24 hours curing at room temperature, disks were further cured in tap water at room temperature. Disks were removed from the plastic vial molds after ten days curing, chipped to pieces with a chisel and further crushed to powder using a robust pharmaceutical stone mortar. Powdered cement stone (1.84 g representing 1 cm^3, containing 10^9 spores) was subsequently slurried and diluted 10-fold by addition of 9 volumes sterile mineral medium. In parallel, the endospore-containing cell suspensions which were kept in a fridge at 4°C were diluted to a spore density of 10^9 ml^{-1} and also 10-fold diluted by addition of 9 ml sterile mineral medium. Cement stone slurries and original spore suspensions were further homogenized by three cycles of vigorous mixing at 2500 rpm and 20 seconds ultrasonic treatment in a Branson 1210, 47 KHz, 80 Watts Ultrasonic bath. Number of viable spores in cement stone slurries and spore suspensions was estimated according to the Most Probable Number (MPN) dilution technique. For this procedure, 8×12 wells sterile microtiter plates were filled with mineral medium, 180 μl per well. Four consecutive wells of the first row were inoculated with 20-μl slurry or control cell suspension aliquots, and these were subsequently serially diluted in ten-fold dilution steps up to the 10^{11} dilution level, leaving the last (12th) row as non-inoculated control to check for medium contamination. Thus, each cement stone slurry and corresponding control cell suspension was serially diluted in four parallel series. During the following incubation period at room temperature, growth occurred in the lower but not in the higher dilution levels due to dilution-to-extinction of the viable cells present in the samples. Growth could easily be determined visually due to increased turbidity of positive wells during the following 2-weeks incubation period. Viable number of cells in cement stone slurries and their corresponding spore suspensions was calculated from the number of positive wells using the MPN computer program of Clarke and Owens (1983).

2.4 Calcite precipitation potential of bacterial concrete

Chips of 10 days cured cement stone samples (see under 2.3) were incubated in rich medium (yeast-extract and peptone based medium) after pasteurizing for 30 min at 70°C. Pasteurization of bacterial and control cement stone chips before incubation was done to inactivate bacteria that potentially came into contact with the cement stone samples during curing period or non-sterile handling of the cement stone samples and chips after curing. As bacterial endospores are not killed by the pasteurization procedure, this treatment ensured that potential differences between control and bacterial concrete samples after incubation were

mediated by added bacteria and not by accidentally introduced contaminants. Rich medium contained 5 g peptone, 3 gram yeast extract and 8.4 g sodium bicarbonate and had a pH of 8.6. Individual chips were incubated aerobically in 100-ml medium aliquots on a shaker table at 100 rpm at 25°C for 12 days. Chips were rinsed with tap water after incubated and stored wet in closed plastic vials until ESEM analysis, what was done within two days after incubation without any further treatment. Chips were mounted on a 1-cm^2 metal support and kept in place with adhesive tape and observed with a Philips XL30 Series Environmental Scanning Electron Microscope.

3 RESULTS

3.1 Cultivation of alkali-resistant spore-forming bacteria

Three out of four strains produced copious spores in mineral medium, except for *S.pasteurii*, which did not grow in this medium. Spore production was considerably less in rich yeast extract- and peptone-containing medium. Percentage of cells with endospores was quantified by microscopic counting, and amounted to 75, 50 and 25% for *B.cohnii*, *B.halodurans* and *B.pseudofirmus* respectively.

3.2 Strength characteristics of bacterial concrete

The 20-ml washed cell suspension of the *S.pasteurii* culture used for the making of bacterial concrete bars contained $3.48 * 10^{12}$ cells, what resulted in a final density of $1.14 * 10^9$ cells cm^{-3} concrete. As the average volume of an *S.pasteurii* cell equals about 2.5 μm^3, the total cell volume amounts to 0.3% of the bacterial concrete volume. Tensile- and compressive strength tests after 3, 7 and 28 days curing revealed no significant difference between control- and bacterial concrete (Fig 1).

3.3 Viability of cement stone-immobilized spores

The number of viable spores in cement stone samples after 10 days curing as well as original spore suspensions, both with a spore density of 10^9 cm^{-3}, were estimated (Table 2). Results revealed that about one percent of the spores in the spore suspensions (10^7ml^{-1}) could be retrieved as viable (Table 2). The number of viable spores in the corresponding cement stone samples appeared significantly lower, i.e. between 10^5 and 10^6 cm^{-3}. Compared to spore suspensions, estimated viable spores in cement stone slurries amounted to 1.9, 7.0 and 2.0% for *B.halodurans*, *B.pseudofirmus* and *B.cohnii* respectively. Number of viable bacteria in control cement stone samples (no bacterial spores added) and tap water used for concrete

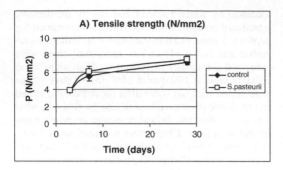

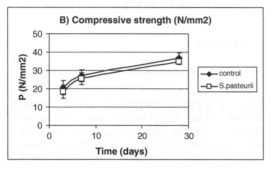

Figure 1. Flexural tensile- (A) and compressive- (B) strength testing, after 3, 7 and 28 days curing, revealed no significant difference between control- and bacterial concrete. The latter contained $1.14 * 10^9$ S.pasteurii cells cm^{-3} concrete.

sample preparation were below detection limit (<500 cells cm^{-3}).

3.4 Calcite precipitation potential of bacterial concrete

ESEM analysis revealed that bacterial cement stone, in contrast to control cement stone samples, precipitated substantial amounts of calcite-like crystals on its surface when incubated in peptone- and yeast extract containing medium. An example is shown in Figure 3, depicting control and B.pseudofirmus endospore-containing cement stone pieces incubated for 12 days.

4 DISCUSSION AND CONCLUSIONS

Application of self-healing concrete, i.e. concrete that is able to repair, seal or plug newly formed cracks autogenously, will not only result in more durable structures but will also save a significant amount of money as labor intensive check and repair can be minimized. In this study we investigated the potential of bacteria-mediated calcium carbonate production as a possible healing mechanism. In contrast to

Table 2. Estimate of number of viable (cultivable) bacterial spores in spore suspension and cement stone in which spores of respective bacterial species were immobilized. In brackets: confidence interval; In square brackets: percentage of number in spore suspension

Spore suspension and tap water (control):	Number cm^{-3}:
B.cohnii	5.73 E7 (1.76–18.58)
B.halodurans	5.63 E6 (1.74–18.17)
B.pseudofirmus	7.98 E6 (2.63–24.24)
Tap water (control)	<500
Cement stone samples:	
B.cohnii	1.15 E6 (3.80–34.80) [2.0]
B.halodurans	1.07 E5 (0.36–3.20) [1.9]
B.pseudofirmus	5.62 E5 (1.74–18.14) [7.0]
Control	<500

some previous studies where bacteria were externally applied for concrete and monument crack repair (Bang et al. 2001; Ramachandran et al. 2001; Dick et al. 2006; Rodriguez-Navarro et al. 2003), we here incorporated bacterial spores, i.e. dormant or resting cells, in the concrete matrix. The results of our study are promising. The estimated number of viable spores retrieved from young cement stone, i.e. after ten days curing, was between 1.9 and 7.0% of the number of viable spores present in the original spore suspension used for the preparation of cement stone samples.

These numbers are substantial, considering the mechanical forces (grinding) needed to liberate and suspend the cement stone-immobilized bacterial spores. Moreover, even if the percentages retrieved reflect truly viable spores, absolute numbers are still high, i.e. between 1.7 and $7.5*10^7$ spores cm^{-3} cement stone, realizing that one viable cell is theoretically enough to start microbial growth and calcite precipitation, providing that suitable conditions prevail. Incubation experiments with 10-days cured cement stone samples demonstrated the mineral precipitation potential of bacterial concrete. We hypothesize that the actual bacterial mineral precipitation mechanism is as follows. Once into contact with copious amounts of water and growth substrates (yeast extract and peptone), bacterial endospores germinate and start to produce CO_2 due to metabolic turnover of growth substrates. CO_2, what can locally reach high concentrations due to rapid metabolic conversion of organic compounds, will chemically react with $Ca(OH)_2$ produced from C_2S and C_3S hydration reactions. The $Ca(OH)_2$ that leaks out of the concrete's pore system reacts with CO_2 and precipitates as calcite or any other calcium carbonate based mineral. The calcite-like crystals found on the surface of bacterial but not on the surface of control cement stone samples support this hypothesis.

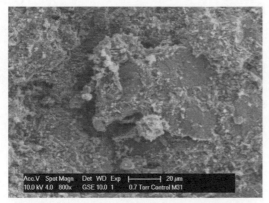

A: Control

B: + Bacteria

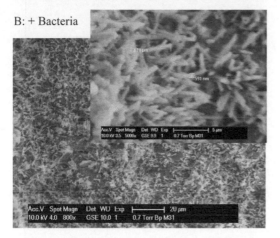

Figure 2. Concrete samples incubated in yeast extract- and peptone-containing medium. A: Control (concrete with no bacteria added) and B: Concrete containing 10^9 cm^{-3} *B.pseudofirmus* endospores. The inset in Figure 2B (5000x magnification) shows a close up of the massive calcite-like crystals formed on the concrete surface.

The experiments done in this study show that alka-liphilic endospore-forming bacteria integrated in the concrete matrix can actively precipitate calcium carbonate minerals. Water, needed for the activation of endospores, can enter the concrete structure through freshly formed cracks. Furthermore, for mineral pre-cipitation, active cells need an organic substrate that can metabolically be converted to inorganic carbon what can subsequently precipitate with free calcium to calcium carbonate. Free calcium is usually present in the concrete matrix, but organic carbon is not. In the present experiments organic carbon was applied externally as a part of the incubation medium, while ideally it should also be part of the concrete matrix.

In that case only external water is needed to activate the concrete-immobilized bacetria which can then con-vert organic carbon present in the concrete matrix to

calcium carbonate and by doing so seal freshly formed cracks. We currently investigate which specific kind of organic compounds are suitable to include in the concrete matrix. This is certainly not trivial as such compounds should be a suitable food source for bac-teria as well as be compatible with concrete. Certain classes of organic compounds are less- or not suitable at all, e.g. compounds such as carbohydrate deriva-tives that are known to inhibit the setting of concrete even at low concentrations. We furthermore presently investigate the long-term viability and potential pos-sibilities to increase the viability of concrete immo-bilized endospores to ensure long-lasting bacterially enhanced self-healing. Other ongoing investigations address the possible decrease in concrete permeability and the change of mechanical characteristics of healed cracked concrete due to bacterial calcite precipitation.

To conclude we can state that the bacterial approach has potential to contribute to the self-healing capacity of concrete. We have shown that bacteria incorporated in high numbers (10^9 cm^{-3}) do not affect concrete strength, that a substantial number of added bacteria remain viable and, moreover, that these viable bacte-ria can precipitate calcium carbonate needed to seal or heal freshly formed cracks.

ACKNOWLEDGEMENTS

Financial support from the Delft Center for Materi-als (DCMat: www.dcmat.tudelft.nl) for this study is gratefully acknowledged.

REFERENCES

Bang, S.S., Galinat, J.K., Ramakrishnan, V. 2001. Calcite pre-cipitation induced by polyurethane-immobilized Bacillus pasteurii. Enzyme and Microbial Technology 28:404–409.

Clarke, T.R. & Owens, N.J.P. 1983. A simple and versatile micro-computer program for the determination of 'most probable number'. J. Microbiol. Meth. 1: 133–137.

De Graef, B., De Windt, W., Dick, J., Verstraete, W., De Belie, N. 2005. Cleaning of concrete fouled by lichens with the aid of Thiobacilli. Materials and Structures 38(284): 875–882.

De la Torre, J.R., Goebel, B.M., Friedmann, E.I., Pace, N.R. 2003. Microbial diversity of cryptoendolithic communi-ties from the McMurdo Dry Valleys, Antarctica. Appl. Environm. Microbiol. 69(7):3858–3867.

Dick, J., De, Windt, W., De Graef, B., Saveyn, H., Van der Meeren, P., De Belie, N., Verstraete, W. 2006. Bio-deposition of a calcium carbonate layer on degraded limestone by Bacillus species. Biodegradation 17(4): 357–367.

Dorn, R.I. & Oberlander, T.M. 1981. Microbial origin of Desert Varnish. Science 213:1245–1247.

Fajardo-Cavazos, P. & Nicholson, W. 2006. Bacillus endospores isolated from granite: Close molecular rela-tionships to globally distributed Bacillus spp. from endolithic and extreme environments. Appl. Environm. Microbiol. 72(4):2856–2863.

Ghosh, P., Mandal, S., Chattopadhyay, B.D., Pal, S. 2005. Use of microorganism to improve the strength of cement mortar. Cement and Concrete Research 35(10):1980–1983.

Jorgensen, B.B. & D'Hondt, S. 2006. A starving majority deep beneath the seafloor. Science 314:932–934.

Pedersen, K., Nilsson, E., Arlinger, J., Hallbeck, L., O'Neill, A. 2004. Distribution, diversity and activity of microorganisms in the hyper-alkaline spring waters of Maqarin in Jordan. Extremophiles 8(2):151–164.

Ramachandran, S.K., Ramakrishnan, V., Bang, S.S. 2001. Remediation of concrete using micro-organisms. ACI Materials Journal 98(1):3–9.

Rodriguez-Navarro, C., Rodriguez-Gallego, M., Ben Chekroun, K., Gonzalez-Munoz, M.T. 2003. Conservation of ornamental stone by Myxococcus xanthus-induced carbonate biomineralization. Appl. Environm. Microbiol. 69(4): 2182–2193.

Sagripanti, J.L. & Bonifacino, A. 1996. Comparative sporicidal effects of liquid chemical agents. Appl. Environm. Microbiol. 62(2):545–551.

Schlegel H.G. 1993. *General microbiology*, 7th edition. Cambridge University Press.

Sleep, N.H., Meibom, A., Fridriksson, T., Coleman, R.G., Bird, D.K. 2004. H-2-rich fluids from serpentinization: Geochemical and biotic implications. PNAS 101(35):12818–12823.

Fracture Mechanics of Concrete and Concrete Structures – High-Performance Concrete,
Brick-Masonry and Environmental Aspects – Carpinteri, et al. (eds)
© 2007 Taylor & Francis Group, London, ISBN 978-0-415-44617-4

Mechanical behavior of self-healed Ultra High Performance Concrete: From experimental evidence to modeling

S.Granger, G.Pijaudier-Cabot & A.Loukili

ERT R&DO, GeM, Ecole Centrale de Nantes, France

ABSTRACT: Self healing of cracks, in an ultra high performance concrete, is investigated in this paper, and especially the role of the phenomenon on mechanical properties. An experimental program is thus developed in order to characterize the mechanical behavior of prismatic specimens, initially cracked and then submitted to self healing, by total immersion in water. The most significant results are a fast recovery of global stiffness and a slight improvement of flexural structural resistance. Microscopic investigations are also proposed to qualify the nature of newly-formed crystals that precipitate in the crack. Then, a first approach of modeling of the mechanical behavior of concrete specimens, including the self healing process, is proposed. A coupling between hydration and elastic-damage models is thus developed and permits to get first qualitative results showing the experimental tendencies.

1 INTRODUCTION

Under special conditions and without any external intervention of repair, the phenomenon of self healing of cracks can appear and act positively on the durability and serviceability problems of concrete structures. The phenomenon is only possible in presence of water (dissolved CO_2 is not always needed) and consists of chemical reactions of compounds present on the crack surfaces. There are two major hypotheses regarding the chemical reactions (Neville, 2002): the hydration of anhydrous cement available in the microstructure of hardened concrete (especially for concrete with low Water to Cement (W/C) ratio), and the precipitation of calcium carbonate $CaCO_3$ (also called calcite) after the dissolution of portlandite (especially for concrete with high W/C ratio).

The majority of research works carried out on this topic highlights the phenomenon by means of water permeability tests. A decrease of flow rate through cracked specimens is the main method to show the self healing of cracks. Such tests have been carried out on concretes with high W/C ratio by Edvardsen (1999) or Hearn & Morley (1997), who show the precipitation of new calcite crystals. The influences of temperature and crack width have also been investigated (Reinhardt & Joos, 2003). The role of the phenomenon on transfer properties (see also, Jacobsen et al., 1996a, for chloride migration) has thus been fully characterized.

Concerning the mechanical impact, a few studies have been conducted. Jacobsen et al. (1996b)

have shown a substantial recovery of frequency resonance on concrete cubes damaged by freeze/thaw cycles and then stored in water, but only a small recovery of compressive strength. SEM investigations have shown the precipitation of new C-S-H crystals (Jacobsen et al., 1995). Pimienta & Chanvillard (2004) also provide some insights about the mechanical properties of healed specimens. The authors reported that the frequency resonance of specimens damaged and then aged in water, tends to recover its initial value. Nevertheless, knowledge about the mechanical properties is scarce, and this contribution aims at providing some new insights about the mechanical behavior of healed concrete specimens.

An experimental program is developed on an ultra high performance concrete. Prismatic notched specimens are cracked under three-point bending (different types of residual crack width) and then totally immersed in water for different ageing times. After this ageing phase, the mechanical behavior of the healed specimens is characterized by means of three-point bending tests, and compared with the mechanical behavior of non healed cracked specimens. This characterization is completed by microscopic investigations in the zone of the pre-existing crack. All these results enable to provide a first approach of modeling of the mechanical behavior of the healed specimens, by coupling hydration and elastic-damage models. Simulations of the threepoint bending tests are thus proposed.

2 THE MECHANICAL CHARACTERIZATION

2.1 Concrete specimens

The experimental program is carried out on an ultra high performance concrete (UHPC). This concrete is characterized by a low W/C ratio, close to 0.2. This implies that the amount of anhydrous clinker in the microstructure is very high, in the order of 50% (Loukili et al., 1998). The composition of this concrete is composed of sand, water, cement, silica fume and a superplasticizer, but no coarse gravels. This UHPC has a quite homogenous microstructure with a high amount of anhydrous clinker, and thus a high potential for self healing by hydration of this cement.

In order to have a localized crack during the mechanical tests, a notch of depth 20 mm and thickness 1.5 mm is performed in each specimen (dimensions 50 × 100 × 500 mm). After casting, the concrete specimens are cured for 2 days at 20°C and 100% relative humidity. A thermal treatment is then applied, so as to accelerate hydration, to activate the pozzolanic reaction, and to get chemically stable concrete. The specimens are placed in a climate chamber with a controlled environment of 90°C and 100% relative humidity during 48 hours.

2.2 Mechanical tests

During the first phase of the mechanical program, specimens are loaded under three-point bending in order to be cracked (Figure 1). The tests are crack opening controlled with a constant rate of 0.05 μm/s. The aim of this first step is to get a controlled cracking of the specimens. Pre-cracking is performed in the post peak regime: after having reached the peak load, specimens are unloaded at different stages in order to get residual crack widths of respectively 10, 20 and 30 μm. This unloading is also crack opening controlled with the same rate as loading.

After this first step, the specimens are stored in specific conditions for ageing. There are two kinds of ageing: in air at 20°C and 50% relative humidity, and in total immersion in tap water, without movement neither renewal. For specimens cracked with residual crack widths of 10 μm, the different periods of ageing are 1, 3, 10, 20 and 40 weeks, in order to analyze the influence of ageing time. The influence of cracked width is studied with the complementary results of specimens cracked at 20 and 30 μm, and aged for 10 and 20 weeks.

After this ageing phase, the last step of the experimental program consists in reloading the specimens under three-point bending, so as to characterize their residual mechanical behavior. Tests are also crack opening controlled and conducted until total failure.

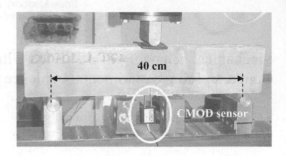

Figure 1. Mechanical test configuration.

2.3 Mechanical behavior of aged specimens

Figure 2 presents the mechanical behavior of specimens that have been aged in water for 1, 3, 10 and 20 weeks, after having been pre-cracked with a residual crack width of 10 μm. These are average curves (three tests for each kind of ageing). The initial value of the crack opening has been shifted to zero, in order to have the same initial state, but in reality there is still the value of the residual crack opening. These results are compared to the average mechanical behavior of specimens stored in air, which is the same as those of non aged cracked specimens (Granger et al., 2006). The curves display the crack opening versus the ratio between load applied and load while unloading in the pre-cracking phase.

These results show the evolution of the mechanical behavior with the time of storage in water. The initial reloading stiffness is not the same as for non aged specimens, and it increases with the ageing time. There is also a slight improvement of flexural strength and a change in stiffness (limit of the elastic phase) in the pre peak regime, which evolves with time, and should be associated with an evolution of the mechanical characteristics of the healed zone of the specimen.

Figure 3 represents the evolution of the ratio between the reloading stiffness with healing, and the reloading stiffness without healing, as a function of the ageing time.

It is noticeable that there is a fast recovery of the structural stiffness by self healing, and this stiffness tends to the one of healthy specimens, which is represented by the straight line on the graph. Figure 4 now shows the evolution of the ratio between peak load after ageing and load measured upon unloading prior to ageing. We can thus notice that there is a slight improvement of the flexural resistance in comparison with the healthy specimens, and that the resistance of initial undamaged specimens can not be achieved after self healing.

The influence of the crack opening has also been investigated. Figure 5 shows the global stiffness during the reloading phase as a function of the residual crack

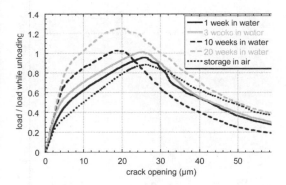

Figure 2. Mechanical behavior of aged cracked specimens.

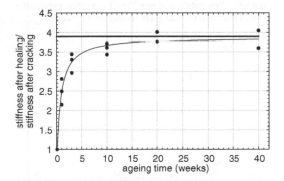

Figure 3. Evolution of the ratio between the stiffness with healing and the stiffness without healing – Comparison with the average ratio of healthy specimens.

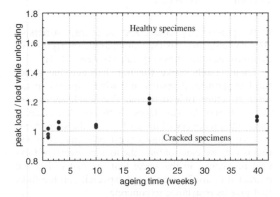

Figure 4. Evolution of the ratio between peak load and load while unloading prior to ageing, for healed specimens, in comparison with cracked and healthy ones.

opening got at the end of the pre-cracking phase. The results for 10 and 20 weeks of ageing are presented.

So, there is a clear influence of the residual crack opening on the stiffness while reloading. We can thus notice, like on figure 3, that, for specimens cracked at

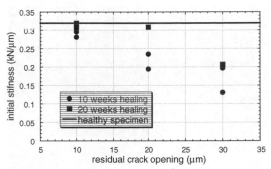

Figure 5. Evolution of the global stiffness for different residual crack openings, compared with the average stiffness of healthy specimens.

10 μm, the recovery of stiffness is complete. In comparison, this is not the case for specimens cracked at 30 μm, even for 20 weeks ageing which is quite a long period for the concrete studied.

2.4 The precipitation of crystals

The two most important mechanical results presented in the previous section, are attributed to the phenomenon of self healing, by precipitation of new crystals in the crack. An acoustic emission analysis of the cracking processes of the specimens, during the reloading phase after ageing, has been conducted (Granger et al., 2006). It shows that the damage of specimens stored in water begins sooner than those of non healed beams, and that the micro-cracks detected are located in the zone of the pre-existing crack. The analysis of the dissipated acoustic energies also shows that the newly formed crystals seems to be less resistant than C-S-H from the first hydration. In order to get more information about their nature, and to make a link with the mechanical experimental results, microscopic investigations are carried out. Figure 6 represents a polished section of the cement paste, containing a crack, so at the end of the pre-cracking phase of the specimens, without healing.

We can thus notice the presence of anhydrous clinker in the microstructure, which appears as white particles on figure 6 (the black ones are sand grains, and the hydration products are in grey). As already said, this illustrates the high potential of this cement for self healing. Moreover, the crack propagates in the cement paste, fracturing anhydrous grain in their whole volume or putting grains available on the crack surface. We can see this in detail on figure 7.

Similar investigations are then carried out on healed specimens. Figures 8 and 9 show that, locally, either in fractured cement grains or in the cement paste, the continuity between the two lips of the crack can be re-established. New crystals have thus precipitated

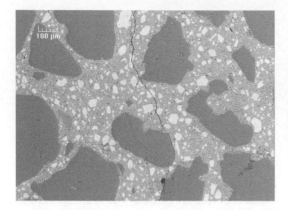

Figure 6. Cement paste of the UHPC with a crack (enlargement ×100).

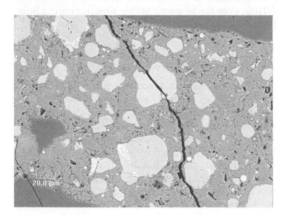

Figure 7. Propagation of the crack in and near cement grains (enlargement ×500).

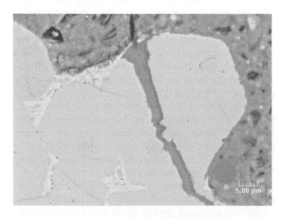

Figure 8. Precipitation of new C-S-H in a crack fracturing an anhydrous clinker grain (enlargement ×2000).

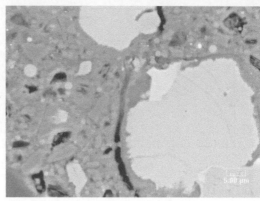

Figure 9. Precipitation of new C-S-H in a crack, outside the zone of fractured cement grain (enlargement ×2000).

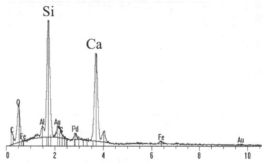

Figure 10. Result of the energy dispersive spectrometry analysis on the new crystals formed in the crack – typical spectrum of C-S-H crystals.

in the pre-existing crack, and an energy dispersive spectrometry analysis (see an example on figure 10) demonstrates that these crystals are new C-S-H.

These results make the link between the structural experimental results and what happens in the microstructure. Indeed, the fact that new links are created between the two faces of the crack enables a substantial recovery of the stiffness of the specimens. Nevertheless, this continuity is established locally, and this is why we can not have the total recovery of flexural resistance, in the sense that the material is not recreated in all the crack, and especially all the links that give its resistance to concrete.

3 A FIRST APPROACH OF MODELING

With all the information given by mechanical tests, acoustic emission analysis, and microscopic investigations, the aim of this part is now to give a first approach of modeling of the mechanical behavior of healed concrete specimens.

3.1 Elastic damage model for healthy concrete

The behavior of healthy concrete is described by an elastic damage model, which is written with the crack opening parameter. This parameter is important if we consider the fact that the crack opening has a real influence on the occurrence of the phenomenon of self healing, as we have seen before. The approach presented here is one-dimensional, but could, of course, be extended.

The model is based on the crack band theory proposed by Bazant & Oh (1983). Fracture of concrete is thus represented by a band where micro-cracks appear, in a dense and distributed way. In one dimension, if we consider a bar subjected to traction, the total strain of the bar is divided in two parts: the elastic part, and the part due to cracking, as follows:

$$\varepsilon = \frac{\sigma}{E} + \frac{w}{L} \tag{1}$$

where w = the crack opening; and L = width of the crack band.

In the case where the total length on the bar is equal to the width of the crack band, the elastic strain is linked to the tensile strength, and Equation 1 becomes as follows:

$$\varepsilon = \frac{f_t}{E} + \frac{w}{L} \tag{2}$$

where f_t = tensile strength.

In order to link the applied stress and the crack opening in the cracking zone (fracture process zone as defined by Hillerborg et al., 1976), a fictitious crack law is used, and put in the expression of the strain. By this way, the crack opening is smeared on the total length on the bar. The parameters of the fictitious crack law (crack opening versus applied stress) are represented on figure 11.

W_c is the critical crack opening when the applied stress between the two lips of the crack vanishes. f_t is the tensile strength, and the two parameters W_{int} and f_{tint} are intermediary ones. The evolution of the stress as a function of the crack opening is thus as follows:

$$\sigma = f_t - \frac{f_t - f_{tint}}{w_{int}} w \quad \text{if} \quad 0 \le w \le w_{int} \tag{3}$$

$$\sigma = \frac{f_{tint} w_{fin}}{w_{fin} - w_{int}} - \frac{f_{tint}}{w_{fin} - w_{int}} w \quad \text{if} \quad w_{int} \le w \le w_{fin} \tag{4}$$

The substitution of equations 3 and 4 in equation 2 leads to the formulation of a classical scalar damage damage law, $\sigma = (1 - D)E\varepsilon$. The definition of the

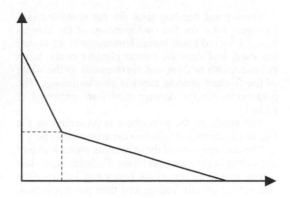

Figure 11. Parameters of the fictitious crack law

damage parameter D is thus given by the following equations:

$$D = 1 - \frac{1}{E\varepsilon} \left[f_t - \frac{f_t - f_{tint}}{w_{int}} L \left(\varepsilon - \frac{f_t}{E} \right) \right]$$

$$\text{if} \quad \frac{f_t}{E} \le \varepsilon \le \frac{w_{int}}{L} + \frac{f_t}{E} \tag{5}$$

$$D = 1 - \frac{1}{E\varepsilon} \frac{f_{tint}}{w_{fin} - w_{int}} \left[w_{fin} - L \left(\varepsilon - \frac{f_t}{E} \right) \right]$$

$$\text{if} \quad \frac{w_{int}}{L} + \frac{f_t}{E} \le \varepsilon \le \frac{w_{fin}}{L} + \frac{f_t}{E} \tag{6}$$

Loading and unloading are described by means of the loading function (in one dimension):

$$f(\varepsilon, \kappa) = \varepsilon - \kappa \tag{7}$$

where κ is a hardening-softening parameter. The initial value of κ is linked to the tensile resistance of concrete:

$$\kappa_0 = \frac{f_t}{E} \tag{8}$$

The evolution of damage is then specified as follows:

– for loading, i.e. for $f(\varepsilon, \kappa) = 0$ and $\dot{f}(\varepsilon, \kappa) = 0$, then $\varepsilon = \kappa$ and D is described by equations 5 and 6.
– for unloading or reloading, i.e. for $f(\varepsilon, \kappa) \le 0$ and $\dot{f}(\varepsilon, \kappa) < 0$ then $\dot{D} = 0$ and $\dot{\kappa} = 0$

Unloading is performed without any residual strain, and this is one of the limitations of this first approach. The model is then implemented, only for uniaxial stress, in a finite element code using layered beam elements (see Bazant & Pijaudier-Cabot, 1987 for the principle, and Granger, 2006 for the implementation).

Three-point bending tests ate the simulated (see Granger, 2006 for the configuration of the simulations). Layered finite beam elements (with 15 layers) are used, and only the central element of the beam (whose width is 2 cm and corresponds to the width of the fracture process zone) is able to damage. The parameters for the damage model are presented in table 1.

The result of the simulation is presented on the figure 12, compared with two experimental tests.

The damage state of the specimens for each step of unloading in the post peak phase. Each damaged state is assumed to be one of the factor that influence the occurrence of self healing and then the recovery of structural mechanical properties.

3.2 Thermodynamics of the mechanical behavior of healed layer

Self healing appears in the damaged zone of the specimen which is represented, according to the previous section, by the damaged layers at the unloading state of the bending test simulation.

The first approach of modeling consists in introducing new mechanical properties in these damaged layers, simulating locally the effects of the self healing. New simulations will then be done considering layered finite beam elements, with layers having different mechanical behaviors (healthy, damaged or healed).

The behavior of healed layer has thus to be described in a correct thermodynamic framework. The principle is to couple an hydration model (proposed by Ulm &

Table 1. Parameters of the damage model for three-point bending tests simulations.

Young modulus	42 GPa
Tensile strength	4.3 MPa
Critical crack opening (W_c)	16 μm
f_{tint}	1.5 MPa
W_{int}	4 μm

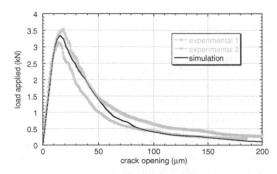

Figure 12. Simulation of three-point bending tests

Coussy, 1995 and 1996), describing the evolution of properties, and a non linear mechanical behavior model, representing the fracture of concrete (the model developed in the previous section). The coupled model is described according to the local state method proposed by Lemaitre & Chaboche (1991).

3.2.1 State variables and thermodynamic potential

The layer which is going to heal is already characterized by a scalar damage variable D_1, which is the damage state when the specimens are unloaded during the pre-cracking phase.

The evolution of healing, which is linked to hydration of anhydrous cement, can be described by the same kind of variable as the one used by Ulm & Coussy, which is called here x.

Damage during the second loading phase, is described by a second and new damage variable called D_2.

Thus, a thermodynamic potential, relative to the healed layer, can be proposed:

$$\rho\psi = \frac{1}{2}(1-D_2)E(x).\varepsilon.\varepsilon + \frac{1}{2}(1-D_1)E.\varepsilon.\varepsilon$$
$$+ \left(\frac{1}{2}\kappa x^2 - A_{x_0} x\right) \quad (9)$$

where κ is a constant variable and A_{x_0} the initial chemical affinity (see Ulm & Coussy, 1995, for these parameters).

3.2.2 State equations

The state equations are then obtained by the derivation of the thermodynamic potential by the state variables:

$$\sigma = \frac{\partial\rho\psi}{\partial\varepsilon} = (1-D_2)E(x).\varepsilon + (1-D_1)E.\varepsilon \quad (10)$$

$$A_x = -\frac{\partial\rho\psi}{\partial x} = -\frac{1}{2}(1-D_2)\frac{\partial E(x)}{\partial x}.\varepsilon.\varepsilon - (\kappa x - A_{x_0}) \quad (11)$$

$$Y_1 = \frac{\partial\rho\psi}{\partial D_1} = -\frac{1}{2}E.\varepsilon.\varepsilon \quad (12)$$

$$Y_2 = \frac{\partial\rho\psi}{\partial D_2} = -\frac{1}{2}E(x).\varepsilon.\varepsilon \quad (13)$$

The first equation is the mechanical behavior law for the healed layer, which takes into account the damage of the initial healthy material (D_1) and the damage of the newly formed material (D_2). The two materials are in parallel. The mechanical properties of the first one do not evolve with healing, while the mechanical ones of the second material evolve with hydration.

Y_1 and Y_2 are the variables associated to damage and A_x the chemical affinity of the reaction of healing (as it has been described for hydration by Ulm & Coussy, 1995, 1996).

The energy dissipation, linked to the appearance of damage and to healing, writes as follows:

$$\varphi = -\dot{D_1}Y_1 - \dot{D_2}Y_2 + A_x\dot{x} \tag{14}$$

Thus, including equations 11 to 13:

$$\varphi = \frac{1}{2}\dot{D_1}E.\varepsilon.\varepsilon + \frac{1}{2}\dot{D_2}E(x).\varepsilon.\varepsilon$$
$$+ \left[A_{x_0} - \kappa x - \frac{1}{2}(1-D_2)\frac{\partial E(x)}{\partial x}.\varepsilon.\varepsilon\right]\dot{x} \tag{15}$$

Neglecting the terms where ε appears in the second order, it comes:

$$\varphi = \left[A_{x_0} - \kappa x\right]\dot{x} \cong A_x\dot{x} \tag{16}$$

The chemical affinity is defined so as to be positive (Ulm & Coussy, 1995). So the energy dissipation is always positive and the Clausius-Duhem principle is checked. So the thermodynamic framework, concerning the re-introduction of mechanical properties, is defined.

3.2.3 The mechanical behavior of healed material

If we now consider the state equation defined by equation 10, the mechanical behavior of the healed layer writes as follows:

$$\sigma = \left[E(1-D_1) + E(x)\right]\varepsilon = E_c\varepsilon \quad \text{if} \quad \varepsilon \leq \frac{f_{tc}}{E_c} \tag{17}$$

$$\sigma = E(1-D_1)\varepsilon + (1-D_2)E(x)\varepsilon \quad \text{if} \quad \varepsilon > \frac{f_{tc}}{E_c} \tag{18}$$

where f_{tc} and E_c are the tensile strength and Young modulus of the healed layer, which depend on the healing variable.

So, the linear elastic part of the mechanical behavior is characterized by the new elastic modulus, depending on x and D_1, and also related to the initial one E. We can thus define a new expression for E_c:

$$E_c = E \times (1 - D_1 + g(x, D_1)) \tag{19}$$

where g is a scalar function to be defined.

Then including equation 19 in equation 18, the mechanical behavior in the non linear part is as follows:

$$\sigma = E\left[(1-D_1) + (1-D_2)g(x,D_1)\right]\varepsilon \tag{20}$$

$$\sigma = \left[(1-D_1 + g(x,D_1))E - D_2g(x,D_1)E\right]\varepsilon \tag{21}$$

$$\sigma = E_c\left[1 - \frac{D_2g(x,D_1)}{1-D_1+g(x,D_1)}\right]\varepsilon \tag{22}$$

The equation 22 can be thus assimilated to the "classical" form of a damage model expression, where the damage variable D_c is defined as follows:

$$D_c = \frac{D_2g(x,D_1)}{1-D_1+g(x,D_1)} \tag{23}$$

So, the mechanical behavior of each layer is characterized by its tensile strength f_{tc}, its Young modulus E_c and the parameters of the damage law defined on figure 11. The evolution of damage is thus described by the equation 23. First, the new material damages, with D_2 varying from 0 to 1. When it reaches the value 1, the new material is totally damaged, and the mechanical behavior becomes the one of the initial damaged layer, with D_1 varying from its initial value after the pre-cracking phase, to 1, when the layer is totally damaged. Finally, the mechanical behavior is defined only by f_{tc}, E_c and the value of D_1. The others parameters are then got by interpolation, in order to reach the initial behavior of the damaged layer, when D_2 is equal to 1.

3.2.4 Simulations of the bending tests

First simulations are proposed for the bending tests after healing. The initial state of damage, before the occurrence of the phenomenon, is given by the numerical result of the bending test proposed on figure 12.

After that, new mechanical properties are given to the damaged layers, and the new mechanical behaviors (damage laws) are defined, according to the section 3.2.3. All the parameters and the configuration of the tests are given in Granger (2006). The results of the simulations for 1, 3, 10 and 20 weeks healing are presented on figure 13, and compared with the experimental results on figure 14.

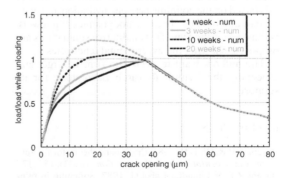

Figure 13. Numerical simulations of bending tests on healed concrete specimens (comparison with the figure 14).

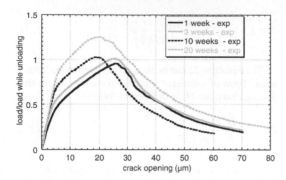

Figure 14. Experimental results of bending tests on healed concrete specimens (comparison with the figure 13).

These first simulations show that the tendencies in the pre-peak regime are quite well represented. The facts that the modeling do not take into account the residual crack opening, and that the recovery of mechanical properties is linked to the state of damage and not to the residual crack width, do not enable to represent the whole mechanical behavior of the healed specimens under three-point bending tests, and especially the post peak regime. Nevertheless, this first approach of modeling has permitted to define the thermodynamic framework and insights in order to simulate the mechanical behavior of healed specimens.

4 CONCLUSION

This research work has provided some new insights about the mechanical behavior of UHPC specimens initially cracked and then healed in water. In particular, a fast recovery of global stiffness and a light improvement of resistance have been highlighted. Experimental results have been completed by microscopic investigations which show that new crystals have precipitated in the crack.

After that, a very first approach of modeling of the mechanical behavior of healed specimens is proposed. It is based on the coupling between hydration and elastic damage models. The thermodynamics framework is thus established and first qualitative simulations are proposed, showing the experimental tendencies.

The authors want to thank the firm Lafarge, and especially the Central Research Laboratory, for its support.

REFERENCES

Bazant Z.P. & Oh B.H. 1983. Crack band theory for fracture of concrete. *Matériaux et Constructions* 16 (93): 155–177
Bazant Z.P. & Pijaudier-Cabot G. 1987. Softening in reinforced concrete beams and frames. *Journal of Structural Engineering* 113 (12): 2333–2347

Edvardsen C. 1999. Water permeability and autogenous healing of cracks in concrete. *ACI Materials Journal* 96 (4): 448–454
Granger S. 2006. Caractérisation expérimentale et modélisation du phénomène d'auto-cicatrisation des fissures dans les bétons. pHD thesis. Ecole Centrale de Nantes (in french)
Granger S., Loukili A., Pijaudier-Cabot G. & Chanvillard G. 2006. Experimental characterization of the self healing of cracks in an ultra high performance cementitious material: mechanical tests and acoustic emission analysis. *Cement and Concrete Research (2006). In press*
Hearn, N. & Morley, C.T. 1997. Self healing property of concrete – Experimental evidence. *Materials and Structures* 30: 404–411
Hillerborg A., Modeer M. & Petersson P.E. 1976. Analysis of crack formation and crack growth in concrete by means of fracture mechanics and finite elements. *Cement and Concrete Research* 6: 773–782
Jacobsen S., Marchand J. & Hornain H. 1995. SEM observations of the microstructure of frost deteriorated and self healed concrete. *Cement and Concrete Research* 25 (8): 55–62
Jacobsen S., Marchand J. & Boisvert L. 1996a. Effect of cracking and healing on chloride transport in OPC concrete. *Cement and Concrete Research* 26 (6): 869–881
Jacobsen S. & Sellevold E.J. 1996b. Self healing of high strength concrete after deterioration by threeze/thaw. *Cement and Concrete Research* 26 (1): 55–62
Lemaitre J. & Chaboche J.L. 1991. Mécanique des matériaux solides. Paris. Dunod
Loukili A., Richard P. & Lamirault J. 1998. A study on delayed deformations of an ultra high strength cementitious material. *ACI Recent Advances in Concrete Technology* 179: 929–949
Neville, A. 2002. Autogenous healing – A concrete miracle? *Concrete International* November: 76–82
Pimienta P. & Chanvillard G. 2004. Retention of the mechanical performances of Ductal® specimens kept in various aggressive environments. *Fib Symposium 2004, Avignon, 26–27 April 2004*
Reinhardt H.W. & Joos M. 2003. Permeability and self healing of cracked concrete as a function of temperature and crack width. *Cement and Concrete Research* 33 (7): 981–985
Ulm F.J. & Coussy O. 1995. Modeling of thermomechanical couplings of concrete at early ages. *Journal of Engineering Mechanics* 121 (7): 785–794
Ulm F.J. & Coussy O. 1996. Strength growth as chemo-plastic hardening in early age concrete. *Journal of Engineering Mechanics* 122 (12): 1123–1132

Concrete damage due to alkali-silica reaction: A new method to determine the properties of the expansive gel

E. Schlangen
Delft University of Technology, Faculty of Civil Engineering and Geosciences, Microlab,
The Netherlands & Intron, Culemborg, The Netherlands

O. Çopuroğlu
Delft University of Technology, Faculty of Civil Engineering and Geosciences, Microlab,
The Netherlands & Ecole des Mines, Douai, France

ABSTRACT: In this study the mechanism of Alkali-Silica Reaction (ASR) is investigated. A combined numerical and experimental research is presented. A meso mechanical model based on lattice theories is used as a starting point. Examples show that the model is able to simulate the damage mechanism in concrete due to ASR. One of the important input parameters in the model, but also one of the key players in the mechanism of ASR, is the amount of expansion of the gel and as a result the internal forces that are generated by this expansion. An experimental set-up is developed to measure the pressure generated during the reaction on a micro scale in order to assess the local pressure developed on each grain by its swelling.

1 INTRODUCTION

1.1 *The ASR phenomenon*

Damage due to alkali-silica reaction (ASR) is a phenomenon that has been observed in many structures all over the world. Numerous studies have been published on this subject (see for example Bödeker (2003)), but the mechanisms of ASR are not yet completely understood. ASR occurs between certain forms of silica present in the aggregates and the hydroxile ions (OH^-) in the pore water of a concrete. The hydroxile ions will attack the siloxene bonds and alkali silicate gel is formed. The formation of the gel itself is not deleterious. However, the gel absorbs water and its subsequent expansion is the start of the deterioration of the concrete structure. If the gel can creep into pores or existing cracks it is probably not doing any damage, but when all the free space is filled up, further expansion of the gel will create internal stresses in the cement matrix, which can lead to cracks propagating radially from the reactive aggregates. Externally, damage in concrete structures due to ASR is visual as random oriented crack patterns, similar to crack patterns known from drying shrinkage. The damage due to ASR reduces the mechanical properties of the concrete (Schlangen & van Breugel 2005) and with that the structural safety of a structure can be lost. Furthermore, cracks formed in concrete structures due

to ASR increase the permeability and the ingress of for instance water and chlorides, which can lead to reinforcement corrosion.

1.2 *What has been done?*

Already a series of 12 international conferences have been organized in the past on the topic of alkali aggregate reaction in concrete. All this research has led to national and international standards, recommendations and procedures describing how to test reactivity of aggregates, how to determine the risk of getting ASR in a certain concrete and methods to determine whether or not it was ASR that caused damage in a structure.

To test the possible swelling in a concrete due to ASR the concrete prism test (CPT) and the (ultra) accelerated mortar bar test (U)AMBT (see for instance Xu et al. 2000 and Grosbois & Fontaine 2000) are now widely accepted and standardized. The result of these tests is an expansion of the concrete in time. The CPT should run for a year, while the UAMBT gives results in about 1 or 2 weeks. If the measured values stay below a certain threshold, the risk for getting ASR in a structure with this concrete is low. Both tests, however, give no explanation for the mechanism that takes place. Furthermore these tests do not tell if the mechanism is possibly different if the deformations are to some extent restrained as is the case in a real structure.

Ferraris et al. (1997) and Binal (2004) developed methods to tests the pressure that is generated in a prism of concrete in which ASR takes place. The deformation of the concrete is restrained and the forces generated measured. This is the structural effect. But on a local scale a similar restraining happens. The cement matrix in between the aggregate particles will give this restraining. If this cement matrix is stronger there will be more restraining. The composition of the concrete determines how the ASR mechanism will evolve. The silica in the aggregates react with the alkali ions in the pore water and a gel is formed. The gel takes up water and wants to swell. This gel is maybe inside the aggregates and could crack the aggregates, if the force generated by the swelling gel is enough to overcome the strength of the aggregate. The same holds for the strength of the interface and the strength of the cement matrix. ASR gel could enter cracks, swell and maybe propagate the cracks (see Figure 1). But in all cases it should be taken into account that the aggregates, interface and cement matrix can not freely expand but are restrained by its surroundings. If this restraining is too large the reaction might even stop the ASR process. An important parameter in the whole mechanism is the pressure that the gel can generate. This is also an important input parameter for models that could help in explaining the ASR mechanism.

Models to simulate ASR in concrete structures on various levels can be found in literature. Ulm et al. (2000), for instance, developed a model on the macro-level to simulate the expansion of concrete due to ASR. In this model a lot of attention is given to the coupling of diffusion taking place through the material, the chemical reaction taking place inside the material and the resulting deformations. Modelling of the reduction of mechanical properties due to ASR is performed by Schlangen & van Breugel (2005) with a 2D meso-level model in which aggregate are explicitly

taken into account. Comby-Peyrot performed similar simulations, however, in this case a 3D model was developed, resulting in more detailed outcome. In the research by Andic-Cakir et al (2007) different types of pressure generated by the gel were discussed on a meso-level, resulting in different expansions of concrete. This paper is a continuation of that research.

1.3 The aim of the present research

The aim of the research presented in this paper is to model ASR mechanism and to develop a test method to measure the input parameters for the model, especially the local expansions and pressures generated by the gel.

2 MODELLING ASR DAMAGE

2.1 The basics of the model

The mechanism of ASR is modelled in this research with a meso-mechanical lattice type model (Delft Lattice model) in which the aggregate structure is taken into account using digitized images of the real material, see Figure 2. In the model, the materials are discretized as a lattice consisting of small beam elements that can transfer normal forces, shear forces and bending moments (Schlangen & Mier, 1992; Schlangen & Garboczi, 1997). The simulation of fracture is realized by performing a linear elastic analysis of the lattice under loading and removing an element from the mesh that exceeds a certain threshold. In the present simulations the normal stress in each element is compared to its strength. Details on the elastic equations as well as the fracture procedure of the model are explained by Schlangen & Garboczi (1997).

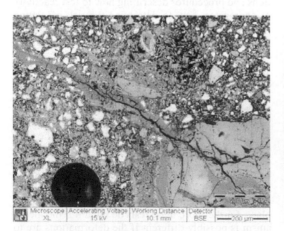

Figure 1. Penetration of ASR gel into crack.

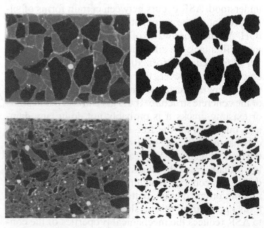

Figure 2. Digitized images of cross sections of M1 (upper row) and M2 (lower row) concrete and threshold binary images.

In this application of the lattice model attention is focused on the simulation of eigenstresses and crack growth that develop in the material as a result of ASR as discussed in Schlangen & van Breugel (2005).

2.2 Experiments that are modelled

Andic-Cakir et al (2007) performed tests in order to determine the effect of aggregate gradation on the concrete microbar expansions. Two different aggregate gradations were studied at constant cement content. Prismatic microbars of $40 \times 40 \times 160$ mm were prepared with aggregate/cement ratio of 1 by weight and water/cement ratio of 0.33 by weight. Although in the original test method proposed by Grattan-Bellew et al. (2003) no fine particles (<4.75 mm) were used in the preparation of microbar specimens; in one of the mixtures tested in this study (M2), fine particles were also added to the mixture. Aggregate gradations and abbreviations of the relevant mixtures are given in Table 1. The concrete microbars were cured in 80°C water for one day and in 80°C 1M NaOH solution for the remaining test period. The average expansion values of 3 microbars were recorded up to 40 days.

The main conclusions from the tests were that the expansion values of the two mixtures tested express a linear time-expansion relationship and that the mixture containing fine aggregate expanded more (about a factor 2.5) than the ordinary microbar sample. This may be due to the increased reaction sites by the addition of fine particles. Note, that the total volume of particles was the same in both mixtures.

2.3 Simulations performed

In the simulations a 2D regular triangular lattice consisting of 20000 beam elements is used. For the implementation part of the binary images of Figure 2 are used. The cross sections in the simulations represent an area of 30×30 mm^2.

The beams that fall inside an aggregate, cement matrix, or just on the boundary of both (interfacial transition zone, ITZ) are assumed to have the properties given in Table 2.

From the experiments it is also clear that the gel enters the cracks and possibly also exerts pressure

inside the cracks (Figure 1). To get some insight in the contribution of the different expansions (inside aggregates, at the ITZ and inside the cracks) three different loading modes are adopted in the simulations:

- only ITZ: All the elements at the boundary of aggregate and cement matrix, forming the ITZ are given an expansion (local strain). From a linear elastic analysis of the lattice the amount of strain is calculated to crack (stress is larger than strength) an element in the lattice. This element is removed from the lattice and the strain is applied again.
- ITZ+crack: The same procedure as the previous one. However, when a crack is formed, the element is not removed in the next step, but it is given the same expansion and properties as the elements in the ITZ.
- Aggregate particle: All the elements in the aggregate particles are given an expansion. A linear elastic analysis determines the element that is removed in the next step.

In the analysis the boundaries of the lattice are not restrained. The analysis in each step is linear elastic. In the present simulations no visco-elastic material behaviour is adopted for both the concrete and the gel.

2.4 Results of the simulations

The results of the simulations are presented as graphs with local versus total global strain (Figure 3) and cracks patterns in the material (Figure 4). In the graph presented in Figure 4 the local strain is the applied strain in the single elements. The total strain is the strain of the complete sample.

From this graph and from the crack patterns the following observations can be made:

- The initial strains for the loading modes ITZ and ITZ+crack are equal. However if the number of cracks increases, the internal pressure in the cracks in case of loading mode ITZ+crack results in a higher total strain. In case of the loading mode ITZ+crack the cracks localize much faster in a single crack. In case of only ITZ loading and especially in case of particle loading first distributed micro-cracks develop, which later coalesce in major cracks.

Table 1. Aggregate gradation of the mixtures.

Sieve aperture mm	Aggregate content (by weight)	
	M1	M2
<4.75	–	60
4.75–9.5	25	20
9.5–12.5	75	20
W/C	0.33	0.33

Table 2. Properties of beam elements in lattice.

Beam location	Strength [MPa]	E-modulus [GPa]
Aggregate	10	70
Matrix	4	25
Interface/ITZ	1	10

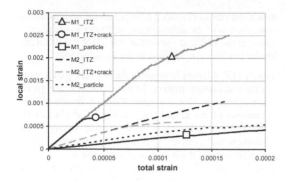

Figure 3. Local strain versus total global strain for M1 and M2 concrete with different loading modes.

- The strain that is obtained for the two concretes M1 and M2 is almost equal in the case of the loading mode particle. Therefore the M1 concrete also has a slightly higher total strain.The volume (area in 2D) of particles is almost equal in both concretes. The M1 has somewhat more particle area in these simulations, because it is not possible to take into account the smaller fine aggregate particles of the M2 concrete for this scenario of expansion.
- In the case of ITZ loading (and also ITZ+crack loading) it can be seen that for a given local strain, the total strain that is obtained in the M2 concrete is about 3 times higher than in the M1 concrete. Explanation for this is that the surface area around the particles (the number of ITZ elements) in case of M2 concrete is a factor 3 higher than for the M1 concrete.
- The number of cracks in case of M2 concrete with smaller particles is larger than in case of M1 concrete.

3 MEASURING EXPANSION AND FORCE

3.1 Description of device

The swelling of the gel is the mechanism that creates the internal damage in the concrete and leads to the expansion of the concrete. As described above there is a need to know the force that is generated by the swelling of the gel, since that is the main missing input parameter for modelling the mechanism of mechanics involved in the ASR process. ASR is a slow process, which means an accelerated test is needed to obtain results in a short period of time. Similar conditions are used as in the accelerated mortar bar tests or micro-concrete tests (Grattan-Bellew et al 2003). The specimens tested have a cross section of $15 \times 15 \text{ mm}^2$ and a length of 20 mm. To make the specimens first an aggregate particle is sawn to the size of $15 \times 15 \times 10 \text{ mm3}$. This specimen is placed in a mould

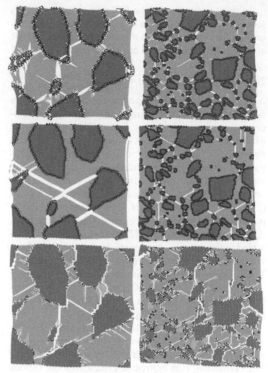

Figure 4. Simulated crack patterns in M1 (left) and M2 (right) concrete. From top to bottom the crack patterns for the different loading modes are shown: ITZ, ITZ+crack, particle.

and the remainder of the length of the final specimen is filled with cement paste (see figure 5). The specimens are cured in the mould for 1 day at 20°C and 99% RH, then they are placed in 80°C water for one day and tested in 80°C 1M NaOH solution after that. To test the specimens they are glued to a stainless steel frame as shown in Figure 5.

The steel frame is attached to a micro tensile-compression testing device (developed by Kammrath & Weiss). The shape of the testing frame is such that the specimen can hang inside a pool with the solution at 80°C and that the loading and measurement-parts are outside this pool, see Figures 5 and 6. The solution in the pool is covered with a layer of oil to prevent evaporation. The deformation of the specimen is measured with a displacement gauge mounted on to the steel frame as shown in Figures 5 and 6. The test can either be run in deformation (zero deformation) or in load control (zero load). In this way it is possible to test the free deformation that will take place due to the ASR formation, but also the stress that is generated if this deformation is restrained. Also different loading regimes are possible, for instance first a restraining of the deformations until a certain stress is reached and after that a free deformation to simulate the situation

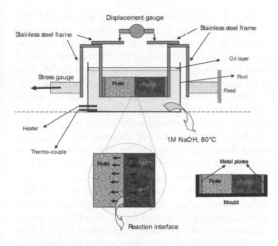

Figure 5.   Principle design of new measuring device.

Figure 6.   New measuring device in tensile testing machine.

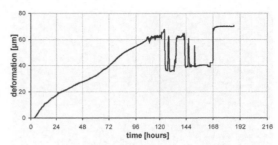

Figure 7.   Measurement of expansion in ASR-test.

3.2   First results

The first tests are performed on basalt aggregates, the same material as in the tests described in paragraph 2.2. As is often the case with the development of a new device a lot of small difficulties have to be tackled before any results are obtained. Until now only a few tests have been performed. In Figure 7 one of the results is shown that were obtained in load control. In the graph the free deformation of the specimen during the test is given. The measurement in the graph starts, a few hours after the specimen has been submerged into the pool containing the solution, from the moment the temperature in the specimen was 80°C and the temperature in the frame and the machine reached a stable value. What can be seen is that the specimen elongates during the first 5 days to a total value of about 60 μm. Then the temperature sensor in the pool started to malfunction and due to that the temperature in the pool decreased to room temperature which can be seen in the graph as a (thermal) shrinkage of the specimen. After 7 days the sensor was replaced, the pool heated again to 80°C and the deformation seems to pick up the same curve.

The tests with restrained deformation are not performed yet, but are still in preparation.

After the test explained above, the specimen was examined in the ESEM. In Figure 8 images are given which are taken from the interface between the basalt particle (on the left side of the image) and the cement paste (on the right side of the image). The band with a width around 70 μm in the middle of the image is ASR-gel. With the ESEM it is also observed that the basalt particle in a band near the interface was very porous and that all the glass phase present in this part of the rock was dissolved. The microstructure of the remainder of the aggregate particle seemed not to be changed.

4   DISCUSSION AND CONCLUSIONS

In this paper a test method and set-up is presented to measure expansions and forces that are generated by

inside a concrete. Here first a stress has to be created to overcome the strength of the material and then the deformation due to the swelling of the gel can take place.

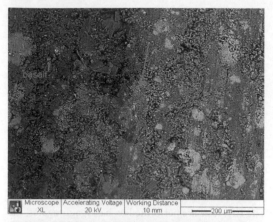

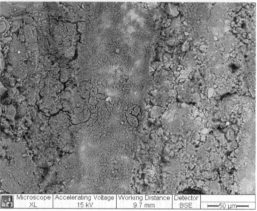

Figure 8. ESEM/BSE microphotographs of ASR gel at interface between aggregate particle (left) and cement paste (right).

the gel that is formed in concrete due to ASR. The background for starting this research is that such values have to be known in order to explain the mechanical action inside concrete when it is attacked by ASR. Restraining on a structural level, but also inside the material itself will have an effect on the ASR mechanism, which can be investigated by numerical modelling.

The Delft lattice model is used to simulate crack patterns and deformation in concrete due to ASR expansions. Although assumptions are made for local properties of the material and the expansive gel, the results that are obtained look promising.

Different loading modes are applied in the simulations. Expansion of ITZ, expansion of ITZ+cracks (gel in cracks) and expansion of particles are used as variables for the two concretes M1 and M2 consisting of large and small particles respectively. It was found that when the expansion of the particles is used as loading mode, almost no difference in total expansion

is found between the two concretes M1 and M2. But it was not possible to taken in account all the small reactive aggregates in the simulation of the M2 concrete. In case of ITZ loading the M2 concrete has a 3 times higher total expansion compared to the M1 concrete. Explanation is that the volume (area) of particles is equal for the two materials, but the surface area is about 3 times higher for the M2 concrete with smaller particles. The higher total deformation and the higher number of cracks in the M2 concrete as founds in the simulations is also found in the experiments on these materials as described in Andic-Cakir et al (2007).

The first tests that are performed in the new test set-up show promising results. It seems to be possible to both measure expansions in a single aggregate-cement matrix connection and also the stresses that are generated when the deformation is restrained. With device it is also possible to study whether the reaction slows down in the case of restrained deformations. This could then answer the question if a better concrete (concrete with a stronger cement matrix or a more optimized packing) or a fibre concrete which does not allow for the opening of cracks can possibly slow down or stop the reaction.

In the device tests can be performed on different materials, both aggregate type and cement matrix (cement type). Specimens will be examined by ESEM after the tests. In this way it can be examined if it is only the interface or also the aggregate that expands. Furthermore, it can be measured how much of the aggregate will dissolve; is it only a rim at the surface, or will it be the complete aggregate? This depends on the type of mineral, but definitely contributes a lot to the damage mechanism due to ASR as discussed in Copuroglu et al (2007).

ACKNOWLEDGEMENT

Discussions on the subject with E. Garcia-Diaz from Ecole des Mines de Douai in France and O. Andic-Cakir from Ege University in Turkey are gratefully acknowledged.

REFERENCES

Andic-Cakir, O., Copuroglu, O., Schlangen, E., & Garcia-Diaz, E. (2007) Combined experimental and modelling study on the expansions of concrete microbars due to ASR. Proc. Consec07, Tours, France.

Binal, A. (2004) A new experimental method and device for measuring alkali silica gel pressure in mortar. Proc. 12th Int. Conf. on Alkali-Aggregate Reaction in concrete, Beijing, pp. 266–272.

Bödeker, W. (2003) Alkali Reaction in Structural Concrete, a View from Practice. Deutscher Ausschuss für Stahlbeton, Vol 539, (in German).

Comby-Peyrot, I. (2006) Development and validation of a 3D computational tool to describe damage and fracture due to alkali silica reaction in concrete structures. PhD-thesis, Ecole des Mines de Paris, France.

Copuroglu, O., Andic-Cakir, O., Broekmans, M.A.T.M., Kuhnel, R. (2007) Mineralogy, geochemistry and expansion testing of an alkali reactive basaltoid from western Anatolia, Turkey. 11th Euroseminar on microscopy applied to building materials, 5–9 June 2007, Porto, Portugal.

Ferraris, C.F., Clifton, J.R., Garboczi, E.J. & Davis, F.L. (1997) 'Stress due to alkali-silica reactions in mortars'. Mechanisms of Chemical Degradation of Cement-Based Systems. Proceedings of the Materials Research Society's Symposium on Mechanisms of Chemical Degradation of Cement-Based Systems. November 27–30, 1995, Boston, MA, E & FN Spon, London, England, Scrivener, K. L. & Young, J. F., Editors, p. 75–82.

Grosbois, M. & Fontaine, E. (2000) Evaluation of the potential alkali-reactivity of concrete aggregates: performance of testing methods and a producer's point of view. Proc, 11th Int. Conf. on Alkali-Aggregate Reaction, Quebec, 267–276.

Grattan-Bellew, P.E., Cybanski, G., Fournier, B. & Mitchell L. (2003) Proposed universal accelerated test for alkali-aggregate reaction the concrete microbar test, Cement, Concrete, and Aggregates, Vol. 25, No.2, 29–34.

Schlangen, E. & van Mier, J.G.M. (1992) Experimental and numerical analysis of micromechanisms of fracture of cement-based composites, Cem. Conc. Composites, Vol. 14, 105–118.

Schlangen, E. & Garboczi, E.J. (1997) Fracture simulations of concrete using lattice models: computational aspects, Eng. Fracture Mech., Vol. 57(2/3), 319–332.

Schlangen, E. & van Breugel, K. (2005) Prediction of tensile strength reduction of concrete due to ASR, in Construction Materials (N. Banthia et. al. eds.), The University of British Columbia, Vancouver.

Ulm, F.-J., Coussy, O., Kefei, L. & Larive, C. (2000) Thermo-chemo-mechanics of ASR expansion in concrete structures, J. Eng. Mech., Vol. 126, No. 3, pp. 233–242.

Xu, Z., Lan, X., Deng, M. & Tang, M. (2000) A new accelerated method for determining the potential alkali-carbonate reactivity, Proc. 11th Int. Conf. on Alkali-Aggregate Reaction, Quebec, 129–138.

Fracture Mechanics of Concrete and Concrete Structures – High-Performance Concrete,
Brick-Masonry and Environmental Aspects – Carpinteri, et al. (eds)
© 2007 Taylor & Francis Group, London, ISBN 978-0-415-44617-4

Mitigation of expansive deterioration processes through crack control

C.P. Ostertag, J. Blunt & J. Grubb
University of California, Berkeley, CA, USA

ABSTRACT: Damage due to corrosion and alkali silica reaction is being mitigated through crack control. The paper discusses how durability can be enhanced by controlling microcracking in close vicinity to the reaction sites, which limits the egress of reaction products away from the reaction site. Mortar specimens with and without microfibers were exposed to either alkaline or corrosive environments. Crack control on the microscale in close vicinity to the reaction site reduced both the alkali silica reaction rate and the corrosion rate. Corrosion current density measurements based on polarization resistance and Tafel measurements indicate that the microfiber-reinforced specimens are more resistant to corrosion than the unreinforced control specimens.

1 INTRODUCTION

The concrete infrastructure in the US and in most industrialized countries is deteriorating at a faster pace than predicted. According to the Civil Engineering Research Foundation, the US has an estimated $1 trillion in deteriorated concrete structures including bridges, highways, piers, wharfs, structures and buildings (Cerf 2004).

There are four main deterioration mechanisms in concrete: Corrosion, Frost Action, Alkali Silica Reaction (ASR) and Sulfate attack (Rostam 2001). The common approach to enhance durability addresses each of the deterioration processes in isolation and recommends different remedies for each deterioration mechanism because each process (i.e. corrosion, frost action, ASR, sulfate attack) involves different reactants. However, all these deterioration processes have one common signature: they are expansive and cause cracking. Cracks allow the ingress of water and aggressive agents into the interior of concrete, thereby accelerating the deterioration process. Hence crack control is paramount in order to mitigate damage in concrete structures. Furthermore, cracks initiate as small microcracks. Therefore, to minimize damage the initiation and propagation of microcracks needs to be controlled and their coalescence into macrocracks significantly delayed.

The effect of crack control from the micro to the macrolevel on mechanical properties is briefly discussed. However, the emphasis of this paper is on the effect of crack control on expansive deterioration processes such as alkali silica reaction (ASR) and corrosion. ASR is a chemical reaction between alkalis from cement and certain forms of silica present in aggregates. The chemical reaction forms an ASR gel that imbibes water resulting in volumetric expansion. This volumetric expansion causes cracking in cement based materials if the expansion pressure exceeds the tensile capacity of the matrix. Corrosion is also an expansive process. When reinforcing steel within concrete corrodes, the rust product applies expansive pressure on the surrounding concrete inducing cracking in the matrix in close vicinity to the reinforcing steel. Since cracks initiate as microcracks in close vicinity to the reaction site, this paper focuses on the effect of crack control on the microscale on alkali silica reaction rate and corrosion rate.

2 MECHANICAL PERFORMANCE ENHANCEMENT DUE TO CRACK CONTROL

Crack control from the micro to the macrolevel enhances the mechanical properties of reinforced concrete as shown in Figure 1. In this case crack control was achieved through fiber hybridization. The hybrid fiber reinforced concrete (HyFRC) composite shown in Figure 1 was developed for the use in bridge approach slabs exposed to severe environmental conditions (Blunt & Ostertag 2007). The composite utilizes two types of fibers, macrofibers (conventional fibers) and microfibers. The microfibers control the microcracks and the macrofibers control and resist propagation of macrocracks. Figure 1 shows that fiber hybridization provides deflection hardening which delays macrocrack formation until more than twice the load levels when compared to control specimens (plain concrete). The HyFRC and control specimens

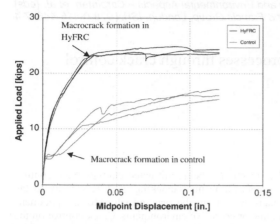

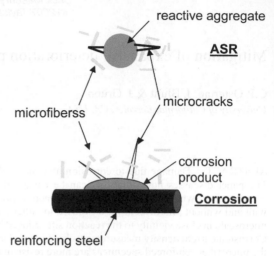

Figure 1. Enhancement in mechanical properties due to crack control.

Figure 2. Microcracks caused by either ASR or corrosion are being controlled by microfibers in close vicinity to the reaction sites.

both contain coarse aggregates and conventional steel reinforcements and were tested in four point bending. The concept of micro and macrofiber hybridization has been used by other researchers (Quian & Stroeven 2000; Banthia & Soleimani 2005) to primarily enhance the mechanical performance of cement based composites. HyFRC composites have not yet been applied to study the effect of crack control on expansive deterioration processes.

Cracks due to expansive deterioration processes initiate as microcracks in close vicinity to the reaction site. The microfibers in HyFRC composites due to their small diameter are able to bridge these microcracks at onset as shown in Figure 2 contrary to macrofibers which are not only too thick but also spaced too far apart to influence these microcracks. Furthermore, microfibers exhibit steep crack growth resistance behavior due to toughening mechanisms associated with crack fiber interactions (Yi & Ostertag 2002, 2007). The focus of this paper is on the effect of crack control on the microscale on expansive deterioration processes such as alkali silica reaction and corrosion.

The difference in crack growth behavior between unreinforced (control) and microfiber reinforced specimens is illustrated schematically in Figure 3. Let's assume a crack initiates at time t_1 at the reaction site (i.e. either reactive aggregate/matrix interface or steel reinforcing bar/matrix interface) in the control specimen. The small resistance to crack extension in the control specimen is shown by the large increase in crack length and crack opening displacement with increasing exposure time to either NaOH or NaCl solution at t_2 and t_3. The lack of crack growth resistance behavior in the unreinforced specimens reduces the driving force for crack extension, and hence smaller tensile stresses are sufficient to increase the crack length and the accompanying crack width. The crack

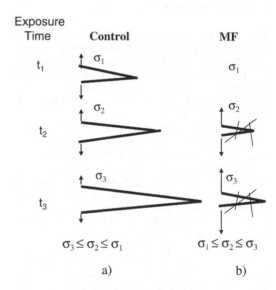

Figure 3. Difference in crack growth behavior of (a) control and (b) microfiber (MF) reinforced specimens with increasing exposure time, t, to either NaOH (to cause ASR) or NaCl (to cause corrosion).

in the microfiber reinforced specimen (Fig. 3b) initiates at a higher expansion stress (i.e. at exposure time t_2 and not t_1) and extends far less with increasing exposure time due to crack fiber interactions such as crack pinning and crack wake bridging processes (Yi & Ostertag 2007). These energy absorbing mechanisms increase with increasing crack length due to the formation of a bridging zone behind the crack tip.

Consequently, not only is a higher expansive pressure (due to ASR or corrosion) required for cracks to initiate and propagate in microfiber reinforced specimens (Fig. 3b) but the ASR gel is mechanically confined (Yi & Ostertag 2005). Furthermore, microcrack control in close vicinity to the reaction site limits the egress of reaction products away from the reaction site. The effect of delay in crack formation and propagation on ASR and corrosion processes and the effect of preventing the reaction products from leaving the reaction site will be discussed in the following sections.

3 EFFECT OF CRACK CONTROL ON ALKALI SILICA REACTION RATE

The effect of crack control on alkali silica reaction rate was studied in close vicinity to the reactive aggregate using plain and steel microfiber reinforced (SMF) mortar specimens. Rod shaped reactive aggregates of constant diameter were used to investigate differences in ASR gel formation between the SMF reinforced specimens and the control specimens. Each specimen contains a Pyrex rod of 5 mm in diameter as reactive aggregate embedded in its center. Prisms of $2.5 \times 2.5 \times 28.12$ cm were cast, cured in 80°C water bath for 1 day and immersed in a 1 N NaOH solution stored at 80°C following the ASTM C-1260 procedure (ASTM 1999). The mortar matrix is reinforced with 0 and 7vol% of steel microfibers (SMF), respectively.

ASR gel formation in SMF reinforced specimens was not only delayed but also the Pyrex rod reacted far less compared to the Pyrex rod embedded in the unreinforced mortar matrices at same exposure times to the NaOH solution. Figure 4a and 4b are backscattered images of the remaining cross-sections of the Pyrex rods for plain and SMF reinforced specimens, respectively, exposed to NaOH solution for 42 days. The reaction starts at the outer surface of the rod and continues towards the center of the Pyrex rod. The dark regions seen between the remaining Pyrex rod and the matrix in Figure 4a and 4b, respectively, will be referred to as the alkali silica reaction rims. In Figure 4a and b, these regions are filled with epoxy used for polishing the specimen surfaces and hence show up dark in the backscattered images. However, these reaction rims are originally filled with solid ASR products and a liquid alkali-silicate solution. Some of the solid ASR products remain but the liquid ASR gel is lost once the samples are sliced off for the sample preparation.

The steel microfibers (show up white in the backscattered image) are evenly distributed and in close proximity to the reactive Pyrex rod. The size of the ASR rim is a function of the dissolution of the rod, the expansion of the matrix due to the gel leaving the reaction sites and swelling in cracks and voids, and the

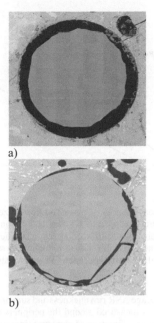

a)

b)

Figure 4. Backscattered (SEM) images of the remaining cross-section of the glass rod after exposure to 1 N NaOH (aq) for 42 days (actual area size: 6 mm × 6 mm; diameter of glass rod: 5 mm); a) specimen without microfibers, b) specimen reinforced with steel microfibers (visible as white regions in mortar matrix).

shrinkage of the ASR gel after being exposed to lower relative humidity. Therefore, only the reacted area of the Pyrex rod normalized by its initial area is plotted in Figure 5 as a function of exposure time to NaOH solution. Not only do we observe a delay in ASR gel formation but also a reduction in reactivity of the Pyrex rod in the SMF reinforced specimen compared to the control specimens.

The difference in ASR gel formation and ASR rate may be related to the difference in crack formation and crack width between the control and SMF reinforced specimens. Radial cracks that formed due to ASR are visible in Figure 4a but are difficult to see in Figure 4b at the same magnification due to their small crack opening displacements. Table 1 presents results on the ASR rim thickness and the sum of crack widths of the radial cracks measured around the periphery of the reactive aggregate as a function of exposure time to NaOH solution for the control and SMF specimens, respectively. For the unreinforced matrix (Table 1a), both the ASR rim thickness and the crack widths increase with increasing exposure time to NaOH solution. On the other hand, in SMF reinforced specimens no ASR gel formation was observed up to 13 days due to the delay in crack initiation. Once cracks initiated in the SMF reinforced specimens, the width of these cracks were too small to be measured accurately

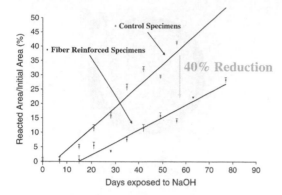

Figure 5. Reacted area of pyrex rod normalized by initial area versus time exposed to 1 N NaOH solution for control and steel microfiber reinforced specimens. The SMF reinforced specimens exhibit a 40% reduction in reaction rate compared to the control specimen (Yi & Ostertag 2005).

Table 1. Average ASR rim thickness and sum of crack width of radial cracks measured around the periphery of the reactive aggregate as a function of exposure time to 1 N NaOH solution for a) unreinforced matrix and b) steel microfiber reinforced matrix

Days	Crack width (Sum), μm	Avg. Rim Thickness, μm	Stand. Dev., Rim, μm
a)			
6	41.9	56.5	39.3
13	103.9	41.9	36.2
20	198.5	142.1	96.0
27	176.3	156.9	84.9
37	287.1	178.5	35.0
41	195.2	249.3	139.9
b)			
6	0	0	
13	0	0	
20	NA	51.4	18.4
27	NA	16.2	30.5
37	39.2	46.6	9.7
41	52.5	74.7	27.9

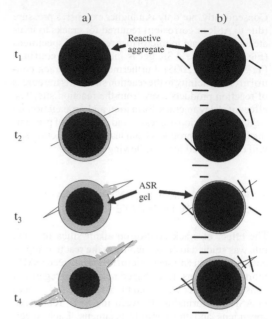

Figure 6. ASR gel formation and reduction in reactive aggregate size with increasing exposure time, t, to NaOH solution; With increasing exposure time the reactive aggregate reacts and decreases in size (dark regions) and the ASR rim thickness increases (i.e. light region between original size of aggregate and remaining aggregate). a) control specimen: ASR gel formation increases and size of original reactive aggregate decreases with increasing exposure time t due to crack formation and increasing crack width. b) SMF reinforced specimen: ASR gel formation and reduction in size of reactive aggregate occur at lower rate due to reduced crack length and reduced crack width associated with crack fiber interactions.

around the Pyrex rod up to exposure time of 27 days in NaOH solution. The ASR gel formation was considerably reduced in SMF reinforced specimens compared to the control specimens due to the delay in crack initiation and reduced crack width.

The lack of crack control in the control specimens increases the amount of gel formation as shown schematically in Figure 6. Gel formation increases with increasing exposure time to NaOH solution once a crack initiates (at t_2) and increases in width and length (t_3-t_4). The gel, now able to leave the reaction site fills the cracks and continues to swell, thereby increasing the width and length of the existing cracks which causes the specimen to expand with increasing exposure time to NaOH solution. Gel formation in the SMF reinforced specimens is delayed up to t_3 due to the delay in crack initiation (see also Table 1b). Furthermore, the small crack widths limit the migration of the ASR gel away from the reaction site into the surrounding matrix (t_4). Since the ASR gel can not leave the reaction site, the ion concentration of the ASR products has to be different in the SMF specimens compared to the control specimens. Indeed, this was the case. A higher Na and Si ion concentration was found in the ASR product extracted from the SMF specimens and analyzed by inductive coupled plasma spectroscopy. Because of the higher Si concentration the dissolution of the reactive aggregate in the SMF reinforced specimens is retarded which reduces the reactivity of the reactive aggregate and hence the gel formation. Viscosity measurements of the liquid ASR products extracted from the SMF and control specimens were conducted using a Rheometrics RMS 800 Rheometer. The liquid ASR gel extracted from the SMF specimens exhibits a 10 fold increase in viscosity compared to the control specimens which further enforces the lack of escape of the reaction products away from the reaction site. The viscosity of alkali silica reaction gels is observed to depend on the Na/Si ratios (Helmuth & Stark 1989).

In our case the ASR gel extracted from the reaction sites had the same ratio (Na/Si = 1) for both the control and SMF specimens, however, they differ in their ion concentration. Both the Na and Si ion concentration in the SMF reinforced specimens is 33% and 45% higher compared to the control specimen. It was observed in sodium silicate solutions for constant Na/Si ratios of >0.25 that the viscosity increased greatly with the increase in Na concentration (Vail 1952). In our experiment, we observed an increase in viscosity by a factor of 10 due to an increase in Na concentrations from 3.46 to 4.59 mol/l.

4 EFFECT OF CRACK CONTROL ON CORROSION

In this study, mortar specimens with and without steel microfiber reinforcements are exposed to a corrosive environment. Microfiber reinforced specimens (4.5% by volume) and control specimens were prepared with water/cement ratios of 0.55 cured for 28 days, and then submerged in aerated 3.5% NaCl solution.

Electrochemical measurements were performed prior to the immersion of the specimens, at 4 weeks of exposure, and at approximately 2-week intervals thereafter up to 7 months. Potentiodynamic polarization measurements were performed with a potentiostat and a three electrode setup. The three electrodes are the working electrode (in this case the steel rebar), the reference electrode and a counter electrode. A stainless steel wire mesh was used as a counter electrode. The specimen preparation and the electrochemical testing procedure is described in detail by Grupp (Grupp et al. 2007). The primary purpose of potentiodynamic tests was to determine the corrosion current (I_{corr}). This is a measure of the rate of charge transfer between the anodic and cathodic reactions at the corrosion potential. The corrosion current density (i_{corr}), denoted by a lower case 'i', is the I_{corr} normalized over the exposed area of steel in the working electrode. The corrosion current density cannot be directly measured, however the anodic/cathodic current differential can. Unfortunately this value is zero at the corrosion potential. Thus various analysis methods were utilized to estimate i_{corr}. These methods included Polarization Resistance, Tafel, and Cyclic Polarization. The same corrosion monitoring cell was used for all tests. The difference

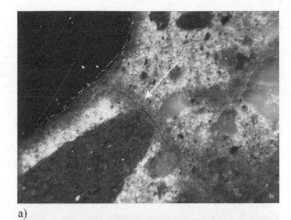

a)

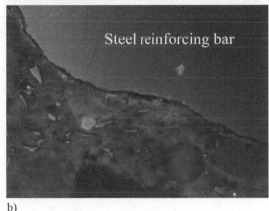

b)

Figure 7. Optical micrographs of a) control and b) microfiber reinforced specimens taken at 5× magnification; radial microcrack in matrix of control specimen is filled with corrosion product; b) no microcracks are observed in the microfiber reinforced specimen. The steel microfibers are in close vicinity to the conventional steel reinforcing bar.

was the range to which the specimen was polarized and how the data from the test was analyzed.

Microscopic analysis was performed on the specimens after exposure to NaCl solution for 22 weeks. Radial cracks were observed in the control specimens emanating from the rebar into the mortar as shown in Figure 7a taken under blue fluorescent light. Small spots of corrosion were seen as small red colored areas on the rebar-mortar interface under regular light. No cracks were observed in the SMF reinforced specimens as shown in Figure 7b. Figure 7b provides evidence that the microfibers were located in close vicinity to the steel rebar.

Crack control due to SMF leads to a reduction in corrosion current density (i_{corr}) and hence corrosion

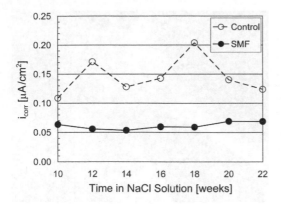

Figure 8. Average Corrosion Current Densities; C stands for control, F for the microfiber specimen.

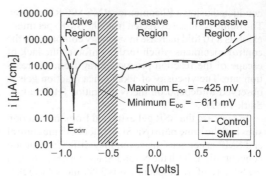

Figure 9. Observed Equilibrium Potential Range Superimposed over 22 Week Forward Polarization Scans; C stands for control specimen, F for steel microfiber specimen.

rate as shown in Figure 8. The corrosion density is calculated using equation 1 (Gamry 2003).

$$i_{corr} = \frac{B}{r_p} \qquad (1)$$

where:
i_{corr} = corrosion current density (units of Current/ Area)
r_p = polarization resistance of the steel ($r_p = \Delta E/\Delta i$ at $E = E_{oc}$)
B = the Stearn-Geary Constant

$$B = \frac{\beta_a \, \beta_c}{2.303 \cdot (\beta_a + \beta_c)} \qquad (2)$$

where B is the Stearn Geary constant

B is calculated from the anodic (β_a) and cathodic (β_c) Tafel slopes which were determined by the non-linear regression across the experimental Tafel data. Rp is the polarization resistance and the polarization resistance values are obtained from the tangent of the net current differential between the anodic and cathodic reaction, I(E) curve, at the corrosion potential, Eoc. As can be seen from Figure 8, the control specimen exhibits a higher corrosion rate compared to the SMF specimens for the duration of the observation period. The difference in corrosion rates is distinct enough to assert that microfibers reduce the corrosion rate of conventional steel reinforcing bars.

Cyclic polarization measurements were only made at 22 weeks with a forward scan range of −1.0 V to +1.0 V. Results on the cyclic polarization measurements, are shown in Figure 9. Observation of the average equilibrium potentials showed that the general tendency was for stabilization within the approximate range of −400 mV to −600 mV. During this period of stabilization, the least negative value was −425 mV while the most negative value was −611 mV. Thus, the corrosion state of the specimens existed at the border between the active and passive regions (Fig. 9). The region of the polarization curves that is indicative of the actual behavior with respect to corrosion is the region of the equilibrium potentials. Within this region the control specimens had notably higher current densities, signifying the control specimens are more susceptible to corrosion than the microfiber-reinforced specimens.

5 CONCLUSION

Crack control on the microscale reduced both the alkali silica reaction rate and the corrosion rate.

Effect of crack control on ASR: Crack control due to steel microfibers leads to a chemo-mechanical confinement of the ASR gel. The resistance in crack propagation and crack opening displacement not only imposes compressive stresses on the expanding ASR gel but also prevents the ASR gel from leaving the reaction site. Preventing the ASR gel from leaving the reaction site increases the ion concentration of the ASR gel. The higher Si ion concentration of the ASR gel in the SMF reinforced specimen retards further dissolution of the reactive aggregate, thereby, reducing the ASR gel production and the ASR rate. The higher viscosity of the ASR gel in SMF reinforced specimens confirms the lack of escape of the reaction products away from the reaction site.

Effect of crack control on corrosion: The fibers may act in much the same way they did in the presence of ASR. In the case of conventional plain concrete, the formation of iron-oxides induces expansive stresses which cause microcracking. Once a crack has formed, the magnitude of expansive stress required to propagate the crack is reduced (see Fig. 3a) and the rate of egress of corrosion products is increased due to the crack opening. Microfibers close to the steel surface provide a source of passive confinement. Cracks can only propagate under increases in the magnitude of expansive stress (see Fig. 3b). In this way, expansive corrosion products that form near the surface of the steel bar remain there and collect. It is postulated that under this confined condition, the solid products formed from the corrosion process will fill surrounding voids and any cracks that may have initiated, locally densifying the cement matrix and cutting off further ingress of deleterious compounds.

REFERENCES

ASTM C 1260-94. 1999. Standard test method for potential alkali reactivity of aggregates (mortar-bar method) in Annual book of ASTM Standards v. 04.02.

CERF 2004, Civil Engineering Research Foundation (CERF) website; http://www.cerf.org/conmat/hotprosp/frp2.htm

Banthia, N. & Soleimani, M. 2005. Flexural response of hybrid fiber-reinforced cementitious composites. ACI Materials Journal 102: 382–389.

Blunt, J. & Ostertag, C.P. 2007. Performance based materials approach to bridge approach slabs. ACI Materials Journal, in press.

Gamry Echem Analyst. 2003. CD-ROM. Warminster, Pennsylvania: Gamry Instruments, Inc. Version 1.30.

Grupp, J.A., Blunt, J., Ostertag, C.P. & Devine, T.M. 2007. Effect of steel microfibers on corrosion of steel reinforcing bars. Cem. Concr. Research, in press.

Helmuth, R. & Stark, D. 1989. Alkali-silica reactivity mechanisms. Mat. Sci. of Concrete III, ed. by J. Skalny & S. Mindess, , Am. C. Soc., Westernville, OH., 131–208.

Ostertag, C.P. & Yi, C.K. 2007. Crack/fiber interaction and crack growth resistance behavior in microfiber reinforced mortar specimens. Materials and Structures, in press.

Qian, C. & Stroeven, P. 2000. Development of hybrid polypropylene steel fiber reinforced concrete. Cem. Conr. Res. 30: 63–69.

Rostam, S. 2001. Design for durability: the great belt link. Concrete Technology: New Trends, Industrial Applications. Ed.by A. Auguado, R. Gettu, and S.P. Shah. RILEM.

Vail, J.G. 1952. Soluble silicates, their properties and uses, Vol1: Chemistry; ACS Monograph No. 116, Reinhold Pub. Corporation

Yi, CK. & Ostertag, C.P. 2002. Strengthening and toughening mechanisms in microfiber reinforced cementitious composites. J. Mat. Sci. 36: 1513–1522.

Yi, CK. & Ostertag, C.P. 2005. Mechanical approach in mitigating alkali silica reaction. Cem. Concr. Res. 35: 67–76.

Fracture Mechanics of Concrete and Concrete Structures – High-Performance Concrete,
Brick-Masonry and Environmental Aspects – Carpinteri, et al. (eds)
© 2007 Taylor & Francis Group, London, ISBN 978-0-415-44617-4

Experimental study on concrete carbonation to work out performance-based specifications

E. Rozière, A. Loukili, & G. Pijaudier-Cabot
Institut de Recherche en Génie Civil et Mécanique, Ecole Centrale de Nantes, Nantes, France

F. Cussigh
GTM Construction, Nanterre, France

ABSTRACT: This study is part of a research programme aimed at designing performance-based specifications for durability of concrete, through the equivalent performance concept. A concrete mixture will be qualified for a given exposure if appropriate tests show that concrete is at least as durable as a concrete mixture which complies with prescriptive specifications from the standard. This study deals with carbonation, as it is a widely spread and relatively well known exposure. 8 concrete mixtures were designed by varying three parameters – water/binder ratio, fly ash content, and aggregate type – and they comply with specified limits. As each concrete mixture underwent two curing conditions, 16 sets of specimens were made. The accelerated test showed good sensitivity and pointed out significant variability of performances. A comprehensive study of properties of porous net including commonly used durability tests suggested that behaviour of concrete exposed to carbonation would not only depend on porosity but on chemical properties of binder.

1 INTRODUCTION

Carbonation of concrete is a chemical reaction between CO_2 from air and hydrates – mainly portlandite $Ca(OH)_2$ – from cimentitious matrix. It results in production of calcite $CaCO_3$ and in a drop of pH in concrete, which may lead to corrosion of reinforcement and spalling of surface concrete. Although it is not the most severe it is a very common degradation and it has been defined as an exposure class – as well as corrosion due to chloride penetration, freeze/thaw cycles, chemical attacks – in new standard NF EN 206-1 (2004). As in most of standards, specified limits are given for maximum water/cement ratio, minimum cement content, and maximum mineral admixture proportions. The specified limits are deemed to ensure aimed durability of concrete, provided that good construction practices are followed. But the same set of specifications may lead to different actual levels of durability, and this prescriptive approach does not take into account advances in concrete research, which might have pointed out more relevant parameters to assess durability resulting of a concrete mixture than composition parameters (Baroghel-Bouny et al. 2004). European standard does not give such durability parameters and criteria but it take into account the

need for performance-based specifications through the equivalent performance concept. In accordance with this concept, a concrete mixture is qualified for a given exposure if its potential durability is at least as good as test performances of the reference concrete mixture, which complies with prescriptive specifications. The purpose of this experimental study is to investigate the variability of performances of concrete mixtures which comply with threshold values of prescriptive requirements. Concrete mixtures have been designed by varying water/cement ratio, binder composition, and aggregate type. The effect of curing condition has been investigated on the eight concrete mixtures. As good sensitivity and repeatability of performance tests are necessary to design a comparative performance-based approach, they are also an interesting result of this study. The performance test is the accelerated carbonation test, according to the French AFPC-AFREM procedure. Carbonation depths have been compared with results from usual durability tests, as porosity, gas permeability, and diffusivity are often considered as relevant parameters to assess potential durability. A study of porous distribution and properties of porous net through mercury intrusion porosimetry has been carried out to confirm results and understand effects of composition and curing on global properties.

Table 1. Mix proportions.

Exposure classes (kg/m^3)	XC1,XC2					XC3, XC4		
	B1	B2	B3	B7	B8	B4	B5	B6
Gravel Boulonnais 12/20	541	561				541	561	
Gravel Graves de mer 12/20			553	553	553			553
Gravel Boulonnais 4/12	416	432				416	432	
Gravel Graves de mer 4/12			446	425	452			446
Sand Boulonnais 0/4	992	915				980	900	
Sand Pilier 0/4			819	880	712			805
Fine sand SIFRACO 0/1					144			
Cement (C) CEMI 52,5 N	260	207	207	260	260	280	223	223
Fly ash (FA) Cordemais (k = 0.6)		89	89				95	95
W$_{eff.}$	163.8	163.8	163.8	163.8	163.8	162.4	162.4	162.4
WRA (Glenium 27)	3.82	3.65	0.97	0.50	0.60	3.67	2.98	1.69
W$_{eff.}$/Binder content	0.63	0.63	0.63	0.63	0.63	0.58	0.58	0.58
Binder content (C + k.FA)	260	260	260	260	260	260	260	260
FA/(FA + C)	0	0.30	0.30	0	0	0	0.30	0.30
Fine elements (<125 μm)	447	468	317	283	289	465	488	339
Volume of paste Vp (L)	318.1	328.3	271.4	256.3	258.4	322.3	339.7	283.5

2 EXPERIMENTAL PROCEDURE

2.1 Materials, mix proportions and curing

Eight concrete mixtures have been designed by choosing two values of each following composition parameters, namely: water/binder ratio, binder type, aggregates, as shown in Table 1. The two water/binder ratios 0.63 and 0.58 were chosen from the maximum values for XC1-XC2 and XC3-XC4 classes (defined in NF EN 206-1, Table NA.F.1). These exposure classes respectively correspond to ordinary and high carbonation risk. Binder contents comply with the same requirements, they are the minimum values given in French standard. The same Portland cement CEM I 52,5 N was used in all concrete mixtures, fly ash content of B1, B2, B5, and B6 concrete mixtures is the maximum mineral admixture content which complies with prescriptions from the standard, that is 30% of binder content. Two different aggregates mixtures were used to investigate the effect of aggregate type and aggregate mixture density. Boulonnais sand and gravel are crushed dense limestone, they have been used in other laboratory studies and their main feature is a high proportion of fine elements. Graves de mer gravels and Pilier sand are sand-lime sea aggregates, they have a low fine elements proportion and a relatively high porosity. Very fine sand was used in B8 concrete to compare the effect of mineral admixture (fly ash) with aggregate mixture density.

Two batches of each concrete were made and cast in 7 × 7 × 28 cm molds (for accelerated carbonation test) and cylindrical Ø 11 × 22 cm molds. After 24 hours of sealed curing, the first set of specimens

was cured under water for 28 days, and will be referred to as "24 h – Water". The second set of specimens was sealed cured until concrete reached 50% of the minimum 28-day characteristic strength from standard, which is 20 MPa for XC1-XC2 classes, and 25 MPA for XC3-XC4 classes. Then they were cured in a room at a constant temperature of 20° C and a constant relative humidity of 50% RH for 28 days. They will be referred to as "50%".

2.2 Durability tests

The 16 previously described materials underwent the same set of durability tests, after 28 days. Each test was done on three samples from the same batch, except the mercury intrusion porosimetry test, which was done on two samples after 90 days.

Compression tests were carried out just after sealed curing and after 28 days, on cylindrical Ø 11 × 22 cm specimens. Gas permeability was measured on cylindrical Ø 11 × 5 cm specimens in accordance with AFPC-AFREM procedure (1997), at a relative pressure of 1.00 bar, to assess apparent permeability, and at five different pressures from 2 bar to 4 bar, to assess intrinsic permeability by Klinkenberg approach.

Accelerated carbonation test was done in accordance with AFPC-AFREM procedure, in a carbonation chamber at 20° C and 65% RH, with 50% CO_2 by Laboratoire Matériaux et Durabilité des Constructions (LMDC) in Toulouse, France. Porosity was also measured in accordance with AFPC-AFREM procedure by LMDC. Chloride diffusivity was assessed from the steady state migration test called LMDC Test.

Table 2. Global properties

	Strength (MPa)		Porosity (%)		Porosity/volume of paste (%)		Apparent gas permeability ($10^{-17}\,m^2$)		Intrinsic gas permeability ($10^{-17}\,m^2$)	
	24 h – W	50%	24 h – W	50%	24 h – W	50%	24 h – W	50%	24 h – W	50%
B1	47.8	36.2	12.7	14.4	40.0	45.2	1.8	11.8	2.6	9.8
B2	40.0	31.2	13.7	15.0	41.6	45.6	8.0	23.0	3.1	21.3
B3	35.4	25.3	14.1	17.3	52.1	63.7	5.2	33.4	4.4	20.2
B7	37.8	31.3	13.6	17.3	52.9	67.4	4.9	24.9	4.3	17.3
B8	37.7	29.4	13.2	15.4	51.1	59.7	4.2	30.7	3.9	18.7
B4	52.1	38.6	14.2	13.9	44.0	43.2	35.6	20.1	6.5	10.2
B5	52.7	38.8	14.2	15.7	41.8	46.3	1.0	26.3	3.1	8.1
B6	37.5	32.5	15.1	15.9	53.1	56.2	14.4	21.4	4.2	12.5

3 RESULTS AND DISCUSSION

3.1 Global properties

Compressive strength, porosity, gas permeability, chloride diffusivity may be considered as global properties. Comprehensive results are given in Table 2. Each value is the mean value, from three tests on different samples.

A 20-MPa characteristic strength was required for B1, B2, B3, B7, and B8 concrete mixtures, and a 25-MPa characteristic strength was required for B4, B5, and B6 concrete mixtures. 28-day compressive strengths were from 50% (B6) to 139% (B1) higher than required. As concrete mixtures comply with maximum Water/Binder ratios and minimum binder contents from the standard, this result may seem surprising. However, the highest compressive strengths were measured on concrete including crushed limestone aggregates (B1, B2, B4, and B5). The sand contained 7% of fine elements (below 80 μm), which are not taken into account in binder content. So fine element contents (given in Table 1) could partly explain this difference. The difference in compressive strength of concrete mixtures including crushed limestone or sea aggregates may also come from densities of aggregate mixtures and actual Water/Binder ratios. Effective water content ($W_{eff.}$) differs from added water by the amount of water which is absorbed by aggregates. This is assessed by WA24 (water which is absorbed after 24 hours), given for each type of sand or gravel. Sea gravels were more porous and had a WA24 of 2%, instead of 0.7% for limestone gravel. But during batching and curing gravels may not absorb so much water, so the real water content of paste is higher for concrete made of sea gravels, and its strength is lower.

Porosity results show the same trend and Porosity/Volume of paste ratios are consistent with strengths. Volume of paste has been defined as $W_{eff.}$ + Volume of fine elements (<125 μm), it is given in Table 1. 125 μm is the maximum size of cement grains;

If it is assumed that porosity of concrete is porosity of paste, Porosity/Volume of paste ratios give porosity of paste (Table 2). It is closely linked to aggregate type, and the significant differences are consistent with what was assumed about real water content of paste to explain differences in strengths. Use of fine sand in B8 concrete mixture reduces concrete porosity and porosity of paste, especially for dry curing ("50%"). Fly ash would not have such a positive effect on density, but would only have a chemical part. But global properties have been assessed after 28 days, and positive effect of pozzolanic reaction might not have been observed yet.

The type of curing seems to have a significant effect, especially on gas permeability, as shown in Table 2. This does not only come from curing conditions but from configuration of samples for durability tests. Compressive strengths of studied concrete mixtures are significantly higher than required characteristic strengths. So strength at the end of sealed curing is less than 50% of real 28-day compressive strength. It actually ranges from 22% (B1 50%) to 31% (B6 50%). So because of early drying hydration rate of surface concrete of "50%" samples is relatively low and this explains the increase of transfer properties, such as gas permeability. Finally, for the "50%" curing condition, the stronger concrete undergoes the worst curing conditions. This discrepancy between required and actual compressive strength and its consequences on actual curing is also to be kept in mind to analyze results from carbonation test. But configuration of samples and testing device should also be taken into account to explain effect of curing conditions on gas permeability. The test is done on cylindrical Ø 11 × 5 cm specimens sawed from Ø 11 × 22 cm samples. As gas flux is transverse, it may flow through surface concrete layer, which is likely to be more permeable, for the "50%" curing condition, because of early drying. In real atmospheric conditions, air or gas gets into concrete from the surface, so configuration and effect of curing conditions are different.

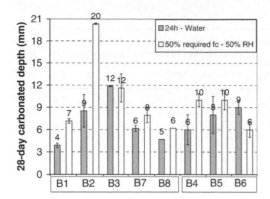

Figure 1. Carbonated depths after 28 days of accelerated test.

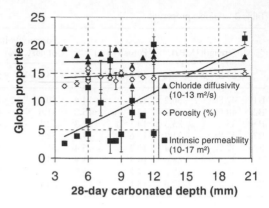

Figure 2. Carbonated depths and global properties.

3.2 Accelerated carbonation test

The main result of the accelerated carbonation test is carbonation depth. It is assessed by a colorimetric method on a cross section of prismatic $7 \times 7 \times 28$ cm samples, at 7, 14, 21, 28, and 56 days.

28-day carbonated depths are given in Figure 1. Standard deviations are given on the graph. In spite of dispersion of some results, significant differences may be observed. Most of carbonation depths ranged from 4 to 12 mm, whereas Water/Binder ratio and Binder content were kept constant. So the accelerated carbonation test is interesting in a performance based approach, as it is sensitive. Moreover classifications of materials seem not to be affected by curing condition, which is of great interest in a comparative approach.

Effects of composition parameters and curing conditions appear on the results of the accelerated test. "Dry" cured ("50%") concrete generally had higher carbonation depth than concrete cured under water, except for B6 50% concrete. It should be noted that this result may not be relevant, as it is not consistent with other properties of the same material (see Table 2) or results for B3 concrete (Fig. 2), which had the same binder and aggregates.

Water/Binder ratio had not systematically the expected effect on carbonation, as B1, B7, and B8 concrete mixtures (W/B = 0.63) shown lower carbonated depth than B4, B5, and B6 concrete mixtures (W/B = 0.58). But the difference between the two W/B ratios is relatively low, and the difference between the water contents – respectively 163.8 and 162.4 L/m^3 – is even more reduced (0.9%), as binder content also varied. So the accelerated test might not have been sensitive enough to show the difference, which also occurred for other properties, such as porosity and gas permeability. The effect of fly ash is clearer and has already been reported (Papadakis 1999). From concrete mixtures which had the same aggregates and W/B ratio, it may be deduced a negative effect of fly ash. That may be explained by a lower portlandite content of the hydration products. Moreover, pozzolanic reaction reduces portlandite content. The effect of aggregates could have been more complicated. Moreover the test may not be sensitive enough to draw conclusions, and "50%" curing may have lead to better curing for sea aggregates concrete mixtures than limestone aggregates concrete mixtures, as shown in 3.1. As far as water cured concrete are concerned, from B1/B7, B2/B3, and B5/B6 concrete mixtures, it may be deduced that the crushed limestone aggregates lead to better resistance against carbonation. This may come from the density of the mix, as the sea aggregates mix lacks fine elements. In B8 concrete mixture, fine sand was added, and the concrete seemed to have a better behavior than B7.

In Figure 2 the second graph results from durability tests are plotted against carbonation depth. The highest sensitivity was shown with gas permeability, but no good correlation between global properties and carbonation may be deduced from these results. For instance the same porosity can be associated with very different carbonation rates. This has already been shown in literature (De Schutter & Audenaert 2004).

3.3 Mercury intrusion

As global properties may be difficult to analyze, mercury intrusion porosimetry (MIP) tests have been carried out, to confirm results and to investigate effects of properties of porous net of concrete. Pore size distribution curves and quantitative results, such as porosities, median and average pore diameters, may be useful to analyze properties of concrete (Roy & al. 1999).

MIP tests were carried out on B4 samples after 28 and 90 days, to study the effect of curing condition, as shown in Figure 3. The porous mode which appears clearly at 28 days for "50%" curing tended to decrease, but it could still be observed after 90 days, and it has an influence on median pore diameter.

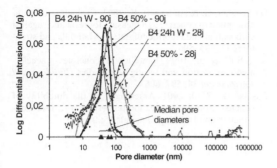

Figure 3. Pore size distribution of B4 concrete.

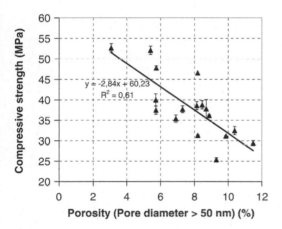

Figure 4. Compressive strength.

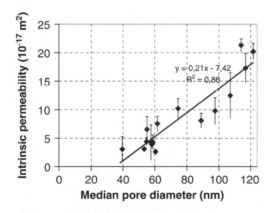

Figure 5. Intrinsic permeability.

Median pore diameter may be used as a quantitative result to be linked to global properties. In Figure 4, compressive strength is plotted against macroporosity, which is defined as porosity for pore diameters 50 nm (Basheer & al. 2001).

Compressive strength decreases with an increase in porosity, which is to be linked to W/B ratio of paste. In Figure 5, intrinsic permeability is plotted against median pore diameter. Equations may be found in literature to link porosity and permeability, through MIP and other data. From the results of the study, pore diameter seems to be a relevant parameter as far as gas permeability is concerned.

4 CONCLUSIONS

This study includes accelerated carbonation test and a comprehensive study of porosity of 16 concrete mixtures. It can be used as a work on standards to translate prescriptions into data on potential durability and as an investigation on durability of concrete exposed to carbonation.

In the standard context, the objective was to investigate performances of reference concrete mixtures for XC exposure classes. Experimental data from accelerated test and other durability tests have shown significant variability of properties and resistance to carbonation. Carbonation depths after 28 days of accelerated test ranged continuously from 4 to 12 mm, whereas the studied concrete mixtures had the same binder content and water/binder ratio, in the standard meaning.

This experimental work gives data to investigate the effect of composition parameters and curing conditions. The effect of binder type was found to be more significant than the effect of aggregates type or water/binder ratio. So, for this set of concrete mixtures carbonation would mainly depend on chemical behavior, rather than density of concrete, and a study on portlandite content of paste and concrete would bring useful data. Saturation degree of concrete has not been assessed but it could explain some trends, as CO_2 diffusivity is much higher in dry concrete than in saturated porosity. Curing condition is a major parameter, but it has to be defined precisely, and one has to pay attention to the configuration of samples and testing device.

ACKNOWLEDGEMENTS

The authors would like to acknowledge the financial support of the Fédération Nationale des Travaux Publics (FNTP), Paris, France. The authors are grateful to Mr. Raphaël Edieux, of LMDC, in Toulouse, France, for his technical support.

REFERENCES

NF EN 206-1 standard, Béton – Partie 1: Spécification, performances, production et conformité, AFNOR, 2004.

Baroghel-Bouny, V & al. 2004. *Conception des bétons pour une durée de vie donnée des ouvrages*. Paris. Association Française de Génie Civil.

Compte-rendu des journées techniques AFPC-AFREM Durabilité des bétons, Méthodes recommandées pour la mesure des grandeurs associées à la durabilité, 11 et 12 décembre 1997, Toulouse, 1997, p. 121–158.

Papadakis, V. G. 2000. Effect of supplementary cementing materials on concrete resistance against carbonation and chloride ingress. *Cement and Concrete Research* 30: 291–299.

De Schutter, G. & Audenaert, K. 2004. Evaluation of water absorption of concrete as a measure for resistance against carbonation and chloride migration. Materials and structures 37: 591–596.

Roy, S.K. & al. 1999.Durability of concrete – accelerated carbonation and weathering studies. *Building and Environment* 34: 597–606.

Basheer, L. & al. 2001. Assessment of the durability of concrete from its permeation properties: a review. Construction and Building Materials 15: 93–103.

Fracture Mechanics of Concrete and Concrete Structures – High-Performance Concrete,
Brick-Masonry and Environmental Aspects – Carpinteri, et al. (eds)
© 2007 Taylor & Francis Group, London, ISBN 978-0-415-44617-4

Reinforced-concrete cover cracking due to the pressure of corrosion products

A. Muñoz & C. Andrade
Institute of Science Construction Eduardo Torroja, Madrid, Spain

A. Torres
Mexican Transport Institute, Queretaro, Mexico

ABSTRACT: Reinforcement corrosion is one of the most important phenomena that reduce the service life of the reinforced concrete structures. Steel corrosion reduces the strength and bond of the reinforcement and the oxides formed cause internal stresses that crack the concrete cover. The prediction of the evolution of these effects is a problem that involves chemical and mechanical aspects. The phenomenon is not well known and quantitative descriptions of development and magnitude of stresses produced by a corroding rebar to concrete are scarce and although there are several models in the literature they do not universally reproduce the experimental results. This work presents a summary of the available experimental evidence on the amount of internal expansion needed at the reinforcement level for concrete cover to crack and formulas to estimate the expansion. Also, results are presented on the pressure needed for concrete cover to crack together with the experimental technique used for the verification of the assumptions and crack propagation analysis.

1 INTRODUCTION

The steel reinforcement is protected from corrosion by passivation due to the high alkaline environment provided by the cement hydration. However, in the marine ambient chloride ions from seawater accumulate on the surface of the concrete and slowly diffuse through the concrete cover to the underlying steel. When the chloride ion concentration at the rebar depth exceeds a critical threshold value, the protective passive layer on the steel surface breaks down and active steel corrosion begins.

The transformation from steel to corrosion products in concrete is only partially understood. Analytical techniques cannot be used in situ to determine the type of corrosion products generated at the embedded rebar surface without exposing the rebar to the exterior. Furthermore corrosion products may oxidize upon exposing the rebar to air. Although discrepancies on the type of corrosion products formed at the steel/concrete interface are still present (Bedu 1993, Fontana 1986, Sagoe-Crentsil & Glasser 1989a, b, 1993), it appears nevertheless well established that these corrosion products have smaller mass densities than steel (Tuutti 1982), resulting in volume expansion and concrete cover cracking.

The mechanical process of the corrosion product expansion due to corrosion is shown in Figure 1. The steel might be considered as a metal cylinder with an initial radius r_0, immersed in a semi-infinite concrete

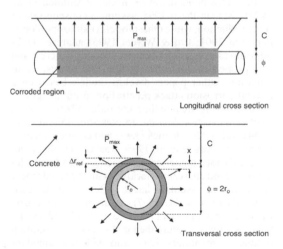

Figure 1. Corrosion process parameters.

medium with a cover C, and undergoing corrosion only in the region L. As corrosion progresses, the radius decreases by an amount x (corrosion penetration). However, corrosion products occupy a volume that is larger than the original metal. The final volume increase corresponds to an increase Δ_{ref}, over the initial rebar radius, for a total value of $r_0 + \Delta_{ref}$. The surrounding concrete is stressed by this effective radial expansion and provokes the concrete-cover cracking and spalling.

In the literature (Atimatay & Ferguson 1973, Bazant 1979a, b, Beeby 1983, Andrade et al. 1993a, Andrade et al. 1996, Alonso et al. 1994, Andrade et al. 1998), some experimental studies, theoretical investigations, and field observations of concrete cracking due to corrosion product expansion have been documented. However, fundamental aspects of the cracking mechanism essential for durability forecasting, remain unsolved.

As explained in the literature (Reinhardt 1984), rebar corrosion induces the development of internal stresses that may crack the concrete. However, quantitative descriptions of development and magnitude of stresses produced by a corroding rebar to concrete are scarce. Thus, further information on the relationship between corrosion expansion and internal pressure is desirable for modeling predictions.

The relation between the crack opening and the quantity of oxide generated by the corrosion expressed as the penetration of the corrosion or loss of diameter of the bars has been the subject of previous works by the authors by means of accelerated and not accelerated corrosion tests. One model (Leung 2001) obtains one upper and lower bound assuming the steel / concrete interface to be perfectly smooth or perfectly bonded. Some models (Andrade et al. 1993b, Martín-Perez 1998) assume a constant rate of rust production, while other models (Pantazopoulou & Papoulia 2001, Liu & Weyers 1995) analyze cracking time as a function of concrete cover, concrete and rust properties controlled by the rate of rust accumulation. Other papers develop models based on a critical corrosion attack penetration to initiate cracking and they relate it to the rebar radius (Torres 1999), steel cross section loss due to corrosion (Vidal et al. 2004) and cover / diameter ratio and concrete characteristics (Andrade et al. 1995, Rasheeduzzafar et al. 1992). Various numerical approaches use a finite element method analyzing cracking with the fixed smear crack model, assuming linear softening of the concrete (Padovan & Jae 1997), assuming linear elastic fracture mechanics and movable mesh placed around the crack tip to capture the local stress concentration (Ohtsu & Yoshimura 1997) and with the boundary element approach (Torres & Sagüés 2000). All calculations and the simulated cracking patterns of the papers are compared with experimental tests. It can be concluded, in general, that the beginning of the cracking depends principally on the relation between concrete cover thickness / diameter of the bars, the quality of the concrete and its tensile strength.

This work contributes to the study of the pressure needed to crack a certain cover and to confirm a predictive model for corrosion penetration taking into account specimen dimensions and fracture mechanic properties of the concrete.

2 EXPERIMENTAL WORK

2.1 Materials and specimens

The concrete mix was made with ordinary portland cement type II and the mix proportions (in kg/m^3) for each specimen are presented in the Table 1.

The steel used was BS-500 having a 16 mm diameter. In the rebars of all specimens, four strain gauges were glued to measure the strain (pressure indirectly) at the steel-concrete interface. The specimen sizes are shown in the Figures 1 and 2. The specimens were cured for 24 hours in the moulds and 28 days in a curing room with 95% RH and 20°C. After the 28 day curing, the specimens were dried for some days

Table 1. Concrete mix proportions.

| Material | Specimens | |
	C1	P1, P2, P3
Cement	320	327
Max. Agg.	650	1016
Fine Agg.	1240	975
Water	200	165

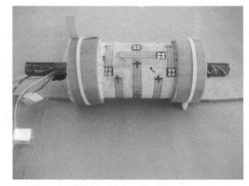

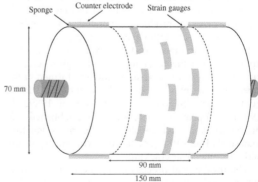

Figure 2. Cylinder C1 specimen.

to allow strain gauges at the concrete surface to reach the cracking moment. These were located as shown in Figures 2 and 3.

2.2 Accelerated corrosion and cracking test

A 90 mm corrosion length was used for the C1 specimen and 230 mm for P1, P2 and P3. In order to activate the corrosion process, a 3% NaCl in cement weight were added to the mix and to accelerate the corrosion process, an electrical current (galvanostatic) was applied to the steel bar.

The accelerated corrosion procedure employed for the specimens consists in a galvanostat that applies a constant current density through the counter electrode placed at the ends of the specimens (Figure 2 and 3). The electric contact between the counter electrode and the concrete surface was provided by sponges maintained moistened by a water dropping system.

Three different current densities were used. Until the first crack appearance, the current densities were 1 (P1), 5 (P2) and 10 (C1 and P3) $\mu A/cm^2$ and after the first crack appearance the current densities were 10 (P1), 50 (P2) and 100 (C1 and P3) $\mu A/cm^2$ to follow the crack evolution.

The test is considered to end when a target cracking size is reached. After this, the specimens

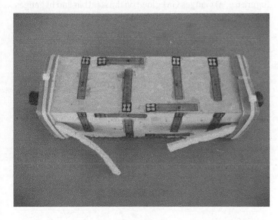

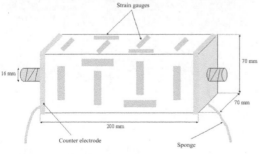

Figure 3. Prisms P1, P2 and P3

weredisconnected from the corrosion equipment and were broken to study the types of oxides, the colors and their spread. Then, the bars were cleaned, dried and weighed to obtain the difference with respect to the initial weight (gravimetrical loss)

The conversion of corrosion rate into radius loss was performed using a formula based on Faraday's law (Andrade et al. 1993b):

$$P_x = 0.0116 \ I_{corr} \ t \tag{1}$$

were x is the attack penetration in microns, I_{corr} is the current density in $\mu A/cm^2$, t is the elapsed time in years since the current was applied, and 0.0116 is a conversion factor of $\mu A/cm^2$ to $\mu m/año$ in the case of homogeneous corrosion.

Preliminary, the radius losses were calculated by means of the expression (1). That is, all the current applied is assumed to be spent in the oxidation of the steel (100% of current efficiency is assumed). The losses so calculated are named "theoretical" steel losses. However, the 100% efficiency of the current was not produced and the "real" steel radius losses in every case were higher than the "theoretical" loss. This fact was verified by comparing the theoretical loss with the gravimetrical loss at the end of the tests.

3 RESULTS

3.1 Expansion evolution

Figure 4 shows the behavior of the strain gauges glued on the steel and the concrete surface of the C1 specimen. The gauge GM1 glued on the steel bar (Figure 4a) shows the most informative steel concrete interface stress behavior (by strain) due to the corrosion process. In the case of the gauges placed on the concrete surface (Figure 4b), it takes longer to show the expansion of gauges G3 and G9 where the first detecting the crack appearance.

For the case of prisms P1, P2 and P3 (Figures 5 to 7) the strain gauges glued on the steel bar (part a of Figures) measured similar behavior to the C1 specimen but the gauges placed on the concrete surface (part b of Figures) detect higher strains attributed to the swelling due to the permanent contact with the chloride solution. After 30 days, the gauges placed on the concrete surface showed more stability.

3.2 Relation Attack penetration/cracking of concrete cover

Previous investigations (Andrade et al. 1993a, Alonso et al. 1998) reported that the amount of corrosion needed to crack the concrete cover was only 15 to 50 microns for specimens with uniform corrosion, while the other author (Rodriguez et al. 1996) reports

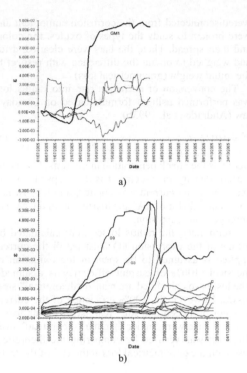

Figure 4.   a) Gauges GM glued on the steel bar and b) gauges G placed on the concrete surface of C1 specimen.

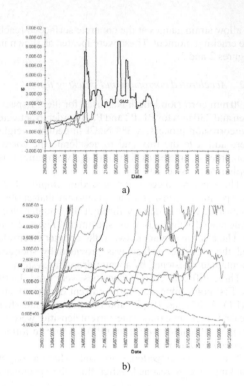

Figure 6.   a) Gauges GM glued on the steel bar and b) gauges G placed on the concrete surface of P2 specimen.

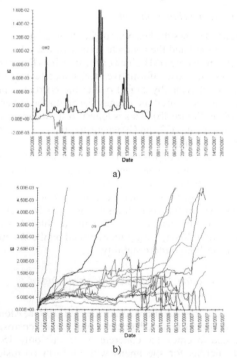

Figure 5.   a) Gauges GM glued on the steel bar and b) gauges G placed on the concrete surface of P1 specimen.

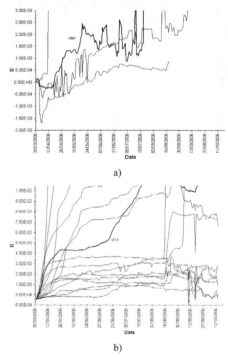

Figure 7.   a) Gauges GM glued on the steel bar and b) gauges G placed on the concrete surface of P3 specimen.

Table 2. Corrosion penetration needed to crack the concrete cover.

Specimen	Attack penetration (μm)
C1	8.51
P1	4.26
P2	19.18
P3	34.78

Table 3. Results of corrosion amount estimated by equations (2) and (3) with the results of this work.

Eq.	Amount of corrosion needed to first crack generation (microns)			
	C1	P1	P2	P3
2	31.87	35.36	35.36	35.36
3	31.24	23.62	23.98	23.77
real	8.51	4.26	19.18	34.78

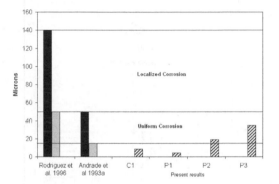

Figure 8. Amount of corrosion or attack penetration needed to crack the concrete cover; results of this work compared and results of other authors.

amounts of 50 to 140 microns for specimens with localized corrosion. For the tests carried out in this work, the corrosion penetration needed to crack the concrete cover of the specimens is shown in Table 2. The attack penetration results of P2 and P3 specimens are similar to the studies for specimens with uniform corrosion (Figure 8). While specimens C1 and P1 needed an even smaller amount of corrosion to crack.

4 DISCUSSION

4.1 Test methodology

The strain gauges glued on the steel bar and placed on the concrete surface is a common technique used to try to detect the concrete cover cracking initiation (Andrade et al 1993a, b and Torres 1999). The results presented in this work confirm that the strain gauge test methodology can give important information to estimate the pressure needed to initiate the concrete cover cracking, but it is important to mention that not all the gauges work correctly. The temperature and humidity can affect the correct behavior of the gauges and the measurements can give erroneous information.

4.2 Amount of corrosion needed to crack the concrete cover

There are some models to estimate the corrosion amount needed to crack the concrete cover in the

bibliography (Rasheeduzzafar et al. 1992, Andrade et al. 1993b, Liu & Weyers 1995, Rodriguez et al. 1996, Ohtsu & Yoshimura 1997, Padovan & Jae 1997, Martin-Perez B. 1998, Torres 1999, Leung 2001, Pantazopoulou & Papoulia 2001, Vidal et al. 2004). Equations (2) and (3) given by Rodriguez et al. 1996 and Torres 1999 respectively are used to compare the theoretical values of the corrosion amount needed to crack the concrete cover with the results obtained in this paper. Equation (2) considers the cross section properties of the specimens and concrete tensile strength while Equation (3) considers only the cross section properties and the corrosion length.

$$P_{xo} = \left(83.8 + 7.4\frac{C}{\phi} - 22.6 f_{ct,sp} \right) \cdot 10^{-3} \tag{2}$$

$$x_{crit} = 0.0111 \frac{C}{\phi} \left(\frac{C}{L} + 1 \right)^{1.95} \tag{3}$$

where C is the concrete cover (mm), ϕ is the steel bar diameter (mm), L is the corrosion length (mm), P_{xo} is the corrosion amount needed to generate the first crack (mm) and $f_{c,sp}$ is the concrete splitting tensile strength (Kg/cm^2).

Table 3 shows the results of the corrosion amount needed to cracking initiation given by Equations (2) and (3) compared with the results obtained in this work.

The attack penetration by corrosion in the P3 specimen estimated with Equation (3) is similar to the real values obtained in the tests although Equation (3) is calibrated with localized corrosion results and the corrosion lengths were between 5 and 6 times greater approximately. The real attack penetration in P3 specimen is greater than the results obtained with Equation (2) but very similar to the results obtained with Equation (3). The attack penetration obtained in the C1 specimen is smaller than the estimated values with Equations (2) and (3) which is attributed to the cylindrical section of C1 and to the low corrosion rate applied (1 μA/cm^2) in the P2 specimen.

Table 4. Results of P_r in kg/cm^2.

Specimen	Periods	Strain gauges	P_r Ec. (4)	P_r Ec. (5)	P_r Ec. (6)
C1	1	GM1	101.2	93.6	111.2
		G3	101.2	93.6	29.0
	2	GM1	101.2	93.6	183.2
		G3	101.2	93.6	115.4
P1	1	GM2	80.7	80.3	81.7
		G9	80.7	80.3	43.6
	2	GM2	80.7	80.3	134.6
		G1	80.7	80.3	136.6
P3	1	GM2	80.7	80.3	81.9
		G1	80.7	80.3	34.2
	2	GM2	80.7	80.3	257.7
		G1	80.7	80.3	298.6
P4	1	GM1	80.7	80.3	81.2
		G14	80.7	80.3	452.8
	2	GM1	80.7	80.3	176.7
		G14	80.7	80.3	799.5

4.3 Pressure needed to concrete cover cracking

To determine the pressure needed to crack the concrete cover P_r Equation (4) (Torres 1999), (5) (Sagüés et al. 1998) and (6), for a thick walled cylinder (Timoshenko 1989), were considered.

$$\frac{P_r}{f_t} = 1.54 \frac{C}{\phi} \left(\frac{C}{L} + 1 \right)^{0.72} \tag{4}$$

$$P_r = 1.5 \left(\frac{C}{\phi} \right)^{0.85} \left(\frac{C}{L} + 1 \right) f_t \tag{5}$$

$$P_r = \frac{E \left(R_2^{\,2} - R_1^{\,2} \right) \varepsilon}{R_2^{\,2} \, 2} \tag{6}$$

Where: f_t is the concrete tensile strength in kg/cm^2, C/ϕ is the cover/diameter ratio, C/L is the cover/length ratio, R_1 y R_2 are the internal radius (steel bar diameter) and external radius (specimen concrete cover) in mm, ε is the strain measured by the strain gauges placed in the concrete surface and E is the Young modulus for the concrete in kg/cm^2.

Table 4 shows the results obtained with Equations (4), (5) and (6) for the 4 specimens tested. The strain values with the GM1 strain gauge glued on the steel bar and the G3 strain gauge placed in the concrete surface were used to make the calculation with Equation (6) because both gauges detect the first crack appearance. The strain gauges GM1 and GM2 (for P3 and for P1 and P2 specimens respectively) were glued on the steel bar and gauges G9, G1 and 14 (for P1, P2 and P3 specimens respectively) placed on the concrete surface there were used too.

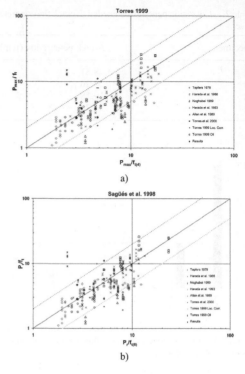

Figure 9. Results P_r/f_t obtained with Equations (4) and (6) for some authors and test specimens.

The strain data obtained with the gauges are considered in two periods as Table 4 shows. After 15 days of accelerated corrosion of specimen C1 the gauges GM1 and G3 measured considerable strain at the steel bar and the concrete surface (Figure 4). Following the gauge GM1 behavior, it was observed that it measured a "cracking" strain (P_r) at 30 days and is when assumes the maximum pressure produced on the concrete to generate the first crack (end of first period). Meanwhile, the gauge G3 continues the strain measure at the concrete surface. A gauge GM1 relaxation was observed after the 30 days indicating that the first crack generation at the interface and the corrosion products cannot maintain the same pressure because they fill the hollow generated by the crack. The second period ended when the first crack at the concrete surface appeared and the gauge G3 measured the maximum strain (after 60 days approximately).

The same procedure was followed for the P1, P2 and P3 specimen with the respective gauges mentioned before.

Two models were chosen to estimate the pressure needed to crack the concrete cover. Those by Torres 1999 and Sagüés et al. 1998 (Equations (4) and (5)).

A comparison between the results of other authors and the results obtained in this paper of P_r/f_t estimated with Equations (4) and (6) are shown in Figure 9.

The behavior of both models is similar but the results obtained with the model proposed by Torres 1999 is less scattered than the results obtained with the model proposed by Sagüés el al. 1998. The test results obtained with glued gauges in the steel bar and at the concrete surface of the specimen's technique and with uniform corrosion lengths remain in general model proposed trends.

5 CONCLUSIONS

The strain gauges glued on the steel bar and on the concrete surface of the test specimens provided good information to detect the concrete cover crack initiation period.

The galvanostatic procedure has proven to be an important tool in maintaining a constant rate of oxidation, although only the final gravimetrical losses can give reliable results.

The corrosion rate has a very significant influence on the limits of attack penetration to generate the first crack. A slower corrosion rate generates earlier cracking with lower attack penetrations.

From the test carried out in this work it is confirmed that the attack penetration or radius losses of $10-50\,\mu\text{m}$ are necessary to generate crack widths of 0.05–0.1 mm.

Finally, the maximum pressure needed to crack the concrete cover can be estimated with the data obtained by the gauges and the models proposed. This pressure reported to be in present assumed 80 to $100\,\text{kg/cm}^2$.

REFERENCES

Allan, M. L. & Cherry, B. W. 1989. Mechanical simulation of corrosion induced cracking in reinforced concrete, *Corrosion/89*, Conference paper No. 377, NACE, Houston, Texas.

Alonso, C., Andrade, C., Rodriguez, J., Casal, J. & Garcia, M. 1994. Rebar corrosion and time to cover cracking. *In concrete across borders international conference*, Odense, Denmark, pp. 301–319.

Alonso, C., Andrade, C., Rodriguez, J. & Diez, J. M. 1998. Factors controlling cracking of concrete affected by reinforcement corrosion. *Materials and Structures*, 31, August–September, pp. 435–441.

Andrade, C., Alonso, C. & Molina, F. J. 1993a. Cover cracking as a function of rebar corrosion: Part I – Experimental test. *Materials and Structures*, 26, pp. 453–464.

Andrade, C., Alonso, C. & Molina, F. J. 1993b. Cover cracking as a function of rebar corrosion: Part II – Numerical model. *Materials and Structures*, 26, pp. 532–548.

Andrade, C., Alonso, C., Rodriguez, J. & Casal, J. M. 1995. Relation between corrosion and concrete cracking. *Internal report* Brite/Euram BE-4062. DG XII, C.E.C.

Andrade, C., Alonso, C., Rodriguez, J. & Garcia, M. 1996. Cover cracking and amount of rebar corrosion: importance of the current applied accelerated tests. *In concrete repair,*

Rehabilitation and protection, R. K. Dhir and M. R. Jones eds., E&FN Spon, London, UK, pp. 263–273.

Atimatay, E. & Ferguson, M. 1973 Early corrosion of reinforced concrete – A test report. *ACI Structural Journal*, 70, 9, pp. 606–611.

Bazant, Z. P. 1979. Physical model for steel corrosion in concrete sea structures – theory. *Journal Structural Division*, ASCE, 105, ST6, pp., 1137–1153.

Bazant, Z. P. 1979 Physical model for steel corrosion in concrete sea structures – application. *Journal Structural Division*, ASCE, 105, ST6, pp., 1155–1166.

Bedu, P. 1993. Volumetric changes of cement paste under exposure to the simulated corrosion products of steel and their influence on cracking susceptibility. M. Sc. Eng. Thesis, Florida Atlantic University, Boca Raton, Florida.

Beeby, A. W. 1983. Cracking, cover and corrosion of reinforcement. *Concrete International*, 5, 2, Pg. 35–40.

Fontana, M. G. 1986. *Corrosion engineering*. McGraw-Hill, 3rd ed., pp. 556, New York, New York, USA.

Harada, T., Idemitsu, T. & Watanabe, A. 1986. Demolition of concrete with expansive demolition agents. *Concrete library of Japanese Society of Civil Eng.*, 3 (360), pp. 63–81.

Harada, T., Soeda, K., Idemitsu, T. & Watanabe, A. 1993. Characteristics of expansive pressure and expansive demolition agent and the development of new pressure transducers, *Conc. Lib. JSCE*, 21 (478), pp. 95–109.

Leung, K. Y. 2001. Modeling of concrete cracking induced by steel expansion. *Journal of Materials in Civil Engineering*, May–June.

Liu, Y. & Weyers, R. E. 1995. Modeling the time to corrosion cracking in chloride contaminated reinforcement concrete structures. *ACI Materials Journal*, 95 (6), pp. 675–681.

Martín-Perez B. 1998. Service life modeling of RC highway structures exposed to chlorides. Ph.D. dissertation, Dept. of Civil Engineering, University of Toronto.

Noghabai, K. 1999. Discrete versus smeared versus element-embedded crack models on ring problem, J. Eng. Mechs., ASCE, 125 (3), pp. 307–315.

Ohtsu, M. & Yoshimura, S. 1997. Analysis of crack propagation and crack initiation due to corrosion reinforcement, Const. And Build. Mat., 11 (7–8), pp. 437–442.

Padovan, J. & Jae J. 1997. FE modeling of expansive oxide induced fracture of rebar reinforced concrete, *Engineering Fracture Mechanics*, 56 (6), pp. 797–812.

Pantazopoulou, S. J. & Papoulia, K. D. 2001. Modeling cover cracking due to reinforcement corrosion in RC structures. *Journal of Engineering Mechanics*. April.

Rasheeduzzafar, S., Al-Saadoun S. & Al-Gahtani, A. S. 1992. Corrosion cracking in relation to bar diameter cover and concrete quality. *Journal of Material in Civil Engineering*, Vol. 4 (4).

Reinhardt, H. W. 1984. *Fracture mechanics of an elastic softening material like concrete*. Heron (ed.). Vol. 29 No. 2.

Rodríguez, J., Ortega, L., Casal, J. & Diez, J. 1996. Corrosion reinforcement and service life of concrete structures, *Durab. Build. Mater. Compon.* 7 (1), pp. 117–126.

Sagoe-Crentsil, K. K. & Glasser, F. O. 1989a. Steel in concrete: Part I. A review of the electrochemical and thermodynamic aspects. *Magazine of Concrete Research*, 41, 149, pp. 205–212.

Sagoe-Crentsil, K. K. & Glasser, F. O. 1989b. Steel in concrete: Part II. Electron microscopy Analysis. *Magazine of Concrete Research*, 41, 149, pp. 213–220.

Sagoe-Crentsil, K. K. & Glasser, F. O. 1993. Constitution of green rust and its significance to the corrosion of steel in Portland cement. *Corrosion*, 49, 6, pp. 457–463.

Sagüés, A, and Torres A. 1998. "Concrete cover cracking and corrosion expansion of embedded reinforcing steel". Proceedings of the third NACE Latin American region corrosion congress on rehabilitation of corrosion damaged infrastructures, Castro, P., Troconis, O. y Andrade, C. eds., pp. 215–229.

Tepfers, R. 1979. Cracking of concrete cover along anchored deformed reinforcing bars, *Mag. Conc. Res.*, 31 (106), pp. 3–12.

Timoshenko, S. 1989. *Strength of materials*, Espasa (ed.), Vol. II: Theory and complex problems, pp. 244.

Torres, A. 1999. Cracking induced by localized corrosion of reinforcement in chloride contaminated concrete. Ph. D. Thesis, University of South Florida, Florida, USA.

Tuutti, K. 1982. *Corrosion of steel in concrete*. Swedish Cement and Concrete Research Institute, Stockholm, Sweden.

Vidal, T., Castel, A. & Francois, R. 2004. Analyzing crack width to predict corrosion in reinforced concrete. *Cement and Concrete Research* 34, pp. 165–174.

Fracture Mechanics of Concrete and Concrete Structures – High-Performance Concrete,
Brick-Masonry and Environmental Aspects – Carpinteri, et al. (eds)
© 2007 Taylor & Francis Group, London, ISBN 978-0-415-44617-4

Monitoring system for chloride content in concrete cover using electromagnetic wave and impedance methods

T. Mizobuchi
Hosei University, Tokyo, Japan

K. Yokozeki, K. Watanabe, M. Hiraishi & R. Ashizawa
Kajima Corporation, Tokyo, Japan

ABSTRACT: Reinforcement corrosion caused by the presence of chloride ions in the neighborhood of the re-bars has been identified as one of the major causes of deterioration of concrete structures. The chlorides could find their way to concrete either as part of constituent materials when sea sand is used, or, by gradual permeation and diffusion as in the case of marine structures, or, cases where deicing salts are used to melt away snow on highways, etc. Thus, determination of chloride content in a concrete structure is an important part of periodic nondestructive testing carried out for structures identified to be vulnerable to chloride induced reinforcement corrosion. In this report, the applicable monitoring system which is possible to estimate the distribution of chloride content included in cover concrete in the existing structures using electromagnetic wave method and impedance method which are non-destructive testing is reported.

1 INTRODUCTION

Reinforcement corrosion caused by the presence of chloride ions in the neighborhood of the re-bars has been identified as one of the major causes of deterioration of concrete structures. The chlorides could find their way to concrete either as part of constituent materials when sea sand is used, or, by gradual permeation and diffusion as in the case of marine structures, or, cases where deicing salts are used to melt away snow on highways, etc. Thus, determination of chloride content in a concrete structure is an important part of periodic nondestructive testing carried out for structures identified to be vulnerable to chloride induced reinforcement corrosion.

On the other hand, a definite understanding about any corrosion of reinforcement is very difficult unless corrosion induced cracks appear on the surface. Thus, in order to detect chloride-induced corrosion at an early stage, chloride content within concrete needs to be investigated using cores drawn from the RC structure, and carrying out chemical analysis. Now, drawing cores could be structurally unacceptable, damage the reinforcement and the repair could be aesthetically unappealing, only very limited sampling can actually be carried out. In addition, drawing cores to estimate the chloride content in concrete could not make it possible to study the changes in chloride content over time (at exactly the same place). On the other hand, it

has been obtained from laboratory tests under limited conditions that chloride content within concrete has could be almost estimated by using electromagnetic waves as one of the non-destructive tests. However, as all specimens have be added sodium chloride during the mixing of concrete in the laboratory tests, the experiments have been not carried out to estimate chloride content using specimens permeated chlorides from the concrete surface after placing. Since experiments to estimate the chloride content in concrete using electromagnetic waves have been carried out to apply existing concrete structures to be a premise, the convenient measuring instrument that the frequency was fixed was used. However, it should be considered that excellence frequency may be changed by water content and chloride content in concrete. In case of the estimation of chloride content using the electromagnetic wave, though it is possible to estimate only average chloride content from the concrete surface to the reinforcing bar, it is not possible to estimate the distribution of chloride content from concrete surface to the reinforcing bar.

Furthermore, in our studies, experiments have been carried out to estimate moisture content and chloride content near concrete surface from resistance value and frequency of concrete using impedance method. From results of the experiments, the frequency band as maximum effective value has tended to be changed by chloride content and water content with age. Also, maximum effective value and frequency has tended to increase according as the measurement point was closer to the concrete surface. In addition, the maximum effective value has tended to increase according to decrease average water content of the concrete. By utilizing results of our studies obtained until now, it is possible that the system in which it is able to carry out the monitoring using these methods of non-destructive testing is developed in respect of grasping the distribution of the chloride content in cover concrete.

In this report, the applicable monitoring system which is possible to estimate the distribution of chloride content included in cover concrete in the existing structures using electromagnetic wave method and impedance method which are non-destructive testing is reported.

2 ESTIMATION OF CHLORIDE CONTENT USING ELECTROMAGNETIC WAVE

2.1 Outline on measurement of chloride content in concrete using electromagnetic waves

The dielectric constant of dry concrete varies in between 4 to 10, and that of wet concrete in the range of 10 to 20. Thus, the dielectric constant of concrete varies depending on the moisture content of concrete. As mentioned above, the dielectric constant is the same for both fresh water and seawater.

In this study, as shown in Figure 1, when the distance to the reflecting surface, i.e. reinforcing bar, was known, it was detected that changes in the properties of the electromagnetic waves such as the dielectric constant are caused by differences in the properties of the intervening medium such as the moisture content, which in this case is concrete. Furthermore, it was also confirmed that changes of dielectric constant in the

electromagnetic waves weren't caused by differences in the chloride content in concrete.

As mentioned above, the conductivity is likely to vary considerably with the amount of chloride ions in concrete. That is to say, when the electrolyte like sodium chloride exists in the concrete, it seems to be changed the electrical properties such as conductivity in comparison with the concrete without the chloride ions. Furthermore, because the amount of electrolyte varies with the difference of chloride ions concentration, the amount of electrolyte also changes depending on the moisture content in the concrete.

In this study, therefore, it was detected that changes in reflected waveform of electromagnetic waves were caused by differences in chloride ions concentration and moisture content in the concrete.

2.2 Method of investigation

Experiments were carried out by dividing them into two steps. In the first step, experiment were carried out to chloride content in concrete from properties of electromagnetic wave waveforms measured using test specimens cast using varying chloride content. Further measurements for attenuation of the output waveforms and the dielectric constant were carried out over a period of 13 weeks, to study the changes in the moisture content in concrete over that period. As shown in Table 1, six types of test specimens in which the chloride content in the test specimens was varied from $0 \, kg/m^3$ to 6 kg/m^3 in steps of 1 kg/m^3 were used to study.

In the second step, the specimens with the seal were made to penetrate in water of chloride ion concentration whose was 3% and 10% for about 3 months.

2.3 Materials used and mix proportion

Table 2 and Table 3 show the materials used and the mix proportion of concrete used in casting the test specimens.

Table 1. Patterns of specimens.

Reinforced concrete	Plain concrete	Content of chloride (kg/m³)
R-0	P-0	0.0
R-1	P-1	1.0
R-2	P-2	2.0
R-3	P-3	3.0
R-4	P-4	4.0
R-5	P-5	5.0
R-6	P-6	6.0

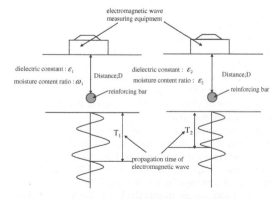

Figure 1. Concept of electromagnetic wave measurement.

2.4 Specimens

In the first step, as shown in Figure 2, two kinds of specimens, R and P, whose size were $100 \times 100 \times 400$ mm were prepared. Specimens R contained one 16 mm diameter deformed bar as shown, whereas specimens P did not contain any reinforcement.

In the second step, Specimens, whose size were $100 \times 100 \times 400$ mm were prepared. The test specimens were sealed all other planes of the specimen to permeate chloride ions from a plane of the specimen.

2.5 Test methods

One day after placement, the concrete specimens were removed and sealed in plastic bags so that the chloride ions in the specimens did not leach. The specimens were cured underwater for seven days, the moisture on the surface of the concrete specimens was wiped off thoroughly before carrying out any measurements. To study the change of the moisture content in concrete, the mass of the specimen, the temperature and humidity in the laboratory were measured during the electromagnetic wave measurement. For measurement of electromagnetic waves, an antenna of about 1.0 GHz with specifications as given in Table 4 was installed on the specimen. As shown Figure 3, the specimens were placed on a steel plate to intensify the reflected electromagnetic waves. As shown in Figure 4, the scale of the monitor was fixed at the level of the gain of the machine when the measurement was started, and the amplitude of the reflected electromagnetic wave estimated using that scale (Full scale = 100%).

Moreover, since the reflected wave from the reinforcement and the reflected wave from the steel plate are likely to interfere with the reflected wave in the case

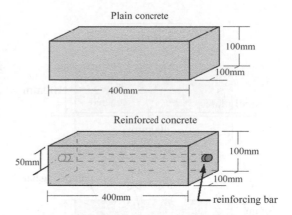

Figure 2. Specimens used electromagnetic wave measurement.

Table 4. Specifications of electromagnetic wave measuring equipment.

Item	Specifications
Rader frequencies	1.0 GHz
Measurement method	Impuls method
Transmission voltage	17Vp-p (at load 50 Ω)
Horizontal resolution	80 mm

of R specimens, the reflected waveform of the P specimen was subtracted from the measured waveform of the R specimen. Thus, as shown in Figure 5, the effect of only the presence of the reinforcement in the case of R specimens could be independently studied.

Table 2. Details of materials used.

Materials	Summary
Cement	Ordinary portland cement, density: 3.16 g/cm^3, fineness: 3320 cm^2/g
Fine aggregate	Hill sand, saturated surface-dry particle density: 2.62 g/cm^3, fineness modulus: 2.57
Coarse aggregate	Crushed stone, saturated surface-dry particle density: 2.65 g/cm^3, solid content: 59.4%
Chemical admixture	Air-entraining and water reducing agent of Libin sulfonic acid compound, density: 1.25 g/cm^3
	Air-entraining agent of denatuted rosinate-based anionic surface active agent

Table 3. Mix proportion of concrete.

Maximum size of coarse aggregate (mm)	Water to cement ratio (%)	Sand content (%)	Unit quantity (kg/m^3)				
			Water	Cement	Sand	Gravel	AEWRA
20	60.0	45.5	165	275	838	1015	0.96

NOTE; AEWRA: Air-entraining and water reducing agent.

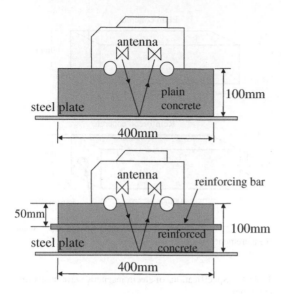

Figure 3. Measuring method.

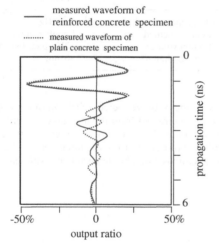

Figure 4. Effect of reinforcing bar in measured waveform.

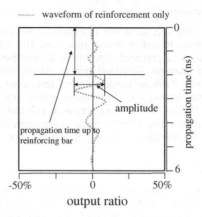

output ratio

Figure 5. Example of waveform with subtraction.

2.6 Test results and discussions

This study confirmed that the amount of moisture in concrete affected the measured amplitude of electromagnetic waves, and this needs to be appropriately considered when making an estimate of the chloride content in concrete. In this study, the latter estimates have been carried out by making the estimates in two ways – without and with considering the effect of water content, and the results are shown in Figure 6 and Figure 7 respectively. The multiple regression coefficients for the cases when moisture content was neglected and was considered was found to be 0.69 and 0.87, respectively, clearly showing the improvement in the estimated values by considering the moisture content.

The above results show that the accuracy of estimation of content of chloride ions using electromagnetic wave was improved in cases when the effect of moisture content was considered. However, as shown in Figure 7, it seemed that the variations of the estimated values for the respective content of chloride ions were large. Thus, to investigate the cause of these variations in the estimated values, multiple regression analysis was carried out separately for each day of measurement, and the results are shown in Figure 8. The multiple regression coefficient was found to be 0.98, indicating substantially improved estimates. In addition, the multiple regression analysis considering amount of moisture in concrete was also carried out for each day of measurement. When the moisture was considered, the estimated values tended to approach closer to the actual values. However, in this method, test-specimens for calibration need to be always prepared, and the calibration needs to be carried out on the day of the measurement. Moreover, partial regression coefficients calculated using the multiple regression analysis for the respective day of the measurement are needed to estimate the content of chloride ions. Thus, this method for estimating the content of chloride ions using electromagnetic waves requires a lot of time and effort. Then, the following were compared: Estimates using all the data and using the data of the different days of measurement.

Figure 9 shows the results in the former case. Figure 10 shows the changes depending upon ages on temperature and humidity in the laboratory during the measurement. As shown in Figure 9, the estimated values tend to have negative slope. As shown in Figure 10, temperature and humidity in the laboratory also tended to have a negative slope as the measurements were carried out from the middle of August to the middle

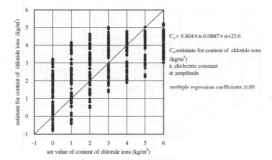

Figure 6. Estimate of chloride content without considering moisture.

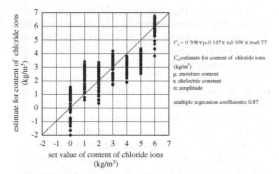

Figure 7. Estimate of chloride content with considering moisture.

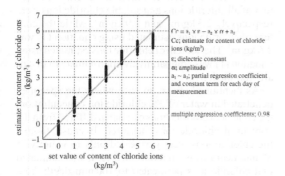

Figure 8. Estimate of chloride content for each day of measurement.

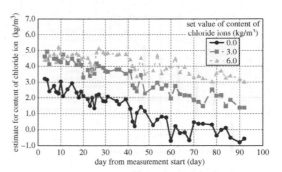

Figure 9. Changes of estimate of chloride content.

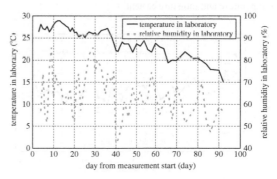

Figure 10. Change of temperature and relative humidity.

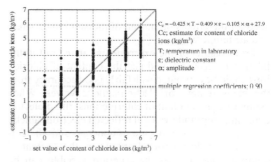

Figure 11. Estimate of chloride content with considering temperature

of November. It was proven that the changes of estimates shown Figure in 9 was similar to the changes of temperature in the laboratory shown in Figure 10. To consider this factor, multiple regression analysis was carried out again taking the temperature in the laboratory as an additional predictor variable.

As shown in Figure 11, the multiple regression coefficients were 0.90 and the estimates were reasonably accurate. When the amount of moisture in concrete was considered, as shown in Figure 12, the regression coefficient was 0.94.

Figure 13 shows the example of the estimated result in the case when the content of chloride ions is $3.0\,kg/m^3$. The result of multiple regression analysis not considering the effect of moisture content and the temperature at the time of measurement, had a negative slope, but when these factors are accounted for, the estimated values become quite close to the actual ones. Thus, it can be concluded that the content of chloride ions can be estimated with good accuracy from the amplitude values of electromagnetic waves, taking into

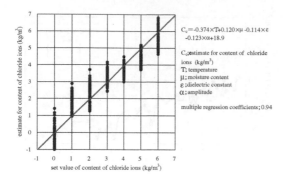

$C_c=-0.374\times T+0.120\times\mu-0.114\times\varepsilon$
$\qquad -0.123\times\alpha+18.9$

C_c;estimate for content of chloride
ions (kg/m³)
T; temperature
μ; moisture content
ε; dielectric constant
α; amplitude

multiple regression coefficients; 0.94

Figure 12. Estimate of chloride content with considering temperature and moisture content.

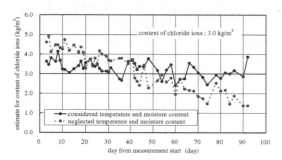

Figure 13. Effect of temperature and moisture content for chloride content.

account the effects of amount of moisture in concrete and air temperature.

From results of the experiments carried out in the laboratory, it was found that the content of chloride ions could be estimated with fairly good accuracy from the measurement of electromagnetic waves. When estimating the content of chloride ions, the dielectric constant, amplitude vales of reflected electromagnetic waves, amount of moisture in concrete and air temperature should be considered. Then, the content of chloride ions, C_c, as the criterion variable can be estimated by the multiple regressions analysis taking temperature in the laboratory T, amount of moisture in concrete μ, the dielectric constant ε and the amplitude value α as the predictor variables, using equation 1 below.

$$C_c=-0.374\times T+0.120\times\mu-0.114\times\varepsilon-0.123\times\alpha+18.9 \quad \text{(kg/m}^3\text{)} \quad (1)$$

Next, two cores were respectively drawn from one test specimen in order to analyze content of all chloride ions and content of soluble chloride ions within each specimen which respectively was penetrated in water of 3% and 10% chloride ion concentration. Figure 14 shows relationship between depth from the concrete surface and content of chloride ions.

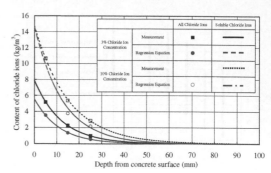

Figure 14. Distribution of content of chloride ions.

In case when the content of chloride ions is estimated using electromagnetic waves, it is seemed that the content of chloride ions estimates the average content to object (steel plate laid under the test specimen in this experiment) reflected the electromagnetic waves from concrete surface. Then, Figure 15 shows respectively the average of content of all chloride ions and soluble chloride ions shown in Figure 14 and the content of chloride ions estimated using equation (1). In case where the specimen penetrated in water of 3% chloride ion concentration, the average of content of all chloride ions was about half content estimated and the average of content of soluble chloride ions was about 1/3 content estimated. On the other hand, in case where the specimen penetrated in water of 10% chloride ion concentration, the average of content of all chloride ions and soluble chloride ions were respectively 2.3 kg/m³ and 1.9 kg/m³, while the content of chloride ions estimated using equation (1) was 2.6 kg/m³. The content of chloride ions estimated using equation (1) was comparatively approximate to that of all chloride ions.

As mentioned above, in case where the specimen penetrated in water of 3% chloride ion concentration, the content of chloride ions estimated differed from the content of chloride ions measured. Because, though the chloride ions were permeated only to vicinity of 50 mm depth from concrete surface, it was estimated that chloride ions permeated to 100 mm depth from concrete surface. In case where the specimen penetrated in water of 10% chloride ion concentration, it seemed that the estimated content of chloride ions became comparatively similar for the average content of chloride ions measured, since the chloride ions permeated to vicinity of 80 mm depth.

If it is assumed that the electromagnetic wave is not almost attenuated within not containing chloride ions, when the content of all chloride ions is averaged within permeating of chloride ions, the average content of chloride ions becomes 1.9 kg/m³ in case where the specimen penetrated in water of 3% chloride ion concentration and becomes 2.6 kg/m³ in case where

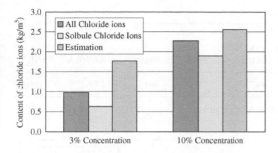

Figure 15. Comparison of content of chloride ions.

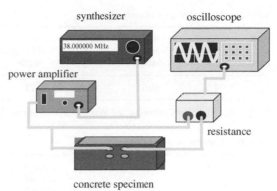

Figure 16. Measuring circuit.

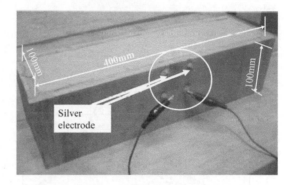

Figure 17. Specimen used measurement.

the specimen penetrated in water of 10% chloride ion concentration. Therefore, in both cases, the estimated content of chloride ions became similar for the average content of chloride ions measured. However, in actual structures, it is proven whether chloride ions have permeated in the neighborhood of the reinforcing bars. Therefore, the content of chloride ions estimated using electromagnetic waves may be estimated larger than that permeated in concrete, when the chloride ions have not permeated to the neighborhood of the reinforcing bars yet.

Therefore, in case of the estimation of chloride content using the electromagnetic wave, though it is possible to estimate only average chloride content from the concrete surface to the reinforcing bar, it is not possible to estimate the distribution of chloride content from concrete surface to the reinforcing bar.

3 ESTIMATION OF CHLORIDE CONTENT IN SURFACE DIVISION OF CONCRETE USING IMPEDANCE METHOD

3.1 Outline on measurement of chloride content in concrete using impedance method

In this study, as shown in Figure 16, alternating current was applied across one pair of the embedded silver electrode from the synthesizer and frequency (MHz) changed and effective values were read from the oscilloscope. The effective value shows the change of the amount of alternating current which passes in the concrete at the voltage difference (mV). The input waveform was made to be amplitude (2Vp-p) and frequency.

3.2 Specimens

As shown in Figure 17, specimen whose size was $100 \times 100 \times 400$ mm was prepared. A reinforcing bar and silver electrodes were embedded in the specimen. As shown in Figure 17, the silver electrode was embedded from the surface to the position of 1 cm and 5 cm and each electrode was embedded in 4 cm interval in

the specimen. In the concrete surface, the silver electrode was fixed in insulating tape, insulation rubber and concrete weight. Table 5 shows the mix proportion of concrete used in casting the test specimens. As shown in Table 6, five types of test specimens in which the chloride content in the test specimens was varied from 0 kg/m^3 to 10 kg/m^3 were used to study.

3.3 Test results and discussions

Figure 18 shows results of the measurements using impedance method. As shown in Figure 18, the frequency that the maximum effective value was obtained decreased depending upon increasing in the distance from concrete surface. The maximum effective value was in the range from 37.0 MHz to 38.5 MHz on each case. Figure 19 shows relationship between the maximum effective values and the content of chloride ions. The maximum effective values tended to decrease with the increase in the content of chloride ions. The maximum effective value at the 1 cm depth is larger than that at the 5 cm depth. Figure 20 shows relationship between the effective values and the average moisture content. The effective values tended to increase with the decrease in the average moisture content.

Table 5. Mix proportion of concrete.

Slump (cm)	Air content (%)	Water to cemnt ratio	Sand content (%)	Unit quantity (kg/m³)				AEWRA	AEA
				Water	Cement	Sand	Gravel		
12	4.5	60.0	45.4	173	347	798	961	1.08	0.0035

NOTE; AEWRA:Air-Entraining and Water Reducing Agent, AEA: Air-Entraining.

Table 6. Patterns of specimens.

Type of specimen	Content of chlorids (kg/m³)
A-0	0
A-1	3
A-2	5
A-3	7
A-4	10

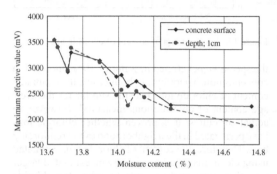

Figure 20. Relationship between moisture content and Maximum effective value.

and frequency tended to increase, as the measurement point was closer to the concrete surface. In addition, the maximum effective value tended to increase when average moisture content in the concrete specimens decreased. However, in present state, as the change of the maximum effective value is smaller than the change of chloride content and measurements of the maximum effective values greatly fluctuate in the concrete surface, it is necessary to pay attention that chloride content in concrete surface is quantitatively estimated using impedance method.

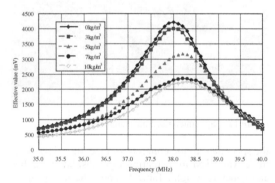

Figure 18. Relationship between frequency and effective value.

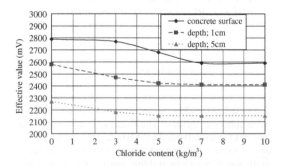

Figure 19. Relationship between chloride content and effective value.

As mentioned above, it was made clear that the frequency band where the maximum effective value was obtained changed by chloride content and moisture content and that the maximum effective value

4 MONITORING SYSTEM OF CHLORIDE CONTENT USING ELECTROMAGNETIC WAVE AND IMPEDANCE METHOD

As mentioned above, since the value estimated using the electromagnetic wave method is the average of chloride content from the concrete surface to the reinforcing bar, it is not possible to evaluate the distribution of chloride content. Therefore, chloride content has excessively been evaluated, when chloride content has not reached the reinforcing bar. In the meantime, if electrode used impedance method is not embedded at the depth beforehand determined, it will be not easy that chloride content in the position is estimated. Then, it seems to be possible that the chloride content in the reinforced concrete is estimated at the good accuracy by utilizing merits of each other and supplementing demerits of each other. The estimation method on distribution of chloride content in reinforced concrete is shown in the following.

First of all, in an object position, current resistance values are measured by installing the electrode in the concrete surface using the impedance method. For the current resistance values, chloride content of concrete surface division is estimated from the relationship between current resistance values and chloride content of concrete surface division on the basis of results gotten in laboratory tests. Next, the average chloride content from concrete surface to reinforcing bar is estimated using the electromagnetic wave method. Figure 21 shows the relationship between results from each measurement mentioned above.

Generally, it is possible to obtain a permeation depth of chloride content using the Fick's law of diffusion. It is possible to obtain chloride content in an arbitrary position and in an arbitrary time by solving the differential equation showed equation (2).

$$\frac{\partial C}{\partial t} = D_c \frac{\partial^2 C}{\partial x^2} \quad (2)$$

$$C(x,t) = C_0 \left(1 - erf\left(\frac{x}{2\sqrt{D_c \cdot t}} \right) \right) \quad (3)$$

Where, C shows chloride content in an arbitrary position (x) and in an arbitrary time (t) (kg/m^3), D_c shows the value of diffusion coefficient of chloride ions into concrete (cm^2/year), C_0 shows chloride content at concrete surface (kg/m^3). erf (s) is error function defined as follow;

$$erf(s) = \frac{2}{\sqrt{\pi}} \int_0^s e^{-\eta^2} d\eta \quad (4)$$

Results from past studies give the value of diffusion coefficient of chloride ions into concrete in equation (4) in following equation.

$$log\, D_c = a\left(\frac{W}{C} \right)^2 + b\left(\frac{W}{C} \right) + c \quad (5)$$

Where, a, b and c are the coefficient determined by the type of cement and W/C is water cement ratio.

It is possible to calculate total chloride content permeated from concrete surface by using chloride content in concrete surface estimated using impedance method and the value of diffusion coefficient into concrete gotten from equation (5).

$$C(t)_{total} = \int C(x,t) dx \quad (6)$$

Where, $C(t)_{total}$ is total chloride content permeated from concrete surface in an arbitrary time.

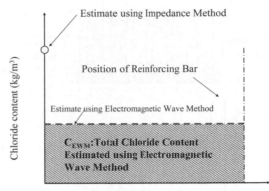

Figure 21. Estimate using electromagnetic wave and impedance method.

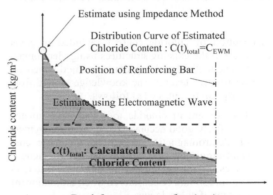

Figure 22. Estimate of distribution curve of chloride content.

As shown Figure 22, by determining permeation depth so that the total chloride content gotten in equation (6) may be equivalent to the average chloride content obtained by the electromagnetic wave method, it is possible to estimate the distribution of the chloride content from concrete surface.

At present, as shown Figure 23, in order to verify the distribution of chloride content estimated from both measurement results, chemical analysis to investigate chloride content within concrete is carried out with cores drawn from the specimens. The results of verification will be reported in the different opportunity.

5 CONCLUSIONS

The present research examined the applicable monitoring system which is possible to estimate the

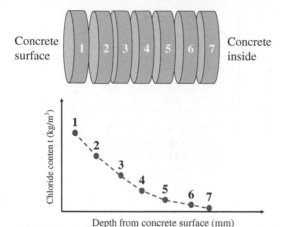

Figure 23. Verification of distribution of chloride content.

distribution of chloride content included in cover con crete in the existing structures using electromagnetic wave method and impedance method which are non-destructive testing. The knowledge gained from this study can be outlined as follows.

(1) The content of chloride ions can be estimated with fairly good accuracy from the measurement of electromagnetic waves. The dielectric constant, amplitude vales of reflected electromagnetic waves, amount of moisture in concrete and air temperature should be considered when estimating the content of chloride ions.

(2) From results of estimating the content of chloride ions using electromagnetic wave in the test specimens permeated chloride ions from concrete surface, however it is necessary to carry out experiments varied permeation depth of the chloride ions in the future, it was found that it is possible to estimate the content of chloride ions from the measurement of electromagnetic waves.

(3) From the result of the measurement of the content of chloride ions in reinforce concrete structure, though it was limited condition, it was possible to estimate the content of chloride ions using electromagnetic waves by applying estimated equation of the content of chloride ions obtained from results of tests in the laboratory.

(4) It was made clear that the maximum effective value was largely affected to the estimation of chloride content and amount of moisture content using impedance method in cover concrete.

(5) As he maximum effective value tended to decrease by supplying chloride ions intermittently, it seemed to find the possibility of estimating migration phenomenon of chloride ions using the impedance method.

(6) In present state, as the change of the maximum effective value is smaller than the change of chloride content and measurements of the maximum effective values greatly fluctuate in the concrete surface, it is necessary to pay attention that chloride content in concrete surface is quantitatively estimated using impedance method.

However, as many problems have been held on application to reinforced concrete structures, further works need to be examined on influences of environmental condition, shape of the structure, moisture condition in concrete, etc.

REFERENCES

Akihiko Y. et al. 1998. Non-destructive tests for diagnosis of concrete structures – electromagnetic wave method, and non-destructive inspection, Vol. 47, No. 10, Japan Society of Non-Destructive Inspection, 712–716. (in Japanese)

Arthur v. H.1954. Dielectric Materials and Applications, Artech House Publishers.

Corrosion & Protection Committee. 1987. Test methods and standards (provisional) related to corrosion and protection of concrete structures, Japan Concrete Institute, 57–58. (in Japanese)

Daisuke H. & Toshiaki M. 2005. Experimental Study on Applicability of Measuring Method of Chloride Content using Electromagnetic Wave in Reinforced Concrete Structures, The Third US-Japan Symposium on Advancing Applications and Capabilities in NDE.

Japan Society of Civil Engineering 2003. Enactment of Test Method for Diffusion Coefficient of Chloride Ion in Concrete and Trend of Test Method of normalization desired, Concrete Engineering Series 55. (in Japanese)

Jun-ichi A. et al. 2002. Research on the measurement of chlorides in reinforced concrete by non-destructive inspection, Annual Proceedings of Concrete Engineering, Vol. 24, No. 1, Japan Concrete Institute, 1515–1520. (in Japanese)

Junnichi A. & Toshiaki M. 2003. Study on Measurement of Chloride Content using Electromagnetic Wave in Reinforced Concrete Structures", Non-Destructive Testing in Civil Engineering.

Kazumasa M. et al. 1999. Relationship between water content and relative dielectric constant in concrete, Japan Society of Non-Destructive Inspection, Proceedings of the 1999 Spring Symposia, 91–94. (in Japanese)

Kyouichi F. et al. 2003. Influence Factors for Measurement of Content of Chloride Ions in Reinforced Concrete Using Electromagnetic Waves, Annual Proceedings of Concrete Engineering, Vol. 25, No. 1, Japan Concrete Institute, 1667–1672. (in Japanese)

Osamu H. 2003. Measurement Method on Material Properties, Morikita Publishers, 30–44.

Satake, N. 2005. Study on measuring method of chloride content in concrete surface division. *Proceedings of the 60th JSCE Annual Meeting*, Vol.5: pp.1199–1200. (in Japanese)

Society of Exploration Geophysicists of Japan 1998. Handbook of Geophysical Exploration, principle edition. (in Japanese)

Taketomo K. et al 2004. Experimental Study on Evaluation of Content of Chloride Ions in Reinforced Concrete Using Electromagnetic Waves, Annual Proceedings of Concrete Engineering, Vol. 25, No. 1, Japan Concrete Institute, 1673–1678. (in Japanese)

Toshiaki M. et al. 2002. Considerations on the measurement of chlorides in reinforced concrete by electromagnetic waves, Annual Proceedings of Concrete Engineering, Vol. 24, No. 1, Japan Concrete Institute, 1509–1514. (in Japanese)

Toshiaki M. & Junnichi A. 2003. Experimental Study on Measurement of Chloride Content using Electromagnetic Wave in Reinforced Concrete Structures, Structural Faults and Repair-2003, 241–248.

Toshiaki M. & Kumiko S. 2005. Experimental Study on Applicability of Measuring Method of Chloride Content using Electromagnetic Wave in Reinforced Concrete Structures, The 11th International Conference on Fracture.

Fracture Mechanics of Concrete and Concrete Structures – High-Performance Concrete,
Brick-Masonry and Environmental Aspects – Carpinteri, et al. (eds)
© 2007 Taylor & Francis Group, London, ISBN 978-0-415-44617-4

Assessment of reinforcement corrosion in concrete façades

J. Lahdensivu & S. Varjonen
Tampere University of Technology, Tampere, Finland

ABSTRACT: Outdoor concrete structures are damaged by several different phenomena whose propagation, again, is influenced by many structural, condition and material factors. In Finnish climate disintegration of concrete and corrosion of reinforcement are the most significant mechanisms creating need to repair concrete facades. The aim of facade condition investigations is to provide the property owner and designers information to help them decide on the repair needs and possibilities of facades and to select the repair methods best suited for each facility. The content of the condition investigation must be such that set goals are met. Generally, a condition investigation is made to determine the remaining service life of examined structures, their need of repair and safety. To achieve that, the investigation must reveal any damage to structures and defects in their performance. This requires determining the existence of damage, its extent, location, degree, impacts and future propagation by damage types.

1 INTRODUCTION

1.1 Background

A lot of concrete facades, over 30 million square metres, have been built in Finland since the 1960s as well as more than half a million concrete balconies. Thousands of other structures have also been built so far of concrete such as bridges, multi-level car parks, industrial plants, etc.

The maintenance and repair of these concrete structures has presented many problems. They are subject to many different types of degradation whose propagation, again, is affected by many structural, condition and material factors. Consequently, their service lives vary a lot in practice. Unexpected and technically and cost-wise significant repair need has occurred in the structures early on – sometimes only 10 years from completion.

Concrete structures have been repaired extensively in Finland since the early 1990s. Repair methods and materials have also been developed simultaneously as well as instructions for determining repair needs. Besides the correct repair method, it is also important to be able to determine the optimal time of repairs both technically and economically.

1.2 Condition investigations

Damage to structures, its degree and extent, due to various degradation phenomena can be determined by a comprehensive systematic condition investigation. A condition investigation involves systematic determination of the condition and performance of a structural element or an aggregate of structural elements (e.g. a facade or balcony) and their repair need with respect to different degradation mechanisms by various research methods such as examining design documents, various field measurements and investigations and sampling and laboratory tests.

The wide variation in the states of degradation of buildings, and the fact that the most significant deterioration is not visible until it has progressed very far, necessitate thorough condition investigation at most concrete-structure repair sites. Evaluation of reinforcement corrosion and the degree of frost damage suffered by concrete are examples of such investigations.

Condition investigation systematics for concrete facades and balconies have been developed in Finland since the mid-1980s. The following is based on the authors' experiences from about 150 condition investigations of concrete structures, long-continued development of condition investigation systematics and national condition investigation instructions (Anon. 2002).

2 GENERAL ASPECTS OF CONDITION INVESTIGATION

Information about the condition of concrete structures needed for the planning and design of repair work can only be acquired by a systematically conducted condition investigation. The main reason for this is that the general degradation mechanisms related to concrete structures normally proceed for relatively long before becoming visible.

The following instructions should be followed in a condition investigation in order to ensure its reliability:

1. The investigator must have sufficient knowledge about the performance and properties of the examined structures. He must also thoroughly understand the degradation mechanisms, defects, deficiencies, and repair methods at issue. Then, he can focus on studying the appropriate problems and collecting information and observations that are relevant to the client's needs.
2. The examined subject must be divided up into structure groups according to performance, type, material and exposure conditions, so that the systematic variation in the state of degradation and properties of structures between the groups can be detected.
3. The investigation must be able to determine the state of all potential degradation mechanisms that endanger the performance and durability of the structure as well as the structural and exposure factors that affect their progress.
4. An effort must be made to examine each issue under scrutiny using several parallel information collection and measurement methods.
5. Observations and measurements describing the state of different degradation mechanisms must be produced in sufficiently representative and large samples, so that the conclusions drawn on the basis of samples are reliable.
6. Sufficiently valid and reliable methods must be used for observations and measurements.
7. The collected information must be analysed carefully, so that the condition and repair need of structure groups can be determined. Conclusions must be clearly based on collected factual information. The existence, extent, degree and reasons for different types of degradation are examined as factors governing the condition of each structure group. Then, the impact of degradation on the structure's performance (e.g. safety), the propagation of degradation, suitable repair methods, and recommended timing of repairs can be assessed.

3 DEGRADATION MECHANISMS

The degradation of concrete structures with age is due primarily to weathering action which deteriorates material properties. Degradation may be unexpectedly quick if used materials or the work performance have been of poor quality or the structural solutions erroneous or non-performing. Weathering action may launch several parallel deterioration phenomena whereby a facade is degraded by the combined impact of several adverse phenomena. Degradation phenomena proceed slowly initially, but as the damage propagates, the rate of degradation generally increases.

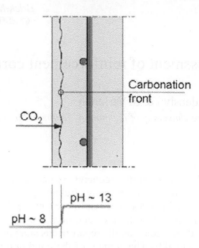

Figure 1. Carbonation propagates as a front into the structure (Varjonen et al. 2006).

The most common degradation mechanisms causing the need to repair concrete facades, and concrete structures in general, are corrosion of reinforcement due to carbonation or chlorides as well as insufficient frost resistance of concrete which leads to, for instance, frost damage (Pentti et al. 1998).

These degradation mechanisms may result in, for instance, reduced bearing capacity or bonding reliability of structures. Experience tells that defective performance of structural joints and connection details generally causes localised damage thereby accelerating local propagation of deterioration.

3.1 Corrosion of steel

Reinforcing bars in concrete are normally well protected from corrosion due to the high alkalinity of the concrete pore water. Corrosion may start when the passivity is destroyed, either by chloride penetration or due to the lowering of the pH in the carbonated concrete.

3.2 Carbonation

Carbonation of concrete is a chemical reaction where atmospheric carbon dioxide diffuses into the pore system of concrete and reacts with the alkaline hydroxides of concrete. Due to carbonation, the pH of the concrete decreases from the level of 13 to the level of 8.5 and the passivity of reinforcing steel is lost. This initiates the active corrosion if there is enough moisture and oxygen available in the concrete.

Carbonation begins at the surface of a structure and propagates as a front at a decelerating rate deeper into the structure. The speed of propagation is influenced foremost by the quality of concrete (proportion of cement and density) as well as rain stress. Heavy rain stress slows down carbonation.

The high alkalinity of concrete protects the reinforcement within from corrosion. When the carbonation front advances in concrete to the depth of the reinforcement, the surrounding concrete neutralises and corrosion of reinforcement can begin. The rate of corrosion clearly depends on the moisture content of concrete and advances significantly only at over 80% RH (Tuutti 1982). Corrosion lowers the tensile and bond strength of reinforcement while the pressure from corrosion products causes the concrete cover around the reinforcement to crack.

3.3 Chloride contamination

It is possible that chlorides have been added to the concrete mix during preparation to accelerate hardening. Chlorides were used mainly in the 1960s in connection with on-site concreting and in prefabrication plants during the cold season when concrete hardens slowly. The amount of salt used as accelerator was generally manyfold compared to steel's corrosion threshold.

Chlorides may also penetrate into hardened concrete if the concrete surface is subjected to external chloride stress, for instance, on bridges in the form of de-icing salts. Strong spotty corrosion is characteristic of chloride corrosion of reinforcement, and it may propagate also in relatively dry conditions. Chloride-induced corrosion becomes highly accelerated when carbonation reaches reinforcement depth whereby the extent of visible damage may increase strongly in a short time.

3.4 Active corrosion

Once the passivity is destroyed either by carbonation or by chloride contamination, active corrosion may start in the presence of moisture and oxygen (Parrott 1987). Corrosion may run for a long time before it can be noticed on the surface of the structure. Because corrosion products are not water soluble, they accumulate on the surface of steel nearby he anodic area (Mattila 1995). This generates an internal pressure, because the volume of the corrosion products induced by carbonation is four to six times bigger than original steelbars (Tuutti 1982).

Internal pressure caused by corrosion products leads to cracking or spalling of the concrete cover. Visible damage appears first on the spots where the concrete cover is smallest.

Due to corrosion, the diameter of steel bars becomes smaller and their tensile capacity is weakened. Thus, besides the aesthetic problems the corrosion may also cause a safety hazard.

3.5 Disintegration of concrete

Concrete is a very brittle material. It can stand only extremely limited tensile strains without cracking.

Internal tensile stresses due to expansion processes inside concrete may result in internal cracking and, therefore, disintegration of concrete. Disintegration of concrete accelerates carbonation and this way also steel corrosion. Concrete may disintegrate as a result several phenomenon causing internal expansion, such as frost weathering, formation of late ettringite or alkali-aggregate reaction.

3.6 Frost resistance of concrete

Concrete is a porous material whose pore system may, depending on the conditions, hold varying amounts of water. As the water in the pore system freezes, it expands about 9% by volume which creates hydraulic pressure in the system (Tuutti 1982). If the level of water saturation of the system is high, the overpressure cannot escape into air-filled pores and thus damages the internal structure of the concrete resulting in its degradation. Far advanced frost damage leads to total loss of concrete strength.

The frost resistance of concrete can be ensured by air-entraining which creates a sufficient amount of permanently air-filled so-called protective pores where the pressure from the freezing dilation of water can escape. Finnish guidelines for the air-entraining of facade concrete mixes were issued in 1976 (Anon. 2002).

Moisture behaviour and environmental stress conditions have an impact on frost stress. For instance, the stress on balcony structures depends on the existence of proper waterproofing.

3.7 Formation of ettringite

In certain conditions crystalline substances may form in concrete pores and take up pore space. In the initial stages this is not generally detrimental to the concrete structure. It becomes problematic if the protective pores start to fill up which lowers the frost resistance of concrete. It is also possible that the crystallising substance itself creates tensions within the concrete that damage it. One substance of this type is ettringite that forms in concrete, for instance, as a result of excessive thermal treatment.

The ettringite reaction is a chemical reaction caused by sulphate minerals that occurs in hydrated cement. It involves strong volume expansion of reaction products, or swelling, since the volume of ettringite is 300-fold compared to the volume of the reactants (Anon. 1989). The forming ettringite mineral crystallises onto the walls of the air-filled pores whereby the volume of protective pores and the frost resistance of concrete decrease. An ettringite reaction may lead to concrete degradation either as a result of frost weathering or as the pressure created by the filling of pores produces cracks in the concrete.

3.8 Alkali reactivity of aggregate

An alkali-aggregate reaction is an expansion reaction in the concrete aggregate due to the alkalinity of hydrated cement, which may degrade concrete. The reaction requires that the cement contains an abundance of alkalis (Na, K), the aggregate includes minerals with low alkali resistance, and the moisture content of the concrete is sufficiently high (Punkki & Suominen 1994).

Alkali-aggregate reactions are generally divided into alkali-silicon, alkali-carbonate and alkali-silicate reactions depending on the reacting aggregate (Punkki & Suominen 1994). Finnish dense deep-seated rocks generally have high chemical resistance which makes the phenomenon rare in Finland.

A concrete structure affected by an alkali-aggregate reaction is typically stained by surface moisture, exhibits irregular pattern cracking and swelling and has a gel-like reaction product oozing out of the cracks (Pentti et al. 1998). The damage caused by the alkali-aggregate reaction resembles the cracking due to frost weathering and both often appear simultaneously.

4 FIELD AND LABORATORY INVESTIGATIONS

The cause and exact extent of corrosion damage can be determined by laboratory and field tests. Initial data for the study of corrosion of reinforcing steel is gathered from the documents and by visual inspection. By visual inspection it is possible to estimate:

- the amount and location of visible damages (spalls, cracks, rust stains or spots)
- the depth of covering concrete in damaged spots
- the moisture behaviour that affects the rate of active corrosion.

The assessment of the amount of steel in active corrosion state at various depths is based on samples of carbonation depths and cover depths of reinforcement. In the field, the distribution of cover depths is measured by covermeter separately from each group of structures. In laboratory it is possible to determine the penetration of carbonation from core samples by using phenolphthalein indicator.

The progress of chloride corrosion can be estimated on the basis of depth of chloride penetration. The critical content is usually considered to be between 0.03–0.07% by weight of concrete (Varjonen et al. 2006). When this content is exceeded at the level of reinforcement, corrosion is initiated. The chloride contamination is measured from drilled powder samples for example by titration.

In condition investigation it is not possible to investigate every facade panel or balcony element one by one, but field examination and taking of samples for laboratory tests must be carried out by sampling. Investigation methods must be cost-effective. Cheaper methods, like visual inspection, can be used for extensive investigation, which are confirmed by more expensive methods like sampling and laboratory tests.

5 ESTIMATION OF SERVICE LIFE CONCERNING STEEL CORROSION

The remaining service life of a concrete structure is estimated on the basis of the expected impact of damage on reinforcement corrosion and related safety of the structure as well as the future rate of damage propagation.

The extent of corroding reinforcement can be estimated by comparing the carbonation or chloride depth distribution of concrete to the cover depth distribution of reinforcement.

5.1 Propagation of reinforcement corrosion

The corrosion state of reinforcement can be estimated by comparing the carbonation distribution of concrete to the cover depth distribution of rebars according to Figure 2.

Future propagation of concrete carbonation can be estimated by the so-called square-root model (Tuutti 1982), according to which carbonation in concrete advances at a decelerating rate as a function of time:

$$x = k \, t^{0.5} \tag{1}$$

where,
  x is carbonation depth [mm]
  k is carbonation coefficient [mm/a$^{0.5}$]
  t is time [a].

This model allows calculating the time when the carbonation front reaches the reinforcement or the share of rebars in corrosion state at each moment in time.

The advancement of chloride corrosion can be estimated by the so-called critical chloride content. Literature considers the critical content to be around 0.03–0.07% of concrete by weight, and in Finnish

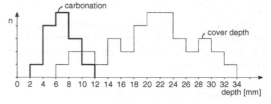

Figure 2. An example of histograms of measured carbonation and cover depth distributions (Pentti 1999).

guideline (Anon. 2002) 0.05% is usually used critical chloride content. When the critical chloride content is exceeded at the level of reinforcement, it starts to corrode.

The amount of chlorides added during mixing of concrete does not increase as a function of time which means that if the chloride threshold is not exceeded, there is no chloride corrosion. In the case of concrete structures exposed to chloride solutions, such as road bridges, the penetration depth of chlorides must be monitored regularly.

5.2 Safety of structures

The estimation of the overall service life of a building or structure requires combining the effects on various structures by different degradation mechanisms. The continued safety of each structure must be assessed and a decision made whether to use the structure to the end of its service life or repair it – appearance of the building must also be a consideration.

The safety of a structure is essentially affected by, for instance, frost damage to reinforcement, various embedded fixtures and anchorage zones of connecting trusses as well as corrosion of reinforcement. Far advanced degradation may also cause spalling of the concrete cover and breaking off of pieces of concrete.

6 CONCLUSIONS

A condition investigation allows evaluating the remaining service life of structures. This involves assessing the service life of a damaged structure which the generally used service-life models normally do not do. Based on the results of the condition investigation, the impacts of degradation can be estimated and the occurrence of future damage predicted when no visible damage yet exists.

Properly timed maintenance measures can often arrest propagation of degradation effectively thereby increasing the service life of a structure.

REFERENCES

Anon. 1989: Guidelines for Durability and Service Life Design of Concrete Structures. Concrete Association of Finland by 32. Jyväskylä. 60 p. (In Finnish)
Anon. 2002: Condition Investigation Manual for Concrete Facade Panels. Helsinki. Concrete Association of Finland BY 42 178 p. (In Finnish)
Mattila Jussi 1995: Realkalisation of Concrete by Cement-based Coatings. Tampere, Tampere University of Technology, Structural Engineering. Licentiate's Thesis. 161 p. (In Finnish)
Parrott L.J. 1987: Review of Carbonation in Reinforced Concrete. Cenebt and Concrete Association. Wexham Springs.
Pentti Matti, Mattila Jussi, Wahlman Jyrki 1998: Repair of Concrete Facades and Balconies, part I: Structures, Degradation and Condition Investigation. Tampere, Tampere University of Technology, Structural Engineering. Publication 87. 157 p. (In Finnish)
Pentti, M. 1999. The Accuracy of the Extent-of-Corrosion Estimate Based on the Sampling of Carbonation and Cover Depths of Reinforced Concrete Façade Panels. Tampere, Tampere University of Technology, Publication 274. 105 p.
Punkki J., Suominen V. 1994: Alkali Reactivity of Aggregate in Norway – and in Finland? Betoni 2/1994. Helsinki. Suomen Betonitieto Oy. pp. 30–32. (In Finnish)
Tuutti K. 1982: Corrosion of Steel in Concrete. Stockholm. Swedish Cement and Concrete Research Institute. CBI Research 4:82. 304 p.
Varjonen Saija, Mattila Jussi, Lahdensivu Jukka, Pentti Matti 2006: Conservation and Maintenance of Concrete Facades. Technical Possibilities and Restrictions. Tampere, Tampere University of Technology, Structural Engineering. Publication 136. 27 p.

Author index

T - #0060 - 071024 - C0 - 244/170/35 [37] - CB - 9780415446174 - Gloss Lamination